APPLIED
MULTIVARIATE
ANALYSIS AND
EXPERIMENTAL DESIGNS

McGRAW-HILL BOOK COMPANY
New York
St. Louis
San Francisco
Auckland
Düsseldorf
Johannesburg
Kuala Lumpur
London
Mexico
Montreal
New Delhi
Panama
Paris
São Paulo
Singapore
Sydney
Tokyo
Toronto

N. KRISHNAN NAMBOODIRI

Department of Sociology
University of North Carolina

LEWIS F. CARTER

Department of Sociology
Washington State University

HUBERT M. BLALOCK, JR.

Department of Sociology
University of Washington

Applied Multivariate Analysis and Experimental Designs

This book was set in Times New Roman.
The editors were Lyle Linder and M. E. Margolies;
the cover was designed by Anne Canevari Green;
the production supervisor was Leroy A. Young.
The drawings were done by Eric G. Hieber Associates Inc.
Kingsport Press, Inc., was printer and binder.

Library of Congress Cataloging in Publication Data
Krishnan Namboodiri, N
    Applied multivariate analysis and experimental
designs

    Includes bibliographical references.
    1. Multivariate analysis. 2. Experimental
design. I. Carter, Lewis F. II. Blalock, Hubert M.
III. Title.
QA278.K72   519.5'3   74-17356
ISBN 0-07-045865-0

**APPLIED
MULTIVARIATE
ANALYSIS AND
EXPERIMENTAL DESIGNS**

234567890 KPKP 798765

To KADAMBARI, JEANNE, and ANN

CONTENTS

PART THREE   SIMULTANEOUS-EQUATION MODELS

PART FOUR   MODELS INVOLVING MEASUREMENT ERRORS

PREFACE

This book is intended primarily for sociologists and political scientists and secondarily for social scientists in other fields. It is directed toward advanced graduate students and professionals who are actively engaged in quantitative research and is intended to be intermediate between general statistics texts, such as Blalock's *Social Statistics*, and more advanced specialized texts on linear models, experimental designs, or simultaneous-equations estimation techniques. It is our fundamental assumption that fields like sociology and political science must continuously strive to upgrade the level of graduate training in applied statistics and that we have reached a point where a textbook at this level is needed. On the other hand, we recognize that the vast majority of professionals in these fields still do not have the necessary backgrounds in mathematical statistics, matrix algebra, and computer programming to enable them to move directly from discussions in the more elementary texts to the more advanced treatments in specialized works. It is our hope that the present work will help to bridge this gap.

The book consists of four parts and an introductory chapter. The introductory chapter is addressed to a number of general questions that students or professional social scientists who have already had a course or two in statistics may ask when confronted with the question of studying more about statistics. Part One has as its focus the general linear model. The goal of this part is to free the researcher from reliance upon a set of isolated computing formulas. It demonstrates that all analysis-of-variance and ordinary least-squares tests and measures are derivable by imposing hypotheses as linear conditions on an appropriate regression model. Nonstandard research questions for which computational rituals are unavailable are emphasized. It is our hope that the reader will learn from Part One how to represent complex theoretical questions in terms of symbolic models which are machine-solvable.

Part Two is devoted to experimental designs. No attempt has been made to cover the field exhaustively. Designs and analytical techniques have been chosen for exposition in terms of their relevance to social-psychological experiments, action research, or evaluation studies and/or in terms of their utility in providing models for the analysis of data from nonexperimental studies. In Part Two many of the tests developed in Part One are developed within the context of experiments. Derivations from probability theory are given more explicit attention. Particular emphasis is paid to the partitioning of sums of squares. Also, Part Two is more explicit in the development of assumptions. It explores test strategies for many of the assumptions, discusses assessment of the impact of violations of these, and proposes remedies for many violations. Finally, Part Two examines the application of the factorial-experiments model to the analysis of multidimensional contingency tables.

Part Three introduces simultaneous-equation models in which reciprocal causation is permitted. In such models, an independent variable for some equations may be represented as the dependent variable in others. It is our hope that this section will permit the reader to formulate and test causal models of realistic complexity. Solutions to such models involve the introduction of lagged models, indirect least squares, two-stage least squares, and block-recursive designs. Complications introduced by such models and simplifying assumptions are developed. In Chapter 10 we have included a brief section on ordinal scales and causal inference.

Part Four deals with problems of measurement. The importance of explicit inclusion of error terms and auxiliary theory in models is illustrated. Models are developed which allow us to account for random measurement errors. The importance of multiple measurement and multiple analytic strategies is emphasized. Alternatives are formulated for testing models in which nonrandom measurement error is present.

We assume that most readers will have already been exposed to some analysis-of-variance and multiple-regression theory, but our treatments of the general linear model and experimental designs are intended to be self-sufficient. We assume essentially no knowledge of mathematics beyond elementary algebra, although we have occasionally used matrix algebra and calculus. Sections or parts of sections that use matrix algebra or calculus are indicated by a footnote and may be skimmed on first reading or omitted entirely. A rather extensive discussion of matrix algebra is provided in Appendix C, which may be of use for those who wish to go back over skimmed portions as well as for those who wish to make the transition to more advanced-level texts. The exposition of linear models in Part One makes use of computer programs. Chapter 2 provides the necessary introduction to Fortran language and computer programming. No prior knowledge of these is required to follow the introduction given in Chapter 2.

There is a shift in notation which may be troublesome as the reader moves from Part One to the other parts. Part One uses the Latin alphabet in all equations and models, whereas the other parts use Greek letters for error components and regression weights. For example, error components are designated by E in Part One and ε in Parts Two to Four. Similarly, regression weights are designated by a's, b's, etc., in Part One and α's, β's, etc., in the other parts. While the shift is minor enough to cause little confusion, it is intentional and requires explanation. Multivariate analysis is sufficiently complex to require computer solutions in most cases. Consequently, Part One emphasizes the development of *representations* of research questions as symbolic models which are machine-solvable. These machine-solvable models are closely integrated with computer instructions for their solution. Fortran and other computer languages use only Latin letters. This convention is followed in our models. To do otherwise would require the reader to shift notation as often as four times per page. Of course, the models in Parts Two to Four would ordinarily be solved by computer. However, detailed machine solutions are not a part of the exposition in these sections, the researcher being expected to generalize the machine skills developed in conjunction with Part One. Further, Parts Two to Four draw more heavily on the written traditions of mathematical statistics and econometrics. These traditions, developed before extensive machine use, have relied heavily on Greek-letter notations. For these reasons, we felt it better to follow machine-usable notation in Part One and the more conventional notation in the other parts.

In any book, decisions must be made not only with respect to level of presentation but also with respect to omitted topics. We have decided not to cover non-parametric or distribution-free approaches because of space limitations and because of the existence of a number of books addressed to the same kind of audience we have in mind. For these same reasons we have also omitted factor analysis, latent-structure analysis, and related topics. It is our hope, of course, that this text will make it rather easy for the reader to study these more specialized topics, as well as to master the materials in more advanced-level texts on experimental designs, applied multivariate analysis, and econometrics.

Carter has been responsible for writing Part One and Appendix B; Namboodiri for Part Two and Appendix C; and Blalock for Parts Three and Four. We all contributed to Chapter One. The computer program CARTROCK shown in Appendix A was developed by Richard C. Rockwell.

The order in which our names appear as authors does not in any sense reflect a distinction between senior and junior authors. Carter and Namboodiri wish to acknowledge, however, that this book would not have materialized had it not been for Blalock's initiation of the project.

We wish to thank the following persons for reading Part One and making valuable suggestions: Richard V. Andree, Marilyn L. Bidnick, Charles E. Bowerman, Jeanne P. Carter, Joe Wilson Floyd, Theodore Greenstein, Ramon E. Henkel, Earl Jennings, Eric Jensen, Cris Kukuk, Sanford Labovitz, Richard Lewis, Patrick J. Marnane, Paula May, S. Dale McLemore, Mae Z. Meidav, Ronald Miller, Randolph Monchick, David Musick, Richard C. Rockwell, Evan D. Rogers, Sally Y. Sedelow, Walter A. Sedelow, and Michael S. Sullivan. Partial support for work on Part One was received from the National Science Foundation Program in Computer Science in

Social and Behavioral Science Education, Department of Computer Science, University of Colorado, directed by Daniel E. Bailey, and from the Departments of Sociology of the University of California at Riverside and the Washington State University. Portions of Part Two were completed while Namboodiri was a visiting faculty member of the Population Studies Center of the University of Michigan. The preliminary versions of Part Two were used in different courses at the University of North Carolina at Chapel Hill. Suggestions received from the students in these courses were of considerable help in improving the presentation. Work on Parts Three and Four was completed at the University of Washington.

The following persons helped in the difficult task of typing the manuscript: Cynthia Baines, Ina Howell, Merri Lewis, Ann Lipscomb, Barbara Parker, Beulah Reddaway, and Carolyn Riply.

We are indebted to the Literary Executor of the late Sir Ronald A. Fisher, to Dr. Frank Yates, and to Longman Group Ltd., London, for permission to reprint Tables III, IV, and V from *Statistical Tables for Biological, Agricultural and Medical Research*, and to E. S. Pearson and H. O. Hartley, editors, and to the *Biometrika* Trustees for permission to reproduce Table 29 from *Biometrika Tables for Statisticians*, vol. 1. We are equally grateful to Richard C. Rockwell for permission to reproduce CARTROCK, and to those other publishers, acknowledged in the appropriate places, who kindly gave permission to reprint in part various journal articles.

N. KRISHNAN NAMBOODIRI

LEWIS F. CARTER

HUBERT M. BLALOCK, JR.

INTRODUCTION

1.1 SIGNIFICANCE OF MULTIVARIATE ANALYSIS

Students or professional social scientists who have already had a course or two in statistics or who are reasonably familiar with cross tabulations, simple correlational analysis, and a reasonable number of significance tests, will very naturally ask: Why should I learn more, given the fact that there are obviously many other things to study? This is a very reasonable question in view of the necessity of mastering a wide variety of substantive materials that cut across several disciplines, to say nothing of the specialties within a single field such as sociology or political science. Since there must inevitably be some kind of division of labor, it might seem that such relatively technical subjects could safely be left to specialists who may then serve as consultants for the remainder. A second kind of question that we must attempt to answer can be put as follows: Given all the imperfections in our data, our inability to meet all the necessary assumptions required in statistical inference, and the complexity of the real world in relation to the adequacy of our theories, how can we justify the use of sophisticated statistical techniques? In effect, this kind of question implies that attempts to use analysis techniques outstripping the quality of one's data or theoretical knowledge will only mislead by giving a false sense of precision and security. Sophisticated multivariate techniques seem to require much better data than we presently have and are therefore often deemed inappropriate except in a very few substantive fields such as world population and ecology or experimental social psychology.

These are both important kinds of questions, and we must attempt to provide a brief response in this introductory chapter, hoping that a more conclusive answer will be given once the reader has completed the book. Let us turn to the first type of question: Why should I study multivariate analysis? The simplest answer is that if our theories recognize reality is complex, and if we are unable to apply rigid controls through laboratory experiments, then our data analyses must also take this complexity into consideration. We cannot, for example, run a series of cross tabulations involving one or two control variables and hope to make valid inferences from our data. It seems safe to say that the overwhelming majority of quantitative studies in sociology, political science, and even psychology and economics make use of data-analysis procedures and assumptions that are far too simplistic for what we commonly assume to be true of the real world. If we are to take ourselves at all seriously, and if we are to develop adequate tests of our theories, we shall need to improve the quality of our data analyses. This, alone, will not be enough, since no degree of sophistication in data analysis will compensate for poor data quality. As we shall see, however, there is a sense in which one may argue that the poorer the quality of one's data, the more complex one's analysis must be in order to correct for such deficiencies. This point is emphasized in Part Four, where we consider the implications of measurement errors for statistical analysis.

But even if this argument were granted, must *all* social scientists learn to master multivariate-analysis techniques? Certainly not, if this is meant to imply that they need to have an active, working knowledge of every detail. But in another sense we must recognize that the average level of statistical literacy in a given discipline is very likely to determine what studies get read and therefore integrated with the main body of literature in the field. This, in turn, is likely to determine what is written, particularly in books that are designed for relatively large audiences. Finally, what one writes is likely to affect even the mode of data analysis. If readers are not likely to be able to understand multiple regression, simultaneous-equation estimation, or a certain kind of factorial design, an investigator may decide not to bother to analyze the data except in the simplest of ways. As a result, many important insights may be lost. In many instances, the conclusions reached may be highly misleading. Presumably, every social scientist is a reader or consumer of social research, and in this role he or she will need to be at least minimally informed about multivariate analysis and other aspects of modern statistical inference. If this does not occur, and if each successive generation of social scientists is not better trained than its predecessors, it is difficult to imagine how cumulation of a valid body of knowledge can be achieved. In short, it is not sufficient for these techniques and methodologies to be known only to a small band of specialists who speak and write only to each other. We must find ways to transmit relatively technical materials to the wider audience of social scientists in such a way that the fundamental ideas and underlying assumptions are reasonably well understood. One of the major purposes of this text is to achieve this objective.

Graduate students face a special kind of difficulty that more established scholars can overcome. A graduate student is expected to be something of a generalist who must master just enough of every major substantive area in a discipline to be able to pass examinations and other requirements. Consequently, the demands on his or her time may become considerable. In view of this, additional courses in statistics and research methodology may seem to be a luxury that only a few specialists can afford. To this dilemma we can only respond that it is usually relatively more difficult to learn technical bodies of literature by informal means and later in life. Most students find it much easier to learn statistics at a time when the mathematics is fresher in their minds, when formal courses are available, and when instructors are present to answer their questions. Many substantive fields of specialization, on the other hand, are sufficiently nontechnical to be mastered by informal means and at a more relaxed pace, through a combination of reading, discussions, teaching, and the active participation in on-going research. It is the authors' position, based on considerable teaching experience, that students who have mastered a technical body of literature are in a much stronger position to study most substantive fields later on than those who first master the substantive fields and then later try to pick up the more technical materials by informal means. Of course students will differ in this respect, and we speak in terms of modalities. Most persons, we believe, will benefit to the extent that they find it possible to include as much quantitative training as they can relatively early in their academic careers.

Let us now turn to the second kind of question, which is of greater fundamental importance because it involves the issue of why *any* social scientist should bother to learn more than very simple statistical tools in view of the numerous inadequacies in our data. This kind of question cannot really be answered on the general level. One must list specific kinds of data inadequacies and pose specific questions in connection with each. But we may put the matter in the following way. Suppose we knew that we were faced with five sources of error. It might be argued that it would be foolish to correct for only one of these because there would still be four remaining ones. Since we could not achieve perfection without correcting for all of them at once, it might even be argued that it would be useless to correct for four of the sources of error if the fifth remained uncorrected.

Clearly, this kind of reasoning would be self-defeating, particularly for the discipline as a whole. It implies an all-or-nothing approach to research; errors are either present or absent. We reject this mode of reasoning on the basis of the assumption that most errors are additive or cumulative in nature and that they vary by degree. Thus, if one does not have a perfectly random sample, this does not mean that the biases need be serious ones. If an error distribution is nonnormal, this does not imply that one should never use a test that requires the normality assumption, since the test may be very "robust" or insensitive to the violation of this particular assumption. If one does not have a perfect measure of x, this does not indicate that one should treat the measure as extremely crude and resort to dichotomies. If the "dependent" variable in an equation is thought to have some very minor effect on one of the independent variables, this does not mean that one should automatically rule out the use of ordinary least squares. But if all these assumptions are violated to a major degree, then we shall be in difficulty. Our approach, then, must be to correct for each kind of deficiency as well as we can, thereby reducing our errors or biases as much as possible.

If we adopt this general strategy, it becomes important to study each potential source of difficulty as carefully as we can, and this will almost inevitably introduce complexity into our analyses. Sometimes this involves substituting cluster-sample formulas for those which presuppose simple random sampling. Sometimes it requires us to introduce additional variables, nonlinearities, or interactive terms into our equations. Or perhaps we cannot conduct a full factorial experiment for one reason or another, in which case we shall need to know how to modify a basic design in specific ways. Or we may need to work with several equations simultaneously, rather than one at a time. *It is because of these compromises with reality that it becomes crucially important to understand what we are doing and the nature of the assumptions that we are making.* In this sense, the realist or the practically minded scientist requires a basic understanding of multivariate analysis and experimental designs because of the fact that we rarely if ever are confronted with simple cases where, for example, "other things" are really equal or where one can control for "all relevant variables." If the reality is complex, so must our data analyses be. Only when we bring our implicit assumptions into the open do we begin to see just how complex these analyses must become in many instances. A basic strategy encouraged in this text is to make our assumptions fully explicit and then to attempt to tackle the unrealistic ones by adding just enough complexities to enable us to cope with reality, while at the same time arriving at defensible answers to our questions.

In the subsections that follow we shall deal very briefly with several specific kinds of complications or imperfections in the data that necessitate the violation of certain assumptions often made in statistical analyses. We alert the reader to problems of significance testing, inadequate measurement, assumptions concerning disturbance terms, and problems of reciprocal causation. These discussions are incomplete since the issues are quite complex. Specific instances of these problems are dealt with in detail at appropriate places in Parts One to Four. Our aim here is to emphasize the main point that the way to handle complexities is to acknowledge their existence and to modify one's analysis accordingly, rather than taking the position that too many complexities imply that one should revert to simpler and perhaps less pretentious modes of analysis. The latter merely hide the problems from view rather than making them disappear.

1.2   SIGNIFICANCE TESTS, RANDOMIZATION, AND GENERALIZATION

There has been an extensive debate in the sociological literature on the appropriateness of significance tests in social research (see especially Selvin, 1957; McGinnis, 1958; Kish, 1959; Morrison and Henkel, 1969; and Winch and Campbell, 1969).[1] Selvin's 1957 article, which sparked the debate, concluded that "conditions under which tests of significance may validly be used are almost impossible of fulfillment in sociological research..." (p. 520). This conclusion was arrived at by reasoning that there are certain "problems of design" and certain "problems of interpretation"

---

[1] References are given in Appendix E.

peculiar to social research which make tests of significance inapplicable. The critical "problem of design" for social scientists, particularly those engaged in survey research, stems from the fact that the experimental procedure of randomization is impossible in most of their research. "Problems of interpretation," on the other hand, have to do with the fact that some of those who apply tests of significance in data analysis do a very poor job of it by confusing statistical significance with substantive importance, by misinterpreting the random-error components, or by engaging in data-dredging exercises. Misinterpretations of the results of statistical analyses are probably as frequent now as when Selvin wrote his 1957 article, but the fact that there are some researchers who misuse significance tests does not provide a compelling argument for abandoning such analytical tools altogether. A more healthy strategy seems to be to try to make sure that research workers understand clearly what significance tests can tell them and when to apply them. In the remaining paragraphs of this section we shall discuss (1) what significance tests are, (2) when such tests can be applied, and (3) what specific research questions such tests can answer.

The Logic of Significance Tests

A test of significance may be broadly described as a method of data analysis which helps to decide whether a given data set is consistent with a given theory or hypothesis. All tests of significance rest on probability laws. In order to apply probability laws to any given data set, we must assume that the data set at hand has been generated by some random process (see below for examples). The analyst may have introduced a random process while collecting the data through random sampling. Alternatively, he may have introduced randomization by his assignment procedures for exposing subjects to conditions. In any event, the analyst must assume that a random process underlies the data set at hand before he can validly apply the theory of probability. The abstract random process and the associated probability law form the mathematical counterpart of the concrete data situation. It may be noted that there may be reason to suspect the validity of the abstract picture with which the analyst represents the concrete data situation. This is true with experiments as well as with observational data. We shall comment on this later. Let us first illustrate how abstract pictures are formulated for given experimental designs.

Design 1

Experimental design 1, portrayed in Table 1.1, has the corresponding abstract representation shown in Table 1.2. Let us imagine that the data in Table 1.1 were

Table 1.1 RANDOM ASSIGNMENT OF SUBJECTS TO TREATMENT AND CONTROL GROUPS

Group	Reaction X present	Reaction X absent	Total
Exposed to treatment	a	b	$a + b$
Control	c	d	$c + d$
Total	$a + c$	$b + d$	$n = a + b + c + d$

obtained in the following manner. A total of n subjects were selected from those who volunteered to participate in an experiment. For each subject, it was decided by tossing a not necessarily unbiased coin whether to expose him to a treatment under investigation or to keep him as a control. It turned out that altogether $a + b$ subjects were exposed to the treatment while $c + d$ subjects were kept as controls. Among those in the treatment group, a subjects showed reaction X while b subjects did not. In the control group, c subjects showed the reaction while d subjects did not.

The research question in this case may be whether the treatment has been "effective." More specifically we ask: Is the tendency to show reaction X independent of whether the subject has been exposed to the treatment or not? The usual test of significance we apply in this case may be described as the test for independence between the rows and columns of the 2×2 table. The appropriate testing procedure, discussed below, for such questions involves the use of the single multinomial model. Subsequently, we shall comment on the application of this same testing procedure to investigations involving other models.

Single Multinomial Model

Imagine an urn containing a very large number of balls, some of them red and others white. Each ball has the letter S (success) or F (failure) marked on it. Let the proportion of each kind of ball be as shown below:

$$
\begin{aligned}
\text{Red balls marked} \quad &\text{S} = p \\
\text{Red balls marked} \quad &\text{F} = q \\
\text{White balls marked} \ &\text{S} = r \\
\text{White balls marked} \ &\text{F} = s
\end{aligned}
\tag{1.1}
$$

Notice that by definition $p + q + r + s = 1$. Suppose the results of n independent draws turn out to be distributed as in Table 1.2. For purposes of statistical analysis we may replace Table 1.1 with its conceptual analog, Table 1.2. Notice that the probability law applicable to the configuration in Table 1.2 is the multinomial with

Table 1.2 DISTRIBUTION OF n BALLS DRAWN FROM AN URN

	S	F
Red	a	b
White	c	d

four classes. Recalling that the cells in Table 1.2 have the following probabilities associated with them

	S	F
Red	p	q
White	r	s

we get the probability associated with the particular configuration in Table 1.2 as

$$P(\text{configuration}) = \frac{n!}{a!\,b!\,c!\,d!}\, p^a q^b r^c s^d \qquad (1.2)$$

where $p + q + r + s = 1$ and $a + b + c + d = n$. The quantity (1.2) may be interpreted as the relative frequency with which we would obtain the particular configuration in Table 1.2 if we were to repeat the urn experiment indefinitely. This indefinite number of repetitions of the urn experiment generates a population of samples. The configuration in Table 1.2 constitutes one sample drawn at random from this population of samples. It will be useful for us to rewrite expression (1.2) as the product of the three quantities L_1, L_2, and L_3

$$P(\text{configuration}) = L_1 L_2 L_3 \qquad (1.3)$$

where

$$L_1 = \frac{n!}{(a + b)!\,(c + d)!}\,(p + q)^{a+b}(r + s)^{c+d}$$

$$L_2 = \frac{n!}{(a + c)!\,(b + d)!}\,(p + r)^{a+c}(q + s)^{b+d}$$

$$L_3 = \frac{(a + c)!\,(b + d)!\,(a + b)!\,(c + d)!}{n!\,a!\,b!\,c!\,d!}$$

$$\times \left[\frac{p}{(p + r)(p + q)}\right]^a \left[\frac{q}{(p + q)(q + s)}\right]^b$$

$$\times \left[\frac{r}{(r + s)(p + r)}\right]^c \left[\frac{s}{(q + s)(r + s)}\right]^d$$

The first two expressions [L_1 and L_2 in (1.3)] give the probability of obtaining the marginals in Table 1.2 from repetitions of the urn experiment, while the third expression [L_3 in (1.3)] gives the conditional probability of the cell frequencies for fixed values of the marginals. In other words, the first two expressions in (1.3) give respectively the relative frequencies with which we may expect to observe among repetitions of the urn experiment the column and the row marginals exactly as in Table 1.2, while the third expression gives the relative frequency with which we may expect to observe the cell frequencies of Table 1.2 among repetitions of the urn experiment that have the row and column marginals exactly as in that table. Notice that in this interpretation of L_1 and L_2 we view the sample at hand, i.e., the particular configuration in Table 1.2, as one among the whole population of samples which would result if we were to repeat the urn experiment described above, while the interpretation given L_3 amounts to viewing the observed configuration as a particular member of a subpopulation of samples, namely, those with the same marginals as in Table 1.2.

We can now translate our hypothesis of independence between rows and columns in Table 1.1 to statements about the parameters in (1.3). For this purpose, we must stipulate that the hypothesis of independence between rows and columns in Table 1.1 is identical to the hypothesis of independence between color and type of balls in the urn experiment under reference. In this urn experiment color and type of balls are said to be independent if the following relationship obtains between the parameters p, q, r, and s:

$$p = (p + r)(p + q) \qquad (1.4)$$

Since $p + q + r + s = 1$, this implies

$$q = (q + s)(p + q) \qquad r = (p + r)(r + s) \qquad s = (q + s)(r + s)$$

Equation (1.4) may be referred to as our null hypothesis H_0. Given that H_0 is true, the third expression in (1.3) reduces to

$$L_3 = \frac{(a + c)!\,(b + d)!\,(a + b)!\,(c + d)!}{n!\,a!\,b!\,c!\,d!} \qquad (1.5)$$

Observe that this expression contains none of the parameters p, q, r, or s. Hence its value can be exactly calculated without knowing p, q, r, or s. It is in this manner that the familiar Fisher's exact test of independence in 2×2 tables is derived. We calculate the value of (1.5) and the corresponding values of configurations which have the same marginals as in Table 1.2 but which are less favorable to H_0. The sum of those values is then compared to some chosen level of significance, for example, .05. If the combined probability of the observed configuration and the more extreme ones is less than the chosen level of significance, we reject H_0. Although we did not fix or determine any of the marginals in our data-collection process, the testing procedure just described is organized around the sampling distribution of cell frequencies within the subpopulation of samples whose row and column marginals are both fixed.

Design 2: An Alternative

It is pertinent to point out that even if we had fixed in advance one set of marginals in data sets such as the one in Table 1.1, it would be appropriate to use the same testing procedure. To illustrate this, imagine that the numbers of subjects allocated to the experimental and control groups were fixed in advance. The difference between this design and design 1 is that in design 1 we let the marginals $a + c$ and $b + d$ be determined by chance whereas in the present case we fix them beforehand. More specifically we allocate exactly $a + b$ subjects to the experimental group and $c + d$ subjects to the control group. Suppose the configuration in Table 1.1 was arrived at in the above manner. Under this design we may be interested in the following hypothesis: Is the proportion of subjects showing reaction X among the treatment group different from the corresponding proportion among the control group?

The following is an urn experiment which may be regarded as the conceptual analog for this data situation. One urn contains a large number of red balls of which $100p_1$ percent are marked S while others are marked F. Another urn contains a large number of white balls of which $100p_2$ percent are marked S while others are marked F. From the first urn we select with replacement $a + b$ balls and count how many of them are marked S and how many are marked F. From the second urn we select $c + d$ balls at random with replacement and count how many among them are marked S and how many marked F. If the results of this experiment were as shown in Table 1.2, the probability associated with the observed configuration would be the product of two binomials

$$P(\text{configuration}) = \left[\frac{(a + b)!}{a!\, b!}\, p_1{}^a(1 - p_1)^b\right]\left[\frac{(c + d)!}{c!\, d!}\, p_2{}^c(1 - p_2)^d\right] \qquad (1.6)$$

The null hypothesis in this case may be H_0: $p_1 = p_2$, which, if true, reduces Eq. (1.6) to

$$P(\text{configuration}) = \frac{(a + b)!\, (c + d)!}{a!\, b!\, c!\, d!}\, p_1{}^{(a+c)}(1 - p_1)^{(b+d)} \qquad (1.7)$$

This, it may be noted, is the probability of the configuration in Table 1.2 for fixed values of $a + b$ and $c + d$. In other words, (1.7) gives the conditional probability $P(a, b, c, d \mid a + b, c + d)$.

Now let us pool the contents of the two urns. If H_0 is true, the probability of observing $a + c$ S's and $b + d$ F's in a random sample of n balls from the pooled universe is

$$\frac{n!}{(a + c)!\, (b + d)!}\, p_1{}^{(a+c)}(1 - p_1)^{(b+d)} \qquad (1.8)$$

The conditional probability of the configuration in Table 1.2 for fixed values of $a + b$, $c + d$, $a + c$, and $b + d$ is therefore expression (1.7) divided by expression (1.8), i.e.,

$$\frac{(a + b)!\, (c + d)!\, (a + c)!\, (b + d)!}{n!\, a!\, b!\, c!\, d!} \qquad (1.9)$$

which is the same as (1.5). This shows that in design 1 as well as design 2 we may use the same testing procedure in which the sample at hand is compared to a sub-population of samples with fixed marginals. What has been presented above allows us to reach certain conclusions about the applicability of significance tests to common research settings.

Hypothetical Populations and Target Populations

Statistical analysis of data sets like that in Table 1.1 need not assume that the subjects involved have been selected at random or on some other probability basis from a fixed universe. As long as we are willing to superimpose a random process and an associated probability law on a given data set, the logic of statistical inference can be applied.

The population to which our inference regarding the relationship under investigation is generalized is a hypothetical population resulting from hypothetical repetitions of the experiment. We imagine repeating the experiment under essentially the same conditions, each repetition involving random reassignment of the subjects already used in the experiment. It is important to recognize that it is frequently impossible to repeat an experiment under the same conditions. For example, "reaction X" may involve producing irreversible changes in the subjects, e.g., change of attitudes or orientation. Nevertheless, many researchers feel comfortable with assuming such repetitions and thus stipulating the probability laws (such as the multinomial in design 1 or the product of the binomials in design 2) to be the mathematical counterpart of the relative frequency distributions associated with those hypothetical repetitions.

It should now be clear that there is nothing intrinsically wrong with applying tests of significance to data including an entire concrete universe such as the population of all counties in the United States. In using such tests, we compare the observed data with a random process and treat the observed distribution as a randomly sampled one from the population of all possible (indefinitely repeated) experiments (cf. Blalock, 1972, pp. 238–239; Winch and Campbell, 1969).

It is essential that we distinguish between the hypothetical population of repeated studies and the "target population" to which the experimenter wishes to generalize the results of a particular experiment. Imagine an experimenter who wants to compare two different teaching methods. Suppose he seeks the cooperation of the schools in an area for conducting an experiment. Assume further that no public schools will participate while the only private school in the area agrees to cooperate. Can the results of the experiment conducted in the private school be generalized to all the schools in the area? (All the schools in the area may be the target population in this case.) The answer to this question depends upon (1) whether the private school differs from the public schools in any relevant characteristics and (2) whether there is *interaction* between those characteristics and the treatment effects. One might think of several characteristics which distinguish the private school from the public schools. Perhaps the average IQ of the children in the private school is higher than that of the children in the public schools. The teachers in the two types of schools may differ in their morale, in their zeal for investigating the effectiveness of different methods of teaching, and so forth. The question is whether these differences have any impact on the treatment comparisons in the experiment under consideration.

The concept of interaction is given detailed attention in Parts One and Two. Suffice it to say here that if the treatment difference tends to be higher (lower) in the higher average IQ situation than in the lower IQ situation, the findings of the experiment in question are not generalizable to the whole universe of schools in the given area, or for that matter to any universe of which the particular school used in the experiment is not a representative. It is pertinent to ask: Would the situation be any different if the private school referred to above was selected on a random basis from among the whole set of schools in the area? Theoretically this situation is different from the one described earlier in which self-selection was allowed to confound the treatment comparison. In practical terms, however, the two situations are not much different from each other, as our discussion below of the limitations of randomization will reveal. With small samples of units (here, one) generalization even from random samples is too risky to be advisable.

Matched Comparisons and Randomization

Suppose it is not possible to randomize subjects to treatment and control groups. For example, suppose we are asked to investigate whether in a particular community the propensity to have high self-esteem is different among Protestants than among Catholics. Suppose we observed the pattern in Table 1.3 based on a random sample of 100 Catholics and a random sample of 100 Protestants in the community. Superficially these data resemble design 2 described earlier. The question arises, however, whether it is valid to assume that the cell frequencies a and b and similarly c and d can be treated as the outcomes of two binomial experiments. It is plausible that the propensity to have high self-esteem is determined by many factors other than religion, e.g., income, education, father's social class. If this were the case, it would be logical to assume that the data in each cell of Table 1.3 result from several binomials each with different p's. If we know the factors affecting the propensity to have high self-esteem, it may be possible to arrive at a more valid specification of the mathematical model by incorporating all those variables as controls. This would result in a large number of tables such as Table 1.3 each of which may be viewed as a matched comparison between Catholics and Protestants. In each matched comparison we may apply the procedure discussed under design 2 above and draw our inferences accordingly. This may be called the method of eliminating the role of confounding factors through the process of matching. But two problems may arise in applying this method: (1) we may not know *all* the factors affecting the propensity to have high self-esteem; (2) our sample may not be large enough to permit the introduction of more than a few factors as controls. Because of these limitations, after controlling for a few factors, we tend to assume that the role of other confounding factors, if any, is negligible. This undoubtedly is a matter of faith in most data situations, and we should be aware of the possibility that such an assumption may be untenable.

We now turn to the question: Would randomization rule out the operation of confounding factors? The usual textbook answer to this question may be outlined as follows. We assume that the observation (response), say y, when a particular treatment is applied to a particular experimental unit (subject) can be expressed thus:

Observed response = a quantity depending on treatment applied

$$+ \text{ a quantity depending on properties of unit} \qquad (1.10)$$

The properties in the second component in (1.10) may belong either to (1) those common for all subjects throughout the experiment or (2) those varying among subjects.

Table 1.3 SELF-ESTEEM AMONG CATHOLICS AND PROTESTANTS

| | Self-esteem | | |
	High	Low	Total
Catholics	a	b	$a + b = 100$
Protestants	c	d	$c + d = 100$

A symbolic representation of (1.10) is

$$y_{ij} = \beta_j + \varepsilon_{ij} \qquad (1.11)$$

where y_{ij} is the response (observation) for the ith subject under treatment j, β_j the quantity associated with the jth treatment (we may call it the effect of the jth treatment), and ε_{ij} the quantity associated with the properties of the ith subject under the jth treatment (we may call this the combined effect of all the properties of the particular subject). Remember that we do not know the values of β_j or ε_{ij}. If subjects have been randomized over treatments, the particular set of ε_{ij}'s appearing under treatment j may legitimately be regarded as a random sample selected without replacement from the universe of all possible ε_{ij}'s. Hence we may apply the mathematical theory of random sampling (without replacement) to the behavior of the ε_{ij}'s associated with our observations. One pertinent result is that the expected value of the sample mean is equal to the population mean. Applying this result, we have

$$E(\bar{\varepsilon}_{.j}) = \mu$$

for all j, and hence

$$E(\bar{y}_{.j}) = \beta_j + \mu$$

where μ represents the population mean of the ε's. Consequently

$$E(\bar{y}_{.j} - \bar{y}_{.l}) = \beta_j - \beta_l$$

In other words, $\bar{y}_{.j} - \bar{y}_{.l}$ is an unbiased estimate of $\beta_j - \beta_l$. It is in this restricted sense that we expect randomization to remove unsuspected biases.

Notice, however, that we may be led to incomplete and often misleading conclusions if the experimental design does not keep under control, before randomization, relevant characteristics of subjects which interact with our experimental manipulations. For example, if we have a large population of students differing in ability levels, random assignment of these students to treatments, e.g., teaching techniques, should result in approximately equal proportions of high-, medium-, and low-ability students being exposed to the different treatments. If some techniques differentially affect students of different ability levels, our conclusions concerning the effects of the techniques will still be incomplete and misleading unless we "control" for ability levels. Uncontrolled findings may "generalize" on the average to other settings with the same talent blends but can tell us nothing about results in settings with different blends. This limitation of randomization is elaborated in Secs. 3.10 and 3.11.

Other Limitations of Randomization

In many experiments the experimenter wishes to distinguish the conceptual treatment from the actual (manipulated) treatment. Miller and Bugelski (1948) reported a study in which the conceptual treatment was "frustration." The study took place in a boys' summer camp. One group of boys was subjected to an impossible test that forced them to miss bank night at the movies, while another group (the control) was spared this manipulation. The objective was to see if the boys who had been frustrated expressed a significant change in their prejudice levels as compared with the control group. Let us denote the manipulated treatment by T, the conceptual treatment (here "frustration") by C, and the response variable as Y. The objective of the experiment may be described as answering the question: What is the causal impact of C on Y, knowing only how Y varies with T? The usual analysis procedure would permit the analyst to infer the causal impact of C on Y from the available data only if the analyst is willing to make the following assumptions: (1) T operates on Y through and only through C, and (2) T is independent of any unaccounted for factors operating on Y independently of C. *Randomization of subjects between the treatment and control groups would not necessarily guarantee the validity of assumptions such as (1) above.* See Costner (1971a) and Miller (1971) for discussions of different possible interpretations of the patterns observed in such data situations as the above. Complications of this sort are discussed in Part Four.

Another limitation of randomization stems from the danger that the randomization process may sometimes yield an unfortunate arrangement of subjects. In other words, the arrangement (layout) obtained as a result of randomization may be one that is inappropriate for the actual conduct of the experiment. To give one example from the field of agricultural experimentation, it might happen that when selecting a Latin square at random the experimenter obtains an arrangement in which the treatments run in regular slanting lines across the field. In this case, although the arrangement has been arrived at by random means, it would not be advisable to carry out such an experiment in which the effects of treatments are highly correlated with possible gradients in soil fertility. To put it more generally, the danger we are referring to is that *randomization may accidently bring about a high association between the treatment effects and the effects of one or more of the variables left uncontrolled.* This danger will be of little consequence if it is possible to repeat the experiment independently several times. If each replication involves a new randomization and our inferences are based on the repeated regularities, we should have no problem. Usually we are less fortunate. Often we have access only to a single experiment. If the inference from a given single experiment is to be taken seriously, we must be assured that tremendous systematic errors have not become confounded with the treatment comparisons. Sir Ronald Fisher, the father of the modern theory of experimental design and the originator of randomization in the design of experiments, was once asked: What would you do if, drawing a Latin square at random for an experiment, you happened to draw a Knut Vik square? Fisher replied that he would discard the selection and draw again and remarked further that ideally a theory explicitly excluding regular squares should be developed (Savage, 1962, p. 88). For work along these lines, see Jones (1958). *Analysts should, in general, avoid random arrangements which are clearly patterned.* As there is no theory to help the investigator mechanically identify such results, the investigator should be wary of blind reliance on randomization.

A major objective of randomization is to guarantee against personal interference by the experimenter or the subjects involved in the experiment. Randomization is expected to help make the data more reasonably convincing to other people. In large experiments in which all factors that look important (relevant) have been physically controlled, it seems reasonable to expect that randomization would almost always help produce a layout that does not arouse excessive suspicion in the minds of any impartial observer. In small experiments, however, this may not be the case. Let us illustrate. Suppose the response variable Y is known to be affected by a large number of factors all of which are dichotomous, e.g., white and nonwhite, Democratic and Republican. Suppose we have only 10 subjects available for the experiment. It is quite possible that any random splitting of these 10 subjects will result in a split along the lines of one or another of the dichotomous independent variables. *In such cases, randomization of the subjects into the treatment and control groups often will not satisfactorily "control" for these extraneous variables.*

In many textbooks one finds the statement that the objective of randomization is to neutralize the combined effect of all uncontrolled variables, i.e., to prevent such variables from confounding the comparison between the average responses of the treatment group and that of the control group. This objective can be more or less attained in a *single* experiment if it involves a very large n. In small experiments this objective cannot be attained if we depend on a *single* experiment. Whether the experiment is large or small, it is true that the expected value of the combined effects of the uncontrolled factors would be the same in the treatment group and the control group. However, the variations around that expected value may be extremely large and troublesome in small experiments. The result may be more than just troublesome where few replications can be anticipated.

1.3 PROBLEMS OF MEASUREMENT

Measurement errors abound in social research. Often theoretical conceptualization is so rudimentary that it is difficult to assess just what kinds of measurement difficulties are involved. Errors which may be considered random typically arise from non-systematic biases in reporting, coding, and processing or from the omission of unrelated and fairly minor causes from theoretical models. More systematic and more serious errors are created when the errors of reporting are related to the values of the variables being measured. This problem arises with classes of variables for which respondents having certain values typically distort more than respondents having other values. Further, there are types of multiplicative (and other interactive models) which produce such correlated errors. Serious measurement error is also introduced when conceptually interval variables, for which effects are specifiable functions, are measured by crude nominal or ordinal categories. Finally, and most seriously, measurement error is produced when any form of indirect measurement is introduced. Indirect measurement is required whenever the researcher has no direct access to the information defining his conceptual variables. Indirect measurement may be necessary for practical reasons, as in studies of whole societies by researchers with finite resources or in secondary analyses of data gathered for other purposes. It sometimes happens that the conceptual variables are explicitly or implicitly defined in ways which make direct measurement impossible, e.g., frustration, anomie, alienation, status, consensus. Finally, indirect measurement often causes serious problems when the researcher has devoted too little attention to the auxiliary theory linking the theoretical concepts to the observations. This occurs most commonly in surveys when indices are substituted for concepts, e.g., income for status, scale scores for attitudes. In experiments, the same problem arises when the researcher assumes that the manipulations may be directly substituted for the conceptual variables. Yet many of the parametric procedures discussed in statistics texts commonly presuppose that one has obtained interval or ratio scales with only negligible measurement errors!

What can one do under the circumstances? One extreme strategy is to move ahead without really challenging the adequacy of one's measures. For example, one may have five or six ordered categories to which arbitrary numerical scores are given, and one may then proceed as though a genuine interval scale has been achieved. At the other extreme, one may decide to reject parametric data analyses altogether and to fall back on procedures that require only ordinal or nominal data. Sometimes investigators may even throw away information by collapsing ordered categories into only two or three categories, not recognizing that in doing so they are merely adding in still more measurement error. The strategy we recommend lies somewhere in between these extremes and consists of a more flexible approach that recognizes the tentative nature of any scientific conclusions.

It is always possible to analyze one's data in *several* different ways and to take note of and to attempt to account for any apparent inconsistencies that are found. Although this text does not deal with ordinal procedures, we would strongly recommend that an investigator who has serious doubts about an interval-scale assumption should supplement the kinds of analyses discussed in this book with those which are more appropriate for ordinal scales. But the latter procedures, too, have certain disadvantages which make it desirable to supplement either type of analysis with the other. For example, transitivity is undefined for nominal variables and is not defined for multivariate analyses with ordinal variables. Consider three ordinal variables X, Y, and Z. Suppose that increases in X cause *increases* in Y while increases in Z cause *decreases* in Y. If a person is higher than another on both X and Z, we have no way of asserting whether he should be higher or lower than the other on Y. Certain of these specific difficulties will be discussed very briefly in Chap. 10. The essential point that we want to stress is that *in situations where one or more assumptions are in doubt, there is nothing to prevent an investigator from analyzing data in several different ways, each of which may have advantages and disadvantages that are mutually complementary.* This has the double advantage of providing additional insurance against errors that may be peculiar to only one of the techniques and also generating new insights whenever one procedure leads to substantially different conclusions from the other.

We can also recommend a much more positive approach to the problem of inadequate measurement. Certain of the techniques discussed in the present text seem very well suited to this purpose. Whenever measurement has been imperfect, we need to ask very specifically about the sources of the measurement errors and how they may be related to the other variables that appear in our theoretical system. In Part Four we shall demonstrate a way to represent strictly random measurement errors through the device of writing a very simple equation linking a measured indicator X' with a true value X. The latter will of course be an unknown, but if we are willing to assume certain things about the nature of this equation and the sources of the measurement error, and if we are able to collect a sufficient number of pieces of empirical information to compensate for the unknowns, we shall find it possible to estimate the true equation linking X to other variables. Furthermore, if we have several different measures of each variable, we may make a series of consistency checks to assess the plausibility of the assumptions about these measurement errors. Then we can go still further and begin to construct models involving different kinds of systematic biases or nonrandom measurement errors. We must pay a price whenever we admit such complications, but if we can obtain a sufficient number of indicators of each variable, we may again utilize consistency checks and derive estimates of the parameters relating the "true values" of each variable.

The value of this general strategy of constructing specific measurement-error models is that it forces us to be very self-conscious about just what assumptions are being made about our measures and their relationships to the theoretical variables of interest. Is "years of schooling" a good measure of "education" as conceptually defined in some manner? Most of us would probably agree that it is not. But having said this, what can we do about it? If we applied the first extreme strategy, we would take years of education as an interval scale and analyze our data by parametric procedures without concern. At the other extreme we might collapse education into three levels and analyze our data using cross tabulations. The first of the two strategies we recommend is to analyze one's data according to *both* these procedures and see whether the conclusions reached are similar. The second and more complete strategy would be to construct a measurement-error model linking "years of schooling" with "education" as conceptually defined, try to locate the variables that might be expected to complicate the relationship between the indicator and the construct, and then to analyze the data in accordance with the implications of this model. This latter strategy cannot be utilized, however, unless one takes advantage of the kinds of statistical tools that will be discussed in the various chapters of this text, and in particular the approaches treated in Part Four.

Problems of measurement illustrate very well the nature of the interplay between theory and research methodology. Some social scientists tend to see theory and methodology as separable and each as something quite apart from substantive research. To be sure, theory is supposed to provide the research hypotheses, and the empirical findings will either confirm or fail to confirm the theory. But in actuality the process is much more complex than this. Unfortunately, it is impossible to discuss theory construction and measurement adequately in a text on applied multivariate analysis and experimental designs. The general point that needs to be stressed, however, is that theoretical assumptions underlie everything we do in statistical analyses, even though it would be tedious to mention them at each point in the analysis (see for example, Carter, 1971; Blalock, Wells, and Carter, 1970). One fundamental assumption that will be made in Parts One to Three is that all independent variables have been perfectly measured; i.e., there is an exact correspondence between a variable as measured or as classified and the conceptual variable of interest. We have already commented on the dangers arising from the fact that most experimental designs assume a treatment to stand for the sum totality of the manipulations made by the experimenter and that no conceptual ambiguity is created by equating these manipulations with the treatment as conceptually defined. This is merely a special case of the more general assumption that whenever we relate some X to Y, we have a clear understanding of what each of these variables means conceptually. Unfortunately, there is nothing in the statistical models and equations that can enable one to evaluate this kind of assumption directly. It is in this sense that we must make theoretical assumptions when employing or discussing statistical techniques. We must strongly emphasize, however, that in any particular application of these techniques it is crucial that the investigator pay careful attention to these matters. This implies a flexible approach to data analysis whenever one or more of these assumptions is in doubt.

1.4 ASSUMPTIONS ABOUT DISTURBANCE TERMS

Throughout the book we shall find it necessary to make a number of assumptions about disturbance terms, designated e_i and E in Part One and ε_i and ε in the remainder. These disturbances represent our ignorance about what may be considered to be errors, unexplained variance, or the effects of causal factors that have not been explicitly identified. They may also include strictly random measurement errors in our dependent variables. As we shall see in Part Four, similar measurement errors in independent variables will create biases in our parameter estimates. Ideally, we might want to allow for such disturbance terms without having to make any assumptions about them. But this will be impossible if we wish to be able to say anything at all about our estimates of parameter values. Generally, we wish to be able to utilize the principles of mathematical statistics that permit one to derive estimators with certain optimal properties, such as those having a minimum variance about the true parameter values (most efficient estimates) or those having expected values exactly equal to the parameters (unbiased estimates) or, at least, those having expected values that approach the parameter values as the sample size increases (consistent estimates). Without making certain assumptions about the disturbance terms, however, we cannot derive any theorems that enable us to say anything definite about these properties. Therefore such assumptions play a crucial role in statistical inference, and we naturally need to know just how sensitive our tests and estimates are to distortions whenever these assumptions are violated in actual practice.

Part One is primarily concerned with laying out the basic features of the linear model and showing that all conventional least-squares tests can be represented as special cases of this model by the imposition of hypotheses as linear restrictions. The reader will find that more attention is paid to the assumptions about disturbance terms and to the implications of their violation in the remaining parts of the book. In Part Two the discussion of experimental designs treats implications of violation of normality and homoscedasticity assumptions and discusses transformations that may in part correct for such departures from these assumptions. Another crucial and commonly made assumption is that the disturbance term is uncorrelated with all the independent variables in a given equation. This assumption will occupy our attention in Part Three, where we shall see that whenever reciprocal causation is presumed to exist, the assumption is untenable. Finally, whenever our variables are subject to measurement errors that become absorbed into these disturbance terms, we must find ways to modify our analyses in the attempt to correct for the biases that are produced. This is the principal topic of Part Four.

In certain (but not all) experimental designs in which the levels of the independent variables are set by the investigator, it would appear that we need not be concerned about correlations between disturbance terms and independent variables. However, as our discussion in Sec. 1.2 indicates, it may be that one's actual manipulations in an experiment cannot be equated with the treatment variable, as conceptually defined. It is entirely possible that certain variables are inadvertently manipulated along with the conceptually defined independent variables. For example, consider a classic study of the effects of severity of initiation on expression of liking for a group (Aronson and Mills, 1959). *Conceptually*, Aronson and Mills sought to manipulate only "severity of initiation." Their *technique* for doing so required that subjects undergoing "severe initiation" read aloud embarrassing and sexually loaded material. It is quite likely that this manipulation had consequences in addition to those intended. Some of these are explored by Gerard and Mathewson (1966). *Our only point, here, is that these unknown variables may appear in the disturbance term which will then be partly confounded with or correlated with what the investigator believes to be the "treatment" variables that are being described in the research report.* Therefore, even in experimental designs one must be cautious about one's assumptions about disturbance terms.

In the nonexperimental situations that characterize a very large proportion of social research the situation is even more problematic. If certain unknown or unmeasured causes of a dependent variable are correlated with any of the independent variables in the equation, the disturbance term will also be correlated with these variables and least-squares procedures will yield biased estimates. Unfortunately, the extent of such biases will be unknown because of our lack of knowledge of the makeup of the disturbance term or the degree of its correlation with each of these independent variables. Nor can we rely on the empirical distribution of the (sample) residuals about our least-squares estimates of the true regression equations or parameter values. This is the case because ordinary least squares *automatically* produce residuals that are uncorrelated with the independent variables, regardless of the truth of the assumption that the true (population) disturbances are uncorrelated with these variables.

As we shall discuss in Chap. 10, the only remedy for this state of affairs is to decompose the disturbance term into components by a process of identifying additional causes of the dependent variable (previously neglected and therefore absorbed into the disturbance term) that may be correlated with the independent variables. This process of course introduces more variables into the equation and adds complexity to the analysis. Then, at some point, one must stop with a finite disturbance term and make assumptions about its properties. In addition to assumptions about normality and homoscedasticity, the crucial assumption will have to be that this reduced disturbance term is uncorrelated with each of the independent variables that appear explicitly in the equation. Anyone who objects to this assumption in a particular context may then proceed to identify remaining causes that implicitly appear in the new disturbance term. If these can be measured, they can be included explicitly in a more complex model and the process repeated. *Thus, any particular formulation may be in error in the sense that it involves incorrect assumptions about disturbances. However, the procedure is sufficiently flexible to permit increasingly refined and complex formulations once we are in a position to locate and measure specific variables that have previously been omitted from the analysis.*

The assumptions we shall need to make about disturbance terms constitute a somewhat less restrictive version of the "other things being equal" assertion that is conventionally tacked onto most theoretical generalizations. In effect, we are assuming that the "other things" do not disturb the patterning of relationships between the variables under study. We assume that their covariances with independent variables are all zero but we are unable to account theoretically for their variances. Our aim is then to reduce these "unexplained variances" as much as possible. This mode of thought rejects as unreasonable arguments to the effect that an explanation must be invalid if it admits of any exceptions or cannot account for all of the variance. The assumption is that if the number of cases or replications is sufficiently large (as compared with parameters to be estimated) there will always be an imperfect fit between the postulated model or explanatory system and the data. *The aim of research is to reduce the unexplained portion of the variance as much as possible while still adhering to a relatively simple explanatory system.*

This implies that reducing the unexplained variance cannot be the sole criterion by which a quantitative study is evaluated. If we increase the number of independent variables sufficiently, or if we allow for a great enough degree of complexity in the functional forms of our equations, we can always reduce this unexplained variance to zero. But we would then probably find that the addition of further cases or replications would again give rise to considerable unexplained variance. Furthermore, certain of the "independent" variables that we may insert into our equations may not be conceptually distinct from the dependent variable under investigation. In this case, reduction of the disturbance term to zero or near zero still fails to provide a causal explanation in spite of the high proportion of "explained variance." For example, one of the "independent" variables may be a past value of the dependent variable, as when one uses grades in high school to predict grades in college, or perhaps population growth between 1960 and 1970 to predict the growth during the following decade. Such instances of autocorrelation do not constitute explanations. *These remarks imply that one must first select a set of independent variables on the basis of sound theoretical reasoning and careful conceptualization.* Only then does it make theoretical sense to invoke the goal of reducing the unexplained variance through the process of adding complexities to a model and comparing the relative explanatory power of alternative models.

1.5 PROBLEMS OF RECIPROCAL CAUSATION

It may be objected that in many kinds of research the investigator is not in a position to designate certain variables as dependent and others as independent. In mathematics, one may simply define the dependent variable to be whichever variable is placed on the left-hand side of an equation, which of course means that one can turn the equation around and solve for any one of the remaining variables. Consequently, the very use of the terms dependent and independent variable is passing from mathematics. However, in a causal sense, X may affect Y and also Y may affect X. Thus not only do motives and goals affect our behaviors, but we commonly recognize that these same behaviors may begin to modify the motives, goals, or other internal states. To be sure, there may be a time lag involved, but our research may not have enabled us to pinpoint the temporal sequences with any degree of precision. On the macro level, industrialization, urbanization, literacy, and political behaviors may all mutually influence one another, so that it makes little sense to select one as dependent on the others.

In situations like these, it might seem reasonable to select one of the variables as dependent upon the others and then to analyze one's data in a straightforward manner. When finished, one might then be tempted to select a second variable as dependent and repeat the analysis until each variable in the set had been explained by all of the remainder. Such a procedure is not legitimate, however, as we shall see in Chap. 11. From the standpoint of the assumptions required by least-squares analysis, this kind of reciprocal causation results in incorrect assumptions about the disturbance terms, which in turn create biases in the estimates of the effects of each variable on the others. Furthermore, there is an even more fundamental analytic problem created by reciprocal causation. In some situations there will be too many unknowns for solution. This implies that it will be impossible to estimate the parameters by *any* means, even where there is no sampling error or measurement error in any of the variables. Such situations require reformulation of one's theory, as will be shown in Chap. 11. Once the theory has been reformulated, we can apply a modified version of least squares to obtain consistent estimates that have negligible biases for large samples.

Many of the questions that are raised concerning the adequacy of any type of quantitative approach to data analysis involve a basic skepticism about whether it is possible to cope analytically with a reality that is continually changing and with individuals each of whom is unique in many distinct ways. How is it possible to break this real world apart and analyze a set of components as though they were separable and operated in relatively closed systems? Is it not true that the whole is greater than (or at least different from) the sum of its parts? These kinds of issues have divided students of human affairs for centuries, and we cannot hope to resolve them here. Nevertheless, a few comments are in order.

Such skepticism reflects an awareness of the complexity of the real world that can be answered only by an effort to incorporate as much of this complexity as possible into our models and data analyses. Knowing that we cannot possibly identify and measure all the variables or factors that may affect a given variable, we try to accommodate some of this complexity by explicitly introducing a disturbance term to allow for our ignorance of the many minor factors that may be at work in any given context. We also attempt to build in complexity by modifying the forms of our equations, e.g., when we introduce nonadditive terms or allow for nonlinearities. In Part One we attempt to free the analyst from ritual reliance upon a set of conventional computing formulas which address conventionalized (and often simplistic) questions. There we show how a generalized model can be used to address realistically complex questions for which no traditional formulas exist. While Part One relies on ordinary least squares, the strategies can be generalized to other procedures. Theoretical formulations which include notions of reciprocal causation force us to allow for the mutual influence of a number of endogenous (or interdependent) variables on each other by studying a number of equations *simultaneously* rather than one at a time. In effect, we introduce another type of complexity when we admit the possibility that *equations* cannot be added together after having been studied in isolation. This need not impel a retreat into mysticism or defeatism. It does imply that certain kinds of questions cannot be answered unless our theories can be specified with considerable care. It is premature in this introductory chapter to discuss such issues further until the single-equation approaches have been studied in Parts One and Two; however, we can say a bit more about the kinds of statistical models that we shall be using to approximate the changing, complex, and interdependent reality that is of interest to the social scientist.

A very simple example of a fixed population is a deck of cards, which has a finite size and essentially three variables: face value, suit, and color. The deck has been constructed in such a manner that face value is unrelated to either suit or color, though the latter two variables are related in the sense that knowing the suit enables one to predict color exactly (though, of course, the reverse is not the case). We may then approximate random sampling, with or without replacement, by thoroughly shuffling the cards and drawing samples of a given size. We may take it as our task to estimate the population parameters on the basis of sample characteristics. We do not imagine that the population is changing while this is going on, nor do we attach any causal significance to the relationship between color and suit. We do not imagine, for example, that a change in color will produce a change in suit, or vice versa.

When we turn our attention to real populations, we need to allow for changes, however, and our problem of prediction becomes much more complex. Not only do we want to estimate population parameters, e.g., the proportion of clubs, from sample statistics, but we may also want to predict what would happen to the other variables if all the cards were to turn into spades or aces. In short, we want to infer something about the *causal processes* that have produced what we previously took to be a fixed population. It is obvious that such processes cannot be inferred from fixed characteristics of any number of samples. In the analysis of social relationships, we might further want to know how such structural changes would affect the character of the games which could be played and the interrelationships of players.

At this point, theoretical *assumptions* must be added to the picture. One important class of such assumptions are those which assert that certain variables or factors do *not* affect the others. Thus if we could assume that the suit of a card does not affect its face value, we could infer that if all cards were to change to spades, this would not affect the distribution of face values. But it *might* affect the distribution of red and black cards unless we were to rule out this possibility as well. Notice that we may allow for causal asymmetry by assuming that a change in suit may change the color but that a change in color cannot change the suit. Of course if we are able to observe the temporal sequences, we may be able to rule out certain possibilities by imposing the assumption that effects cannot precede causes. Note, however, that even if we were to observe the same temporal order in the sequences—say suit changes followed by color changes—this could not establish a causal connection because of the possibility of what we term a *spurious association*. That is, some unknown factor might be producing a change in both suit and color in accord with this particular temporal sequence.

Thus when we wish to move from statistical estimates of population parameters to substantive interpretations concerning the processes that have generated population characteristics, we must supplement our statistical procedures with a priori assumptions, only some of which can be subjected to direct or indirect test. We imagine populations which are only temporarily fixed and which are subjected to exogenous events that may shift the levels of one or more of the variables or even change the values of the parameters in our equations. Thus, if we write the equation

$$Y = \alpha + \beta X + \varepsilon$$

we recognize that not only are the values of X and Y subject to change, but this will even be true of the parameters α and β that have been taken as constants for purposes of simplification. In a more refined formulation we might have taken the slope coefficient β as a variable that is influenced by one or more other variables in our theoretical system, including even X and Y themselves. But for purposes of arriving at an approximation we have assumed it to be a constant for the particular population under study. When we compare across populations, one of our principal aims is to study the behaviors of these "constants" to see whether or not they have the same numerical values in a diversity of populations. If they do not, we must then turn our attention to explaining these differences. Presumably, we infer that these parameters are subject to change according to ascertainable principles which, when understood, will enable us to modify the original formulation by making it more complex.

The import of these rather general remarks is that the kinds of simplifications that we make in statistical analyses are not intended to imply rigidity or stability in the real world. In effect, we imagine that populations remain fixed for long enough periods to enable us to collect data and infer their characteristics at a single point in time. If we were to stop with a single population at one point in time, this exercise would be theoretically trivial, though it might still serve some practical purpose such as the short-run estimation of characteristics of interest to policy makers. Whenever we are interested in the processes or causal laws that have presumably generated the population parameters of interest, our focus of attention must be on *comparisons* of populations or across time. But even these comparisons are not ends in themselves unless they can shed light on the underlying processes at work. For example, it is of little value to state that the median income in Chicago is $13,000 whereas that in San Francisco is $11,000 unless this finding can help us infer the mechanisms producing such differences. Similarly, it is of little theoretical interest to note that the median income in Chicago has increased by $2,000 during a given decade unless this information can be used to provide information concerning the causes of this shift.

All of this does not imply that empirical information about populations is not needed in order to test hypotheses concerning the underlying dynamics. If our theories are carefully enough specified, we can make a number of predictions, only some of which will be compatible with the factual state of affairs in the real world.

In other words, we may postulate a number of theoretical models that may be consistent with the facts that pertain to a particular population. But many others will imply predictions that will *not* hold up, and we may therefore utilize our population data to reject or modify those explanations which do not fit the facts. In Part Three we shall see how this can be done in a way permitting formulation, testing, and modification of a series of models which admit as much complexity as any verbal theories intended to account for real-world processes involving change and reciprocal causation.

PART ONE

Multiple Regression
and Related Techniques

2

DATA REPRESENTATION AND TRANSFORMATION

Multiple regression is the most versatile and powerful of the quantitative inferential frameworks. Unfortunately, it is often presented as a special-purpose technique of rather limited application. A variety of techniques which can be derived from the same framework are developed as if they were isolated tests without common foundation.[1] As we shall see in this section, multiple regression includes such diverse topics as tests of differences between means and proportions, simple and complex analysis of variance, least-squares correlation, multiple correlation, partial correlation, multiple partial correlation, path analysis, dummy-variable analysis, and analysis of covariance. Additionally the inferential foundations of the technique are similar to those employed in the development of chi square, discriminant-function analysis, latent-structure analysis, and probit analysis.

Inferential tests associated with all these techniques can be reduced to a single sampling distribution, the F ratio. Definition of this F ratio will be shown to involve the definition of two symbolic least-squares models: one which involves the algebraic imposition of the hypotheses to be tested and another which does not include these hypotheses. *The F ratio will always be defined as the proportionate error increase*

---

[1] Earlier, though partial, recognition of this criticism may be found in Cohen (1968), Gujarati (1970a,b), Ward (1969), Mendenhall (1968), Jennings (1967), Li (1964), Bottenberg and Ward (1963), Ward (1962), Scheffé (1959), Anderson and Bancroft (1952), and Mood (1950). An excellent discussion of the repeated rediscovery of the commonalities among apparently diverse least-squares techniques may be found in Rockwell (1974).

attributable to the imposition of our hypotheses. It will develop that all conventional tests associated with these techniques conform to this definition. Of more interest is that just *what* is being tested becomes much more obvious within this framework. This is especially true for the tests associated with the various multivariate techniques. Finally, the linear-model approach permits straightforward technically correct formulations of many substantively interesting nonstandard tests. These will be shown to include many otherwise complicated interaction questions, some powerful replication questions, and questions which do not conform to idealized models because of necessary research compromises.

2.1 INTRODUCTION TO VECTOR ALGEBRA

The few notions from vector algebra necessary to understand these applications of least-squares theory will be summarized briefly below.[1] This elementary treatment, while not exhaustive of vector algebra, will suffice for our purposes. Hand computation of results would be inefficient and undesirable, if not impossible, save for the most superficial examples. Consequently, each topic will be treated in terms of conventional vector notation as well as Fortran notation, also summarized below. It will develop that all computations save the calculation of the appropriate F ratio can be performed with any standard multiple-regression program, one of which is presented in Appendix A.

2.2 DEFINITION AND CHARACTERISTICS OF VECTORS

In Chaps. 2 to 4, we shall conventionally represent a vector by a boldface capital letter such as **A** or **X**. This letter represents an ordered number set or an ordered set of elements. The elements will have subscripts to indicate that they are the first, second, third, ..., nth element of the vector. Thus, the following vectors are defined as indicated:

$$\mathbf{A} = \begin{bmatrix} a_1 \\ a_2 \\ \vdots \\ a_n \end{bmatrix} \qquad \mathbf{X} = \begin{bmatrix} x_1 \\ x_2 \\ \vdots \\ x_n \end{bmatrix}$$

The elements may be defined symbolically, or they may have numeric values. Thus if $a_1 = 5$, $a_2 = 3$, and $a_n = 1$:

$$\mathbf{A} = \begin{bmatrix} 5 \\ 3 \\ \vdots \\ 1 \end{bmatrix}$$

[1] The reader's attention is called to the discussion of notation in general and shifts in notation in different parts of the text to be found in the Preface.

Notice that the elements are *ordered* in the sense that the value for a_1 is first, for a_2 is second, etc. They are *not* necessarily ordered in terms of the magnitude of the values.

Type

Vectors may be of two types, row or column. If **B** is a column vector and **C'** is a row vector, they are written

$$\mathbf{B} = \begin{bmatrix} b_1 \\ b_2 \\ \vdots \\ b_n \end{bmatrix} \qquad \mathbf{C'} = \begin{bmatrix} c_1 & c_2 & \cdots & c_n \end{bmatrix}$$

By convention column vectors will be written with boldface capital letters, for example, **A**, **B**, **X**, and row vectors with primed boldface capital letters, for example, **A'**, **B'**, **X'**.

Transpose

The transpose of a vector **A** is designated as **A'** (read A transpose). In the general case, **A'** is a row vector if **A** is a column vector and **A'** is a column vector if **A** is a row vector. The elements of **A'** are each equivalent to the corresponding elements of **A**. That is, $a_i' = a_i$. Thus,

$$\mathbf{A'} = \begin{bmatrix} 5 & 4 & 2 & 6 \end{bmatrix} \qquad \text{if} \qquad \mathbf{A} = \begin{bmatrix} 5 \\ 4 \\ 2 \\ 6 \end{bmatrix}$$

Although it is a violation of general vector notation, we shall conventionally refer to column vectors without the transpose sign and to row vectors with the transpose sign. It should be obvious that the transpose of the transpose of a vector is the original vector; i.e.,

$$(\mathbf{A'})' = \mathbf{A}$$

Dimension

The number of elements in a vector is called its *dimension*. Thus, in the following examples, **A'** is a row vector of dimension 4, **B** is a column vector of dimension 3, **C'** is a row vector of dimension n.

$$\mathbf{A'} = \begin{bmatrix} 1 & 3 & 6 & 2 \end{bmatrix} \qquad \mathbf{B} = \begin{bmatrix} b_1 \\ b_2 \\ b_3 \end{bmatrix} \qquad \mathbf{C'} = \begin{bmatrix} c_1 & c_2 & \cdots & c_n \end{bmatrix}$$

Special Vectors

We shall commonly employ five special vectors in our explanations: null vectors, unit vectors, orthogonal binary vectors, orthogonal score vectors, and scalars (vectors of dimension 1). All these vectors have useful properties which will assist in defining some rather complex notations. They are merely defined here.

A *null vector* is a vector each of whose elements is zero. It will usually be designated N or N'. Thus,

$$N = \begin{bmatrix} 0 \\ 0 \\ \vdots \\ 0 \end{bmatrix} \qquad N' = \begin{bmatrix} 0 & 0 & \cdots & 0 \end{bmatrix}$$

A *unit vector* is a vector each of whose elements is unity. It will usually be designated U or U'. Thus,

$$U = \begin{bmatrix} 1 \\ 1 \\ \vdots \\ 1 \end{bmatrix} \qquad U' = \begin{bmatrix} 1 & 1 & \cdots & 1 \end{bmatrix}$$

Orthogonal binary vectors are any set of vectors satisfying three conditions, (1) each vector contains only ones or zeros; (2) each vector must contain at least one zero element; (3) the ith elements may have the value 1 in one and only one vector of the set. Thus, the pair of vectors A and B are orthogonal binary vectors, where

$$A = \begin{bmatrix} 1 \\ 1 \\ 1 \\ 0 \\ 0 \\ 0 \end{bmatrix} \qquad B = \begin{bmatrix} 0 \\ 0 \\ 0 \\ 1 \\ 1 \\ 1 \end{bmatrix}$$

Also, the triplet of vectors C, D, and E are orthogonal binary vectors, where

$$C = \begin{bmatrix} 1 \\ 1 \\ 0 \\ 0 \\ 0 \\ 0 \end{bmatrix} \qquad D = \begin{bmatrix} 0 \\ 0 \\ 1 \\ 1 \\ 0 \\ 0 \end{bmatrix} \qquad E = \begin{bmatrix} 0 \\ 0 \\ 0 \\ 0 \\ 1 \\ 1 \end{bmatrix}$$

Orthogonal score vectors are any set of vectors satisfying two conditions: (1) each vector must contain at least one zero element, and (2) the ith element must have a nonzero value in one and only one vector of the set. Thus, a pair of vectors G and H are orthogonal score vectors where

$$G = \begin{bmatrix} 16 \\ 17 \\ 23 \\ 0 \\ 0 \\ 0 \end{bmatrix} \qquad H = \begin{bmatrix} 0 \\ 0 \\ 0 \\ 14 \\ 17 \\ 18 \end{bmatrix}$$

The general notion of orthogonality is discussed later. The conditions defining orthogonal binary and score vectors satisfy the generic definition, though they constitute a special limited case of that definition.

Scalars are vectors of dimension 1. Otherwise stated, a scalar is a vector with only one element. Thus, if $\mathbf{A}' = a_i' = a$, $\mathbf{A}'$ is a scalar. Conventionally scalars are designated by lowercase unsubscripted letters.

2.3 VECTOR RELATIONS AND OPERATIONS

The equivalence relationship, vector addition, scalar products, and vector products are necessary notions for our presentation. Additionally, they will serve to define three very important notions: linear independence, length, and orthogonality.

Any vector $\mathbf{A}$ is said to be *equivalent* to any vector $\mathbf{B}$ if and only if $\mathbf{A}$ is of the same type (row or column) as $\mathbf{B}$ and every element of $\mathbf{A}$ has the same value as the corresponding element of $\mathbf{B}$. Thus, $\mathbf{A} = \mathbf{B}$ if and only if $a_i = b_i$ for all i's. Implicit, of course, is that $\mathbf{A}$ and $\mathbf{B}$ are of the same dimension. Thus for the following example vectors, only $\mathbf{A}$ and $\mathbf{B}$ are equivalent. However, $\mathbf{A}$ (or $\mathbf{B}$) is equivalent to the unwritten vector $\mathbf{E}$, which is $(\mathbf{E}')'$.

$$\mathbf{A} = \begin{bmatrix} 5 \\ 4 \\ 6 \\ 1 \end{bmatrix} \quad \mathbf{B} = \begin{bmatrix} 5 \\ 4 \\ 6 \\ 1 \end{bmatrix} \quad \mathbf{C} = \begin{bmatrix} 5 \\ 5 \\ 6 \\ 1 \end{bmatrix} \quad \mathbf{D} = \begin{bmatrix} 5 \\ 4 \\ 6 \\ 1 \\ 2 \end{bmatrix} \quad \mathbf{E}' = \begin{bmatrix} 5 & 4 & 6 & 1 \end{bmatrix}$$

The *sum* of a pair (or set) of vectors is a vector each element of which is the sum of the corresponding elements in the pair (or set). Thus, $\mathbf{A} + \mathbf{B} = \mathbf{C}$ may be defined, symbolically, as

$$\begin{bmatrix} a_1 \\ a_2 \\ \vdots \\ a_n \end{bmatrix} + \begin{bmatrix} b_1 \\ b_2 \\ \vdots \\ b_n \end{bmatrix} = \begin{bmatrix} a_1 + b_1 \\ a_2 + b_2 \\ \vdots \\ a_n + b_n \end{bmatrix} = \begin{bmatrix} c_1 \\ c_2 \\ \vdots \\ c_n \end{bmatrix}$$

A single numerical example should suffice:

$$\mathbf{A} = \begin{bmatrix} 1 \\ 5 \\ 9 \end{bmatrix} \quad \mathbf{B} = \begin{bmatrix} 6 \\ 2 \\ -2 \end{bmatrix} \quad \mathbf{C} = \mathbf{A} + \mathbf{B} = \begin{bmatrix} 7 \\ 7 \\ 7 \end{bmatrix}$$

Obviously, the vectors to be summed must be of the same dimension and type. Thus, under our conventions, $\mathbf{A} + \mathbf{B}'$ is not a defined operation.

A *scalar product* is a vector $\mathbf{A}$ which results from the multiplication of any vector $\mathbf{B}$ times any scalar d. Each of the elements of the product vector $\mathbf{A}$ is the product of d and the corresponding element of $\mathbf{B}$. Symbolically, $d\mathbf{B} = \mathbf{A}$ can be defined as follows:

$$d \begin{bmatrix} b_1 \\ b_2 \\ \vdots \\ b_n \end{bmatrix} = \begin{bmatrix} db_1 \\ db_2 \\ \vdots \\ db_n \end{bmatrix} = \begin{bmatrix} a_1 \\ a_2 \\ \vdots \\ a_n \end{bmatrix}$$

Where $d = 3$ and $C' = [1 \quad 5 \quad 2 \quad 6]$, the operation $E' = dC'$ produces

$$E' = [3 \quad 15 \quad 6 \quad 18]$$

A vector of either type and any dimension can be multiplied by any scalar. The resulting product vector will be of the same type and dimension as the initial vector.

A set of vectors is said to be *linearly independent* if and only if no subset of them can be weighted (by scalars) and summed to produce a vector equivalent to one of the other vectors. Thus, among the following vectors, several linear dependencies exist.

$$A = \begin{bmatrix} 1 \\ 2 \\ 3 \end{bmatrix} \quad B = \begin{bmatrix} 2 \\ 4 \\ 6 \end{bmatrix} \quad C = \begin{bmatrix} 1 \\ 0 \\ 1 \end{bmatrix} \quad D = \begin{bmatrix} 0 \\ 2 \\ 2 \end{bmatrix} \quad E = \begin{bmatrix} 0 \\ 4 \\ 4 \end{bmatrix}$$

including the following:

1 A and B are linearly dependent, as $A = .5B$.

2 D and E are linearly dependent, as $D = .5E$.

3 A, C, and D are linearly dependent, as $A = C + D$.

4 A, C and E are linearly dependent, as $A = C + .5E$.

However, **A**, **D**, and **E** are *not* linearly dependent, as no weighted combination of two of them will result in a vector equivalent to the third. For a formal solution to questions of linear independence, see Noble (1969), pp. 97–104. The average social scientist need not master these technical solutions, however, but should always avoid *definitional* dependence between independent variables. Where linear dependencies exist in models, program CARTROCK will notify the user that the model cannot be solved.

We shall define a *vector product* as the product of a row vector (premultiplier) times a column vector (postmultiplier). This operation produces a scalar. It is performed by taking the sum of the products of the corresponding elements of the two vectors. Of course, the vectors must be of the same dimension for these operations to be possible. Symbolically,

$$c = A'E$$

$$= [a_1 \quad a_2 \quad \cdots \quad a_n] \begin{bmatrix} e_1 \\ e_2 \\ \vdots \\ e_n \end{bmatrix}$$

$$= a_1 e_1 + a_2 e_2 + \cdots + a_n e_n = \sum_{i=1}^{n} a_i e_i$$

It should be obvious that several familiar operations can be represented as the result of vector multiplication. Let vector **A** represent the scores for a sample on the variable age

$$A = \begin{bmatrix} 14 \\ 15 \\ 12 \\ 16 \\ 13 \end{bmatrix}$$

If n is the dimension of A, the mean age can be represented as $(1/n)U'A$, which is

$$\tfrac{1}{5}\begin{bmatrix} 1 & 1 & 1 & 1 & 1 \end{bmatrix}\begin{bmatrix} 14 \\ 15 \\ 12 \\ 16 \\ 13 \end{bmatrix} = \tfrac{1}{5}(1 \times 14 + 1 \times 15 + 1 \times 12 + 1 \times 16 + 1 \times 13)$$

$$= \tfrac{1}{5}(70) = 14$$

The *length* of a vector A is defined as $\sqrt{A'A}$. Of course, in our example, $A'A$ is the same as the sum of squares of the variable age. In general,

$$A'A = \begin{bmatrix} a_1 & a_2 & \cdots & a_n \end{bmatrix}\begin{bmatrix} a_1 \\ a_2 \\ \vdots \\ a_n \end{bmatrix}$$

$$= a_1{}^2 + a_2{}^2 + \cdots + a_n{}^2 = \sum_{i=1}^{n} a_i{}^2$$

While we shall use this concept of length directly, it is worthy of note that the variance for a variable with mean $= 0$ can be defined, in vector terms, as $(1/n)A'A$.

The reader might profitably persuade himself that where A and B are vectors of standard scores, i.e., with mean $= 0$ and standard deviation $= 1$, and n is their dimension, the correlation between variables A and B is

$$r = \frac{1}{n}A'B$$

Two vectors X and Y are *orthogonal* if and only if $X'Y = 0$. (This is sometimes written $X \perp Y$.) It should be clear that the special orthogonal binary and score vectors referred to above satisfy this condition. The following two illustrations should suffice on this point:

$$A = \begin{bmatrix} 1 \\ 1 \\ 0 \\ 0 \end{bmatrix} \qquad B = \begin{bmatrix} 0 \\ 0 \\ 1 \\ 1 \end{bmatrix} \qquad A'B = 1 \times 0 + 1 \times 0 + 0 \times 1 + 0 \times 1 = 0$$

$$C = \begin{bmatrix} 15 \\ 16 \\ 0 \\ 0 \end{bmatrix} \qquad D = \begin{bmatrix} 0 \\ 0 \\ 12 \\ 17 \end{bmatrix} \qquad C'D = 0$$

The notion of orthogonality must *not* be confused with correlation, though both are sometimes loosely defined in terms of independence. In the example above, $A'B = 0$, and therefore A and B are orthogonal. However, A and B have a perfect negative correlation in the Pearson product-moment sense.

While it is obvious that what we have called special orthogonal binary and score vectors satisfy the orthogonality criterion that their vector product equals zero, many vectors are orthogonal without conforming to the rather rigid definitions of these vectors. Consider the vectors

$$A = \begin{bmatrix} -3 \\ 2 \\ -2 \\ 1 \\ -1 \\ 3 \end{bmatrix} \quad B = \begin{bmatrix} 6 \\ 8 \\ 8 \\ 1 \\ 1 \\ 6 \end{bmatrix} \quad C = \begin{bmatrix} 1 \\ 2 \\ 3 \\ 4 \\ 5 \end{bmatrix} \quad D = \begin{bmatrix} -5 \\ -4 \\ 0 \\ 2 \\ 1 \end{bmatrix}$$

$$E = \begin{bmatrix} 0 \\ -11 \\ 3 \\ 2 \\ 1 \end{bmatrix} \quad N = \begin{bmatrix} 0 \\ 0 \\ 0 \\ 0 \\ 0 \\ 0 \end{bmatrix} \quad U = \begin{bmatrix} 1 \\ 1 \\ 1 \\ 1 \\ 1 \\ 1 \end{bmatrix}$$

Notice that $A'B = 0$, $C'D = 0$, $C'E = 0$, $A'U = 0$, $A'N = 0$, $B'N = 0$. When the question of orthogonality is important, then, the appropriate check by vector multiplication should always be performed. If X and Y are vectors and $X'Y = 0$, it will always be true that $Y'X = 0$.

Utilization of Vectors

Variables for a sample are commonly represented as a column vector. Thus, for a sample of couples, X might be the man's age, Y the woman's age, and Z the employment status of the wife (coded 1 if she is employed and coded 0 otherwise).

$$X = \begin{bmatrix} 25 \\ 42 \\ 50 \\ 22 \\ 17 \end{bmatrix} \quad Y = \begin{bmatrix} 22 \\ 40 \\ 47 \\ 21 \\ 17 \end{bmatrix} \quad Z = \begin{bmatrix} 0 \\ 0 \\ 1 \\ 1 \\ 1 \end{bmatrix}$$

The notion that these are *ordered sets* is important. It means that the first, second, third, . . . , nth element in X can be meaningfully paired with the corresponding element in Y and Z. This means that *the* man whose age is 50, x_3, is married to *the* woman whose age is 47, y_3, and *that* woman is employed, as indicated by the code 1 for element z_3. Consequently, we can define for these data, five row vectors A', B', C', D', and E', the elements of which are the measured characteristics for each of the couples. For the first couple, the vector would be

$$\mathbf{A}' = [x_1 \quad y_1 \quad z_1] = [25 \quad 22 \quad 0]$$

For the last couple, it would be

$$\mathbf{E}' = [x_5 \quad y_5 \quad z_5] = [17 \quad 17 \quad 1]$$

For each of the variables, we can define some conventional and elementary operations in terms of vector operations. The examples will be for the mean and the percent. The mean age for the males is $\frac{1}{5}\mathbf{U}'\mathbf{X}$

$$\frac{1}{5}[1 \quad 1 \quad 1 \quad 1 \quad 1] \begin{bmatrix} 25 \\ 42 \\ 50 \\ 22 \\ 17 \end{bmatrix} = \frac{1}{5}(1 \times 25 + 1 \times 42 + 1 \times 50 + 1 \times 22 + 1 \times 17)$$

$$= \frac{1}{5}(156) = 31.2$$

The percentage of the females who are employed is $\frac{1}{5}\mathbf{U}'\mathbf{Z}$

$$\frac{1}{5}[1 \quad 1 \quad 1 \quad 1 \quad 1] \begin{bmatrix} 0 \\ 0 \\ 1 \\ 1 \\ 1 \end{bmatrix} = \frac{1}{5}(1 \times 0 + 1 \times 0 + 1 \times 1 + 1 \times 1 + 1 \times 1)$$

$$= \frac{1}{5}(3) = .60$$

In general, a column vector will represent a single variable for all units in a sample, and a row vector will represent all data (variables) for a single unit. Analytically, we shall develop symbolic models allowing us to test whether, how, and to what degree some variables may be said to influence others. Ordinarily, computation required for such tests will be provided by computer. We shall have to transmit row vectors (all data for each case) of the proper form to a computer program. At least, we shall have to tell a programmer how those vectors are to be defined. All the ideas that we have discussed for vector notation can be translated into Fortran notation. Some familiarity with computer conventions is thus desirable.

2.4 INTRODUCTION TO COMPUTER NOTATION AND UTILIZATION

At their least technical level, computer programs perform a set of operations on the current contents of a collection of memory locations and report the results of those operations. For example, a program named AMEAN might function by taking the mean for the contents of locations $a_1, a_2, a_3, \ldots, a_n$ and reporting that mean. All that an applied computer user would have to know to use such a program is how to start program AMEAN and how to see to it that a vector $\mathbf{A}$ of dimension n contains the distribution for which the mean is desired. One of the great advantages of using already prepared programs is that no more knowledge is required for a program which calculates many statistics for the distribution than for one which calculates only the mean.

A program is provided in Appendix A which performs all necessary operations for testing any of the statistical models presented here. As each analytic model is developed, instruction for use of this program will be provided. The most casual user of computers should be able to use this program with the aid of these instructions. However, it may be useful here to give a brief summary of program organization, Fortran notation, and some basic computer procedures for data manipulation and transformation.

Program Organization

A computer program usually consists of two kinds of information, a program segment and a control/data segment. The program segment contains sequenced instructions which bring data into specific locations, computes analytic results, and prints those results. The program segment for all analytic techniques treated in Chaps. 2 to 4 will consist principally of the program found in Appendix A. Typically, a program segment will be in the form of a sequenced deck of IBM cards, with a one-line instruction on each card.[1] The control/data segment will usually contain sequenced cards of two types: (1) control codes tell the program how many variables are to be used, where they are located on cards, how many cases are involved, and which variables are to be used in each of any set of analyses; (2) the raw data themselves will be contained in sequenced cards in the control/data segment. The data may also be stored on other media, but it is most convenient for the beginner to use data on cards which he can see and feel. Much mystery is thus avoided. In practice, we shall usually use the same program with different control/data segments. We shall think of both as being stored on cards and as physically adjacent, as shown below. However, either segment can be stored on any medium, and they may be continents apart physically. If telephone lines connected three computers, we could use the job-control language to seize the electrical representation of a program segment from California and the electrical representation of a data segment from Florida and put the two together in a computer in Minnesota, printing the results on a printer in New York.

Each of the two segments will be delimited by (preceded and followed by) job-control statements, called *JCL cards*. These statements enable the computer to identify and utilize the two segments properly. Since the job-control requirements vary slightly from installation to installation, the job-control statements given here (Table 2.1) are merely illustrative. Consult your local computation center or your instructor concerning job-control cards. Table 2.1 and its explanation will help to visualize program organization.[2] Technical details of how computer centers process jobs are ignored because they vary greatly from center to center and depend largely on the hardware available. This schema is essentially accurate from the naïve user's point of view.

---

[1] Technically, the program may be in the form of logical records stored on magnetic tapes, magnetic disks, or other media. Additionally, these segments *may* be longer than an IBM card. It will be convenient, however, to equate such records with a one-line instruction per card.

[2] *Note:* In actual computer printouts, all symbols (including letters and numbers) occupy equal spaces; vertical alignment is exact. However, we have not followed that system in this text. Underlining in the table is for emphasis only.

A set of cards sequenced as in Table 2.1 constitutes a program and a data set to be operated on by that program. All job-control cards contain statements beginning in column 1 and terminating at or before column 80. All statements within the program segment begin in column 7 and may extend through column 72. For these statements, columns 1 to 6 are reserved for special functions discussed later. Cards in the control/data segment may include punches in any position depending on the format instructions used for bringing data into the computer.

When such a package (program and control/data segment) is submitted to a computer center, the following occurs:

1 The first instruction card, JCL 1, provides a program name for internal referencing, an account number for billing, a programmer name for billing, and assorted control parameters such as time limit, page limit, etc. The underlined portions *must* be varied from user to user. Each computation center has different control options.

2 The second instruction card provides for a particular compiler (translating program), which translates the program segment into machine-usable codes. This card also prepares the machine to transfer control to the readied program after it has been translated.

3 The cards designated JCL 3 and JCL 4 delimit the physical boundaries of the program segment; that is, JCL 3 alerts the computer that what follows is to be considered a program and sent to a compiler for translation, and JCL 4 signals the computer that the last program statement has been received. All of this segment is transmitted at a single time for translation into a language the computer can use.

4 The cards designated JCL 5 and JCL 6 delimit the physical boundaries of the control/data segment. This segment is set aside until after the program is

Table 2.1 PROGRAM ORGANIZATION

Illustrative card content	Type of card
//PROGI JOB (199029,1), CARTER, MSGLEVEL=1	JCL 1
// EXEC FORTGCLG	JCL 2
//FORT.SYSIN DD *	JCL 3
DIMENSION X(100), Y(100)	Program 1
READ (5,10) N	Program 2
10 FORMAT (1I5)	Program 3
. . . .	
END	Program *n*
/*	JCL 4
//GO.SYSIN DD *	JCL 5
50	Control 1
102463	Data 1
104572	Data 2
114521	Data 3
. . . .	
114261	Data *m*
/*	JCL 6

translated and then made available to the program as it calls for needed control cards or portions of the data.

5   Control is passed to the program; the statements of the program segment are then executed in the order indicated. The program statements cause the computer to read and process the data in a predetermined fashion and to print the results.

For the program in Appendix A, the user will typically need to know how to set up the control/data segment for particular analyses. Some data sets may require knowledge of how to modify format statements in the program segment. These statements specify where (in which columns, etc.) certain data are to be found. In many cases, the user will also have to transform data vectors internally to make them conform to theoretical models of interest. We shall find in Chaps. 3 and 4, for example, that analysis of variance is conveniently defined in terms of operations on a set of orthogonal binary vectors and analysis of covariance in terms of operations on a set of orthogonal score vectors.

2.5   COMPONENTS OF THE FORTRAN LANGUAGE

To facilitate such solutions, a summary of Fortran notation and terminology follows. Much of the elegance and many of the options of the Fortran language are omitted. Only that subset of Fortran which is almost universally accepted by different machines is presented. Where several programming approaches exist for a problem, only the most straightforward is developed. A good complete treatment of Fortran IV which is compatible with Chaps. 2 to 4 can be found in Veldman (1967).

It is convenient to consider the Fortran language as composed of the following types of elements, in increasing order of complexity:

1   Constants
2   Variables
3   Operators
4   Statements
5   Programs and subprograms

2.6   CONSTANTS

Constants are numeric values which can be used in Fortran arithmetic statements. We typically add or subtract constants from the value stored in a location, multiply or divide the value stored in a location by constants, or raise the value stored in a location to the power of the constant. The notion of a constant is equivalent to the vector notion of a scalar of definite (numerical) value. Thus the scalar *a is not* a Fortran constant, but the scalar 1.5 *is* a Fortran constant. In Fortran, constants may be either *integer numbers* (also called fixed-point numbers) or *real numbers* (also called floating-point numbers).

Integer numbers, commonly called whole numbers, have no decimal part.

Examples of integer numbers are

$$1 \quad 999 \quad -20 \quad 0 \quad -104 \quad +16$$

The number may be written with or without the sign if positive, but the sign is required if negative. In Fortran, integers are used when only whole numbers have meaning. Universally, this is the case for specifying the dimension of a vector, the subscript of an element in a vector, or the number of times an operation or string of operations are to be performed. Usually, contents of locations to be raised to powers will be raised to integer powers.

Real numbers are numbers treated as having a decimal part, even if that part has the value zero. Examples of real numbers are

$$1.0 \quad 1. \quad 999.0 \quad 999. \quad -20.0 \quad -20. \quad 0.0 \quad 0.$$
$$-104.0 \quad -10. \quad +16.0 \quad +16.$$
$$1.333 \quad .92 \quad .090 \quad -0.5 \quad -.5$$

The number may be written with or without the sign if positive, but the sign is required if negative. Leading zeros to the left of the decimal point are optional. Trailing zeros to the right of the decimal point are optional. Thus, in Fortran, 0201., 201., 201.0, and 201.00 are exactly equivalent, and 0.0090, .0090, .009, and 0.009 are exactly equivalent. Notice that *all* real numbers have the decimal point represented even if the decimal part is zero. Its representation may never be omitted; in the numbers 1., 0., 205., −15. the decimal point is essential for the computer to differentiate them from the integer quantities

$$1 \quad 0 \quad 205 \quad -15$$

Thus it is important always to distinguish integers from real numbers by including a decimal point in all real numbers; otherwise, integer arithmetic may be employed in computations involving these numbers.[1] Integer arithmetic is arithmetic involving *only* whole numbers. While integer arithmetic has legitimate and valuable uses, it will not reliably give the results expected of conventional arithmetic. For example, the division of an integer by another integer results in an integer which is the whole number of times the second can be divided into the first. Thus, in Fortran integer arithmetic

1 divided by 2 equals 0 (exactly).

2 divided by 2 equals 1 (exactly).

3 divided by 2 equals 1 (exactly).

4 divided by 2 equals 2 (exactly).

5 divided by 2 equals 2 (exactly).

If you are careless about the decimal notation when analyzing a set of proportions, all proportions will be zero (exactly).

---

[1] Most computers have a limited capability for handling mixed-mode arithmetic expressions. They are programmed to assume that the programmer did not really mean the integer quantities to be integers. Reliance on such (usually unknown) programmed assumptions is extremely risky and should be avoided. Further *some* machines have no such capabilities, and mixing of modes can become a pernicious habit.

2.7 VARIABLES

In Fortran variables are *names of locations* which may contain different values at different times. The expression A=3. tells the computer "take the real number value 3. (or 3.000000) and store that value in the location *named* A." The expression B=A tells the computer "take the real-number value currently to be found in the location named A and store that value in the location named B." Storing a value in a location permanently destroys whatever value was previously stored there. (Technically, this is called *destructive read-in*.) However, retrieving a value from a location does not destroy the storage of that value in that location. (Technically, this is called *nondestructive read-out*.) Consider the statements in Table 2.2 performed in the sequence indicated. The question marks indicate that we have no way of knowing the contents of a location before placing some quantity there. The contents will be whatever was left there by the previous machine user. Failure initially to define locations results in the same type of errors that will result from using a desk calculator without clearing from it the numbers left there by the previous user.[1]

The name of a location in Fortran may be any combination of one to five[2] alphanumeric (letters A to Z, numerals 0 to 9) characters, but it must begin with an alphabetic character. Thus the following names are valid:

A	A15	J2XI
X	BIP	MOMEN
I	LAM	FIELD

The following names are *not* valid for the reasons indicated:

APONDE	More than five characters
1SP12	Begins with a numeral
A12*	Contains a nonalphanumeric character

---

[1] *Usually*, and in *most* machines, the residual numbers in uninitialized locations will in fact be zeros. However, this is not reliably the case, and failure to initialize locations has cost many tears as well as much time and money.

[2] The GE-235 time-sharing system allows variable names of up to 15 characters. Some IBM machines allow names as long as 8 characters. No major machine employs rules *more* restrictive than those specified here. If these rules are followed, one should be safe.

Table 2.2 DESTRUCTIVE READ-IN AND
NONDESTRUCTIVE READ-OUT

Time	Statement	Value in A	Value in B	Value in C
1	A=3.	3.	?	?
2	C=A	3.	?	3.
3	B=A	3.	3.	3.
4	C=4.	3.	3.	4.
5	A=C	4.	3.	4.

Mode

Like the constants, variables may be designated either in the fixed-point mode or the floating-point mode. If a number is stored in a fixed-point location, it will be stored and used as an integer; if it is stored in a floating-point location it will be stored and used as a real number.

By convention,[1] any variable name beginning with the letter I, J, K, L, M, or N is a fixed-point location. Any number placed in such a location will be represented as an integer. Any fractional part will simply be truncated rather than rounded. While such truncation is sometimes desirable and has important technical uses, ordinarily it should be avoided. For example, placing a set of proportions into a set of integer locations will result in each location being filled with the number 0 (exactly). Integer arithmetic is ordinarily used in operations involving such locations.

By convention, any variable name beginning with the letters A through H or letters O through Z is a floating-point location. Any number placed in such a location will be represented as a real number with a decimal part. While this decimal part may be .000000, it will be represented and will influence computations. The beginning programmer should *not* ordinarily combine real and integer locations in an arithmetic operation. The results of such combinations are totally predictable but not always obvious. Assume that we give the computer the instruction string

A = 3.

B = 1.

I = B/A

It will store the real-number quantity 3. in location A and the real-number quantity 1. in location B. It will divide the value in B by the value in A, yielding 0.33333, which will be stored as an integer quantity 0 in location I.

Dimension

In the discussion so far variables have been names of *single* storage locations. Such single location variables are equivalent to the vector notion of a scalar, with its numerical value defined by the current contents of the location. If A is the name of a physical storage location [   ] , every time we direct the computer to store some value in A, that value is physically placed in the same location. Also, each time we call for the contents of A, the machine retrieves the value from the same physical location. For this reason, we shall find that statements of the form A = A + 1. make sense in Fortran. The statement means "take the current contents of location A, add 1. to that value, and store the new value in location A." In more conventional notations, this operation is represented as $a_{t+1} = a_t + 1$. At any point in time the name of a single location (also referred to as an *unsubscripted variable*) has the same function as the vector notion of a scalar whose value may change from time to time. If we multiply the values in any locations by location A,

---

[1] On many machines, these conventions can be changed by the user, with specialized Fortran instructions which specify which names (or which beginning letters) are to be integer and which real. In the absence of explicit user-provided conventions, computers will follow those given here.

this is the same as multiplying the values in those locations by the scalar *a*. The obvious is labored because our vector notation will use small unsubscripted letters to designate scalars while our Fortran notation will use capital unsubscripted letters to refer to single locations which are used as scalars. Finally, notice that no analytic distinction is made in vector notation between the scalar *i* and the scalar 1. However, in Fortran, I is an integer *variable* while 1 is an integer constant.

Fortran also permits us to reference a series or string of locations with the same name and appended subscripts. Thus a single subscripted variable, A(I), is the name of a sequentially ordered string of locations:

A(1) ☐

A(2) ☐

A(3) ☐

. . . .

A(N) ☐

A(1) is the first location in this string, A(3) is the third location, and A(N) is the *n*th location. The single subscripted variable corresponds to our notion of a vector. A(1) is the first element of the vector a_1; A(2) the second element a_2; and A(I) refers to any element a_i. However, in Fortran, I is itself the name of a location and must have a numerical value stored in it before we can use the contents of location A(I). We can place values in any specific location in the vector. A(3) = 6. means "place a real number 6. in the third location in the vector string **A**." Further, where the variable used as a subscript has a meaningful current value, we can *sequentially* perform operations on each element of vector **A**.

The string of operations in Table 2.3 is equivalent to the vector notation of multiplying a vector times a scalar. In vector notation it would be represented as **B** = c**A**. Two special features of the Fortran process may need clarification. First the sign * is the Fortran multiplication sign, discussed later under operators. Second, steps *b*, *c*, and *e* provide the necessary iteration. This is achieved by initializing the index I, incrementing the index I, and counting the number of elements which have been processed. This feature is discussed later under transfer of control. While the

Table 2.3   SEQUENTIAL PROCESSING OF AN ARRAY OR VECTOR

Step	Instruction	Interpretation
a	C = 3.	Store a 3. in location C
b	I = 0	Store a 0 in location I
c	I = I + 1	Increase I by 1 and store result in I
d	B(I) = A(I)*C	Multiply A(I) by 3. and store result in B(I)
e	Return to step *c* unless all A's have been processed	

vector notation looks initially much less cumbersome, it should be remembered that the operations implied by that notation are as follows (where $C = 3$):

$$\begin{bmatrix} b_1 \\ b_2 \\ \vdots \\ b_n \end{bmatrix} = \begin{bmatrix} 3a_1 \\ 3a_2 \\ \vdots \\ 3a_n \end{bmatrix}$$

The reader should verify that these are the same operations performed by the Fortran sequence. Other parallels between Fortran and vector notation will be provided when the necessary ideas have been developed.

Fortran also provides for the possibility of doubly subscripted variables. The variable $A(I, J)$ is the name of a *structured string of locations* such that one can access (for storage or retrieval) specific locations as if one of the subscripts represented the row and the other subscript represented the column in a matrix. It is useful and permissible to *think* of such a variable as a matrix, as follows, though the *physical* storage arrangements are quite different:

A(1,1),A(1,2),A(1,3),...,A(1,N)

A(2,1),A(2,2),A(2,3),...,A(2,N)

A(3,1),A(3,2),A(3,3),...,A(3,N)

. . . .

A(M,1),A(M,2),A(M,3),...,A(M,N)

For such a variable, the statement $A(2,3) = 5.$ would store a real number 5. in a location which is conceptually in the second row and third column of the matrix **A**. Raw data for multivariate analyses will usually be conceptualized as such a matrix, where each row contains elements which are the values on each variable of interest for one case and each column contains the distribution values for every case on one variable. As with elements of vectors, elements of matrices can be referenced with variable subscripts if the subscripts are themselves defined. Thus the sequence of operations

I = 3

J = 6

A(I,J) = 7.

will store a real number 7. in the cell of A defined by the third row and sixth column. Of course, considerable power can be gained by systematic manipulation of these subscripts. On some computers, variables may have as many as 15 independently manipulated subscripts. However, we shall have need for only one, two, or three subscripts for our purposes. Whenever a subscripted variable (conceptually a vector, matrix, or space) is to be used in a program, the *first* statement in the program must be a DIMENSION statement (discussed later) which sets aside a set of adjacent locations. These statements include the name of the variable and its maximum dimensions.

2.8   OPERATORS: REPLACEMENT, LOGICAL, ARITHMETIC, FUNCTIONS, GROUPING

The Fortran operators specify how constants and variables are to be combined, compared, and transformed. As we shall see, most of these operators are familiar ones, drawn from arithmetic, algebra, symbolic logic, and trigonometry. The operators are grouped into priority levels which determine which operators are to be performed in what order. The function of each of the operators and the application of the priority-level principle is discussed in detail later. There are eight levels of operators in Fortran. These notions are very useful whenever data are to be manipulated. They are listed in Table 2.4 in the order of their precedence.

Any of these operators can be combined in strings including other operators as well as variables and constants to be operated upon. As Table 2.4 gives the operators in their order of priority, the grouping operation will be performed before any others, library functions before any save the grouping operation, ..., and the replacement operation is performed last. If a string includes two operations on the same level of priority, execution proceeds from left to right on most IBM, Control Data, and General Electric computers. It is prudent to check out this rule with your local computer-center staff, especially if you are using a Univac, RCA, PDP, or Xerox computer. The definition of operators on each level is given below. We shall deal only briefly with the logical operators and relations, as the other operators permit us to perform any operations which can be done with these. We shall begin with the simplest of the operators, deferring the notion of grouping until last, as this operator forces us to address the rather complicated notion of priority of operations directly.

Replacement

This operator is always used in the form

Name of recipient storage location = expression to be evaluated

Table 2.4   **TYPES OF FORTRAN OPERATORS AND LEVEL OF PRIORITY**

Level	Type of operator	Symbolic representation
1	Grouping	( )
2	Library functions (examples)†	SQRT, ALOG, SIN, ABS, etc
3	Exponentiation	**
4	Multiplication, division	*,/
5	Addition, subtraction	+,−
6	Logical relations	.EQ.,.NE.,.GT.,.GE.,.LT.,.LE.
7	Logical operators	.AND., .OR., .NOT.
8	Replacement ("store")	=

† Usually, more than 50 such library functions are available at an installation. The functions we discuss are available at all installations. If you need additional functions, the computer-center personnel can give you a manual listing all the library functions available and specifying how they are used.

The operator means "evaluate the expression to the right of the = symbol and store that value in the location designated to the left of the symbol." It is usually said that the contents of the recipient storage location are replaced by the evaluation of the expression. The recipient storage location may be an unsubscripted or a subscripted variable with constant or variable subscripts. Notice that subscripts *must* always be integers, whether constants or variables. All the following are valid examples for placing a real number 3. in the indicated location:

A = 3.

B(5) = 3.

C(2,4) = 3.

D(2,5,1) = 3.

B(I23) = 3.   if location I23 contains a proper value

C(KCHCK,MNO5) = 3.   if locations KCHCK and MNO5 contain proper values

D(I,J,K) = 3.   if locations I, J, and K contain proper values

Also, the expression to be evaluated may be quite simple or a complex string. We shall explore variations of such strings later. For the moment, let us say that the simple possibilities include constants and other unsubscripted or subscripted variables (locations) (Table 2.5). It should be obvious that some powerful transformations can be performed by systematically manipulating the subscripts of the source and recipient storage locations. As a simple example, if we perform the instruction

A(J,I) = B(I,J)

for all *i*'s and *j*'s, the matrix **A** will be the transpose of the matrix **B**. In matrix notation **A** = **B**′ is the equivalent of this Fortran notation. (See Appendix C for a definition and discussion of the transpose of a matrix.)

Logical Operators

There are three logical operators .AND., .NOT., and .OR.. Their use assumes that locations contain either a 0 (false) or a 1 (true). The expression A = B.AND.C finds the values in locations B and C. If *both* are true (1s), a true (1) is stored in

Table 2.5 ILLUSTRATION OF STORAGE

Expression	Meaning
A = 3.	Store 3. in A
A = B	Take the value of B and store it in A
A = C(5,2)	Take the value of the fifth row and second column of C and store it in A
A = D(I,J,K)	Find the values of I, J, and K; use them to designate a row, column, and depth in the space D; retrieve the value from that cell and store it in A

location A. If *either or both* are false (0), a false (0) is stored in location A. This is equivalent to making A the *product* of B and C. The expression D = .NOT.E finds the value of E. If it is true (1), false (0) is stored in D. If it is false (0), true (1) is stored in D. The expression A = B.OR.C finds the values in locations B and C. If either or both are true (1), a true (1) is stored in A. If both are false (0), a false (0) is stored in A. This expression is equivalent to making A the product of .NOT.B and .NOT.C. These operators may be combined in any sequence. The interpretations in Table 2.6 may be helpful. (The expression "iff" will be used henceforth to mean "if and only if.")

Two things should be noticed about the priority of operations at this point: (1) the logical operators are utilized *before* the replacement operator; (2) as all logical operators are on the same level, execution proceeds from left to right. Thus the following two statements are quite different:

A = B.AND.C.OR.D

A = B.OR.C.AND.D

In the first, only D being true is sufficient for A to be true. In the second, D must be true as well as at least one of the others for A to be true.

Logical Relations

There are six logical relations in Fortran. They have the interpretations indicated in Table 2.7. Since the logical relations take precedence over the logical operators,

Table 2.6   EVALUATION OF COMBINED LOGICAL OPERATORS

Expression	Interpretation
A = B.AND.C.AND.D	Store a "true" in A iff B, C, *and* D contain "trues"
A = B.OR.C.OR.D	Store a "true" in A iff one or more of B, C, or D contain "trues"
A = B.AND.C.OR.D	Store a "true" in A iff one or more of the following statements are "true": *a.* D is "true" *b.* Both B and C are "true"

Table 2.7   DEFINITION OF LOGICAL RELATIONS

Expression	Meaning
A = B.EQ.C	Store "true" in A iff B=C, "false" otherwise
A = B.NE.C	Store "true" in A iff B≠C, "false" otherwise
A = B.GT.C	Store "true" in A iff B>C, "false" otherwise
A = B.GE.C	Store "true" in A iff B≥C, "false" otherwise
A = B.LT.C	Store "true" in A iff B<C, "false" otherwise
A = B.LE.C	Store "true" in A iff B≤C, "false" otherwise

they are evaluated before any logical operators in a sequence. Thus, the statement $A = B.EQ.D.OR.E.LT.F$ is executed in the following way:

1 The computer encounters a replacement, two logical relations, and one logical operator. The logical relations are performed first. Since they are on the same level, the leftmost is performed before the rightmost.
2 B is compared with D. If they are equivalent, the expression $B.EQ.D$ is assigned a 1 (true) otherwise a 0 (false). Let us call this value X.
3 E is compared with F. If E is less than F, the expression $E.LT.F$ is assigned a 1 (true) otherwise a 0 (false). Let us call this value Y.
4 X and Y are linked by the .OR. operation. If either or both are true, a 1 (true) is stored in A. If both are false, a 0 (false) is stored in A.

As with all other operators, the operands may be constants or unsubscripted or subscripted variables. Also the recipient location may be an unsubscripted or subscripted variable.

Addition, Subtraction

The addition and subtraction operators are the same as in ordinary arithmetic, and the same symbols are used. $A = B + C$ means "find the contents of B, add to that the contents of C, and store the result in location A." Similarly, $A = B - C$ means "find the contents of B, subtract from that the contents of C, and store the result in location A."

Of course, operands which are constants or subscripted variables may be substituted for B and C. This simple notion is made more powerful when we utilize subscripted variables. If a computer is made to perform the instruction $A(I) = B(I) + C(I)$ for all I's from 1 to n, it has carried out the vector addition operation. This notation is equivalent to our vector representation

$$A = B + C = \begin{bmatrix} a_1 \\ a_2 \\ \vdots \\ a_n \end{bmatrix} = \begin{bmatrix} b_1 + c_1 \\ b_2 + c_2 \\ \vdots \\ b_n + c_n \end{bmatrix}$$

Multiplication and Division

The multiplication and division operators are the same as in ordinary arithmetic, though notation differs slightly. $A = B * C$ means "find the contents of B, multiply that by the contents of C, and store the result in A." $A = B/C$ means "find the contents of B, divide that by the contents of C, and store the result in A." Operands which are constants or subscripted variables may be substituted for A, B, and C. While we can write expressions like $A(I) = X(I) * C(I)$, such expressions produce the *outer product* of two vectors, a vector operation for which we shall have no use in these chapters. Notions of scalar products and matrix products cannot be easily handled until we have developed the later topic of *transfer of control*. However, the scalar multiplication operation can readily be written as A(I) = B $*$ C(I) if performed for all i's from 1 to n. This is equivalent to our vector representation

$$
\mathbf{A} = b\mathbf{C} = \begin{bmatrix} a_1 \\ a_2 \\ \vdots \\ a_n \end{bmatrix} = \begin{bmatrix} bc_1 \\ bc_2 \\ \vdots \\ bc_n \end{bmatrix}
$$

Exponentiation

The exponentiation operation is the same as in conventional arithmetic, though the symbol used is a double asterisk. A = B $**$ 2 means "raise the contents of B to the second power and store the result in A." The power to which a number is raised may be either an integer or a real number. The above statement yields the same result as A = B $**$ 2.. In the first, the integer 2 was used; in the second the real number 2.. Square roots may be taken by using the fractional power as in the statement E = F $**$ .5, which will store the square root of the contents of location F in location E. Numbers can also be raised to variable powers, as in the expression X = Y $**$ I, which raises the contents of location Y to the power stored in location I and stores the result in location X. In this expression Y or I may be unsubscripted variables or subscripted variables.

Library Functions

Useful in many transformation operations on data are the *library functions*. These preprogrammed miniprograms can be used by simply calling their names and giving them a location upon which to operate. People at your local computer center can tell you which of these are available at your installation. All installations have the following illustrative functions:

The statement A = SQRT(B) causes the contents of location B to be accessed, the square root of that value to be computed, and that square root to be stored in location A.

Social scientists commonly find log transformations of data to be useful. The laborious chore of transforming each observation's value to its logarithm can readily be performed by computer. For example, if we perform the following statements for all i's:

B(I) = ALOG(B(I)) stores the logarithm (base e) of all numbers in the vector B back in the appropriate locations in vector B.

B(I) = ALOG10(B(I)) stores the logarithm (base 10) of all numbers in the vector B back in the appropriate locations in vector B.

The ability to transform and combine data easily in theoretically meaningful ways is built into the regression program provided. It gives great flexibility and power to the researcher.

Grouping Operator and the Hierarchy of Operations

Where any portion of a string of operations is enclosed with a *group* (indicated by opening and closing parentheses), computers are designed to give priority to the portion within that group. The necessity for this operator and its use are illustrated below. Let us say that location A contains the number of like-ordered pairs of data on two variables and location B contains the number of unlike-ordered pairs. Further, we wish to compute and store in location G a statistic which is defined as

$$G = \frac{A-B}{A+B}$$

In Fortran we could *not* perform the desired operations by the simple coding

G = A − B/A + B

The hierarchy of operations would assign priority to the division operator, dividing B by A and saving the result. This result would then be subtracted from A and saved since the minus and plus operators are on the same level of priority and the minus is leftmost. The computer would then add B to this result and store the final result in G. However, G would now contain a quantity which is conceptually

$$G = A - \frac{B}{A} + B$$

We can modify the conventional hierarchy, however, with the grouping operator. Consider the following Fortran coding:

G = (A − B)/(A + B)

The computer now encounters, in order of conventional priority, two grouping operators, a division operator, a subtraction operator, an addition operator, and a replacement operator. However, the subtraction and addition will be performed *before* the division because they are *enclosed* within the priority groupings. The execution of this statement by a computer will proceed as follows:

1. Satisfy the leftmost group by subtracting B from A.
2. Satisfy the rightmost group by adding B to A.
3. Divide the first result (above) by the second result (above).
4. Store the result of this division in location G.

A functionally equivalent alternative to the use of the grouping operation can always be obtained by using a sequence of several statements. For the computation discussed here the same results could be obtained by executing the following statements in the order indicated:

C = A − B

D = A + B

G = C/D

While perfectly satisfactory, such sequences are often more cumbersome than parentheses (the grouping operator). Also, where subscripted locations are involved, waste of storage locations can be considerable.

Grouping operators may contain within them any legitimate operator, including other groups. The rule to be followed is that the *innermost* group must be satisfied first. One then works out systematically. The process is rather like peeling an onion from the inside out. For example, $A = ((B+C)/B**(I+J))+D$ is performed *as if* it were the sequence shown in Table 2.8 (X, K, Y, and Z do not appear in the statement but are used to show intermediate storage of results). Placing a grouping operation inside of a grouping operation is referred to as *nesting*. On most computer systems, groups may be nested 50 deep.

2.9 STATEMENTS: ARITHMETIC, SPECIFICATION, CONTROL, TRANSFER OF CONTROL

Statements are executed in series, from first to last, unless this order of events is altered by one of the transfer-of-control statements. We shall discuss five classes of statements with which you can do almost any computer work. There are many specialized statements with which specific work can be performed more easily, but all work needed for our purposes can be performed with the statements discussed. These may be conveniently classified under the following types:

1 Arithmetic
2 Specification
3 Control
4 Transfer of control
5 Input/output/formats

In general a Fortran statement begins in column 7 of a card and ends at or before column 72. Column 6, usually blank, is used for indicating that a statement is a continuation from a previous line (card). Columns 1 to 5 are reserved for statement label numbers, which are used when a transfer-of-control statement passes control to a statement out of the ordinary sequence. This will be discussed under transfer of control. One and only one statement should appear on each line (card).

Arithmetic

We have already discussed the arithmetic statement in the treatment of operators. Most of the work of computers is done with these statements. These are statements

Table 2.8 COMPUTATION USING A NESTED GROUP

Operation	Reason for priority
$X = B+C$	+ has highest priority *within* leftmost *inner* group
$K = I+J$	+ has highest priority *within* rightmost *inner* group
$Y = B**K$	** has highest priority *within outer* group
$Z = X/Y$	/ has next highest priority *within outer* group
$A = Z+D$	+ has highest priority *outside* the *outer* group, followed finally by the lowest priority =

giving a recipient storage location, a replace sign ($=$), and some other location or expression to be evaluated. The statement evaluates the expression to the right of the replace sign and then stores that value in the designated location. Examples are

$A = 5.$

$A = B$

$C(I) = D(I)**.5$

$D(I,J) = (5.*B(I,J) + C**I)*6.2$

An arithmetic statement must contain one and only one replace sign. The only thing which may appear to the left of the replace sign is the name of a location and its subscripts, if any.

Specification

The only specification statement which we shall need is the DIMENSION statement, which reserves strings or structures of locations to be used as vectors or matrices. The first statement in a program segment is the DIMENSION statement, and it must include the names and *maximum* dimensions of any *subscripted* variables used in the program. For example:

DIMENSION X(1000) means "save 1,000 adjacent locations to be referenced by the name X, followed by a single subscript which is greater than zero and equal to or less than 1,000."

DIMENSION X(10), A(50,50) means "save 10 adjacent locations to be referenced by the name X, followed by a single subscript which is positive and equal to or less than 10 *and* save 2,500 adjacent locations to be referenced by the name A, followed by two subscripts each of which is to be positive and less than or equal to 50."

You will notice that the first statement in the program and in each subroutine in Appendix A is a DIMENSION statement. Ordinarily, we shall not need to modify these statements. However, knowledge of what they mean will allow you to change them should you need to add a vector or change the maximum dimensions.

Control

Any program or subprogram (discussed later) should have as its *last* card the statement END. This simply marks the physical end of the program so that the computer can know to stop translating a segment at that point.

Ordinarily, the next to the last statement in a program will be the statement STOP, which means "stop executing this program and return control to the computer monitor for billing." The next to the last statement in a subprogram is the statement RETURN, which means "return control to the program utilizing the subprogram."

Additionally, there is a statement of the form

statement label number CONTINUE

This statement does nothing except to provide a location to which control can be transferred. We shall see the use for this statement later. Thus far you should understand that the organization of a program is as shown in Table 2.9.

We shall now discuss how the remaining statements organize the use of data/control information from that segment and control the sequenced use of arithmetic statements to produce a functioning program.

Transfer of Control

It was noted earlier that arithmetic statements are performed in order of their appearance. One of the most powerful features of computers is that this ordering can be altered by contingencies which arise in the processing of data; i.e., a sequence of statements may be repeated before proceeding outside the sequence, *or* alternative statements may be performed, depending on the data encountered. How useful this is will appear later. First, we must discuss the transfer-of-control statements which make this possible. Any executable[1] statement may have a *label* (called also a statement label number) in columns 1 to 5. This must be an integer number right-justified in column 5. It provides a label (or tag) for that statement so that we can direct the computer to proceed out of the usual sequence to one of these labeled (numbered) statements. While many kinds of statements can be used to achieve this alteration of usual sequence, any work which we shall do can be performed by statements of the following three forms:

GO TO n

IF (arithmetic expression) n_1, n_2, n_3

DO n I$=m_1, m_2, m_3$

The specific functions and examples of the usefulness of these statements are discussed below.

---

[1] Executable statements include all statements which we shall cover *except* DIMENSION and FORMAT statements. Even these two are, in a sense, executed (or performed). The DIMENSION statement is executed *only* at the beginning of the program, when it saves adjacent storage spaces. It may never again be executed, and therefore control may not be transferred to it. The FORMAT statement (Sec. 2.11) is executed only when it is used *with* another statement, which must be a READ or WRITE statement. It may, therefore, never have control passed directly to it. By convention, we refer to statements of these types as *nonexecutable*.

Table 2.9 SKELETAL OUTLINE OF
PROGRAM ORGANIZATION

Job-control language (as needed)
　　DIMENSION list of locations (if needed)
　　remaining program segment
　　STOP
　　END
Job-control language (as needed)
　　data/control segment
/*

GO TO n

Here n is a statement label number. This sends control to the statement labeled n (out of ordinary sequence) then (in the ordinary sequence) to the statement after the statement labeled n.

IF (arithmetic expression) n_1, n_2, n_3

Here n_1, n_2, n_3 are three possibly different statement label numbers. This statement evaluates the arithmetic expression. If that value is negative, the computer moves next to the statement indicated by the label number in the first position (n_1); if the value is zero, the computer moves to the statement indicated by the label number in the second position (n_2); if the value is positive, the computer moves to the statement indicated by the label number in the third position (n_3). Thus, the computer has an ability to make simple decisions and to modify the sequencing of steps based on these decisions. This capability is the principal characteristic distinguishing a computer from a calculator.

By using the GO TO, IF, and CONTINUE statements, it is possible to show how a computer can do work of some import. Assume that we have dimensioned variables (vectors) A, B, and C, each of dimension 10,000. Imagine that B and C already contain data values and that we seek some A each element of which is the sum of the corresponding elements in B and C. This can be done for all 10,000 cells by the simple coding

```
. . . . . . . . . . . . . . . . . .

     I = 0
8    I = I + 1
     A(I) =  B(I) + C(I)
     IF  (I − 10000)8,2,2
2    CONTINUE

. . . . . . . . . . . . . . . . . .
```

This sequence of statements will take each of the 10,000 b_i's, add it to the corresponding c_i, store each sum in the corresponding a_i, and finally transfer control to the statement labeled 2 so that any subsequent programming operations can be performed. The result is achieved by structuring the statements so that the sequence of the second, third, and fourth statements is repeated 10,000 times *before* control moves to the fifth statement. This is achieved by setting the *counter location* I to zero (equivalent to clearing a calculator's counting register), incrementing the counter by 1 before performing each addition, and then checking the counter to see if its terminal value is reached before repeating the process. Specifically, these five statements may be considered to generate the computer *execution* of the following 30,002 steps:

Step 1   The location I is set to 0.
Step 2   The contents of location I are increased by 1 (becoming *currently* 1).

Step 3   B(I) is added to C(I), and the result stored in A(I). (*Currently*, b_1 is added to c_1 and stored in a_1.)

Step 4   The maximal value[1] (10,000) is subtracted from the value of I, producing a negative quantity (*currently*, $-9{,}999$), and thus control moves to the statement labeled 8, the label number indicated in the first (negative) position following the arithmetic portion of the IF statement.

Step 5   Location I is increased by 1 (becoming *currently* 2).

Step 6   B(I) is added to C(I) and the result stored in A(I) (*currently* b_2 is added to c_2 and stored in a_2).

Step 7   The maximum value (10,000) is subtracted from the value of I, producing a negative number (*currently*, $-9{,}998$), and thus control moves again to the statement labeled 8, the first (negative position) statement number in the list of alternatives.

. . . .

This process will continue, with repetition of steps 2, 3, and 4 until:

. . . .

Step 29,999   Location I is increased by 1 (becoming *currently* 10,000).

Step 30,000   B(I) is added to C(I) and the result stored in A(I) (*currently*, b_{10000} is added to c_{10000} and stored in a_{10000}).

Step 30,001   The maximum value (10,000) is subtracted from the value of I, producing a zero, and thus control passes to the statement labeled 2, which is the label number in the second (or zero) position in the list of alternative destination statements.

Step 30,002   The statement labeled 2 is executed.

The reader should note that the counter controlling structure involves three elements. The counter is initialized at some value (here 0), incremented with each execution of the sequence by some value (here 1), and finally checked against a terminal value (here 10,000) before the reiteration of the sequence is terminated and control passes to the indicated statement outside the sequence. The current value of the counter at each stage is used for indexing (subscripting) purposes for the three vectors. Notice the efficiency of producing the execution of 30,000 steps with only three of the five statements.

An additional simple illustration will demonstrate some of the powerful potential of this structured sequence of statements. Assume that the dimensioned variable D references (is the name of) 10,000 locations and that data, e.g., the ages

[1] Notice especially that large numbers like 10,000 are written *without the comma* in Fortran; thus the number is 10000. The reason for this will be obvious, if you reflect that the computer has no way to distinguish the subscript (1,100), as the 1100th location, from the pair of subscripts designating the 1st row and 100th column. Thus, no number however large may include embedded commas. Also, in structures like the looping structure described here, the maximal value is the largest value which the index is allowed to obtain before switching to a new sequence of operations. This maximum may be an integer constant, unsubscripted integer value, subscripted integer value, or some complicated computed integer value.

of 10,000 people, are already stored in the D(I) locations. Consider the following sequence for computing the mean of the variable age:

··························

      DMEAN=0.

      I=0

12 I=I+1

      DMEAN=DMEAN+D(I)

      IF (I−10000)12,13,13

13 DMEAN=DMEAN/10000.

··························

From our previous discussion it should be obvious that the sequence comprising the third, fourth, and fifth statements will be executed 10,000 times. Specifically, the execution sequence is:

Step 1 The sum box DMEAN is zeroed.

Step 2 The counter I is zeroed.

Step 3 Location I is increased by 1 to the value 1.

Step 4 The location DMEAN is increased by the value of d_1. (*Currently*, DMEAN $= 0. + d_1$.)

Step 5 The counter is checked, and since the terminal value has not been achieved, control moves back to the statement labeled 12.

Step 6 Location I is increased by 1 to the value 2.

Step 7 The location DMEAN is increased by the value of d_2. (*Currently*, DMEAN $= 0. + d_1 + d_2$.)

··

The process continues until:

··

Step 30,000 Location I is increased by 1 to the value 10,000.

Step 30,001 The location DMEAN is increased by the value of d_{10000}. (*Currently*, DMEAN $= 0. + d_1 + d_2 + d_3 + d_4 + \cdots + d_{10000}$.) Thus, *at this time*, DMEAN is the sum of the ages stored in D.

Step 30,002 The counter is checked, by subtracting the terminal value. This having been achieved, a zero results and control is shifted to the statement labeled 13.

Step 30,003 The value stored in DMEAN (conceptually, the sum of the ages originally stored in D) is divided by 10,000 (conceptually, the number of cases) and the result (conceptually, the mean of the ages) is stored in DMEAN.

Modifying this program sequence so that it calculates not only the mean but the variance for a distribution is quite simple. A brief exploration of the modification should serve to clarify the function of the *looping* structure employed to process all cases and cumulate sums. The modification would require only three statements:

1   A new sum box called (for instance) VAR would have to be created. The statement VAR=0., inserted after DMEAN=0., would initialize (clear) a location for this purpose.

2   The sum of squares of the d_i's would have to be cumulated. The statement VAR=VAR+D(1)**2, inserted after DMEAN=DMEAN+D(I), would cumulate the sum of squares of the ages into the location VAR just as the sum of the scores were cumulated into DMEAN.

3   The variance would have to be calculated from the sum of squares and the mean. The statement VAR=VAR/10000. − DMEAN**2, inserted after DMEAN=DMEAN/10000., would achieve this result.

Before proceeding the reader should persuade himself that these modifications would result in the calculation of the mean and variance of D. If necessary, the reader should go back, insert the statements where indicated and analyze the functioning of the new program sequence. Of more interest to the user (as opposed to the writer) of the program should be the fact that *no* new input or control of the program is necessary. It will develop that the remaining transfer-of-control statement to be considered is merely a shorthand way of achieving this iterative structure for processing cases serially.

DO *m* I = n_1, n_2, n_3

Here, *m* is a statement label number, n_1 is the initial value for a counter I, n_2 is the terminal value for the counter I, and n_3 is the amount by which I is incremented with each execution. (Limitations on these values are $n_1 \geq 0$, $n_2 \geq n_1$, $n_3 \geq 0$.) This statement means "perform the sequence or string of statements from this one through and including the statement labeled *m*. Continue repeating this sequence *until* some counter (begun at the value n_1 and increased each time by the value n_3) exceeds the value n_2." Thus, a simpler program sequence written as follows would function exactly in the same way as the earlier sequence which calculated a mean:

```
         . . . . . . . . . . . .
         DMEAN=0.
         DO  12  I=1,10000,1
   12    DMEAN=DMEAN+D(I)
         DMEAN=DMEAN/10000
         . . . . . . . . . . . . . . . . . . . . . .
```

The statements in this segment mean the following:

1   Store a 0. in location DMEAN, which will become a sum box for cumulation.
2   Execute all statements from the DO down through and including the statement labeled 12 (in this case the sequence from the DO through the referrent

statement has only the one referent statement in it). Perform this (10,000 times) for $I = 1, 2, 3, 4, \ldots, 10000$. This will cause DMEAN to contain

$$0. + d_1 \qquad \text{where } I = 1$$
$$0. + d_1 + d_2 \qquad \text{where } I = 2$$
$$\ldots\ldots\ldots\ldots\ldots\ldots\ldots\ldots\ldots\ldots\ldots\ldots\ldots$$
$$0. + d_1 + d_2 + \ldots + d_{10000} \qquad \text{where } I = 10000$$

3 After the loop is satisfied (the looping index having achieved its terminal value), proceed to execute the next statement following the loop, which is DMEAN=DMEAN/10000.. At this point DMEAN will contain the quantity which is, conceptually, the mean of the vector D.

2.10 RELATIONSHIP OF FORTRAN STATEMENTS TO VECTOR OPERATIONS

We can now illustrate how, in Fortran, the vector operations can be performed for vectors of any dimension. Assume that we have saved strings of locations for the vectors A, B, C, D with the statement

DIMENSION A(10000),B(10000),C(10000),D(10000)

Further assume that we have stored data (values) in vectors **A** and **B** for N (any number of) cases (where $1 \le N \le 10000$). Finally, assume that the number of cases is stored in location N and a scalar is stored in location E. To produce a vector **C** which is the sum of the vectors **A** and **B** we need insert only the following two statements:

```
         ....................
      DO 18 K = 1,N,1
  18  C(K) = A(K) + B(K)
         ......................
```

Notice that these two statements will process vectors of any dimension depending on the values stored in location N. This is limited only by the maximum dimensions assigned to A, B, and C in the DIMENSION statement. The efficiency of being able to process many thousands of cases with only two statements is obvious.

To produce a vector **D** which is the product of the scalar E and the vector **A** we need insert only the following two statements:

```
         ....................
      DO 19 L = 1,N,1
  19  D(L) = E*A(L)
         ....................
```

To produce a scalar F which is the vector product of the vectors **A** and **B** we need insert only the following three statements:

```
 . . . . . . . . . . .
     F = 0.
     DO 55 I = 1,N,1
 55  F = F+A(I)*B(I)
 . . . . . . . . . . . . . . . . . .
```

In our vector notation, this is conceptually the product of **A′B**. However, in Fortran, no explicit distinction is made between row and column vectors. The programmer must keep track of which is conceptually row and which column. To have the computer check for orthogonality of **A** and **B**, we need only follow these three statements with

```
     IF (F)1,2,1
```

where the statement labeled 2 is the subsequent statement we wish to perform if **A** and **B** are orthogonal (written **A** ⊥ **B**) and the statement labeled 1 is the subsequent statement we wish to perform if **A** and **B** are not orthogonal.

The usefulness of the DO statement can be appreciated by noting that it permits us to perform all these things at one time. With the following six statements we can produce (1) the vector **C**, which is the sum of vectors **A** and **B**, (2) the vector **D**, which is the product of the scalar E and the vector **A**, (3) the scalar F, which is the product of the vectors **A** and **B**, and (4) we can check to see whether **A** ⊥ **B**.

```
 . . . . . . .
     F = 0.
     DO 20 I = 1,N,1
     C(I) = A(I)+B(I)
     D(I) = E*A(I)
 20  F = F+A(I)*B(I)
     IF (F)1,2,1
 . . . . . . . . . . . . . . .
```

Often, simple transformations of data are required for substantive or theoretical reasons. Most commonly used are the square and cubic functions (to permit testing of models involving equations not of the first degree) and the logarithmic functions. While such transformations may have great theoretical import, they are laborious by hand. However, these and other transformations are very easy to do with a computer, using preprogrammed library functions, discussed earlier under operators. The following five statements will perform four such transformations for any number of cases, limited only by the value stored in N and the maximum dimensions of A, B, C, and D.

```
.................
        DO 90 I=1,N,1
        A(I) = A(I)**3
        B(I) = SQRT(B(I))
        C(I) = ALOG(C(I))
   90   D(I) = ALOG10(D(I))
.......................
```

These statements will (1) replace each element in A with the cube of the original value, (2) replace each element in B with the square root of the original value, (3) replace each element in C with the logarithm (base e) of the original value, and (4) replace each element in D with the logarithm (base 10) of the original value. Transformation possibilities are as varied as the researcher's imagination. Many more than these illustrative possibilities have already been programmed as library functions and are available on most computers. It should be obvious that the transfer-of-control operations permit great flexibility and efficiency in computer manipulation of data. We now turn to the question of how we get data into the computer and results out of it.

2.11 INPUT/OUTPUT/FORMATS

Input statements are specialized statements which cause the computer to take information from some external source (punched cards, magnetic tapes or disks, keyboards, paper tapes) and store those data in specified locations. These locations are then used for control purposes (as with the location N, above) or analysis purposes (as with the vectors **A**, **B**, **C**, and **D**, above). We shall concern ourselves principally with taking information from punched cards located in the control/data segment, although trivial modifications will allow retrieving data from other storage media (tapes, disks, etc.).

Input Statements

The most generally useful input statement, and the only one to be discussed here, is of the form

READ (u,f) list

where u is a unit number, indicating whether the data are to come from the card-reader unit, a specified disk or tape unit, or some other specified input device. Which unit number designates the card reader varies from installation to installation. The unit designations can also be changed with job-control language, not treated here. We shall assume unit number 5 to be the card reader for our examples.

The f is a statement label number for a FORMAT statement which designates exactly where (in which columns) the data are to be found and what their form (mode and decimal placement) is to be. Every READ statement has associated with it one and only one FORMAT statement. These are discussed below.

The *list* is a string of location names (separated by commas) into which the retrieved values are to be stored. Consider the READ statement

READ (5,12) N,A(1),A(2),A(3),B,C(N)

This statement says:

1 Take the next (the first, if this is the first READ statement in the program) card from the data segment (because we have assumed unit number 5 to be the card reader).

2 Use the FORMAT statement numbered 12 to determine what parts of the card to read.

3 Store the first number in location N, the second through fourth numbers in locations A(1) through A(3), the fifth number in location B, and the sixth number in location C(N), the subscript for which must be defined by the number already stored in location N.

FORMAT Statements

At some point within the program segment[1] there must appear a FORMAT statement labeled 12 which describes the form and location on cards of these six numbers. No other statement in the program segment may be labeled number 12.[2] The general form for FORMAT statements is

n FORMAT (specification)

where *n* is the statement label number (a one- to five-digit positive integer right-justified in columns 1 to 5) and the specification portion is a set of codes specifying the form and location of each number to be read from cards and stored in the location specified by the READ statement. While there are many specification codes, only five types will commonly be used in FORMATs associated with READ statements. These may be combined in any order, separated by commas.

1 Integer numbers are specified by the form
  nIw where the I indicates that an integer number is to be read, the w is

---

[1] FORMAT statements may be placed in any order and any position between the DIMENSION statement and the END statement. They are nonexecutable and therefore do *not* affect the flow of operations. Whatever their placement, they are used only when the READ statement referencing them is being executed.

[2] Strictly, no two statements within the *same* program or subroutine (discussed later) may have the same statement label number, as the computer has no way of ascertaining which one the programmer really meant. However, a statement in one subroutine may have the same label number as a statement in a different subroutine (though both are in the same program segment). Variable names and statement label numbers are said to be *locally defined*, i.e., defined only *within* a particular program or subprogram.

an integer specifying the field width (number of adjacent columns, including one for the sign, if necessary, which constitute the number), and the n is an integer (usually 1) specifying the number of adjacent numbers following identical specifications.

2 Real numbers are specified by the form

$nFw.d$ where the F indicates that a real number is to be read, the w is an integer specifying the field width (total number of adjacent columns including one for the sign and one for a decimal, if necessary, which constitute the number), the d (following the decimal point) is an integer specifying how many of the columns are to be considered to be to the right of the decimal point, and the n is an integer specifying the number of adjacent numbers following identical specifications. The entire number found in the field (columns specified) will be brought into the machine. If no decimal point is punched in the data field, the decimal point will be placed to the left of d digits before the number is stored. If, however, a decimal point is punched in the data, this will override the d specification and the number will be stored *exactly* as punched.

3 Alphanumeric codes (not to be used in arithmetic operations) are specified by the form

nAw where the A indicates that an A-mode quantity (a string of any legitimate alphanumeric codes, including blanks) is to be read, the w is an integer specifying field width, and the n is an integer specifying the number of adjacent quantities following identical specifications. A-mode fields may contain any legitimate alphanumeric characters. There are technical storage reasons why an A-mode field may never exceed a four-column (four-character) field. Such fields are used only for labeling purposes in generalized programs. The neophyte need not be terribly concerned by them.

4 Skipping of columns is designated by the form

nX where the X indicates that a column is to be skipped (ignored) in the reading operation and the n is an integer indicating how many such skips are to be made of adjacent columns.

5 Skipping to a new record (card, line, etc.) is designated by the form

/   where the / indicates that a skip to a new card (line if one is writing rather than reading) is to occur. This is commonly used when one seeks to read data from more than one card for each individual. If, for any reason, several successive skips are wanted, they can be achieved by repeating the / symbol. Thus, /// would skip to the third card (record) after the one currently being processed.

2.12   READING IN FORTRAN

The reading operation can be understood only by the simultaneous consideration of the READ statement, the associated FORMAT statement, and the portion of the data segment to which these refer. It should be noted that the first time a READ statement is encountered in the execution of a program, it causes the first card in the data segment to be read and its contents to be stored as indicated. That card is then discarded and the next card in the data segment is available for reading. Each subsequent time that a READ statement is encountered (or when a / symbol is encountered in a FORMAT being used) the next card is taken for reading. Let us now trace the operation of several READ/FORMAT pairs with associated data segments. The program and data features necessary for this exposition are

```
. . . . . . . . . . . . . . . . . . . . . . . . . . . . .
        DIMENSION A(3),C(3)
. . . . . . . . . . . . . . . . . . . . . . .
        READ (5,12) N,A(1),A(2),A(3),B,C(N)
      1 2 FORMAT (1I1,3X,1F2.0,1F2.1,1F2.2,9X,2F3.0)
. . . . . . . . . . . . . . . . . . . . . . . . . . . . . . . . . . . . . . . . . . . . . . . . .
        END
/*
//G O .SYSIN DD *
2___ 333333_____2312.3
. . . . . . . . . . . . . . . . . . . . . . . . . . . .
```

In general, the first element in the READ list is associated with the first variable specification in the FORMAT list and the first portion of the data line to be read. Then the next element is paired with the next, etc. In highly generalized programs, it is possible to develop general FORMATs which may allow portions of the FORMAT to be reused (and portions to be ignored). The rules for this are exceedingly complex and machine-specific. Most scientific users need never know of the possibilities. Most of the basics of reading data can be illustrated by tracing the computer actions generated by the structure above. That READ/FORMAT/data triplet will cause the following steps.

STEP 1   Location N is to be filled with an integer of one column's width (1I1). This is to be taken from the first column of the first data card, because no specification in the FORMAT list precedes the 1I1 and no READ statement precedes the one considered. *Here* location N will be filled with an integer quantity (2).

STEP 2   Location A(1) is to be filled with a real number of two columns' width with neither column to be considered to the right of the decimal point (1F2.0). This is to be taken from columns 5 and 6 of the first data card because column 1 has been used and the (3X) specifies that the next three columns (2, 3, 4) are to be skipped. Thus, (1F2.0) will access columns 5 and 6. *Here* location A(1) will be filled with a real-number quantity (33.).

STEP 3   Location A(2) is to be filled with a real number of two columns' width with one column (the rightmost) to be considered to the right of the decimal point (1F2.1). This is to be taken from columns 7 and 8 of the card because columns 1 through 6 have already been used. *Here* location A(2) will be filled with a real-number quantity (3.3).

STEP 4   Location A(3) is to be filled with a real number of two column's width with both columns to be considered to the right of the decimal point (1F2.2). This is to be taken from columns 9 and 10 because columns 1 through 8 have already been used. *Here* location A(3) will be filled with a real-number quantity (.33). Notice that because of the varying *d* specification in the *nFw.d* form, the numbers stored in A(1), A(2), A(3) are quite different even though their card representations were identical. Specifically:

Location	Number in card	Format	Number stored
A(1)	33	1F2.0	33.
A(2)	33	1F2.1	3.3
A(3)	33	1F2.2	.33

STEP 5   Location B is to be filled with a real number of three columns' width with none of the columns to be considered to the right of the decimal point (2F3.0). This is to be found in columns 20, 21, and 22 because columns 1 through 10 have already been used and the (9X) specifies that the next nine columns are to be skipped. *Here* location B will be filled with a real-number quantity (231.). It is important to note that expressions like (2F3.0) could be rewritten, with no change in meaning, (1F3.0,1F3.0). Notice that B would be paired only with the first of these; C(N) would be paired with the second. This shorthand convention can be used only when the specifications are identical in all respects and refer to adjacent fields.

STEP 6   Location C(N) is to be filled with a real number of three columns' width none of which are to be considered to be to the right of the decimal point (2F3.0). This is to be found in columns 23, 24, and 25 because columns 1 through 22 have already been used. Notice, however, that these columns contain the quantity 2.3, the decimal point being punched in the data card. Where the decimal point is explicitly represented, the FORMAT specification (here no columns to the right of it) is overridden and the datum is stored as punched. Here location C(3) will be filled with the real-number quantity (2.3). Notice that C(N) is paired with the second half of the (1F3.0,1F3.0) implied by the shorthand specification (2F3.0). Notice further that one digit is treated as being to the right of the decimal point, though the *d* specification says that none should be. This is because a punched decimal point will always override the FORMAT specification of decimal placement.

Output Statements

Output statements are specialized statements which take information from storage locations (usually computed results) and place them on some external medium, e.g., printed page, punched card, punched tape, magnetic tape, magnetic disk. The computer program in Appendix A, like most packaged programs, automatically prints all the information which one ordinarily needs. However, it is often quite useful to know enough about output statements to be able to modify them slightly, as packages frequently compute all the desired information but print only most of it. For example, many programs calculate variance and standard deviation but print only one of them. Also, some programs calculate correlations but not slopes, though all the information to calculate slopes is available. A *single* new line of Fortran and an addition to the WRITE list and FORMAT would provide both.

Fortunately, output statements are so similar to input statements that they can be explained readily. We shall concern ourselves only with output to the printed page, although change of the unit number would allow output to any other medium. The most generally useful output statement, and the only one to be discussed here, is of the form

WRITE (u, f) list

where u is a unit number indicating whether the data are to go to the printer, a specified magnetic disk or tape unit, the card punch, or some other specified output device. The unit number designating the printer varies from installation to installation. Job-control language allows the experienced user to define unit numbers at will. The inexperienced user should ask a programmer which is the standard unit number for the printer at his installation. We shall assume that unit number 6 is the printer for our example. The f is a statement label number for a FORMAT statement which designates exactly where (in which columns) the data are to be placed and in what form (mode and decimal placement). Every WRITE statement has associated with it a FORMAT statement. The *list* is a string of location names (separated by commas) from which data are to be retrieved for printing. Consider the following WRITE statement:

WRITE (6,22) SUM,SMEAN,N

This statement says:

1 Take the contents of locations SUM, SMEAN, and N and write them on the line printer (unit number assumed to be 6).
2 Use the FORMAT statement numbered 22 to determine how and where these quantities are to appear on the printed page.

The FORMAT statement associated with this statement follows the same form and rules as the FORMAT statement used for input purposes. Briefly, the form is

n FORMAT (specification)

where n is the statement label number and the specification portion is a set of codes, separated by commas, specifying the form and location in which the variables are to be represented. All the codes discussed with the READ/FORMAT pair are permissible and have only slightly modified meanings. There are several additional FORMAT elements which are important when the visual quality of the output is a concern. The beginning programmer has far too many other concerns to become preoccupied with aesthetics. Therefore, only minimal output possibilities are discussed. They can be summarized as follows:

nIw where I indicates printing of an integer, which is w columns wide (including a column for a sign) and n (usually 1) specifies the number of adjacent variables following identical specifications.

nFw.d   where *F* indicates printing of a real number, which is *w* columns wide (including a column for a sign and a column for a decimal point). The *d* indicates how many digits are to be represented to the right of the decimal point. The *n* specifies the number of identical adjacent specifications. It is especially critical on output that the programmer provide enough columns for (1) a sign, (2) as many digits to the *left* of the decimal as will ever develop, (3) a decimal point, (4) as many positions as desired to the right of the decimal point. If you try to print a number that is too large (too many digits to the left of the decimal), most machines will merely give a line of asterisks ****** which fill the field. When this happens, you should expand the field width.

nAw   where *A* indicates printing of an A mode (alphanumeric string, usually for labeling), which is *w* columns wide. The *n* specifies the number of identical adjacent specifications.

nX   where *X* indicates skipping (leaving blank for spacing purposes) a column on the printer and *n* indicates the number of adjacent skips.

/   where / indicates skipping to a new line on the printer and multiple skips may be achieved by repeated use of this symbol, e.g., /// · · · /.

One additional type of code not discussed with reading operations will be found especially useful for output operations. This code is of the form:

'label'   called a *literal*. The pair of single quotation marks will cause the alphanumeric expression between them (here designated by 'label') to be printed exactly. This is especially useful for identifying output. While the literal is only one of several ways to accomplish labeling, it is the easiest to use and the only one needed by the beginner.

2.13   WRITING IN FORTRAN

The writing operation can be understood only by a simultaneous treatment of (1) the contents of the referenced locations, (2) the WRITE statement, (3) the associated FORMAT statement, and (4) the appearance of data on the printed page. Assume that the location SUM contains the sum of a set of scores, for example, $-120.$, the location SMEAN contains the mean for the scores, for example, -2.4, and the location N contains the number of scores, for example, 50. Further, assume a program including the following WRITE/FORMAT pair placed at some point after completion of the computation of the sum of scores and the mean but before the END statement:

```
. . . . . . . . . . . . . . . . . . . . . . . . . . . . . . . .
      WRITE (6,22) SUM,SMEAN,N
  22  FORMAT (1X,'SUM=',1F8.2, 'MEAN=',1F8.2,'N=',1I4)
. . . . . . . . . . . . . . . . . . . . . . . . . . . . . . . . . . . . . . . . . . .
```

The line printed on the printer will appear as follows (_ means blank):

SUM=−120.00_MEAN=___−2.40_N=__50

Let us analyze why this is the case.

The WRITE statement causes the contents of location SUM to be paired with the first variable specification (1F8.2). Before this can be printed, the elements preceding it in the FORMAT statement must be satisfied. Thus, the computer will first "print" a blank in column 1 to satisfy the skip specification (1X). (The first print position is reserved for specialized instructions which control the carriage on the printer. The beginner should not worry about these but should *always* skip the first column on the printer. Failure to do so occasionally results in the printer's spewing forth 3,000 or 4,000 blank pages before it can be stopped.) Next the computer will satisfy the literal specification ('SUM =') by printing the expression SUM = in columns 2 through 5. Next it will right-justify the contents of location SUM (− 120.) in the eight-column field 6 through 13, representing the quantity to two decimal places of accuracy. This results in the printing of _ − 120.00 in columns 6 through 13.

The WRITE statement causes the contents of location SMEAN to be paired with the second variable specification (1F8.2). Before this can be printed the yet unsatisfied element preceding it must be satisfied. Thus, the computer will print the expression _ MEAN = in columns 14 through 19. Next, it will right-justify the contents of location SMEAN in the eight-column field 20 through 27, representing the quantity to two decimal places of accuracy. This results in the printing of ___ − 2.40 in columns 20 through 27.

The WRITE statement causes the contents of location N to be paired with the third variable specification (1I4). Before this can be printed the yet unsatisfied element preceding it must be satisfied. Thus, the computer will now print the expression _N= in columns 28 through 30. Next, it will right-justify the contents of location N in the four-column field in columns 31 through 34. This results in the printing of __50 in columns 31 through 34.

The computer will typically skip to a new line (1) when another WRITE statement is encountered, (2) when the same WRITE statement is encountered another time (because it is embedded in a DO loop), or (3) when a / symbol is encountered in a FORMAT.[1] Input and output statements become far more powerful when we take advantage of the iterative capability provided by the DO statement. We shall illustrate these capabilities only for input statements because we shall need them more often. Further, the functioning of input and output statements is so similar that the generalization is quite straightforward. Our illustration (Table 2.10) involves a general program for computing and reporting the mean and variance for any data. The job-control language has been omitted, where indicated with dots.

We can summarize much that we have covered concerning constants, variables, operators, statements, and structures by detailing the operation of this (complete) program segment and associated data/control segment. While it is a meticulous job, it will organize much that the reader has learned.

STEP 1   Statement 1 is the first encountered READ statement. Thus, it causes the machine to read the first data card (statement 19) as specified in the FORMAT labeled 31 (statement 12) and store the quantity in location N. The FORMAT specifies that the

---

[1] Several other subtle conditions can cause skipping to a new line, but the beginner is not likely to encounter them. The experienced programmer can use carriage-control characters to suppress the usually automatic spacing discussed here.

first three columns of the card are to be read and treated as an integer. Thus the integer 5 is read and stored in location N. This number is to be, conceptually, the number of cases. Notice that the location N is later used to control a loop which repeatedly reads numbers and sums them and their squares. By changing the first data card to 100 (or any number) and following it with 100 (or any specified number of) data cards, this program would perform the same operations for any data (with only the possible change of the FORMAT which reads the data).

STEP 2   Statement 2 stores the contents of location N (integer) in location T (real). Both N and T now contain, conceptually, the number of cases. The integer will be used to control looping and the real number for division in subsequent computations, avoiding integer arithmetic.

STEP 3   Statements 3 and 4 clear (store zeros in) the locations SUM and SUMSQ. These locations must be cleared (zeroed) because they are to be used as sum boxes to cumulate the sum of scores and sum of squares of scores.

STEP 4   Statement 5 sets up a DO loop which causes the string of statements down through and including the statement labeled 1 to be executed N times. For these

Table 2.10   SAMPLE PROGRAM FOR COMPUTING THE MEAN AND VARIANCE

Fortran statement	Statement number
READ (5,31) N	1
T=N	2
SUM=0.	3
SUMSQ=0.	4
DO 1 I=1,N,1	5
READ (5,32) X	6
SUM=SUM+X	7
1 SUMSQ=SUMSQ+X**2	8
SUM=SUM/T	9
SUMSQ=SUMSQ/T−SUM**2	10
WRITE (6,33) SUM,SUMSQ,N	11
3 1 FORMAT (1I3)	12
3 2 FORMAT (3X,1F2.0)	13
3 3 FORMAT (1X,2F8.2,1I3)	14
STOP	15
END	16
/*	17
// GO. SYSIN DD *	18
_ 5	19
_ _ 2 1	20
_ _ 2 2	21
_ _ 2 3	22
_ _ 2 0	23
_ _ 1 8	24
/*	25

data (N = 5), this will cause the sequence of statements 6 to 8 to be executed five times. (If the first data card had contained the number 100, the sequence would be performed 100 times. Of course, then, 100 cards should follow the first data card rather than 5.) Notice below, that the entire string, i.e., statement 6, then 7, then 8, is performed in sequence, starting again at the first of the string for each iteration. Beginning programmers sometimes overlook this structural feature (unique to computer languages) and *mistakenly* imagine that the first statement is performed N times, then the second, etc. Much of the power of Fortran is attributable to this structural possibility, which has no analog in conventional mathematics.

STEP 5   The sequence of statements 6 to 8 will be performed five times:

First pass   Statement 6 is, at this time, the second READ statement encountered. Thus, it will cause the reading of the second data card (statement 20) according to the FORMAT labeled 32 (statement 13) and the number on that card to be stored in X. As the FORMAT specifies a real number in columns 4 and 5 (3X,1F2.0) with no places considered to be to the right of the decimal point, the number 21. will be stored in X. Statement 7 will add the *current* contents of X to SUM and store the result in SUM. Conceptually at this time, $SUM = 0. + x_1$. Numerically at this time, $SUM = 0. + 21. = 21.$. Statement 8 will add the square of the *current* contents of X to SUMSQ and store the results back in SUMSQ. Conceptually now, $SUMSQ = 0. + x_1^2$. Numerically, $SUMSQ = 0. + 441. = 441.$.

Second pass   Statement 6 is, at this time, the third READ statement encountered. Thus it will cause the reading of the third data card under the associated FORMAT and store the datum (22.) in location X. Statement 7 will add the *current* contents of X to SUM and store the result in SUM. Conceptually now, $SUM = 0. + x_1 + x_2$. Numerically, $SUM = 0. + 21. + 22. = 43.$. Statement 8 will add the square of the current contents of X to SUMSQ and store the result in SUMSQ. Conceptually now, $SUMSQ = 0. + x_1^2 + x_2^2$. Numerically, $SUMSQ = 0. + 441. + 484. = 925.$.

. . . .

Fifth pass   Statement 6 will cause the datum (18.) from the sixth data card to be stored in X [in the general case, this will be the *n*th pass and the $(n + 1)$st data card]. After execution of statement 7, $SUM = 0. + x_1 + x_2 + x_3 + x_4 + x_5$. Numerically, $SUM = 0. + 21. + 22. + 23. + 20. + 18. = 104.$. (In the general case, SUM will contain the SUM of the *n* scores after *n* passes.) After execution of statement 8, $SUMSQ = 0. + x_1^2 + x_2^2 + x_3^2 + x_4^2 + x_5^2$. Numerically, $SUMSQ = 0. + 441. + 484. + 529. + 400. + 324. = 2178.$. (In the general case, SUMSQ will contain the SUM of the squares of the *n* scores after *n* passes.)

STEP 6   The DO loop having been satisfied, control passes to statement 9, which divides SUM (the sum of scores) by T (the number of cases), producing the mean of the scores, which is then stored in location SUM. Numerically, this quantity is 20.8 for these data. Statement 10 divides SUMSQ (the sum of squares) by T (the number of cases) and subtracts SUM squared (the mean squared), producing the variance of the scores which is stored in SUMSQ. Numerically, this quantity is 2.96 for these data.

Notice that a location may contain values with different conceptual meaning at different points in the program. Only immediately after the loop has been satisfied will the contents of SUM be interpretable as the sum of the scores. After statement 9 has been executed, the contents of SUM will be interpretable as the mean of the scores. We must constantly be aware of the difference in a location, its numerical contents at any time, and the conceptual meaning of those contents *at any time.*

STEP 7   Statement 11 will print the contents of the locations SUM, SUMSQ, and N (conceptually, the mean, the variance of the scores, and the number of cases) under the FORMAT labeled 33 (statement 14). Notice that the *same* WRITE statement would *not* give us the mean and variance had it been positioned earlier in the program. The output on the printer will look as follows:

   ____20.80____2.96__5

Notice that we provided no labeling in the FORMAT. Consequently, to interpret these numbers one would have to know that the program printed the mean and variance or would have to analyze the program to make this determination.

STEP 8   The computer would encounter the STOP statement as the FORMATs are used only in connection with the READ and WRITE statements containing their label number. It would transfer control from the program to the master control system, effectively stopping work with this program and data segment.

2.14   READING AND WRITING WITH THE IMPLIED DO LOOP

One final elaboration of the READ statement should be presented. While there are many specialized forms of the Fortran READ statement, those discussed here will suffice. It will often be necessary to read several variables from a single card (or set of cards) for each case. In complex multivariate analyses we might commonly need to read eight (or more) variables for each case. These are usually conceptualized as a row vector, VAR(I), where VAR(1) is the first variable, VAR(2) is the second variable, ..., and VAR(8) is the eighth variable. We *could* use a READ statement like

   READ (5,7) VAR(1),VAR(2),VAR(3),VAR(4),VAR(5),VAR(6),VAR(7),VAR(8)

However, this is an unnecessarily awkward way of achieving the desired result. A briefer statement which will function in *identically* the same way is

   READ (5,7) (VAR(I),I=1,8,1)

This statement involves an *implied DO loop.* It means "Set I=1. Read VAR(I) [that is, VAR(1)]. Increase I by 1. Read VAR(I) [that is, VAR(2)]. Continue doing this until I exceeds 8." This form of the statement becomes almost essential where even more variables are to be read. Often we might need to read as many as 25 variables, later to be employed in several analyses. This is possible through the statement

   READ (5,7) (VAR(I),I=1,25,1)

This will function exactly as if we had written out all the location names from VAR(1) through VAR(25).

It is possible to have many subscripts for variables in READ statements. One or several of these can be manipulated by (external) explicit DO loops and one or several manipulated by these implied DO loops. Incredibly complicated reading (and writing) operations can thus be defined, but we shall have no need for such complexity in this book. Our use of the implied DO loop will be limited to its time-saving, space-saving function as a substitute for long variable lists in the READ statement.

2.15 DEFINITION AND UTILIZATION OF SUBROUTINES

A subroutine is a (usually) small program which can be repeatedly used by another operating program. The iterated use of one program by another is achieved by turning over control to the first statement in the subroutine in the same way that the master control system ordinarily turns over control to the first statement in a program. Instead of ending with a STOP statement, followed by an END statement, a subroutine usually terminates with a RETURN statement, followed by an END statement. When a RETURN statement is reached, control is returned to the calling program using the subroutine at the statement immediately following the one which used the subroutine. It is as if the entire subroutine had been physically substituted into the calling program at the physical location where it was called. Where the subroutine is fairly complex, e.g., a matrix-inversion subroutine, and the operations it performs are needed at several points in a program, a considerable amount of time and effort are saved by writing the complex of operations only once in a subroutine and calling the subroutine when needed. In moderately complicated programs, e.g., the one in Appendix A, subroutines may even call (use) subroutines, which call (use) subroutines. Fortunately, these labyrinthine matters need not be mastered by the ordinary user of such programs. We shall not examine the rather intricate rules for setting up such subroutines. It will be helpful, however, to note how the use of subroutines changes our notion of program organization. Additionally, the reader will have to modify one subroutine in the program we shall use. This is called SUBROUTINE DATA, and it simply reads data from cards and performs any transformations required before analyses are performed. Programs involving the use of subroutines are organized as shown in Table 2.11.

2.16 SUBROUTINE DATA

The last subroutine in the program package (Appendix A) is SUBROUTINE DATA. It is within this subroutine that the user must write the Fortran statements to bring data into the computer from the data cards. For this purpose, we must know how to design a READ statement and FORMAT statement. The READ list must treat these variables as a row vector named VAR(I) and of maximum dimension 25. Additionally within this subroutine, the user must perform any necessary transformations which are to be applied to the input data before analyses begin and store the (possibly) transformed variables to be analyzed in a row vector named VEC(I) of maximum dimension 100. As the program is currently set up, no more than 500 cases can be used, though this may be changed. The subroutine should be set up to read and transform data for each case. The computer will repeatedly use the subroutine until all cases have been processed. The simplest use of this program would involve analyses in which all data are read directly from cards and these (untransformed) data are analyzed. In this case, the subroutine would be set up as follows:

```
      SUBROUTINE DATA

      DIMENSION VAR(25),VEC(100)
      COMMON NVAR,NVEC,VAR,VEC

      READ (5,19) (VAR(I),I=1,8,1)
 19   FORMAT (details depend on data positioning in cards)
      DO 3 I=1,8,1
  3   VEC(I)=VAR(I)

      RETURN
      END
```

Table 2.11   REPRESENTATION OF PROGRAM
ORGANIZATION INCLUDING
SUBROUTINES

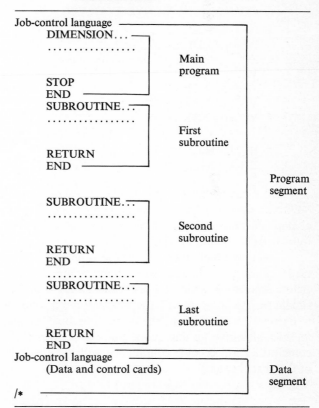

The first three and last two statements must always be used for this subroutine. Between these groups is the user-supplied portion of the subroutine. The four example statements would read eight variables from data cards for *one* case (given an appropriate FORMAT specification), store those values in the vector VAR, and transfer them to the vector VEC. The main program would do this repeatedly until all cases are processed. The last step is, of course, trivial where the variables are to be used in the same form in which they appear on cards. It is necessary, however, that the program allow the option of transforming the data brought from cards (VARs) in ways that make analytic sense before they are placed in other locations for analysis (VECs). A simple example of such a transformation follows. Let us assume that we believe it worthwhile to examine the possibility that a variable Y is related by an equation of the third degree to some variable X. We shall see later (Chap. 4) that this will require four vectors: the Y scores, the X scores, the square of the X scores, and the cube of the X scores. We *could*, of course, punch all these vectors into cards and use a subroutine like the one presented above. An easier approach involves reading only Y and X from cards as VAR(1) and VAR(2) and letting the machine construct the additional requisite values. Such a subroutine might look like following:

```
SUBROUTINE DATA

DIMENSION VAR(25),VEC(100)
COMMON NVAR,NVEC,VAR,VEC

     READ (5,33) VAR(1),VAR(2)
33   FORMAT (depends on data locations)
     VEC(1)=VAR(1)
     VEC(2)=VAR(2)
     VEC(3)=VAR(2)**2
     VEC(4)=VAR(2)**3


     RETURN
     END
```

After having been performed for all cases, this subroutine would produce four VECs for analysis. The first would contain all values for the y_i's, the second would contain all values for the x_i's, the third would contain all values for the x_i^2's, and the fourth would contain all values for the x_i^3's.

The only other steps involved in running the program are writing two control cards, which are explained, where appropriate, in Chaps. 3 and 4 and Appendix A. Before proceeding to Chap. 3, the reader (whether novice or experienced programmer) should run the program with the sample subroutine, data, and analysis cards provided in Exercise 2.9. In this way the user can become familiarized with the mechanics and logistics of setting up and submitting the program. If this is not done, mechanical complexities will become confounded with the theoretical and analytic complexities of the next chapter.

EXERCISES

2.1 Assume the following:

$$A = \begin{bmatrix} 1 \\ 2 \\ 3 \\ 4 \end{bmatrix} \quad B = \begin{bmatrix} 4 \\ 3 \\ 2 \\ 1 \end{bmatrix} \quad C = \begin{bmatrix} -1 \\ -2 \\ 2 \\ 6 \end{bmatrix} \quad D = \begin{bmatrix} 2 \\ 4 \\ 0 \\ 5 \\ 1 \end{bmatrix} \quad E = \begin{bmatrix} 12 \\ 3 \\ 4 \\ -7 \\ -1 \end{bmatrix}$$

$F' = \begin{bmatrix} 3 & 4 & 2 & 1 \end{bmatrix}$ $\quad G' = \begin{bmatrix} 5 & 1 & -2 & -3 & -1 \end{bmatrix} \quad H' = \begin{bmatrix} 1 & 1 & 1 & 1 & 1 \end{bmatrix}$
$x = 3 \quad y = 8$

(*a*) Find A', G, $(A')'$, $[(C')']'$, x'.

(*b*) Where possible, find the following (where not possible, indicate why):

$$\begin{array}{cccccccc} A+B & D+E & C+D & F'+G' & G'+H' & A+F' & A'+F' \\ D+G & x+A & x+y & A+B-C+F \end{array}$$

(*c*) Where possible, find the following (where not possible, indicate why):

$$\begin{array}{cccccccc} A'B & AB' & H'D & D'H & xD & xy & H'A & U'A & N'A \\ DE & D'EA \end{array}$$

(*d*) Show that

A, B, and U are a linearly dependent set of vectors.

$$C \perp B \quad B \perp C \quad D \perp E \quad G \perp H$$

(*e*) Find and interpret

$$H'H \quad H'D \quad \tfrac{1}{5}H'D \quad A'A \quad A'B$$

2.2 Indicate which of the following are invalid variable names (for valid variable names indicate the type).

A	A(2)	X(I,J)
I	I(3)	Y(1,2)
ABLOT	A(I)	C(1.,3.)
I12I	A(B)	I(I,1)
XMEANS	I(J)	A(C,D)
A1254	I(I)	P(K,L,M2)
A*2	A(I124)	B(1,3,I)
2B	A(I3/)	L(1,3,K,L)
B(	A(I12345)	K12(I,3,I2)

2.3 Perform the following Fortran operations and indicate the results.
(*a*) Constants (remember integer arithmetic):

$$\begin{array}{lll} 1+2 = & 1/2 = & 1.+2. = \\ 1./2. = & 1.*2. = & 2**2 = \\ 2**.5 = & 2.**.5 = & 4.**4 = \\ 3*2 = & & \end{array}$$

2.4 Variables (remember integer arithmetic). Assume that locations A, B, C, I, and J; vectors **D** and **E**; and matrices **G** and **H** contain the following:

A	B	C	I	J	D	E
5.1	8.	−2.3	1	3	1.3	−1.
					2.1	0.
					3.8	1.

G				H		
1.	2.	3.		1.	4.	7.
4.	5.	6.		2.	5.	8.
7.	8.	9.		3.	6.	9.

Assume also, an empty vector **F** (of dimension 3) and an empty matrix **P** (of dimensions 3,3).

(*a*) Perform the indicated Fortran operations and give the result:

Q=A+B−C	Q now contains _____?
F(1)=D(1)+B	F(1) now contains _____?
F(J)=E(2)+E(I)	F(J) now contains _____?
(and which F is it?)	
R=A**2	R now contains _____?
S=B**J	S now contains _____?
T=B**E(I)	T now contains _____?
U=SQRT(G(2,1))	U now contains _____?
K=J**2	K now contains _____?
L=I*J	L now contains _____?
M=I/J	M now contains _____?
V=A/B	V now contains _____?
P(I,J) = G(I,J)+H(I,J)	P(I,J) now contains _____?
(and which P is it?)	
P(3,J)=G(I,1)+H(1,I)	P(3,J) now contains _____?
(and which P is it?)	

(*b*) For each of the following expressions outline the order in which the operations are performed and indicate what the resultant number is and where it is stored.

Q=A+B*C	R=(A+B)*C	S=A*C+B*C
T=A+B**2/C	U=(A+B**2)/C	V=(A+B)**2/C
W=(A+B)**(2./E(I))	Y=SQRT((A+B)**2)	Z=4.**.5

2.5 Write a program which will read a number and then write it.

2.6 Write a program which will read two numbers, sum them, and write the sum.

2.7 Write a program which will read two vectors, calculate the sum of the two, and write the sum.

2.8 Set up the program segment in Table 2.10, and add to it any necessary job-control statements, and run it. Change the data set and run the program.

2.9   Set up and run program CARTROCK by:

(a) Inserting the following Fortran statements into SUBROUTINE DATA:

          . . . .

          READ (5,19) (VAR(I),I=1,5)

    19   FORMAT (5F3.0)

          DO 3 I=1,5

     3   VEC(I)=VAR(I)

          VEC(6)=1.

          . . . .

(b) Inserting the following main card, control/data cards, and model cards into the data segment. See Appendix A (_ means a blank space).

          . . . .

          //GO. S YSIN_DD_*

          TEST_NO. _1 _ _ _1 0____5____6____7

          __1_10 __1__5 _ _1

          __2__9 __2__4 _ _4

          __3__8 __3__3 _ _9

          __4__7 __4__2 _16

          __5__6 __5__1 _25

          __6__5 __5__1 _36

          __7__4 __4__2 _49

          __8__3 __3__3 _64

          __9__2 __2__4 _81

          _10__1 __1__5100

          TEST_A___ __ _ _1____2____6____2

          TEST_B ___ __ _ _1____2____6____3

          TEST_C___ __ _ _1____2____6____4

          TEST_D___ __ _ _1____2____6____5

          TEST_E ___ __ _ _1____1____6

          TEST_F ___ __ _ _3____2____6____4

          TEST_G___ __ _ _3____2____6____1

          /*

3

LINEAR MODELS AND THE ANALYSIS OF VARIANCE

3.1 INTRODUCTION TO MODELS AND THEIR USES IN HYPOTHESIS FORMULATION

Hypothesis testing through regression is usually based on the criterion of *least squares*. We shall first approach least-squares theory through discussion of the logically simplest single-equation[1] model and the logically most complete one. The simplest we shall call the *random-variation model*, as it implicitly assumes that all deviations from the expected value for a variable are due to chance or random forces and may be safely ignored. The most complete model we shall call the *idiographic model*, as it treats each observation (value on a variable) as the result of forces which are unique in important respects. Neither of these models will be useful for explanatory or causal analytic purposes, but the exposition will help to clarify what is meant by a model and by the criterion of least squares. Analytically useful models, the real concern of this chapter, will fall logically between these two (trivial) extreme cases.

---

[1] In Part One, we shall restrict ourselves to ordinary least-squares single-equation models. In a single-equation model one variable is viewed as dependent in all equations and never as an independent variable. In Part Three, there is some discussion of multiple-equation models, where the same variable may at one time be viewed as causally independent and at another as causally dependent.

Random-Variation Model

If we believe that some variable Y has a single characteristic expected value for a population and that all deviations from that value are produced by random factors (which can be safely ignored), we can solve for that characteristic with the system of equations implied by the model

$$\mathbf{Y} = b\mathbf{U} + \mathbf{E}$$

In this mathematical statement, $\mathbf{Y}$ is the dependent variable, b is the characteristic expected value which we seek, $\mathbf{U}$ is a unit vector (called the independent or predictor variable), and $\mathbf{E}$ is the vector of residuals (errors of prediction) assumed here to be random.[1] Of course, we know how to solve directly for b in this case, as it is $\overline{Y}$. However, it is useful to illustrate what is meant by a model and by the least-squares criterion with a familiar and simple case before proceeding to more powerful and complex cases. The elements of this model can be represented more explicitly by writing out the vectors

$$\begin{bmatrix} y_1 \\ y_2 \\ \vdots \\ y_n \end{bmatrix} = b \begin{bmatrix} 1 \\ 1 \\ \vdots \\ 1 \end{bmatrix} + \begin{bmatrix} e_1 \\ e_2 \\ \vdots \\ e_n \end{bmatrix}$$

In more conventional notation, the model can be represented as the following set of simultaneous linear equations:

$$y_1 = b \times 1 + e_1$$
$$y_2 = b \times 1 + e_2$$
$$\cdots\cdots\cdots\cdots\cdots$$
$$y_n = b \times 1 + e_n$$

In both cases, it is explicit that each person's (unit's) score is y_i and the error of predicting that score from b is e_i. Note that we have symbolized the assumption that the dependent variable has a single characteristic expected value by using predictor equations of identical *form* for all values of the dependent variable. The n equations are said to be of the same form in that they all include the same unknown. Analytically useful models will usually *not* involve identical equations for all values of the dependent variable. If $\mathbf{Y}$ were a vector containing the ages of five people, the model could be represented numerically as

$$\begin{bmatrix} 25 \\ 42 \\ 50 \\ 22 \\ 17 \end{bmatrix} = b \begin{bmatrix} 1 \\ 1 \\ 1 \\ 1 \\ 1 \end{bmatrix} + \begin{bmatrix} e_1 \\ e_2 \\ e_3 \\ e_4 \\ e_5 \end{bmatrix}$$

[1] For all tests discussed in Chaps. 3 and 4, the dependent variable Y must be *conceptually* interval. This is because we are treating errors of prediction as varying in magnitude and summing them. The notion of distance is critical to the least-squares criterion. However, we need not fret too much about this if our dependent variable is many-valued and not strictly rank ordered (see McNemar, 1962, pp. 374–376). Indeed, it can be maintained that the designation of data as interval is a substantive and judgmental decision (see Carter, 1971, especially n. 9). Also, the dependent variable is said to be *conceptually* continuous but in practice need only be many-valued. After all, continuity in the mathematical sense is not an empirical notion and is never observable.

In the absence of constraints, this model (these equations) has no unique solution; b may be arbitrarily defined as any constant, and (if the elements of **E** are free to take any values) all equations will balance. This is because the six quantities $(b, e_1, e_2, e_3, e_4, e_5)$ are not determined by five equations. A unique solution is possible if we impose the restriction that the solution for the constant b minimize the length of the error vector ($\mathbf{E'E} = \min$). This is the same as saying that $\sum_{i=1}^{n} e_i^2$ must be a minimum. This is the simplest statement of the least-squares criterion. When a solution minimizing $\mathbf{E'E}$ has been found, the error vector **E** will be orthogonal to the predictor vector **U** ($\mathbf{E} \perp \mathbf{U}$, or $\mathbf{E'U} = \mathbf{U'E} = 0$). This is a way of saying that the errors of prediction will be independent of the predictor vectors. In more useful and complicated models, involving many binary orthogonal predictor vectors, **E** will be orthogonal to every predictor (independent) vector.

A direct solution for the proper constant b can be provided. It is

$$b = \frac{1}{n}\mathbf{U'Y}$$

In our numerical example, this solution will be

$$b = \tfrac{1}{5}\begin{bmatrix} 1 & 1 & 1 & 1 & 1 \end{bmatrix}\begin{bmatrix} 25 \\ 42 \\ 50 \\ 22 \\ 17 \end{bmatrix} = 31.2$$

The example, including the least-squares solution for b and values for the vector **E**, will then become

$$\begin{bmatrix} 25 \\ 42 \\ 50 \\ 22 \\ 17 \end{bmatrix} = 31.2 \begin{bmatrix} 1 \\ 1 \\ 1 \\ 1 \\ 1 \end{bmatrix} + \begin{bmatrix} -6.2 \\ 10.8 \\ 18.8 \\ -9.2 \\ -14.2 \end{bmatrix}$$

That $\mathbf{E'E}$ is a minimum with this value of b can be verified by showing that $\mathbf{U'E} = 0$; that is, $\mathbf{U} \perp \mathbf{E}$. Here

$$\begin{bmatrix} 1 & 1 & 1 & 1 & 1 \end{bmatrix}\begin{bmatrix} -6.2 \\ 10.8 \\ 18.8 \\ -9.2 \\ -14.2 \end{bmatrix} = -6.2 + 10.8 + 18.8 - 9.2 - 14.2 = 0$$

We shall later show how to use the computer program to get solutions which minimize the length of the error vector for models of any degree of complexity. We shall also have considerable interest in the quantity $\mathbf{E'E}$ (or $\sum e_i^2$), the error variation associated with our assumptions of the predictor model. In the present case, this quantity is the error variation associated with the assumption that the same form of equation (involving as an unknown the same expected value) reasonably typifies the value of Y for all cases. Notice that this least-squares estimator is the mean.

We shall later examine the reasonableness of such an assumption (for any data) by comparing the magnitude of this error sum of squares (SS) with the error SS associated with alternative views.

Idiographic Model

Momentarily, assume that important, unique forces produce the n observed values on the variable Y and that no overriding generic or summary laws may be resorted to with safety. Such a view leads to the logically most complex least-squares model. This state of affairs can be summarized by the model

$$\mathbf{Y} = b_1\mathbf{X}_1 + b_2\mathbf{X}_2 + b_3\mathbf{X}_3 + \cdots + b_n\mathbf{X}_n + \mathbf{E}$$

where $\mathbf{Y}$ is the dependent variable of dimension n, the $b_1, b_2, b_3, \ldots, b_n$ are the unknown weights to be determined by solving the model by the least-squares criterion, the $\mathbf{X}_1, \mathbf{X}_2, \mathbf{X}_3, \ldots, \mathbf{X}_n$ constitute a set of binary orthogonal vectors such that $\mathbf{X}_i$ is 1 for the ith case and 0 for all others, and the $\mathbf{E}$ is a vector of residuals (errors of prediction). When the elements are defined, this model becomes

$$\begin{bmatrix} y_1 \\ y_2 \\ y_3 \\ y_4 \\ \vdots \\ y_n \end{bmatrix} = b_1 \begin{bmatrix} 1 \\ 0 \\ 0 \\ 0 \\ \vdots \\ 0 \end{bmatrix} + b_2 \begin{bmatrix} 0 \\ 1 \\ 0 \\ 0 \\ \vdots \\ 0 \end{bmatrix} + b_3 \begin{bmatrix} 0 \\ 0 \\ 1 \\ 0 \\ \vdots \\ 0 \end{bmatrix} + b_4 \begin{bmatrix} 0 \\ 0 \\ 0 \\ 1 \\ \vdots \\ 0 \end{bmatrix} + \cdots + b_n \begin{bmatrix} 0 \\ 0 \\ 0 \\ 0 \\ \vdots \\ 1 \end{bmatrix} + \begin{bmatrix} e_1 \\ e_2 \\ e_3 \\ e_4 \\ \vdots \\ e_n \end{bmatrix}$$

Also, to labor our notations a bit until they become familiar, the model implies the following set of simultaneous linear equations:[1]

$$y_1 = b_1 + 0 + 0 + 0 + \cdots + 0 + e_1$$
$$y_2 = 0 + b_2 + 0 + 0 + \cdots + 0 + e_2$$
$$y_3 = 0 + 0 + b_3 + 0 + \cdots + 0 + e_3$$
$$y_4 = 0 + 0 + 0 + b_4 + \cdots + 0 + e_4$$
$$\cdots\cdots\cdots\cdots\cdots\cdots\cdots\cdots\cdots\cdots\cdots$$
$$y_n = 0 + 0 + 0 + 0 + \cdots + b_n + e_n$$

These equations have no unique solution. However, we can solve for the unique solution for the b_i's which minimize $\mathbf{E}'\mathbf{E}$ (or $\sum e_i{}^2$). For this solution $\mathbf{E}$ will be orthogonal to all $\mathbf{X}$'s. The least-squares solution produces $b_i = y_i$, for all i's, and every $e_i = 0$. For this solution $\mathbf{E}'\mathbf{E} = 0$ and $\mathbf{E} \perp \mathbf{X}_i$ for all i's. Obviously, this solution involves predicting every element's score on $\mathbf{Y}$ from its score on $\mathbf{Y}$. Also, the vector $\mathbf{E}$ satisfies the orthogonality criterion because a null vector is orthogonal to any vector.

---

[1] The equations have been simplified by eliminating any terms multiplied by zero. Thus, the complete representation for the first would be

$$y_1 = b_1 \times 1 + b_2 \times 0 + b_3 \times 0 + b_4 \times 0 + \cdots + b_n \times 0 + e_1$$

Thus this limiting case of the mathematics of single-equation least squares is trivial because a unique equation is permitted for every observation. All useful cases of single-equation models involve the imposition of linear conditions which group observations and use identical equations for predicting several different (conceptually grouped) observations. We are using the term prediction in its conditional sense rather than the temporal sense. We are specifying the best guess for a value conditioned on the distribution of values (or, later, conditioned on the relationships of that distribution with other distributions). We are *not* predicting the value for some future point in time.

These two extreme models are closely related to each other. Notice that imposing the following conditions on the idiographic produces the random-variation model:

$$b_1 = b_2 \quad (= \text{some general } b)$$
$$b_1 = b_3 \quad (= \text{some general } b)$$
$$b_1 = b_4 \quad (= \text{some general } b)$$
$$\cdots\cdots\cdots\cdots\cdots\cdots\cdots\cdots\cdots\cdots$$
$$b_1 = b_n \quad (= \text{some general } b)$$

Thus the model becomes

$$\mathbf{Y} = b\mathbf{X}_1 + b\mathbf{X}_2 + b\mathbf{X}_3 + b\mathbf{X}_4 + \cdots + b\mathbf{X}_n + \mathbf{E}_1$$
$$\mathbf{Y} = b(\mathbf{X}_1 + \mathbf{X}_2 + \mathbf{X}_3 + \mathbf{X}_4 + \cdots + \mathbf{X}_n) + \mathbf{E}_1$$
$$\mathbf{Y} = b\mathbf{U} + \mathbf{E}_1$$

The reader should verify that the definition of $\mathbf{X}_1, \mathbf{X}_2, \mathbf{X}_3, \mathbf{X}_4, \ldots, \mathbf{X}_n$ implies that $\sum_{i=1}^{n} \mathbf{X}_i = \mathbf{U}$. In general, the sum of a set of orthogonal binary vectors is a unit vector.

We have shifted from the subscripted b_i to some general b to make explicit that the b_i's are least-squares weights for one model (their solution will be $b_i = y_i$) and the b is the least-squares weight for another (its solution will be $b = \overline{Y}$). Also, we have changed the subscript for the error vector. That vector is $\mathbf{E}$ in the *full model* (idiographic model), and *here* $\mathbf{E}_1$ in the derived *restricted model* (random-variation model). The shift makes explicit that the two error vectors *may* be different. They *will* be different unless all cases have the same value; i.e., the hypotheses are exactly true.

Hypothesis Formulation

Regression models are most usefully employed in pairs, one designated the full model and the other the restricted model. The full model will represent a set of equations in which the hypotheses of interest are allowed to be either true or false. In the restricted model these hypotheses are forced to be true. In fact, it will appear that all regression hypotheses are derivable from direct algebraic impositions of linear conditions on appropriate full models. An example should help clarify the difference between these two models.

Assume that we believe that the determinants of the votes of Democratic and Republican representatives are different in important respects; i.e., they produce distinctly different expected values for some measure of liberalism (such as ADA ratings) for Democrats and Republicans. Statistical evidence for this belief could be provided by comparing the errors of prediction produced by an appropriate pair of regression models. These are developed below:

Full model:
$$Y = b_1 X_1 + b_2 X_2 + E_F$$

where Y is the dependent variable for the representatives, i.e., each element is a single representative's ADA rating; X_1 and X_2 are a pair of binary orthogonal vectors such that an element of X_1 is 1 if the representative is a Democrat (0 otherwise) and an element of X_2 is a 1 if the representative is a Republican (0 otherwise); E_F is a vector each element of which is the error of prediction for a single representative's ADA rating; and b_1 and b_2 are the regression weights which minimize the length of the error vector (produce a minimum $E_F'E_F$).

If we write out the elements of this model with nd the number of Democrats and nr the number of Republicans, we get

$$
\begin{bmatrix} y_1 \\ y_2 \\ \vdots \\ y_{nd} \\ y_{nd+1} \\ y_{nd+2} \\ \vdots \\ y_{nd+nr} \end{bmatrix}
= b_1 \begin{bmatrix} 1 \\ 1 \\ \vdots \\ 1 \\ 0 \\ 0 \\ \vdots \\ 0 \end{bmatrix}
+ b_2 \begin{bmatrix} 0 \\ 0 \\ \vdots \\ 0 \\ 1 \\ 1 \\ \vdots \\ 1 \end{bmatrix}
+ \begin{bmatrix} e_1 \\ e_2 \\ \vdots \\ e_{nd} \\ e_{nd+1} \\ e_{nd+2} \\ \vdots \\ e_{nd+nr} \end{bmatrix}
$$

Notice that this model generates N, that is, $nd + nr$, equations, which are of two forms. The equations for Democrats are all of the form

$$y_i = b_1 \times 1 + b_2 \times 0 + e_i$$

(There will be nd of these.) The equations for Republicans are all of the form

$$y_i = b_1 \times 0 + b_2 \times 1 + e_i$$

(There will be nr of these.)

These equations have unique solutions for both b_1 and b_2 if we use the least-squares criterion. Notice that the Republican ADA ratings (and errors) have no effect on the solution for b_1 because b_1 is weighted 0 for predicting Republican ADA ratings. Similarly, the Democratic ADA ratings (and errors) have no effect on the estimates for b_2 because b_2 is weighted 0 for Democratic predictions. Thus, in seeking the solution for b_1, we can concern ourselves *only* with the simplified equations of the form

$$y_i = b_1 \times 1 + e_i \qquad \text{for } i = 1, 2, \ldots, nd$$

The least-squares solution for b_1 for systems of equations of this form, as we saw when discussing the random-variation model, is the mean of the observations on the dependent variable. This is, of course, the mean ADA rating for Democrats ($b_1 = \overline{Y}_D$).

Similarly, in seeking the solution for b_2, we can concern ourselves *only* with the simplified equations of the form

$$y_i = b_2 \times 1 + e_i \quad \text{for } i = nd + 1, nd + 2, \ldots, nd + nr$$

Of course, this least-squares solution produces $b_2 = \overline{Y}_R$ (the mean ADA rating for Republicans). Notice also that the error SS is made up of the familiar quantities

$$e_i = \begin{cases} y_i - b_1 = y_i - \overline{Y}_D & \text{for Democrats} \\ y_i - b_2 = y_i - \overline{Y}_R & \text{for Republicans} \end{cases}$$

Consequently,

$$\sum_{i=1}^{nd} e_i{}^2 = \frac{\text{variation of Democrats' ADA ratings around}}{\text{Democratic mean ADA rating}}$$

$$\sum_{i=nd+1}^{nd+nr} e_i{}^2 = \frac{\text{variation of Republicans' ADA ratings around}}{\text{Republican mean ADA rating}}$$

The error SS is the sum of these two sums, or $\sum_{i=1}^{nd+nr} e_i{}^2$. As such, it is exactly equivalent to the notion of within-group variation (the sum of the squared deviations of each observation from the group mean for the group in which the observation is embedded).

Recall that our initial hypothesis involved the question: Do Democratic and Republican representatives differ with respect to expected ADA rating? Our first task was to construct a full model which allowed this to be true or false. The model outlined above satisfies this condition; b_1 is the expected (least-squares) value for Democrats' ADA rating, and b_2 is the expected (least-squares) value for Republicans' ADA rating. It is possible for the least-squares solutions for this model to produce weights such that $b_1 = b_2$ or $b_1 \neq b_2$. Statistical evaluation of the hypothesis requires that we formulate our hypothesis explicitly in the form of linear conditions which can be imposed on the full model. Specifically, the null hypothesis to be tested is that the error of prediction $\mathbf{E'E}$ will not increase (significantly) if we impose the research hypothesis as a linear condition.[1] The proper linear condition for our question is that $b_1 = b_2$ (= some general b which reasonably typifies *all* representatives regardless of party). If we substitute b for b_1 and b_2 in the full model, we get a restricted model which "forces" the hypothesis to be true and gives the best fit under the *assumption* of the hypothesis. This will be

Restricted model:

$$\mathbf{Y} = b\mathbf{X}_1 + b\mathbf{X}_2 + \mathbf{E}_R = b(\mathbf{X}_1 + \mathbf{X}_2) + \mathbf{E}_R = b\mathbf{U} + \mathbf{E}_R$$

---

[1] The null hypothesis is sometimes taken to be that a pair of means are equal. In fact the null hypothesis can be stated: The difference in a pair of means differs from some expected value d by zero. This is equivalent to our usage. In all cases our null hypothesis will be that our research hypotheses result in zero (null) increase in the error SS $\mathbf{E'E}$. A *linear* condition is one which expresses any weight b_i as a weighted sum of other weights, $b_j, b_k, \ldots$. Thus, the following hypotheses are linear conditions:

$$b_1 = b_2 \qquad b_1 = 2b_2 \qquad b_1 = b_2 + 2 \qquad b_1 = b_2 + b_3$$

However, a hypothesis of the form $b_1 = b_2 b_3$ is not a linear condition. We shall later discover that hypotheses of this form can be imposed by conversions to logarithms.

Of course, this is the random-variation model, and b will be $\overline{Y}_T$. Also, the error SS will be the total variation.

Notice that if the hypothesis ($b_1 = b_2$) were (exactly) true, the weight b would be equal to b_1 and b_2. Therefore, the least-squares predictions for y_i would be the same for all cases in both the full and restricted models. Consequently, the errors of prediction would be the same in both models (since $e_{i_F} = e_{i_R}$). Thus, if the hypothesis were (exactly) true, $\mathbf{E}'_F\mathbf{E}_F$ would equal $\mathbf{E}'_R\mathbf{E}_R$. There would be *no* increase of error caused by the imposition of our hypothesis. In general, the difference in $\mathbf{E}'_F\mathbf{E}_F$ and $\mathbf{E}'_R\mathbf{E}_R$ (or $\sum e_{i_F}^2$ and $\sum e_{i_R}^2$) is a measure of the degree to which our hypothesis is (or hypotheses are) false. The hypothesis (hypotheses) will always be imposed algebraically on an appropriate full model to produce the restricted model. Notice that to this point no assumptions are necessary save that the *dependent* variable be conceptually interval-scaled, as the notion of distance is necessary for errors to be treated as varying in magnitude. No assumptions are made about the form of the distribution of the dependent variable, and no assumptions whatsoever are made about the independent variable(s).

3.2   TESTING HYPOTHESES WITH THE F RATIO

Usually it is not possible to impose our hypotheses without producing some increase in the error SS. Thus, the next question of interest is how much the error SS must be increased before we are forced to consider our hypothesis (hypotheses) to be false. To answer this familiar question of statistical inference we shall always resort to the F distribution. Several traditional assumptions become necessary for rigorous interpretation of the F ratio; they are discussed below. The robustness literature cited at that point suggests that these assumptions need not overly trouble the careful researcher. For our purposes we shall consider the F statistic in the form

$$F = \frac{s_H^2}{s_A^2} = \frac{\text{error-variance estimate attributable to hypotheses}}{\text{error-variance estimate attributable to assumptions}}$$

By *assumptions* we mean here those conditions which are imposed without test on our full model. In the example above, the full model uses a single equational form for predicting ADA rating for all Democrats (also a single form for all Republicans). *Within* party it is assumed by the model that a random-variation model is adequate. It is quite possible, indeed true, that other factors, e.g., region, constituency, or ideology, significantly influence votes *among* Democrats (or Republicans). However, the full model presented will not allow us to test these notions. More complicated models, discussed later, will permit such tests. The error variance attributable to the assumptions will be the full-model error SS divided by the appropriate degrees of freedom. Specifically,

$$s_A^2 = \frac{\mathbf{E}'_F\mathbf{E}_F}{\text{df}_2} = \frac{\sum e_{i_F}^2}{\text{df}_2}$$

By *hypotheses* we mean those conditions imposed algebraically on the full model to produce the restricted model. In the example above, the only hypothesis is that $b_1 = b_2$, that is, that the mean ADA rating is the same for Democrats and Republicans. Notice that the assumptions are present and produce error variance in both models, while the hypothesis (hypotheses) is (are) present only in the restricted model and thus can produce error variance in that model only. The error variance attributable to the hypotheses will be a function of the *difference* in the full-model error SS (without the hypotheses) and the restricted-model error SS (with the hypotheses). Specifically,

$$s_H{}^2 = \frac{\mathbf{E}_R'\mathbf{E}_R - \mathbf{E}_F'\mathbf{E}_F}{\mathrm{df}_1} = \frac{\sum e_{i_R}{}^2 - \sum e_{i_F}{}^2}{\mathrm{df}_1}$$

The F ratio can now be rewritten in one of the forms which we shall most commonly use for hypothesis testing

$$F = \frac{(\sum e_{i_R}{}^2 - \sum e_{i_F}{}^2)/\mathrm{df}_1}{\sum e_{i_F}{}^2/\mathrm{df}_2} \tag{3.1}$$

Where $F < 1$, we shall be unable to reject the hypotheses we have imposed. Where $F > 1$, we shall have to refer to the F distribution at df_1 and df_2 degrees of freedom. If F exceeds one of the critical values in a standard F table, we shall be able to reject our hypotheses at the .05, .01, or .001 level of significance.

The only terms which we have yet to define are df_1 and df_2. Precise definition of these terms is rather involved. Fortunately, as we shall see below, rules for determining their values are rather simple. df_2 is the number of degrees of freedom associated with the error vector in the full model. df_1 is the *difference* in the number of degrees of freedom associated with the full- and restricted-model error vectors. Before exploring the implications of these definitions we must briefly explicate the general notion of degrees of freedom.

Degrees of Freedom

When linear conditions are imposed on a set of observations, some of the observations are thereby constrained, their values being fixed given the values of the remaining observations. The most familiar linear condition is that the set of observations sums to some quantity. For example, the computation of a mean for a set of N observations requires summing the observations. Deviation scores are then defined by subtracting this mean from each score. Thus, a set of N deviation scores is constrained by the fact that the sum of the set must be zero. This constraint means that the set of N scores has $N - 1$ degrees of freedom. Notice that with the set of N scores $X_1, X_2, X_3, X_4, \ldots, X_{N-1}, X_N$ any values may be assigned any subset of $N - 1$ of them and the set may still sum to zero. The excluded element *must* be the negative of the sum of the subset for this to be true. Its value is not free to vary. It is in this sense that the imposition of the condition that the elements sum to zero is said to reduce the freedom of the set by 1. If we treat all elements as initially free to vary (N degrees of freedom), this constraint produces a system with $N - 1$ degrees of freedom.

The error terms in our regression equations are a system of deviation scores ($e_i = y_{i,\mathrm{obs}} - y_{i,\mathrm{pred}}$). In the random-variation model the predicted score for all cases is the mean. In this case, the error terms are simple deviation scores, and all that has been said above applies directly to the errors for that model. Specifically, the errors must sum to zero, and therefore the model has $N - 1$ degrees of freedom, where N is the number of observations.

The situation is only slightly complicated by more comprehensive models. For example, we observed that the full model discussed above has two *types* of predictor equations, one type for Democrats and another for Republicans. Let us talk of the error vector as being partitioned into the Democratic subset and the Republican subset. *Within* each subset, we can reduce the predictor equation to a random-variation model. Thus by our earlier argument, *within* each subset the errors must sum to zero. Consequently, the Democratic subset has $nd - 1$ degrees of freedom (where nd is the number of Democrats), and the Republican subset has $nr - 1$ degrees of freedom (where nr is the number of Republicans). The total number of degrees of freedom for the error vector $\mathbf{E}_F$ is $N - 2 = (nd - 1) + (nr - 1)$, where $N = nd + nr$. Were there k partitions or groups, 1 degree of freedom would be lost *within* each group and the degrees of freedom would be $N - k$.

While this discussion should give the reader some grasp of the notion of degrees of freedom, their actual determination for a model can be made more simply. The number of degrees of freedom associated with a regression model is N minus the number of *linearly independent* unknown weights to be determined by the least-squares criterion. Thus, if no linear dependencies are present in the model, the number of degrees of freedom of the model is N minus the number of vectors in the model (*exclusive* of the dependent variable and the error vector).

Linearly independent weights are weights associated with linearly independent vectors, discussed in Sec. 2.3. The computer package provided will warn the user if linear dependencies exist. A formal treatment of the topic can be found in Noble (1969, pp. 97–104). Each linear dependency will reduce the number of weights to be counted for this purpose by 1. The dependent variable is excluded from counting as it has known weight 1, and the error vector is excluded from counting as it is a linear combination of the dependent and all independent variables. If df_F is the number of degrees of freedom for the full model and df_R is the number of degrees of freedom for the restricted model, we can define the requisite degrees of freedom for the F ratio rather simply as

$$df_1 = df_R - df_F \qquad df_2 = df_F$$

It may be useful to note that df_1 and df_2 can be defined in several alternative ways which are, of course, equivalent.

Definitions of df_1

1 df_1 is the difference in the number of degrees of freedom associated with the restricted and full models.

2 df_1 is the difference in the number of unknown weights (associated with *linearly independent* vectors) in the full and restricted models.

3 df_1 is the number of *independent*[1] conditions (hypotheses) imposed on the full model to produce the restricted model.

---

[1] Often we shall impose conditions, for example, $b_1 = b_2$, $b_1 = b_3$, which formally imply other equivalences, for example, $b_2 = b_3$. These implied conditions are not to be counted, as they are already imposed algebraically by the imposition of the conditions from which they may be derived.

Definitions of df_2

1 df_2 is the number of degrees of freedom associated with the full model. (Notice that the idiographic has zero degrees of freedom.)

2 df_2 is N minus the number of unknown weights (associated with *linearly independent* vectors) in the full model. Notice that the idiographic has N unknown weights. Thus, df_2 is the difference in the number of (*independent*) unknown weights in the idiographic and the full model.

3 df_2 is the number of *independent* conditions which would have to be imposed on the idiographic in order to produce the full model. This is another way of saying that df_2 is the minimal number of linear assumptions embodied in the full model.

Alternative Form of the F Ratio

Where, as is usually the case,[1] the *dependent variable* is the same for the full and restricted model and there is an explicit or implicit unit vector, we can compute the F statistic from the multiple R^2's provided by most standard regression programs. In the regression program provided, this is labeled the *coefficient of determination*. We shall later derive this quantity and show it to be the proportion of variance in the dependent variable which is accounted for by the string of independent variables. It is also the square of the correlation between the predicted and observed values of Y. As such, R^2 can be related directly to the error SS (used in the definition of F, above) by the following formulas, making no assumptions save that Y is conceptually interval-scaled (see McNemar, 1962, pp. 129–132; Blalock, 1972, pp. 389–393):

$$\sum e_{i_F}^2 = \sum y_{i_F}^2 (1 - R_F^2) \qquad \sum e_{i_R}^2 = \sum y_{i_R}^2 (1 - R_R^2)$$

where $\sum e_i^2$ = error SS in full or restricted model, as indicated

$\sum y_i^2$ = variation (sum of squared deviation scores) of Y (the dependent variable) for full or restricted model, as indicated

R^2 = coefficient of determination (multiple R^2, proportion of variance accounted for) for full or restricted model, as indicated

Recall the F ratio, given in Eq. (3.1)

$$F = \frac{(\sum e_{i_R}^2 - \sum e_{i_F}^2)/df_1}{\sum e_{i_F}^2/df_2}$$

If we substitute the definitions given above for the error SS, we get

$$F = \frac{[\sum y_{i_R}^2(1 - R_R^2) - \sum y_{i_F}^2(1 - R_F^2)]/df_1}{\sum y_{i_F}^2(1 - R_F^2)/df_2}$$

Notice that where the *dependent* variable Y is the same for the full and restricted models, $\sum y_{i_R}^2 = \sum y_{i_F}^2 = \sum y_i^2$, and the unit vector is in the predictor spaces

---

[1] We shall discuss exceptions to this later. Usually, where generalization or replication questions are being tested, the dependent variable will *not* be the same for full and restricted models. In such cases, we must use the error SS formulation for the F ratio.

for both models, we may substitute the quantity $\sum y_i^2$ for both these quantities, producing

$$F = \frac{[\sum y_i^2(1 - R_R^2) - \sum y_i^2(1 - R_F^2)]/df_1}{\sum y_i^2(1 - R_F^2)/df_2}$$

which can be simplified to

$$F = \frac{(R_F^2 - R_R^2)/df_1}{(1 - R_F^2)/df_2} \qquad (3.2)$$

where R_F^2 = coefficient of determination in (proportion of variance accounted for by) full model

R_R^2 = coefficient of determination in (proportion of variance accounted for by) restricted model

df_1 = difference in number of (independent) unknown weights in full and restricted models

df_2 = N minus number of (independent) unknown weights in full model

It will always be possible to define F in terms of the full- and restricted-model error SS. Often, our exposition will be easier if we use the alternative formulation employing full and restricted model R^2's. We shall always note cases for which this alternative formulation is inappropriate because of a change in the definition of the dependent variable or the absence of a unit vector.

Assumptions and Robustness of the F Ratio

The statistic which we have defined above in terms of proportionate differences in error variance attributable to hypotheses and assumptions (and the alternative R^2 formulation where appropriate) will be distributed according to F, where:

1 The dependent variable Y is conceptually interval-scaled (see Carter, 1971; Senders, 1958, p. 50; McNemar, 1962, pp. 374–376).

2 The elements y_i of the dependent variable are independent of each other; e.g., the elements of the **Y** vector do not include matched pairs such as observations on the same people at two points in time or observations on linked cases such as husbands and wives. This assumption will not preclude time-series analyses or analysis of linked individuals; rather we shall simply have to deal with change and difference scores for dependent variables.

3 The measurement of each element's value on the dependent variable y_i is randomly sampled from the possible measurements for *that* case on *that* variable; i.e., we have not *systematically* overestimated some values and underestimated others. Problems with this assumption can generally be avoided if we are cautious. For example, we must not measure the performance of a set of students if some are up, some straight, and some down unless the bio-psycho-chemical state of the organism is a variable which is treated analytically.

4 The distribution of possible observations for each case on the dependent variable is normal around the "true" value for *that* case on *that* variable. Where groups to be compared have equal numbers of cases, the F ratio is quite robust with respect to this assumption (see Box and Watson, 1962, and Norton, 1952, p. 252). In Part Two we develop cautions concerning the effects of heteroscedasticity where groups have unequal n's.

5 The variances of the distributions of possible observations on the dependent variable around the "true" value for *that* case on *that* variable are equal. The F ratio is quite robust with respect to this assumption (see Norton, 1952).

6 Our hypotheses imposed for test are true.

If the F is quite large (larger than one would expect by chance), we shall conclude that *at least one* of the above is false. It will appear that with some diligence we can avoid serious distortion of our inference from the first three assumptions above (interval dependent variable, random selection of measured values for each case, independence of elements of the dependent variable). Also, as indicated in the citations above, inferences based on the F statistic are not seriously distorted by violations of assumptions 4 and 5 (normality and homoscedasticity of distributions of possible observations for cases around "true" values for the case on that variable). Consequently, most of the distortion of the F statistic will ordinarily be produced by the imposition of false hypotheses. Thus, when F is large, we shall usually be safe in rejecting the hypotheses which we have imposed for test.

3.3 A PARADIGM FOR HYPOTHESIS FORMULATION AND TESTING

In the remainder of Part One, we shall develop all the conventional questions (and some nonconventional ones of special use to social researchers) posed by analysis of variance, multiple correlation, partial correlation, regression, and analysis of covariance. These topics will be organized in terms of the following scheme:

A *Natural-language statement* Each test will be prefaced with a brief natural-language (nonmathematical) statement of what hypotheses are to be tested by the strategy.

B *Full model* An appropriate symbolic additive linear model which permits the hypotheses to be either true or false will be developed at three levels of abstraction:
  1 Algebraic formulation
  2 Vector representation
  3 Simultaneous linear equation system

C *Hypotheses* Symbolic representation of the natural-language hypotheses will be developed, requiring:
  1 Solution for independent unknowns
  2 Substitution in the full model
  3 Expansion of the resultant model
  4 Simplification and collection of terms in the resultant model

D *Restricted model* The above steps will produce an appropriate restricted model which forces the hypotheses to be true. This model will be explored at the following levels:

1   Algebraic formulation

2   Vector representation

3   Simultaneous linear equation system

E   *Inference*   The inferential steps for making decisions about the hypotheses will be detailed, including:

1   Use of coefficients of determination (or $\sum e_i^2$'s)

2   Determination of degrees of freedom

3   Calculation of the F ratio

4   Interpretation of the F ratio with appropriate cautions about possible violations of assumptions

F   *Computer instructions*   Instructions will be provided for computer solutions.

G   *Numerical example*   Usually a numerical example will be provided. In some cases, hypothetical data will be used. For some of the more complicated analyses, it is too cumbersome to provide data.

H   *Generalization*   Where possible, i.e., with conventional questions, general cases will be developed symbolically.

I   *Traditional equivalence*   For all conventional questions, discussion will show the solutions to be equivalent to more traditional presentations.

J   *Nonstandard questions*   Often we shall formulate questions for which there is no standard formula but which are interesting.

Not all topics will be presented for every case. Where several closely allied topics are presented at a time, some redundancy can be avoided.

3.4   CATEGORICAL INDEPENDENT VARIABLES: THE TWO-GROUP CASE

t tests and one-, two-, and *n*-way analyses of variance all involve a single dependent variable which is conceptually interval-scaled. Cases are then divided into two or more *groups*, determined by values on substantively interesting categorical variables. The notion is entertained that the expected value for each group (defined by a unique configuration of properties) may be unique and produced by importantly different causal forces. At least initially, this notion will be symbolically represented in most full models by allowing a unique categorical (binary) vector for each group. The least-squares solution will be the mean. All such vectors will be orthogonal to each other. All traditional tests employed within these frameworks will be shown to involve hypotheses about equivalences (or relationships) between these expected values.

Assume that we have randomly assigned 30 children from a school to two class-rooms employing quite different teaching techniques which we wish to compare. The dependent variable of interest will be the percentage improvement by each child on a standard achievement test. To ascertain whether technique 1 or 2 is better we should proceed as follows.

A Natural-Language Statement

On the average, do randomly assigned students improve more when taught by one technique than when taught by the other?

B Full Model

$$Y = b_1 X_1 + b_2 X_2 + E_F$$

where Y = percent improvement for each child
X_1 = 1 for those taught by technique 1, 0 for those taught by technique 2
X_2 = 1 for those taught by technique 2, 0 for those taught by technique 1
b_1 = least-squares estimate of improvement associated with group 1
b_2 = least-squares estimate of improvement associated with group 2
E_F = error of prediction for each child
In terms of vectors this would be represented

$$
\begin{bmatrix} y_1 \\ y_2 \\ \vdots \\ y_{15} \\ y_{16} \\ y_{17} \\ \vdots \\ y_{30} \end{bmatrix}
= b_1 \begin{bmatrix} 1 \\ 1 \\ \vdots \\ 1 \\ 0 \\ 0 \\ \vdots \\ 0 \end{bmatrix}
+ b_2 \begin{bmatrix} 0 \\ 0 \\ \vdots \\ 0 \\ 1 \\ 1 \\ \vdots \\ 1 \end{bmatrix}
+ \begin{bmatrix} e_1 \\ e_2 \\ \vdots \\ e_{15} \\ e_{16} \\ e_{17} \\ \vdots \\ e_{30} \end{bmatrix}
$$

This system of vectors implies two *types* of equations. For children exposed to technique 1 (y_1 to y_{15}), the predictor equations are all of the form

$$y_i = b_1 \times 1 + b_2 \times 0 + e_i = b_1 + e_i \qquad i = 1, 2, \ldots, 15$$

As we have seen earlier, the least-squares solution for b_1 produces $b_1 = \bar{Y}_1$ (where $\bar{Y}_1$ is the mean improvement for children exposed to technique 1). For children exposed to technique 2 (y_{16} to y_{30}), the predictor equations are all of the form

$$y_i = b_1 \times 0 + b_2 \times 1 + e_i = b_2 + e_i \qquad i = 16, 17, \ldots, 30$$

As we have seen earlier, the least-squares solution for b_2 produces $b_2 = \bar{Y}_2$ (where $\bar{Y}_2$ is the mean improvement for children exposed to technique 2).

That this is an appropriate full model is apparent from the fact that it *allows* different estimators for the improvement of members of each of the two groups. For possible data, of course, the improvement *could* be the same, that is, $\bar{Y}_1 = \bar{Y}_2$ and $b_1 = b_2$. Thus the model will allow our hypothesis to be either true or false.

C Hypothesis

$$b_1 = b_2 \qquad (= \text{some general } b)$$

The only null hypothesis suggested by our natural-language statement is this equivalence. We shall substitute b for both b_1 and b_2 in the full model to produce a proper restricted model:

$$Y = bX_1 + bX_2 + E_R = b(X_1 + X_2) + E_R = bU + E_R$$

D   Restricted Model

$$\mathbf{Y} = b\mathbf{U} + \mathbf{E}_R$$

This hypothesis produces a restricted model which is a random-variation model, since $\mathbf{X}_1 + \mathbf{X}_2 = \mathbf{U}$. In vector notation the model would be

$$\begin{bmatrix} y_1 \\ y_2 \\ \vdots \\ y_{30} \end{bmatrix} = b \begin{bmatrix} 1 \\ 1 \\ \vdots \\ 1 \end{bmatrix} + \begin{bmatrix} e_1 \\ e_2 \\ \vdots \\ e_{30} \end{bmatrix}$$

Notice that for this model (under the hypothesis that $b_1 = b_2$) all equations are of the form

$$y_i = b + e_i \qquad i = 1, 2, \ldots, 30$$

As we have seen earlier, such a system produces a solution for b which is $b = \overline{Y}_T$ (where $\overline{Y}_T$ is the mean improvement for all children regardless of which technique they were exposed to). It should be obvious that this restricted model forces the hypothesis ($b_1 = b_2$) to be true. The question of interest is whether the imposition of this hypothesis (significantly) increases the error SS (or decreases the R^2).

E   Inference

For our models, the least-squares solutions will be $b_1 = \overline{Y}_1, b_2 = \overline{Y}_2, b = \overline{Y}_T$. The error SS for the full model will be

$$\sum e_{i_F}^2 = \sum_{i=1}^{15} (y_i - \overline{Y}_1)^2 + \sum_{i=16}^{30} (y_i - \overline{Y}_2)^2$$

This is, of course, equivalent to the traditional within-group SS (see McNemar, 1962, pp. 253–255; Blalock, 1972, pp. 322–323). The error SS for the restricted model will be

$$\sum e_{i_R}^2 = \sum_{i=1}^{30} (y_i - \overline{Y}_T)^2$$

This is, of course, equivalent to the traditional total SS (see McNemar, 1962, p. 254; Blalock, 1972, p. 326).

As there are two unknown weights in the full model and one unknown weight in the restricted model,

$$\mathrm{df}_2 = N - 2$$
$$\mathrm{df}_1 = 2 - 1 = 1$$

Thus, the F ratio becomes a special case of Eq. (3.1):

$$F = \frac{(\sum e_{i_R}^2 - \sum e_{i_F}^2)/1}{\sum e_{i_F}^2/(N - 2)}$$

which is equivalent to the traditional

$$F = \frac{(\text{total SS} - \text{within SS})/df_1}{(\text{within SS})/df_2} = \frac{s_b^2}{s_w^2}$$

(see McNemar, 1962, pp. 267–269; Blalock, 1972, p. 327).

F Computer Instructions

While this analysis could readily be done by hand, subsequent examples will be too complex for hand tabulation. It will therefore be of value for the reader to familiarize himself with the appended program package by using it for solutions. To do so requires the following:

1 Writing an appropriate set of Fortran statements for SUBROUTINE DATA which will produce the vectors VEC(1) defined as X_1, above; VEC(2) defined as X_2, above; VEC(3) defined as U, above; and VEC(4), defined as Y, above. The data provided in the numerical example below should be used with this program.
2 A main control card which includes a label, the number of cases, the number of input variables, the number of vectors, and the number of models (see Appendix A).
3 Two analysis control cards, each indicating the subscript for the dependent variable, the number of independent variables, and their subscripts.

By convention, let the data cards be defined in the following way (see numerical example below):

Columns 1–3   Case identification number
Columns 4–5   Percentage improvement on achievement test
Column 6   Type of teaching technique (1 or 2)

With the conventions in mind, we can indicate the statements in SUBROUTINE DATA which would produce the proper vectors. The statements which should be embedded between the COMMON statement and the RETURN statement are

```
. . . . . . . . . . . . . .
      VEC(1)=0.
      VEC(2)=0.
      VEC(3)=1.
      READ (5,11) VAR(1),VAR(2)
      VEC(4)=VAR(1)
      K=VAR(2)
      VEC(K)=1.
   11 FORMAT (3X,1F2.0,1F1.0)
. . . . . . . . . . . . . . . . . . . . . . . . . . .
```

When the program has automatically executed these statements for each case, it will have produced:

VEC(1), which is 1. for all students exposed to technique 1 and 0. for all others. [Notice that statements 1 and 2 zero VEC(1) and VEC(2) for each case. In statement 4, the group number, 1. or 2., is stored in VAR(2). In statement 6, the group number, 1 or 2, is stored in location K. Statement 7 changes VEC(K), *either* VEC(1) or VEC(2), to a 1. depending on which group the person was in.]

VEC(2), which is 1. for all students exposed to technique 2 and 0. for all others. [Note that statements 1, 2, 4, 6, and 7 do this as they did for VEC(1).]

VEC(3), which is 1. for all cases, i.e., a unit vector. (Note that statement 3 does this.)

VEC(4), which is percentage improvement for each case, that is, Y. [Note that statement 4 places this information in VAR(1) and statement 5 transfers it to VEC(4).]

The main control card: the first card in the data segment would be set up as follows (_ means a blank space, and N is assumed to be 30):

2ND_TEST_____30____2____4____2

Columns 1 to 10 contain a label for the run. Columns 11 to 15 contain 30 (the number of cases). Columns 16 to 20 contain 2 (the number of VARs to be read in from cards). Columns 21 to 25 contain 4 (the number of VECs to be created by SUBROUTINE DATA). Columns 26 to 30 contain 2 (the number of regressions to be performed, i.e., one for the full model and one for the restricted model).

The main control card is to be followed by the 30 data cards, one for each student, set up as detailed above. Immediately following the last of these, there must be an analysis control card for each of the two models. These should be set up as follows:

FULL_MODEL____4____2____1____2
REST_MODEL____4____1____3

For the first card, columns 1 to 10 contain the label, columns 11 to 15 contain the subscript of the dependent variable [a 4 to designate VEC(4)], columns 16 to 20 contain the number of independent variables (2), the next two five-column fields contain the subscripts for those independent variables [1 for VEC(1) and 2 for VEC(2)]. Similarly, the second card has a label, designates the dependent variable as VEC(4), requires one independent variable and designates it as VEC(3). The reader should verify for himself that these models correspond to the full and restricted models discussed above. It is extremely important that the reader run the computer program with the numerical example below to familiarize himself with the process on this fairly simple example. This will allow us to concentrate on more substantive and technical issues in subsequent discussions.

G   Numerical Example

Assume the data shown in Table 3.1. The question of interest is whether techniques 1 and 2 differ in the amount of change characteristically produced in students' achievement. The models defined in the discussion above are:

Full model: $\qquad$ $Y = b_1X_1 + b_2X_2 + E_F$

Hypothesis: $\qquad$ $b_1 = b_2 (= b)$

Restricted model: $\qquad$ $Y = bU + E_R$

The relevant computer solutions for these models will produce the following results (comparing them with *your* computer printouts will greatly assist you in later interpretations):

Full model	Restricted model
$b_1 = 13.3333$	$b = 13.6667$
$b_2 = 14.0000$	
$\sum e_i^2 = 619.3$	$\sum e_i^2 = 622.7$
$R^2 = .0054$	$R^2 = 0$

Applying Eq. (3.1) produces

$$F = \frac{(622.7 - 619.3)/(2 - 1)}{619.3/(30 - 2)} \approx .15$$

Entering the F table (Appendix Table D.3) at 1 and 28 degrees of freedom, we find that the F would have to exceed 4.20 for us to reject the hypothesis $b_1 = b_2$ at the .05 level of significance. As our obtained F is .15, we cannot reject the hypothesis that

Table 3.1   DATA FOR IMPROVEMENT OF 30
STUDENTS BY TWO TECHNIQUES

Technique 1		Technique 2	
Case	Improvement, %	Case	Improvement, %
001	15	016	20
002	14	017	17
003	19	018	13
004	17	019	16
005	15	020	14
006	19	021	15
007	17	022	17
008	15	023	21
009	15	024	18
010	14	025	19
011	9	026	5
012	6	027	10
013	7	028	8
014	6	029	10
015	12	030	7

the two teaching techniques produce substantially the same results. Conclusions like these are warranted only where we have considered relevant control variables and where our assumptions of the full model (*within*-technique variation is random) are valid. We shall have an opportunity to examine some of these later under two-way analysis of variance.

We should note here that the F ratio could have been computed using $R_F{}^2$ and $R_R{}^2$ (the coefficients of determination) because the dependent variable is the same for the full and restricted models. Numerically, of course, the result would be the same if we had used Eq. (3.2):

$$F = \frac{(.0054 - 0)/(2 - 1)}{(1 - .0054)/(30 - 2)} \approx .15$$

I Traditional Equivalence

The alert reader will have noticed that we have asked the equivalence-of-means question normally approached through the t test. This simple case is presented first to familiarize the reader with our research paradigm for questions he already understands. In this way, more complicated questions can be treated more readily. It is worth noting that our approach yields an F which is equivalent to t^2. This is proved below. Recall that for this test

$\sum e_{i_R}{}^2$ = sum of squared deviations from total mean

$\sum e_{i_F}{}^2$ = sum of squared deviations for each case from its group mean

Thus, where $N_1 = N_2 = N/2$,

$$\sum e_{i_R}{}^2 = \sum_1^N (y_i - \bar{Y}_T)^2 = \sum_1^N y_i{}^2 - \frac{\left(\sum_1^N y_i\right)^2}{N}$$

(the proof that $F = t^2$ is easier where $N_1 = N_2$, but this is not essential) and

$$\sum e_{i_F}{}^2 = \sum_1^{N/2} (y_i - \bar{Y}_1)^2 + \sum_{N/2+1}^N (y_i - \bar{Y}_2)^2$$

$$= \sum_1^N y_i{}^2 - \frac{2\left(\sum_1^{N/2} y_i\right)^2}{N} - \frac{2\left(\sum_{N/2+1}^N y_i\right)^2}{N}$$

By substituting these terms in our formula for F and simplifying we shall find that for this test

$$F = \frac{\bar{Y}_1{}^2 - 2\bar{Y}_1\bar{Y}_2 + \bar{Y}_2{}^2}{2(s_1{}^2 + s_2{}^2)/(N - 2)} = \frac{(\bar{Y}_1 - \bar{Y}_2)^2}{(s_1{}^2 + s_2{}^2)/(N/2 - 1)}$$

$$= \frac{(\bar{Y}_1 - \bar{Y}_2)^2}{s_1{}^2/(N/2 - 1) + s_2{}^2/(N/2 - 1)}$$

And since, by assumption, $N_1 = N_2 = N/2$,

$$F = \frac{(\bar{Y}_1 - \bar{Y}_2)^2}{s_1{}^2/(N_1 - 1) + s_2{}^2/(N_2 - 1)} \quad \text{and} \quad \sqrt{F} = \frac{\bar{Y}_1 - \bar{Y}_2}{\sqrt{s_1{}^2/(N_1 - 1) + s_2{}^2/(N_2 - 1)}}$$

which is the formula for the t test (see McNemar, 1962, p. 268; Blalock, 1972, p. 224).

3.5 NONSTANDARD QUESTIONS: REPLICATION AND GENERALIZATION

Replication of a Specified Difference

As social science develops a tradition of replication, it will often be of interest to test hypotheses of greater precision than the conventional ones like the hypothesis of no difference between means. Let us assume that some previous researcher found that technique 2 produced an average improvement which was 5 percentage points higher than technique 1. We might be interested in seeing whether our findings are reasonably in agreement with this.[1] Although discussion of the t test is usually not such as to permit the formulation of such questions, the generalization is straightforward from our regression perspective. An appropriate strategy for such a test follows:

Full model: $$Y = b_1X_1 + b_2X_2 + E_F$$

where all definitions are the same as in the previous example.

Hypothesis: $$b_1 = b_2 - 5$$

The symbolic statement of the hypothesis is that the mean change expected for technique 2 is 5 units more than for technique 1. Under this hypothesis, if $b_1 = b$, $b_2 = b + 5$.

Restricted model: $Y = bX_1 + (b + 5)X_2 + E_R = bX_1 + bX_2 + 5X_2 + E_R$

$$= b(X_1 + X_2) + 5X_2 + E_R = bU + 5X_2 + E_R$$

$$Y - 5X_2 = bU + E_R$$

The last step requires comment. We subtracted $5X_2$ from both sides. This was necessary because the right side of the model may have only vectors with unknown weights (to be fitted by the least-squares criterion) and an error vector. Notice that we are required to construct a new vector VEC(5) to be used as the dependent variable. Given our solution for the t test problem, we would only need the following modifications for this problem:

1 Addition of the statement

VEC(5) = VEC(4) − 5.∗VEC(2)

immediately before the RETURN statement in SUBROUTINE DATA
2 Changing the master control card to permit five VECs and three regressions
3 Adding a model card for this alternative restricted model

ALT._REST. ____5____1____3

It is quite important to note that *the dependent variable is not the same in the full and restricted models*. In the full model, the dependent variable is Y, while in the restricted model it is a new variable Z(where $Z = Y - 5X_2$). For this reason, we

---

[1] Tests to demonstrate agreement with previous findings can be viewed as either replication tests or generalization tests. They are replication tests where the same observational procedures are used under the same circumstances. If the circumstances (conditions, populations, etc.) are changed, the same formal tests become generalization questions.

must calculate F using the two error SS. We *cannot* use the coefficients of determination (R^2's). As $y_{i_F} \neq y_{i_R}$, the dependent variable being different in the two models, $s_{y_F}^2 \neq s_{y_R}^2$ and our earlier equivalence does not hold. Using the R^2's under these conditions will always yield an incorrect F ratio. Sometimes the F so calculated will be negative. Solutions for the two error SS will be

$$\sum e_{i_F}^2 = 619.3 \qquad \sum e_{i_R}^2 = 760.2$$

Consequently

$$F = \frac{(760.2 - 619.3)/(2 - 1)}{619.3/(30 - 2)} = \frac{140.9}{22.12} \approx 6.37$$

As this exceeds the value of 4.20 associated with F at the .05 level of significance at 1 and 28 degrees of freedom, we would *reject* our hypothesis that technique 2 produced 5 percentage points more improvement than technique 1. Substantively, we would reject the idea that we had replicated (or generalized) the finding of our hypothetical previous researcher. This is because imposition of such an hypothesis produces a significant increase in error.

In the general case, one can determine whether the finding of a specified difference c has been replicated by analysis of the following models. This is a much more powerful hypothesis than the usual $\overline{Y}_1 - \overline{Y}_2 = 0$ hypothesis. This possibility is implied in many elementary and intermediate statistics books, but is not really developed (see, for example, Blalock, 1972, p. 224, formula 13.1):

Full model: $\qquad\qquad \mathbf{Y} = b_1 \mathbf{X}_1 + b_2 \mathbf{X}_2 + \mathbf{E}_F \qquad$ defined as above

Hypothesis: $\qquad\qquad b_1 = b_2 - c$

where c is a definite value determined by previous research. Thus, if $b_1 = b$, $b_2 = b + c$ and we may substitute b for b_1 and $b + c$ for b_2.

Restricted model: $\qquad\qquad\qquad\qquad \mathbf{Y} = b\mathbf{X}_1 + (b + c)\mathbf{X}_2 + \mathbf{E}_R$

$$= b\mathbf{X}_1 + b\mathbf{X}_2 + c\mathbf{X}_2 + \mathbf{E}_R$$

and since c is *not* an unknown weight but a definite constant,

$$\mathbf{Y} - c\mathbf{X}_2 = b\mathbf{U} + \mathbf{E}_R \qquad \text{and} \qquad \mathbf{Z} = b\mathbf{U} + \mathbf{E}_R$$

where $\mathbf{Z}$ is a *new* dependent variable equal to $\mathbf{Y} - c\mathbf{X}_2$.

The F ratio *must* be computed using error SS rather than R^2's, and $df_1 = 1$, $df_2 = N - 2$.

Replication Involving a Single Characteristic Value[1]

Assume that some previous researcher found that technique 2 typically produces an average improvement of 15 percentage points, ignoring completely the improvement produced by technique 1. It might be of interest to us to decide whether our results with technique 2 differ significantly from this. An appropriate strategy for answering such a question would be

Full model: $\qquad\qquad\qquad\qquad\qquad\qquad\qquad \mathbf{Y} = b\mathbf{U} + \mathbf{E}_F$

---

[1] This limited special case is treated here, rather than earlier as a one-group case, because its solution requires additional computer programming as no least-squares solutions are to be determined.

where Y = dependent variable for 15 cases exposed to technique 2

$\quad\quad U$ = unit vector of dimension 15

$\quad\quad E_F$ = the error vector

Of course, this yields a solution for $b = \bar{Y}$, and $E'_F E_F$ will be the variation for the dependent variable. (For these data, $b = 14$ and $E'_F E_F = 348$.)

Hypothesis: $\hfill b = 15$

Restricted model: $\hfill Y = 15U + E_R$

and since 15 is not an unknown weight but a constant,

$$Y - 15U = E_R$$

Notice that for this restricted model, i.e., for the hypothesis that $b = 15$, $e_i = y_i - 15$. Though our regression program *cannot* solve such a restricted model, there being no weights to fit in a deterministic model, the solution for $\sum e_i^2$ is straightforward. By hand (or computer) we simply subtract the hypothesized value from each y_i, getting the vector of e_i's. We then get the sum of the squared e_i's. If Y is a vector stored in the computer, the requisite Fortran follows (N contains the number of cases, C contains the hypothesized value, and SSQ will cumulate $\sum e_i^2$):

```
. . . . . . . . . . .
    SSQ=0.
    DO 5 I=1, N
  5 SSQ=SSQ+(Y(I)-C)**2
. . . . . . . . . . . . . . . . . . . . . . .
```

For our example data for group 2 (above),

$$\sum e_{i_F}^2 = 348 \quad\quad \sum e_{i_R}^2 = 363$$

$$df_1 = 1 - 0 = 1$$

there being one unknown weight in the full model and none in the restricted,

$$df_2 = 15 - 1 = 14$$

there being one unknown weight in the full model. Thus

$$F = \frac{(363 - 348)/(1 - 0)}{348/(15 - 1)} \approx .60$$

As this is less than the value (at 1 and 14 degrees of freedom) of 4.60 required for significance at the .05 level, we are unable to reject the hypothesis that $b = 15$.

Replication Involving Two Specified Values

We can, of course, simultaneously test whether we have replicated the findings of previous research with respect to the expected improvement for each technique. Let c_1 and c_2 be the (definite) mean improvement found in previous research for techniques 1 and 2, respectively. The replication hypotheses can then be tested as follows:

Full model: $\qquad\qquad Y = b_1X_1 + b_2X_2 + E_F \qquad$ defined as above

Hypotheses: $\qquad\qquad b_1 = c_1 \qquad b_2 = c_2$

Restricted model: $\qquad\quad Y = c_1X_1 + c_2X_2 + E_R$

and since c_1 and c_2 are constants rather than unknown weights,

$$Y - c_1X_1 - c_2X_2 = E_R$$

$\sum e_{i_F}^2$ is (as before) within-group variation

$$\sum e_{i_R}^2 = \sum_{i=1}^{N} (y_i - c_1x_{1_i} - c_2x_{2_i})^2$$
$$\mathrm{df}_1 = 2 - 0 = 2$$

Two hypotheses were imposed, or, alternatively, this is the difference in the number of unknown weights in the two models.

$$\mathrm{df}_2 = N - 2$$

there being two unknown weights in the full model.

Alternative nonstandard hypotheses such as replication of a specified difference, replication involving a single characteristic value, and replication involving two specified values are possible for any data for which two groups are defined. They are of *interest* only within a well-developed replication (generalization) tradition or within a very precise theoretical tradition.

3.6 CATEGORICAL INDEPENDENT VARIABLES: THE THREE-GROUP CASE

Assume that we randomly assigned 15 students to each of three groups, each group being taught by a different technique. We might well be interested in whether the techniques differ in their effects on students. This interest would lead us to the following analysis.

A Natural-Language Question

Do techniques 1, 2, and 3 produce different average improvement in students' performances?

B Full Model

$$Y = b_1X_1 + b_2X_2 + b_3X_3 + E_F$$

where Y = dependent variable (improvement)

$\quad X_1$ = 1 for members of group 1, 0 otherwise

$\quad X_2$ = 1 for members of group 2, 0 otherwise

$\quad X_3$ = 1 for members of group 3, 0 otherwise

$\quad E_F$ = the error-of-prediction vector

By this point you should be able to write out these vectors, which must be created for the computer analysis. It should also be apparent that this model generates three types of equations; specifically, for those taught by technique 1, the prediction will be

$$y_i = b_1 + e_i \qquad \text{for } i = 1, 2, \ldots, 15$$

for those taught by technique 2

$$y_i = b_2 + e_i \qquad \text{for } i = 16, 17, \ldots, 30$$

for those taught by technique 3

$$y_i = b_3 + e_i \qquad \text{for } i = 31, 32, \ldots, 45$$

Also, it should be obvious that the least-squares solutions will be $b_1 = \bar{Y}_1$, $b_2 = \bar{Y}_2$, and $b_3 = \bar{Y}_3$. Further, the error vector will comprise the deviations of each case from the mean of the group of which it is a part, and $\mathbf{E}_F'\mathbf{E}_F$ (or $\sum e_{i_F}^2$) will be what is conveniently called within-group variation. As there are now three partitions and the errors *within* each partition are deviation scores (summing zero), 3 degrees of freedom will be lost for this model ($\mathrm{df}_2 = N - 3 = 45 - 3 = 42$). The appropriateness of this full model is obvious from the fact that it allows our hypotheses (equivalence of all b_i's) to be either true or false.

C   Hypotheses

Our natural-language question generates the following *two* hypotheses:

$$b_1 = b_2 \,(= b) \qquad b_1 = b_3 \,(= b)$$

Also, implicitly there is the hypothesis that $b_2 = b_3 (= b)$. However, this is not an *independent* hypothesis as the truth of the first two implies (formally) the truth of the third. It is in this sense that df_1 was said to be the number of *independent* hypotheses to be imposed. Any two of these three hypotheses constitute a symbolic representation of our natural-language question. Given two of them, the third is redundant. The truth of any two of the three allow us to substitute b for b_1, b_2, and b_3.

D   Restricted Model

$$\mathbf{Y} = b\mathbf{X}_1 + b\mathbf{X}_2 + b\mathbf{X}_3 + \mathbf{E}_R = b(\mathbf{X}_1 + \mathbf{X}_2 + \mathbf{X}_3) + \mathbf{E}_R$$
$$= b\mathbf{U} + \mathbf{E}_R$$

Notice that the restricted model forces our hypotheses ($b_1 = b_2$, $b_1 = b_3$) to be true. Under these hypotheses our model reduces to the random-variation model. Thus b will be $\bar{Y}_T$, and $\mathbf{E}_R'\mathbf{E}_R$ will be what is conventionally called *total variation*. Comparison of the two models, of course, tells us how much more error (if any) is attributable to the imposition of the hypotheses.

E   Inference

For our models, the least-squares solutions will be $b_1 = \bar{Y}_1$, $b_2 = \bar{Y}_2$, $b_3 = \bar{Y}_3$, $b = \bar{Y}_T$. The error SS for the full model will be

$$\sum_{i=1}^{N} e_{i_F}^2 = \sum_{j=1}^{3} \sum_{i=1}^{n_j} (y_{ij} - \bar{Y}_j)^2$$

where y_{ij} = observation (dependent variable) for ith person in jth group

$\quad \overline{Y}_j$ = mean for jth group

$\quad n_j$ = number of people in jth group

It should be obvious that this quantity is what is traditionally called the within-group SS (or within-group variation). The error SS for the restricted model will be

$$\sum_{i=1}^{N} e_{i_R}^{2} = \sum_{j=1}^{3} \sum_{i=1}^{n_j} (y_{ij} - \overline{Y}_T)^2$$

where y_{ij} = observation for ith person in jth group

$\quad \overline{Y}_T$ = mean for combined groups

$\quad n_j$ = number of people in jth group

It should be obvious that this quantity is what is traditionally called the total SS (or total variation). As there are three unknown weights in the full model and one unknown in the restricted model,

$$df_2 = N - 3 \quad \text{and} \quad df_1 = 3 - 1 = 2$$

Thus, the F ratio becomes a special case of our Eq. (3.1) at 2 and $N - 3$ degrees of freedom.

F Computer Instructions

Computer solution requires the following modifications for program CARTROCK:

1 Fortran statements embedded in SUBROUTINE DATA which will produce the VEC(1) defined as X_1, above; VEC(2) defined as X_2, above; VEC(3) defined as X_3, above; VEC(4) defined as U, above; and VEC(5) defined as Y, above. Assume that data are provided in the following form:

Columns 1–3   Case identification number

Columns 4–5   Percentage improvement on achievement test

Column 6   Type of teaching technique (1, 2, or 3)

The Fortran statements to be inserted in the SUBROUTINE would then be

```
. . . . . . . . . . . . . . .
     VEC(1)=0.
     VEC(2)=0.
     VEC(3)=0.
     VEC(4)=1.
     READ (5,11) VAR(1),VAR(2)
     VEC(5)=VAR(1)
     K=VAR(2)
     VEC(K)=1.
  11 FORMAT (3X,1F2.0,1F1.0)
. . . . . . . . . . . . . . . . . . . . . . . . . . .
```

2 The main control card (the first card in the data segment) would be

3RD_Test_____45____2____5____2

The 45 data cards follow immediately after this card.

3 Following the data cards, the two analysis control cards would be

FULL_MODEL____5____3____1____2____3

REST_MODEL____5____1____4

The first card specifies dependent variable number 5 or Y, three independent variables which are to be numbers 1 or X_1, 2 or X_2, and 3 or X_3. The second card specifies dependent variable number 5 or Y, one independent variable which is to be number 4 or U.

G   Numerical Example

Assume the data shown in Table 3.2. The question of interest is whether techniques 1, 2, and 3 produce characteristically different changes in students' achievement. The models as defined in the discussion above will permit test of such a question.

Full model: $\quad\quad\quad\quad\quad Y = b_1X_1 + b_2X_2 + b_3X_3 + E_F$
Hypotheses: $\quad\quad\quad\quad b_1 = b_2(= b) \quad\quad b_1 = b_3(= b)$
Restricted model: $\quad\quad Y = bU + E_R$

The relevant computer solutions for these models will produce the following results:

Full model	Restricted model
$b_1 = 13.3333$	$b = 13.7778$
$b_2 = 14.0000$	
$b_3 = 14.0000$	
$\sum e_i^2 = 973.3$	$\sum e_i^2 = 977.8$
$R^2 = .0045$	$R^2 = 0$

Table 3.2   **DATA FOR IMPROVEMENT OF 45 STUDENTS BY THREE TECHNIQUES**

Technique 1		Technique 2		Technique 3	
Case	Score	Case	Score	Case	Score
001	15	016	20	031	17
002	14	017	17	032	16
003	19	018	13	033	12
004	17	019	16	034	19
005	15	020	14	035	16
006	19	021	15	036	9
007	17	022	17	037	12
008	15	023	21	038	6
009	15	024	18	039	6
010	14	025	19	040	7
011	9	026	5	041	20
012	6	027	10	042	21
013	7	028	8	043	17
014	6	029	10	044	16
015	12	030	7	045	16

Calculating F by formula (3.1) produces

$$F = \frac{(977.8 - 973.3)/(3 - 1)}{973.3/(45 - 3)} \approx .10$$

Entering the F table at 2 and 42 degrees of freedom (actually at 2 and 40 degrees of freedom to provide a slightly conservative test), we find that the F would have to exceed 3.23 for us to reject the hypotheses at the .05 level of significance. As our obtained F is .10, we cannot reject the hypotheses that the three teaching techniques produce substantially the same results. This depends on the validity of our full-model assumptions (within-group variation is random).

As the dependent variable is the same for the two models, we could have calculated F by the alternative formula (3.2):

$$F = \frac{(.0045 - 0)/(3 - 1)}{(1 - .0045)/(45 - 3)} \approx .09$$

It should be noted that while we proved that F defined in this way is mathematically identical to F defined in terms of the error SS, it is computationally slightly different. This is caused by differential rounding errors in the third significant digit produced when the computer calculates the error SS and the R^2's. (Notice the difference in the number of significant digits reported on the computer printout.) In general, rounding errors in the third significant digit will have *no* consequences for the conclusions reached in social science research and such differences can therefore be safely ignored. Where greater accuracy is essential, e.g., in simulations or population projections, it can be provided by double-precision computations in Fortran.

3.7   GENERALIZATION: THE k-GROUP CASE

If we have k groups of individuals and we wish to ask whether the k groups differ from each other with respect to some dependent variable, we may do so by analysis of the following models:

Full model: $\mathbf{Y} = b_1\mathbf{X}_1 + b_2\mathbf{X}_2 + b_3\mathbf{X}_3 + b_4\mathbf{X}_4 + \cdots + b_k\mathbf{X}_k + \mathbf{E}_F$ (3.3)

where $\mathbf{Y}$ = dependent variable
  $\mathbf{X}_j$ = 1 for members of group j, 0 for all others
  b_j = least-squares weight associated with each $\mathbf{X}_j$ (it will be the $\overline{Y}_j$, the mean on $\mathbf{Y}$ for group j)
  $\mathbf{E}_F$ = vector of errors of prediction

Hypotheses: $b_1 = b_2 (= b)$
        $b_1 = b_3 (= b)$
        $b_1 = b_4 (= b)$ (there will be $k - 1$ of these)
        . . . .
        $b_1 = b_k (= b)$

Restricted model: $\mathbf{Y} = b\mathbf{X}_1 + b\mathbf{X}_2 + b\mathbf{X}_3 + b\mathbf{X}_4 + \cdots + b\mathbf{X}_k + \mathbf{E}_R$

$\qquad\qquad\quad = b(\mathbf{X}_1 + \mathbf{X}_2 + \mathbf{X}_3 + \mathbf{X}_4 + \cdots + \mathbf{X}_k) + \mathbf{E}_R$

$\qquad\qquad\quad = b\mathbf{U} + \mathbf{E}_R$

For such models

$$F = \frac{(\sum e_{i_R}^2 - \sum e_{i_F}^2)/\mathrm{df}_1}{\sum e_{i_F}^2/\mathrm{df}_2}$$

$$\mathrm{df}_1 = k - 1$$

because there will always be k unknown weights, one for each group, in the full model and one unknown weight, for the unit vector, in the restricted model.

$$\mathrm{df}_2 = N - k$$

because there will always be k unknown weights in the full model (see McNemar, 1962, p. 254; Blalock, 1972, p. 327).

For computer solutions we shall find it convenient to make the first through the kth VEC equivalent to $\mathbf{X}_1$ through $\mathbf{X}_k$, the $(k + 1)$st VEC equivalent to $\mathbf{U}$, and the $(k + 2)$nd VEC equivalent to the dependent variable. This will involve minimal changes in the computer instructions employed for the three-group case, above. Specifically, we must zero the first k VECs rather than the first 3. The master control card will contain a label followed by the number of cases (N), the number of VARs to be read in (usually 2), the number of VECs to be constructed ($k - 2$), and the number of regressions (2). This will be followed by the N data cards.

The full-model card will contain a label followed by the subscript for the dependent variable ($k + 2$), the number of independent variables (k), and their subscripts ($1, 2, 3, 4, \ldots, k$). If these will not all fit on a single card, they may be continued on other cards as described in Appendix A.

The restricted-model card will contain a label followed by the subscript of the dependent variable ($k + 2$), the number of independent variables (1), and its subscript ($k + 1$).

3.8   TRADITIONAL EQUIVALENCE: ONE-WAY ANALYSIS OF VARIANCE

We observed that for models of this type $\sum e_{i_F}^2$ is equivalent to within-group sum of squares (within SS) and $\sum e_{i_R}^2$ is equivalent to total sum of squares (total SS). From traditional discussions of analysis of variance we know that between-group sum of squares (between SS) is equal to total SS – within SS. Substituting these equivalences for the error SS and the quantities developed in the general case above for df_1 and df_2, we get

$$F = \frac{(\text{total SS} - \text{within SS})/(k - 1)}{(\text{within SS})/(N - k)} = \frac{(\text{between SS})/(k - 1)}{(\text{within SS})/(N - k)} = \frac{s_b^2}{s_w^2}$$

where s_b^2 is the between-group variance estimate and s_w^2 is the within-group variance estimate. This, of course, is the formula developed for the F test for equivalence of means in traditional one-way analysis of variance (see McNemar, 1962, pp. 265–267; Blalock, 1972, pp. 327–328). The development here, however, makes the hypotheses being tested, the rationale for the degrees of freedom, and the interpretation of the SS more explicit. Further, the way is opened for several possibly interesting substantive tests for which there are no conventional computing formulas. Of course, statisticians can and do develop tests for such questions. However, the computing formulas usually given in introductory texts do not permit such tests. Our goal is to make rigorous solutions for such problems available to the novice.

3.9 NONSTANDARD QUESTIONS: REPLICATION, GENERALIZATION, PARTITIONING

Our tack so far has been to develop sensible natural-language research questions, to show how least-squares theory can be applied to them, and to demonstrate that *these* are the questions addressed by routine computing formulas. However, we intend the reader to gain more general use of least-squares theory. In fact, our purpose is to equip the reader to formulate almost *any* sensible natural-language research question in a way permitting a rigorous answer whether the question fits a convenient formula or not. Consequently, we cannot give an exhaustive accounting of the possibilities. However, at this point we shall illustrate three useful types of nonconventional questions which can be asked about sets of means. These are (1) determining whether *replication of findings of specific means* has been achieved, (2) determining whether *replication of findings of specific relationships between sets of means* has been achieved, (3) *partitioning our groups into substantively interesting subsets* and testing for equivalences within subsets.

QUESTION 1   Assume that we have four groups, each exposed to different teaching techniques, and that previous researchers have found the techniques to produce the following mean improvement:

$$\bar{Y}_1 = 14 \qquad \bar{Y}_2 = 16 \qquad \bar{Y}_3 = 12 \qquad \bar{Y}_4 = 5$$

We could test to see whether we have replicated this with the following models:

Full model: $$\mathbf{Y} = b_1\mathbf{X}_1 + b_2\mathbf{X}_2 + b_3\mathbf{X}_3 + b_4\mathbf{X}_4 + \mathbf{E}_F$$

where $\mathbf{Y}$ = dependent variable
$\quad \mathbf{X}_j$ = 1 for members of group j, 0 otherwise
$\quad b_j$ = least-squares weight for each $\mathbf{X}_j$
$\quad \mathbf{E}_F$ = error vector

Hypotheses: $\qquad b_1 = 14 \qquad b_2 = 16 \qquad b_3 = 12 \qquad b_4 = 5$

Restricted model: $\qquad \mathbf{Y} = 14\mathbf{X}_1 + 16\mathbf{X}_2 + 12\mathbf{X}_3 + 5\mathbf{X}_4 + \mathbf{E}_R$

Moving all constants to the left side gives

$$\mathbf{Y} - 14\mathbf{X}_1 - 16\mathbf{X}_2 - 12\mathbf{X}_3 - 5\mathbf{X}_4 = \mathbf{E}_R$$

We could solve for $\sum e_{i_F}^2$ with our program, and we could get $\sum e_{i_R}^2$ directly, as, for this model, $e_i = y_i - 14x_{1i} - 16x_{2i} - 12x_{3i} - 5x_{4i}$. Also, for this test $df_1 = 4$, and $df_2 = N - 4$.

If we were to reject our hypotheses, we could test any subset of them merely by imposing only that subset on our full model.

QUESTION 2   Assume that we have three groups, each exposed to different teaching techniques, and that previous researchers have found that technique 3 produces an average of 5 units more change than technique 2 and 10 units more change than technique 1. We could test whether we have replicated this with the following:

Full model: $$\mathbf{Y} = b_1\mathbf{X}_1 + b_2\mathbf{X}_2 + b_3\mathbf{X}_3 + \mathbf{E}_F$$

the definitions being the same as for the three-group case, above.

Hypotheses: $$b_3 = b_2 + 5 \qquad b_3 = b_1 + 10$$

Thus, if $b = b_3, b - 5 = b_2, b - 10 = b_1$, we have

Restricted model:
$$\mathbf{Y} = (b - 10)\mathbf{X}_1 + (b - 5)\mathbf{X}_2 + b\mathbf{X}_3 + \mathbf{E}_R$$
$$= b\mathbf{X}_1 - 10\mathbf{X}_1 + b\mathbf{X}_2 - 5\mathbf{X}_2 + b\mathbf{X}_3 + \mathbf{E}_R$$
$$\mathbf{Y} + 10\mathbf{X}_1 + 5\mathbf{X}_2 = b\mathbf{X}_1 + b\mathbf{X}_2 + b\mathbf{X}_3 + \mathbf{E}_R$$
$$= b(\mathbf{X}_1 + \mathbf{X}_2 + \mathbf{X}_3) + \mathbf{E}_R$$
$$= b\mathbf{U} + \mathbf{E}_R$$
$$\mathbf{Z} = b\mathbf{U} + \mathbf{E}_R$$

where $z_i = y_i + 10x_{1i} + 5x_{2i}$.

If we construct vectors $\mathbf{Y}, \mathbf{X}_1, \mathbf{X}_2, \mathbf{X}_3, \mathbf{Z}$, and $\mathbf{U}$, we can get solutions for both models with our program. Of course, we would have to use the $\sum e_i^2$'s (rather than R^2's) because the dependent variables (Y and Z) are different; $df_1 = 2$, and $df_2 = N - 3$.

QUESTION 3   Imagine that we have a measure of political liberalism for each representative in each of the 50 states. A conventional question would be to ask whether the states differ with respect to mean political liberalism. This could be answered by comparing the models below:

Full model: $$\mathbf{Y} = b_1\mathbf{X}_1 + b_2\mathbf{X}_2 + b_3\mathbf{X}_3 + \cdots + b_{50}\mathbf{X}_{50} + \mathbf{E}_F$$

where $\mathbf{Y} =$ each representative's political liberalism score
  $\mathbf{X}_j = 1$ for individuals in state j, 0 otherwise
  $b_j =$ least-squares weight
  $\mathbf{E}_F =$ error vector

Hypotheses: $$b_1 = b_2 \,(= b)$$
$$b_1 = b_3 \,(= b) \qquad \text{(there are 49 of these)}$$
$$\dots\dots\dots\dots$$
$$b_1 = b_{50} \,(= b)$$

Restricted model:

$$\mathbf{Y} = b\mathbf{X}_1 + b\mathbf{X}_2 + b\mathbf{X}_3 + \cdots + b\mathbf{X}_{50} + \mathbf{E}_R$$
$$= b\mathbf{U} + \mathbf{E}_R$$

notice that $df_1 = 49$ and $df_2 = N - 50$.

It would be remarkable if representatives from all of the states did not differ significantly on any reasonable scale of political liberalism. Hypotheses of such scope will ordinarily increase the error SS to a significant degree and therefore be deemed false. However, the creative, curious analyst will not stop here with the rather pedestrian report that "representatives from different states differ in liberalism." The thoughtful analyst may suspect that the error is increased because some, but not all, of the hypotheses are false. The problem of deciding just which hypotheses are false is straightforward if analysis is informed by theory or previous research. A political theorist would tell us that it is likely that a relatively small number of regions can be delimited with common interests and that *within* these regions most representatives are fairly homogeneous.[1] For illustrative purposes, let us imagine the following regions:

Region	States
Northeast	1–11
Midwest	12–24
Mountain	25–32
Pacific	33–37
South	38–50

Having already concluded that representatives differ in liberalism, depending on their state, we could proceed to test the more interesting regional hypothesis. We can do this by using the same full model and writing out the specific algebraic hypotheses implied by the regional hypothesis, conditioned, of course, by our definition of region. While this is meticulous, it is worth the effort in its substantive payoff. The 45 hypotheses are

$$\left. \begin{array}{l} b_1 = b_2 \\ b_1 = b_3 \\ \cdots\cdots \\ b_1 = b_{11} \end{array} \right\} \quad (= b_1') \qquad \left. \begin{array}{l} b_{12} = b_{13} \\ b_{12} = b_{14} \\ \cdots\cdots \\ b_{12} = b_{24} \end{array} \right\} \quad (= b_2')$$

$$\left. \begin{array}{l} b_{25} = b_{26} \\ b_{25} = b_{27} \\ \cdots\cdots \\ b_{25} = b_{32} \end{array} \right\} \quad (= b_3') \qquad \left. \begin{array}{l} b_{33} = b_{34} \\ b_{33} = b_{35} \\ \cdots\cdots \\ b_{33} = b_{37} \end{array} \right\} \quad (= b_4')$$

$$\left. \begin{array}{l} b_{38} = b_{39} \\ b_{38} = b_{40} \\ \cdots\cdots \\ b_{38} = b_{50} \end{array} \right\} \quad (= b_5')$$

[1] Some political scientists may object that the critical variable, party affiliation, is not considered. Later we develop ways to handle more than two variables at a time.

It is worth taking the time to reflect on the substantive interpretation which we may attach to the reasonableness of the conventional and alternative hypotheses.

At the algebraic level, the conventional hypothesis $b_1 = b_j$ $(j = 2, 3, \ldots, 50)$ says that the mean liberalism for representatives from any state is the same as for any other state. Were these 49 hypotheses reasonable, we would arrive at the substantive inference that "state makes no difference in the liberalism of representatives." Statistically, there is a single expected value for representatives' liberalism, and we can safely ignore state.

At the algebraic level, it is best first to discuss the alternative hypotheses in groups. The first 10 hypotheses say that Northeastern states are homogeneous, with a single expected value b_1'; the next 12 hypotheses say that the Midwestern states are homogeneous with the single expected value b_2'; the next 7 hypotheses say that the Mountain states are homogeneous with the single expected value b_3'; the next 4 hypotheses say that the Pacific states are homogeneous with the single expected value b_4'; and the last 12 hypotheses say that the Southern states are homogeneous with the single expected value b_5'. If the 45 hypotheses were deemed reasonable, after test, we would arrive at the substantive inference that "*within* region, state makes no difference in the liberalism of representatives."

If, as seems likely, we rejected the conventional 49 hypotheses but were unable to reject the 45 alternative hypotheses, we would have some evidence for the substantive conclusion that "liberalism or conservatism of United States' representatives is determined largely by which of a set of five large homogeneous regions they were elected from," though there are alternative interpretations.

Just how one addresses such broad questions is made rather more obvious by our paradigm. Even within this perspective, the task is tedious, as we still must impose these hypotheses on the full model.

Restricted model:

$$\mathbf{Y} = b_1'\mathbf{X}_1 + \cdots + b_1'\mathbf{X}_{11} + b_2'\mathbf{X}_{12} + \cdots + b_2'\mathbf{X}_{24} + b_3'\mathbf{X}_{25}$$
$$+ \cdots + b_3'\mathbf{X}_{32} + b_4'\mathbf{X}_{33} + \cdots + b_4'\mathbf{X}_{37} + b_5'\mathbf{X}_{38}$$
$$+ \cdots + b_5'\mathbf{X}_{50} + \mathbf{E}_R$$

$$= b_1'(\mathbf{X}_1 + \cdots + \mathbf{X}_{11}) + b_2'(\mathbf{X}_{12} + \cdots + \mathbf{X}_{24})$$
$$+ b_3'(\mathbf{X}_{25} + \cdots + \mathbf{X}_{32}) + b_4'(\mathbf{X}_{33} + \cdots + \mathbf{X}_{37})$$
$$+ b_5'(\mathbf{X}_{38} + \cdots + \mathbf{X}_{50}) + \mathbf{E}_R$$

$$= b_1'\mathbf{Z}_1 + b_2'\mathbf{Z}_2 + b_3'\mathbf{Z}_3 + b_4'\mathbf{Z}_4 + b_5'\mathbf{Z}_5 + \mathbf{E}_R$$

where $\mathbf{Z}_j$ is 1 for members of region j and 0 otherwise.

Of course, $df_1 = 45$, and $df_2 = N - 50$. The vectors discussed here ($\mathbf{Y}, \mathbf{X}_1, \mathbf{X}_2, \ldots, \mathbf{X}_{50}, \mathbf{Z}_1, \mathbf{Z}_2, \mathbf{Z}_3, \mathbf{Z}_4, \mathbf{Z}_5$) are not difficult to generate in Fortran, though they are arduous to write out in detail by hand. The full and restricted model cards are simple extensions of previous examples.

3.10 TWO-WAY ANALYSIS OF VARIANCE

Often the effect of one variable on another cannot be properly understood without the introduction of other variables which condition that effect. Consider the data in Table 3.3. We have already used these data, but we shall see in this section that the addition of the third variable will change our inferences. This provides a good illustration of the importance of multivariate analyses (as opposed to bivariate analyses). Two rather obvious questions which one can address with these data are: On the average does technique make a difference in amount of improvement? and On the average does ability level make a difference in amount of improvement? While we asked the first of these in the bivariate case, introduction of a third variable will seriously change our conclusions. We shall treat these in detail later under main effects. What is less obvious is that answers to neither of these questions is interpretable if students with different ability levels are differentially responsive to different techniques. Such a result is not at all uncommon. It is termed *statistical interaction*, as many readers will already know.

3.11 INTERACTION IN TWO-WAY ANALYSIS OF VARIANCE

Interaction is said to exist when the effect of one variable on another differs for different values of some third variable.[1] In one sense, a finding of interaction is often extremely interesting from a theoretical point of view. For example, the finding that

---

[1] Strictly speaking, this is a definition of *first-order interaction*. Higher-order interaction involves the ways in which combinations of variables condition the effect of one variable on another. These higher-order questions are difficult to describe other than symbolically.

Table 3.3 PERCENTAGE IMPROVEMENT IN PERFORMANCE OF STUDENTS BY TECHNIQUE AND ABILITY LEVEL

Children	Technique 1	Technique 2	Technique 3
	15	20	17
	14	17	16
Gifted	19	13	12
	17	16	19
	15	14	16
	19	15	9
	17	17	12
Normal	15	21	6
	15	18	6
	14	19	7
	9	5	20
	6	10	21
Retarded	7	8	17
	6	10	16
	12	7	16

education affects ideology differently for different occupational and income levels has generated the extensive research tradition in status inconsistency. From another point of view, however, the existence of interaction can be a nuisance. Where it exists, we can meaningfully say very little *in general* about the average effect of one variable upon another. In any event, the cautious researcher must always check diligently for interaction before proceeding to seek general effects. A brief illustration will convey the dangers of ignoring this concern.

Assume that in a certain educational setting approximately one-third of the students are gifted, one-third normal, and one-third retarded. Further, assume that students are randomly assigned to each of three teaching techniques. Thus, within each technique, approximately one-third of the children would be in each ability category. Imagine that high-ability students respond well to technique 3 only, normal-ability students to technique 2 only, and low-ability students to technique 1 only. Such a result is illustrated by Table 3.4.

If we simply look at the average improvement produced by each technique, it is 5 percent. (Also, each category of student improves by an average of 5 percent.) If we restricted ourselves to considering two variables at a time (improvement and teaching or improvement and ability level), we would arrive at the following misleading conclusions:

1 The three techniques produce substantially the same improvement, *on the average*. It will develop that we made this error in Sec. 3.6.
2 Students of all ability levels improve at about the same rate, *on the average*.

These conclusions are warranted, *on the average*. They are termed misleading rather than false because they do not address the interaction question. Where one variable conditions the effect of another on the variable of interest, tests of overall average effects mask the important insights.

For most data, of course, patterns of interaction are not as blatant as this one. Consequently, we must develop a test for the presence of interaction. We have said that interaction exists when the effect of one variable on another differs for different values of some third (or fourth . . .) variable. To translate this notion into algebraic expressions, let us reference the means for the cells in the example as shown in Table 3.5.

It can readily be seen that the differential effects of techniques 1 and 2 for gifted students can be measured by $b_1 - b_2$. For normal students, this effect is measured by $b_4 - b_5$. For retarded students, this effect is $b_7 - b_8$. If there is no interaction, these

Table 3.4 HYPOTHETICAL MEAN PERCENTAGE IMPROVEMENT FOR THREE TYPES OF STUDENTS AND THREE TECHNIQUES

Students	Technique 1	Technique 2	Technique 3
Gifted	0	0	15
Normal	0	15	0
Retarded	15	0	0

technique effects should be the same for the three ability levels. Thus, the hypothesis of no interaction implies:

$$H_1: \quad b_1 - b_2 = b_4 - b_5 \qquad H_2: \quad b_1 - b_2 = b_7 - b_8$$

and, of course, *implicitly*, $b_4 - b_5 = b_7 - b_8$.

The differential effects of techniques 1 and 3 are similarly measured by $b_1 - b_3$ (for the gifted), $b_4 - b_6$ (for the normal), and $b_7 - b_9$ (for the retarded). The hypothesis of no interaction also implies that these effects will be the same for all ability levels, or

$$H_3: \quad b_1 - b_3 = b_4 - b_6 \qquad H_4: \quad b_1 - b_3 = b_7 - b_9$$

and, of course, *implicitly*, $b_4 - b_6 = b_7 - b_9$.

If all four of these independent hypotheses are true, no interaction exists among the variables and we may safely proceed to ask questions about overall average effects. Before proceeding to set up the testing paradigm, we should note that "no interaction" is a symmetrical hypothesis, i.e., we could as easily have asked whether ability effects were the same for all techniques. Fortunately, it will appear that either natural-language approach to the question produces equivalent algebraic expressions.

Notice that the gifted-normal ability effect is measured by $b_1 - b_4$ (for technique 1), $b_2 - b_5$ (for technique 2), and $b_3 - b_6$ (for technique 3). The no-interaction hypothesis implies that these will be equivalent; thus:

Alternate H_1: $\qquad\qquad\qquad\qquad\qquad\qquad b_1 - b_4 = b_2 - b_5$

Subtracting b_2 from both sides and adding b_4 to both sides makes this equivalent to H_1.

Alternate H_3: $\qquad\qquad\qquad\qquad\qquad\qquad b_1 - b_4 = b_3 - b_6$

Subtracting b_3 from both sides and adding b_4 to both sides makes this equivalent to H_3.

The gifted-retarded ability effect is measured by $b_1 - b_7$ (for technique 1), $b_2 - b_8$ (for technique 2), and $b_3 - b_9$ (for technique 3). The no-interaction hypothesis implies:

Alternate H_2: $\qquad\qquad\qquad\qquad\qquad\qquad b_1 - b_7 = b_2 - b_8$

Table 3.5 CONVENTIONS FOR REFERENCING MEANS BY TECHNIQUE AND ABILITY LEVEL

Students	Technique 1	Technique 2	Technique 3
Gifted	$b_1 = \bar{Y}_1$	$b_2 = \bar{Y}_2$	$b_3 = \bar{Y}_3$
Normal	$b_4 = \bar{Y}_4$	$b_5 = \bar{Y}_5$	$b_6 = \bar{Y}_6$
Retarded	$b_7 = \bar{Y}_7$	$b_8 = \bar{Y}_8$	$b_9 = \bar{Y}_9$

which can readily be shown to be equivalent to H_2, and

Alternate H_4:
$$b_1 - b_7 = b_3 - b_9$$

which can readily be shown to be equivalent to H_4.

Thus we have found that the no-interaction hypothesis can be stated either as "the differential effects of the techniques on improvement are the same for all ability levels" or as "the differential effects of the ability levels on improvement are the same for all techniques." Either statement implies the same four independent algebraic hypotheses. Algebraically, then, these are equivalent natural-language questions.

We are now ready to see how our paradigm allows us to perform a statistical test for interaction.

A Natural-Language Question

Does technique affect improvement to the same degree for all ability levels? (Or, does ability affect improvement to the same degree for all techniques?)

B Full Model

$$\mathbf{Y} = b_1\mathbf{X}_1 + b_2\mathbf{X}_2 + b_3\mathbf{X}_3 + b_4\mathbf{X}_4 + b_5\mathbf{X}_5 + b_6\mathbf{X}_6 + b_7\mathbf{X}_7 + b_8\mathbf{X}_8$$
$$+ b_9\mathbf{X}_9 + \mathbf{E}_F$$

where $\mathbf{Y}$ = dependent variable (improvement)
$\quad \mathbf{X}_j$ = 1 for members of group j (cell$_j$), 0 otherwise
$\quad b_j$ = least-squares weight for each X_j
$\quad \mathbf{E}_F$ = error vector

This is an appropriate full model because it allows our hypotheses (about equivalences in differences in means) to be either true or false.

C Hypotheses

We found, above, that the general no-interaction hypothesis implies the following four independent hypotheses:

$$H_1: \ b_1 - b_2 = b_4 - b_5 \qquad H_2: \ b_1 - b_2 = b_7 - b_8$$
$$H_3: \ b_1 - b_3 = b_4 - b_6 \qquad H_4: \ b_1 - b_3 = b_7 - b_9$$

and, implicitly, $b_4 - b_5 = b_7 - b_8$ and $b_4 - b_6 = b_7 - b_9$.

Before imposing these conditions on the full model, we must solve each in terms of one unknown. Further, each unknown solved for must appear in only one of the hypotheses.[1] It will therefore be necessary to solve for b_5, b_8, b_6, and b_9. Such solutions permit us to rewrite the four hypotheses as follows:

$$H_1: \ b_5 = b_4 + b_2 - b_1 \qquad H_2: \ b_8 = b_7 + b_2 - b_1$$
$$H_3: \ b_6 = b_4 + b_3 - b_1 \qquad H_4: \ b_9 = b_7 + b_3 - b_1$$

[1] We solve for unknowns appearing in only one equation so that we can eliminate them from the restricted model.

We now substitute the indicated expressions for b_5, b_8, b_6, b_9 in the full model to produce the restricted model. If the resultant model produces no increase (or little increase) in the error SS, we shall conclude that the hypotheses are reasonable and that no (or negligible) interaction exists.

D   Restricted Model

$$\mathbf{Y} = b_1\mathbf{X}_1 + b_2\mathbf{X}_2 + b_3\mathbf{X}_3 + b_4\mathbf{X}_4 + (b_4 + b_2 - b_1)\mathbf{X}_5$$
$$+ (b_4 + b_3 - b_1)\mathbf{X}_6 + b_7\mathbf{X}_7 + (b_7 + b_2 - b_1)\mathbf{X}_8$$
$$+ (b_7 + b_3 - b_1)\mathbf{X}_9 + \mathbf{E}_R$$

expanding[1] gives

$$\mathbf{Y} = b_1\mathbf{X}_1 + b_2\mathbf{X}_2 + b_3\mathbf{X}_3 + b_4\mathbf{X}_4 + b_4\mathbf{X}_5 + b_2\mathbf{X}_5 + b_1(-\mathbf{X}_5)$$
$$+ b_4\mathbf{X}_6 + b_3\mathbf{X}_6 + b_1(-\mathbf{X}_6) + b_7\mathbf{X}_7 + b_7\mathbf{X}_8 + b_2\mathbf{X}_8$$
$$+ b_1(-\mathbf{X}_8) + b_7\mathbf{X}_9 + b_3\mathbf{X}_9 + b_1(-\mathbf{X}_9) + \mathbf{E}_R$$

Collecting common terms leads to

$$\mathbf{Y} = b_1(\mathbf{X}_1 - \mathbf{X}_5 - \mathbf{X}_6 - \mathbf{X}_8 - \mathbf{X}_9) + b_2(\mathbf{X}_2 + \mathbf{X}_5 + \mathbf{X}_8)$$
$$+ b_3(\mathbf{X}_3 + \mathbf{X}_6 + \mathbf{X}_9) + b_4(\mathbf{X}_4 + \mathbf{X}_5 + \mathbf{X}_6)$$
$$+ b_7(\mathbf{X}_7 + \mathbf{X}_8 + \mathbf{X}_9) + \mathbf{E}_R$$

To make explicit that these b_j's are merely unknown weights for these vectors and are no longer means for the nine cells, we shall substitute c_k's for them:[2]

$$\mathbf{Y} = c_1(\mathbf{X}_1 - \mathbf{X}_5 - \mathbf{X}_6 - \mathbf{X}_8 - \mathbf{X}_9) + c_2(\mathbf{X}_2 + \mathbf{X}_5 + \mathbf{X}_8)$$
$$+ c_3(\mathbf{X}_3 + \mathbf{X}_6 + \mathbf{X}_9) + c_4(\mathbf{X}_4 + \mathbf{X}_5 + \mathbf{X}_6)$$
$$+ c_5(\mathbf{X}_7 + \mathbf{X}_8 + \mathbf{X}_9) + \mathbf{E}_R$$

For clarity, we shall define new vectors $\mathbf{Z}_i$

$$\mathbf{Y} = c_1\mathbf{Z}_1 + c_2\mathbf{Z}_2 + c_3\mathbf{Z}_3 + c_4\mathbf{Z}_4 + c_5\mathbf{Z}_5 + \mathbf{E}$$

where $\mathbf{Y}$ = dependent variable
$\mathbf{Z}_1 = \mathbf{X}_1 - \mathbf{X}_5 - \mathbf{X}_6 - \mathbf{X}_8 - \mathbf{X}_9$
$\mathbf{Z}_2 = \mathbf{X}_2 + \mathbf{X}_5 + \mathbf{X}_8$
$\mathbf{Z}_3 = \mathbf{X}_3 + \mathbf{X}_6 + \mathbf{X}_9$
$\mathbf{Z}_4 = \mathbf{X}_4 + \mathbf{X}_5 + \mathbf{X}_6$
$\mathbf{Z}_5 = \mathbf{X}_7 + \mathbf{X}_8 + \mathbf{X}_9$
c_k = least-squares weights

---

[1] Notice that when we expand, we shift from $-b_1\mathbf{X}_5$ to $+b_1(-\mathbf{X}_5)$. This form will be useful in later steps.
[2] Note that b_j does not have the same value in the restricted and full models. In either case, the b_j's are least-squares weights for the vectors of the model, but the vectors are different in the two models. To make this explicit and to avoid confusion of the two sets of weights, we shifted from the b_j terms in the full model to c_k terms in the restricted model.

The vectors of the full model are easily described; for example, X_j is 1 for members of group j. Note that the vectors of this restricted model no longer have easily summarized definitions. For example, Z_1 is 1 for members of group 1, -1 for members of groups 5, 6, 8, 9, and 0 for all others. These vectors cannot be interpreted other than algebraically. They are merely a set of vectors which will produce no increase in error if our no-interaction hypotheses are true. Further, the c_k's are not interpretable as means for any definable groups. They are merely least-squares weights associated with the no-interaction restricted model.

E Inference

The traditional interaction test can now be reduced to our familiar F distribution. Notice that $df_2 = N - 9$ (as there are nine unknown weights in the full model) and $df_1 = 9 - 5 = 4$ (as there are five unknown weights in the restricted model). Thus

$$F = \frac{(\sum e_{i_R}^2 - \sum e_{i_F}^2)/4}{\sum e_{i_F}^2/(N - 9)}$$

F Computer Instructions

Computer solution requires the following modifications for program CARTROCK:

1 Fortran statements to read data and construct: VECs 1 through 9 defined as X_1 through X_9, VECs 10 through 14 defined as Z_1 through Z_5, VEC 15 defined as Y. Assume that the data are provided in the following form:

Columns 1–3   Case identification number
Columns 4–5   Percentage improvement on achievement test
Column 6   Type of teaching technique (1, 2, 3)
Column 7   Ability grouping (1 = gifted, 2 = normal, 3 = retarded)
Column 8   Group number (1 through 9 as indicated in Table 3.5)

The Fortran statements for SUBROUTINE DATA could then be

```
. . . . . . . . . . . . . . .
      DO 5 I=1,9
   5  VEC(I)=0.
      READ (5,11) VAR(1),VAR(2)
      VEC(15)=VAR(1)
      K=VAR(2)
      VEC(K)=1.
      VEC(10)=VEC(1)-VEC(5)-VEC(6)-VEC(8)-VEC(9)
      VEC(11)=VEC(2)+VEC(5)+VEC(8)
      VEC(12)=VEC(3)+VEC(6)+VEC(9)
      VEC(13)=VEC(4)+VEC(5)+VEC(6)
      VEC(14)=VEC(7)+VEC(8)+VEC(9)
  11  FORMAT (3X,1F2.0,2X,1F1.0)
. . . . . . . . . . . . . . . . . . . . . . . . . . .
```

2 A main control card with a label, the number of cases (use 45 for numerical example), the number of VARs (2), the number of VECs (15), and the number of regressions (2). It would be

INTER_TEST___45____2___15____2

3 Two analysis control cards following the 45 data cards. They would be

FULL_MODEL___15____9____1____2____3____4___5____6____7____8____9

Specifying the dependent variable to be VEC (15), or Y, and nine independent variables with subscripts 1 through 9, or X_1 through X_9.

REST_MODEL___15____5___10___11___12___13___14

Specifying the dependent variable to be VEC (15), or Y, and five independent variables with subscripts 10 through 14, or Z_1 through Z_5.

G Numerical Example

We shall use the data from Table 3.3 for our example. These are the same data which were used for the three-group case of one-way analysis of variance. The reader will already have punched these on cards for the example in that section. The only additions which should be made to those cards are:

1 In column 7, a code should be added to indicate ability grouping. This code should be 1 (gifted) for cases 1 through 5, 16 through 20, and 31 through 35. The code should be 2 (normal) for cases 6 through 10, 21 through 25, and 36 through 40. The code should be 3 (retarded) for cases 11 through 15, 26 through 30, and 41 through 45.

2 In column 8 a code should be added to indicate the group number (or cell number) following the convention in Table 3.5. Thus, the codes should be

Code	1	2	3	4	5	6	7	8	9
Cases	1–5	16–20	31–35	6–10	21–25	36–40	11–15	26–30	41–45

We can test these data for interaction between ability and technique (as they affect improvement) by the procedure outlined above, i.e., by comparing the following models:

Full model:

$$Y = b_1X_1 + b_2X_2 + b_3X_3 + b_4X_4 + b_5X_5 + b_6X_6 + b_7X_7$$
$$+ b_8X_8 + b_9X_9 + E_F$$

Hypotheses:

$$b_1 - b_2 = b_4 - b_5 \qquad b_1 - b_3 = b_4 - b_6$$
$$b_1 - b_2 = b_7 - b_8 \qquad b_1 - b_3 = b_7 - b_9$$

Restricted model: $Y = c_1Z_1 + c_2Z_2 + c_3Z_3 + c_4Z_4 + c_5Z_5 + E_R$

where $Z_1 = X_1 - X_5 - X_6 - X_8 - X_9$ $Z_2 = X_2 + X_5 + X_8$

$Z_3 = X_3 + X_6 + X_9$ $Z_4 = X_4 + X_5 + X_6$

$Z_5 = X_7 + X_8 + X_9$

The computer solutions will be:

Full model	Restricted model
$b_1 = 16.0000$	$c_1 = 15.5556$
$b_2 = 16.0000$	$c_2 = 16.2222$
$b_3 = 16.0000$	$c_3 = 16.2222$
$b_4 = 16.0000$	$c_4 = 13.5556$
$b_5 = 18.0000$	$c_5 = 10.8889$
$b_6 = 8.0000$	
$b_7 = 8.0000$	
$b_8 = 8.0000$	
$b_9 = 18.0000$	
$\sum e_i^2 = 200.0$	$\sum e_i^2 = 808.9$
$R^2 = .7955$	$R^2 = .1727$

$$F = \frac{(\sum e_{i_R}^2 - \sum e_{i_F}^2)/df_1}{\sum e_{i_F}^2/df_2} = \frac{(808.9 - 200.0)/(9 - 5)}{200.0/(45 - 9)} \approx 27.4$$

or

$$F = \frac{(R_F^2 - R_R^2)/df_1}{(1 - R_F^2)/df_2} = \frac{(.7955 - .1727)/(9 - 5)}{(1 - .7955)/(45 - 9)} \approx 27.4$$

As this F is well in excess of the critical value 7.05 required at 4 and 36 degrees of freedom for the .001 level of significance, we must reject our no-interaction hypotheses. The finding that interaction exists should never be merely reported: it requires interpretation. Recall that interaction, in its most general sense, means that one variable's effect on another is different for different values of some third variable. As such, interaction may take many forms. Just which form can only be determined by examining the table of means. For our data this is presented in Table 3.6. These results can be interpreted as follows:

1 For gifted children, each of the three techniques produce the same amount of improvement.
2 For normal children technique 2 is slightly better than technique 1 and much better than technique 3.
3 For retarded children technique 3 is much better than either technique 1 or 2.

The importance of the interaction finding can be seen clearly if we imagine that we are giving advice to a school system based on our evaluation of these teaching techniques. If we simply looked for average differences in improvement produced by the techniques, we would say that there is no significant difference between them.

Table 3.6   MEAN PERCENTAGE IMPROVEMENT IN PERFORMANCE BY TECHNIQUE AND ABILITY LEVEL

Children	Technique 1	Technique 2	Technique 3
Gifted	$b_1 = \bar{Y}_1 = 16$	$b_2 = \bar{Y}_2 = 16$	$b_3 = \bar{Y}_3 = 16$
Normal	$b_4 = \bar{Y}_4 = 16$	$b_5 = \bar{Y}_5 = 18$	$b_6 = \bar{Y}_6 = 8$
Retarded	$b_7 = \bar{Y}_7 = 8$	$b_8 = \bar{Y}_8 = 8$	$b_9 = \bar{Y}_9 = 18$

We arrived at this conclusion in the three-group case, Sec. 3.6. Were this true, we should advise a school system to choose among the techniques on other grounds, e.g., cost, style, ease. However, our recommendations will be quite different if we have introduced other relevant variables and found interaction. Here we should advise a school system that the choice of technique is unimportant for the gifted, technique 1 or 2 may reasonably be chosen for the normals, and technique 3 must be chosen for the retarded. As sociologists, we would probably further be obligated to recommend either:

1 Ability grouping to permit different techniques for different children or

2 A decision between technique 3 and the others based on whether the system is more willing to write off normals (with technique 3) or retardates (with techniques 1 and 2)

3.12 GENERALIZATION: THE $r \times c$ CASE OF INTERACTION

If we have a row variable with r values (designated $1, 2, 3, \ldots, r$) and a column variable with c values (designated $1, 2, 3, \ldots, c$) we can test for interaction with the following strategy. (The first subscript refers to the row variable, the second to the column.)

B   Full Model

$$Y = b_{11}X_{11} + b_{12}X_{12} + \cdots + b_{1c}X_{1c} + b_{21}X_{21} + b_{22}X_{22} + \cdots + b_{2c}X_{2c}$$
$$+ \cdots + b_{rc}X_{rc} + E_F \qquad (3.4)$$

This allows a unique mean for each cell and there are rc cells.

C   Hypotheses

$$\left. \begin{array}{l} b_{11} - b_{12} = b_{21} - b_{22} \\ b_{11} - b_{12} = b_{31} - b_{32} \\ \cdots\cdots\cdots\cdots\cdots\cdots \\ b_{11} - b_{12} = b_{r1} - b_{r2} \end{array} \right\}$$ This makes the column difference for column values 1 and 2 equal for all rows

$$\left. \begin{array}{l} b_{11} - b_{13} = b_{21} - b_{23} \\ b_{11} - b_{13} = b_{31} - b_{33} \\ \cdots\cdots\cdots\cdots\cdots\cdots \\ b_{11} - b_{13} = b_{r1} - b_{r3} \end{array} \right\}$$ This makes the column difference for column values 1 and 3 equal for all rows

$$\cdots\cdots\cdots\cdots\cdots\cdots$$

$$\left. \begin{array}{l} b_{11} - b_{1c} = b_{21} - b_{2c} \\ b_{11} - b_{1c} = b_{31} - b_{3c} \\ \cdots\cdots\cdots\cdots\cdots\cdots \\ b_{11} - b_{1c} = b_{r1} - b_{rc} \end{array} \right\}$$ The process is continued for all columns compared with the first

A verbal description of these hypotheses may be helpful. The difference between the paired means for the first and second columns is set equal for all rows. Next, the difference between the paired means for the first and third columns is set equal for all rows. Finally, this process is continued for the differences between the paired means for the first column and each remaining (fourth, fifth, ..., cth) column. There will be $(r - 1)(c - 1)$ such independent hypotheses. After they have been specified, we must solve for an independent unknown from each. The appropriate unknowns will be the rightmost ones in the equations, that is, $b_{22}, b_{32}, \ldots, b_{r2}, b_{23}, b_{33}, \ldots, b_{r3}, \ldots, b_{2c}, b_{3c}, \ldots, b_{rc}$. The resulting expressions can then be substituted directly in the full model to produce the proper restricted model.

D Restricted Model

After the hypotheses are substituted into the full model, the resultant restricted model must be expanded and common terms collected. Then the character of the necessary vectors will be made obvious. A general restricted model would be unnecessarily complex, but the process for arriving at a proper restricted model for any set of data should be clear.

F ratio:
$$F = \frac{(\sum e_{i_R}^2 - \sum e_{i_F}^2)/[(r - 1)(c - 1)]}{\sum e_{i_F}^2/(N - rc)}$$

(For equivalences, see McNemar, 1962, p. 306 and Blalock, 1972, p. 342.) Note that $df_1 = (r - 1)(c - 1)$, as this is the number of independent conditions imposed; also, $df_2 = N - rc$, as there are rc unknown weights in the full model.

3.13 NONSTANDARD QUESTIONS: PARTITIONS AND FORMS OF INTERACTION, CONTROLS

We have shown the hypotheses which must be imposed to address the standard interaction question. If there were no interaction, the analyst could then proceed immediately to the standard main-effect questions. As this is often impossible, we shall now develop three types of useful nonstandard questions. The first is directed at isolating the sources of interaction; the second will be useful where we seek to demonstrate a specific *form* of interaction; and the third provides an answer to the question: What is the difference in the effect of technique (or ability) *controlling for* ability (or technique)?

QUESTION 1   Suppose we believe (on theoretical or ad hoc bases) that the interaction can be restricted to certain portions of our table. We can test such a notion by imposing only a subset of the no-interaction hypotheses. In the current example, it seems reasonable to conclude that technique 3 is the only technique which interacts with ability.[1] It produces unusually low improvement for normals and unusually high

---

[1] We shall later exclude technique 3 from our analyses of main effects; however, this may be the most interesting technique as it is the only one differentiating the three types of students and the only one producing large changes for the retarded. If we were doing a study in retardation, we might elect to exclude techniques 1 and 2 from further analysis.

improvement for retardates. If these speculations are true, we can proceed to make general statements about ability and technique for the first two techniques. A test of this notion follows.

A   Natural-Language Question

Is the difference in the effects of techniques 1 and 2 the same for all ability groups? (Or, is the effect of ability the same for techniques 1 and 2?)

B   Full Model

$$Y = b_1X_1 + b_2X_2 + b_3X_3 + b_4X_4 + b_5X_5 + b_6X_6 + b_7X_7 + b_8X_8$$
$$+ b_9X_9 + E_F$$

C   Hypotheses

$$b_1 - b_2 = b_4 - b_5 \qquad b_1 - b_2 = b_7 - b_8$$

Solving for independent weights b_5 and b_8 gives

$$b_5 = b_4 + b_2 - b_1 \qquad b_8 = b_7 + b_2 - b_1$$

D   Restricted Model

Substituting for b_5 and b_8, we have

$$Y = b_1X_1 + b_2X_2 + b_3X_3 + b_4X_4 + (b_4 + b_2 - b_1)X_5 + b_6X_6$$
$$+ b_7X_7 + (b_7 + b_2 - b_1)X_8 + b_9X_9 + E_R$$

Expanding gives

$$Y = b_1X_1 + b_2X_2 + b_3X_3 + b_4X_4 + b_4X_5 + b_2X_5 + b_1(-X_5) + b_6X_6$$
$$+ b_7X_7 + b_7X_8 + b_2X_8 + b_1(-X_8) + b_9X_9 + E_R$$

Collecting common terms leads to

$$Y = b_1(X_1 - X_5 - X_8) + b_2(X_2 + X_5 + X_8) + b_3X_3 + b_4(X_4 + X_5)$$
$$+ b_6X_6 + b_7(X_7 + X_8) + b_9X_9 + E_R$$

To make explicit that the weights are no longer necessarily the same as the weights of the full model we write

$$Y = c_1(X_1 - X_5 - X_8) + c_2(X_2 + X_5 + X_8) + c_3X_3 + c_4(X_4 + X_5)$$
$$+ c_5X_6 + c_6(X_7 + X_8) + c_7X_9 + E_R$$

and for simplicity

$$Y = c_1Z_1 + c_2Z_2 + c_3X_3 + c_4Z_3 + c_5X_6 + c_6Z_4 + c_7X_9 + E_R$$

where
$$Z_1 = X_1 - X_5 - X_8$$
$$Z_2 = X_2 + X_5 + X_8$$
$$Z_3 = X_4 + X_5$$
$$Z_4 = X_7 + X_8$$

We have retained the X_j notation for vectors X_3, X_6, and X_9, which are unchanged from the full to the restricted models, as these vectors are for technique 3, for which no hypotheses are imposed.

This alternative partial-interaction test would require the creation of the following vectors, in addition to those developed in Sec. 3.10 for the traditional interaction test:

$$VEC(16) = VEC(1) - VEC(5) - VEC(8)$$
$$VEC(17) = VEC(2) + VEC(5) + VEC(8)$$
$$VEC(18) = VEC(4) + VEC(5)$$
$$VEC(19) = VEC(7) + VEC(8)$$

The model card could be written

PART.INTER___15____7____3____6____9___16___17___18___19

This specifies dependent variable VEC(15) or Y, 7 independent variables which are VEC(3) or X_3, VEC(6) or X_6, VEC(9) or X_9, VEC(16) or Z_1, VEC(17) or Z_2, VEC(18) or Z_3, and VEC(19) or Z_4. The relevant computer results are:

	Full model	Restricted model
	$b_1 = 16.0000$	$c_1 = 15.6667$
	$b_2 = 16.0000$	$c_2 = 16.3333$
	$b_3 = 16.0000$	$c_3 = 16.0000$
	$b_4 = 16.0000$	$c_4 = 16.6666$
	$b_5 = 18.0000$	$c_5 = 8.0000$
	$b_6 = 8.0000$	$c_6 = 7.6667$
	$b_7 = 8.0000$	$c_7 = 18.0000$
	$b_8 = 8.0000$	
	$b_9 = 18.0000$	
	$\sum e_i^2 = 200.0$	$\sum e_i^2 = 206.7$
	$R^2 = .7955$	$R^2 = .7886$

$$F = \frac{(\sum e_{i_R}^2 - \sum e_{i_F}^2)/df_1}{\sum e_{i_F}^2/df_2} = \frac{(206.7 - 200.0)/(9 - 7)}{200.0/(45 - 9)} \approx .60$$

As this F is considerably less than the value of 3.32 required at 2 and 36 degrees of freedom for significance at the .05 level, we are unable to reject our hypotheses. Consequently it seems reasonable to assume no interaction between the techniques (1 and 2) and ability level.[1] An alternative formulation for this partial-interaction hypothesis could be obtained by simply excluding all cells for technique 3 (cells 3, 6, 9) from consideration in the full and restricted models. This would produce the following analysis:

Full model:

$$Y = b_1X_1 + b_2X_2 + b_4X_4 + b_5X_5 + b_7X_7 + b_8X_8 + E_F$$

Notice that we have retained our convention for designating the cells in our table. Technique 3 is excluded by omitting X_3, X_6, and X_9 from the full (and therefore from the restricted) model.

Hypotheses:
$$b_1 - b_2 = b_4 - b_5 \quad \text{or} \quad b_5 = b_4 + b_2 - b_1$$
$$b_1 - b_2 = b_7 - b_8 \quad \text{or} \quad b_8 = b_7 + b_2 - b_1$$

---

[1] As we wish to *accept* the no-interaction hypotheses, it would be prudent to reject them if the error increase is significant at the .10 (or .30, .50, ...) level. Requiring the traditional .05 level may put us in the position of assuming no interaction in situations where we will be wrong 19 of 20 times.

These are the hypotheses which we developed above to assert that the differential effects of techniques 1 and 2 are the same for all ability levels.

Restricted model:

$$
\begin{aligned}
\mathbf{Y} &= b_1\mathbf{X}_1 + b_2\mathbf{X}_2 + b_4\mathbf{X}_4 + (b_4 + b_2 - b_1)\mathbf{X}_5 + b_7\mathbf{X}_7 \\
&\qquad\qquad\qquad + (b_7 + b_2 - b_1)\mathbf{X}_8 + \mathbf{E}_R \\
&= b_1\mathbf{X}_1 + b_2\mathbf{X}_2 + b_4\mathbf{X}_4 + b_4\mathbf{X}_5 + b_2\mathbf{X}_5 + b_1(-\mathbf{X}_5) + b_7\mathbf{X}_7 \\
&\qquad\qquad\qquad + b_7\mathbf{X}_8 + b_2\mathbf{X}_8 + b_1(-\mathbf{X}_8) + \mathbf{E}_R \\
&= b_1(\mathbf{X}_1 - \mathbf{X}_5 - \mathbf{X}_8) + b_2(\mathbf{X}_2 + \mathbf{X}_5 + \mathbf{X}_8) + b_4(\mathbf{X}_4 + \mathbf{X}_5) \\
&\qquad\qquad\qquad + b_7(\mathbf{X}_7 + \mathbf{X}_8) + \mathbf{E}_R \\
&= c_1(\mathbf{X}_1 - \mathbf{X}_5 - \mathbf{X}_8) + c_2(\mathbf{X}_2 + \mathbf{X}_5 + \mathbf{X}_8) + c_3(\mathbf{X}_4 + \mathbf{X}_5) \\
&\qquad\qquad\qquad + c_4(\mathbf{X}_7 + \mathbf{X}_8) + \mathbf{E}_R \\
&= c_1\mathbf{Z}_1 + c_2\mathbf{Z}_2 + c_3\mathbf{Z}_3 + c_4\mathbf{Z}_4 + \mathbf{E}_R
\end{aligned}
$$

where
$$
\begin{aligned}
\mathbf{Z}_1 &= \mathbf{X}_1 - \mathbf{X}_5 - \mathbf{X}_8 \\
\mathbf{Z}_2 &= \mathbf{X}_2 + \mathbf{X}_5 + \mathbf{X}_8 \\
\mathbf{Z}_3 &= \mathbf{X}_4 + \mathbf{X}_5 \\
\mathbf{Z}_4 &= \mathbf{X}_7 + \mathbf{X}_8
\end{aligned}
$$

These are the same vectors which we used in the previous example as VEC(16), VEC(17), VEC(18), and VEC(19). We have already developed the Fortran to produce VEC(1) through VEC(9) as $\mathbf{X}_1$ through $\mathbf{X}_9$. We can use these statements with the 30 cases subject to techniques 1 and 2 to produce our requisite $\mathbf{X}_1$ as VEC(1), $\mathbf{X}_2$ as VEC(2), $\mathbf{X}_4$ as VEC(4), $\mathbf{X}_5$ as VEC(5), $\mathbf{X}_7$ as VEC(7), and $\mathbf{X}_8$ as VEC(8). For these 30 cases, those statements will produce, additionally, three null vectors VEC(3), VEC(6), and VEC(9), which we shall not use in these analyses. The model control cards would appear as follows:

ALT.FULL _____15____6____1____2____4____5____7____8

ALT.REST. ____15____4___16___17___18___19

The relevant computer results would be:

	Alternative full model	Alternative restricted model
	$b_1 = 16.0000$	$c_1 = 15.6667$
	$b_2 = 16.0000$	$c_2 = 16.3333$
	$b_4 = 16.0000$	$c_3 = 16.6666$
	$b_5 = 18.0000$	$c_4 = 7.6667$
	$b_7 = 8.0000$	
	$b_8 = 8.0000$	
	$\sum e_i^2 = 126.0$	$\sum e_i^2 = 132.7$
	$R^2 = .7976$	$R^2 = .7869$

$$
F = \frac{\left(\sum e_{i_R}^2 - \sum e_{i_F}^2\right)/\mathrm{df}_1}{\sum e_{i_F}^2/\mathrm{df}_2} = \frac{(132.7 - 126.0)/(6 - 4)}{126.0/(30 - 4)} \approx .64
$$

As this F is less than the value of 3.40 required at 2 and 24 degrees of freedom at the .05 level of significance, we are unable to reject our hypotheses. Thus, we conclude that it is reasonable to assume no interaction.

Ordinarily, it will not matter in terms of practical inference which of these two approaches we choose for demonstrating that, while interaction exists, it is restricted to certain subcells of our table. Notice that the error increase attributable to the (partial) no-interaction hypotheses is the same ($\sum e_{i_R}^2 - \sum e_{i_F}^2 = 6.7$) for either of these approaches. This is because in the first approach the error associated with vectors $\mathbf{X}_3$, $\mathbf{X}_6$, and $\mathbf{X}_9$ (excluded in the second approach) appears in both full and restricted models. Thus, as two hypotheses are imposed in each case, the numerator is the same. The fuller (nine-group) approach has a larger $\sum e_{i_F}^2$ in the denominator. However, as the degrees of freedom are also larger, the net difference is negligible. The full model with six groups is, of course, the traditional interaction test for the restricted data set (technique 3 excluded). The advantage of the full model using all nine groups is that it allows partial no-interaction hypotheses which have no conventional analog. This possibility is elaborated below. The effects of this choice are discussed under pooling, Sec.3.21.

QUESTION 2   As we have noted, interaction findings may be the most substantively interesting results of an analysis. Increasingly, social scientists are developing more precise theoretical formulations and traditions of replication. Either of these may permit us to formulate rather precise interaction hypotheses which have considerably more payoff than the more exploratory no-interaction hypotheses. Although traditional computing formulas exist for such hypotheses, tests of many interaction hypotheses of specific forms can be developed quite straightforwardly through the application of least-squares theory. Since specific interactional forms are as varied as the imaginations of researchers, we cannot be exhaustive about this. A single illustration must suffice. Assume that either theory or previous research leads us to believe that the following differences are likely:

1 The differential improvement for technique 2 over technique 1 will be 5 percentage points more for the gifted than for normals but 10 percentage points less for the retarded than for normals.

2 The differential improvement for technique 3 over technique 1 will be 5 percentage points less for the gifted than for normals but 5 percentage points more for the retarded than for normals.

These are interaction hypotheses in that the effects of one variable are said to depend on the value of another. They can be tested by comparing the following models, in which these specific hypotheses are algebraically expressed and directly imposed:

Full model:

$$\mathbf{Y} = b_1\mathbf{X}_1 + b_2\mathbf{X}_2 + b_3\mathbf{X}_3 + b_4\mathbf{X}_4 + b_5\mathbf{X}_5 + b_6\mathbf{X}_6 + b_7\mathbf{X}_7$$
$$+ b_8\mathbf{X}_8 + b_9\mathbf{X}_9 + \mathbf{E}_F$$

Hypotheses:

$b_5 - b_4 = b_2 - b_1 - 5$ $b_5 - b_4 = b_8 - b_7 + 10$

$b_6 - b_4 = b_3 - b_1 + 5$ $b_6 - b_4 = b_9 - b_7 - 5$

Solving for independent unknowns gives

$$b_2 = b_5 - b_4 + b_1 + 5 \qquad b_8 = b_5 - b_4 + b_7 - 10$$
$$b_3 = b_6 - b_4 + b_1 - 5 \qquad b_9 = b_6 - b_4 + b_7 + 5$$

Restricted model:

$$\begin{aligned}
\mathbf{Y} = {} & b_1\mathbf{X}_1 + (b_5 - b_4 + b_1 + 5)\mathbf{X}_2 + (b_6 - b_4 + b_1 - 5)\mathbf{X}_3 + b_4\mathbf{X}_4 \\
& + b_5\mathbf{X}_5 + b_6\mathbf{X}_6 + b_7\mathbf{X}_7 + (b_5 - b_4 + b_7 - 10)\mathbf{X}_8 \\
& \qquad\qquad\qquad + (b_6 - b_4 + b_7 + 5)\mathbf{X}_9 + \mathbf{E}_R
\end{aligned}$$

$$\begin{aligned}
= {} & b_1\mathbf{X}_1 + b_5\mathbf{X}_2 + b_4(-\mathbf{X}_2) + b_1\mathbf{X}_2 + 5\mathbf{X}_2 + b_6\mathbf{X}_3 + b_4(-\mathbf{X}_3) \\
& + b_1\mathbf{X}_3 - 5\mathbf{X}_3 + b_4\mathbf{X}_4 + b_5\mathbf{X}_5 + b_6\mathbf{X}_6 + b_7\mathbf{X}_7 + b_5\mathbf{X}_8 \\
& + b_4(-\mathbf{X}_8) + b_7\mathbf{X}_8 - 10\mathbf{X}_8 + b_6\mathbf{X}_9 + b_4(-\mathbf{X}_9) + b_7\mathbf{X}_9 \\
& \qquad\qquad\qquad\qquad\qquad\qquad\qquad\qquad + 5\mathbf{X}_9 + \mathbf{E}_R
\end{aligned}$$

$$\begin{aligned}
\mathbf{Y} - 5\mathbf{X}_2 + 5\mathbf{X}_3 + 10\mathbf{X}_8 - 5\mathbf{X}_9 = {} & b_1(\mathbf{X}_1 + \mathbf{X}_2 + \mathbf{X}_3) \\
& + b_4(\mathbf{X}_4 - \mathbf{X}_2 - \mathbf{X}_3 - \mathbf{X}_8 - \mathbf{X}_9) \\
& + b_5(\mathbf{X}_5 + \mathbf{X}_2 + \mathbf{X}_8) + b_6(\mathbf{X}_6 + \mathbf{X}_3 + \mathbf{X}_9) \\
& \qquad\qquad + b_7(\mathbf{X}_7 + \mathbf{X}_8 + \mathbf{X}_9) + \mathbf{E}_R
\end{aligned}$$

$$\mathbf{W} = c_1\mathbf{Z}_1 + c_2\mathbf{Z}_2 + c_3\mathbf{Z}_3 + c_4\mathbf{Z}_4 + c_5\mathbf{Z}_5 + \mathbf{E}_R$$

where $\mathbf{W} = \mathbf{Y} - 5\mathbf{X}_2 + 5\mathbf{X}_3 + 10\mathbf{X}_8 - 5\mathbf{X}_9$

$\mathbf{Z}_1 = \mathbf{X}_1 + \mathbf{X}_2 + \mathbf{X}_3$

$\mathbf{Z}_2 = \mathbf{X}_4 - \mathbf{X}_2 - \mathbf{X}_3 - \mathbf{X}_8 - \mathbf{X}_9$

$\mathbf{Z}_3 = \mathbf{X}_5 + \mathbf{X}_2 + \mathbf{X}_8$

$\mathbf{Z}_4 = \mathbf{X}_6 + \mathbf{X}_3 + \mathbf{X}_9$

$\mathbf{Z}_5 = \mathbf{X}_7 + \mathbf{X}_8 + \mathbf{X}_9$

Our test then would require the construction of vectors $\mathbf{Y}$, $\mathbf{X}_1$ through $\mathbf{X}_9$, $\mathbf{W}$, and $\mathbf{Z}_1$ through $\mathbf{Z}_5$. The two models have different dependent variables, and therefore we must calculate F from the $\sum e_i^2$'s. Notice that $df_1 = 4$ and $df_2 = N - 9$. If the F ratio is large, we reject these specific interaction hypotheses. If it is small, we are unable to reject them. When we find interaction with the traditional general test, it is a good idea to try to discern pattern in the interaction. If some regular pattern makes good theoretical sense, it is probably worthwhile to document the patterning by using the approach discussed here. Of course, the resulting hypotheses will be post hoc and should be reported as such. However, in subsequent research such patterns can then be predicted in the strict sense of the word.

QUESTION 3   If interaction is found, whether general or neatly patterned, we shall be unable to say anything about the overall average effect of either of the variables. Although no traditional computation is available, we can nevertheless test to see whether either variable has statistically significant effects on the dependent variable, *controlling for the other variable*. This can be done by imposing relational conditions on the means for one of the variables *within categories of the other*. Recall our data matrix representation:

	Technique 1	Technique 2	Technique 3
Gifted children	b_1	b_2	b_3
Normal children	b_4	b_5	b_6
Retarded children	b_7	b_8	b_9

If we hypothesize no technique effects for the first row, that is, $b_1 = b_2, b_1 = b_3$, these hypotheses will not be contaminated by interaction effects from ability level since ability level is treated as a constant for these hypotheses. Similarly, tests of hypotheses of no technique effects within the second row ($b_4 = b_5, b_4 = b_6$), and the third row ($b_7 = b_8, b_7 = b_9$) cannot be influenced by interaction with ability level. Thus, we can ask whether the techniques produce significantly different levels of improvement *controlling for ability* by comparing the following models:

Full model:

$$\mathbf{Y} = b_1\mathbf{X}_1 + b_2\mathbf{X}_2 + b_3\mathbf{X}_3 + b_4\mathbf{X}_4 + b_5\mathbf{X}_5 + b_6\mathbf{X}_6 + b_7\mathbf{X}_7$$
$$+ b_8\mathbf{X}_8 + b_9\mathbf{X}_9 + \mathbf{E}_F$$

Hypotheses:

$$\left. \begin{array}{l} b_1 = b_2 \,(= c_1) \\ b_1 = b_3 \,(= c_1) \end{array} \right\} \quad \text{no technique effect for the gifted}$$

$$\left. \begin{array}{l} b_4 = b_5 \,(= c_2) \\ b_4 = b_6 \,(= c_2) \end{array} \right\} \quad \text{no technique effect for normals}$$

$$\left. \begin{array}{l} b_7 = b_8 \,(= c_3) \\ b_7 = b_9 \,(= c_3) \end{array} \right\} \quad \text{no technique effect for the retarded}$$

Restricted model:

$$\mathbf{Y} = c_1\mathbf{X}_1 + c_1\mathbf{X}_2 + c_1\mathbf{X}_3 + c_2\mathbf{X}_4 + c_2\mathbf{X}_5 + c_2\mathbf{X}_6 + c_3\mathbf{X}_7 + c_3\mathbf{X}_8$$
$$+ c_3\mathbf{X}_9 + \mathbf{E}_R$$
$$= c_1(\mathbf{X}_1 + \mathbf{X}_2 + \mathbf{X}_3) + c_2(\mathbf{X}_4 + \mathbf{X}_5 + \mathbf{X}_6)$$
$$+ c_3(\mathbf{X}_7 + \mathbf{X}_8 + \mathbf{X}_9) + \mathbf{E}_R$$
$$= c_1\mathbf{Z}_1 + c_2\mathbf{Z}_2 + c_3\mathbf{Z}_3 + \mathbf{E}_R$$

where $Z_1 = X_1 + X_2 + X_3 = \begin{cases} 1 & \text{for all gifted children} \\ 0 & \text{for all others} \end{cases}$

$Z_2 = X_4 + X_5 + X_6 = \begin{cases} 1 & \text{for all normal children} \\ 0 & \text{for all others} \end{cases}$

$Z_3 = X_7 + X_8 + X_9 = \begin{cases} 1 & \text{for all retarded children} \\ 0 & \text{for all others} \end{cases}$

Notice that these hypotheses, taken together, say that *within ability level*, the three techniques produce substantially the same average improvement. Of course, the c_j's are the overall row means. Computer solution requires the construction of Y, X_1 through X_9 (as before) and Z_1 through Z_3 (as defined above). The full-model card should designate Y as the dependent variable and X_1 through X_9 as the independent variables. The restricted-model card should designate Y as the dependent variable and Z_1 through Z_3 as the independent variables. The relevant computer results will be:

Full model	Restricted model
$b_1 = 16.0000$	$c_1 = 16.0000$
$b_2 = 16.0000$	$c_2 = 14.0000$
$b_3 = 16.0000$	$c_3 = 11.3333$
$b_4 = 16.0000$	
$b_5 = 18.0000$	
$b_6 = 8.0000$	
$b_7 = 8.0000$	
$b_8 = 8.0000$	
$b_9 = 18.0000$	
$\sum e_i^2 = 200.0$	$\sum e_i^2 = 813.3$
$R^2 = .7955$	$R^2 = .1682$

$$F = \frac{(\sum e_{i_R}^2 - \sum e_{i_F}^2)/df_1}{\sum e_{i_F}^2/df_2} = \frac{(813.3 - 200.0)/(9 - 3)}{200.0/(45 - 9)} \approx 18.4$$

The critical value for F at 6 and 36 degrees of freedom (actually at 6 and 30 degrees of freedom for a conservative test) for the .001 level of significance is 5.12. As the obtained F is considerably larger than this, we reject our hypothesis that technique has no effect on improvement controlling for ability.

The parallel analysis of the effect of ability on improvement controlling for technique is now quite straightforward:

Full model:

$$Y = b_1X_1 + b_2X_2 + b_3X_3 + b_4X_4 + b_5X_5 + b_6X_6 + b_7X_7 + b_8X_8$$
$$+ b_9X_9 + E_F$$

Hypotheses:

$$\left.\begin{array}{l} b_1 = b_4 (= c_1) \\ b_1 = b_7 (= c_1) \end{array}\right\} \text{ no ability effect for technique 1}$$

$$\left.\begin{array}{l} b_2 = b_5 (= c_2) \\ b_2 = b_8 (= c_2) \end{array}\right\} \text{ no ability effect for technique 2}$$

$$\left.\begin{array}{l} b_3 = b_6 (= c_3) \\ b_3 = b_9 (= c_3) \end{array}\right\} \text{ no ability effect for technique 3}$$

Restricted model:

$$\mathbf{Y} = c_1\mathbf{X}_1 + c_1\mathbf{X}_4 + c_1\mathbf{X}_7 + c_2\mathbf{X}_2 + c_2\mathbf{X}_5 + c_2\mathbf{X}_8 + c_3\mathbf{X}_3 + c_3\mathbf{X}_6$$
$$+ c_3\mathbf{X}_9 + \mathbf{E}_R$$
$$= c_1(\mathbf{X}_1 + \mathbf{X}_4 + \mathbf{X}_7) + c_2(\mathbf{X}_2 + \mathbf{X}_5 + \mathbf{X}_8)$$
$$+ c_3(\mathbf{X}_3 + \mathbf{X}_6 + \mathbf{X}_9) + \mathbf{E}_R$$
$$= c_1\mathbf{Z}_1 + c_2\mathbf{Z}_2 + c_3\mathbf{Z}_3 + \mathbf{E}_R$$

where $\mathbf{Z}_1 = \mathbf{X}_1 + \mathbf{X}_4 + \mathbf{X}_7$ $\qquad \mathbf{Z}_2 = \mathbf{X}_2 + \mathbf{X}_5 + \mathbf{X}_8$ $\qquad \mathbf{Z}_3 = \mathbf{X}_3 + \mathbf{X}_6 + \mathbf{X}_9$

The computer setup should by now be routine. It will produce the following results:

Full model	Restricted model
$b_1 = 16.0000$	$c_1 = 13.3333$
$b_2 = 16.0000$	$c_2 = 14.0000$
$b_3 = 16.0000$	$c_3 = 14.0000$
$b_4 = 16.0000$	
$b_5 = 18.0000$	
$b_6 = 8.0000$	
$b_7 = 8.0000$	
$b_8 = 8.0000$	
$b_9 = 18.0000$	
$\sum e_i^2 = 200.0$	$\sum e_i^2 = 973.3$
$R^2 = .7955$	$R^2 = .0045$

$$F = \frac{(\sum e_{i_R}^2 - \sum e_{i_F}^2)/\mathrm{df}_1}{\sum e_{i_F}^2/\mathrm{df}_2} = \frac{(973.3 - 200.0)/(9 - 3)}{200.0/(45 - 9)} \approx 23.2$$

As this F exceeds the critical value of 5.12 for the .001 level of significance, we reject our hypothesis that ability has no effect on improvement controlling for technique. As interaction is not at all uncommon, tests of this type are often quite useful. There is no traditional analog for them.

3.14 GENERALIZATION: EFFECTS OF ONE VARIABLE CONTROLLING FOR ANOTHER

The question which is addressed is, in general: Does the column (or row) variable significantly affect the dependent variable controlling for the row (or column) variable? If the first subscript refers to the row values and the second to the column values, the general form for the column effect (controlling for row values) is

Full model:

$$\mathbf{Y} = b_{11}\mathbf{X}_{11} + b_{12}\mathbf{X}_{12} + \cdots + b_{1c}\mathbf{X}_{1c} + b_{21}\mathbf{X}_{21} + b_{22}\mathbf{X}_{22}$$
$$+ \cdots + b_{2c}\mathbf{X}_{2c} + \cdots + b_{rc}\mathbf{X}_{rc} + \mathbf{E}_F$$

Hypotheses:

$b_{11} = b_{12}$ $\quad b_{11} = b_{13} \quad \cdots \quad b_{11} = b_{1c}$ $\quad = c_1$, setting all b's equal in first row

$b_{21} = b_{22}$ $\quad b_{21} = b_{23} \quad \cdots \quad b_{21} = b_{2c}$ $\quad = c_2$, setting all b's equal in second row

....

$b_{r1} = b_{r2}$ $\quad b_{r1} = b_{r3} \quad \cdots \quad b_{r1} = b_{rc}$ $\quad = c_r$, setting all b's equal in last row

Restricted model:

$$Y = c_1X_{11} + c_1X_{12} + \cdots + c_1X_{1c} + c_2X_{21} + c_2X_{22} + \cdots + c_2X_{2c}$$
$$+ \cdots + c_rX_{rc} + E_R$$
$$= c_1(X_{11} + X_{12} + \cdots + X_{1c}) + c_2(X_{21} + X_{22} + \cdots + X_{2c})$$
$$+ \cdots + c_r(X_{r1} + X_{r2} + \cdots + X_{rc}) + E_R$$
$$= c_1Z_1 + c_2Z_2 + \cdots + c_rZ_r + E_R \tag{3.5}$$

where $Z_1 = X_{11} + X_{12} + \cdots + X_{1c}$

$\qquad Z_2 = X_{21} + X_{22} + \cdots + X_{2c}$

$\qquad \cdots\cdots\cdots\cdots\cdots\cdots\cdots\cdots\cdots$

$\qquad Z_r = X_{r1} + X_{r2} + \cdots + X_{rc}$

For the test statistic, $df_1 = r(c - 1)$, and $df_2 = N - rc$, as there are rc unknown weights in the full model and $r(c - 1)$ conditions are imposed.

3.15 MAIN EFFECTS IN TWO-WAY ANALYSIS OF VARIANCE

Where it is ascertained that the effect of each independent variable is the same for all values of the other (no interaction), traditional treatments of analysis of variance proceed to ask whether those effects produce statistically significant differences. This is termed the *main effect* of a variable. As we found interaction between our three techniques and ability, we have seen that we cannot sensibly proceed to these questions about overall average effects. However, we found no interaction between ability and the first two techniques. Therefore, we shall illustrate the analysis of main effects with technique 3 excluded. This exclusion necessitates removing cases 31 through 45 from the data deck. We shall retain the numbering convention for the cells, as before, to permit the use of the data-group numbers as previously designated. Thus, we shall be concerned with a table of the form of Table 3.7.

Having found no interaction for this reduced set of data, we can proceed to ask whether the two techniques produce different average amounts of improvement and whether different ability levels produce different average amounts of improvement.

Table 3.7 MEAN IMPROVEMENT BY THREE ABILITY LEVELS AND TWO TECHNIQUES

Children	Technique 1	Technique 2
Gifted	$b_1 = 16$	$b_2 = 16$
Normal	$b_4 = 16$	$b_5 = 18$
Retarded	$b_7 = 8$	$b_8 = 8$

3.16 COLUMN MAIN EFFECTS

A Natural-Language Question

In general, on the average, do the two techniques produce different amounts of improvement? This question is equivalent to asking whether it is reasonable to hypothesize that the overall mean for technique 1 ($\overline{Y}_{T1}$) is equal to the overall mean for technique 2 ($\overline{Y}_{T2}$).

B Full Model

$$Y = b_1X_1 + b_2X_2 + b_4X_4 + b_5X_5 + b_7X_7 + b_8X_8 + E_F$$

Notice that the full model allows different amounts of improvement for each cell. Also, note that we have represented the exclusion of technique 3 by omitting X_3, X_6, and X_9.

C Hypothesis

Since $b_i = \overline{Y}_i = \sum_{i=1}^{n_i} y_i/n_i, \quad n_i b_i = \sum_{i=1}^{n_i} y_i.$

Thus, the hypothesis that $\overline{Y}_{T1} = \overline{Y}_{T2}$ can be represented as

$$\frac{n_1 b_1 + n_4 b_4 + n_7 b_7}{n_1 + n_4 + n_7} = \frac{n_2 b_2 + n_5 b_5 + n_8 b_8}{n_2 + n_5 + n_8}$$

Where the number of cases is the same for all cells, that is, $n_1 = n_2 = \cdots = n_i$; this hypothesis can be simplified to

$$b_1 + b_4 + b_7 = b_2 + b_5 + b_8$$

Solving for b_8 gives $b_8 = b_1 + b_4 + b_7 - b_2 - b_5$.

D Restricted Model

$$\begin{aligned}
Y &= b_1X_1 + b_2X_2 + b_4X_4 + b_5X_5 + b_7X_7 \\
&\quad + (b_1 + b_4 + b_7 - b_2 - b_5)X_8 + E_R \\
&= b_1X_1 + b_2X_2 + b_4X_4 + b_5X_5 + b_7X_7 + b_1X_8 + b_4X_8 + b_7X_8 \\
&\quad + b_2(-X_8) + b_5(-X_8) + E_R \\
&= b_1(X_1 + X_8) + b_2(X_2 - X_8) + b_4(X_4 + X_8) + b_5(X_5 - X_8) \\
&\quad + b_7(X_7 + X_8) + E_R \\
&= c_1(X_1 + X_8) + c_2(X_2 - X_8) + c_3(X_4 + X_8) + c_4(X_5 - X_8) \\
&\quad + c_5(X_7 + X_8) + E_R \\
&= c_1Z_1 + c_2Z_2 + c_3Z_3 + c_4Z_4 + c_5Z_5 + E_R
\end{aligned}$$

where $Z_1 = X_1 + X_8$
$Z_2 = X_2 - X_8$
$Z_3 = X_4 + X_8$
$Z_4 = X_5 - X_8$
$Z_5 = X_7 + X_8$

E Inference

The F ratio is defined in the usual way with $df_1 = 1$, as one hypothesis was imposed, and $df_2 = N - 6$, as there are six unknown weights in the full model.

F Computer Instructions

The following Fortran statements will create VEC(1), VEC(2), VEC(4), VEC(5), VEC(7), and VEC(8) as X_1, X_2, X_4, X_5, X_7, and X_8. [For these data, they also create VEC(3), VEC(6), and VEC(9) as null vectors which we do not use in any analysis.] They create VEC(10) as Y and VECs 11 through 15 as Z_1 through Z_5:

```
. . . . . . . . . . . . . . .

      DO 5 I=1,9

   5  VEC(I)=0.

      READ (5,11) VAR(1),VAR(2)

      VEC(10)=VAR(1)

      K=VAR(2)

      VEC(K)=1.

  11  FORMAT (3X,1F2.0,2X,1F1.0)

      VEC(11)=VEC(1)+VEC(8)

      VEC(12)=VEC(2)-VEC(8)

      VEC(13)=VEC(4)+VEC(8)

      VEC(14)=VEC(5)-VEC(8)

      VEC(15)=VEC(7)+VEC(8)

. . . . . . . . . . . . . . . . . . . . . . . . . . .
```

The main control card should specify 30 cases, 2 input variables, 15 constructed vectors, including a dependent variable, and 2 regressions:

    COL. MAIN_____30____2___15____2

The full-model card should specify dependent variable 10 and six independent variables, which are 1, 2, 4, 5, 7, 8:

    FULL_MODEL___10____6____1____2____4____5____7____8

The restricted-model card should specify dependent variable 10 and five independent variables, which are 11, 12, 13, 14, 15:

    REST. MODEL___10____5___11___12___13___14___15

G   Numerical Example

Computer results for these data will be:

Full model	Restricted model
$b_1 = 16.0000$	$c_1 = 16.3333$
$b_2 = 16.0000$	$c_2 = 15.6667$
$b_4 = 16.0000$	$c_3 = 16.3333$
$b_5 = 18.0000$	$c_4 = 17.6667$
$b_7 = 8.0000$	$c_5 = 8.3333$
$b_8 = 8.0000$	
$\sum e_i^2 = 126.0$	$\sum e_i^2 = 129.3$
$R^2 = .7976$	$R^2 = .7923$

$$F = \frac{(\sum e_{i_R}^2 - \sum e_{i_F}^2)/df_1}{\sum e_{i_F}^2/df_2} = \frac{(129.3 - 126.0)/(6 - 5)}{126.0/(30 - 6)} \approx .63$$

The critical value for F at the .05 level at 1 and 24 degrees of freedom is 4.20. As the obtained F is considerably smaller, we cannot reject the hypothesis that techniques 1 and 2 produce substantially the same results.

3.17   GENERALIZATION: THE c-COLUMN CASE

If we have a row variable with r values (designated $1, 2, 3, \ldots, r$) and a column variable with c values (designated $1, 2, 3, \ldots, c$) we can test for column main effects with the following (the first subscript refers to the row variable, the second to the column):

Full model:

$$\mathbf{Y} = b_{11}\mathbf{X}_{11} + b_{12}\mathbf{X}_{12} + \cdots + b_{1c}\mathbf{X}_{1c} + b_{21}\mathbf{X}_{21} + b_{22}\mathbf{X}_{22}$$
$$+ \cdots + b_{2c}\mathbf{X}_{2c} + \cdots + b_{rc}\mathbf{X}_{rc} + \mathbf{E}_F$$

Hypotheses: Where equal cell n's are assumed

$$b_{11} + b_{21} + b_{31} + \cdots + b_{r1} = b_{12} + b_{22} + b_{32} + \cdots + b_{r2}$$
(col. 1 mean = col. 2 mean)

$$b_{11} + b_{21} + b_{31} + \cdots + b_{r1} = b_{13} + b_{23} + b_{33} + \cdots + b_{r3}$$
(col. 1 mean = col. 3 mean)

. . . .

$$b_{11} + b_{21} + b_{31} + \cdots + b_{r1} = b_{1c} + b_{2c} + b_{3c} + \cdots + b_{rc}$$
(col. 1 mean = col. c mean)

The proper restricted model for any case can be derived by solving these hypotheses for independent unknowns, for example, $b_{r2}, b_{r3}, \ldots, b_{rc}$, and substituting for them in the full model. Notice that there are $c - 1$ of these hypotheses and there are rc unknown weights in the full model. Therefore, $df_1 = c - 1$, and $df_2 = N - rc$.

3.18 ROW MAIN EFFECTS

A Natural-Language Question

In general, on the average, do the three ability levels produce different amounts of improvement? This question amounts to asking whether it is reasonable to hypothesize that the overall mean for the gifted $\overline{Y}_G$ is equal to the overall mean for the normals $\overline{Y}_N$ and also equal to the overall mean for the retarded $\overline{Y}_R$.

B Full Model

$$Y = b_1 X_1 + b_2 X_2 + b_4 X_4 + b_5 X_5 + b_7 X_7 + b_8 X_8 + E_F$$

Technique 3 is excluded because we have omitted vectors X_3, X_6, and X_9.

C Hypotheses

By our earlier arguments, we can show that the hypotheses $\overline{Y}_G = \overline{Y}_N$ and $\overline{Y}_G = \overline{Y}_R$ can be represented as

$$\frac{n_1 b_1 + n_2 b_2}{n_1 + n_2} = \frac{n_4 b_4 + n_5 b_5}{n_4 + n_5} \qquad \frac{n_1 b_1 + n_2 b_2}{n_1 + n_2} = \frac{n_7 b_7 + n_8 b_8}{n_7 + n_8}$$

Where $n_1 = n_2 = \cdots = n_i$, these hypotheses can be simplified to

$$b_1 + b_2 = b_4 + b_5$$

$$\text{(solving for } b_5 \text{ gives } b_5 = b_1 + b_2 - b_4)$$

and
$$b_1 + b_2 = b_7 + b_8$$

$$\text{(solving for } b_8 \text{ gives } b_8 = b_1 + b_2 - b_7)$$

D Restricted Model

$$\begin{aligned}
Y &= b_1 X_1 + b_2 X_2 + b_4 X_4 + (b_1 + b_2 - b_4) X_5 + b_7 X_7 \\
&\qquad\qquad + (b_1 + b_2 - b_7) X_8 + E_R \\
&= b_1 X_1 + b_2 X_2 + b_4 X_4 + b_1 X_5 + b_2 X_5 + b_4(-X_5) \\
&\qquad\qquad + b_7 X_7 + b_1 X_8 + b_2 X_8 + b_7(-X_8) + E_R \\
&= b_1(X_1 + X_5 + X_8) + b_2(X_2 + X_5 + X_8) \\
&\qquad\qquad + b_4(X_4 - X_5) + b_7(X_7 - X_8) + E_R \\
&= c_1(X_1 + X_5 + X_8) + c_2(X_2 + X_5 + X_8) \\
&\qquad\qquad + c_3(X_4 - X_5) + c_4(X_7 - X_8) + E_R \\
&= c_1 Z_1 + c_2 Z_2 + c_3 Z_3 + c_4 Z_4 + E_R
\end{aligned}$$

where $\mathbf{Z}_1 = \mathbf{X}_1 + \mathbf{X}_5 + \mathbf{X}_8$
$\mathbf{Z}_2 = \mathbf{X}_2 + \mathbf{X}_5 + \mathbf{X}_8$
$\mathbf{Z}_3 = \mathbf{X}_4 - \mathbf{X}_5$
$\mathbf{Z}_4 = \mathbf{X}_7 - \mathbf{X}_8$

E   Inference

The F ratio is defined in the usual way with $df_1 = 2$, as two hypotheses were imposed, and $df_2 = N - 6$, as there are six unknown weights in the full model.

F   Computer Instructions

In addition to the vectors created in the last problem, we now need VEC(16) through VEC(19), defined as $\mathbf{Z}_1$ through $\mathbf{Z}_4$, above. This can be achieved by adding the following statements to those used in the analysis above:

. . . .

VEC(16) = VEC(1) + VEC(5) + VEC(8)
VEC(17) = VEC(2) + VEC(5) + VEC(8)
VEC(18) = VEC(4) − VEC(5)
VEC(19) = VEC(7) − VEC(8)

. . . .

The main control card from the previous problem should be changed to permit 19 VECs and 3 regressions. A third model control card should be added for this restricted model:

REST2 _ _ _ _ _ _ _ _ 10 _ _ _ _ 4 _ _ _ 16 _ _ _ 17 _ _ _ 18 _ _ _ 19

G   Numerical Example

Computer results for these data would be:

	Full model	Restricted model
	$b_1 = 16.0000$	$c_1 = 13.6667$
	$b_2 = 16.0000$	$c_2 = 13.6667$
	$b_4 = 16.0000$	$c_3 = 12.6667$
	$b_5 = 18.0000$	$c_4 = 13.6667$
	$b_7 = 8.0000$	
	$b_8 = 8.0000$	
	$\sum e_i^2 = 126.0$	$\sum e_i^2 = 612.7$
	$R^2 = .7976$	$R^2 = .0161$

$$F = \frac{(\sum e_{i_R}^2 - \sum e_{i_F}^2)/df_1}{\sum e_{i_F}^2/df_2} = \frac{(612.7 - 126.0)/(6 - 4)}{126.0/(30 - 6)} \approx 46.4$$

The critical value for F at the .001 level and at 2 and 24 degrees of freedom is 9.34. Therefore, we shall have to reject our hypotheses at the .001 level of significance. The different ability levels produce different amounts of improvement.

3.19 GENERALIZATION: THE r-ROW CASE

If we have a row variable with r values $(1, 2, 3, \ldots, r)$ and a column variable with c values $(1, 2, 3, \ldots, c)$, we can test for row main effects with the following (the first subscript refers to the row variable, the second to the column):

Full model:

$$\mathbf{Y} = b_{11}\mathbf{X}_{11} + b_{12}\mathbf{X}_{12} + \cdots + b_{1c}\mathbf{X}_{1c} + b_{21}\mathbf{X}_{21} + b_{22}\mathbf{X}_{22}$$
$$+ \cdots + b_{2c}\mathbf{X}_{2c} + \cdots + b_{rc}\mathbf{X}_{rc} + \mathbf{E}_F$$

Hypotheses: Assuming equal n's in all cells

$$b_{11} + b_{12} + b_{13} + \cdots + b_{1c} = b_{21} + b_{22} + b_{23} + \cdots + b_{2c}$$
(row 1 mean = row 2 mean)

$$b_{11} + b_{12} + b_{13} + \cdots + b_{1c} = b_{31} + b_{32} + b_{33} + \cdots + b_{3c}$$
(row 1 mean = row 3 mean)

. . . .

$$b_{11} + b_{12} + b_{13} + \cdots + b_{1c} = b_{r1} + b_{r2} + b_{r3} + \cdots + b_{rc}$$
(row 1 mean = row r mean)

The proper restricted model for any case can be derived by solving these hypotheses for independent unknowns, for example, b_{2c}, b_{3c}, b_{rc}, and substituting for these in the full model. Notice that there are $r - 1$ of these hypotheses and there are rc unknown weights in the full model. Therefore, $\mathrm{df}_1 = r - 1$, and $\mathrm{df}_2 = N - rc$.

3.20 TRADITIONAL EQUIVALENCES AND SUMMARY FOR TWO-WAY ANALYSIS OF VARIANCE

The interaction and main-effect hypotheses developed above are identical to those implicitly tested with traditional computing formulas. However, we believe that the nature of the hypotheses is more clearly manifested through the least-squares formulation. For the interested reader, we can demonstrate equivalences between our F ratios and the components of traditional tests. Recall that for these tests we have employed comparisons of three restricted models to a single full model. In the six-group (two-technique) case these models were

Full model: $\mathbf{Y} = b_1\mathbf{X}_1 + b_2\mathbf{X}_2 + b_4\mathbf{X}_4 + b_5\mathbf{X}_5 + b_7\mathbf{X}_7 + b_8\mathbf{X}_8 + \mathbf{E}_F$

The error SS for this model is the sum of the squared deviations of cases in each subclass from the subclass mean. As such this quantity is equivalent to the within-subclass SS (sometimes called *the* error SS[1]). It is the error attributable to the assumption that within-subcell variation is random. The model has associated with it $N - 6$

---

[1] This conventional usage ignores that *all* the quantities are *error* SS but errors from different models. However, conventional computing approaches never make explicit what models are being compared.

(in general, $N - rc$) degrees of freedom. The error SS from this model divided by its degrees of freedom is the denominator for all of our F ratios and is identical to the denominator for the traditional formulas. For our data, $\sum e_{i_F}^2 = 126.0 =$ within-subclass SS.

Restricted model 1: Where no interaction hypotheses are imposed

$$\mathbf{Y} = c_1(\mathbf{X_1} - \mathbf{X_5} - \mathbf{X_8}) + c_2(\mathbf{X_2} + \mathbf{X_5} + \mathbf{X_8}) + c_3(\mathbf{X_4} + \mathbf{X_5}) \\ + c_4(\mathbf{X_7} + \mathbf{X_8}) + \mathbf{E}_{R1}$$

The error SS for this model is the error attributable to the assumptions of the full model (within-subcell variation is random) *plus* the error attributable to our hypotheses (no interaction) as *both* the assumptions and hypotheses are imposed on this model. If we subtract from this combined error the error attributable to the assumptions ($\sum e_{i_F}^2$, above), we have left only the error attributable to our no-interaction hypotheses. This difference will have 2 degrees of freedom [in general, $(r - 1)(c - 1)$ degrees of freedom]. For our data,

$$\sum e_{i_{R1}}^2 - \sum e_{i_F}^2 = 132.7 - 126.0 = 6.7 = \text{interaction SS}$$

Restricted model 2: No column main-effects hypotheses are imposed

$$\mathbf{Y} = c_1(\mathbf{X_1} + \mathbf{X_8}) + c_2(\mathbf{X_2} - \mathbf{X_8}) + c_3(\mathbf{X_4} + \mathbf{X_8}) + c_4(\mathbf{X_5} - \mathbf{X_8}) \\ + c_5(\mathbf{X_7} + \mathbf{X_8}) + \mathbf{E}_{R2}$$

The error SS for this model is the error attributable to the assumptions of the full model *plus* the error attributable to our hypotheses (no column main effects). Subtracting the error attributable to the assumptions from this combined error, we have remaining only the error attributable to the no-column-main-effect hypotheses. This difference will have 1 degree of freedom (in general, $c - 1$ degrees of freedom). For our data,

$$\sum e_{i_{R2}}^2 - \sum e_{i_F}^2 = 129.3 - 126.0 = 3.3 = \text{between-column SS}$$

Restricted model 3: No row main-effects hypotheses are imposed

$$\mathbf{Y} = c_1(\mathbf{X_1} + \mathbf{X_5} + \mathbf{X_8}) + c_2(\mathbf{X_2} + \mathbf{X_5} + \mathbf{X_8}) + c_3(\mathbf{X_4} - \mathbf{X_5}) \\ + c_4(\mathbf{X_7} - \mathbf{X_8}) + \mathbf{E}_{R3}$$

The error SS for this model is the error attributable to the assumptions of the full model *plus* the error attributable to our hypotheses (no row main effects). Subtracting the error attributable to the assumptions from this combined error, we have remaining only the error attributable to the no-row-main-effect hypotheses. This difference will have 2 degrees of freedom (in general, $r - 1$ degrees of freedom). For our data,

$$\sum e_{i_{R3}}^2 - \sum e_{i_F}^2 = 612.7 - 126.0 = 486.7 = \text{between-rows SS}$$

These components are the only ones used in the traditional tests. However, we can construct the remaining entries in the traditional analysis-of-variance table from these components (within-subclass, interaction, between-columns, between-rows SS).

Specifically, the between-subclass SS can be obtained by adding together interaction, between-column, and between-row SS. Also, the total SS can be obtained by adding together the within-subclass and the between-subclass SS. This is done in Table 3.8 (see McNemar, 1962, p. 306; Blalock, 1972, p. 342).

The general linear-model approach also makes explicit the rationale for the various degrees of freedom. In this regard, see page 118 for within-subclass degrees of freedom, page 118 for interaction degrees of freedom, page 130 for between-column degrees of freedom, and page 133 for between-row degrees of freedom. The between-subclass degrees of freedom is the sum of the degrees of freedom for its three components (lines 2 to 4). The total degrees of freedom is the sum of the degrees of freedom for its two components (lines 1 and 5). The total degrees of freedom is also the degrees of freedom for a random-variation model (see pages 84 to 86).

3.21 POOLING INTERACTION AND WITHIN-SUBCLASS-ERROR SUMS OF SQUARES

It is sometimes argued that the interaction SS should be combined with the within-class SS for the denominator for the F ratio when interaction is found to be statistically insignificant (see Blalock, 1972, pp. 342–347). This is done by adding the two SS together. The proper degrees of freedom then become $(N - rc) + (r - 1)(c - 1) = N - r - c + 1$, or the sum of the degrees of freedom for the two error components. While the mechanics of pooling are straightforward, it is difficult to present a compelling general case for the desirability of pooling or not pooling. Most defenses of pooling follow one or both of two arguments: the interaction error SS is random variation, *or* pooling provides a "conservative" test (see McNemar, 1962, pp. 338–339). The first argument is questionable, and the second is not true in any general sense. It is readily demonstrated that under certain conditions pooling can either decrease or increase the F ratio (see Paull, 1950, pp. 539–556). Further, F ratios based on pooling do not have the same interpretation as those in which pooling is avoided.

The best general advice is to do whatever minimizes the chances of an analysis coming out the way the researcher desires. Thus, where one seeks to find a statistically significant effect, the cautious researcher will minimize the F ratio. When replication or generalization questions are being asked, the conservative tactic is to maximize the

Table 3.8 TRADITIONAL EQUIVALENCES FOR TWO-WAY ANALYSIS OF VARIANCE

Line	Source of error	Derived from models	Value for data	General df
1	Within subclasses	Full	126.0	$N - rc$
2	Interaction	Full − restricted 1	6.7	$(r - 1)(c - 1)$
3	Between columns	Full − restricted 2	3.3	$c - 1$
4	Between rows	Full − restricted 3	486.7	$r - 1$
5	Between subclasses	Sum of lines 2 to 4	496.7	$rc - 1$
6	Total	Sum of lines 1 and 5	622.7	$N - 1$

F ratio and thus to provide a difficult criterion for replication. For the best detailed treatment of the desirability and effects of pooling, see Paull (1950, pp. 539–556).

Also, it should be noted that the pooling decision will ordinarily have little effect on robust substantive findings. Finally, we can only note that the substantive inference which is deemed true or false because we pooled or failed to do so is hardly the type of information upon which one rests a science or a reputation.

3.22 TWO-WAY ANALYSIS OF VARIANCE WITH UNEQUAL SUBCLASSES

In nonlaboratory research the number of cases in each subclass commonly is not equal. This poses problems with conventional computing formulas, which are based on the simplifying assumption of equal subclass n's. These problems are not critical, how-ever, from the least-squares perspective we have taken. Indeed, the interaction test and tests for the effects of one variable controlling for another can be performed in exactly the same way with equal or unequal subclass n's. The main-effects tests, however, require some modification. We shall illustrate this with our two-technique column main-effects test. Where the subclass n's are unequal, we can still use the same full model. However, the hypotheses and restricted model will differ.

B Full Model

$$\mathbf{Y} = b_1\mathbf{X}_1 + b_2\mathbf{X}_2 + b_4\mathbf{X}_4 + b_5\mathbf{X}_5 + b_7\mathbf{X}_7 + b_8\mathbf{X}_8 + \mathbf{E}_F$$

C Hypotheses

We noted that the column main-effects hypothesis for these data was that $\bar{Y}_{T1} = \bar{Y}_{T2}$ (Sec. 3.16). It developed that this hypothesis is equivalent to

$$\frac{n_1 b_1 + n_4 b_4 + n_7 b_7}{n_1 + n_4 + n_7} = \frac{n_2 b_2 + n_5 b_5 + n_8 b_8}{n_2 + n_5 + n_8}$$

This is true whether the subclass n's are equal or not. In developing the traditional (equal-n) case for column main effects we simplified this hypothesis to $b_1 + b_4 + b_7 = b_2 + b_5 + b_8$ before solving and substituting in our full model: If we omit this simplifying assumption and use our more general (unequal-n) hypothesis, we have an appropriate restricted model for testing column main effects with unequal n's (see Jennings, 1967, pp. 95–108). Let us solve for b_8 from the general hypothesis, above:

$$\frac{n_2 + n_5 + n_8}{n_1 + n_4 + n_7}(n_1 b_1 + n_4 b_4 + n_7 b_7) = n_2 b_2 + n_5 b_5 + n_8 b_8$$

$$n_8 b_8 = \frac{n_2 + n_5 + n_8}{n_1 + n_4 + n_7}(n_1 b_1 + n_4 b_4 + n_7 b_7) - n_2 b_2 - n_5 b_5$$

Simplifying our notation, let $n' = (n_2 + n_5 + n_8)/(n_1 + n_4 + n_7)$. Then

$$n_8 b_8 = n'(n_1 b_1 + n_4 b_4 + n_7 b_7) - n_2 b_2 - n_5 b_5$$

$$b_8 = \frac{n'}{n_8}(n_1 b_1 + n_4 b_4 + n_7 b_7) - \frac{n_2 b_2}{n_8} - \frac{n_5 b_5}{n_8}$$

$$= \frac{n' n_1 b_1}{n_8} + \frac{n' n_4 b_4}{n_8} + \frac{n' n_7 b_7}{n_8} - \frac{n_2 b_2}{n_8} - \frac{n_5 b_5}{n_8}$$

Substituting in the full model gives the restricted model.

D   Restricted Model

$$\mathbf{Y} = b_1 \mathbf{X}_1 + b_2 \mathbf{X}_2 + b_4 \mathbf{X}_4 + b_5 \mathbf{X}_5 + b_7 \mathbf{X}_7$$

$$+ \left(\frac{n' n_1 b_1}{n_8} + \frac{n' n_4 b_4}{n_8} + \frac{n' n_7 b_7}{n_8} - \frac{n_2 b_2}{n_8} - \frac{n_5 b_5}{n_8} \right) \mathbf{X}_8 + \mathbf{E}_R$$

After expanding and collecting terms, this will simplify to

$$\mathbf{Y} = b_1 \left(\mathbf{X}_1 + \frac{n' n_1}{n_8} \mathbf{X}_8 \right) + b_2 \left(\mathbf{X}_2 - \frac{n_2}{n_8} \mathbf{X}_8 \right) + b_4 \left(\mathbf{X}_4 + \frac{n' n_4}{n_8} \mathbf{X}_8 \right)$$

$$+ b_5 \left(\mathbf{X}_5 - \frac{n_5}{n_8} \mathbf{X}_8 \right) + b_7 \left(\mathbf{X}_7 + \frac{n' n_7}{n_8} \mathbf{X}_8 \right) + \mathbf{E}_R$$

or for clarity

$$\mathbf{Y} = c_1 \mathbf{Z}_1 + c_2 \mathbf{Z}_2 + c_3 \mathbf{Z}_3 + c_4 \mathbf{Z}_4 + c_5 \mathbf{Z}_5 + \mathbf{E}_R$$

where $\mathbf{Z}_1 = \mathbf{X}_1 + \dfrac{n' n_1}{n_8} \mathbf{X}_8$

$$\mathbf{Z}_2 = \mathbf{X}_2 - \frac{n_2}{n_8} \mathbf{X}_8$$

$$\mathbf{Z}_3 = \mathbf{X}_4 + \frac{n' n_4}{n_8} \mathbf{X}_8$$

$$\mathbf{Z}_4 = \mathbf{X}_5 - \frac{n_5}{n_8} \mathbf{X}_8$$

$$\mathbf{Z}_5 = \mathbf{X}_7 + \frac{n' n_7}{n_8} \mathbf{X}_8$$

Vectors $\mathbf{Z}_1$ through $\mathbf{Z}_5$ are no more difficult to create in Fortran than any of our earlier restricted-model vectors. The only difference is that VEC(8) is weighted by (multiplied by) some constant before being added to or subtracted from the other vectors. If $\mathbf{Z}_1$ through $\mathbf{Z}_5$ are VEC(11) through VEC(15) and $\mathbf{Y}$ is VEC(10), the restricted-model control card is

UNEQUAL_N____10____5___11___12___13___14___15

The full-model control card is the same as with the equal-n test. The use of the error SS and the computation of df_1 and df_2 are the same as with the equal-n test.

As this procedure is quite straightforward, there is really no excuse for social researchers to continue to assume equal n's where they have grossly disproportionate subgroups. Where the disproportionality is minor, of course, the weighting factors n' and n_i/n_j will approach 1 and can be ignored in favor of the somewhat simpler equal-n main-effect hypotheses. It is somewhat surprising that the unequal-n case is termed "incredibly complex" by most elementary and intermediate statistics books. The logic is straightforward and the algebra, while meticulous, is quite elementary.

3.23 DUMMY-VARIABLE ANALYSIS: A REPRESENTATIONAL ALTERNATIVE

We have shown that all analysis-of-variance questions can be asked by imposing linear conditions on a model of the form [see Eqs. (3.3) and (5.16b)]

$$\mathbf{Y} = b_1\mathbf{X}_1 + b_2\mathbf{X}_2 + \cdots + b_k\mathbf{X}_k + \mathbf{E}$$

Recall from Sec. 3.7 that $\mathbf{Y}$ is the dependent variable, $\mathbf{X}_i$ is defined as 1 for members of group i and 0 otherwise, and $\mathbf{E}$ is the error vector. Each unique least-squares estimate b_i for such a model is the mean on $\mathbf{Y}$ for all members of group i. Because the b_i's have this interpretation, hypothesis testing about values of means, equality of means, and relationships between means can be represented as direct translations of natural-language questions; e.g., the hypothesis that all categories of $\mathbf{X}$ have the same expected value on $\mathbf{Y}$ becomes $b_i = b_j$ for all j's ($j \neq i$).

In Sec. 5.7, we point out that another appropriate model [see Eq. (5.16a)] is

$$\mathbf{Y} = b_0\mathbf{U} + b_1\mathbf{X}_1 + b_2\mathbf{X}_2 + \cdots + b_k\mathbf{X}_k + \mathbf{E}$$

We show that unique estimates are possible for the b_i's if a side condition, for example, $b_0 = 0$, is imposed. This has the effect of making the grand mean the reference point for our system. Such a side condition eliminates $\mathbf{U}$ (there $\mathbf{X}_0$) from the model producing Eq. (3.3). We also show that comparisons (differences) between the b_i's, expected values for all y_j's, and the errors are unique for both models. In fact, the models define the same predictor space because they are identical save for the inclusion of the unit vector $\mathbf{U}$. Recall that $\mathbf{U}$ is linearly dependent on the $\mathbf{X}_i$'s, that is,

$$\mathbf{U} = \sum_{i=1}^{k} \mathbf{X}_i$$

Consequently, the inclusion or exclusion of the unit vector causes no change in expected values or error terms. Nor will degrees of freedom be modified as b_0 is not counted in the model, for it is not an independent (of the other b_i's) unknown weight. We can elect as a side condition that any other weight or combination of weights be 0. If we choose to impose that $b_k = 0$, the following model defines the same predictor space as our model (3.3):

$$\mathbf{Y} = b_0\mathbf{U} + b_1\mathbf{X}_1 + b_2\mathbf{X}_2 + \cdots + b_{k-1}\mathbf{X}_{k-1} + \mathbf{E}$$

Such a model would produce the following codings for solution:

$$\mathbf{U} = 1 \quad \text{for all cases}$$

$$\mathbf{X}_i = \begin{cases} 1 & \text{for members of group } i \\ 0 & \text{otherwise} \end{cases}$$

Proponents of dummy-variable analysis, in fact, analyze such models. The kth category is said to be *suppressed*. Another way of saying this is that unique codings can be produced for members of any group without $\mathbf{X}_k$, as $\mathbf{X}_k$ (1 for members of group k, 0 otherwise) is linearly dependent on $\mathbf{U}$ and $\mathbf{X}_1$ through $\mathbf{X}_{k-1}$. Similarly, the codings employed in dummy-variable analysis for interaction and contrast effects can be shown to define the same space as our comparable models. Consequently, *no mathematical virtue* can be claimed for either representation. Selection of either side condition (or of another) produces the same expected values y_j, the same errors e_j, and the same error SS.

With Bottenberg (1963), Jennings (1967), Rockwell (1974), and Ward (1969), we have chosen the side condition that $b_0 = 0$ for pedagogic rather than mathematical reasons. In dummy-variable analysis b_0 is interpretable as the grand mean on $\mathbf{Y}$ and the other b_i's as deviations of group means from that mean. Equivalent hypotheses about differences among the means (equivalence, no interaction, main effects, etc.) can be imposed as validly on one model as the other. However, these are difficult to state in natural-language statements where the side condition is that $b_k = 0$. The unpersuaded reader should try to summarize the dummy-variable codings for the no-interaction test in a straightforward fashion. Our choice of the side condition that $b_0 = 0$ allows us to translate many complex analytic questions into obvious algebraic relations between the means, b_i's. Dummy-variable codings are just as valid (indeed equivalent), merely less obvious. Consequently, many students learn the codings by rote and cannot innovate rigorously when faced with unconventional research questions.

Unfortunately, some regression programs are designed so that the elimination of the unit vector (side condition $b_0 = 0$) is not practical. In some programs, the unit vector is implicit because the program operates only on deviation scores. Thus, for such programs, our representation will involve a linear dependency, and no solutions will be given. For this reason, we strongly recommend the program in Appendix A or other programs which operate on raw scores and do not require unit vectors.

3.24 SUMMARY FOR CONTINUOUS DEPENDENT AND CATEGORICAL INDEPENDENT VARIABLES

We have shown that single-equation regression models with a continuous dependent variable and discrete independent variables subsume, as special cases, the t test and all traditional one- and two-way analysis-of-variance questions. The regression perspective makes explicit just which hypotheses are tested by traditional computing formulas. Further, the determination of degrees of freedom is removed from the realm of faith (or formula, which is perhaps the same thing). A brief summary of our paradigm may give some perspective to this:

1 Formulate a natural-language question(s) about what patterning you expect or seek to demonstrate in your data.
2 Formulate a full model which *allows* that patterning to exist or to be absent.
3 Translate your natural-language question(s) into linear conditions which can be imposed on the full model, producing a restricted model.
4 Compare the error SS in the two models to see whether your hypotheses increase the error.
5 Use the *F* ratio for a decision rule where

$$F = \frac{s_H^2}{s_A^2} = \frac{(\sum e_{i_R}^2 - \sum e_{i_F}^2)/df_1}{\sum e_{i_F}^2/df_2}$$

This can always be interpreted as the ratio of the error variance attributable to your hypotheses to the error variance attributable to your assumptions. If *F* is large, you should reject your hypotheses. Otherwise, the hypotheses will be regarded as tenable. df_1 is always the number of independent hypotheses imposed, and df_2 is *N* minus the number of unknown weights in the full model.

Of more interest than the traditional equivalences is the power which the general linear model gives for formulating and testing specific, precise, and substantively interesting hypotheses. We have merely illustrated some of these possibilities. You can use the approach developed so far to test almost any hypotheses about the values of means or patterned relationships between means. We have restricted ourselves to questions about means as these are the simplest cases of hypothesis testing with respect to expected values. In all our full models, the expected value (least-squares solution) for each case has been the mean of the group in which the case is embedded. We now move to a discussion of continuous *independent* variables, and this limited view of expected value will be discarded.

EXERCISES

All exercises refer to data in Appendix B. You will find it convenient to use the standard READ *and* FORMAT *statements given at the end of that appendix. Then the necessary vectors can be created by referencing the variables by their subscripts* [VAR(1), VAR(2), *etc.*].

3.1 Set up program CARTROCK to give means and variations for husband's liberalism, VAR(8); wife's liberalism, VAR(12); husband's father's liberalism, VAR(17); and husband's mother's liberalism, VAR(20). For each random-variation model, the mean will be the weight for the unit vector, and the variation will be the error SS (as well as the total SS).
3.2 Develop a full model, hypotheses, and restricted model to test the hypothesis that husband's income VAR(6) is the same for all five regions (Northeast, Midwest, Mountain, Pacific, South) defined by VAR(1).
   (*a*) Set up program CARTROCK to provide solutions for the two models.
   (*b*) Calculate, report, and interpret the *F* ratio as a test of these hypotheses.
   (*c*) Report and interpret the total mean, regional means and error SS.

3.3 Decide which of the regions might reasonably be grouped together on either substantive or empirical grounds. Develop a full model, hypotheses, and restricted model to test the hypothesis that husband's income is the same for "comparable" regions.

(*a*) Set up program CARTROCK to provide solutions for the two models.

(*b*) Calculate, report, and interpret the *F* ratio as a test of these hypotheses.

(*c*) Report and interpret the total mean and the *grouped* regional means.

(*d*) Compare these results to those in Exercise 3.2.

3.4 Construct a full model for determining within-subcell means and the within-subcell SS where the dependent variable is husband's liberalism and the independent variables are region of residence and political-party preference. Conceptualize the problem in terms of a table of the form

	Northeast	Midwest	Mountain	Pacific	South
Republicans	b_1	b_2	b_3	b_4	b_5
Democrats	b_6	b_7	b_8	b_9	b_{10}

(Note that the subcell frequencies will be different.)

(*a*) Set up the hypotheses and restricted model to test the no-interaction hypothesis.

(*b*) Calculate, report, and interpret the *F* ratio as a test of this hypothesis.

(*c*) By comparing and interpreting differences in the means describe the interaction (if any).

3.5 If the interaction *is not* significant, ask (1) whether party makes a difference in liberalism *on the average* and (2) whether region makes a difference in liberalism *on the average*. If the interaction *is* significant, ask (1) whether party makes a difference in liberalism *controlling for region* and (2) whether region makes a difference in liberalism *controlling for party*.

(*a*) For both situations (no interaction *and* interaction), set up the full models, hypotheses, and restricted models.

(*b*) For the appropriate case (no interaction *or* interaction), set up program CARTROCK to give solutions for your models.

(*c*) For the appropriate case, calculate, report, and interpret the *F* ratios as tests of your hypotheses.

(*d*) Give a substantive summary of your findings.

4

LINEAR MODELS, CORRELATION, AND THE ANALYSIS OF COVARIANCE

In Chap. 3 we developed ways to test hypotheses about how categorical independent variables may condition or determine the relationships between expected values (means) for interval dependent variables. We examined how teaching technique conditions improvement in performance and how party affiliation conditions liberalism. For categorical variables, we are warranted to say only that two expected values are different, e.g., the expected value for technique 1 is not that for technique 2, nor is either expected value that for technique 3; Republicans are not the same as Democrats. We may say that categorical independent variables are *related* to our dependent variable whenever different dependent variable means tend to occur for different values of our categorical independent variables, e.g., when there exist different levels of improvement for the different techniques and/or ability categories or different levels of liberalism for representatives affiliated with different parties. However, we cannot specify on a priori grounds the magnitudes of the differences in categorical variable values on any scale, metric, or dimension. Often, social scientists are fortunate enough to have independent variables which are conceptually interval-scaled. By interval-scaled we simply mean that we can give substantively meaningful interpretation to differences in the magnitudes of values for such variables; i.e., it *makes sense* to say that the distance is the same from value i to $i + 1$ as it is from value $i + 1$ to $i + 2$ on some substantive dimension. Where this property is plausible, we shall find that we can arrive at very powerful and complex insights into the social forces producing our observations. Let us begin with a simple case.

If two interval variables are linearly related, we can use a single simple equation $Y = aU + bX$ to map from one of them to the other. Where the values of Y are produced by the values of X *and* minor random (unsystematic) forces which can be ignored, the equation should be written

$$Y = aU + bX + E$$

where Y = dependent variable
  U = unit vector
  X = independent variable
  E = error vector
  a = Y intercept for least-squares line
  b = slope for least-squares line for predicting Y from X

In Chap. 3 the expected value for Y was typically *the point* which was the mean for some category of X. In the current formulation, the expected value for Y is *any point on the line* determined by the equations generated by this model. These are of the form

$$y_i = a \times 1 + bx_i + e_i$$

The point which constitutes the predicted value for y_i depends on the value of x_i. If such a formulation is reasonable, we shall be able to use the *same form* of equation for all y_i's. In earlier formulations, we typically had to use a different form of equation for predicting Y values for each category of X. Before extolling the analytic power permitted by such linear relationships, we should always determine whether the assumptions of such models are reasonable.

4.1 TEST OF THE LINEARITY ASSUMPTION

The appropriate full model for testing the linearity assumptions is one which treats the values of the independent variable X as merely categorical. This permits the y_i expected values for each value of x_i to fall on a straight line (conform to a first-degree equation) or not.

A Natural-Language Question

Do the expected values for Y increase (or decrease) by a constant amount as the values of X increase by a constant amount? Alternatively, can the relationship between Y and X be reasonably represented by a straight line?

B Full Model

$$Y = c_1Z_1 + c_2Z_2 + c_3Z_3 + c_4Z_4 + c_5Z_5 + \cdots + c_kZ_k + E_F \qquad (4.1)$$

where $\mathbf{Y}$ = dependent variable

$\mathbf{Z}_1$ = 1 if the case has the first value on $\mathbf{X}$, 0 otherwise

$\mathbf{Z}_2$ = 1 if the case has the second value on $\mathbf{X}$, 0 otherwise

. . . .

$\mathbf{Z}_k$ = 1 if the case has the kth value on $\mathbf{X}$, 0 otherwise[1]

c_j = least-squares weights

$\mathbf{E}_F$ = error vector

This model is equivalent to the one used for the general case of one-way analysis of variance [Eq. (3.3)]. Thus, the solution for any c_j will be the mean of the Y scores for those cases having an X score of j (in the jth group). $\mathbf{E}_F'\mathbf{E}_F$ will be the variation of all cases around their category means (within-group variation). Under this model, the category means may occur in any pattern. In effect, we allow the $\mathbf{X}$ scores to determine category membership on $\mathbf{Z}$.

C   Hypotheses

The linearity hypothesis implies that these category means will be separated by a constant (will fall on a straight line). Algebraically, this can be developed as follows. Let c_1 (the Y mean for cases having the first value of X) have the value a. Let the distance between c_1 and c_2 have the value b. By these conventions $c_1 = a$ and $c_2 = c_1 + b = a + b$. The linearity assumption implies that all other adjacent c_j's (means) will be separated by this same constant, b. Thus, the implied hypotheses are

$$c_3 = c_2 + b \qquad \text{or} \quad c_3 = c_1 + 2b \qquad \text{or} \quad c_3 = a + 2b$$

$$c_4 = c_3 + b \qquad \text{or} \quad c_4 = c_1 + 3b \qquad \text{or} \quad c_4 = a + 3b$$

$$c_5 = c_4 + b \qquad \text{or} \quad c_5 = c_1 + 4b \qquad \text{or} \quad c_5 = a + 4b$$

. . . .

$$c_k = c_{k-1} + b \qquad \text{or} \quad c_k = c_1 + (k-1)b \qquad \text{or} \quad c_k = a + (k-1)b$$

There are $k - 2$ of these hypotheses.

D   Restricted Model

We can now impose these linearity conditions on the full model to produce a proper restricted model. This is done by substituting the above equivalences for c_1 and c_2 (by our conventions) and for $c_3, c_4, c_5, \ldots, c_k$ (by our hypotheses). Note that $c_1 = a = a + 0 \times b$, which we shall use in the substitution below. The new model is

$$\mathbf{Y} = (a + 0b)\mathbf{Z}_1 + (a + 1b)\mathbf{Z}_2 + (a + 2b)\mathbf{Z}_3 + (a + 3b)\mathbf{Z}_4$$
$$+ (a + 4b)\mathbf{Z}_5 + \cdots + [a + (k-1)b]\mathbf{Z}_k + \mathbf{E}_R$$

---

[1] The set of vectors $\mathbf{Z}$ should be treated as binary orthogonal vectors representing all *possible* values of X. While any particular X score may not be observed in a sample, it must nonetheless be considered in determining distances.

which can be shown to be equivalent to

$$\mathbf{Y} = a(\mathbf{Z}_1 + \mathbf{Z}_2 + \mathbf{Z}_3 + \mathbf{Z}_4 + \mathbf{Z}_5 + \cdots + \mathbf{Z}_k) + b[0\mathbf{Z}_1 + 1\mathbf{Z}_2 \\ + 2\mathbf{Z}_3 + 3\mathbf{Z}_4 + 4\mathbf{Z}_5 + \cdots + (k-1)\mathbf{Z}_k] + \mathbf{E}_R$$

In the full model, the $\mathbf{Z}_j$'s were defined as binary orthogonal vectors. Therefore, their sum is a unit vector. Thus, the model becomes

$$\mathbf{Y} = a\mathbf{U} + b[0\mathbf{Z}_1 + 1\mathbf{Z}_2 + 2\mathbf{Z}_3 + 3\mathbf{Z}_4 + 4\mathbf{Z}_5 + \cdots + (k-1)\mathbf{Z}_k] + \mathbf{E}_R$$

If we call the parenthetic component $\mathbf{W}$, the model simplifies to

$$\mathbf{Y} = a\mathbf{U} + b\mathbf{W} + \mathbf{E}_R$$

The vector $\mathbf{W}$ has the following properties. The value of an element w_i is 0 for those with the first value on X, 1 for those with the second value on X, 2 for those with the third value on $X, \ldots, k - 1$ for those with kth value on X. If we let d be the first X value, then $x_i = w_i + d$ and we can rewrite the model as

$$\mathbf{Y} = a\mathbf{U} + b\mathbf{X} + \mathbf{E}_R \qquad (4.2)$$

where $\mathbf{Y}$ = dependent variable
$\quad\quad \mathbf{X}$ = independent variable
$\quad\quad \mathbf{U}$ = unit vector
$\quad\quad a$ = intercept (least-squares weight)
$\quad\quad b$ = slope (least-squares weight)
$\quad\quad \mathbf{E}_R$ = error vector

The elements of the vector $\mathbf{X}$ have the value d for the first X value, $d + 1$ for the second X value, $d + 2$ for the third X value, $\ldots$, and $d + k - 1$ for the kth X value.

In vector representation this becomes

$$
\begin{bmatrix} y_1 \\ y_2 \\ y_3 \\ y_4 \\ y_5 \\ y_6 \\ y_7 \\ y_7 \\ y_8 \\ y_9 \\ y_{10} \\ \vdots \\ y_{n-1} \\ y_n \end{bmatrix}
= a \begin{bmatrix} 1 \\ 1 \\ 1 \\ 1 \\ 1 \\ 1 \\ 1 \\ 1 \\ 1 \\ 1 \\ 1 \\ \vdots \\ 1 \\ 1 \end{bmatrix}
+ b \begin{bmatrix} d \\ d \\ d+1 \\ d+1 \\ d+2 \\ d+2 \\ d+3 \\ d+3 \\ d+3 \\ d+4 \\ d+4 \\ \vdots \\ d+k-1 \\ d+k-1 \end{bmatrix}
+ \begin{bmatrix} e_1 \\ e_2 \\ e_3 \\ e_4 \\ e_5 \\ e_6 \\ e_7 \\ e_7 \\ e_8 \\ e_9 \\ e_{10} \\ \vdots \\ e_{n-1} \\ e_n \end{bmatrix}
$$

where each X value $(d, d + 1, d + 2, \ldots, d + k - 1)$ is represented by two cases. In general, then, the model can be represented

$$
\begin{bmatrix} y_1 \\ y_2 \\ y_3 \\ y_4 \\ y_4 \\ y_5 \\ \vdots \\ y_n \end{bmatrix} = a \begin{bmatrix} 1 \\ 1 \\ 1 \\ 1 \\ 1 \\ 1 \\ \vdots \\ 1 \end{bmatrix} + b \begin{bmatrix} x_1 \\ x_2 \\ x_3 \\ x_4 \\ x_4 \\ x_5 \\ \vdots \\ x_n \end{bmatrix} + \begin{bmatrix} e_1 \\ e_2 \\ e_3 \\ e_4 \\ e_4 \\ e_5 \\ \vdots \\ e_n \end{bmatrix}
$$

For each case, the expected value will be determined by the equation

$$ y_i = a + bx_i + e_i \qquad (4.3) $$

The prediction will thus be determined by the best-fit straight line mapping from the X values to the Y values. In the previous chapter, the predictions were the best-fit points (means) for the X categories. The errors were deviations from those points (means). In the current formulation the errors will be deviations from the least-squares line.

E   Inference

If the error sum of squares (SS) is not significantly increased by the imposition of the $k - 2$ linearity hypotheses, we shall conclude that a straight line (or first-degree equation) provides a reasonable representation of the relationship between X and Y. We can then proceed to ask questions about the parameters of that equation. The decision about the linearity hypotheses will, as usual, be based on the F ratio, where

$$ F = \frac{(\sum e_{i_R}^2 - \sum e_{i_F}^2)/\mathrm{df}_1}{\sum e_{i_F}^2/\mathrm{df}_2} = \frac{(R_F^2 - R_R^2)/\mathrm{df}_1}{(1 - R_F^2)/\mathrm{df}_2} $$

Here $\mathrm{df}_1 = k - 2$, where k is the number of X values used, as there are k unknown weights in the full model and two unknown weights in the restricted (also, there are $k - 2$ hypotheses imposed), and $\mathrm{df}_2 = N - k$, where N is the number of cases and k is the number of X values used, as there are k unknown weights in the full model.

F   Computer Instructions

We shall test the hypothesis that spouses' years of education are related to each other by a linear function. For this purpose we shall use the data in Appendix B, where we have suggested a standard READ and FORMAT statement which should be used with these data. These statements permit us to refer to the variables by their subscript only. According to these statements, the wife's education is VAR(10) and the husband's education is VAR(4). For our full model, we shall need to construct VEC(1) through VEC(13), defined as Z_1 through Z_k above ($k = 13$ for these data), and VEC(14), defined as Y above. Note that for these data the husbands' education ranges from

7 through 19 years *with no intermediate values missing*[1]. Thus, the first value of X is $d = 7$. Also, Z_1 above will be 1 where the husband's education is 7 and 0 otherwise; Z_2 will be 1 where the husband's education is 8 and 0 otherwise; ... ; Z_{13} (or Z_k) will be 1 where the husband's education is 19 and 0 otherwise. For our restricted model, we shall need to construct VEC(14), defined as Y above; VEC(15), defined as U above; and VEC(16), defined as X above.

1 Fortran statements The statements to be embedded in SUBROUTINE DATA will be

. . . .

(standard READ statement)

(standard FORMAT statement)

DO 8 I=1,13

8   VEC(I)=0.

    I=VAR(4)−6.

    VEC(I)=1.

    VEC(14)=VAR(10)

    VEC(15)=1.

    VEC(16)=VAR(4)

. . . .

2 Main control card The main control card should contain a label, the number of cases (200), the number of VAR's (for these data we shall use a standard 20), the number of VECs(16), and the number of regressions (2). The card should look like

LIN.TEST_____200___20___16_____2

3 Model cards The full-model card should contain a label, designate VEC(14) or Y as the dependent variable, and designate 13 independent variables whose subscripts will be 1 to 13 or Z_1 through Z_k, where $k = 13$. As all this cannot be accommodated on a single card (see Appendix A), we shall continue the *last* subscript to the next card. The full-model *cards* should look like

LIN.FULL_____14___13____1____2____3____4____5____6____7____8____9___10___11___12

----------------13

---

[1] Any values which do not occur in the sample *must* be excluded from the full model (and therefore the restricted model). If, for example, the values 10, 14, and 15 years of education were missing, Z_4, Z_8, and Z_9 [Eq. (4.1)] would be eliminated, as would the hypotheses restricting the weights for these vectors. As we are postulating a continuum including these as *possible* values, the hypotheses restricting the weights of other variables remain the same, however. Thus, even where Z_8 and Z_9 are missing, Z_{10} is still three steps above Z_7 and nine steps above Z_1, that is, $Z_{10} = a + 9b$. To leave empirically missing values in the model would produce individual equations (row vectors) of the following type: $y_i = c_1 \times 0 + c_2 \times 0 + \cdots + c_z \times 0 + e_i$. Such null vectors cannot be solved by the least-squares criterion.

The restricted-model card should contain a label, designate VEC(14) or **Y** as the dependent variable, and designate two independent variables whose subscripts are 15 or **U** and 16 or **X**. The restricted-model card should look like

LIN.REST.____14____2___15___16

G   Numerical Example

The question of interest is whether a linear function may reasonably be said to typify the relationship between the education of spouses. We have shown how to set up the computer program using data from Appendix B.

Full model: $Y = c_1 Z_1 + c_2 Z_2 + c_3 Z_3 + \cdots + c_{13} Z_{13} + E_F$

(By convention, we set $c_1 = a$ and $c_2 = a + b$.)

Hypotheses:

$$c_3 = a + 2b$$
$$c_4 = a + 3b$$
$$\cdots\cdots\cdots\cdots$$
$$c_{13} = a + 12b$$

$$c_j = a + (j-1)b$$

Restricted model: $Y = aU + bX + E_R$

where $x_i = w_i + d$ and d is the first X value, here 7 [see text following Eq. (4.1) for details of derivation]. The relevant computer solutions for these models will produce the following results:

	Full model	Restricted model
$c_1 =$	7.5000 ($X =$ 7)	$a = 5.1428$
$c_2 =$	12.0000 ($X =$ 8)	$b = .5883$
$c_3 =$	11.0000 ($X =$ 9)	
$c_4 =$	13.0000 ($X =$ 10)	
$c_5 =$	13.5556 ($X =$ 11)	
$c_6 =$	11.7059 ($X =$ 12)	
$c_7 =$	11.7391 ($X =$ 13)	
$c_8 =$	11.4667 ($X =$ 14)	
$c_9 =$	13.3333 ($X =$ 15)	
$c_{10} =$	14.3636 ($X =$ 16)	
$c_{11} =$	15.5000 ($X =$ 17)	
$c_{12} =$	16.8571 ($X =$ 18)	
$c_{13} =$	17.3500 ($X =$ 19)	
$\sum e_i^2 =$	799.0	$\sum e_i^2 = 1{,}061.0$
$R^2 =$	.4916	$R^2 = .3250$

$$F = \frac{(\sum e_{i_R}^2 - \sum e_{i_F}^2)/df_1}{\sum e_{i_F}^2/df_2} = \frac{(1061.0 - 799.0)/(13 - 2)}{799.0/(200 - 13)} \approx 5.57$$

or $$F = \frac{(R_F^2 - R_R^2)/df_1}{(1 - R_F^2)/df_2} = \frac{(.4916 - .3250)/(13 - 2)}{(1. - .4916)/(200 - 13)} \approx 5.57$$

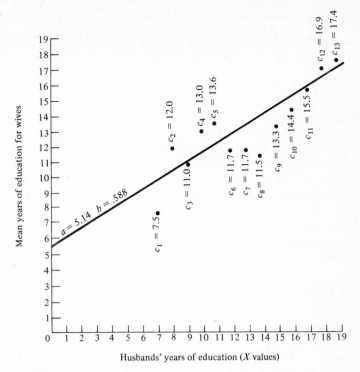

FIGURE 4.1
Average education of wives with husbands of specified educational levels.

The critical value for F at the .001 level of significance at 11 and 187 degrees of freedom (actually at 8 and 120 degrees of freedom for a conservative test) is 3.55. As the obtained F exceeds this value, we must reject the linearity hypotheses at the .001 level of significance. A graphic representation of the means computed for our full model should clarify this. For the husbands' education $X = 7$ through $X = 19$, the mean values for the wives' education are c_1 through c_{13}. These are portrayed in Fig. 4.1.

Husbands with little education tend to have wives with more education than they do, while college-educated husbands tend to have wives with less education than they do. We therefore have found it unreasonable to typify the relationship between spouses' education as a linear one, in spite of the fact that the correlation between spouses' education is .57 ($R_R^2 = .3250$). It is quite likely that many small to moderate correlations between social variables are the result of computing linear correlations for variables which are not linearly related. The square of these correlations may still be interpreted as the proportion of variance accounted for, as we shall see later. However, it is likely that the correlations would be much stronger if we were dealing with the proper functional form, e.g., a second- or third-degree equation. We shall explore some of these possibilities later.

I   Traditional Equivalence

The traditional linearity test is given by the formula (see Blalock, 1972, pp. 408–413; McNemar, 1962, pp. 275–278):

$$F = \frac{(\varepsilon^2 - r^2)/(k - 2)}{(1 - \varepsilon^2)/(N - k)} \qquad (4.4)$$

where ε^2 = squared *correlation ratio*, or the proportion of variance accounted for by Y means for X categories

r^2 = squared correlation coefficient for linear regression or proportion of variance accounted for by a straight-line equation

N = number of cases

k = number of X categories

As ε^2 is defined equivalently to $R_F{}^2$ and r^2 is defined equivalently to $R_R{}^2$, our obtained F ratio [Eq. (3.2)] is equivalent to this traditional test. However, the specific hypotheses being tested are here made explicit, and the degrees of freedom are rationalized. Also, it should be noted that $\sum e_{i_F}{}^2 = (1 - \varepsilon^2)\sum \mathbf{y}^2$ and $\sum e_{i_R}{}^2 = (1 - r^2)\sum \mathbf{y}^2$. Therefore, we can demonstrate the equivalence to the traditional test of our F ratio defined in terms of error SS [Eq. (3.1)].

4.2   NONSTANDARD QUESTIONS: HIGHER-DEGREE EQUATIONS, ISOLATING NONLINEARITY

Where nonlinearity is disclosed, we have open two general extensions of our analysis. The first involves determining whether some alternative simple relational form, e.g., second- or third-degree equation, is appropriate. The second is to determine whether our variables are linearly related *within some range of interest*.

Higher-Degree Equations

Some social variables are related to each other by second-degree equations. For example, the literature on status liability implies that violation of rules will be tolerated for persons of moderate group status but is likely to evoke sanctions if the violator is either very low or very high in group status. Also, for some attitudes college-educated persons may be similar to grade-school-educated persons and both groups dissimilar from high-school-educated persons, the grade-school-educated never having been exposed to American high schools and the college-educated having unlearned some of their high-school lessons. In either of these examples, we are suggesting that the value of one variable goes up as the value of the other variable goes up (down), but only to a certain point, after which a decline (increase) occurs. Such substantive assertions can be represented by second-degree equations of the form

$$\mathbf{Y} = a\mathbf{U} + b_1\mathbf{X}_1 + b_2\mathbf{X}_1{}^2 + \mathbf{E}$$

where $\mathbf{Y}$ = dependent variable

$\quad\mathbf{X_1}$ = independent variable

$\quad\mathbf{X_1}^2$ = vector whose elements are squares of corresponding elements for $\mathbf{X_1}$

$\quad\mathbf{E}$ = error vector

a, b_1, b_2 = least-squares weights

Occasionally, social variables are related to each other by third- (or higher-) degree equations. Notice that if only two *values* (regardless of the number of *cases* distributed between them) of a variable occur, the variable will exactly determine a least-squares first-degree equation and cannot be fitted to a least-squares second-degree equation. If only three values occur, no third-degree equation can be fitted. In general, the number of values occurring should be considerably greater than the degree of any equation fitted, and there should be at least several cases per value.

Decisions about the functional form relating two variables constitute important substantive findings. The procedure for deciding among first-, second-, and third-degree equations is straightforward. Simply, we ascertain the error (or R^2) associated with the best-fitting higher-degree equation; we hypothesize that the highest-order term may be weighted zero (eliminated), reducing our model to a lower-order equation; and we compare the error (or R^2) for the higher- and lower-order equations. We illustrate this with the data for spouses' educations in Table 4.1.

For these tests, we need to construct VEC(1) equal to $\mathbf{Y}$, VEC(2) equal to the unit vector, VEC(3) equal to $\mathbf{X}$, VEC(4) equal to $\mathbf{X}^2$, and VEC(5) equal to $\mathbf{X}^3$. This can be done with wives' educations [VAR(10)] being $\mathbf{Y}$ and husbands' educations [VAR(4)] being $\mathbf{X}$ by using the following Fortran in SUBROUTINE DATA:

Table 4.1 TEST PROCEDURE FOR COMPARISONS OF FIRST-, SECOND-, AND THIRD-DEGREE EQUATIONS

	Comparison of second- and first-degree equations	Comparison of third- and second-degree equations
Natural-language question	Does a first-degree equation describe the relationship between Y and X as well as a second-degree equation?	Does a second-degree equation describe the relationship between Y and X as well as a third-degree equation?
Full model	$\mathbf{Y} = a_1\mathbf{U} + b_1\mathbf{X} + b_2\mathbf{X}^2 + \mathbf{E}_F$	$\mathbf{Y} = a_1\mathbf{U} + b_1\mathbf{X} + b_2\mathbf{X}^2 + b_3\mathbf{X}^3 + \mathbf{E}_F$
Hypothesis	$b_2 = 0$	$b_3 = 0$
Restricted model	$\mathbf{Y} = a_1\mathbf{U} + b_1\mathbf{X} + \mathbf{E}_R$	$\mathbf{Y} = a_1\mathbf{U} + b_1\mathbf{X} + b_2\mathbf{X}^2 + \mathbf{E}_R$
F	$\dfrac{(R_F^2 - R_R^2)/(3 - 2)}{(1 - R_F^2)/(N - 3)}$	$\dfrac{(R_F^2 - R_R^2)/(4 - 3)}{(1 - R_F^2)/(N - 4)}$

....

(standard READ statement)

(standard FORMAT statement)

VEC(1) = VAR(10)

VEC(2) = 1.

VEC(3) = VAR(4)

VEC(4) = VAR(4)**2

VEC(5) = VAR(4)**3

....

The master control card would be

DEGREE_TST __ 200___ 20_____5_____3

The three requisite model cards would be

3RD_DEGREE_____1_____4_____2_____3_____4_____5

2ND_DEGREE_____1_____3_____2_____3_____4

1ST_DEGREE_____1_____2_____2_____3

The relevant computer output for these tests is:

	First-degree model	Second-degree model	Third-degree model
	$a = 5.1428$	$a = 19.6563$	$a = -4.5625$
	$b_1 = .5883$	$b_1 = -1.5122$	$b_1 = 4.1406$
		$b_2 = .0733$	$b_2 = -.3491$
			$b_3 = .0102$
	$\sum e_i^2 = 1,061.0$	$\sum e_i^2 = 982.1$	$\sum e_i^2 = 971.0$
	$R^2 = .3250$	$R^2 = .3752$	$R^2 = .3822$

For the comparison of the first- and second-degree equations

$$F = \frac{(R_F^2 - R_R^2)/(3 - 2)}{(1 - R_F^2)/(200 - 3)} = \frac{(.3752 - .3250)/1}{(1. - .3752)/197} \approx 15.8$$

The critical value for F at the .001 level of significance at 1 and 197 degrees of freedom (actually, for a conservative test, at 1 and 120 degrees of freedom) is 11.38. As our obtained F exceeds this, we must reject the hypothesis that $b_2 = 0$; that is, we reject the hypothesis that a first-degree equation fits as well as a second-degree equation. For the comparison of the second- and third-degree equations

$$F = \frac{(R_F^2 - R_R^2)/(4 - 3)}{(1 - R_F^2)/(200 - 4)} = \frac{(.3822 - .3752)/1}{(1. - .3822)/196} \approx 2.2$$

The critical value for F at the .05 level of significance at 1 and 197 degrees of freedom (actually 1 and 120 degrees of freedom) is 3.92. As our obtained F is less than this value, we cannot reject the hypothesis that $b_3 = 0$. Thus we cannot reject the notion that the second-degree equation describes the relationship as well as a third-degree equation. No equation of higher order than the second degree will significantly reduce the error SS.

4.3   GENERALIZATION: COMPARING kTH-DEGREE AND $(k - 1)$ST-DEGREE EQUATIONS

The general case for determining what degree of equation provides the most reasonable description of a relationship is quite straightforward. The question of interest is always whether a (simpler) equation of lesser degree can be used without significantly increasing the errors of prediction. This question can be stated: Can an equation of degree $k - 1$ be fitted to the data almost as well as an equation of degree k? For the general case, the following may be used:

Full model:

$$\mathbf{Y} = a\mathbf{U} + b_1\mathbf{X} + b_2\mathbf{X}^2 + b_3\mathbf{X}^3 + \cdots + b_{k-1}\mathbf{X}^{k-1} + b_k\mathbf{X}^k + \mathbf{E}_F \qquad (4.5)$$

Hypothesis: $\qquad\qquad\qquad\qquad\qquad\qquad\qquad\qquad\qquad b_k = 0$

Restricted model:

$$\mathbf{Y} = a\mathbf{U} + b_1\mathbf{X} + b_2\mathbf{X}^2 + b_3\mathbf{X}^3 + \cdots + b_{k-1}\mathbf{X}^{k-1} + \mathbf{E}_R \qquad (4.6)$$

$$F = \frac{(R_F{}^2 - R_R{}^2)/\mathrm{df}_1}{(1 - R_F{}^2)/\mathrm{df}_2}$$

where $\mathrm{df}_1 = 1$, as one hypothesis is imposed, and $\mathrm{df}_2 = N - (k + 1) = N - k - 1$, as there are $k + 1$ unknown weights in the full model (one weight for each of the k $\mathbf{X}$'s and one for the unit vector).

Linearity Within Range of Interest

It is sometimes the case that we are concerned with only a limited range of a variable. If, for example, we were principally interested in educational determinants for the college-educated population, we could test only those linearity assumptions which involve the X (husband's) values 13 through 19 years of education or VEC(7) through VEC(13). Without going into the details of the derivation, which are somewhat more complex than for the general test, we can specify the models which would allow such a partial linearity test. They are

Full model: $\mathbf{Y} = c_1\mathbf{Z}_1 + c_2\mathbf{Z}_2 + c_3\mathbf{Z}_3 + c_4\mathbf{Z}_4 + c_5\mathbf{Z}_5 + \cdots + c_{13}\mathbf{Z}_{13} + \mathbf{E}_F$

[defined as in Eq. (4.1)]. By convention, we could set $c_7 = a + 0b$ and $c_8 = a + 1b$. Then, paralleling our earlier derivation, the hypotheses would involve all higher values of X.

Hypotheses:

$$c_9 = c_8 + b \quad \text{or} \quad c_9 = a + 2b$$

$$c_{10} = c_9 + b \quad \text{or} \quad c_{10} = a + 3b$$

$$c_{11} = c_{10} + b \quad \text{or} \quad c_{11} = a + 4b$$

$$c_{12} = c_{11} + b \quad \text{or} \quad c_{12} = a + 5b$$

$$c_{13} = c_{12} + b \quad \text{or} \quad c_{13} = a + 6b$$

And where $d = 7$ (or the first X value), the restricted model can be shown to follow from direct substitution.

Restricted model:

$$\mathbf{Y} = c_1\mathbf{Z}_1 + c_2\mathbf{Z}_2 + c_3\mathbf{Z}_3 + c_4\mathbf{Z}_4 + c_5\mathbf{Z}_5 + c_6\mathbf{Z}_6 + a\mathbf{U}_p + b\mathbf{X}_p + \mathbf{E}_R \qquad (4.7)$$

where $\mathbf{Z}_1 = 1$ when husband's education is 7 years, 0 otherwise
$\qquad \mathbf{Z}_2 = 1$ when husband's education is 8 years, 0 otherwise
$\qquad \cdots\cdots\cdots\cdots\cdots\cdots\cdots\cdots\cdots\cdots\cdots\cdots\cdots\cdots\cdots\cdots\cdots\cdots\cdots\cdots\cdots\cdots\cdots$
$\qquad \mathbf{Z}_6 = 1$ when husband's education is 12 years, 0 otherwise
$\qquad \mathbf{U}_p = $ *partial unit vector*, i.e., a vector whose elements are 1 if husband's education is in range of interest (13 through 19 years) and 0 otherwise
$\qquad \mathbf{X}_p = $ *partial score vector*, i.e., a vector whose elements are X score (husband's years of education) if X value is within the range of interest (13 through 19 years) and 0 otherwise
$\qquad c_j = $ least-squares weight which is mean of wife's education for each j value of husband's education ($j = 1$ to 6, where $x_j = 7$ to 12)
$\qquad a = $ least-squares weight which is intercept of least-squares line describing relationship between *college-educated* spouses' educations
$\qquad b = $ least-squares weight which is slope of least-squares line describing relationship between *college-educated* spouses' educations
$\qquad \mathbf{E}_R = $ error vector

Notice that by these definitions, the full-model predictor equations will be (with all zero terms omitted)

$$y_i = c_j \times 1 + e_i \qquad (4.8)$$

where c_j will be the mean on Y for all cases having the jth X value. In the restricted model these equations are unchanged for cases in which the husband has 12 years of education or less. However, the restricted model generates equations of the following form for couples with college-educated males (with all zero terms omitted)

$$y_i = a \times 1 + bx_i + e_i \qquad (4.9)$$

where a is the intercept and b the slope of a least-squares line representing the (hypothesized) linear relationship between spouses' education.

The following Fortran will generate VEC(1) through VEC(13) as Z_1 through Z_{13}, VEC(14) as Y, VEC(15) as U_p, and VEC(16) as X_p:

```
        ............................
        (standard READ statement)
        (standard FORMAT statement)
        DO 8 I=1,16
    8   VEC(I)=0.
        I=VAR(4)-6.
        VEC(I)=1.
        VEC(14)=VAR(10)
        IF(VAR(4)-12.)10,10,9
    9   VEC(15)=1.
        VEC(16)=VAR(4)
    10  CONTINUE
        ............................
```

The main control card should read

PARLINTEST__200___20___16____2

The model cards should read

FULL.PLT._____14__13____1____2____3___4____5___6____7____8____9___10___11___12

_____13
REST.PLT._____14____8____1____2____3____4____5____6___15___16

The relevant computer results will be:

Full model	Restricted model
$c_1 = 7.5000$	$c_1 = 7.5000$
$c_2 = 12.0000$	$c_2 = 12.0000$
$c_3 = 11.0000$	$c_3 = 11.0000$
$c_4 = 13.0000$	$c_4 = 13.0000$
$c_5 = 13.5556$	$c_5 = 13.5556$
$c_6 = 11.7059$	$c_6 = 11.7059$
$c_7 = 11.7391$	$a = -1.9417$
$c_8 = 11.4667$	$b = 1.0191$
$c_9 = 13.3333$	
$c_{10} = 14.3636$	
$c_{11} = 15.5000$	
$c_{12} = 16.8571$	
$c_{13} = 17.3500$	
$\sum e_i^2 = 799.0$	$\sum e_i^2 = 816.2$
$R^2 = .4916$	$R^2 = .4808$

$$F = \frac{(R_F^2 - R_R^2)/df_1}{(1 - R_F^2)/df_2}$$

Note that $df_1 = 5$, as 5 hypotheses were imposed, and $df_2 = 200 - 13$, as there are 13 unknown weights in the full model.

$$F = \frac{(.4916 - .4808)/(13 - 8)}{(1. - .4916)/(200 - 13)} \approx 0.79$$

The critical value for F at the .05 level of significance at 5 and 187 degrees of freedom (actually 5 and 120 degrees of freedom for the conservative test) is 2.29. As our obtained F is less than this value, we shall not be able to reject our limited-linearity hypotheses. (In fact, where F is less than 1.0, as it is here, we need not consult an F table.) On the basis of our two linearity tests (the traditional one and this limited one) we find it reasonable to conclude that a straight line describes the relationship between the years of education for college-educated spouses quite well but that this linear relationship cannot be generalized to other educational levels. We might retain all cases in subsequent analysis but impose questions of a linear nature for the college-educated and analogous questions about differences in means for the other educational levels. There are no traditional equivalences to such a procedure.

4.4 PROPORTION OF VARIANCE ACCOUNTED FOR BY A SIMPLE LINEAR MODEL

The square of the (linear)[1] correlation coefficient is often said to be the "proportion of variation in Y which is accounted for by a knowledge of variation in X." This interpretation requires *no* assumptions about the shape of distributions (normality) or about equality of the variation of Y for fixed values of X (homoscedasticity).[2] The sense of this interpretation is developed below. Consider a random-variation model

$$\mathbf{Y} = a\mathbf{U} + \mathbf{E}_R$$

For such a model, the expected value for each y_i is determined by equations of the form

$$y_i = a + e_i$$

We found earlier that the least-squares solution for a system of such equations produced $a = \bar{Y}$. Thus, $e_i = y_i - \bar{Y}$ and $\sum e_{i_R}^2 = \sum (y_i - \bar{Y})^2$, which is the variation of the Y scores around the Y mean. In all cases save those involving trivially homogeneous variables (where $y_i = \bar{Y}$ for all observations) we seek to reduce the magnitude of these errors of prediction by finding other variables whose values condition, i.e., are related to or cause, the variation in the values of Y. If we use some other (fuller) model, e.g., a linear model (any model more complicated than the

---

[1] We shall see that the proportion of variance accounted for is more general than this. It applies whether relationships are linear or not. However, *certain* tests, about to be developed, implicitly test for variance accounted for by linear predictions.

[2] *Other* uses and interpretations of r^2 and r, for example, slopes, within-array accuracy of prediction, and bivariate normal surface, require these assumptions at points in their derivations. However, we shall not here be concerned with these interpretations (see McNemar, 1962, chap. 9 and especially pp. 134–135).

random-variation model would serve our illustrative function as well as the linear model),

$$\mathbf{Y} = a\mathbf{U} + b\mathbf{X} + \mathbf{E}_F$$

the expected value for each y_i is determined by equations of the form

$$y_i = a + bx_i + e_i$$

We found earlier that the least-squares solution for a system of such equations produced a as the intercept and b as the slope of a straight line to be used for predicting Y from X. Thus, for such a model $e_i = y_i - a - bx_i$ and $\sum e_{i_F}^2 = \sum (y_i - a - bx_i)^2$, which is the variation of the Y scores around the least-squares line.

The difference in these two error SS, $\sum e_{i_R}^2 - \sum e_{i_F}^2$, is the amount by which our predictive model reduces the error associated with a random-variation model, or it is the amount of error found in the random-variation model which is eliminated (accounted for) by our more complicated model. If we divide this quantity by the amount of error in the random-variation model $\sum e_{i_R}^2$, we have an expression which is the proportion of the initial error (for a random-variation model) eliminated by the more complicated (here linear) model. This quantity is[1]

$$\text{Proportion of error variation in restricted model accounted for by full model} = \frac{\sum e_{i_R}^2 - \sum e_{i_F}^2}{\sum e_{i_R}^2} \qquad (4.10)$$

In Eq. (3.2), we noted that[2]

$$\sum e_{i_F}^2 = \sum \mathbf{y}_{i_F}^2(1 - R_F^2) \qquad \text{and} \qquad \sum e_{i_R}^2 = \sum \mathbf{y}_{i_R}^2(1 - R_R^2)$$

Also, we noted that where the dependent variable is the same for both models, as it is here, $\sum \mathbf{y}_{i_F}^2 = \sum \mathbf{y}_{i_R}^2 = \sum \mathbf{y}_i^2$. Through the same algebra used there, we can show that the following equivalence holds:

$$\text{Proportion of error variation in restricted model accounted for by full model} = \frac{R_F^2 - R_R^2}{1 - R_R^2} \qquad (4.11)$$

However, where our restricted model is a random-variation model, $\sum e_{i_R} = \sum \mathbf{y}_i^2$ and therefore $R_R^2 = 0$. Substituting this in the above equation provides

$$\text{Proportion of error variation in random-variation model accounted for by full model} = \frac{R_F^2 - 0}{1 - 0} = R_F^2 \qquad (4.12)$$

In general, then, R_F^2 is the proportion of error variation in the random-variation model which is accounted for (eliminated by) the predictive model. Where the predictive model is the simple (first-degree) linear model, this is traditionally designated r^2 (see McNemar, 1962, pp. 130–131). For the tests involving spouses' years of education

---

[1] Notice that the denominator refers to the error produced by the restricted model as the base for calculating the *amount* of proportionate error reduction. In all our tests for determining the *significance* of the difference in error, the full model provides the proper base for the denominator.

[2] Recall that $\mathbf{y}_i$ designates *deviation* scores.

(results in Fig. 4.1), the full model (permitting a different Y mean for each X category) accounts for 49.16 percent of the variation in Y while the restricted (linear) model accounts for 32.50 percent of the variation in Y.

Test of Significance for r

The most common statistical test involving two variables asks the question: Does knowledge of one of the variables significantly reduce the error in predicting the other? This is equivalent to asking whether the different values of the independent variable are systematically related to different values of the dependent variable. (Such a relationship is a necessary but insufficient condition for the imputation of causality.) In Chap. 3, this question was asked by imposing the conditions that the dependent variable had the same expected value for all categories of the independent variable (group means were all equal.) If these conditions increased error, we said that the independent (categorical) variable was significantly related to the dependent variable. In this chapter we have discussed a conceptually continuous and interval independent variable. Consider the linear model

$$Y = aU + bX + E_F$$

For N cases, this generates N equations of the form

$$y_1 = a + bx_1 + e_1$$
$$y_2 = a + bx_2 + e_2$$
$$y_3 = a + bx_3 + e_3$$
$$\cdots\cdots\cdots\cdots\cdots$$
$$y_N = a + bx_N + e_N$$

Each of these equations may be different, as they depend on the value x_i, which may be unique to each equation. Notice that where the full model is of this form, two special assumptions are already embedded in the model. Interval data are assumed, since the differences in the x_i's would otherwise be uninterpretable. Linearity is assumed, as this is the restricted model which resulted from imposing the linearity conditions. If we want to determine whether the variable X makes a difference in our prediction for the variable Y, we should simply impose a condition (hypothesis) which would force the same equation for predicting all y_i values. This condition is that $b = 0$, reducing the model to a random-variation model of the form $y_i = a + e_i$. If this condition significantly increases error, we shall reject it and conclude that Y and X are related, i.e., that the values of X significantly condition the values of Y.

A Natural-Language Question

Do the values of X condition (influence) the values of Y?

B Full Model $Y = aU + bX + E_F$

C Hypothesis $b = 0$

D   Restricted Model $\mathbf{Y} = a\mathbf{U} + \mathbf{E}_R$

E   Inference

(Note the return of $R_F{}^2$ to the denominator, as we are now dealing with an F ratio rather than a measure of variation accounted for.)

$$F = \frac{(\sum e_{i_R}{}^2 - \sum e_{i_F}{}^2)/df_1}{\sum e_{i_F}{}^2/df_2} = \frac{(R_F{}^2 - R_R{}^2)/df_1}{(1 - R_F{}^2)/df_2}$$

where $df_1 = 1$ as one hypothesis is imposed, and $df_2 = N - 2$ as there are two unknown weights in the full model.

   If the error is increased (R^2 is decreased) significantly, we conclude that Y and X are related. If the hypothesis produces no or little increase in error, we conclude that there is *no or little linear* (!) *relationship between Y and X*. We cannot conclude from this test that there is *no* pattern of relationship, as our full model *assumes* that any relationship about which we construct hypotheses, for example, $b = 0$, is of the linear form; e.g., the perfect second-degree relationship $Y = X^2$ will produce a $b = 0$ (and $r^2 = 0$) for our full model which assumes a first-degree equation.

F   Computer Instructions

We shall first use husband's age as $\mathbf{Y}$ and wife's age as $\mathbf{X}$ for our example. These are VAR(3) and VAR(9), respectively, for the data in Appendix B, when the standard READ and FORMAT statements are used. We shall need to construct VEC(1) as $\mathbf{Y}$, above; VEC(2) as $\mathbf{U}$, above; and VEC(3) as $\mathbf{X}$, above. The following Fortran statements will do this:

```
     .....................
     (standard READ statement)
     (standard FORMAT statement)
     VEC(1)=VAR(3)
     VEC(2)=1.
     VEC(3)=VAR(9)
     ..............
```

The main control card should read

```
     B=0_TEST_____200___20_____3_____2
```

The model cards should read

```
     B=0_FULL_____1____2_____2_____3
     B=0_REST. _____1____1____2
```

G   Numerical Example

The question of interest is whether the ages of spouses are related to a significant degree by a linear model. Arbitrarily, the husband's age is taken as dependent and the wife's age as independent.

Full model:	$Y = a\mathbf{U} + b\mathbf{X} + \mathbf{E}_F$
Hypothesis:	$b = 0$
Restricted model:	$\mathbf{Y} = a\mathbf{U} + \mathbf{E}_R$

The relevant computer results are:[1]

Full model	Restricted model
$a = 7.1926$	$a = 36.3550$
$b = .7850$	
$\sum e_i^2 = 909.3$	$\sum e_i^2 = 19{,}750$
$R^2 = .9540$	$R^2 = 0$

$$F = \frac{(R_F^2 - R_R^2)/\mathrm{df}_1}{(1 - R_F^2)/\mathrm{df}_2} = \frac{(.9540 - 0)/(2 - 1)}{(1 - .9540)/(200 - 2)} \approx 4{,}106$$

We shall not enter the F table for this test but shall reject (as a matter of professional judgment) the hypothesis that $b = 0$ at the .001 level of significance. Having found that spouses' ages are significantly related by a linear model, we should turn our attention to interpretation of the parameters of that model. The following interpretations of a and b may be of substantive interest. In the restricted model, a is the mean of the dependent variable (husband's age). Thus, for this sample the average of the husbands' ages is 36.3550. In the full model a is the intercept and b the slope of the least-squares line predicting husbands' ages from wives' ages. This prediction is $y_i' = a + bx_i$. The parameters produce the predictions in Table 4.2. The likely

---

[1] Some explanation of the printout that will result from these models is in order. All the information for this test *can* be obtained from the printout for the *full model*. This is true whenever the restricted model is a random-variation model. The *full-model printout* gives mean of dependent variable, which is a_R, and total variation, which is $\sum e_{i_R}^2$; R_R^2 is always zero for a random-variation model. However, on the *restricted-model* printout we get more significant digits for a_R (36.3550 rather than 36.35). Notice that for a random-variation model error SS = total SS. A minor computational complication sometimes gives $R^2 = -0.0000$ for random-variation models. Disregard the $-$. Where numbers are too large for the standard output FORMAT, this program automatically shifts to E (exponential) mode to avoid loss of accuracy. Thus, for $\sum e_{i_R}^2$, the computer reports .1975E 05. The E 05 means shift the decimal point five places to the right, producing 19,750. The number .1975E 06 would be 197,500. A negative number after the E means shift the decimal point that many spaces to the left. Thus, the number .1975E $-$ 05 is .000001975. Exponential mode reports numbers in scientific form, that is, $aEb = a \times 10^b$.

Table 4.2 HUSBAND'S AGE
PREDICTED FROM
WIFE'S AGE

Wife's age	Predicted husband's age
20	22.9
30	30.7
40	38.6
50	47.0
60	54.3

import of this can be summarized by saying that young women tend to marry men about 3 years older than themselves. However, the sex differences in mortality and the cumulative effects of divorce (or widowhood) and remarriage produce older couples in which the man is about 6 years younger than the woman. Any such interpretation, of course, assumes proper sampling from some population of interest, as well as cultural information.

I Traditional Equivalence

The traditional t test for the hypothesis that $r = 0$ is in fact the square root of our F test for the hypothesis that $b = 0$ in a linear model. The formula for the traditional t test (see Blalock, 1972, pp. 397–400; McNemar, 1962, p. 138) is

$$t = \frac{r\sqrt{N - 2}}{\sqrt{1 - r^2}} \tag{4.13}$$

$$t^2 = \frac{r^2(N - 2)}{1 - r^2} = \frac{r^2}{(1 - r^2)/(N - 2)} \tag{4.14}$$

Our formulation of the ratio was, as always,

$$F = \frac{(R_F^2 - R_R^2)/df_1}{(1 - R_F^2)/df_2}$$

However, for these models, R_F^2 is r^2, or the proportion of variation accounted for by a linear model. R_R^2 is 0. $df_1 = 1$, as one hypothesis ($b = 0$) was imposed, and $df_2 = N - 2$ as the full model has two unknown weights (a and b). Substituting these special case equivalences in our general F ratio produces the formula for t^2, above.

4.5 NONSTANDARD QUESTIONS: REPLICATION, RELIABILITY, THEORETICAL FORMS

While this test for the hypothesis that $b = 0$ is the most commonly used test in social science analyses, it is also one of the weakest hypotheses available to researchers. Where N's are quite large, trivially weak relationships are shown to be *statistically significant*, i.e., the variables are not totally unrelated to each other. The almost ritual use of the computing formula for this test has caused much social analysis to stop when a "significant" relationship is found rather than proceeding to uncover underlying relational structures. Consider, for example, our findings above that spouses' educations are related by a second-degree equation (a first-degree equation being appropriate only for the college-educated). If we had performed no linearity tests (and considered no alternative forms for our predictor equations), we would have been likely to arrive at the following defensible but simplistic analysis.

Question: Are spouses' educations related (by a linear model, *assumed*)?

Full model: $$Y = aU + bX + E_F$$

where Y = wife's education
    X = husband's education
    U = unit vector
    E_F = error vector
    a, b = least-squares weights

Hypothesis: $$b = 0$$

Restricted model: $$Y = aU + E_R$$

[The computer analysis is identical to that in the previous section except that the variables used for Y and X are changed. Thus, VEC(1) = VAR(10) and VEC(3) = VAR(4).] The computer results are:

	Full model	Restricted model
	$a = 6.7129$	$a = 14.1550$
	$b = .5525$	
	$\sum e_i^2 = 996.4$	$\sum e_i^2 = 1{,}476.0$
	$R^2 = .3250$	$R^2 = 0$

$$F = \frac{(R_F^2 - R_R^2)/\mathrm{df}_1}{(1 - R_F^2)/\mathrm{df}_2} = \frac{(.3250 - 0.)/(2 - 1)}{(1. - .3250)/(200 - 2)} \approx 95.3$$

This F is sufficiently large to force us to reject our hypothesis of no relationship ($b = 0$) at the .001 level of significance. At this point, much analysis stops, having rejected the simplistic no-relationship hypothesis at an extreme level of statistical significance. To have done so, however, would mean that we would never have found that a second-degree equation is the more reasonable form for this relationship (Sec. 4.2) as we would not have asked the question. Further, we would never have discovered that the linearity assumption is reasonable, but only for the college-educated (Sec. 4.3). The reason for this oversight would be that our model *assumes* rather than *hypothesizes* a linear relationship and that where relationships are quite strong (and N's large) even the *wrong* model will often be significantly better than a random one.

In this section we shall develop several nonstandard alternatives which are preferable to the simple no-relationship question addressed by the traditional computing formula. Specifically, we shall deal with questions of relevance to (1) replication or generalization, (2) reliability, and (3) certain theoretical relational constructs. These examples are not, of course, exhaustive, as we have indicated that hypotheses may be as varied as the theories which generate them. There also follow some cautions about interpretations associated with certain tests used with grouped data.

Replication or Generalization

To determine whether one has replicated a previous finding about how much difference a unit of change in one variable makes in another variable, one simply imposes the previously determined b as an hypothesis on a full model which allows the b to be determined by one's data. Formally, the procedure is the same for the question of whether a relationship (slope between two variables) is the same across ranges or in a new context. The analysis is developed below:

Full model: $$\mathbf{Y} = a\mathbf{U} + b\mathbf{X} + \mathbf{E}_F$$

Hypothesis: $$b = d$$

some constant determined by previous research

Restricted model: $$\mathbf{Y} = a\mathbf{U} + d\mathbf{X} + \mathbf{E}_R$$

and since d is a constant rather than a least-squares weight,

$$\mathbf{Y} - d\mathbf{X} = a\mathbf{U} + \mathbf{E}_R \qquad (4.15)$$

Letting $\mathbf{Z} = \mathbf{Y} - d\mathbf{X}$, we have

$$\mathbf{Z} = a\mathbf{U} + \mathbf{E}_R$$

Notice that the new dependent variable Z is not the same as the full-model dependent variable Y. It can easily be defined in SUBROUTINE DATA, however. Consequently, the F ratio *must* be defined in terms of error SS rather than R^2's:

$$F = \frac{(\sum e_{i_R}{}^2 - \sum e_{i_F}{}^2)/\mathrm{df}_1}{\sum e_{i_F}{}^2/\mathrm{df}_2}$$

where $\mathrm{df}_1 = 1$, as one hypothesis is imposed, and $\mathrm{df}_2 = N - 2$, as the full model has two unknown weights.

Where it makes theoretical sense, we can easily test to see whether both parameters are replicated (or generalized). This can be done with the following models:

Full model: $$\mathbf{Y} = a\mathbf{U} + b\mathbf{X} + \mathbf{E}_F$$

Hypotheses: $$\begin{array}{ll} a = c & \text{some previously determined constant} \\ b = d & \text{some previously determined constant} \end{array}$$

Restricted model: $$\mathbf{Y} = c\mathbf{U} + d\mathbf{X} + \mathbf{E}_R$$

Since c and d are constants rather than least-squares weights,

$$\mathbf{Y} - c\mathbf{U} - d\mathbf{X} = \mathbf{E}_R \qquad (4.16)$$

Letting $\mathbf{Z} = \mathbf{Y} - c\mathbf{U} - d\mathbf{X}$ gives

$$\mathbf{Z} = \mathbf{E}_R$$

F must be defined in terms of error SS rather than R^2's as the dependent variable is different for the two models. Also, such a model cannot be solved by our computer program as there are no unknown weights to be determined. However, our program will give us $\sum e_{i_F}{}^2$, and we can readily solve for $\sum e_{i_R}{}^2$ as $e_{i_R}{}^2 = (y_i - c \times 1 - dx_i)^2$.

The following Fortran structure will provide this computation (this is to be an independent program, *not* part of program CARTROCK):

```
............................
       C=(hypothesized intercept)
       D=(hypothesized slope)
       N=(number of cases)
       ERSUM=0.
       DO 8 I=1,N
  9    READ(5,9)Y,X
       FORMAT(some appropriate specifications)
  8    ERSUM=ERSUM+(Y-C-D*X)**2
       WRITE(6,10)ERSUM
  10   FORMAT(some appropriate specification)
............................
```

$$F = \frac{(\sum e_{i_R}^2 - \sum e_{i_F}^2)/\mathrm{df}_1}{\sum e_{i_F}^2/\mathrm{df}_2}$$

where $\mathrm{df}_1 = 2$, as two hypotheses were imposed (or the difference in the number of unknown weights is $2 - 0 = 2$), and $\mathrm{df}_2 = N - 2$, as there are two unknown weights in the full model. The fact that the restricted model has zero unknown weights is of no concern. It simply means that the model has $N - 0$, or N, degrees of freedom.

Reliability

We do not intend to address the general issue of reliability in this volume, as that is a massive and technical topic. Instead we intend to develop preferable alternatives to the common practice of asserting interrater, and test-retest reliability on the basis of an r which is significantly different from zero. With as few as 200 cases, an $r = .14$ will be significantly different from 0 at the .05 level. Obviously, this is not a test of what we mean by interrater reliability. In the interrater-reliability context, the test that $r = 0$ ($b = 0$) is a test of the hypothesis that two organisms are randomly responding with respect to each other. This would not be tenable for very large numbers of cases even if our judges were pigeons. It would be preferable practice to test the hypothesis that the r is not smaller than (one-tailed test) some arbitrarily large standard, for example, .7 or .8 or .9 (for such a test, consult Blalock, 1972, pp. 400–407, or McNemar, 1962, pp. 139–143). The discussions of Fisher's Z transformation there should be of use in this regard.

Better still would be to decide what we mean by ideal reliability and design a test to determine to what degree it is approximated. Even an $r = 1.0$ between judges' ratings does not imply total reliability. We could have one judge scoring a set of observations (1, 2, 3, 4, 5) and another scoring the same observations (9, 10.5, 12,

13.5, 15) and the $r = 1.0$. If our scale ranged from 1 to 15, the interpretation of such judgments would be quite different. What we do mean by interrater (or test-retest, or, for that matter, split-half) reliability is that one set of ratings is equivalent to another set of ratings. A simple but precise test for such a question follows.

Question: Do two raters (ratings) provide the *same* values for the same set of observations?

Full model: $$\mathbf{Y} = a\mathbf{U} + b\mathbf{X} + \mathbf{E}_F$$

where $\mathbf{Y}$ = one set of ratings
$\quad\ \mathbf{X}$ = other set of ratings
$\quad\ \mathbf{U}$ = unit vector
$\quad\ \mathbf{E}_F$ = error vector
$\quad a, b$ = least-squares weights

Hypotheses: $\qquad\qquad\qquad\qquad\qquad\qquad a = 0 \qquad b = 1$

Restricted model: $\qquad\quad \mathbf{Y} = 0\mathbf{U} + 1\mathbf{X} + \mathbf{E}_R \quad \mathbf{Y} = \mathbf{X} + \mathbf{E}_R \qquad$ (4.17)

and since $\mathbf{X}$ has no unknown weight associated with it,

$$\mathbf{Y} - \mathbf{X} = \mathbf{E}_R \qquad (4.18)$$

and where $\mathbf{Z} = \mathbf{Y} - \mathbf{X}$

$$\mathbf{Z} = \mathbf{E}_R$$

Since the dependent variable $\mathbf{Z}$ is not the same as the dependent variable $\mathbf{Y}$, we must define F in terms of error SS:

$$F = \frac{(\sum e_{i_R}{}^2 - \sum e_{i_F}{}^2)/df_1}{\sum e_{i_F}{}^2/df_2}$$

where $df_1 = 2$, as two hypotheses were imposed, and $df_2 = N - 2$, as the full model has two unknown weights.

This is a very precise and stringent test for reliability, but then, reliability is a very stringent notion. Some of the restrictiveness of the test can be avoided by some relaxation of our probability criteria for the test.

Theoretical Relational Constructs

Certain theoretical constructs in social science are formally equivalent to the reliability problem. Notions of concensus in political science and social psychology and the homogamy principle (likes marry likes) in mate-selection literature are two cases in point. In its simplest and most general form, the homogamy principle is that spouses tend to have the same attitudes, values, demographic characteristics (except sex, of course), and backgrounds. The main tests of the homogamy principle (and the competing complementary-needs principle) have involved tests of r's (correlations between characteristics of spouses). A more direct test of the homogamy principle is that outlined above for reliability. The example below uses spouses' ages, but any other characteristics can be substituted.

Question: Do women tend to marry men of the same age (other characteristic) as their own?

Full model: $$Y = aU + bX + E_F$$

where Y is the wife's age (characteristic), X is the husband's age (characteristic), and all other terms are defined as before.

Hypotheses: $$a = 0 \qquad b = 1$$

Restricted model:
$$Y = 0U + 1X + E_R$$
$$Y = X + E_R$$
$$Y - X = E_R$$
$$Z = E_R$$

where $Z = Y - X$.

We developed the Fortran and findings for this full model and these data in Sec. 4.4. The restricted model involves no least-squares weights and therefore cannot be solved by program CARTROCK. However, the following Fortran will provide the error SS for this model, as

$$\sum e_{i_R}^2 = \sum (y_i - x_i)^2$$

. . . .

```
      ERSUM=0.
      DO 8 I=1,200
      (standard READ statement)
      (standard FORMAT statement)
    8 ERSUM=ERSUM+(VAR(9)−VAR(3))**2
      WRITE (6,9) ERSUM
    9 FORMAT (1X,1F10.1)
```

. . . .

For these data, the following findings were obtained:

	Full model	Restricted model
$a =$	7.1926	
$b =$	.7850	
$\sum e_i^2 =$	909.3	$\sum e_i^2 = 2449.0$
$R^2 =$	.9540	R^2 (undefined)

F must be defined in terms of error SS, as the dependent variable is different for the two models:

$$F = \frac{(\sum e_{i_R}^2 - \sum e_{i_F}^2)/df_1}{\sum e_{i_F}^2/df_2} = \frac{(2449.0 - 909.3)/(2 - 0)}{909.3/(200 - 2)} \approx 167.6$$

As this far exceeds the critical value for F at 2 and 198 degrees of freedom at the .001 level of significance, we shall reject the age-homogamy hypotheses at the .001 level of significance. These findings taken with those in Sec. 4.4 indicate that ages of spouses are strongly related but the relationship is not one of equivalence. Unfortunately, much of the homogamy literature has treated these two notions as equivalent.

It is, of course, easy to develop regression models which are precise tests of any theoretical statements about slopes and intercepts. If we believe, for example, that women marry men 2 years younger than themselves, the hypotheses should be $a = -2; b = 1$. The literature on achievement and IQ of siblings of varying degrees of relationship could well profit by development of models which test exact specifications of relationships given competing genetic or cultural theories.

Cautions on the Use of Grouped Data

It has been suggested in some of the operant conditioning and experimental literature that "the proportion of variance accounted for" can be greatly increased by using the means for groups of observations rather than the observations themselves. While this is true, its value may be illusory. Basically, the strategy is to use the Y mean for all cases with X value i and use this mean as the expected value for all such Y cases. This technique *assumes* that variation around those means is random and can be safely ignored. Further, it ignores all variation around the regression line save that produced by the *means* falling off the line. Tests based on such measurements are more akin to our linearity test than to our tests for the proportion of variation in one variable accounted for by another. R^2's using such data are *not* the proportion of variance one variable (on a case-by-case basis) accounts for in another. Rather, R's for such tests are the proportion of variation in the category *means* accounted for by the linear regression with the other variable. The R's tend to be quite large because the main source of error variation (within-category variation) is eliminated by assumption and without test.

Special caution must be exercised when comparing findings based on individual versus such grouped data. Also, the number of cases (and consequently the degrees of freedom) for such analyses is not based on N (the number of observations) but k (the number of categories).

4.6   PROPORTION OF VARIANCE ACCOUNTED FOR BY COMPLEX MODELS: MULTIVARIATE REGRESSION

Only in the most highly controlled laboratory setting can a social variable reasonably be treated as being influenced or caused by a single variable. We shall first discuss multiple correlation in its simplest case involving two independent variables and straight-line relationships. The logic of the technique is unchanged by generalization to include k independent variables and relationships of any specified form.

If we predict husband's education from a linear function of father's education, we shall find that some of the variation in husband's education is accounted for (eliminated) by our knowledge of father's education. The correlation between the two is .75 and the proportion of variation accounted for is $r^2 = .56$. Reference to Fig. 4.2 may now be helpful. Recall that we are using a linear model. The best-fit straight line is determined by least-squares solutions to the equation $\mathbf{Y} = a\mathbf{U} + b\mathbf{X} + \mathbf{E}_R$.

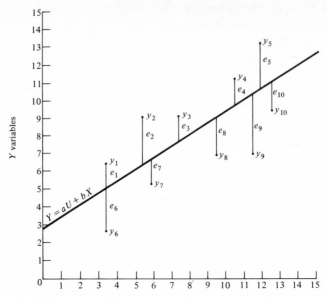

FIGURE 4.2
Graphic representation of error terms for regression with two variables.

Notice that the Y scores do not fall exactly on the regression line. The regression line defines predictions for each Y score given the X score. These are $y_i' = a + bx_i$. The errors, $e_i = y_i - y_i' = y_i - a - bx_i$, are defined by the distance from each y_i to the regression line measured perpendicularly to the X axis. Recall from Sec. 3.1 that the least-squares solutions produce an error vector which is independent of (orthogonal to) all predictor vectors. Thus $\mathbf{E} \perp \mathbf{U}$ ($\mathbf{E}$ is orthogonal to $\mathbf{U}$), and $\mathbf{E} \perp \mathbf{X}$. This can be verified by showing that $\mathbf{E'U} = 0$ and $\mathbf{E'X} = 0$. If we can find a variable which is related to (accounts for variation in) these e_i's, we can further reduce our errors of prediction of Y. Such a variable, of course, must not be perfectly correlated with X. If it were, it would also be independent of E and would give us no information not already provided by X. Where we believe that two variables each have some independent effects on a dependent variable, we can determine the magnitude of those effects by a model of the following form:

$$\mathbf{Y} = a_1\mathbf{U} + b_1\mathbf{X}_1 + a_2\mathbf{U} + b_2\mathbf{X}_2 + \mathbf{E}_F = (a_1 + a_2)\mathbf{U} + b_1\mathbf{X}_1 + b_2\mathbf{X}_2 + \mathbf{E}_F$$

and where $a_1 + a_2 = a$

$$\mathbf{Y} = a\mathbf{U} + b_1\mathbf{X}_1 + b_2\mathbf{X}_2 + \mathbf{E}_F$$

Such a model will allow variable $\mathbf{X}_1$ (or $\mathbf{X}_2$) to account for all of the variation in $\mathbf{Y}$ that it can and then the other independent variable to account for all of the variation in the residuals (e_i's, errors of prediction) that it can. We found earlier [Eq. (4.11)] that

$$\frac{R_F{}^2 - R_R{}^2}{1 - R_R{}^2}$$

is the proportion of error variation in a restricted model accounted for by a (less restricted) full model. Also, we noted that where our restricted model was a random-variation model, $R_R{}^2 = 0$. Substituting this in the above and simplifying produces $R_F{}^2$ as the proportion of error variation accounted for by our full model. Where our full model involves two independent variables, X_1 and X_2, $R_F{}^2$ is the proportion of error variation in Y accounted for by X_1 and X_2 considered together. In conventional notation this is designated $R^2_{Y \cdot X_1 X_2}$ and is called a *multiple R square*. Where our full model employs k independent variables,

$$\mathbf{Y} = a\mathbf{U} + b_1\mathbf{X}_1 + b_2\mathbf{X}_2 + \cdots + b_k\mathbf{X}_k + \mathbf{E}_F \qquad (4.19)$$

the $R_F{}^2$ is the proportion of error variation in Y accounted for by $X_1, X_2, \ldots, X_k$ considered together. In conventional notation this is designated $R^2_{Y \cdot X_1 X_2, \ldots, X_k}$ and also called a *multiple R square*. We shall have two uses for R^2's from models of this type. First, we shall be directly concerned with them where we are interested in pure prediction problems, i.e., where we want to see whether and to what degree a set of variables can be used to predict some dependent variable. In Sec. 4.9, we shall use such R^2's to define partial correlation coefficients which can be used in testing causal and other hypotheses. First, we should explore the direct utility of such models. The question addressed by the conventional test for the multiple R is developed below.

A   Natural-Language Question

Do the values of variables X_1 and X_2 give any information about values of the dependent variable Y which allows us to predict those values with more accuracy than we would get from $\overline{Y}$ alone?

B   Full Model $\qquad\qquad\qquad \mathbf{Y} = a\mathbf{U} + b_1\mathbf{X}_1 + b_2\mathbf{X}_2 + \mathbf{E}_F$

C   Hypotheses $\qquad\qquad\qquad\qquad b_1 = 0 \qquad b_2 = 0$

D   Restricted Model $\qquad \mathbf{Y} = a\mathbf{U} + 0\mathbf{X}_1 + 0\mathbf{X}_2 + \mathbf{E}_R = a\mathbf{U} + \mathbf{E}_R$

E   Inference $\qquad\qquad\qquad\qquad R^2_{Y \cdot X_1 X_2} = R_F{}^2$

$$F = \frac{(R_F{}^2 - R_R{}^2)/\mathrm{df}_1}{(1 - R_F{}^2)/\mathrm{df}_2}$$

where $\mathrm{df}_1 = 2$, as two hypotheses were imposed, and, $\mathrm{df}_2 = N - 3$, as there are three unknown weights in the full model.

F   Computer Instructions

In our numerical example below, we shall use husband's education, VAR(4), as $\mathbf{Y}$; father's education, VAR(13), as $\mathbf{X}_1$; and mother's education, VAR(18), as $\mathbf{X}_2$. Thus we shall need to construct a VEC(1), defined as VAR(4) or $\mathbf{Y}$; a VEC(2), defined as $\mathbf{U}$; a VEC(3), defined as VAR(13) or $\mathbf{X}_1$; and a VEC(4), defined as VAR(18) or $\mathbf{X}_2$.

1 To achieve this, we need the following Fortran statements embedded in SUBROUTINE DATA:

. . . .

(standard READ statement)

(standard FORMAT statement)

VEC(1) = VAR(4)

VEC(2) = 1.

VEC(3) = VAR(13)

VEC(4) = VAR(18)

. . . .

2 The main control card would be

MULT.TEST____200____20_____4_____2

3 The model cards would be

MULT.FULL_____1____3_____2_____3_____4

MULT.REST. ____1_____1_____2

G Numerical Example

The question of interest is whether the education of one's parents significantly determines one's own education.

Full model: $\qquad\qquad\qquad\qquad\qquad$ $\mathbf{Y} = a\mathbf{U} + b_1\mathbf{X}_1 + b_2\mathbf{X}_2 + \mathbf{E}_F$

Hypotheses: $\qquad\qquad\qquad\qquad\qquad$ $b_1 = 0 \qquad b_2 = 0$

Restricted model: $\qquad\qquad\qquad\qquad$ $\mathbf{Y} = a\mathbf{U} + \mathbf{E}_R$

The relevant computer results are

	Full model	Restricted model
	$a = 0.9034$	$a = 14.15$
	$b_1 = 0.5340$	
	$b_2 = 0.5465$	
	$\sum e_i^2 = 115.1$	$\sum e_i^2 = 1{,}476$
	$R^2 = 0.9220$	$R^2 = 0$

$$F = \frac{(R_F^2 - R_R^2)/\mathrm{df}_1}{(1 - R_F^2)/\mathrm{df}_2} = \frac{(.9220 - 0)/(3 - 1)}{(1. - .9220)/(200 - 3)} \approx 1{,}105$$

With an F as large as this, we must reject our hypotheses at the .001 level of significance. Notice that the parents' educations account for 92.2 percent of the variation in the son's education. We should interpret the parameters of a model which fits this well. In the restricted model, $a = 14.15$ and is the mean education for all sons. In the full model a, b_1, and b_2 are least-squares weights providing predicted values for sons' educations given educations of their parents: $Y = .9034 + .5340X_1 + .5465X_2$. The model provides the predictions given in Table 4.3. Generally, the finding can be interpreted by saying that the son's education is very *nearly* the average $(.5X_1 + .5X_2)$ of his parents' educations plus a constant. At lower levels of parents' education the equation produces an expected son's education of about 1.5 years more than his parents. At higher levels of parents' education the expected son's education is about 2.2 years greater than his parents.

4.7 GENERALIZATION: THE k-INDEPENDENT-VARIABLE CASE

The general test for the hypothesis that a multiple $R^2 = 0$ is quite straightforward.

A Natural-Language Question

Do the independent variables taken together give us any information about the values of the dependent variable?

B Full Model

$$Y = aU + b_1X_1 + b_2X_2 + \cdots + b_kX_k + E_F \qquad \text{from Eq. (4.19)}$$

C Hypotheses

$$b_1 = 0, \qquad b_2 = 0, \qquad \ldots, \qquad b_k = 0 \qquad \text{There are } k \text{ of these.}$$

D Restricted Model

$$Y = aU + E_R$$

$$F = \frac{(R_F{}^2 - R_R{}^2)/\text{df}_1}{(1 - R_F{}^2)/\text{df}_2}$$

where $\text{df}_1 = k$, as there are k hypotheses imposed, and $\text{df}_2 = N - (k + 1) = N - k - 1$, as there are $k + 1$ unknown weights in the full model.

Table 4.3 EXPECTED SON'S EDUCATION GIVEN PARENTS' EDUCATIONS

Father's education, years	Mother's education, years		
	8	12	16
8	9.5	11.7	13.9
12	11.7	13.9	16.1
16	13.8	16.0	18.2

I   Traditional Equivalence

For these models, $R_R{}^2 = 0$, $R_F{}^2$ is conventionally designated $R^2_{Y \cdot X_1 X_2, \ldots, X_k}$, $df_1 = k$, and $df_2 = N - k - 1$. Substituting these quantities in our definition for F above yields

$$F = \frac{(R^2_{Y \cdot X_1, X_2, \ldots, X_k})/k}{(1 - R^2_{Y \cdot X_1, X_2, \ldots, X_k})/(N - k - 1)} \qquad (4.20)$$

which is the conventional formula for the test for significance of multiple correlation (see Blalock, 1972, p. 465).

4.8   NONSTANDARD QUESTIONS: REPLICATION, NONLINEARITY, INTERACTION

All that was said about the weaknesses of the hypothesis that $r^2 = 0$ in the bivariate case (Sec. 4.5) applies a fortiori to the hypothesis that $R^2 = 0$ in the multivariate case. Especially where N's are quite large the finding that the X_k independent variables are significantly related to the dependent variable is often quite trivial. We shall briefly develop five types of nonstandard hypotheses which are considerably more powerful than the no-relationship hypotheses associated with the standard test. These will include questions of relevance to (1) replication or generalization, (2) multiple regression where linearity is not tenable, (3) certain theoretical relationships, (4) two special cases which will detect special types of interaction between the independent variables, and (5) true multiplicative models.

Replication or Generalization

Occasionally we want to determine whether we have found the same relational structure as some previous researcher (replication) or the same structure found under different conditions (generalization). These questions are developed quite straightforwardly by imposing the previous findings as conditions on our full model.

Full model: $\mathbf{Y} = a\mathbf{U} + b_1\mathbf{X}_1 + b_2\mathbf{X}_2 + \cdots + b_k\mathbf{X}_k + \mathbf{E}_F$

Hypotheses: $b_1 = c_1$

$b_2 = c_2$

. . . .

$b_k = c_k$

where the c_j's are constants determined by previous research rather than unknown weights to be fitted to data.

Restricted model: $\mathbf{Y} = a\mathbf{U} + c_1\mathbf{X}_1 + c_2\mathbf{X}_2 + \cdots + c_k\mathbf{X}_k + \mathbf{E}_R$

Since the c_j's are constants,

$$\mathbf{Y} - c_1\mathbf{X}_1 - c_2\mathbf{X}_2 - \cdots - c_k\mathbf{X}_k = a\mathbf{U} + \mathbf{E}_R \qquad (4.21)$$

Setting $Z = Y - c_1X_1 - c_2X_2 - \cdots - c_kX_k$ gives

$$Z = aU + E_R$$

As the dependent variables are not the same in the full and restricted models, we must define F in terms of error SS:

$$F = \frac{(\sum e_{i_R}^2 - \sum e_{i_F}^2)/df_1}{\sum e_{i_F}^2/df_2}$$

where $df_1 = k$, as k hypotheses are imposed, and $df_2 = N - (k + 1) = N - k - 1$, as there are k unknown weights in the full model.

We could, of course, impose hypotheses on only some of the b_j's, leaving the others free to vary. In such a case df_2 would remain the same, but df_1 would be the number of independent hypotheses imposed.

Multiple Regression with Nonlinearity

Occasionally it may be useful to combine in a model predictor variables some of which are thought to be related to the dependent variable by a first-degree equation and others of which are thought to be related by some other relational form. We cannot handle this exhaustively as the possibilities are unlimited, but we shall discuss a limited example. Logical extensions should be rather obvious to the user who has followed our analyses to this point.

Suppose we believe that when two independent variables are operating together, one of them, X, influences a dependent variable in a linear fashion and the other, Z, influences the dependent variable in a fashion best described by a second-degree equation. If neither of these effects can be observed when isolated from the other, the following analytic strategy would be appropriate.

The analysis should proceed in three stages: (1) we should test whether higher-degree forms than those posited will give better prediction; (2) we should test to see whether simpler forms will do as well; (3) if our general theory seems tenable, we should formulate tests and interpretations of the parameters of our preferred model.

QUESTION 1   Does a first-degree equation for the variable X and a second-degree equation for the variable Z give us as good a prediction of Y as equations of the next higher degrees?

Full model 1: $Y = aU + b_1X + b_2X^2 + c_1Z + c_2Z^2 + c_3Z^3 + E_F$

Hypotheses 1: $b_2 = 0$ $c_3 = 0$

Restricted model 1: $Y = aU + b_1X + c_1Z + c_2Z^2 + E_R$

$$F_1 = \frac{(R_F^2 - R_R^2)/df_1}{(1 - R_F^2)/df_2}$$

where $df_1 = 2$, as two hypotheses are imposed, and $df_2 = N - 6$, as there are six unknown weights in the full model.

QUESTION 2   Does a first-degree equation for the variable X and a second-degree equation for the variable Z give us a better fit than equations of the next lower degree?

Full model 2:
$$\mathbf{Y} = a\mathbf{U} + b_1\mathbf{X} + c_1\mathbf{Z} + c_2\mathbf{Z}^2 + \mathbf{E}_F$$

Hypotheses 2:
$$b_1 = 0 \qquad c_2 = 0$$

Restricted model 2:
$$\mathbf{Y} = a\mathbf{U} + c_1\mathbf{Z} + \mathbf{E}_R$$

$$F_2 = \frac{(R_F{}^2 - R_R{}^2)/df_1}{(1 - R_F{}^2)/df_2}$$

where $df_1 = 2$, as two hypotheses are imposed, and $df_2 = N - 4$, as there are four unknown weights in the full model.

Our theory is tenable if we cannot reject the first set of hypotheses but can reject the second set. The specific example has limited application, but the strategy which it illustrates is quite general.

Theoretical Relationships

Any linear theoretical relationships which can be specified between the slopes of the independent variables can be tested by the simple algebraic substitution of these in the full model. For example, let us say that we have the notion that the father's and mother's educations are equally important in determining the educational achievement of their son. Another way of stating this is that the son's education will be the mean of his parents' educations plus a constant. To test this, we could take

Full model:
$$\mathbf{Y} = a\mathbf{U} + b_1\mathbf{X}_1 + b_2\mathbf{X}_2 + \mathbf{E}_F$$

where $\mathbf{Y}$ = son's education
$\quad \mathbf{X}_1$ = father's education
$\quad \mathbf{X}_2$ = mother's education
$\quad \mathbf{U}$ = unit vector
$\quad \mathbf{E}_F$ = error vector
a, b_1, b_2 = least-squares weights
If the variabilities of father's and mother's educations are comparable and there is no interaction, we can translate our question into

Hypothesis:
$$b_1 = b_2 \, (= b)$$

Restricted model:

$$\mathbf{Y} = a\mathbf{U} + b\mathbf{X}_1 + b\mathbf{X}_2 + \mathbf{E}_R = a\mathbf{U} + b(\mathbf{X}_1 + \mathbf{X}_2) + \mathbf{E}_R$$

and where $\mathbf{Z} = \mathbf{X}_1 + \mathbf{X}_2$

$$\mathbf{Y} = a\mathbf{U} + b\mathbf{Z} + \mathbf{E}_R$$

This has the effect of converting our prediction for the son's education into the simple weighted sum (average) of the father's and mother's educations plus a constant, a. The two models can be evaluated by

$$F = \frac{(R_F^2 - R_R^2)/df_1}{(1 - R_F^2)/df_2}$$

where $df_1 = 1$, as one hypothesis is imposed, and $df_2 = N - 3$, as there are three unknown weights in the full model.

Special Cases of Interaction

We shall discuss interaction in detail later under the topic of analysis of covariance. Here, we discuss two rather common techniques which have been developed to reveal interactions of very special types. One strategy involves inclusion of a term which is the product of the independent variables; the other involves inclusion of a term which is their absolute difference. Here we shall explore these models and determine exactly *what kinds* of interaction can be detected by the inclusion and exclusion of such terms in our regression models.

Multiplicative Terms

Models commonly used for these tests are of the following form:

Full model: $$\mathbf{Y} = a\mathbf{U} + b_1\mathbf{X}_1 + b_2\mathbf{X}_2 + b_3\mathbf{Z} + \mathbf{E}_F \qquad (4.22)$$

where $z_i = x_{1i}x_{2i}$ and all other terms are defined as above.

Recall that interaction means that the effect of one independent variable on the dependent variable is different for different values of another independent variable. Let us assume that all the unknown weights in the above model are positive. While unnecessary, this assumption will facilitate the exposition. This would mean

1 As X_1 increases by 1 unit, Y increases by b_1 units.
2 As X_2 increases by 1 unit, Y increases by b_2 units.
3 As X_1 increases by 1 unit, Y increases by b_3X_2 units (or as X_2 increases by 1 unit, Y increases by b_3X_1 units).

Because of the multiplicative term Z, the effect of X_1 is *increased* by higher values of X_2 (and the effect of X_2 is *increased* by higher values of X_1). Such a finding is consistent with our general definition of interaction. *However*, the absence of a significant b_3 does *not* preclude interaction. It merely precludes interaction of the multiplicative form, i.e., interaction such that each variables' effect is a simple increasing function of the other variable. A substantive context in which such a model might make sense follows. Let us assume we have a theory which makes the following suggestions:

Proposition 1   The higher the prestige of an occupation, the more the income which is produced for its incumbents.

Proposition 2   The higher the education a person has, the more income he will command.

Proposition 3   The more education a person has, the more he will be able to mobilize the claims to more income inherent in his occupation. (Or the higher the occupational prestige of a person, the more he will be able to utilize his education to produce income.)

A test of this third proposition can be developed by comparing a model including a multiplicative term with one which excludes the multiplicative term.

Full model: $$\mathbf{Y} = a\mathbf{U} + b_1\mathbf{X}_1 + b_2\mathbf{X}_2 + b_3\mathbf{Z} + \mathbf{E}_F$$

Hypothesis: $$b_3 = 0$$

Restricted model: $$\mathbf{Y} = a\mathbf{U} + b_1\mathbf{X}_1 + b_2\mathbf{X}_2 + \mathbf{E}_R$$

where
  $\mathbf{Y}$ = husband's income [VAR(6)]
  $\mathbf{X}_1$ = husband's years of education [VAR(4)]
  $\mathbf{X}_2$ = husband's occupational prestige [VAR(5)]
  $\mathbf{Z}$ = product of X_1 and X_2
  $\mathbf{U}$ = unit vector
  $\mathbf{E}_F$ = error vector
a, b_1, b_2, b_3 = least-squares weights

Using the data in Appendix B, the following Fortran will generate the requisite vectors VEC(1) as $\mathbf{Y}$, VEC(2) as $\mathbf{U}$, VEC(3) as X_1, VEC(4) as $\mathbf{X}_2$, and VEC(5) as $\mathbf{Z}$:

. . . .

(standard READ statement)

(standard FORMAT statement)

VEC(1)=VAR(6)

VEC(2)=1.

VEC(3)=VAR(4)

VEC(4)=VAR(5)

VEC(5)=VAR(4)*VAR(5)

. . . .

The main control card would be

        MULTSTATUS__200____20_____5_____2

The model cards would be

        FULL_MULT._____1_____4_____2_____3_____4_____5

        REST. MULT._____1_____3_____2_____3_____4

The relevant computer results would be:[1]

	Full model	Restricted model
	$a = -1,275.0000$	$a = -5,187.6250$
	$b_1 = 186.3750$	$b_1 = 494.5195$
	$b_2 = 130.7422$	$b_2 = 197.1616$
	$b_3 = 4.9900$	
	$\sum e_i{}^2 = 3,550,000,000$	$\sum e_i{}^2 = 3,568,000,000$
	$R^2 = .6569$	$R^2 = .6551$

$$F = \frac{(R_F{}^2 - R_R{}^2)/\mathrm{df}_1}{(1 - R_F{}^2)/\mathrm{df}_2} = \frac{(.6569 - .6551)/(4 - 3)}{(1. - .6569)/(200 - 4)} \approx 1.1$$

The critical value for F at 1 and 194 degrees of freedom (actually 1 and 120 degrees of freedom) at the .05 level of significance is 3.92. As our F is considerably less than this, we shall be unable to reject the no-(multiplicative)-interaction hypothesis. Strictly, we have found that the multiplicative-interaction term adds very little information for predicting income over what is given by education and occupational prestige alone.

Though not common practice, it is often informative to perform the parallel test which asks whether the statuses (education and occupational prestige) add anything to what we know from the multiplicative-interactive variable. This can be asked in the following way:

Full model: $\quad\quad\quad\quad \mathbf{Y} = a\mathbf{U} + b_1\mathbf{X}_1 + b_2\mathbf{X}_2 + b_3\mathbf{Z} + \mathbf{E}_F$
Hypotheses: $\quad\quad\quad b_1 = 0 \quad\quad b_2 = 0$
Restricted model: $\quad\quad \mathbf{Y} = a\mathbf{U} + b_3\mathbf{Z} + \mathbf{E}_F$

This requires only the substitution of the following model card:

ALT.REST._ _ _ _ _ _1_ _ _ _2_ _ _ _2_ _ _ _5

The relevant computer results are

	Full model	Restricted model
	$a = -1,275.0000$	$a = 2,359.6523$
	$b_1 = 186.3750$	$b_3 = 12.6224$
	$b_2 = 130.7422$	
	$b_3 = 4.9900$	
	$\sum e_i{}^2 = 3,550,000,000$	$\sum e_i{}^2 = 3,681,000,000$
	$R^2 = .6569$	$R^2 = .6442$

$$F = \frac{(R_F{}^2 - R_R{}^2)/\mathrm{df}_1}{(1 - R_F{}^2)/\mathrm{df}_2} = \frac{(.6569 - .6442)/(4 - 2)}{(1. - .6569)/(200 - 4)} \approx 3.7$$

[1] This is a good example of why we must always keep in mind the range and variability of independent variables. The extremely large error SS are produced because they are sums of squared deviations of income in dollars (range is 0 to 40,000). If we use income in thousands of dollars, $\sum e_{i_F}{}^2 = 3,550$ and $\sum e_{i_R}{}^2 = 3,568$. However, their relative magnitudes and tests based on them would lead to identical F ratios. The magnitudes of the b's cannot be directly compared or interpreted for similar reasons. *In part*, b_1 and b_2 are larger than b_3 because b_3 refers to a product term X_1X_2 the range for which is much larger than the range for X_1 or X_2.

The critical value for F at 2 and 196 degrees of freedom (actually 2 and 120 degrees of freedom) at the .05 level of significance is 3.07. As our F exceeds this, we must reject the hypotheses implying that the statuses add nothing in predicting income over that provided by the interaction. While it is comforting when only one of these parallel tests is negative, it is by no means always the case and both should always be performed.

Absolute-Difference Terms

One other special-case approach to interaction has appeared in social science literature with some frequency. Sociologists working with the notion of status inconsistency wished to demonstrate that inconsistency between statuses has an effect which is independent of the statuses themselves. Early suggestions for analysis included models of the form

$$\mathbf{Y} = a\mathbf{U} + b_1\mathbf{X}_1 + b_2\mathbf{X}_2 + b_3\mathbf{Z} + \mathbf{E}_F$$

where all terms except Z are defined as above and $Z = X_1 - X_2$ or the difference in the two status variables. Unfortunately, such models have no solutions, as Z, X_1, X_2 are a linearly dependent set. Subsequently, two very similar solutions began to appear in the literature. One involved $Z = (X_1 - X_2)^2$, the square of the differences in the statuses, and the other involved $Z = |X_1 - X_2|$, the absolute value of the differences in the statuses. These are termed similar because $|X_1 - X_2| = \sqrt{(X_1 - X_2)^2}$. In both cases, implicitly or explicitly an interaction term is introduced into the equation which involves the multiplicative term X_1X_2. Consequently, tests based on such models produce results quite similar to those based on models involving the simple product term. These models differ from the one in the previous section only in that they also include (implicitly or explicitly) the terms X_1^2 and X_2^2, which are forced (assumed) to be weighted equally and forced (assumed) to be weighted one-half as much as the product term. Direct interpretation of such models is difficult. However, their similarity to the model involving the product term means that their use constitutes an approximate, albeit muddier, test of the same hypothesis.

True Multiplicative Models

Where one is dealing with true ratio scales (which must satisfy all interval criteria *and* have a *theoretically* grounded zero point), some theories involve propositions of the multiplicative form, for example,

$$Y = X_1^{b_1}X_2^{b_2} \tag{4.23}$$

Such theories can be tested with linear regression by conversion of these nonlinear propositions to a linear form. This can be done by taking the logarithm (any base) of each side of the equation. Thus, the above model becomes

$$\log Y = \log (X_1^{b_1}X_2^{b_2}) = \log X_1^{b_1} + \log X_2^{b_2} \tag{4.24}$$

$$\log Y = b_1\log X_1 + b_2\log X_2 \tag{4.25}$$

and for data involving random error

$$\log Y = b_1 \log X_1 + b_2 \log X_2 + E \qquad (4.26)$$

Most regression packages use computing routines which operate on deviations from the mean for each variable (or Z scores). For this reason, those solutions *assume* a unit vector. For technical reasons, program CARTROCK does not assume a unit vector, and this vector must be made explicit in all regression model cards. In general, solutions to models such as the above will be more satisfactory if we also insert a unit vector, converting the model to

$$\log Y = a\mathbf{U} + b_1 \log X_1 + b_2 \log X_2 + E$$

We can treat such a model as a full model and impose on it any linear conditions suggested for test by our theory; e.g., if the independent variables are not related to the dependent variable in this form, it would imply that $b_1 = 0$ and $b_2 = 0$. It may be of interest to note that the following model produces an $R^2 = .6656$:

$$\log \mathbf{Y} = a\mathbf{U} + b_1 \log \mathbf{X_1} + b_2 \log \mathbf{X_2} + \mathbf{E}$$

where $\mathbf{Y}$ = husband's income

$\quad \mathbf{X_1}$ = husband's years of education

$\quad \mathbf{X_2}$ = husband's occupational prestige

$\quad \mathbf{U}$ = unit vector

$\quad \mathbf{E}$ = error vector

This is very slightly higher than the $R^2 = .6569$ for the model including as independent variables education, occupation, and their product. We cannot use the F ratio to compare the two models as neither of them results from the imposition of linear conditions on the other.

4.9 PARTIAL CORRELATION COEFFICIENTS

It is often extremely important to be able to assess the relationship between two variables *controlling for the effects of other variables*. This is critical with nonexperimental research in which we cannot usually manipulate either the frequency of certain values of variables or the combinations in which these values occur. For research in nonlaboratory settings causal inferences are quite impossible without some way of controlling for the effects of confounding variables. All the current *causal analytic techniques* available to social science are at some level based on the notion of controlling for the effects of extraneous variables. For example, there are gross disparities in the socioeconomic status distributions of ethnic and racial groups in the United States. Some or all of the race and ethnic differences on any variable will be caused by the related socioeconomic differences; some or all of the differences will be caused by cultural experiential differences; and some or all of the differences will be caused by genetic, selective-migration, and other differences. Even tentative sorting out of such causes is beyond us without some partialing technique to determine the effects of some variables controlling for the effects of others. We *cannot* in many cases sample or select respondents who have all combinations of attributes in sufficient numbers to permit *controlling by constancy*.[1]

Unless all relationships are additive and neatly patterned and all independent variables can be manipulated, some partialing technique (or partialing analog) must even be used in experimental research. Recall the problem in Chap. 3 of sorting out the effects of teaching technique and ability levels of students on student improvement. Note that randomization offers no answer here. With balanced (naturally occurring) numbers of gifted, normal, and retarded students, perfect randomization would tell us that there are no differences in the three techniques (averaged across our random-ized groups). Only when some sort of control was introduced which allowed us to talk about technique effects controlling for ability were we able to isolate the differ-ences in the three techniques (see Secs. 3.11 and 3.13).

Fortunately, determining the effects of one variable on a dependent variable controlling for the effects of another (or others) is quite simple with continuous independent variables. Consider the following three models in which $\mathbf{Y}$, $\mathbf{X}_1$, and $\mathbf{X}_2$ are continuous variables, $\mathbf{U}$ is a unit vector, and $\mathbf{E}_{Rj}$ is the error vector in each case:

Full model: $\mathbf{Y} = a\mathbf{U} + b_1\mathbf{X}_1 + b_2\mathbf{X}_2 + \mathbf{E}_F$

Restricted model 1: $\mathbf{Y} = a\mathbf{U} + b_1\mathbf{X}_1 + \mathbf{E}_{R1}$

Restricted model 2: $\mathbf{Y} = a\mathbf{U} + b_2\mathbf{X}_2 + \mathbf{E}_{R2}$

Recall that in Secs. 4.4 and 4.6

$\mathbf{E}'_{R1}\mathbf{E}_{R1}$ = error SS associated with predicting Y from X_1 alone
$R_{R1}{}^2$ = proportion of variation in Y accounted for by X_1 alone
$\mathbf{E}'_F\mathbf{E}_F$ = error SS associated with predicting Y from X_1 and X_2 together
$R_F{}^2$ = proportion of variation in Y accounted for by X_1 and X_2 together

It is axiomatic that $\mathbf{E}'_{R1}\mathbf{E}_{R1} - \mathbf{E}'_F\mathbf{E}_F$ is the *amount* of squared error eliminated by addition of X_2 to the model utilizing X_1 only.

Further

$$\frac{\mathbf{E}'_{R1}\mathbf{E}_{R1} - \mathbf{E}'_F\mathbf{E}_F}{\mathbf{E}'_{R1}\mathbf{E}_{R1}}$$

is the *proportion of the error* for predicting Y from X_1 *which is eliminated by additional consideration of* X_2 *with* X_1. We have already developed [Eq. (3.2)] the algebraic approach for showing that such an expression is equal to

$$\frac{R_F{}^2 - R_{R1}{}^2}{1 - R_{R1}{}^2}$$

---

[1] Even with very large samples many events of interest are sufficiently rare to preclude control by constancy. For example, studies of cross-ethnic mate selection have been plagued by this problem. It would be of special interest to study Japanese-Black couples, as both groups have been discriminated against on racial grounds but in all other terms constitute polar extremes in the United States stratification system. However, with a random sample of 200,000 cases, only 2 such pairings occur. Absolute numbers of cases may even preclude some experimental controls. Certain very important perceptual questions can best be studied by comparing normals with individuals whose brain commissures have been severed (to alleviate disabling epilepsy). There are fewer than a dozen such people alive today.

which is the *proportion of variation in Y accounted for by X_2 over and above that accounted for by X_1*. Alternatively, it is the proportion of variation in Y accounted for by X_2 *controlling for* X_1. This is also called the partial $r^2_{YX_2 \cdot X_1}$. Its square root $r_{YX_2 \cdot X_1}$ is the partial R for predicting Y from X_2 controlling for X_1.

By similar argument, we can show that

$$\frac{R_F{}^2 - R_{R2}{}^2}{1 - R_{R2}{}^2}$$

is the proportion of variation in Y accounted for by X_1 *controlling for* X_2. This is also called the partial $r^2_{YX_1 \cdot X_2}$. Its square root $r_{YX_1 \cdot X_2}$ is the partial R for predicting Y from X_1 controlling for X_2.

Notice that the hypothesis $b_2 = 0$ will produce restricted model 1 from the full model. When this hypothesis is true, $R_F{}^2 = R_{R1}{}^2$ and therefore $r_{YX_2 \cdot X_1} = 0$. Conversely, whenever $r_{YX_2 \cdot X_1} = 0$, $b_2 = 0$.[1] Consequently, for such models, the test that $b_2 = 0$ *is* the test that the partial $r_{YX_2 \cdot X_1} = 0$.

This is, of course,

$$F = \frac{(R_F{}^2 - R_{R1}{}^2)/\mathrm{df}_1}{(1 - R_F{}^2)/\mathrm{df}_2}$$

where $\mathrm{df}_1 = 1$, as one hypothesis was imposed, and $\mathrm{df}_2 = N - 3$, as the full model has three unknown weights. Similarly, the hypothesis $b_1 = 0$ will produce restricted model 2 from the full model, and this hypothesis is equivalent to the hypothesis that the partial $r_{YX_1 \cdot X_2} = 0$. The test is, of course,

$$F = \frac{(R_F{}^2 - R_{R2}{}^2)/\mathrm{df}_1}{(1 - R_F{}^2)/\mathrm{df}_2}$$

where df_1 and df_2 are the same as above.

Partialing should usually involve performing all parallel analyses. For example, with two independent variables X_1 and X_2, one should usually not compute $r_{YX_2 \cdot X_1}$ without computing $r_{YX_1 \cdot X_2}$ because where X_1 and X_2 are themselves highly correlated, it is likely that *neither* will account for much variation if the other is controlled. Doing only half of the tests can lead to seriously erroneous conclusions.

A   Natural-Language Question

Does X_2 give us any additional information for predicting Y in addition to what is given by X_1 and if so how much? (For the alternative question, interchange X_1 and X_2.)

B   Full Model

$$\mathbf{Y} = a\mathbf{U} + b_1\mathbf{X}_1 + b_2\mathbf{X}_2 + \mathbf{E}_F$$

C   Hypothesis

$$b_2 = 0 \qquad \text{Alternative hypothesis:} \quad b_1 = 0$$

---

[1] The one exception to this is the determinate case where X_1 accounts for all of the variation in Y. Thus, $r_{YX_1} = 1$, and the partial $r_{YX_2 \cdot X_1}$ is undefined.

D   Restricted Model

Alternative restricted model:

$$Y = aU + b_1X_1 + E_{R1} \qquad (4.27)$$

$$Y = aU + b_2X_2 + E_{R2} \qquad (4.28)$$

E   Inference

$$F = \frac{(R_F{}^2 - R_{Ri}{}^2)/df_1}{(1 - R_F{}^2)/df_2}$$

where $df_1 = 1$, as one hypothesis is imposed, and $df_2 = N - 3$, as there are three unknown weights in the full model.

F   Computer Instructions

We shall need to develop VEC(1) as **Y** or husband's education, VAR(4); VEC(2) as **U** or the unit vector; VEC(3) as X_1 or father's education, VAR(13); and VEC(4) as X_2 or mother's education, VAR(18). The following Fortran will do this:

```
........................
(standard READ statement)
(standard FORMAT statement)
VEC(1)=VAR(4)
VEC(2)=1.
VEC(3)=VAR(13)
VEC(4)=VAR(18)
...............
```

The main control card would be

```
PARTIALS____200___20____4____3
```

The three model cards would be

```
FULL_PART_____1____3____2____3____4
REST1_PART ____1____2____2____3
REST2_PART ____1____2____2____4
```

G   Numerical Example

The questions to be answered, using the data from Appendix B, are whether and to what degree (1) mother's education gives us information for predicting son's education in addition to that given by father's education and (2) father's education gives us information for predicting son's education in addition to that given by mother's education. These questions can be answered by comparing the following models:

Full model: $\qquad\qquad\qquad\qquad Y = aU + b_1X_1 + b_2X_2 + E_F$

Hypothesis: $\qquad\qquad\qquad\qquad b_2 = 0 \qquad$ or $\qquad b_1 = 0$

Restricted model: $\qquad\qquad\qquad Y = aU + b_1X_1 + E_{R1}$

or $\qquad\qquad\qquad\qquad\qquad\quad Y = aU + b_2X_2 + E_{R2}$

The relevant computer results are

	Full model	Restricted model 1	Restricted model 2
	$a = .9034$	$a = 3.2400$	$a = 5.6306$
	$b_1 = .5340$	$b_1 = .8708$	
	$b_2 = .5465$		$b_2 = .7104$
	$\sum e_i^2 = 115.1$	$\sum e_i^2 = 646.8$	$\sum e_i^2 = 368.0$
	$R^2 = .9220$	$R^2 = .5618$	$R^2 = .7507$

$$r_{YX_2 \cdot X_1}^2 = \frac{R_F^2 - R_{R1}^2}{1 - R_{R1}^2} = \frac{.9220 - .5618}{1. - .5618} \approx .8219$$

Thus mother's education accounts for 82.19 percent of the error variance remaining after son's education is predicted from father's education.

$$r_{YX_1 \cdot X_2}^2 = \frac{R_F^2 - R_{R2}^2}{1 - R_{R2}^2} = \frac{.9220 - .7507}{1. - .7507} \approx .6871$$

thus father's education accounts for 68.71 percent of the error variance remaining after son's education is predicted from mother's education.

$$F_1 = \frac{(R_F^2 - R_{R1}^2)/df_1}{(1 - R_F^2)/df_2} = \frac{(.9220 - .5618)/(3 - 2)}{(1. - .9220)/(200 - 3)} \approx 909.8$$

We must reject the hypothesis that $b_2 = 0$ ($r_{YX_2 \cdot X_1} = 0$) at the .001 level of significance.

$$F_2 = \frac{(R_F^2 - R_{R2}^2)/df_1}{(1 - R_F^2)/df_2} = \frac{(.9220 - .7507)/(3 - 2)}{(1. - .9220)/(200 - 3)} \approx 432.7$$

We must reject the hypothesis that $b_1 = 0$ ($r_{YX_1 \cdot X_2} = 0$) at the .001 level of significance. As each X_1 and X_2 accounts for a large and statistically significant amount of the variation in Y, even controlling for the other, we can conclude that father's and mother's educations each have independent effects on son's education.

4.10 GENERALIZATION: kTH-ORDER PARTIAL CORRELATION COEFFICIENTS

Where we want to know whether a variable X_k has an *independent* effect on dependent variable Y, that is, X_k is still related to Y when we control for other relevant variables $X_1, X_2, \ldots, X_{k-1}$, we can use the following models:

Full model: $\mathbf{Y} = a\mathbf{U} + b_1\mathbf{X}_1 + b_2\mathbf{X}_2 + \cdots + b_{k-1}\mathbf{X}_{k-1} + b_k\mathbf{X}_k + E_F$ (4.29)

Hypothesis: $b_k = 0$

Restricted model:

$$\mathbf{Y} = a\mathbf{U} + b_1\mathbf{X}_1 + b_2\mathbf{X}_2 + \cdots + b_{k-1}\mathbf{X}_{k-1} + E_R \qquad (4.30)$$

The partial for Y and X_k, controlling for all other $k - 1$ variables, will be

$$r^2_{YX_k \cdot X_1 X_2, \ldots, X_{k-1}} = \frac{R_F^2 - R_R^2}{1 - R_R^2} \qquad (4.31)$$

and the test for the hypothesis that this partial is zero (or equivalently, that $b_k = 0$) will be

$$F = \frac{(R_F^2 - R_R^2)/\mathrm{df}_1}{(1 - R_F^2)/\mathrm{df}_2}$$

where $\mathrm{df}_1 = 1$, as one hypothesis is imposed, and $\mathrm{df}_2 = N - (k + 1) = N - k - 1$, as the full model has $k + 1$ unknown weights.

Unless a very precise theory is being tested, it is prudent to let each variable take the place of X_k in the above models. In this way, the *independent effect* of each variable controlling for the others can be isolated. Some cautions in this regard are developed later in Sec. 4.12.

Traditional computing formulas are typically provided for first-, second,- or third-order partials, as the general case is extremely complicated within traditional developments (see, for example, Blalock, 1972, pp. 433–440; McNemar, 1962, pp. 164–180). Within the linear-regression perspective we have taken, the general case of the partial and the F test for its significance from zero are readily translated into conventional notation:

$$r^2_{YX_k \cdot X_1 X_2, \ldots, X_{k-1}} = \frac{R^2_{Y \cdot X_1 X_2, \ldots, X_k} - R^2_{Y \cdot X_1 X_2, \ldots, X_{k-1}}}{1 - R^2_{Y \cdot X_1 X_2, \ldots, X_{k-1}}} \qquad (4.32)$$

$$F = \frac{(R^2_{Y \cdot X_1 X_2, \ldots, X_k} - R^2_{Y \cdot X_1 X_2, \ldots, X_{k-1}})/1}{(1 - R^2_{Y \cdot X_1 X_2, \ldots, X_k})/(N - k - 1)} \qquad (4.33)$$

Blalock (1972, pp. 466–467) does provide a general F test for the nth-order partial which is equivalent to this, though no general formula is provided for the nth-order partial itself. Thus, we can demonstrate the equivalence of the two formulations for the first-order partial only. The general traditional test is

$$F = \frac{r^2_{YX_k \cdot X_1 X_2, \ldots, X_{k-1}}/1}{(1 - r^2_{YX_k \cdot X_1 X_2, \ldots, X_{k-1}})/(N - k - 1)} \qquad (4.34)$$

For the F test for the first-order partial, $r_{YX_2 \cdot X_1}$, our general formula becomes

$$F = \frac{(R^2_{Y \cdot X_1 X_2} - R^2_{YX_1})/1}{(1 - R^2_{Y \cdot X_1 X_2})/(N - k - 1)} \qquad (4.35)$$

The numerator (disregarding df_1, which is the same in both formulations) is the proportion of variance explained by X_2 (the referent variable) in addition to that explained by X_1 (the control variable). In Blalock's development, this quantity is $r^2_{YX_2 \cdot X_1}(1 - r^2_{YX_1})$

The denominator (disregarding df_2, which is common to both formulations) is the proportion of variation unexplained by X_1 or X_2 taken together. In Blalock's development, this quantity is $(1 - r^2_{YX_2 \cdot X_1})(1 - r^2_{YX_1})$.

Substituting these representations of these two conceptually defined quantities in our formula produces

$$F = \frac{r_{YX_2 \cdot X_1}^2 (1 - r_{YX_1}^2)/1}{(1 - r_{YX_2 \cdot X_1}^2)(1 - r_{YX_1}^2)/(N - k - 1)} \qquad (4.36)$$

$$F = \frac{r_{YX_2 \cdot X_1}^2/1}{(1 - r_{YX_2 \cdot X_1}^2)/(N - k - 1)} \qquad (4.37)$$

which is the traditional F test formulation for the first-order partial $r_{YX_2 \cdot X_1}$. Notice that our general definitions of the partial and its F test involve only two terms, both of which are multiple R^2's. These are $R_{Y \cdot X_1 X_2, \ldots, X_k}^2$, which is the R_F^2 for the (full) model including the variable of interest and all control variables, and $R_{Y \cdot X_1 X_2, \ldots, X_{k-1}}^2$ which is the R_R^2 for the (restricted) model including only the control variables.

So defined, the nth-order partial is no more difficult to understand than the first-order partial. With program CARTROCK it is no more difficult to compute a partial of any degree than to compute a first-order partial. The reason the general case of the partial is so difficult to express in conventional treatments is that it is developed in terms of zero-order r's and lower-order partials rather than multiple R's. We have shown that two multiple R's are sufficient for this purpose regardless of how many variables are controlled. If partials are defined in terms of zero-order r's, the number of such r's which must be dealt with increases as a geometric function of the number of control variables.

4.11 NONSTANDARD QUESTIONS: REPLICATION AND NONLINEARITY

The partial correlation coefficient is an extremely powerful and flexible statistical device. By contrast with our treatment of earlier techniques, we can add little to its power with nonstandard questions; however, we should mention two possible extensions, one involving replication or generalization questions and the other involving nonlinear relationships.

Replication or Generalization

It is unlikely that current social science researchers would desire to replicate (or generalize) a finding constituted by a particular partial r. As these are determined by the effect of a specific variable operating in the context of other variables, their statistical properties are extremely complex. Further, some of the most useful contexts for using partials are those in which the effects of contaminating or perturbing variables (which are likely to differ across naturally occurring research settings) must be controlled. Should the same configuration of variables be of interest, replication *can* be assessed by imposing as hypotheses all (or part) of the b's derived in previous research. This is done directly by algebraic substitution as in previous discussions of replication and generalization. The infrequency of appropriate research settings for such a procedure is the reason for not developing it in detail here.

Nonlinear Relationships

More commonly, we may need to use variables (as controls or referents) which are related to the dependent variable by equations not of the first degree. This is easily done, but it requires slight modification of the formulas for partials and the F ratio. First, the variable and its square (cube, higher power) must be included in the full model. Thus, if Y is related to X_1 by a straight line but the residuals are related to X_2 by a second-degree equation and those residuals to X_3 by a first-degree equation, the proper full model is

Full model:
$$\mathbf{Y} = a\mathbf{U} + b_1\mathbf{X}_1 + c_1\mathbf{X}_2 + c_2\mathbf{X}_2{}^2 + d_1\mathbf{X}_3 + \mathbf{E}_F \qquad (4.38)$$

If we seek the relationship of Y and X_3 controlling for the other variables, the hypothesis is, as before,

Hypothesis:
$$d_1 = 0$$

The restricted model, of course, retains the squared term:

Restricted model:
$$\mathbf{Y} = a\mathbf{U} + b_1\mathbf{X}_1 + c_1\mathbf{X}_2 + c_2\mathbf{X}_2{}^2 + \mathbf{E}_R \qquad (4.39)$$

The desired partial is

$$R^2_{YX_3 \cdot X_1 X_2 X_2{}^2} = \frac{R_F{}^2 - R_R{}^2}{1 - R_R{}^2} \qquad (4.40)$$

The F-ratio is

$$F = \frac{(R_F{}^2 - R_R{}^2)/\mathrm{df}_1}{(1 - R_F{}^2)/\mathrm{df}_2}$$

where $\mathrm{df}_1 = 1$, as before, as one hypothesis is imposed, and $\mathrm{df}_2 = N - 5$, as there are five unknown weights in the full model.

Notice that where k is the number of variables, here 3, there are $k + 2$ unknown weights, one for each variable, one for the unit vector, and one for the higher-order term $X_2{}^2$. A general formula for this would be quite complex, as any number of variables could be expressed in any degrees, and we shall have no trouble if we simply count the unknown weights. Another way of looking at this problem is simply to count each higher-order term as a new variable for degrees-of-freedom purposes.

While these changes in the full model and df_2 are minor, they must not be overlooked. Slight additional modification is necessary where the variable with the higher-order equation is the referent variable rather than a control. The full model is, as before,

Full model:
$$\mathbf{Y} = a\mathbf{U} + b_1\mathbf{X}_1 + c_1\mathbf{X}_2 + c_2\mathbf{X}_2{}^2 + d_1\mathbf{X}_3 + \mathbf{E}_F$$

If we seek the relationship between X_2 and Y controlling for the remaining variables, we must take *both* weights for vectors involving X_2 out by hypothesis.

Hypotheses:
$$c_1 = 0 \qquad c_2 = 0$$

Restricted model:
$$\mathbf{Y} = a\mathbf{U} + b_1\mathbf{X}_1 + d_1\mathbf{X}_3 + \mathbf{E}_R \qquad (4.41)$$

The desired partial is

$$R^2_{Y(X_2 X_2{}^2) \cdot X_1 X_3} = \frac{R_F{}^2 - R_R{}^2}{1 - R_R{}^2} \tag{4.42}$$

The F ratio is

$$F = \frac{(R_F{}^2 - R_R{}^2)/\mathrm{df}_1}{(1 - R_F{}^2)/\mathrm{df}_2}$$

where $\mathrm{df}_1 = 2$, as *two* hypotheses were imposed, and $\mathrm{df}_2 = N - 5$, as before, as there are five unknown weights in the full model.

4.12 CAUSAL INFERENCES USING PARTIAL CORRELATIONS: SOME CAUTIONS

All rigorous techniques for making causal inferences in nonexperimental settings are based on some partialing technique. We shall not develop this notion in any depth. However, we shall briefly mention two illustrations of the simplest ways in which partials may aid in causal inference. Where two variables are related but neither causes the other (or causes anything else which causes the other), the relationship is said to be *spurious*. This commonly occurs when values of the two variables are caused by some common antecedent factor. Thus, if Z causes X and Z causes Y but neither X nor Y causes variation in the other, all the variables will be related. A diagram of the situation is

In such a situation, all zero-order r's (r_{ZX}, r_{ZY}, and r_{XY}) will be greater than zero. However, only some partials ($r_{ZY \cdot X}$ and $r_{ZX \cdot Y}$) will be greater than zero. The partial $r_{XY \cdot Z}$ will be zero because the relationship between X and Y depends solely on the common antecedent Z, which, if controlled, will eliminate any relationship between X and Y.

Another common causal structure involves a Z which causes X which causes Y. This can be diagrammed

In this case, X is termed an *intervening variable*. All zero-order r's will again be greater than zero. However, only some partials ($r_{ZX \cdot Y}$ and $r_{XY \cdot Z}$) will be greater than zero. The partial $r_{ZY \cdot X}$ will be zero because Z can influence Y only by influencing X, which, if controlled, will eliminate the relationship between Z and Y.

These two cases (common antecedent and intervening variable) are but the simplest cases in which partial correlations are of use. In general, when a partial is not zero, we can say that *some causal path* exists between the two variables which does not involve the variables being controlled. In Part Three, we develop some complex extensions of these ideas.

Nontheoretical Partialing

Partial correlation coefficients should be calculated only when required by theory (or at least when required by specific hypotheses). This is not mere pious dogma. Atheoretical partialing can, and often does, result in incredibly erroneous conclusions. Imagine that an investigator has a collection of some 20 potential independent variables. If any of them are multiply or redundantly measured, e.g., several indices of status, they will be correlated with nothing else when one controls for the alternative measures of the same thing. Further, one should note that in most research those variables which are considered most important are also the variables which are most likely to be multiply measured. Consequently, for most sets of data, mindlessly "doing all of the partials" is, paradoxically, likely to result in elimination of the most important variables (those measured several times) and retention of the (probably ancillary) variables which are measured only once.[1]

Selection by Zero-Order Correlations

A fairly common practice is to select independent variables for inclusion in an analysis on the basis of their zero-order correlations with the dependent variable. Many step-wise-regression packages automatically select the items for inclusion on the basis of such a decision rule. Such a practice is nonsense, as it is theoretically possible for a variable to have an $r = 0$ with the dependent variable and yet to be highly correlated with the residuals of the dependent variable predicted from some other variable. McNemar (1962, p. 186) gives an example of this where $r_{YX_1} = .400$, $r_{YX_2} = .000$, and $r_{X_1X_2} = .707$. Intuition and many computer programs might tell us that X_2 cannot possibly add to our prediction of Y. However, the multiple $r_{1 \cdot 23} = .566$. Thus, predicting Y from X_1 alone accounts for about 16 percent of the variation in Y, while predicting Y from X_1 and X_2 accounts for about 32 percent of the variation in Y. This is true in spite of the zero correlation between Y and X_2.

Substantive examples of this phenomenon are not at all rare. They most commonly occur when two potential independent variables each highly predictive of the dependent variable are themselves highly related. The addition of either to the other will not bring much improvement in prediction, as they both account for the *same* variance in the dependent variable. Often some variable less correlated with the dependent variable will yield more *new* information for predicting the residuals. An illustration of this can be provided by the variables husband's income, occupational prestige, education, and age in Appendix B. We shall take income as the dependent variable.

Predicting income from occupation:　　　　　　　　$R^2 = .6332$

Predicting income from education:　　　　　　　　　$R^2 = .3648$

Predicting income from age:　　　　　　　　　　　　$R^2 = .0295$

---

[1] This problem and solutions to it are developed extensively by Nirannanilathu Lalu in an unpublished master's thesis, University of North Carolina, 1970. (See also Gordon, 1968.)

If we incorporate predictor variables in the order of their zero-order relationships with income, we next get the multiple using occupation and education. This is $R^2 = .6551$. Had we chosen the (nearly) uncorrelated variable age, the multiple (for income predicted from occupation and age) would be $R^2 = .7442$. Using all three independent variables produces $R^2 = .8165$, but by the usual decision rule we probably never would have included age in the equation at all. However, informed social speculation suggests that age is somehow related to income. The conclusion is that one should trust theory (and perhaps even intuition) more than the inadequate decision rules built into most canned "analysis" packages.

4.13 MULTIPLE-PARTIAL CORRELATION COEFFICIENTS

It is often desirable to find the multiple relationship between Y and a cluster of variables $(Z_1, Z_2, Z_3, \ldots, Z_k)$ controlling for another cluster of variables $(X_1, X_2, X_3, \ldots, X_j)$. The desired quantity is the proportion of variance in Y accounted for by the Z_i cluster after the X_i cluster has been allowed to account for all that it can. This quantity is obtained by comparing a full model which includes both clusters with a restricted model which includes only the control cluster X_i. For three variables in each cluster, the resulting quantity is[1]

$$R^2_{Y(Z_1Z_2Z_3) \cdot X_1X_2X_3} = \frac{R_F{}^2 - R_R{}^2}{1 - R_R{}^2} \tag{4.43}$$

$$= \frac{R^2_{Y \cdot Z_1Z_2Z_3X_1X_2X_3} - R^2_{Y \cdot X_1X_2X_3}}{1 - R^2_{Y \cdot X_1X_2X_3}} \tag{4.44}$$

This can be read "the proportion of variation in Y accounted for by the Z cluster controlling for the X cluster." Its square root is called the multiple-partial correlation coefficient. The multiple partial has been little used in social science, principally because its computation and definition have been rather complicated. This need not be the case within the general regression framework, as we shall see. Further, certain very important kinds of questions require such a procedure. Consider the following problems.

Controlling for Powerful Variables Which Obscure Analyses

Blalock (1972, pp. 458–459) notes that an interest in the determinants of family size Y might require examination of the variation accounted for by theoretically interesting (possibly more manipulable) variables Z_1, Z_2, Z_3 after the effects of obvious but less interesting variables X_1, X_2, X_3, for example, duration of marriage, age of wife at marriage, have been removed.

---

[1] This derivation can be developed exactly as the simple partial was with the substitution of $X_1X_2, \ldots, X_j$ for X_1 and the substitution of $Z_1Z_2, \ldots, Z_k$ for X_2.

Multiple Indicators of Independent Variables[1]

Assume that we want to determine whether achieved or ascribed status has more influence on some measure of ideology Y controlling for the effects of the other. As is often the case, we probably would want more than one indicator of achieved status and more than one indicator of ascribed status, assuming that each indicator taps a slightly different dimension of our construct and is subject to somewhat different measurement error. The theoretical model might be

But with multiple indicators, the model tested would be

Obviously we would want the effects of the Z *cluster* controlling for the X *cluster* and the parallel effects of the X *cluster* controlling for the Z *cluster*.

Given appropriately valid measurement, this strategy could be employed to answer powerful questions like: What are the relative effects of heredity clusters and environmental clusters on achievement? What are the relative effects of ideological clusters, legal-system clusters, and technological clusters on economic development? To what degree is early or late socialization more critical in the formation of political ideology? Such broad questions can usually be rigorously addressed only with multiple partials.

A   Natural-Language Question

Do the three indicators of construct Z account for variation in Y controlling for the three indicators of construct X and if so, to what degree? The parallel question concerning the effects of the X *cluster* controlling for the Z cluster can be asked by simply exchanging X and Z in this development.

B   Full Model

$$Y = aU + b_1X_1 + b_2X_2 + b_3X_3 + c_1Z_1 + c_2Z_2 + c_3Z_3 + E_F \qquad (4.45)$$

C   Hypotheses

$$c_1 = 0 \qquad c_2 = 0 \qquad c_3 = 0$$

D   Restricted Model

$$Y = aU + b_1X_1 + b_2X_2 + b_3X_3 + E_R \qquad (4.46)$$

[1] It is possible to have a *dependent* variable W which is also multiply measured by W_1, W_2, W_3. This extension of our framework would then involve canonical correlation.

E   Inference

The *degree* of Z influence controlling for X influence is the multiple partial

$$R^2_{Y(Z_1 Z_2 Z_3) \cdot X_1 X_2 X_3} = \frac{R_F^2 - R_R^2}{1 - R_R^2} \qquad (4.47)$$

The significance test for the hypothesis that this multiple partial is zero is the same as the test of the hypotheses ($c_1 = 0$, $c_2 = 0$, $c_3 = 0$). It is

$$F = \frac{(R_F^2 - R_R^2)/\mathrm{df}_1}{(1 - R_F^2)/\mathrm{df}_2}$$

where $\mathrm{df}_1 = 3$, as three hypotheses are imposed, and $\mathrm{df}_2 = N - 7$, as there are seven unknown weights in the full model.

F   Computer Instructions

In our illustration, we shall explore a question of this type. Using the data in Appendix B, we shall seek to account for husband's liberalism Y by considering his status characteristics X_1, X_2, X_3 (education, occupation, income) and his father's characteristics Z_1, Z_2, Z_3 (education, occupation, income). Thus we shall need to define the following vectors:

VEC(1) = husband's liberalism	or	Y	[VAR(8)]
VEC(2) = unit vector	or	U	
VEC(3) = husband's education	or	X_1	[VAR(4)]
VEC(4) = husband's occupational prestige	or	X_2	[VAR(5)]
VEC(5) = husband's income	or	X_3	[VAR(6)]
VEC(6) = father's education	or	Z_1	[VAR(13)]
VEC(7) = father's occupational prestige	or	Z_2	[VAR(14)]
VEC(8) = father's income	or	Z_3	[VAR(15)]

This can be achieved with the following Fortran:

```
.......................
(standard READ statement)
(standard FORMAT statement)
VEC(1)=VAR(8)
VEC(2)=1.
VEC(3)=VAR(4)
VEC(4)=VAR(5)
VEC(5)=VAR(6)
VEC(6)=VAR(13)
VEC(7)=VAR(14)
VEC(8)=VAR(15)
................
```

The main control card would be

MULT.PART.___200___20____8____3

The model cards would be (we are including the restricted model, above, and its parallel model controlling for the Z variables)

FULL_____1____7____2____3____4____5____6____7____8

REST_1_____1____4____2____3____4____5

REST_2_____1____4____2____6____7____8

G   Numerical Example

The questions of interest are: (1) Do the father's statuses significantly influence the son's liberalism controlling for the son's statuses (and to what degree)? and (2) Do the son's statuses significantly affect his liberalism controlling for the father's statuses (and to what degree)?

Full model:

$$\mathbf{Y} = a\mathbf{U} + b_1\mathbf{X}_1 + b_2\mathbf{X}_2 + b_3\mathbf{X}_3 + c_1\mathbf{Z}_1 + c_2\mathbf{Z}_2 + c_3\mathbf{Z}_3 + \mathbf{E}_F$$

Hypotheses 1: $c_1 = 0 \qquad c_2 = 0 \qquad c_3 = 0$

Restricted model 1: $\mathbf{Y} = a\mathbf{U} + b_1\mathbf{X}_1 + b_2\mathbf{X}_2 + b_3\mathbf{X}_3 + \mathbf{E}_{R1}$

Hypotheses 2: $b_1 = 0 \qquad b_2 = 0 \qquad b_3 = 0$

Restricted model 2: $\mathbf{Y} = a\mathbf{U} + c_1\mathbf{Z}_1 + c_2\mathbf{Z}_2 + c\mathbf{Z}_3 + \mathbf{E}_{R2}$

The relevant computer results are:

Full model†	Restricted model 1	Restricted model 2
$a = 34.5193$	$a = 41.8387$	$a = 46.9900$
$b_1 = 2.1448$	$b_1 = 3.0296$	
$b_2 = .0342$	$b_2 = .0308$	
$b_3 = -.0018$	$b_3 = -.0018$	
$c_1 = 1.6710$		$c_1 = 1.3232$
$c_2 = .0153$		$c_2 = -.0815$
$c_3 = -.0003$		$c_3 = .0004$
$\sum e_i{}^2 = 34{,}070$	$\sum e_i{}^2 = 35{,}070$	$\sum e_i{}^2 = 52{,}790$
$R^2 = .3817$	$R^2 = .3634$	$R^2 = .0418$

† No direct comparisons of the b's can be made as they refer to variables of quite disparate range. Education ranges from 0 to 20; occupation from 0 to 100, and income from 0 to 40,000.

$$F_1 = \frac{(R_F{}^2 - R_{R1}{}^2)/\mathrm{df}_1}{(1 - R_F{}^2)/\mathrm{df}_2} = \frac{(.3817 - .3634)/(7 - 4)}{(1 - .3817)/(200 - 7)} \approx 1.9$$

The critical value for $F\,(.05)$ at 3 and 193 degrees of freedom (actually 3 and 120 degrees of freedom) is 2.68. As our obtained F is less than this, we cannot reject our hypotheses that $c_1 = 0$, $c_2 = 0$, $c_3 = 0$. As these hypotheses are tenable, we can eliminate fathers' statuses and say that fathers' statuses have little or no impact on the sons' liberalism controlling for sons' statuses.

$$F_2 = \frac{(R_F^2 - R_{R2}^2)/\mathrm{df}_1}{(1 - R_F^2)/\mathrm{df}_2} = \frac{(.3817 - .0418)/(7 - 4)}{(1 - .3817)/(200 - 7)} \approx 35.4$$

The critical value for $F\,(.001)$ at 3 and 193 degrees of freedom (actually 3 and 120 degrees of freedom) is 5.79. As our obtained F exceeds this value, we must reject our hypotheses that $b_1 = 0$, $b_2 = 0$, $b_3 = 0$ at the .001 level of significance. Consequently, we cannot eliminate sons' statuses from the model including sons' and fathers' statuses without significantly increasing error. Thus, sons' statuses significantly affect their liberalism, controlling for fathers' statuses. The magnitude of these effects can be determined by the partials. For question 1

$$R^2_{Y(Z_1Z_2Z_3) \cdot X_1X_2X_3} = \frac{R_F^2 - R_{R1}^2}{1 - R_{R1}^2} = \frac{.3817 - .3634}{1 - .3634} \approx .0287$$

Fathers' statuses account for 2.8 percent of the variation left unaccounted for by sons' statuses. For question 2

$$R^2_{Y(X_1X_2X_3) \cdot Z_1Z_2Z_3} = \frac{R_F^2 - R_{R2}^2}{1 - R_{R2}^2} = \frac{.3817 - .0418}{1 - .0418} \approx .3547$$

Sons' statuses account for 35.47 percent of the variation left unaccounted for by fathers' statuses.

4.14 GENERALIZATION: jTH-ORDER CLUSTERS WITH kTH-ORDER CONTROLS

To ask whether any theoretically defined cluster of variables Z_i accounts for a significant amount of the variation in Y controlling for any other set of variables X_i, the following analysis may be used:

Full model:

$$\mathbf{Y} = a\mathbf{U} + b_1\mathbf{X}_1 + b_2\mathbf{X}_2 + \cdots + b_k\mathbf{X}_k + c_1\mathbf{Z}_1 + c_2\mathbf{Z}_2 + \cdots + c_j\mathbf{Z}_j + \mathbf{E}_F \quad (4.48)$$

Hypotheses: $\qquad\qquad c_1 = 0, \qquad c_2 = 0, \qquad \ldots, \qquad c_j = 0$

Restricted model: $\qquad \mathbf{Y} = a\mathbf{U} + b_1\mathbf{X}_1 + b_2\mathbf{X}_2 + \cdots + b_k\mathbf{X}_k + \mathbf{E}_R \qquad (4.49)$

The significance of the otherwise unaccounted for variation accounted for by the cluster Z_i is given by

$$F = \frac{(R_F^2 - R_R^2)/\mathrm{df}_1}{(1 - R_F^2)/\mathrm{df}_2}$$

(no traditional equivalence is ordinarily supplied for this general case), where $\mathrm{df}_1 = j$, as j hypotheses are imposed, and $\mathrm{df}_2 = N - (k + j + 1) = N - k - j - 1$, as there are $k + j + 1$ unknown weights in the full model (k of these for the X variables, j of these for the Z variables, and one for the unit vector).

The proportion of otherwise unexplained variation which the Z cluster accounts for is given by the multiple partial

$$R^2_{Y(Z_1Z_2,\ldots,Z_j)\cdot X_1X_2,\ldots,X_k} = \frac{R_F{}^2 - R_R{}^2}{1 - R_R{}^2} \qquad (4.50)$$

In our notation, the traditional formula for the multiple partial (see Blalock, 1972, pp. 458–459) is

$$r^2_{Y(Z_1Z_2,\ldots,Z_j)\cdot X_1X_2,\ldots,X_k} = \frac{R^2_{Y\cdot Z_1Z_2,\ldots,Z_jX_1X_2,\ldots,X_k} - R^2_{Y\cdot X_1X_2,\ldots,X_k}}{1 - R^2_{Y\cdot X_1X_2,\ldots,X_k}} \qquad (4.51)$$

Notice that the first term in the numerator is our $R_F{}^2$ and the second term is our $R_R{}^2$. The second term in the denominator is our $R_R{}^2$. Substituting these equivalences produces our formulation above.

4.15   NONSTANDARD QUESTIONS: NONLINEARITY, MATCHED VARIABLES ACROSS CLUSTERS

Even with a technique as powerful and flexible as the multiple partial, important substantive questions can arise the answers to which require the formulation of non-standard questions. We shall deal with two of these: (1) nonlinearity and (2) questions of whether comparable variables in different clusters have similar effects. The first is a relatively minor extension; the second is of more substance and can open the way to extremely powerful findings. Of course, any other questions generating linear restrictions and implied by one's theory can be imposed.

Nonlinearity with Multiple-Partial Models

Where one or more variables in a cluster should be related to Y by a higher-degree equation, we need simply add variable(s) to the cluster for the square (cube, etc.) of that variable. When we impose conditions on such a variable, we must take care to impose them on all terms relating to the variable. Further, each term for each such variable must be counted as a separate variable for computing degrees of freedom (see Sec. 4.11).

Matched Variables across Clusters: A Special Case of Interaction

Imagine that we had found that both the sons' statuses and the fathers' statuses had *independent* effects on the sons' liberalism, i.e., neither multiple partial was near zero. It might then have been of interest to ask whether the father's education and son's education affected the son's liberalism *in the same way*. One way of asking this would be to impose the hypothesis $b_1 = c_1$ on the full model. Or, we might have asked whether the status variables for the father and son affected liberalism *in the same ways*. This would have generated the hypotheses that $b_j = c_j$ ($j = 1$ to 3). The analysis would then take the following form:

Question: Do the fathers' statuses and sons' statuses affect the sons' liberalism in the same ways?

Full model: $\mathbf{Y} = a\mathbf{U} + b_1\mathbf{X}_1 + b_2\mathbf{X}_2 + b_3\mathbf{X}_3 + c_1\mathbf{Z}_1 + c_2\mathbf{Z}_2 + c_3\mathbf{Z}_3 + \mathbf{E}_F$

Hypotheses: $\qquad b_1 = c_1 \, (= d_1) \qquad b_2 = c_2 \, (= d_2) \qquad b_3 = c_3 \, (= d_3)$

Restricted model:

$$\mathbf{Y} = a\mathbf{U} + d_1\mathbf{X}_1 + d_2\mathbf{X}_2 + d_3\mathbf{X}_3 + d_1\mathbf{Z}_1 + d_2\mathbf{Z}_2 + d_3\mathbf{Z}_3 + \mathbf{E}_R$$
$$= a\mathbf{U} + d_1(\mathbf{X}_1 + \mathbf{Z}_1) + d_2(\mathbf{X}_2 + \mathbf{Z}_2) + d_3(\mathbf{X}_3 + \mathbf{Z}_3) + \mathbf{E}_R$$

and where $\mathbf{W}_j = \mathbf{X}_j + \mathbf{Z}_j$

$$\mathbf{Y} = a\mathbf{U} + d_1\mathbf{W}_1 + d_2\mathbf{W}_2 + d_3\mathbf{W}_3 + \mathbf{E}_R$$

To test the hypotheses relating such a pair of models requires only the creation of $\mathbf{Y}$, $\mathbf{U}$ and the three clusters $\mathbf{X}_j$, $\mathbf{Z}_j$, $\mathbf{W}_j$. The F ratio would be

$$F = \frac{(R_F{}^2 - R_R{}^2)/\mathrm{df}_1}{(1 - R_F{}^2)/\mathrm{df}_2}$$

where $\mathrm{df}_1 = 3$, as three hypotheses are imposed, and $\mathrm{df}_2 = N - 7$, as there are seven unknown weights in the full model.

4.16 ANALYSIS OF COVARIANCE WITH REGRESSION MODELS

In analysis of variance, our predictor equations produced expected values which were means within categorically defined groups. In correlational and regression analysis our predictor equations produced expected values which were defined by least-squares lines in simple cases (planes and spaces in more complex ones). All tests for these techniques, whether traditional or not, were developed by the imposition of linear conditions on the weights in equations producing these expected values. For analysis of covariance, our predictor equations will produce expected values which are defined by least-squares lines (planes, spaces) within categorically defined groups. In a very strict sense, analysis of covariance involves *regression analyses within categorically defined groups*. Thus it is a combination of analysis of variance (expected values *within* categories) and regression (expected values defined by least-squares lines, planes, spaces).[1] All traditional, and some other, tests will be developed as linear conditions imposed on the weights in these equations.

For the simplest case of covariance, we seek a framework which allows us to examine the relationship between two continuous variables Y and X for separate categories of some categorical variable Z. Thus, we might desire to examine the relationship between liberalism or conservatism (Y) and age (X) (both conceptualized as continuous) with the provision that this relationship might be different for Republicans

---

[1] The approach taken here is *logically* the same as that taken by users of dummy-variable analysis, though some *parameters* have obvious interpretations in our approach but only mathematical interpretations in the other tradition (see Sec. 3.23).

and Democrats (Z). We seek then an analytic scheme which permits possibly different regression lines $Y = aU + bX$ for members of the two parties. Let us designate the regression line for Democrats $Y_d = aU_d + bX_d + E_d$ and the regression line for Republicans $Y_r = aU_r + bX_r + E_r$. We can readily develop a full model which will generate both these equations and allow us to test relationships between them. It is

Full model: $$Y = a_1 U_d + b_1 X_d + a_2 U_r + b_2 X_r + E_F \qquad (4.52)$$

where Y = dependent variable, liberalism for all cases
  U_d = 1 if person is a Democrat, 0 otherwise = *partial* unit vector
  X_d = age if person is a Democrat, 0 otherwise = *partial* score vector
  U_r = 1 if person is a Republican, 0 otherwise = *partial* unit vector
  X_r = age if person is a Republican, 0 otherwise = a *partial* score vector
  E_F = error vector
a_1, a_2, b_1, b_2 = least-squares weights
  In vector representation this model becomes

$$
\begin{bmatrix} y_1 \\ y_2 \\ \vdots \\ y_{nd} \\ y_{nd+1} \\ y_{nd+2} \\ \vdots \\ y_{nd+nr} \end{bmatrix}
= a_1 \begin{bmatrix} 1 \\ 1 \\ \vdots \\ 1 \\ 0 \\ 0 \\ \vdots \\ 0 \end{bmatrix}
+ b_1 \begin{bmatrix} x_1 \\ x_2 \\ \vdots \\ x_{nd} \\ 0 \\ 0 \\ \vdots \\ 0 \end{bmatrix}
+ a_2 \begin{bmatrix} 0 \\ 0 \\ \vdots \\ 0 \\ 1 \\ 1 \\ \vdots \\ 1 \end{bmatrix}
+ b_2 \begin{bmatrix} 0 \\ 0 \\ \vdots \\ 0 \\ x_{nd+1} \\ x_{nd+2} \\ \vdots \\ x_{nd+nr} \end{bmatrix}
+ \begin{bmatrix} e_1 \\ e_2 \\ \vdots \\ e_{nd} \\ e_{nd+1} \\ e_{nd+2} \\ \vdots \\ e_{nd+nr} \end{bmatrix}
$$

The model partitions the sample of nd Democrats and nr Republicans into two subsets. For Democrats, the expected value for Y is generated by

$$y_i = a_1 \times 1 + b_1 x_i + a_2 \times 0 + b_2 \times 0 + e_i = a_1 + b_1 x_i + e_i \qquad (4.53)$$

For Republicans the expected value for Y is generated by

$$y_i = a_1 \times 0 + b_1 \times 0 + a_2 \times 1 + b_2 x_i + e_i = a_2 + b_2 x_i + e_i \qquad (4.54)$$

Thus, for our full model, the following definitions hold:

  a_1 = intercept and b_1 = slope for least-squares line relating liberalism and age for Democrats
  a_2 = intercept and b_2 = slope for least-squares line relating liberalism and age for Republicans

$\displaystyle\sum_{i=1}^{nd} e_i^2$ = sum of squared deviations of the Democrats' liberalism scores from the least-squares prediction (line) for Democrats

$\displaystyle\sum_{i=nd+1}^{nd+nr} e_i^2$ = sum of squared deviations of the Republicans' liberalism scores from least-squares prediction (line) for Republicans

$\displaystyle\sum_{i=1}^{nd+nr} e_i^2$ = within-group error SS for model

R_F^2 = proportion of variation in Y accounted for by model

4.17   INTERACTION IN THE ANALYSIS OF COVARIANCE

Reasonable questions to ask of such a model might include: Does age affect liberalism and how? Or does party affect liberalism and how? First we should recall that the development of two-way analysis of variance showed that such questions cannot reasonably be answered where there is interaction. Interaction was defined as existing when the effect of one independent variable is different for different values of the other independent variable.[1] In a covariance analysis, the effect of the continuous independent variable is measured by its slope (the amount of change in Y associated with a unit change in X). If this slope is the same for both (all) values of Z (the categorical variable), no interaction exists. Thus, in the two-group case the no-interaction hypothesis is that $b_1 = b_2$. Let us consider interaction from the point of view of the categorical variable. The effect of the categorical variable is measured by the difference in liberalism associated with the two values, Democratic and Republican. For this difference to be the same for all values of X, the two least-squares lines must be parallel. That will be the case only where $b_1 = b_2$. Whichever point of view we take, we arrive at the conclusion that only when the slopes are equal across categories may we conclude that no interaction exists. Graphs of four possible relationships between our two equations are shown in Fig. 4.3.

CASE I   Age and liberalism are positively related for Democrats and unrelated for Republicans. The effect of age is different for the two parties, *or* the effect of party is different for different ages. $b_1 \neq b_2$, and interaction exists.

CASE II   Age and liberalism are positively related for Democrats and negatively related for Republicans. The effect of age is different for the two parties, *or* the effect of party is different for different ages. $b_1 \neq b_2$, and interaction exists.

CASE III   Age and liberalism are positively related to the same degree for each party. The effect of age is the same for both parties, *or* the effect of party is the same for all ages. Even though different ages have different amounts of liberalism and the two parties differ in liberalism, $b_1 = b_2$ and there is no interaction.

CASE IV   The Republican and Democratic relationships between age and liberalism are represented by the *same* line. There is no interaction, as $b_1 = b_2$, nor does party affect liberalism.

Only for cases III and IV could we proceed to ask questions about the *general overall* effects of age and party. We can now formulate the interaction test for the analysis of covariance.

A   Natural-Language Question

Does age affect liberalism in the same way for both parties? Or does party affect liberalism in the same way for all ages?

---

[1] In a sense, the *analysis* of covariance is a search for interaction. We know that two variables, X and Y, covary, and we seek to ascertain to what degree this covariance is general across categories of some other variable.

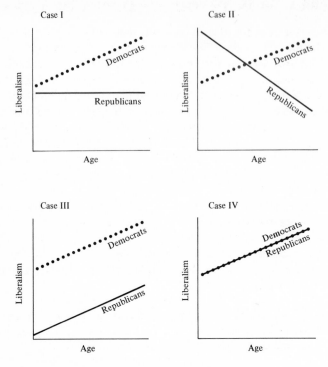

FIGURE 4.3
Four possible relationships for regression lines for two groups.

B   Full Model

$$\mathbf{Y} = a_1 \mathbf{U}_d + b_1 \mathbf{X}_d + a_2 \mathbf{U}_r + b_2 \mathbf{X}_r + \mathbf{E}_F$$

C   Hypothesis

$$b_1 = b_2 \ (= b)$$

D   Restricted Model

$$\mathbf{Y} = a_1 \mathbf{U}_d + b \mathbf{X}_d + a_2 \mathbf{U}_r + b \mathbf{X}_r + \mathbf{E}_R = a_1 \mathbf{U}_d + a_2 \mathbf{U}_r + b(\mathbf{X}_d + \mathbf{X}_r) + \mathbf{E}_R \quad (4.55)$$

and where $\mathbf{X} = \mathbf{X}_d + \mathbf{X}_r$

$$\mathbf{Y} = a_1 \mathbf{U}_d + a_2 \mathbf{U}_r + b \mathbf{X} + \mathbf{E}_R \quad (4.56)$$

E   Inference

$$F = \frac{(R_F{}^2 - R_R{}^2)/\mathrm{df}_1}{(1 - R_F{}^2)/\mathrm{df}_2}$$

where $\mathrm{df}_1 = 1$, as one hypothesis is imposed, and $\mathrm{df}_2 = N - 4$, as there are four unknown weights in the full model.

F   Computer Instructions

We must construct a VEC(1), defined as $\mathbf{Y}$ above; VEC(2), defined as $\mathbf{U}_d$ above; VEC(3), defined as $\mathbf{X}_d$ above; VEC(4), defined as $\mathbf{U}_r$ above; VEC(5), defined as $\mathbf{X}_r$ above; and VEC(6), defined as $\mathbf{X}$ above. Additionally, we shall later need a unit vector $\mathbf{U} = \mathbf{U}_d + \mathbf{U}_r$. We shall define a VEC(7) which is $\mathbf{U}$. The following Fortran statements will define these for the data in Appendix B [notice that VAR(7), party preference, is used to define these partitions]:

```
. . . . . . . . . . . . . . . . . . . . . . . . .
      (standard READ statement)
      (standard FORMAT statement)
      VEC(1) = VAR(8)
      VEC(6) = VAR(3)
      VEC(7) = 1.
      DO 1 I = 2,5
 1    VEC(I) = 0.
      IF(VAR(7))3,2,3
 2    VEC(2) = 1.
      VEC(3) = VAR(3)
      GO TO 4
 3    VEC(4) = 1.
      VEC(5) = VAR(3)
 4    CONTINUE
. . . . . . . . . . . . .
```

The main control card would be

```
COVARIANCE__200___20____7____2
```

The model control cards would be

```
FULL_____1____4____2____3____4____5
INTER_____1____3____2____4____6
```

G   Numerical Example

The question of interest is whether there is interaction between the variables liberalism, age, and party preference.

Full model: $\mathbf{Y} = a_1\mathbf{U}_d + b_1\mathbf{X}_d + a_2\mathbf{U}_r + b_2\mathbf{X}_r + \mathbf{E}_F$

Hypothesis: $b_1 = b_2 \; (= b)$

Restricted model: $\mathbf{Y} = a_1\mathbf{U}_d + a_2\mathbf{U}_r + b\mathbf{X} + \mathbf{E}_R$

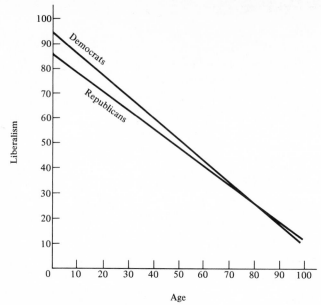

FIGURE 4.4
Least-squares lines for age and liberalism for the two parties.

The relevant computer results are:

	Full model	Restricted model
	$a_1 = 96.6611$	$a_1 = 94.7251$
	$b_1 = -.8595$	$a_2 = 89.4722$
	$a_2 = 87.7536$	$b = -.8052$
	$b_2 = -.7582$	
	$\sum e_i{}^2 = 40{,}340$	$\sum e_i{}^2 = 40{,}390$
	$R^2 = .2678$	$R^2 = .2669$

$$F = \frac{(R_F{}^2 - R_R{}^2)/df_1}{(1 - R_F{}^2)/df_2} = \frac{(.2678 - .2669)/(4 - 3)}{(1 - .2678)/(200 - 4)} \approx .24$$

As the F is less than 1.0, we shall be unable to reject our hypothesis that $b_1 = b_2$. Notice that the lines are not exactly parallel. Each year of age produces a bit more movement from liberalism for Democrats than for Republicans, though the Democrats are in general more liberal than Republicans. Still, the difference in slopes is small enough to be attributable to chance most of the time. The relationships are portrayed in Fig. 4.4. Notice that by about age 80, party makes no difference in liberalism.

4.18 GENERALIZATION: INTERACTION IN k-GROUP COVARIANCE MODELS

The question is whether the effect of each independent variable is the same for all values of the other independent variable. We have shown that this question can be reduced to the hypothesis that the slopes are equal for all categories.

Full model:

$$\mathbf{Y} = a_1\mathbf{U}_1 + b_1\mathbf{X}_1 + a_2\mathbf{U}_2 + b_2\mathbf{X}_2 + a_3\mathbf{U}_3 + b_3\mathbf{X}_3 + \cdots + a_k\mathbf{U}_k + b_k\mathbf{X}_k + \mathbf{E}_F$$

(4.57)

Hypotheses:
$$\left.\begin{aligned} b_1 &= b_2 \ (= b) \\ b_1 &= b_3 \ (= b) \\ &\cdots\cdots\cdots \\ b_1 &= b_k \ (= b) \end{aligned}\right\} \text{ there are } k - 1 \text{ of these}$$

Restricted model:

$$\mathbf{Y} = a_1\mathbf{U}_1 + b\mathbf{X}_1 + a_2\mathbf{U}_2 + b\mathbf{X}_2 + a_3\mathbf{U}_3 + b\mathbf{X}_3 + \cdots + a_k\mathbf{U}_k + b\mathbf{X}_k + \mathbf{E}_R$$

(4.58)

$$= a_1\mathbf{U}_1 + a_2\mathbf{U}_2 + a_3\mathbf{U}_3 + \cdots + a_k\mathbf{U}_k + b(\mathbf{X}_1 + \mathbf{X}_2 + \mathbf{X}_3 + \cdots + \mathbf{X}_k) + \mathbf{E}_R$$

$$\mathbf{Y} = a_1\mathbf{U}_1 + a_2\mathbf{U}_2 + a_3\mathbf{U}_3 + \cdots + a_k\mathbf{U}_k + b\mathbf{X} + \mathbf{E}_R \tag{4.59}$$

where $\mathbf{X} = \mathbf{X}_1 + \mathbf{X}_2 + \mathbf{X}_3 + \cdots + \mathbf{X}_k$, that is, x_i is the value on the variable X regardless of group.

$$F = \frac{(R_F{}^2 - R_R{}^2)/\mathrm{df}_1}{(1 - R_F{}^2)/\mathrm{df}_2}$$

where $\mathrm{df}_1 = k - 1$, as $k - 1$ hypotheses are imposed, and $\mathrm{df}_2 = N - 2k$, as there are $2k$ unknown weights in the full model (one slope and one intercept for each of the k variables).

We have shown that where the dependent variable is common to the full and restricted models (as is the case here),

$$\frac{(R_F{}^2 - R_R{}^2)/\mathrm{df}_1}{(1 - R_F{}^2)/\mathrm{df}_2} = \frac{(\sum e_{i_R}{}^2 - \sum e_{i_F}{}^2)/\mathrm{df}_1}{\sum e_{i_F}{}^2/\mathrm{df}_2}$$

which is our F test.

$\sum e_{i_F}{}^2$ is the error SS for our full model, which allows interaction, as the slopes are free to vary. It is also what is traditionally termed *error variation* or *within-group variation* around the regression line. $\sum e_{i_R}{}^2$ is the error SS for our restricted model, which allows no interaction, as the slopes are forced to be equal. Therefore the difference in these two error SS is the amount of variation attributable to (explained by) interaction. Thus our F test becomes

$$F = \frac{(\text{variation explained by interaction})/(k - 1)}{(\text{error variation})/(N - 2k)} \tag{4.60}$$

which is the traditional test (see Blalock, 1972, p. 487).

4.19 NONSTANDARD QUESTIONS: COVARIANCE MODELS WITH INTERACTION

Where interaction is found, traditional analyses stop. However, there are some interesting questions which may be asked in theoretically grounded analyses. Where we are not dealing with a binary categorical variable, we may be able to isolate the interaction to a few categories and proceed to analyze the data for groups between which there is no interaction. We can determine whether this is reasonable by imposing only part of the no-interaction hypotheses delimited in the general case above.

Where we have a really elegant theory, we may be able to specify a priori the differences in slopes. Test of such a theory only requires imposing such differences as linear conditions.

Finally, we should reiterate that interaction findings are often the most substantively interesting results of an analysis. They require explanation, generating new theory. Often contradictions in social science literature can be resolved by noting that interaction exists and that one researcher had a preponderance of people in one category and his critic a preponderance of people in another.

Where no (or minimal) evidence of interaction is found, we can proceed to ask questions about the general overall effects of the independent variables. These questions will be of the form: On the average, does age (party) significantly affect liberalism? Just how we perform the analysis will differ for the interval and categorical variable. Our discussion of the two must diverge at this point.

4.20 EFFECTS OF THE INTERVAL VARIABLE CONTROLLING FOR THE NOMINAL VARIABLE

Recall that the no-interaction hypothesis $b_1 = b_2 (= b)$ could not be rejected. This means that our restricted model produced approximately as much error as our full model. As the restricted model is simpler, we can elect to utilize it as a full model for further hypothesis testing.[1] We can then ask whether the interval variable, here age, significantly affects the dependent variable, here liberalism. This is outlined as

Full model: $\quad\quad\quad\quad\quad\quad\quad \mathbf{Y} = a_1\mathbf{U}_d + a_2\mathbf{U}_r + b\mathbf{X} + \mathbf{E}_F$

Hypothesis: $\quad\quad\quad\quad\quad\quad\quad b = 0$

Restricted model: $\quad\quad\quad\quad\quad \mathbf{Y} = a_1\mathbf{U}_d + a_2\mathbf{U}_r + \mathbf{E}_R$

$$F = \frac{(R_F{}^2 - R_R{}^2)/\mathrm{df}_1}{(1 - R_F{}^2)/\mathrm{df}_2}$$

where $\mathrm{df}_1 = 1$, as one hypothesis is imposed, and $\mathrm{df}_2 = N - 3$, as there are three unknown weights in the full model.

---

[1] To do so pools the within-category error SS with the interaction SS, as no interaction is now assumed by the model. The resulting index is equivalent to the average within-class correlation.

In the general case, we have the following models:

Full model: $$\mathbf{Y} = a_1 \mathbf{U}_1 + a_2 \mathbf{U}_2 + \cdots + a_k \mathbf{U}_k + b\mathbf{X} + \mathbf{E}_F \qquad (4.61)$$

Hypothesis: $\qquad\qquad\qquad b = 0$

Restricted model: $$\mathbf{Y} = a_1 \mathbf{U}_1 + a_2 \mathbf{U}_2 + \cdots + a_k \mathbf{U}_k + \mathbf{E}_R \qquad\qquad (4.62)$$

F is defined as above, and $df_1 = 1$, as one hypothesis is imposed, and $df_2 = N - (k + 1) = N - k - 1$, as there are $k + 1$ unknown weights in the full model (one partial unit vector for each of the k variables and one vector for the pooled interval variable).

This is equivalent to the traditional test for the *significance of the average within-class correlation* (see Blalock, 1972, p. 490). The magnitude of the effect of X controlling for the means is

$$R^2_{YX \cdot U_d U_r} = \frac{R_F^2 - R_R^2}{1 - R_R^2} \qquad (4.63)$$

This quantity is the proportion of variation (left unexplained by the categorical variable) which is explained by the interval variable (using the weighted-average slope). Its *square root is the average within-class correlation.*

4.21   EFFECTS OF THE NOMINAL VARIABLE CONTROLLING FOR THE INTERVAL VARIABLE

Where we have found no interaction, we have seen that we may reasonably pool the X's and use a full model involving the average within-class slope:

Full model: $$\mathbf{Y} = a_1 \mathbf{U}_d + a_2 \mathbf{U}_r + b\mathbf{X} + \mathbf{E}_F$$

This produces two parallel lines, one for Democrats and one for Republicans. The difference in a_1 and a_2 (the intercepts) is the same as the difference in any two points having the same X value because the lines are parallel. Thus, $a_1 - a_2$ is interpretable as the expected difference in liberalism for Republicans and Democrats *of the same age.* By imposing the condition that $a_1 = a_2$ we may then ask: What are the effects of party on liberalism controlling for age?

Hypothesis: $\qquad a_1 = a_2 \, (= a)$

Restricted model: $\mathbf{Y} = a\mathbf{U}_d + a\mathbf{U}_r + b\mathbf{X}_r + \mathbf{E}_R \qquad$ and $\qquad \mathbf{U} = \mathbf{U}_d + \mathbf{U}_r$

$$\mathbf{Y} = a\mathbf{U} + b\mathbf{X} + \mathbf{E}_R$$

$$F = \frac{(R_F^2 - R_R^2)/df_1}{(1 - R_F^2)/df_2}$$

where $df_1 = 1$, as one hypothesis is imposed, and $df_2 = N - 3$, as there are three unknown weights in the full model.

In the general case the analysis becomes

Full model:
$$Y = a_1U_1 + a_2U_2 + \cdots + a_kU_k + bX + E_F \qquad (4.64)$$

Hypotheses:

$$a_1 = a_2 (= a), \qquad a_1 = a_3 (= a), \qquad \ldots, \qquad a_1 = a_k (= a)$$

Restricted model:

$$Y = aU_1 + aU_2 + \cdots + aU_k + bX + E_R = a(U_1 + U_2 + \cdots + U_k) + bX + E_R$$
$$(4.65)$$

$$U = U_1 + U_2 + \cdots + U_k$$

$$Y = aU + bX + E_R$$

$$F = \frac{(R_F^2 - R_R^2)/df_1}{(1 - R_F^2)/df_2}$$

where $df_1 = k - 1$, as there are $k - 1$ hypotheses imposed, and $df_2 = N - (k + 1) = N - k - 1$, as the full model has $k + 1$ unknown weights (one for each partial unit vector and one for the interval variable). This is equivalent to the traditional test for the *significance of differences among adjusted means* (see Blalock, 1972, p. 496).

4.22   MULTIVARIATE ANALYSIS OF COVARIANCE

There is no reason why within- and between-category regression analyses should be restricted to models involving first-degree equations as they are in traditional treatments of analysis of covariance. It is far too simple to treat attitudinal constructs like liberalism as being caused by a single independent variable like age. We shall develop an extension of the above covariance analysis in which the principal status characteristics (education, income, occupational prestige) and age are allowed to account for variation in liberalism. Of course, such a model produces equations of the form

$$Y = aU + b_1X_1 + b_2X_2 + b_3X_3 + b_4X_4 + E$$

However, we wish to allow for the possibility that these variables operate differently for Democrats and Republicans. This can be achieved with the following model:

Full model:

$$Y = a_1U_d + b_1X_{d1} + b_2X_{d2} + b_3X_{d3} + b_4X_{d4} + a_2U_r + b_5X_{r1}$$
$$+ b_6X_{r2} + b_7X_{r3} + b_8X_{r4} + E_F \qquad (4.66)$$

where
$$U_d = 1 \text{ for Democrats, } 0 \text{ otherwise}$$
$$X_{d1} = \text{age for Democrats, } 0 \text{ otherwise}$$
$$X_{d2} = \text{education for Democrats, } 0 \text{ otherwise}$$
$$X_{d3} = \text{occupation for Democrats, } 0 \text{ otherwise}$$
$$X_{d4} = \text{income for Democrats, } 0 \text{ otherwise}$$
$$U_r = 1 \text{ for Republicans, } 0 \text{ otherwise}$$
$$X_{r1} = \text{age for Republicans, } 0 \text{ otherwise}$$

X_{r2} = education for Republicans, 0 otherwise
X_{r3} = occupation for Republicans, 0 otherwise
X_{r4} = income for Republicans, 0 otherwise
E_F = the error vector
a_1, b_1, b_2, b_3, b_4 = parameters (least-squares weights) for Democrats
a_2, b_5, b_6, b_7, b_8 = parameters (least-squares weights) for Republicans

A general case of no interaction would imply that the *forms* of these multiple-regression equations would be the same, i.e., all paired weights (except the intercepts, a_1 and a_2) should be equal:

Hypotheses:

$$b_1 = b_5 (= c_1) \qquad b_2 = b_6 (= c_2) \qquad b_3 = b_7 (= c_3) \qquad b_4 = b_8 (= c_4)$$

Restricted model:
Substituting and collecting terms gives

$$Y = a_1 U_d + a_2 U_r + c_1 X_1 + c_2 X_2 + c_3 X_3 + c_4 X_4 + E_R \qquad (4.67)$$

where $X_1 = X_{d1} + X_{r1}$ = age for all cases
$X_2 = X_{d2} + X_{r2}$ = education for all cases
$X_3 = X_{d3} + X_{r3}$ = occupation for all cases
$X_4 = X_{d4} + X_{r4}$ = income for all cases
The relevant computer results are:

Full model	Restricted model
$a_1 = 63.6125$	$a_1 = 82.5837$
$b_1 = -.7079$	$a_2 = 88.6133$
$b_2 = 2.7278$	$c_1 = -.6413$
$b_3 = -.3253$	$c_2 = 1.7382$
$b_4 = .0006$	$c_3 = -.2040$
$a_2 = 114.9309$	$c_4 = -.0010$
$b_5 = -.5269$	
$b_6 = .5723$	
$b_7 = -.3177$	
$b_8 = -.0012$	
$\sum e_i^2 = 22{,}130$	$\sum e_i^2 = 30{,}860$
$R^2 = .5983$	$R^2 = .4398$

$$F = \frac{(R_F{}^2 - R_R{}^2)/df_1}{(1 - R_F{}^2)/df_2} = \frac{(.5983 - .4398)/(10 - 6)}{(1 - .5983)/(200 - 10)} \approx 18.7$$

We must reject the no-interaction hypotheses at the .001 level of significance. The variables are not related in the same ways for Democrats and Republicans. We cannot now pursue the traditional main-effect analogs as they assume no interaction. Nevertheless, we can ask many interesting substantive questions of the full model. For example, we can ask: Does each status variable operate to produce liberalism in the same way for members of the two parties in the context of the others?

Answers to such questions can be provided by *serially* imposing the condition that b's referring to the same variable are equivalent rather than imposing these all at a time. We shall not outline the details of these as they only involve the direct substitution of algebraic equivalences in our full model. Results of such a series of analyses[1] indicate that:

1 Increasing age decreases liberalism for both Democrats and Republicans, but to a greater degree for Democrats.
2 Increasing education increases liberalism for both Democrats and Republicans but to a greater degree for Democrats.
3 Increasing occupational prestige decreases liberalism for both Democrats and Republicans but more so for Republicans than Democrats. (Note that this is *not* inconsistent with the small and reversed order of b_3 and b_7 above. In the general analysis testing for interaction, b_3 and b_7 are simple, uncontrolled weights. The conclusions drawn from the serial imposition of equivalence between b's referring to the same variable assess the similarity of the independent effect of each variable controlling for all others.)
4 Increasing income decreases liberalism for both Democrats and Republicans but much more so for Republicans. Note that the small *unstandardized* b's (b_4 and b_8) above cannot be compared directly with the others as they are associated with income which has a very large range (0 to 40,000) compared with the range of the variable liberalism (0 to 100) or the ranges of the other independent variables: 0 to 100 for age, 0 to 20 for education, 0 to 100 for occupational prestige.

4.23   SUMMARY AND CONCLUDING REMARKS

We have shown that the variance, regression, and covariance analytic schemes need not be treated as distinct traditions with different formulas, questions, and significance tests. Instead all traditional questions (and others) in the three schemes can be reduced to special cases of multiple linear regression. All these questions involve the algebraic imposition of linear conditions (consistent with interesting substantive questions) on models generating the expected values for dependent variables. Viewed in this way, the single statistical distribution $F = s_H^2/s_A^2$ is sufficient for all tests (where s_H^2 is the error variance associated with our hypotheses and s_A^2 is the error variance associated with our assumptions). From this perspective, the definition of degrees of freedom is no longer a given, to be memorized, but is reduced to a counting operation with an appropriate and obvious theoretical rationale.

The differences in the variance, regression, and covariance traditions are simply that (1) different levels of measurement are assumed at certain points and (2) that the expected values are produced by slightly different assumptions, specifically:

1 In analysis of variance, full-model expected values are points which are *within-category* means.

---

[1] Readers are cautioned that these analyses are based on the *simulated data* in Appendix B.

2 In traditional regression, full-model expected values are generated by lines (in simple cases) and planes or spaces (in complex cases).

3 In analysis of covariance, full-model expected values are generated *within categories* by lines (planes or spaces).

In all cases, the full-model expected values are produced by least-squares weights for systems of simultaneous linear equations which embody the assumptions we are willing to make (without test) at any point. Restricted-model expected values are produced by least-squares weights for systems of simultaneous linear equations which *additionally* embody the hypotheses we seek to test (these having been imposed on our full model to produce our restricted model).

This perspective makes possible the statement of general cases of many traditional tests which are extremely awkward to develop within traditional frameworks, e.g., two-way analysis of variance with unequal subcell *n*'s, interaction, linearity tests, tests of degree, *k*th-order partials, and *k*th-order multiple partials. Finally, the perspective frees the researcher from the constraints of (possibly inappropriate) traditional tests and allows one to formulate many theoretically generated hypotheses for which there are no conventional computing formulas, e.g., replication or generalization questions, partial-interaction questions, partial-linearity questions, controlling questions where interaction is found, and *any* other questions which can be translated into linear conditions. From this point of view substantive theory generates the questions, and least-squares theory generates the answer. Too often, with specialized computing formulas, least-squares theory generates both the questions and the answers.

EXERCISES

All exercises, below, refer to data in Appendix B. You will find it convenient to use the standard READ *and* FORMAT *statements given at the end of that appendix. Then the necessary vectors can be created by referencing the variables by their subscripts:* VAR(1), VAR(2), *etc.*

4.1 Test the hypothesis that mother's years of education as the independent variable *X* is linearly related to father's years of education as the dependent variable *Y*. Mother's education VAR(18) ranges from 5 through 19 years with no missing intermediate values. Father's education is VAR(13).

(*a*) Formulate an appropriate full model, hypothesis, and restricted model.

(*b*) Solve the two models with program CARTROCK.

(*c*) Compute and interpret the *F* ratio.

(*d*) Interpret the parameters of the models.

(*e*) Graph the path of means.

4.2 Determine whether the husband's education VAR(4) and income VAR(6) are better related by a first, second, third, or fourth-degree equation.

(*a*) Formulate an appropriate collection of full models, hypotheses, and restricted models.

(*b*) Solve the models with program CARTROCK.

(*c*) Compute and interpret the *F* ratios.

(*d*) Summarize your findings.

4.3   Test the hypothesis that husband's education has no effect on husband's income, i.e., a linear model using education as the independent variable accounts for none of the variation in income.

(a) Formulate an appropriate full model, hypothesis, and restricted model.

(b) Solve the models with program CARTROCK.

(c) Compute and interpret the F ratio.

(d) Interpret your findings, here. Then compare your conclusions with those from Exercise 4.2.

4.4   Test the following hypotheses: (1) that husband's liberalism is unrelated to wife's liberalism and (2) that husbands and wives have the *same* liberalism.

(a) Formulate appropriate full models, hypotheses, and restricted models.

(b) Solve all models with program CARTROCK.

(c) Compute and interpret the F ratios.

(d) Summarize your conclusions.

4.5   Evaluate the hypothesis that the wives' status characteristics (age, education) account for none of the variation in their liberalism.

(a) Formulate the appropriate full model, hypotheses, and restricted model.

(b) Solve the models using program CARTROCK.

(c) Compute and interpret the F ratio.

4.6   Evaluate the hypothesis that the father's education and occupation interact significantly (in a multiplicative fashion) to produce the father's liberalism. See whether the interaction adds anything to what we know from the statuses *and* whether the statuses add anything to what we know from the interaction.

(a) Formulate appropriate full models, hypotheses, and restricted models.

(b) Solve the models using program CARTROCK.

(c) Compute and interpret the F ratios.

(d) Summarize your findings.

4.7   Evaluate the independent effects of the wives' statuses (age, education) and the husbands' statuses (age, education, occupation, income) on the wives' liberalism; i.e., find the magnitude and significance of the effect of each cluster on liberalism, controlling for the other cluster.

(a) Formulate appropriate full models, hypotheses, and restricted models.

(b) Solve the models using program CARTROCK.

(c) Compute and interpret the F ratio.

(d) Summarize your findings.

4.8   Evaluate the hypothesis that the husbands' age *and* education add nothing to the prediction of the wives' liberalism over what we know from the wives' age and education and the husbands' occupation and income; i.e., find the magnitude and significance of the effect of husbands' age and education controlling for wives' age and education and husbands' occupation and income.

(a) Formulate an appropriate full model, hypotheses, and restricted model.

(b) Solve the models using program CARTROCK.

(c) Compute and interpret the F ratio.

(d) Summarize your findings.

4.9   Evaluate your findings in Exercises 4.7 and 4.8 and develop some substantively interesting extension of those analyses.

4.10 Consider the relationship between education and liberalism for Republicans and Democrats.

(a) Ask whether education affects liberalism in the same way for Republicans and Democrats.

(1) Formulate an appropriate full model, hypothesis, and restricted model.

(2) Solve the models using program CARTROCK.

(3) Compute and interpret the F ratio.

(4) Explicate the nature of the interaction (if any).

(*b*) Assess the effects of education on liberalism controlling for party.

(*c*) Assess the effects of party on liberalism controlling for education.

(*d*) Summarize your conclusions from parts (*a*) to (*c*).

4.11 Take a model in which husband's liberalism is predicted (separately for Democrats and Republicans) from

(*a*) *Husband's* age, education, occupation, income

(*b*) *Father's* education, occupation, income

(*c*) *Wife's* age, education, liberalism

(*d*) *Mother's* education

Formulate and test three substantively interesting questions to be answered by the imposition of linear conditions on such a model.

PART TWO

Experimental Designs

5

ONE-FACTOR EXPERIMENTS

It is well known that the bulk of the work on experimental design was contributed two to four decades ago by statisticians who were confronted by design problems in agricultural and biological sciences. It is perhaps less well known that more recently in diverse fields such as political sociology, demography, medical sociology, and industrial sociology, not to speak of small-group studies, experimental designs have found applications. It is also often overlooked that in analyzing data from non-experimental studies, we invariably superimpose experimental models of one kind or another on the concrete data set at hand.

The focus of attention in Part Two is the relatively simpler designs and the associated statistical models. The choice of the designs for exposition in Part Two was guided by the consideration that they must have applicability in social sciences, either in experimental or quasi-experimental studies or in data analysis in nonexperimental research. Most designs expounded herein can be described as single-response designs, i.e., those in which there is only one dependent variable measured on each subject (experimental unit). The analysis of covariance and repeated-measures designs are the exceptions. In the analysis of covariance, one response variable and one or more concomitant variables are measured on each subject. The response variable and the concomitant variables are interrelated, but if we make the assumption that only the response variable is affected by the treatments, the situation becomes very similar to the single-response models. Situations in which the concomitant variables are affected by the treatments are not given detailed attention in Part Two.

However, models are presented in Part Three which would help handle such situations. The repeated-measures designs form another category in which multiple-interdependent responses are present. In these designs the same subject receives two or more treatments in succession and yields one observation under each treatment. The analysis therefore has to be multivariate in structure. In the models presented in Part Two, however, highly restrictive assumptions are made about the interdependence between observations on the same subject, thereby making the associated analytical procedure somewhat similar to those employed for single-response designs.

As already indicated, (see Preface), the notation in this part of the book differs from that introduced in Part One and that employed in Part Three. This has been necessitated by a wish to maintain similarity between the notations used herein and those usually found in the literature on experimental designs.

5.1   GENERAL STRUCTURE OF EXPERIMENTAL MODELS

Underlying all the models that we shall be concerned with in this book is the assumption that any phenomenon has a multiplicity of causes and that an observed value y of the phenomenon can be expressed as the sum of the effects on y of its causal factors. We shall not discuss here how to decide whether a given factor is causally connected to another. That question will be given attention in Part Three. Here we assume that that question has been settled.

In some models all the explicitly recognized causal factors are assumed to have fixed effects, while in others, the effects of some, if not all, of them are assumed to be random variables. The meaning of fixed effects and random effects will become clear in a moment. Assuming that such a distinction is worthwhile, we can write the underlying structure of experimental models as follows:

$$\text{Observed value} = \begin{matrix} \text{sum of effects of} \\ \text{accounted for factors} \\ \text{that are assumed to} \\ \text{have fixed effects} \end{matrix} + \begin{matrix} \text{sum of effects of} \\ \text{accounted for factors} \\ \text{that are assumed to} \\ \text{have random effects} \end{matrix} + \begin{matrix} \text{the effects of} \\ \text{other factors} \end{matrix}$$

$$(5.1)$$

An effect is said to be accounted for if the corresponding causal factor is explicitly introduced into the above model. For example, a boy's age, race, socioeconomic status, and a host of other factors may be known to influence the score he gets in an intelligence test. These factors can be explicitly introduced into the model by representing each of them by a set of parameters or random variables as the case may be. To represent race, for example, two parameters may be employed, β_1 for whites and β_2 for nonwhites. If the subject is white, the equation representing his intelligence score contains the parameter β_1, and if he is nonwhite, the parameter β_2 is used instead. In either case the parameter represents the effect of the subject's race on the score he gets in the test. Now, in most practical situations, the analyst may not be able to identify and exhaustively enumerate the factors affecting the phenomenon taken for study. If there is a causal factor the analyst is unable to identify, its effect is assumed to be part of the residual component in (5.1). There is, however, another set of effects that forms part of the residual. That set consists of the effects of constant factors, whether deliberately held constant by the experimenter or remaining constant without his knowledge. To facilitate further discussion of model (5.1) and the assumptions underlying it, let us consider a hypothetical example.

Example A researcher hypothesized that, other things remaining the same, the more irregular the shape of a line, the longer it would appear to the human eye. To test this hypothesis, let us suppose he conducted the following experiment. Three lines of the same length, but differing in irregularity, were drawn and marked *A*, *B*, and *C*. The students in a class were randomly divided into three groups, and group 1 was presented with line *A*, group 2, with line *B*, and group 3, with line *C*. Each student was asked to estimate by visual inspection the length of the line he was presented with.

Suppose the study yielded the data shown in Table 5.1. We note that only one factor was explicitly manipulated here, namely, the shape of the line. Hence, the appropriate model will be of the following form:

$$
\begin{array}{l}
\text{Observed value, i.e., estimate} \\
\text{given by a subject}
\end{array}
=
\begin{array}{l}
\text{term representing effect of} \\
\text{appropriate treatment,} \\
\text{i.e., shape of line}
\end{array}
+
\begin{array}{l}
\text{combined effect of} \\
\text{all factors other} \\
\text{than treatment}
\end{array}
$$

$$(5.2)$$

For the first quantity on the right-hand side of (5.2), we have a choice of using a fixed number (parameter) or a random variable. It is appropriate to represent it by a random variable if we consider the three shapes used in the experiment (lines *A*, *B*, and *C*) as a random sample from a collection of all possible shapes and if the aim of the experiment is to make inferences concerning the variance of the effects of the different shapes included in that collection. On the other hand, if we consider the three shapes as three specific shapes, then the appropriate representation will be by means of a fixed quantity. We shall use Greek letters with appropriate suffixes to denote the effects of treatments if we decide to consider them as fixed quantities, and lowercase Latin letters, with appropriate suffixes, if our decision is to view them as random variables.

Table 5.1 STUDENTS' ESTIMATES (BY EYE INSPECTION) OF THE LENGTHS OF THREE LINES IN UNITS

Line *A*	Line *B*	Line *C*
4.5	3.8	3.6
4.8	4.6	4.0
5.0	4.5	4.4
5.8	4.2	4.2
5.2	4.0	3.8
4.6	4.8	4.6
	3.6	

The second term on the right-hand side of (5.2) represents the combined effect of all factors other than the treatment. It is useful to note that some of these factors may remain constant during the experiment. Several of these might have been deliberately held constant by the experimenter, e.g., the subject's age. But there may be other constant factors whose operation is unknown to the experimenter. The combined effect of all constant factors may be designated by a constant, say μ. What remains in the second term on the right-hand side of (5.2) constitutes the combined effect of all variable but unaccounted for factors. The particular configuration of these factors is not likely to be the same for all observations, even for those under the same treatment, since the experimenter exerts no control over them. It is appropriate therefore to designate this variable but unaccounted for effect by a symbol that has two suffixes, for example, ε_{ij}, the second suffix, j, representing the treatment under which the observation occurs and the first one, i, identifying the observation within those under the same treatment. A convenient notation for the corresponding observed value is y_{ij}, the suffixes having the same meaning as described above for ε_{ij}.

We thus have two ways of specifying the model (5.2):

$$y_{ij} = \mu + \beta_j + \varepsilon_{ij} \qquad (5.3)$$

and

$$y_{ij} = \mu + b_j + \varepsilon_{ij} \qquad (5.4)$$

Model (5.3), in which we represent treatment effects by means of parameters, is referred to as model I in the literature, and (5.4), in which treatment effects are considered as random variables, as model II.

The question immediately arises which model to use in a specific situation. Obviously, we should use the one that represents the situation more closely. The choice should also depend on the aim of the investigation. Model I is appropriate if our aim is to test hypotheses concerning specific differences between the treatment effects, e.g., all treatment effects are equal. If, however, the aim is to test hypotheses concerning the variability among a large number of possible treatment effects, then model II is the appropriate one. We shall examine now how to apply each of these models in specific situations.

Completely Randomized Design

The simplest design we shall discuss is the one known as the completely randomized design. The experiment described [in the above example provides an illustration. Another example follows.

Example Twenty-four subjects, all with the same amount of schooling and of the same age level, were divided randomly into four groups of six each. The groups were then assigned to four different rooms. Each subject was then given a fixed set of arithmetic problems to solve. No consultation was allowed. The rooms varied in the intensity of distracting noises. The research question was whether the intensity of distracting noises influenced the performance of the subjects in solving the problems. The scores received by the subjects are shown in Table 5.2. Suppose that we wish to test the hypothesis that these particular four treatment effects are not different from one another. We shall use model I. Accordingly, we shall represent the observation by

$$y_{ij} = \mu + \beta_j + \varepsilon_{ij} \qquad (5.5)$$

in which β_1 represents the effect of treatment 1, β_2 that of treatment 2, and so on. If the treatment effects are all equal, however, the above equation becomes

$$y_{ij} = \mu + \beta_0 + \varepsilon_{ij} \qquad (5.6)$$

where $\beta_0 = \beta_1 = \beta_2 = \beta_3 = \beta_4$. In other words, Eq. (5.6) is the appropriate representation of the observations under the null hypothesis that the treatment effects are all equal, while Eq. (5.5) is the appropriate one if the null hypothesis is not true. We want to decide which of the above two representations is closer[1] to the real data. The answer to that question will then tell us whether the null hypothesis is true or not. But how do we infer from the available data which of the two representations given above is closer to the actual situation? This is done by means of the analysis of variance, which compares the consequences, as reflected in the data, of the representation in (5.6) with the corresponding consequences of representation (5.5).

We shall now outline the steps in the analysis of variance.

5.2 THE ANALYSIS OF VARIANCE

Assumptions

First we make the following assumptions. Why we need them will become clear later on.

---

[1] One may take the position that model (5.5) is always closer to reality, since strict equality seldom holds. From that point of view the analyst's position will be that of preferring the simpler model, namely (5.6), if departures of β_j's from equality are negligible from a practical point of view.

Table 5.2 SCORES RECEIVED BY THE SUBJECTS IN A PSYCHOLOGICAL EXPERIMENT: AN EXAMPLE OF AN EXPERIMENT USING A COMPLETELY RANDOMIZED DESIGN

	Treatments (noise level)				
	1	2	3	4	Total
	2	3	6	9	
	1	4	5	8	
	1	2	7	9	
	3	5	8	8	
	6	4	7	7	
Sum	13	18	33	41	105
Mean	2.6	3.6	6.6	8.2	5.25
Number of observations	5	5	5	5	20

ASSUMPTION 1   The ε_{ij}'s, that is, the variable but unaccounted for components of the observed values, are uncorrelated both within each treatment group and across all treatment groups. To understand what this means, imagine that you are able to repeat the experiment described above an indefinite number of times under the same conditions. From each repetition, suppose you single out two *observed values*, say the first and the second ones under treatment 1. You will have thus before your mind's eye an infinite number of pairs of values. The first member of each pair represents the first observation under treatment 1 and the other, the second observation under treatment 1, in a particular repetition of the experiment. Imagine now that you calculate the product-moment correlation from that set of paired values. The assumption we have described above states that the correlation you get should be zero. (You may have noted that the correlation between the observed values in those pairs is the same as that between the corresponding ε_{ij}'s since ε_{ij}'s differ in each case from the observation by the same constant, $\mu + \beta_1$.) The assumption that the ε_{ij}'s across treatment groups should be uncorrelated can be understood in a similar manner. Note that in the context of repeated experiments we are viewing each y_{ij} as a single realization of a random variable. From this point of view the observed y values are seen as one observation each on as many random variables as there are observations. This perspective permits us to associate with each y value properties such as sampling distribution, mean, and variance and with each pair of y values properties such as covariance and correlation.

ASSUMPTION 2   The ε_{ij}'s are normally distributed. Again, to understand what this means, imagine repeating the experiment as described above, and note that if the random variable y_{11}, the first observation under treatment 1, has a sampling distribution that coincides with the theoretical normal probability distribution, so will the corresponding random variable ε_{11} because, as already pointed out, in repeated experiments the first observation under treatment 1 differs only by a constant, $\mu + \beta_1$, from the corresponding ε.

ASSUMPTION 3   The probability distribution of each ε_{ij} has mean, i.e., expectation, zero.

ASSUMPTION 4   The probability distributions of all ε_{ij}'s have the same variance.
    Note that assumptions 3 and 4 simply specify explicitly certain characteristic properties of the probability distribution described in assumption 2.
    Let us express the above assumptions in shorthand notation to facilitate future reference.

$$\Omega = \begin{cases} 1 & \text{the } \varepsilon_{ij}\text{'s are uncorrelated} \\ 2 & \text{each } \varepsilon_{ij} \text{ is normally distributed} \\ 3 & E(\varepsilon_{ij}) = 0 \text{ for all } i \text{ and } j \\ 4 & \text{var } \varepsilon_{ij} = \sigma^2 \text{ for all } i \text{ and } j \end{cases}$$

We shall hereafter refer to these assumptions by the symbol Ω and sometimes by the numbers attached to each.

Sums of Squares

We are now ready to examine the consequences of representations (5.5) and (5.6). Let us consider the latter first. In our discussion, we shall take for granted the Ω assumptions. Later we shall see that some of these assumptions are not needed for some purposes.

Now, if Eq. (5.6) is the appropriate representation, i.e., if the treatment effects are equal, we can consider the entire set of values in Table 5.2 as constituting a random sample from a normal population with mean $\mu + \beta_0$ and standard deviation σ. (Note that if ε_{ij} is a random observation from a normal distribution with mean 0 and variance σ^2, then $\mu + \beta_0 + \varepsilon_{ij}$ is a random observation from a normal population with mean $\mu + \beta_0$ and variance σ^2.) This permits us to estimate σ^2 in three different ways, all of which should yield estimates of more or less the same magnitude if the Ω assumptions and model (5.6) are valid.

METHOD 1   Since we are assuming that all 20 observations have come from the same population, we can compute the sum of squared deviations of all the observations from the grand mean and from that sum of squares (SS) obtain an estimate of σ^2. The SS of deviations from the grand mean is

$$2^2 + 1^2 + 1^2 + \cdots + 7^2 - \frac{105^2}{20} = 683 - 551.25 = 131.75$$

which is the same as $(2 - \frac{105}{20})^2 + \cdots + (7 - \frac{105}{20})^2$. This is known as the *total sum of squares* (*total SS*) in the literature on the analysis of variance. To understand the properties of the total SS, imagine that you are able to repeat the experiment an indefinite number of times under constant conditions. Suppose also that the Ω assumptions hold true and that you know the value of σ^2. In each repetition of the experiment imagine that you calculate the total SS and divide it by the value of σ^2, which we have assumed is known to you. If you form the relative frequency distribution of the ratios thus obtained, that distribution will be exactly the same as a theoretical probability distribution known as the chi-square distribution with 19 degrees of freedom (1 less than the total number of observations in the experiment). In other words, the sampling distribution of (total SS)/σ^2, in the present case, is a chi square with 19 degrees of freedom. The average of a variate with a chi-square distribution is equal to the degrees of freedom of the corresponding chi-square distribution. Hence, using the notation $E(u)$ for the average of the sampling distribution of u, we have in the present case $E[(\text{total SS})/\sigma^2] = 19$. Under the Ω assumptions therefore, we can associate with the total SS a number known as its degree of freedom. The above result is true irrespective of the particular value of σ^2 and irrespective of the total number of observations in the experiment. Hence we may write in general

$$E\left(\frac{\text{total SS}}{\sigma^2}\right) = \text{df of total SS}$$

provided, of course, that the Ω assumptions hold true. For a more complete exposition of these points see Sec. 5.3.

From the above general result you must have guessed that if σ^2 is not known to you, you may take the ratio

$$\frac{\text{Total SS}}{\text{df of total SS}}$$

as an estimate of σ^2. Your guess is correct for

$$E\left(\frac{\text{total SS}}{\text{df of total SS}}\right) = \sigma^2$$

Any SS, in the analysis of variance, when divided by the corresponding degrees of freedom is known as the mean sum of squares (MSS). The total MSS is our first estimate of σ^2.

METHOD 2   For the second estimate we note that if the entire set of 20 observations can be considered as a random sample from a common population, the observations under each treatment group can also be considered the same way. We can thus get from each treatment group an estimate of σ^2 and construct a pooled estimate from them. By the same logic as the one applied above for the total SS, we get from the observations under treatment group 1

$$2^2 + 1^2 + \cdots + 6^2 - \frac{13^2}{5} = 51 - 33.8 = 17.2$$

a sum of squares with 4 degrees of freedom giving 17.2/4 as an estimate of σ^2. Similar estimates can be obtained from the other groups. Pooling them, we get an SS with 16 degrees of freedom:

$$\left(2^2 + 1^2 + \cdots + 6^2 - \frac{13^2}{5}\right)$$
$$+ \left(3^2 + 4^2 + \cdots + 4^2 - \frac{18^2}{5}\right)$$
$$+ \left(6^2 + 5^2 + \cdots + 7^2 - \frac{33^2}{5}\right)$$
$$+ \left(9^2 + 8^2 + \cdots + 7^2 - \frac{41^2}{5}\right)$$
$$= 30.40$$

When this SS is divided by the corresponding degrees of freedom (16), we have 1.90 as our second estimate of σ^2. This pooled SS is known as the *within-treatments sum of squares* (within SS), and the corresponding MSS is referred to as *within MSS*.

METHOD 3   For the third estimate, we consider the means under the four treatment groups: 2.6, 3.6, 6.6, and 8.2. If the individual observations have come as a random sample from a normal population with mean $\mu + \beta_0$ and variance σ^2, these means can be considered as a random sample from a normal population with mean $\mu + \beta_0$

and variance $\sigma^2/5$, since they are means of five independent observations. The SS of their deviations from their mean is

$$2.6^2 + 3.6^2 + 6.6^2 + 8.2^2 - \frac{21.00^2}{4} = 20.27$$

with 3 degrees of freedom. Hence $20.27/3 = 6.76$ is an estimate of $\sigma^2/5$, from which to get an estimate of σ^2 all we need do is to multiply by 5. We can do the whole calculation more directly by taking

$$5\left(2.6^2 + 3.6^2 + 6.6^2 + 8.2^2 - \frac{21.00^2}{4}\right)$$

which is equivalent to the following computationally simpler form:

$$\frac{13^2}{5} + \frac{18^2}{5} + \frac{33^2}{5} + \frac{41^2}{5} - \frac{105^2}{20} = 101.35$$

This is known as the *between-treatments sum of squares* (between SS). The corresponding MSS ($= 101.35/3$) is referred to as *between MSS*, which is our third estimate of σ^2.

The computations described so far can be carried out in the following simple steps:

1 Arrange the y values by subclasses (treatment groups).
2 Calculate the subclass (treatment group) totals and their grand total.
3 Square each subclass total and divide the squared value by the number of cases in the corresponding subclass.
4 Square the grand total and divide the squared value by the total number of observations, i.e., the number of cases in all subclasses together. The resulting quantity is referred to in the literature as the *correction factor*, *correction for the mean*, or the *correction term*.
5 Subtract the correction term from the sum of the quantities obtained in step 3. The resulting figure is the between SS. Its degree of freedom is 1 less than the number of subclasses (treatment groups).
6 Square each individual observation, sum the squares, and subtract the correction term from the sum. This gives the total SS. Its degree of freedom is 1 less than the total number of observations.
7 Obtain the within SS by subtraction:

$$\text{Total SS} - \text{between SS} = \text{within SS}$$

The corresponding degree of freedom is also similarly obtained by subtraction.

This last step is valid because of the following identity:

$$\sum_i \sum_j (y_{ij} - \text{grand mean})^2 = \sum_i \sum_j (j\text{th subclass mean} - \text{grand mean})^2$$

$$+ \sum_i \sum_j (y_{ij} - j\text{th subclass mean})^2$$

that is, $\text{Total SS} = \text{between SS} + \text{within SS}$

Notice that

$$\sum_i \sum_j (j\text{th subclass mean} - \text{grand mean})^2 = \sum_j r_j (j\text{th subclass mean} - \text{grand mean})^2$$

where r_j represents the number of cases in the jth subclass. This SS is the same as the between SS computed according to the formula described in step 5. Expressing total SS as the sum of between SS and within SS is known as the *decomposition (partitioning) of the total SS*.

5.3 SAMPLING DISTRIBUTION OF THE SUMS OF SQUARES

The Chi-Square Distribution

We mentioned above several sums of squares (total SS, between SS, within SS) and associated each of them with a certain number called the degrees of freedom. In this section we shall expound the logic underlying the assignment of degree of freedom to sums of squares in the analysis of variance. In order to do this we need to know something about a probability distribution known as the chi-square distribution.

The chi-square distribution is a one-parameter distribution confined wholly to the positive side of the horizontal axis. The chi-square curve has the equation

$$f(x) = Cx^{(v-2)/2}e^{-x/2} \qquad x > 0$$

where C is a constant depending on the value of the parameter v, commonly referred to as the degrees of freedom of the chi square. Later we shall discuss the appropriateness of this label for the parameter v. In this book we shall be concerned only with cases in which v is a positive integer. Figure 5.1 shows the shape of the chi-square curve for different values of v.

It is possible to prove, although we shall not attempt to do it here, that the sampling distribution of the square of a standard normal deviate is the chi-square distribution with $v = 1$. This result is often stated in the following simpler manner: the square of a standard normal deviate is a chi square with $v = 1$.

The chi-square distribution possesses an interesting property known as *additivity*. The sum of two independent variates each of which is a chi square, one with $v = n$ and the other with $v = m$, is itself a chi square with $v = n + m$. In particular, the SS of two independent normal deviates is a chi square with $v = 2$; the SS of n independent normal deviates is a chi square with $v = n$; and so on.

Another important property of the chi-square distribution is that the mean of the distribution is equal to v.

Sampling Distribution of $\sum (y - \bar{y})^2$

In most empirical situations associated with the analysis of variance the SS of interest are of the form $\sum (y_i - \bar{y})^2$, where $\bar{y}$ is the mean of the observed values $y_1, y_2, \ldots, y_n$ and the y's are assumed to be drawn at random from a normal universe with mean μ

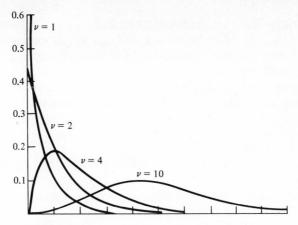

FIGURE 5.1
Chi-square curve for different values of v.

and variance σ^2. It can be shown that $\sum (y_i - \bar{y})^2/\sigma^2$, although it is an SS of n quantities, is a chi square with $v = n - 1$. A direct way of demonstrating this is to prove the following theorem:

> *If $y_1, y_2, \ldots, y_n$ are a random sample from a normal universe with mean μ and variance σ^2, then $\sum (y - \bar{y})^2/\sigma^2$ can be expressed as the SS of $n - 1$ independent standard normal deviates.*

We shall outline below a proof of this important theorem. The reader may find it helpful to work out in full all steps for a particular case such as for $n = 3$. In proving this theorem there is no loss of generality if we assume that the parent universe from which the sample has been drawn is one with mean 0 instead of mean $\mu \neq 0$. If μ is different from 0, all we need do is to replace in the arguments that follow y_1 by $y_1 - \mu$, y_2 by $y_2 - \mu$, and so on. We define n quantities $z_1, z_2, \ldots, z_n$, each a linear combination of the y's:

$$z_1 = \frac{y_1 - y_2}{\sqrt{1^2 + 1^2}}$$

$$z_2 = \frac{y_1 + y_2 - 2y_3}{\sqrt{1^2 + 1^2 + 2^2}}$$

$$\cdots\cdots\cdots\cdots\cdots\cdots\cdots\cdots\cdots\cdots\cdots$$

$$z_{n-1} = \frac{y_1 + \cdots + y_{n-1} - (n - 1)y_n}{\sqrt{\underbrace{1^2 + \cdots + 1^2}_{n-1 \text{ terms}} + (n - 1)^2}}$$

$$z_n = \frac{y_1 + \cdots + y_n}{\sqrt{n}}$$

Note (1) that the sum of the coefficients of y's for each z is zero, (2) that the sum of the product of the coefficients of like terms in z_i and z_j, $i \neq j$, is zero, and (3) that the SS of the coefficients of y's in the numerator for each z is equal to the quantity within the square-root sign in the denominator.

Because of these properties the following results hold true in the framework of repeated sampling, i.e., considering each z_i as a random variable:

1 $z_1, z_2, \ldots, z_n$ each has expectation 0 and variance σ^2.
2 $E(z_i z_j) = 0$ for all $i \neq j$.
3 When the above two results are put together, z_i and z_j, for all $i \neq j$, are uncorrelated.
4 The sampling distribution of each z_i is normal, since each z_i is a linear combination of normally distributed y's.

From this we conclude that $z_1/\sigma, \ldots, z_n/\sigma$ are normally distributed with mean 0 and variance 1. They are also uncorrelated with each other. Given these properties, it can be shown that $z_1/\sigma, \ldots, z_n/\sigma$ are independent of each other (see Rao, 1965, p. 147).

It is not difficult to verify that

$$z_1^2 + z_2^2 + \cdots + z_n^2 = y_1^2 + y_2^2 + \cdots + y_n^2$$

We notice also that

$$z_n^2 = n\bar{y}^2$$

Hence

$$\sum (y_i - \bar{y})^2 = \sum y_i^2 - n\bar{y}^2 = z_1^2 + \cdots + z_n^2 - z_n^2$$
$$= z_1^2 + \cdots + z_{n-1}^2$$

Consequently

$$\frac{\sum (y_i - \bar{y})^2}{\sigma^2} = \frac{z_1^2}{\sigma^2} + \cdots + \frac{z_{n-1}^2}{\sigma^2}$$

Thus $\sum (y_i - \bar{y})/\sigma^2$ is SS of $n - 1$ independent standard normal deviates. Hence it is a chi square with $v = n - 1$.

Having demonstrated that $\sum (y_i - \bar{y})^2/\sigma^2$, an SS of n quantities, is a chi square with $v = n - 1$ rather than one with $v = n$, it is easy to see why the parameter v is referred to as the degree of freedom of the SS in question. What determines the sampling distribution of a given SS is not how many squares are involved in the sum but how many of the terms squared are independent of each other. For $\sum (y - \bar{y})^2/\sigma^2$ only $n - 1$ of the n terms whose squares are involved in the sum are independent of each other or are free to vary independently of each other, since given any $n - 1$ of the n quantities, $(y_i - \bar{y})/\sigma^2$, $i = 1, 2, \ldots, n$, the other is automatically determined by virtue of the identity $\sum (y_i - \bar{y}) = 0$. From this point of view the parameter v that determines the sampling distribution of a given SS may appropriately be called the degrees of freedom of the SS. Obviously if there is no relationship connecting the terms squared and summed, the degrees of freedom of the SS in question will be equal to the number of terms in the SS if each square is a chi square with $v = 1$.

Sampling Distributions of Total SS, Between SS, and Within SS

On the basis of the theorem expounded above it should be clear why we associate with the ratio (total SS)/σ^2, in the analysis of the data in Table 5.2, 19 degrees of freedom. Under the Ω assumptions and model (5.6) the observations in Table 5.2 can be regarded as a random sample from a normal universe with mean $\mu + \beta_0$ and variance σ^2. Consequently, according to the theory expounded above, the ratio (total SS)/σ^2, which is of the form $\sum_i \sum_j (y_{ij} - \bar{y})^2/\sigma^2$, involving altogether 20 terms is a chi square with $20 - 1$ degrees of freedom.

As for the ratio (between SS)/σ^2, we notice that under the Ω assumptions and model (5.6) the subclass means in Table 5.2 can be considered as a random sample of size 4 from a normal universe with mean $\mu + \beta_0$ and variance $\sigma^2/5$. Hence

$$\frac{\sum (\text{subclass mean} - \text{grand mean})^2}{\sigma^2/5}$$

which is the same as

$$\frac{\text{Between SS}}{\sigma^2}$$

is a chi square with 3 degrees of freedom.

Now turning to the ratio (within SS)/σ^2, we notice that under the Ω assumptions and model (5.6) the observations within each subclass can be considered as a random sample of size 5 from a normal universe with mean $\mu + \beta_0$ and variance σ^2. Hence for each subclass

$$\frac{\sum (\text{individual observation} - \text{subclass mean})^2}{\sigma^2}$$

is a chi square with 4 degrees of freedom. Since these chi squares from different subclasses are independent of each other, their sum is also a chi square, with $4 + 4 + 4 + 4 = 16$ degrees of freedom by virtue of the additive property of chi squares.

5.4 CONSEQUENCES OF TREATMENT DIFFERENCES

In Sec. 5.2 we mentioned that if the Ω assumptions hold, and if model (5.6) is valid, we can estimate the residual variance, i.e., the variance of ε's, in three different ways. The three estimates obtained from Table 5.2 were

Estimate 1: $\qquad$ Total MSS $= \dfrac{\text{total SS}}{19} = \dfrac{131.75}{19} = 6.93$

Estimate 2: $\qquad$ Within MSS $= \dfrac{\text{within SS}}{16} = \dfrac{30.40}{16} = 1.90$

Estimate 3: $\qquad$ Between MSS $= \dfrac{\text{between SS}}{3} = \dfrac{101.35}{3} = 33.78$

These estimates differ so much that we feel uncomfortable in regarding them all as estimates of one and the same parameter. Could it be that while one of these MSS is a valid estimate of σ^2, others are not necessarily so? We now turn to a consideration of this question.

Let us begin with a simulated experiment. Table 5.3 contains six random observations from a normal population with mean 5 and variance 1. The observations are put into three groups of two each to simulate an experiment in which model (5.6) holds.

Table 5.4 was constructed from Table 5.3 by subtracting 1.0 from each value under treatment 1 and adding 3.0 to each value under treatment 3. Table 5.4 thus simulates an experiment to which model (5.5) applies.

The analyses of variance of the data from Tables 5.3 and 5.4 are also shown in the respective tables. Comparing the results in the two tables we notice (1) that the within MSS is the same in both the tables and (2) that the between MSS is higher in Table 5.4. These results indicate that if the treatment effects differ, i.e., if model (5.5) is valid, we may expect the between MSS to be greater than it would be otherwise. This is algebraically demonstrated below. Some readers may want to skip to the concluding part of this section where the results derived below are summarized.

Expected Values of MSS

Expected values of MSS play an important role in the analysis of experiments. We shall indicate how to obtain the expectations of between MSS and within MSS under model (5.5) for the design in Table 5.2. We shall then generalize the results for one-factor experiments with t treatments and r_j observations under the jth treatment. The corresponding results under model (5.6) can be obtained as particular cases of those under model (5.5). We shall not be concerned with the expected value of total

Table 5.3 A SIMULATED EXPERIMENT:
TREATMENT EFFECTS $\beta_1 = \beta_2 = \beta_3$

| | Treatments | | | |
	1	2	3	Total
	4.4	3.0	6.1	
	5.0	4.5	4.0	
Sum	9.4	7.5	10.1	27.0

Analysis of Variance			
Source	df	SS	MSS
Between treatments	2	1.81	.90
Within treatments	3	3.51	1.17
Total	5	5.32	

MSS, simply because for hypotheses testing E (total MSS) is of no use, as will become clear later on.

In deriving the expected values of the between and within MSS we assume that the residual terms ε satisfy the Ω assumptions.

Now model (5.5) implies that for the design in Table 5.2

$$j\text{th subclass mean} = \frac{1}{5} \sum_{i=1}^{5} y_{ij} = \frac{1}{5} \sum_{i=1}^{5} (\mu + \beta_j + \varepsilon_{ij})$$

$$= \mu + \beta_j + \frac{1}{5} \sum_{i=1}^{5} \varepsilon_{ij}$$

Here we introduce the dot-and-bar notation. A dot subscript means that summation has been performed over the subscript the dot replaces. Thus $y_{.j} = \sum_i y_{ij}$. A total with a bar, for example, $\bar{y}_{.j}$, represents the corresponding mean. According to this notation,

$$j\text{th subclass mean} = \bar{y}_{.j} = \mu + \beta_j + \bar{\varepsilon}_{.j}$$

and

$$\text{Grand mean} = \bar{y}_{..} = \mu + \bar{\beta}_{.} + \bar{\varepsilon}_{..}$$

Hence

$$\text{Between MSS} = \frac{\sum (\bar{y}_{.j} - \bar{y}_{..})^2}{\frac{1}{5}(3)}$$

$$= \tfrac{5}{3} \sum (\beta_j - \bar{\beta}_{.} + \bar{\varepsilon}_{.j} - \bar{\varepsilon}_{..})^2$$

$$= \tfrac{5}{3} \Big[\sum (\beta_j - \bar{\beta}_{.})^2 + \sum (\bar{\varepsilon}_{.j} - \bar{\varepsilon}_{..})^2$$

$$+ 2 \sum (\beta_j - \bar{\beta}_{.})(\bar{\varepsilon}_{.j} - \bar{\varepsilon}_{..}) \Big]$$

Table 5.4 **A SIMULATED EXPERIMENT: TREATMENT EFFECTS DIFFERENT; β_1 LESS THAN β_2, WHICH IN TURN IS LESS THAN β_3**

| | Treatments | | | |
	1	2	3	Total
	3.4	3.0	9.1	
	4.0	4.5	7.0	
Sum	7.4	7.5	16.1	31.0

Analysis of Variance

Source	df	SS	MSS
Between treatments	2	24.94	12.47
Within treatments	3	3.51	1.17
Total	5	28.45	

Taking expected values, we have

$$E \text{ (between MSS)} = \tfrac{5}{3}E\sum (\beta_j - \bar{\beta}_.)^2 + \tfrac{5}{3}E\sum (\bar{\varepsilon}_{.j} - \bar{\varepsilon}_{..})^2$$
$$+ 2\left[\tfrac{5}{3}E\sum (\beta_j - \bar{\beta}_.)(\bar{\varepsilon}_{.j} - \bar{\varepsilon}_{..})\right] \qquad (5.7)$$

We shall examine separately each term on the right-hand side of (5.7).

RESULT 1   Since β_j's are all constants, and since the expected value of any constant is the constant itself,

$$E\sum (\beta_j - \bar{\beta}_.)^2 = \sum E(\beta_j - \bar{\beta}_.)^2 = \sum (\beta_j - \bar{\beta}_.)^2$$

RESULT 2   Since under the Ω assumptions ε_{ij}'s are NID$(0,\sigma^2)$, that is, normally and independently distributed with 0 mean and variance σ^2, $\bar{\varepsilon}_{.j}$'s each of which is the mean of five independent ε_{ij}'s are NID$(0,\sigma^2/5)$. Furthermore $\bar{\varepsilon}_{..}$ is the mean of $\bar{\varepsilon}_{.j}$'s. Hence by the results of Sec. 5.3

$$\frac{\sum\limits_{j=1}^{4} (\bar{\varepsilon}_{.j} - \bar{\varepsilon}_{..})^2}{\sigma^2/5}$$

is a chi square with 3 degrees of freedom. But the expected value of a chi-square variate is the degrees of freedom of the chi square. Hence

$$E\,\frac{\sum (\bar{\varepsilon}_{.j} - \bar{\varepsilon}_{..})^2}{\sigma^2/5} = 3$$

Consequently

$$\tfrac{5}{3}E\sum (\bar{\varepsilon}_{.j} - \bar{\varepsilon}_{..})^2 = \sigma^2$$

RESULT 3   Since $(\beta_j - \bar{\beta}_.)$'s are constants,

$$E\sum (\beta_j - \bar{\beta}_.)(\bar{\varepsilon}_{.j} - \bar{\varepsilon}_{..}) = \sum (\beta_j - \bar{\beta}_.)E(\bar{\varepsilon}_{.j} - \bar{\varepsilon}_{..})$$

But the sampling distributions of $\bar{\varepsilon}_{.j}$'s all have zero mean. The same is true of $\bar{\varepsilon}_{..}$. Hence $E(\bar{\varepsilon}_{.j}) = 0, j = 1, 2, 3, 4$, and $E(\bar{\varepsilon}_{..}) = 0$. It follows from the above that

$$E \sum (\beta_j - \bar{\beta}_.)(\bar{\varepsilon}_{.j} - \bar{\varepsilon}_{..}) = 0$$

Inserting results 1 to 3 in (5.7) gives

$$E \text{ (between MSS)} = \tfrac{5}{3} \sum (\beta_j - \bar{\beta}_.)^2 + \sigma^2$$

In the general case with t treatments and r observations under each treatment

$$E \text{ (between MSS)} = \frac{r}{t-1} \sum_{1}^{t} (\beta_j - \bar{\beta}_.)^2 + \sigma^2 \qquad (5.8)$$

If the numbers of observations differ between treatments, for example, r_j observations under treatment j, then

$$E \text{ (between MSS)} = \frac{1}{t-1} \sum_{1}^{t} r_j(\beta_j - \bar{\beta})^2 + \sigma^2 \qquad (5.9)$$

It is easy to see that when β_j's are all equal, i.e., when model (5.6) holds,

$$E \text{ (between MSS)} = \sigma^2 \qquad (5.10)$$

irrespective of whether the numbers of observations differ between treatments.
Let us turn to the expected value of within MSS. Under model (5.5)

$$y_{ij} = \mu + \beta_j + \varepsilon_{ij} \qquad \text{and} \qquad \bar{y}_{.j} = \mu + \beta_j + \bar{\varepsilon}_{.j}$$

Hence for the design in Table 5.2

$$\text{Within MSS} = \frac{1}{4} \sum_{j=1}^{4} \frac{\sum_{i=1}^{5} (y_{ij} - \bar{y}_{.j})^2}{5 - 1} = \frac{1}{4} \sum_{j=1}^{4} \frac{\sum_{i=1}^{5} (\varepsilon_{ij} - \bar{\varepsilon}_{.j})^2}{5 - 1} \qquad (5.11)$$

Now since under the Ω assumptions ε_{ij}'s are all NID$(0,\sigma^2)$, we can apply the results of Sec. 5.3 to the sums of squares

$$\sum_{i=1}^{5} \frac{(\varepsilon_{ij} - \bar{\varepsilon}_{.j})^2}{\sigma^2} \qquad j = 1, 2, 3, 4$$

Each of these sums of squares is seen to be a chi square with 4 degrees of freedom. Since these chi squares are all independent of each other, their sum is also a chi square with its degree of freedom determined by the additive property of chi squares. Thus

$$\sum_{j=1}^{4} \sum_{i=1}^{5} \frac{(\varepsilon_{ij} - \bar{\varepsilon}_{.j})^2}{\sigma^2}$$

is a chi square with $4 + 4 + 4 + 4 = 16$ degrees of freedom. Taking expectations, we get

$$E \left[\sum_{j=1}^{4} \sum_{i=1}^{5} \frac{(\varepsilon_{ij} - \bar{\varepsilon}_{.j})^2}{\sigma^2} \right] = 16$$

This implies that

$$E \left[\sum_{j=1}^{4} \sum_{i=1}^{5} \frac{(\varepsilon_{ij} - \bar{\varepsilon}_{.j})^2}{16} \right] = \sigma^2 \qquad (5.12)$$

From (5.11) and (5.12) it immediately follows that for the design in Table 5.2 model (5.6) implies

$$E \text{ (within MSS)} = \sigma^2 \qquad (5.13)$$

If there are t treatments and r_j observations under the jth treatment, the above result generalizes to

$$E \left[\frac{1}{t} \sum \frac{\sum_{i=1}^{r_j} (y_{ij} - \bar{y}_{.j})^2}{r_j - 1} \right] = \sigma^2 \qquad (5.14)$$

Thus irrespective of whether all treatments are applied on an equal number of subjects, the expected value of within MSS is equal to σ^2 under the Ω assumptions.

This result holds true regardless of whether the treatment effects differ between themselves because the treatment effects cancel out when we calculate the SS within each treatment group. Hence (5.14) is valid under model (5.5) as well as under (5.6).

To summarize, for the case of t treatments (subclasses) with r_j observations under treatment (subclass) j we have the following results. Under model (5.5), that is, $y_{ij} = \mu + \beta_j + \varepsilon_{ij}$

$$E \text{ (between MSS)} = \sigma^2 + \frac{1}{t-1} \sum_{j=1}^{t} r_j(\beta_j - \bar{\beta}.)^2$$

$$E \text{ (within MSS)} = \sigma^2$$

Under model (5.6), that is, $y_{ij} = \mu + \beta_0 + \varepsilon_{ij}$

$$E \text{ (between MSS)} = \sigma^2 \qquad E \text{ (within MSS)} = \sigma^2$$

The Variance Ratio: F

The above results suggest that the ratio

$$\frac{\text{Between MSS}}{\text{Within MSS}}$$

is likely to be a good criterion of whether the treatment effects differ between themselves. If the treatment effects do not differ between themselves this ratio should be close to 1 except for sampling fluctuations. If, on the other hand, the treatment effects differ between themselves, this ratio is likely to exceed 1 beyond what we may attribute to sampling fluctuations.

Of course one can think of other test criteria, e.g., the difference (between MSS) − (within MSS). It turns out that under the Ω assumptions the ratio (between MSS)/(within MSS) provides the most powerful test[1] of the hypothesis that the treatment effects do not differ between themselves.

Under the Ω assumptions it can be shown that if the hypothesis that the treatment effects do not differ is valid, the ratio (between MSS)/(within MSS) has a sampling distribution called the F distribution. This is a particular case of the following more general theorem.

If U_1 and U_2 are two independent chi squares with n_1 and n_2 degrees of freedom, respectively, then the ratio

$$V = \frac{U_1/n_1}{U_2/n_2}$$

is distributed as an F with (n_1, n_2) degrees of freedom.

In our present case recall that (between SS)/σ^2 is a chi square with 3 degrees of freedom and (within SS)/σ^2 is a chi square with 16 degrees of freedom. It can also be

---

[1] Power of a test, it may be recalled, is judged in terms of the relative frequency with which the null hypothesis if not true would be rejected in repeated testing under constant conditions.

shown that these two chi squares are independent of each other (see below). Hence the general theorem stated above can be applied. We thus have

$$\frac{\frac{1}{3}[(\text{between SS})/\sigma^2]}{\frac{1}{16}[(\text{within SS})/\sigma^2]} = \frac{\frac{1}{3} \text{ between SS}}{\frac{1}{16} \text{ within SS}} = \frac{\text{between MSS}}{\text{within MSS}}$$

which is distributed as an F with $(3, 16)$ degrees of freedom.

Appendix Table D.3 shows the *theoretical* 5, 1, and .1 percent points of the F distribution. The degrees of freedom shown across the top of the table correspond to that of the MSS in the numerator and those at the left to that of the denominator.

Analysis-of-Variance Table

For the data in Table 5.2 we have

$$\frac{\text{Between MSS}}{\text{Within MSS}} = \frac{101.35/3}{30.40/16} = \frac{33.78}{1.90} = 17.78$$

The 5 and 1 percent values of F with $(3, 16)$ degrees of freedom are 3.24 and 5.29, respectively. Thus if the null hypothesis is true, i.e., if the treatment effects are equal to one another, there is less than 1 chance in 100 of getting a ratio of (between MSS)/(within MSS) larger than 5.29 in repetitions of the experiment under constant conditions. The calculated ratio being greater than 5.29, we infer that the null hypothesis is not true and interpret the result to mean that the treatment effects are different.

A complete analysis-of-variance table shows the different MSS as well as the calculated F ratio. Table 5.5 presents the complete analysis of variance of the data in Table 5.2.

Between SS and Within SS Are Independent

In the remainder of this section we shall demonstrate that (between SS)/σ^2 and (within MSS)/σ^2 are independent chi squares in any one-factor design, provided the treatment effects are not different from each other and the Ω assumptions hold. Some readers may want to skip to the next section.

If the treatment effects do not differ between themselves, the ith observation under the jth treatment can be represented as

$$y_{ij} = \mu + \beta_0 + \varepsilon_{ij}$$

Table 5.5   COMPLETE ANALYSIS-OF-
VARIANCE TABLE

Source	df	SS	MSS	Calculated F ratio
Between	3	101.35	33.78	17.78
Within	16	30.40	1.90	
Total	19	131.75		

There is no loss of generality if we set $\mu + \beta_0 = 0$. If it is desired $\mu + \beta_0$ can be retained, reckoning it to be $\neq 0$; the arguments that follow are still valid if one replaces y_{ij} by $y_{ij} - \mu - \beta_0$ for all i and j.

Setting $\mu + \beta_0 = 0$, we have $y_{ij} = \varepsilon_{ij}$ for all i and j. We confine our attention for the moment to the y values under one treatment, say the jth treatment. Under the Ω assumptions we can apply the results of Sec. 5.3. Accordingly we can write

$$\sum_{i=1}^{r_j} \frac{y_{ij}^2}{\sigma^2} = \frac{z_1^2}{\sigma^2} + \frac{z_2^2}{\sigma^2} + \cdots + \frac{z_{r_j-1}^2}{\sigma^2} + \frac{r_j \bar{y}_{.j}^2}{\sigma^2}$$

where r_j is the number of observations under treatment j; each term on the right-hand side is independent of all the others; in particular, z_1^2/σ^2, z_2^2/σ^2, $z_{r_j-1}^2/\sigma^2$ are all independent of $r_j\bar{y}_{.j}^2/\sigma^2$, and hence $z_1^2/\sigma^2 + \cdots + z_{r_j-1}^2/\sigma^2$ is independent of $r_j\bar{y}_{.j}^2/\sigma^2$. But

$$\frac{z_1^2 + z_2^2 + \cdots + z_{r_j-1}^2}{\sigma^2} = \sum_{i=1}^{r_j} \frac{y_{ij}^2}{\sigma^2} - \frac{r_j\bar{y}_{.j}^2}{\sigma^2} = \sum_{i=1}^{r_j} \frac{(y_{ij} - \bar{y}_{.j})^2}{\sigma^2}$$

Hence $\sum_{i=1}^{r_j} (y_{ij} - \bar{y}_{.j})^2/\sigma^2$ is independent of $r_j\bar{y}_{.j}^2/\sigma^2$. Summation with respect to j shows therefore that

$$\sum_{j=1}^{t} \sum_{i=1}^{r_j} \frac{(y_{ij} - \bar{y}_{.j})^2}{\sigma^2}$$

or the (within SS)/σ^2 is independent of

$$\sum_{j=1}^{t} \frac{r_j\bar{y}_{.j}^2}{\sigma^2}$$

Applying the same logic, we find that

$$\sum_{j=1}^{t} \frac{r_j\bar{y}_{.j}^2}{\sigma^2}$$

has two independent parts

$$\sum_{j=1}^{t} \frac{r_j(\bar{y}_{.j} - \bar{y}_{..})^2}{\sigma^2}$$

or (between SS)/σ^2 and

$$\frac{\bar{y}_{..}^2}{\sigma^2/\sum r_j}$$

or the correction for the mean divided by σ^2. Thus (within SS)/σ^2 is independent of (between SS)/σ^2, which in turn is independent of the correction for the mean divided by σ^2.*

---

* Since between SS and within SS are both parts of total SS, it is obvious that the last one is not independent of either of the first two. Hence ratios of the type (between MSS)/(total MSS) or (within MSS)/(total MSS) are not distributed as F. For this reason, we do not make use of E (total MSS) in the analysis of variance.

5.5   TREATMENT DIFFERENCES

Estimation versus Testing Significance

Let us pursue the consequences of representation (5.5) a little further. Averaging over the r_j observations under treatment j, we get

$$\bar{y}_{.j} = \mu + \beta_j + \bar{\varepsilon}_{.j}$$

Similarly, for treatment l, we have

$$\bar{y}_{.l} = \mu + \beta_l + \bar{\varepsilon}_{.l}$$

Subtraction yields

$$\bar{y}_{.j} - \bar{y}_{.l} = (\beta_j - \beta_l) + (\bar{\varepsilon}_{.j} - \bar{\varepsilon}_{.l}) \qquad (5.15)$$

Note that in getting the above result we have not made use of any of the assumptions Ω about ε_{ij}'s. If, however, we make the assumption that $E(\bar{\varepsilon}_{.j}) = k$, for all j, Eq. (5.15) immediately gives

$$E(\bar{y}_{.j} - \bar{y}_{.l}) = (\beta_j - \beta_l)$$

which tells us that $\bar{y}_{.j} - \bar{y}_{.l}$ is an unbiased estimate of $\beta_j - \beta_l$. Hence, for estimation purposes we need not assume anything about the ε_{ij}'s other than that $E(\bar{\varepsilon}_{.j}) = k$ for all j. The other assumptions, namely, about normality, noncorrelation, and homoscedasticity, become relevant only if we want to conduct a formal testing of the hypothesis concerning the treatment differences or be concerned with the efficiency of our estimates.

Treatment Contrasts

Sometimes we may be interested in parametric functions other than simple differences of the type $\beta_1 - \beta_2$. For example, suppose in an experiment two new methods and a traditional method of teaching are the three treatments taken for comparison. Let β_1 and β_2 stand for the effects of the two new methods and β_3 for that of the traditional one. The objective may be to estimate (1) the average effect of the two new methods as contrasted with the traditional method:

$$\frac{\beta_1 + \beta_2}{2} - \beta_3$$

and (2) the difference between the effects of the two new methods: $\beta_1 - \beta_2$.

If $\bar{y}_{.1}$, $\bar{y}_{.2}$, and $\bar{y}_{.3}$ are the means of observations under treatments 1, 2, and 3, respectively, by following the reasoning in Eq. (5.15) it can be seen that

$$\frac{\bar{y}_{.1} + \bar{y}_{.2}}{2} - \bar{y}_{.3}$$

is an unbiased estimate of

$$\frac{\beta_1 + \beta_2}{2} - \beta_3$$

and $\bar{y}_{.1} - \bar{y}_{.2}$ is an unbiased estimate of $\beta_1 - \beta_2$.

We may consider both the parametric functions described above as particular instances of a general linear combination of the three parameters $l_1\beta_1 + l_2\beta_2 + l_3\beta_3$.

In $[(\beta_1 + \beta_2)/2] - \beta_3$ we have $l_1 = \frac{1}{2}$, $l_2 = \frac{1}{2}$, and $l_3 = -1$, and in $\beta_1 - \beta_2$, $l_1 = 1, l_2 = -1$, and $l_3 = 0$. In both instances, we further note that the sum of the coefficients of the parameters is 0. Thus,

$$\tfrac{1}{2} + \tfrac{1}{2} + (-1) = 0 \qquad \text{and} \qquad 1 + (-1) + 0 = 0$$

Linear combinations of the above kind are called *treatment contrasts*. To give a formal definition, referring to the experiment in which there are t treatments, if β_1, $\beta_2, \ldots, \beta_t$ are the parameters representing the treatment effects and $l_1, l_2, \ldots, l_t$ are known constants whose sum is zero, then $l_1\beta_1 + l_2\beta_2 + \cdots + l_t\beta_t$ is called a *treatment contrast*.

The following points are worth emphasizing in this connection:

1 If assumption 3 of Ω is true, i.e., if $E(\varepsilon_{ij}) = 0$ for all i and j, $l_1\bar{y}_{.1} + l_2\bar{y}_{.2} + \cdots + l_t\bar{y}_{.t}$ is an unbiased estimate of $l_1\beta_1 + l_2\beta_2 + \cdots + l_t\beta_t$. This holds if $E(\varepsilon_{ij}) = k$, for all i and j, where k may not necessarily be equal to zero. This is a consequence of the relation $\sum l_j = 0$.

2 If assumptions 1 and 4 of Ω are true, i.e., if ε_{ij}'s are independently distributed and are homoscedastic, the sampling variance of $l_1\bar{y}_{.1} + l_2\bar{y}_{.2} + \cdots + l_t\bar{y}_{.t}$, where $\bar{y}_{.1}, \bar{y}_{.2}, \ldots$ are based on $r_1, r_2, \ldots, r_t$ observations, respectively, is given by

$$\left(\frac{l_1^2}{r_1} + \frac{l_2^2}{r_2} + \cdots + \frac{l_t^2}{r_t}\right)\sigma^2$$

and an estimate thereof by

$$\left(\frac{l_1^2}{r_1} + \frac{l_2^2}{r_2} + \cdots + \frac{l_t^2}{r_t}\right)(\text{within MSS})$$

3 If assumptions 1 to 4 of Ω are true, i.e., if ε_{ij}'s are independently normally distributed with a common mean, usually assumed to be zero, and are homoscedastic, the following test statistic can be considered as an F with $(1, n - t)$ degrees of freedom, where $n = \sum r_j$ and its square root a Student's t with $n - t$ degrees of freedom for testing the hypothesis that the treatment contrast is equal to zero:

$$\frac{(l_1\bar{y}_{.1} + l_2\bar{y}_{.2} + \cdots + l_t\bar{y}_{.t})^2}{(\text{Within MSS})(l_1^2/r_1 + l_2^2/r_2 + \cdots + l_t^2/r_t)}$$

Multiple Comparisons

It is important to recognize that if the null hypothesis that the treatment effects are not different is rejected, on the basis of the results of the analysis of variance, it does not necessarily follow that each treatment effect is different from the others. Therefore, once the null hypothesis is rejected, the analyst is usually interested in investigating which treatment effects are different and which are equal. This problem has been approached from different points of view by different statisticians. We shall mention only a few here.

The least-significant-difference (LSD) method Consider the experiment described in Table 5.2. We have seen that the analysis of variance has led to the rejection of the null hypothesis. Suppose now we want to compare each treatment with all other treatments. Let β_j and β_l be the two treatment effects to be compared. Let $\bar{y}_{.j}$ and $\bar{y}_{.l}$ be the observed mean scores (based on r_j and r_l cases, respectively, say) under the two treatments. Then, according to the results just described, given the Ω assumptions,

$$\frac{\bar{y}_{.j} - \bar{y}_{.l}}{\sqrt{(\text{Within MSS})(1/r_j + 1/r_l)}}$$

can be used as a Student's t for testing whether $\beta_j = \beta_l$. If the value of the above statistic exceeds the tabulated value of Student's t for the appropriate degrees of freedom (which is the same as that of the within MSS) and the chosen level of significance, we reject the hypothesis that $\beta_j = \beta_l$. The product

$$\begin{array}{c}\text{Appropriate tabulated} \\ \text{value of Student's } t\end{array} \sqrt{(\text{within MSS})\left(\frac{1}{r_j} + \frac{1}{r_l}\right)}$$

is known as the least significant difference (LSD). Note that if the numbers of cases under all treatments are the same, the LSD need be calculated only once. The same value can be applied in all pairwise comparisons of treatment means.

Let us apply the above procedure to the results of Table 5.2. From Table 5.5 we have within MSS = 1.90. We also note that the numbers of observations under all treatments are the same (= 5). Choosing the level of significance as 5 percent, we get from Appendix Table D.1, 2.120 as the value of Student's t for 16 degrees of freedom. Hence we have

$$\text{LSD} = 2.120\sqrt{1.90 \times \tfrac{2}{5}} = 1.84$$

Let us now write down all possible paired comparisons between the treatment means in Table 5.2. These are presented in Table 5.6. Of the six possible comparisons, we find only four exceeding the calculated LSD (= 1.84). Treatments 3 and 4 are both found to be different from all the others, while there is no difference between treatments 1 and 2, the same being the case between treatments 3 and 4.

Table 5.6 **PAIRWISE COMPARISONS OF ALL TREATMENT EFFECTS IN TABLE 5.2**

Treatment number		1	2	3	4
	Treatment mean	$\bar{y}_{.1} = 2.6$	$\bar{y}_{.2} = 3.6$	$\bar{y}_{.3} = 6.6$	$\bar{y}_{.4} = 8.2$
1	$\bar{y}_{.1} = 2.6$	—	1.0	4.0	5.6
2	$\bar{y}_{.2} = 3.6$	—	—	3.0	4.6
3	$\bar{y}_{.3} = 6.6$	—	—	—	1.6
4	$\bar{y}_{.4} = 8.2$	—	—	—	—

It should be noted that the LSD tests may lead to the anomalous situation in which the overall F ratio is not significant while one or more of the pairwise comparisons are significant. The reason is simply that when you make a large number of comparisons of means, it is almost certain that some of them will turn out to be significant even when the corresponding parametric contrasts are zero. For this reason, it is recommended that the LSD method not be applied if the overall F ratio is found not to be significant.

Honestly-significant-difference (HSD) method   Tukey (1953) proposed for multiple comparison of means a test statistic similar to the one used in the LSD method. We shall call it the *honestly-significant-difference* (HSD) method. It is defined as

$$HSD = Q \sqrt{\frac{\text{within MSS}}{r}}$$

where Q is the appropriate value obtained from the table for the Studentized range and r denotes the number of observations under each treatment.[1] See Appendix Table D.4 for 5 and 1 percent values of the Studentized range. To use the table you need two values: the degrees of freedom of the within MSS used in the calculation of HSD and the number of treatments in the experiment. For the results in Table 5.2, we have degrees of freedom of the within MSS = 16, and the number of treatments = 4. Choosing 5 percent level of significance, we find from Appendix Table D.4 that the appropriate value of Q is 4.05. Hence,

$$HSD = 4.05 \sqrt{\frac{1.90}{5}} = 2.51$$

Referring back to Table 5.6, we find that the only pairwise differences that are significant are the ones involving the treatment means $\bar{y}_{.3}$ or $\bar{y}_{.4}$. The inference is that treatment effects β_1 and β_2 form one set while β_3 and β_4 form another set. The members of each set differ from those of the other but do not differ between themselves.

The HSD has the property that if we test some or all of the pairs of means, the probability that no erroneous claim of significance will be made will be greater than or equal to 1 minus the significance level used. It has also the property that the probability that the interval $(\bar{y}_{.j} - \bar{y}_{.i}) \pm HSD$ includes the true difference $\beta_j - \beta_i$ will be equal to 1 minus the significance level used. The price paid, of course, is that, compared with the LSD method, fewer of the differences that really exist will be detected and the confidence interval is wider.

Scheffé's method   If the analysis of variance leads to the rejection of the null hypothesis, a method suggested by Scheffé may be used to set confidence intervals for any complex comparison. (See Scheffé, 1959, pp. 70–71.) Let

$$\Psi = l_1 \beta_1 + l_2 \beta_2 + \cdots + l_t \beta_t$$

---

[1] If the number of cases under the treatments differ, an approximate test is possible by using for r

$$\tilde{r} = \frac{t}{1/r_1 + \cdots + 1/r_t}$$

where t is the number of treatments and r_j is the number of cases under treatment j.

be a comparison of interest, and let

$$\hat{\Psi} = l_1 \bar{y}_{.1} + l_2 \bar{y}_{.2} + \cdots + l_t \bar{y}_{.t}$$

be an estimate thereof, in which the means $\bar{y}_{.1}, \bar{y}_{.2}, \ldots, \bar{y}_{.t}$ are based on $r_1, r_2, \ldots, r_t$ cases, respectively. Then the confidence interval suggested by Scheffé is

$$\hat{\Psi} - S \le \Psi \le \hat{\Psi} + S$$

where

$$S = \sqrt{(t-1)F} \sqrt{(\text{within MSS})\left(\frac{l_1^2}{r_1} + \cdots + \frac{l_t^2}{r_t}\right)}$$

the quantity F being the tabulated value in the F distribution corresponding to the following:

$$\text{df of numerator} = t - 1$$

$$\text{df of denominator} = \text{df of within MSS}$$

Significance level $= 1 -$ chosen confidence coefficient for comparison of interest

Note that if what is needed is a test of significance, a comparison $\hat{\Psi}$ in order to be significant must be larger than S. To apply this procedure to test the significance of the comparisons in Table 5.6, we calculate

$$S = \sqrt{(4-1)(3.24)} \sqrt{\tfrac{2}{5} \times 1.90} = 2.72$$

Using the above value of S, we note that only four out of the six possible pairwise comparisons listed in Table 5.6 are significant.

A sequential variation of the LSD method   Another method, originally due to Newman (1939) and Keuls (1952), that can be used for multiple comparisons is a modified version of the LSD method. This can best be described in terms of an example.

We first arrange the treatment means in ascending order of magnitude. For the data in Table 5.2, the arrangement is

Treatment number	1	2	3	4
Treatment mean	2.6	3.6	6.6	8.2

Treatments 1 and 4 being the farthest apart, we first test whether they are significantly different from each other. Then test the next maximum difference, and so on. For testing these differences, we define treatment 4 as four steps removed from treatment 1 in the above arrangement, treatment 3 as three steps removed from treatment 1, and so on. In a test of significance of any pairwise comparison, the number of steps between the two means involved in that comparison is to be taken into account. The value that must be exceeded by the observed difference between means in order to be declared significant is then calculated as follows:

$$N_v = Q_v \sqrt{\frac{\text{within MSS}}{r}}$$

where $r =$ number of cases each of means is based on
$v =$ number of steps between means involved in comparison
$Q =$ value obtained from Studentized range table

The following three values are used to enter the table:

1 The value of v
2 The chosen level of significance
3 The degrees of freedom of the within MSS

The problem of choosing among the methods   Several procedures for multiple comparisons have been described and illustrated above. This does not by any means exhaust the available methods.[1] You may at this point wonder how to choose among the different methods available. In many situations, it may not matter which method is chosen; all methods may lead to identical conclusions. But not infrequently, some methods may detect more differences than others. The protection against wrong claims of significance can be gained only by decreasing the ability to detect real differences. Hence in critical situations, the analyst must try to judge the relative costs of the two kinds of mistakes and be guided by these costs. In routine situations, either the LSD method or the Newman-Keuls method should be satisfactory.

5.6   EFFECTS OF VIOLATIONS OF ASSUMPTIONS

In all our discussions, we have taken for granted that assumptions Ω hold true whenever the t test or the F test is used. The same set of assumptions underlies the use of other statistics, too, e.g., Scheffé's S or Newman and Keuls' N. The analyst would obviously feel concerned about whether the assumptions are satisfied by his data. He may also be interested in the related question: What difference does it make if the assumptions are violated? The question whether a given set of data violates one or more of the assumptions Ω is relatively easy to answer. For one thing, there are few real data that do not violate the assumptions. For another, more often than not, one could find out whether a given assumption is violated by applying rather simple statistical tests. The other question What does it matter if the assumptions are violated? is more difficult to answer. For the answer depends upon how and to what degree a given assumption is violated. And indeed a given assumption can be violated in numerous ways. For example, nonnormality may be due to any one of the following, or similar, conditions.

1 The dependent variable may be one that takes small positive integers as its values. In that case the underlying probability distribution may be Poisson rather than normal.
2 The values of dependent variable may be proportions very small in magnitude. In this case the underlying probability distribution is binomial.
3 The treatment effect may be multiplicative rather than additive. Model (5.5) assumes that the effect of treatment j is to add a quantity β_j to the overall effect due to the other causal factors. If, instead, the treatment effect is to multiply the overall effect due to all the other factors, e.g., to increase by 40 percent, then the underlying distribution may be *log normal* rather than normal.

---

[1] Reference may be made to Kirk (1968) for a more complete treatment of methods of multiple comparison.

Nonnormality due to one of the above causes may have one set of implications while that due to some other cause may have another set of implications. Thus, if the underlying distribution is Poisson, the variance varies with the means of the treatment groups. In a binomial distribution, the variance varies with $p_j(1 - p_j)$, where p_j is the proportion characteristic of the jth treatment group, and so on. Depending upon which of these is actually the case, the implications for the inferences differ. The remedy for the violation also differs according to the nature of the situation.

Some general points, however, can be made regarding the effects of violations of the assumptions.

EFFECT 1   As a rule, if an assumption is violated, the significance level as well as the sensitivity of the F and t tests will be affected. With respect to the significance level, what may happen is the following. When an analyst thinks that he is using a 1 percent level of significance, he may actually be using a 4 percent level. In other words, the analyst is led to declare more tests as significant than if he had actually been using the significance level he thought he was using. Because of this possibility it will be wise *not* to regard the chosen level of significance level as sacred. In other words, when a level such as 5 percent is chosen and a test is declared significant, the analyst must be prepared to accept that verdict even if it were true only at a significance level for say, 9 percent, or some such level different from the one he thought he was using.

Violation of an assumption may also result in a loss of sensitivity of the test or a loss of accuracy of the estimates of the treatment comparisons. If one knows in what particular way and to what extent an assumption has been violated, it may be possible (1) to construct a more powerful test than the F and t tests and (2) to obtain more accurate estimates of the treatment comparisons.

EFFECT 2   Investigations of the effects of violation of assumptions have taken essentially two forms, mathematical and empirical. For large samples, some investigators have examined mathematically whether the effects of given violations (nonnormality, heteroscedasticity, and correlation between residuals) are serious or negligible. It may be noted that an effect found to be serious for large samples is bound to be serious for small samples too, but an effect found to be negligible for large samples may not necessarily be negligible for small samples. Small-sample situations have been investigated using sampling experiments in a number of studies. Scheffé (1959) has reported results of mathematical as well as empirical investigations. Reference may be made to Boneau (1960) and Baker et al. (1966) for results from more recent empirical investigations covering very small to moderately large numbers of cases under each treatment. Judging from the results of these and other studies, it is clear that as far as inferences about treatment means are concerned, nonnormality has little effect, even when the number of cases under each treatment is as small as 5.

EFFECT 3   Heteroscedasticity also has been found to have negligible effect on inferences about differences between treatment means as long as the means involved are based on equal or nearly equal numbers of cases (whether few or large). This calls for equalizing the cell frequencies (number of observations in each group). In controlled experiments this is relatively easy to accomplish. But not so in nonexperimental research.

EFFECT 4   When, however, heteroscedasticity occurs with unequal numbers of cases under different treatments, inferences about differences between treatment means may be seriously invalidated.

EFFECT 5   The violation of the assumption that residuals associated with the observations are independent between themselves can have very serious consequences as far as inferences about differences between treatments are concerned. In the simplest case, nonindependence between residuals is represented mathematically by supposing that there is an intraclass correlation ρ between any pair of residuals within each treatment (see Sec. 5.10 for a discussion of intraclass correlation). In the absence of real treatment differences, the expected value of between MSS can be shown to be $\sigma^2[1 + (n - 1)\rho]$, where n is the number of observations under each treatment, σ^2 is the variance of the residuals, and ρ the intraclass correlation. The corresponding expected value of within MSS can be shown to be $\sigma^2(1 - \rho)$. Hence the F ratio can be regarded in repeated samples as fluctuating around $[1 + (n - 1)\rho]/(1 - \rho)$. If ρ is positive, which is most frequently the case, this ratio can be considerably larger than 1. For example, if $\rho = 0.25$ and $n = 5$, this ratio takes the value 2.67. Supposing that there were six treatments each with five observations, we would be comparing the above value 2.67 with the tabulated F value of 2.62 (at 5 percent level). It is thus clear that even when there are no real treatment differences, the F test may lead to the rejection of the null hypothesis quite frequently if there is correlation between the residuals.

More complex situations of correlated residuals have not been adequately studied, but there is reason to believe that whatever the nature of the correlations among the residuals may be, they can invalidate inferences about treatment differences quite seriously.

The Principle of Randomization

Many textbooks emphasize that in experiments it is possible[1] to make the observations independent by *randomization*. A few words about how randomization accomplishes this may be in order. The principle of randomization is simply the following. When subjects are assigned to treatments or treatment groups, the assignment should be done at random so that each subject will have the same chance of being assigned to one treatment as to another.

Of course, even when randomization is carried out according to this principle, the particular allocation of experimental units which is made in a given experiment may still work in favor of particular treatments (treatment combinations). Thus, for example, it may happen by chance in the experiment in Table 5.2 that a disproportionately large number of subjects with relatively higher intelligence gets allocated to the less intense noise-level (treatment) group. The excess in the observed mean score for one treatment group over the other groups may then be due at least in part to

---

[1] See in this connection Blalock (1964, pp. 22–26). Blalock makes the point that only properties of subjects can be randomized; the factors that get manipulated by the experimenter when he manipulates the treatments cannot be randomized. If the experimenter is aware of such factors, he should collect data on them too and incorporate them in the analysis. This calls for the use of more complex models. See in this connection Costner (1971a).

causal factors not included in the experiment and not solely to the treatment differences. The experimenter however, would attribute the observed difference to the treatment difference, on the basis of the assumption that he has randomized out all factors, e.g., intelligence, not included in the experiment. This is justifiable only on the following grounds. The experimenter's generalization applies to a hypothetical population generated by an infinite number of imaginary repetitions of the experiment under the same conditions. This hypothetical population includes every possible pattern of allocation of experimental units which randomization could possibly have produced. Within this population, the factors not explicitly or inadvertently manipulated in the experiment cannot show any favor to those explicitly incorporated into the structure, and hence the experimenter's inference is free from bias. It is, however, the responsibility of the experimenter to see whether the hypothetical population just described coincides with his target population, i.e., the population to which he actually wants to generalize his results. Only insofar as the particular sample of subjects used in the experiment can be considered as a random sample from the target population may he generalize the results of the analysis of the data at hand to the target population. Whether the data at hand can be considered as a random sample from the target population is often a matter of personal judgment on the part of the experimenter.

In nonexperimental research, randomization of subjects to treatment groups is not possible. Hence even if the subjects chosen for the study can be considered as a random sample from the target population, there is no guarantee that when all possible random samples are considered together one will get every possible pattern of allocation of subjects to treatment groups. Therefore, violation of assumption 1 of Ω is likely to be a constant threat in nonexperimental research.

5.7 REGRESSION ANALYSIS OF ONE-FACTOR EXPERIMENTS

The relationship between the analysis of designed experiments and regression analysis was recognized more than 35 years ago. Several books on experimental designs, e.g., Kempthorne (1952) and Scheffé (1959), have used the general linear model, i.e., the multiple-regression model, as their point of departure. Chapter 3 has shown how to apply linear models to the analysis of data from experiments. In this section we shall go over the same ground with a view to highlighting certain points which were not explicitly brought out in Chap. 3.

Let us start with the representation

$$y_{ij} = \mu + \beta_j + \varepsilon_{ij} \qquad \begin{array}{l} i = 1, 2, 3, 4, 5 \\ j = 1, 2, 3, 4 \end{array}$$

which was introduced earlier [see Eq. (5.5)] for the one-factor design in Table 5.2. Recall that μ represents the combined effect of all factors that remain constant, β_j the effect of the jth treatment, and ε_{ij} the residual effect, i.e., the combined effect of all variable but unaccounted for factors. A minimal assumption which is always made on the residuals ε_{ij} is that their expected values are zero. This assumption necessitates the introduction of μ to represent the combined effect of constant factors.

The above equation can be rewritten as

$$y = \mu x_0 + \beta_1 x_1 + \beta_2 x_2 + \beta_3 x_3 + \beta_4 x_4 + \varepsilon \qquad (5.16a)$$

where x_1, x_2, x_3, and x_4 are a set of regressors such that for any y_{ij}, $x_j = 1$ and $x_k = 0$ if $k \neq j$, while $x_0 = 1$ always. See Table 5.7, in which the y values are arranged in a standard order, those under treatment 1 followed by those under treatments 2, 3, and 4 in that order.

Notice that according to the convention introduced in Chap. 3, we would use the following representation (the full model) instead of (5.16a).

$$y = \beta_1' x_1 + \beta_2' x_2 + \beta_3' x_3 + \beta_4' x_4 + \varepsilon \qquad (5.16b)$$

The difference between (5.16a) and (5.16b) lies in the presence of μ in the former and its absence in the latter. We have already mentioned the reason for inserting μ in (5.16a). Once we define β as representing exclusively the effect of the manipulated treatment *and* stipulate that the residuals should minimally satisfy the condition of zero expectations, we have no choice but to introduce μ to represent the combined effect of constant factors. We may, however, define a set of coefficients β', as in (5.16b), in such a way that they represent the effect of the manipulated treatment *plus* the combined effect of constant factors. This procedure is also compatible with the assumption that the residuals have zero expectations. It is this latter convention, i.e., defining the regression weight as representing the treatment effect plus the combined

Table 5.7 PREPARATION OF THE DATA SET IN TABLE 5.2 ACCORDING TO THE REGRESSION MODEL (5.16a)

y	x_0	x_1	x_2	x_3	x_4
2	1	1	0	0	0
1	1	1	0	0	0
1	1	1	0	0	0
3	1	1	0	0	0
6	1	1	0	0	0
3	1	0	1	0	0
4	1	0	1	0	0
2	1	0	1	0	0
5	1	0	1	0	0
4	1	0	1	0	0
6	1	0	0	1	0
5	1	0	0	1	0
7	1	0	0	1	0
8	1	0	0	1	0
7	1	0	0	1	0
9	1	0	0	0	1
8	1	0	0	0	1
9	1	0	0	0	1
8	1	0	0	0	1
7	1	0	0	0	1

effect of constant factors, that underlies the exposition in Chap. 3. The connection between (5.16a) and (5.16b) is thus obvious: β'_j of (5.16b) is equal to $\beta_j + \mu$ of (5.16a) for $j = 1, 2, 3, 4$.

It was mentioned in Chap. 3 that the least-squares procedure, i.e., the procedure that minimizes the SS of the residuals, would provide unique estimates of the regression weights in equations such as (5.16b). Before examining whether the same is true of equations such as (5.16a), it is interesting to note that under very broad conditions the least-squares estimates have lower variance than any other estimates of the same kind, namely, estimates which can be expressed as weighted sums of the observed y values. Such estimates are called *linear estimates*. The grand mean, $\bar{y}_{..}$, the subclass means, $\bar{y}_{.j}$, and the difference between subclass mean and grand mean, $\bar{y}_{.j} - \bar{y}_{..}$, are all examples of weighted sums of the observed y values. The sample median is an example of a nonlinear statistic, i.e., a quantity that cannot be expressed as a weighted sum of the sample values. With reference to Eq. (5.16b) we can say that any unbiased estimate of β'_j of the form $\sum \sum c_{ij} y_{ij}$, where c_{ij} are known constants, will have its variance greater than or equal to that of the least-squares estimate of β'_j. This property of least-squares estimates of the regression weights in Eq. (5.16b) holds even when *the residuals are not normally distributed* but are independent and homoscedastic, and have expectation zero. Thus the least-squares estimates possess certain attractive properties that make them preferable to other estimates of the same kind.

Normal Equations

Now let us turn to Eq. (5.16a) and examine where the application of the least-squares procedure leads us. The mechanics involved in obtaining the least-squares estimates consist in solving what are known as the *normal equations*. To obtain the normal equations we equate to zero the partial derivatives with respect to the unknowns of the sum of squares of the residuals. Thus with reference to Eq. (5.16a) we equate to zero the partial derivatives with respect to μ, β_1, β_2, β_3, and β_4 of

$$Q = \sum \varepsilon^2 = \sum (y - \mu x_0 - \beta_1 x_1 - \beta_2 x_2 - \beta_3 x_3 - \beta_4 x_4)^2$$

Those unfamiliar with the procedure for obtaining partial derivatives may get the normal equations by setting equal to zero each of the following sums of cross products: $\sum \varepsilon x_0$, $\sum \varepsilon x_1$, $\sum \varepsilon x_2$, $\sum \varepsilon x_3$, and $\sum \varepsilon x_4$, where $\varepsilon = y - \mu x_0 - \beta_1 x_1 - \cdots - \beta_4 x_4$.

Notice that, in the terminology introduced in Chaps. 2 and 3, when we set $\sum \varepsilon x_j$ equal to zero, we are making the ε vector orthogonal to the x_j vector in Table 5.7. There is another way of describing what $\sum \varepsilon x_j = 0$ means. Notice that $\sum \varepsilon x_0 = 0$ means simply that $\sum \varepsilon = 0$, since $x_0 = 1$ for each ε. But if and when $\sum \varepsilon = 0$, the equation $\sum \varepsilon x_j = 0$ is equivalent to stating that the sample covariance between ε and x_j is equal to zero. This is so because, given $\sum \varepsilon = 0$, we have $\sum (\varepsilon - \bar{\varepsilon})(x_j - \bar{x}_j) = \sum \varepsilon(x_j - \bar{x}_j) = \sum \varepsilon x_j - \bar{x}_j \sum \varepsilon = \sum \varepsilon x_j$. It follows therefore that the procedure for writing down the normal equations that yield the least-squares estimates of the unknowns in equations such as (5.16a) is to set equal to zero the sum of the residuals and also the sample covariance between the residuals and each of the regressors.

Applying the above procedure to Eq. (5.16a), we arrive at the normal equations

$\sum \varepsilon x_0 = 0$: $\quad \sum (y - \hat{\mu}x_0 - \hat{\beta}_1 x_1 - \hat{\beta}_2 x_2 - \hat{\beta}_3 x_3 - \hat{\beta}_4 x_4) \quad = 0$

$\sum \varepsilon x_1 = 0$: $\quad \sum (y - \hat{\mu}x_0 - \hat{\beta}_1 x_1 - \hat{\beta}_2 x_2 - \hat{\beta}_3 x_3 - \hat{\beta}_4 x_4)x_1 = 0$

$\sum \varepsilon x_2 = 0$: $\quad \sum (y - \hat{\mu}x_0 - \hat{\beta}_1 x_1 - \hat{\beta}_2 x_2 - \hat{\beta}_3 x_3 - \hat{\beta}_4 x_4)x_2 = 0$

$\sum \varepsilon x_3 = 0$: $\quad \sum (y - \hat{\mu}x_0 - \hat{\beta}_1 x_1 - \hat{\beta}_2 x_2 - \hat{\beta}_3 x_3 - \hat{\beta}_4 x_4)x_3 = 0$

$\sum \varepsilon x_4 = 0$: $\quad \sum (y - \hat{\mu}x_0 - \hat{\beta}_1 x_1 - \hat{\beta}_2 x_2 - \hat{\beta}_3 x_3 - \hat{\beta}_4 x_4)x_4 = 0$

The hat over the unknowns indicates that the reference is to estimates of the unknowns. When we multiply the terms within parentheses by the factor outside and then sum term by term, the above equations reduce to

$\sum y \ = \hat{\mu} \sum x_0 + \hat{\beta}_1 \sum x_1 + \hat{\beta}_2 \sum x_2 + \hat{\beta}_3 \sum x_3 + \hat{\beta}_4 \sum x_4$

$\sum x_1 y = \hat{\mu} \sum x_0 x_1 + \hat{\beta}_1 \sum x_1^2 + \hat{\beta}_2 \sum x_2 x_1 + \hat{\beta}_3 \sum x_3 x_1 + \hat{\beta}_4 \sum x_4 x_1$

$\sum x_2 y = \hat{\mu} \sum x_0 x_2 + \hat{\beta}_1 \sum x_1 x_2 + \hat{\beta}_2 \sum x_2^2 + \hat{\beta}_3 \sum x_3 x_2 + \hat{\beta}_4 \sum x_4 x_2$

$\sum x_3 y = \hat{\mu} \sum x_0 x_3 + \hat{\beta}_1 \sum x_1 x_3 + \hat{\beta}_2 \sum x_2 x_3 + \hat{\beta}_3 \sum x_3^2 + \hat{\beta}_4 \sum x_4 x_3$

$\sum x_4 y = \hat{\mu} \sum x_0 x_4 + \hat{\beta}_1 \sum x_1 x_4 + \hat{\beta}_2 \sum x_2 x_4 + \hat{\beta}_3 \sum x_3 x_4 + \hat{\beta}_4 \sum x_4^2$

From Table 5.7 we have

$$\sum x_0 = 20$$
$$\sum x_j = 5 \quad j = 1, 2, 3, 4$$
$$\sum x_0 x_j = \sum x_j = 5 \quad j = 1, 2, 3, 4$$
$$\sum x_i x_j = 0 \quad i \neq j$$
$$\sum x_j^2 = 5 \quad j = 1, 2, 3, 4$$
$$\sum y = \text{grand total}$$
$$\sum x_j y = j\text{th subclass total}$$

Hence for the design in Table 5.2 (Table 5.7) the normal equations are (using the dot notation introduced earlier)

Grand total:	$y_{..} = 20\hat{\mu} + 5\hat{\beta}_1 + 5\hat{\beta}_2 + 5\hat{\beta}_3 + 5\hat{\beta}_4$	
First subclass total:	$y_{.1} = \ 5\hat{\mu} + 5\hat{\beta}_1$	
Second subclass total:	$y_{.2} = \ 5\hat{\mu} \qquad\quad + 5\hat{\beta}_2$	(5.17)
Third subclass total:	$y_{.3} = \ 5\hat{\mu} \qquad\qquad\qquad + 5\hat{\beta}_3$	
Fourth subclass total:	$y_{.4} = \ 5\hat{\mu} \qquad\qquad\qquad\qquad + 5\hat{\beta}_4$	

It is obvious from these equations that the first equation is the sum of the last four. Hence we have only four independent equations here. But there are five unknowns. Hence unique solutions of the unknowns are not possible. In fact any one of a number of solution sets can be easily verified to satisfy the above equations. Some examples are

Solution set 1:

$$\hat{\mu} = 0 \quad \hat{\beta}_1 = \bar{y}_{.1} \quad \hat{\beta}_2 = \bar{y}_{.2} \quad \hat{\beta}_3 = \bar{y}_{.3} \quad \hat{\beta}_4 = \bar{y}_{.4}$$

Solution set 2:

$$\hat{\mu} = \bar{y}_{..} \qquad \hat{\beta}_1 = \bar{y}_{.1} - \bar{y}_{..}$$

$$\hat{\beta}_2 = \bar{y}_{.2} - \bar{y}_{..} \qquad \hat{\beta}_3 = \bar{y}_{.3} - \bar{y}_{..} \qquad \hat{\beta}_4 = \bar{y}_{.4} - \bar{y}_{..}$$

Solution set 3:

$$\hat{\mu} = \bar{y}_{.4}$$

$$\hat{\beta}_1 = \bar{y}_{.1} - \bar{y}_{.4} \qquad \hat{\beta}_2 = \bar{y}_{.2} - \bar{y}_{.4} \qquad \hat{\beta}_3 = \bar{y}_{.3} - \bar{y}_{.4} \qquad \hat{\beta}_4 = 0$$

Solution set 4:

$$\hat{\mu} = \bar{y}_{.1}$$

$$\hat{\beta}_1 = 0 \qquad \hat{\beta}_2 = \bar{y}_{.2} - \bar{y}_{.1} \qquad \hat{\beta}_3 = \bar{y}_{.3} - \bar{y}_{.1} \qquad \hat{\beta}_4 = \bar{y}_{.4} - \bar{y}_{.1}$$

Before examining the common properties of these solution sets it may be useful to describe briefly how they were arrived at. The solutions of set 1 resulted from first stipulating arbitrarily that $\hat{\mu} = 0$ and then using this stipulation to solve for $\hat{\beta}_1, \hat{\beta}_2, \hat{\beta}_3$, and $\hat{\beta}_4$ from the last four equations in (5.17). It may be noticed that these are the least-squares estimates of $\beta_1', \beta_2', \beta_3'$, and β_4' in (5.16b). The solutions in set 2 resulted from first stipulating arbitrarily that $\hat{\beta}_1 + \hat{\beta}_2 + \hat{\beta}_3 + \hat{\beta}_4 = 0$. It is not difficult to verify that this stipulation leads to the solution $\hat{\mu} = \bar{y}_{..}$ from the first equation in (5.17). Inserting this solution for $\hat{\mu}$ into the remaining equations gives the solutions for $\hat{\beta}_j, j = 1, 2, 3, 4$, in set 2. As for the solution set 3, it was first arbitrarily stipulated that $\hat{\beta}_4 = 0$, and for the set 4, the arbitrary starting point was the stipulation $\hat{\beta}_1 = 0$.

Obviously one can find a large number of other solution sets in the same fashion. First arbitrarily introduce one stipulation, or one *side condition* as it is referred to in the literature. This, together with the normal equations, will lead to one complete set of solutions. (In more complex designs two or more side conditions may be necessary.)

Notice that in solution set 1 the side condition $\hat{\mu} = 0$ may be interpreted to mean that the solutions are expressed as deviations from $\hat{\mu}$, the combined effects of all constant factors. Similarly in solution set 2 the side condition $\hat{\beta}_1 + \hat{\beta}_2 + \hat{\beta}_3 + \hat{\beta}_4 = 0$ may be interpreted to mean that the solutions are expressed as deviations from the arithmetic mean of $\hat{\beta}_j$'s. The side condition thus specifies a reference point. For the solution set 3 the reference point is $\hat{\beta}_4$, and for the set 4 it is $\hat{\beta}_1$.

Let us now examine the common properties of these and similar solution sets.

Certain Properties of Least-Squares Estimates

PROPERTY 1 *Irrespective of the reference point (side condition) used, the least-squares estimates of comparison between β_j's are unique.* Suppose we want to estimate the comparison $\beta_1 - \beta_2$. The least-squares estimate of $\beta_1 - \beta_2$ is the corresponding comparison $\hat{\beta}_1 - \hat{\beta}_2$, in which $\hat{\beta}_1$ and $\hat{\beta}_2$ are taken from any solution set satisfying the normal equations. It is easy to verify that irrespective of the solution set used, $\hat{\beta}_1 - \hat{\beta}_2$ is equal to $\bar{y}_{.1} - \bar{y}_{.2}$ for the design under consideration. This result can be generalized as follows: the least-squares estimate of any treatment contrast $\psi = \sum l_j \beta_j$, where $\sum l_j = 0$, is $\hat{\psi} = \sum l_j \hat{\beta}_j$, in which $\hat{\beta}_j$'s are taken from any solution set satisfying the normal equations, $\hat{\psi}$ being the same irrespective of the solution set to which $\hat{\beta}_j$'s belong.

PROPERTY 2 *The side conditions used do not affect the predictions of individual y values.* Thus for the individual y values under treatment 1 in Table 5.7 we have the predicted y value

$$\hat{y}_{i1} = \hat{\mu} + \hat{\beta}_1 x_1 + \hat{\beta}_2 x_2 + \hat{\beta}_3 x_3 + \hat{\beta}_4 x_4 = \hat{\mu} + \hat{\beta}_1 = \bar{y}_{.1}$$

This predicted value is the same irrespective of the solution set employed. Similarly the predicted y values yielded by all solution sets for the observations under treatment 2 are $\bar{y}_{.2}$, for those under treatment 3, $\bar{y}_{.3}$, and for those under treatment 4, $\bar{y}_{.4}$.

PROPERTY 3 *A consequence of what we have just described is that the sum of squares due to regression, i.e., the SS of the deviations of the predicted values from their mean, does not differ by the particular solution set used.* In the present illustrative case this SS is

$$5 \sum (\bar{y}_{.j} - \bar{y}_{..})^2$$

which is the same as the between SS in the analysis of variance.

PROPERTY 4 *Since the predicted values do not differ by the particular solution set employed, neither will the difference between the observed and the corresponding predicted value vary from one solution set to another.* For the ith observation under treatment j in Table 5.7 the difference between the observed and the predicted value is

$$y_{ij} - \bar{y}_{.j}$$

The SS due to deviation from regression is thus seen to be the SS of $(y_{ij} - \bar{y}_{.j})$'s. This SS, that is, $\sum_i \sum_j (y_{ij} - \bar{y}_{.j})^2$, is the within SS in the analysis of variance.

The Least-Squares Theory

We are now ready to summarize the main results of the theory of least squares. Let us first define the concept of *estimable functions.* Any linear function of the parameters in linear (multiple-regression) models is called an *estimable function* if it has an unbiased linear estimate. As we have mentioned, by linear estimate is meant an estimate that can be expressed as a weighted sum of the observed y values, using known weights.

From what we have seen above with reference to (5.16a) it is clear that β_j is not estimable but linear functions such as $\beta_1 - \beta_2$, $\beta_1 + \beta_2 - 2\beta_3$, or in general $\sum l_j \beta_j$, where $\sum l_j = 0$, are all estimable. β_j is not estimable because we can estimate it uniquely only up to an additive constant, the value of the constant being dependent on the side condition(s) used in solving the normal equations.

We have mentioned that the least-squares estimate of any linear combination of the unknowns, for example, $\sum l_j \beta_j$, is the corresponding linear combination of the solutions of the normal equations. Thus, the least-squares estimate of $\beta_1 - \beta_2$ is $\hat{\beta}_1 - \hat{\beta}_2$, where $\hat{\beta}_1$ and $\hat{\beta}_2$ are taken from any solution set satisfying the normal equations. This means that $\hat{\beta}_1$ and $\hat{\beta}_2$ are dependent on one set of side conditions or another.

An important result in the least-squares theory is the Gauss-Markoff theorem, which may be stated as follows:

If the residuals in a linear model are independent and homoscedastic and have expectations zero, the least-squares estimate of every estimable function is unique and has smaller variance than any other linear unbiased estimate.

We shall not attempt to prove this theorem here but simply emphasize the following points:

1 The theorem holds even when the residuals are *not normally distributed.*
2 The theorem guarantees that all estimable functions and in particular all treatment contrasts (cf. Sec. 5.5) can be uniquely estimated by the least-squares method. *The least-squares estimates of all treatment contrasts are unbiased and have minimum variance.*

It must be emphasized that the Gauss-Markoff theorem deals only with estimates that are both linear and unbiased. There may exist in certain situations nonlinear estimates that are better, i.e., have smaller variance, than linear estimates. Thus in the case of certain parent universes that are thick-tailed, e.g., the Cauchy universe with the equation $p(x) = [\pi(1 + x^2)]^{-1}$ for the probability curve, it can be shown that the sample median, which is a nonlinear estimate, may be a better estimate of the population mean than the sample mean, which is the least-squares estimate.

Before leaving this section it may be of some interest to point out that under the Ω assumptions, which include normality, the maximum-likelihood estimates of estimable functions coincide with the corresponding least-squares estimates. For a discussion of the maximum-likelihood method of estimation and the attractive properties of the estimates provided by it, see textbooks such as Hoel (1962) or Rao (1965). The maximum-likelihood procedure can be applied only if the form of the distribution of each observation is specified. Under the Ω assumptions the observations are independently drawn from normal populations with common variance. Given these assumptions, the maximum-likelihood method can be shown to lead to the best (minimum-variance) estimates of estimable functions. In fact, in most practical situations, the maximum-likelihood estimate can be shown to possess a number of desirable large-sample properties. The equivalence of the least-squares method and the maximum-likelihood method thus puts the former in good company, providing another justification for using the least-squares method.

5.8 RANDOM-EFFECTS MODEL (MODEL II)

Let us now turn our attention to Eq. (5.4), in which we have assumed that the effects of the treatments in the experiment described in Table 5.2 are random variables. This amounts to considering the four treatments used in that experiment as a random sample from a universe of possible treatments. That is, we consider the four particular noise levels used in that experiment as a random sample from a universe of possible noise levels.

Perhaps before proceeding it may be helpful if we mention one practical situation in which the above model has been used in social research. In the survey-research literature, it is often contended that in interview surveys the interviewer may influence the response to certain questions. For example, consider a person's preferred family size. It is plausible that this is determined by the number of children already born, duration of marriage, age, knowledge of contraception, and a number of such factors. Assuming that at any particular time a person's preferred family size is a fixed number, we may imagine that this number has a certain relative frequency distribution in the population under study. If μ is the average of this distribution, any given person's preferred family size may be represented as $\mu + \varepsilon$, where ε has zero expectation. If interviewers tend to influence responses to interview questions, it seems reasonable to assume that any given respondent's preferred family size recorded in the interview instrument may differ from $\mu + \varepsilon$ by a quantity which we may call the *interviewer effect*. If the same interviewer interviews several persons, it is reasonable to assume that the interviewer effect will remain the same from one respondent to another, as far as the responses to a given question is concerned, unless there are interviewer-interviewee interaction. We shall assume that such an interaction is absent. From this perspective the ith observed value, e.g., the preferred family size of the ith respondent, of the jth interviewer may be represented as

$$y_{ij} = \mu + b_j + \varepsilon_{ij}$$

where b_j is the interviewer effect associated with the jth interviewer.

Let us suppose that the interviewers used in a particular study can be considered as a random sample from a universe of interviewers. Then the interviewer effect for any particular interviewer can be thought of as a random variable, since in repetitions of the study we imagine that at each repetition we select a fresh random sample of interviewers. Here lies the main difference between model I and model II. If we apply model I to the interviewer-effect problem, we imagine that the same group of interviewers will be used in all (hypothetical) repetitions of the experiment, whereas according to model II, the interviewers may differ from one (hypothetical) repetition of the experiment to another.

This difference between model I and model II leads to a corresponding difference in the expected value of the between MSS. We have seen earlier that for model I, for designs involving t treatments and r observations under each treatment

$$E \text{ (between MSS)} = \sigma^2 + \frac{r \sum (\beta_j - \bar{\beta}.)^2}{t - 1}$$

The analogous result for model II is

$$E \text{ (between MSS)} = \sigma^2 + r\sigma_b^2$$

where σ_b^2 is the variance of the treatment effects in the universe of treatments.[1]

This result can be derived by following an extension of the procedure expounded in Sec. 5.4. We replace β's by b's and regard b's as random variables. An easy way to derive the result is to carry out the derivation in two stages. First concentrate on repeated experiments using the same set of treatments, e.g., noise levels or interviewers,

---

[1] The balance of this discussion may be skimmed or omitted on first reading.

and then let the treatments vary. The average of the between SS over the repetitions using the same set of treatments will be called the *first-stage expectation* E_1 (between SS), where the subscript 1 indicates that the result pertains to the end of the first stage of the derivation. At the second stage we let the treatments vary, and the second-stage average of the first-stage expectations is denoted by $E_2 [E_1 \text{ (between SS)}]$. The reasonings of Sec. 5.4 apply as such to the derivation of E_1 (between SS), since b_j's remain constant. This gives us, for t treatments with r observations per treatment,

$$E_1 \text{ (between SS)} = \sigma^2 + \frac{r \sum (b_j - \bar{b})^2}{t - 1}$$

We now introduce the additional assumption that b_j's and ε's are uncorrelated with each other. Under this assumption, the first term on the right-hand side of the first-stage expectation just obtained will be σ^2 irrespective of the particular set of b_j's involved in the given experiment. Taking the second-stage expectation of E_1 (between SS), we get

$$E_2 [E_1 \text{ (between SS)}] = \sigma^2 + rE_2 \left[\frac{\sum (b_j - \bar{b})^2}{t - 1} \right]$$

We notice that when we treat b_j's as random variables in the second stage, $\sum (b_j - \bar{b})^2 / (t - 1)$ should be viewed as the sample variance of a random sample of size t of b's. Applying the well-known result that the expected value of $\sum (b_j - \bar{b})^2 / (t - 1)$ is the population variance of b's we have

$$E_2 \left[\frac{\sum (b_j - \bar{b})^2}{t - 1} \right] = \sigma_b^2$$

This proves that

$$E \text{ (between SS)} = \sigma^2 + r\sigma_b^2$$

It is not difficult to show by following the reasoning of Sec. 5.4 that under model II also

$$E \text{ (within MSS)} = \sigma^2$$

From these results it follows that

$$\hat{\sigma}_b^2 = \frac{\text{between MSS} - \text{within MSS}}{r}$$

is an unbiased estimate of σ_b^2 and the within MSS is an unbiased estimate of σ^2. The quantities σ_b^2 and σ^2 are called the *components of variance*, the former being the component due to treatments.

Note that the null hypothesis of equality between the treatment effects is identical to the null hypothesis stating that $\sigma_b^2 = 0$, since if $\sigma_b^2 = 0$, all treatment effects must be identical.

To test $\sigma_b^2 = 0$ we use the ratio (between MSS)/(within MSS) and consider it as an F under the Ω assumptions supplemented with the assumption that the treatment effects are normally distributed with variance σ_b^2.

The test statistic and the test procedure are thus seen to be the same in model I and model II for the one-way classification.

A Simulated Experiment

It is easy to illustrate model II by means of a simulated experiment. Let us simulate a set of interview responses. We shall assume that there are four interviewers and that they have been selected at random from a universe of interviewers. We shall also assume that the respondents have been selected at random from a universe of potential respondents and that the respondents have been randomly allocated to the interviewers.

To simulate the responses we recall that each response is assumed to have two additive components, the interviewer effect and the net effect of all the other relevant factors. Both these parts are assumed to be random variables. Let us assume both to be normally distributed. Let the interviewer effects be distributed as a $N(5,5^2)$(a normal distribution with mean 5 and standard deviation 5) and the second component be a $N(10,7^2)$. We shall further assume, in order to keep our simulation simple, that there are four interviewers and that each take five interviews. To simulate the responses we first take one observation at random from a $N(5,5^2)$; this represents the interviewer effect of one interviewer. We then select five observations at random from a $N(10,7^2)$. To the observed interviewer effect we add each of the other observations in turn to get five observations for the interviewer in question.[1] The whole procedure is repeated four times to give the results for four interviewers. The data shown in Table 5.8 have been obtained in the above fashion.

[1] An alternative procedure is to select random observations from $N(0,5^2)$ and $N(0,7^2)$ and then add 15 to their sum.

Table 5.8   SIMULATED DATA ON INTERVIEWER EFFECT

Interviewer number (1)	Observation number (2)	Interviewer effect $N(5,5^2)$ (3)	Error $N(10,7^2)$ (4)	Observation, (3) + (4) (5)
1	1	14.43	14.76	29.19
	2	14.43	21.86	36.29
	3	14.43	3.16	17.59
	4	14.43	13.52	27.95
	5	14.43	7.28	21.71
2	1	9.68	4.37	14.05
	2	9.68	15.88	25.56
	3	9.68	11.34	21.02
	4	9.68	11.15	20.83
	5	9.68	6.80	16.48
3	1	9.11	−6.75	2.36
	2	9.11	.07	9.18
	3	9.11	3.20	12.31
	4	9.11	9.51	18.62
	5	9.11	14.27	23.38
4	1	9.77	11.31	21.08
	2	9.77	4.48	14.25
	3	9.77	6.56	16.33
	4	9.77	7.25	17.02
	5	9.77	17.75	27.52

To analyze the data we calculate, as usual, the between SS and the within SS, which, along with the corresponding degrees of freedom and the MSS, are shown in Table 5.9.

Using formulas derived in the preceding subsection we obtain the estimates

$$\hat{\sigma}_b^2 = 21.60 \qquad \hat{\sigma}^2 = 41.55$$

To test for significance of $\hat{\sigma}_b^2$ we calculate the ratio of between MSS to within MSS and compare it with the F tabulated value for the appropriate degrees of freedom, here 3 and 16. Since the observed ratio is greater than the 5 percent tabulated F (2.85), we declare that the hypothesis $\sigma_b^2 = 0$ is not consistent with the data, and accordingly we reject it.

If we now wish to obtain interval estimates for σ_b^2, we may proceed in the following manner. We first get four values from the tabulated F distribution. These values are specified in terms of the following parameters:

Number of observations per treatment $= r = 5$ here

Degrees of freedom of between SS $= f_1 = 3$ here

Degrees of freedom of within SS $= f_2 = 16$ here

Confidence coefficient $= 90\%$, say

If we make use of the tabled F values for 5 percent significance levels, the confidence coefficients associated with the interval estimate will be $100 - 10$ percent. The values required from the F tables are

$$F_1 = \text{tabulated } F \text{ value with } (f_1, f_2) \text{ df} = 3.24$$

$$F_2 = \text{tabulated } F \text{ value with } (f_1, \infty) \text{ df} = 2.60$$

$$F_3 = \text{tabulated } F \text{ value with } (f_2, f_1) \text{ df} = 8.69$$

$$F_4 = \text{tabulated } F \text{ value with } (\infty, f_1) \text{ df} = 8.53$$

$$F = \text{observed } F \text{ value} = 3.5988$$

Table 5.9 ANALYSIS OF VARIANCE

Source	SS	df	MSS
Between	448.582	3	149.527
Within	664.795	16	41.550
Total	1,113.377	19	

Computation
$\hat{\sigma}_b^2 = \dfrac{149.527 - 41.550}{5} = 21.60 \qquad \hat{\sigma}^2 = 41.55$

The boundaries of the confidence interval for σ_b^2 are then

$$\text{Lower boundary} = \frac{(F - F_1)(F + F_1 - F_2)}{FF_2}\frac{\hat{\sigma}^2}{r} = 1.35$$

$$\text{Upper boundary} = \left(FF_4 - 1 + \frac{F_3 - F_4}{FF_3^2}\right)\frac{\hat{\sigma}^2}{r} = 246.78$$

In other words, the interval estimate for σ_b^2 is

$$1.35 \le \sigma_b^2 \le 246.78$$

with a confidence coefficient of 90 percent attached to it. It happens to be in this case a very wide interval!

Effects of Violations of Assumptions

In Sec. 5.1 it was mentioned that random-effects models are used when the inference problem is concerned with variances. Examples of inferences about variances are confidence interval for residual variance, for a variance component, e.g., the one due to treatments, or for a ratio of one variance component to another, and testing the hypothesis of equality of variances between a given number of groups. The assumptions on which such inferences are made, in the case of a one-factor experiment, may be summarized as follows:

$$y_{ij} = \mu + b_j + \varepsilon_{ij}$$

where b_j's are independently normally distributed with mean 0 and variance σ_b^2, ε_{ij}'s are independently normally distributed with mean 0 and variance σ^2, and b_j's and ε_{ij}'s are independent of each other.

It can easily be seen that we make more assumptions for random-effects models than fixed-effects models. We saw earlier that in fixed-effects models the inferences (about means) are not seriously invalidated by nonnormality. The story is not the same for inferences from random-effects models. Even for very large samples (numbers of observations under each treatment), departure from normality can have very serious consequences for the validity of inferences about variances.

Considering a one-factor design with r observations under each of t treatments Scheffé (1959) came to the following conclusions. For large r, if the kurtosis (= 3 less than the ratio of the fourth central moment to the square of the variance) of the universe of treatment effects is zero, nonnormality in these effects or in the residuals does not invalidate the inferences about the ratio σ_b^2/σ^2. But if the kurtosis is different from zero, confidence coefficients and the probabilities of errors of the first and second kinds are seriously affected, except that in testing the hypothesis that $\sigma_b^2 = 0$, the probability of type I errors is not affected. If the kurtosis is less than zero, for confidence $1 - \alpha$ and significance level α, the true α will be less than the one the analyst thinks it is, while the opposite will be true if the kurtosis is greater than zero.

The effects of violations other than nonnormality on the inferences about variances have not been adequately investigated, but there is reason to believe that any violation tends to have more serious effects on inferences about variances than on those about means.

5.9 TESTS OF HOMOGENEITY OF VARIANCE

Since heteroscedasticity is likely to have a serious impact on the inferences drawn using model II, it may be wise to test whether heteroscedasticity prevails in the data taken for analysis. The following test due to Bartlett (1937) may be made use of for that purpose.

When we discussed the analysis of variance under model I, we mentioned that if homoscedasticity can be assumed, we can estimate the variance of the residual components separately from the observations under each of the treatment groups. Thus if $y_{1j}, y_{2j}, \ldots, y_{rj}$ are r observations under treatment j, then

$$\frac{1}{r-1} \sum_{i=1}^{r} (y_{ij} - \bar{y}_{.j})^2$$

is an unbiased estimate of σ^2, the variance of the residual components. If there are t treatments, we can have in this way t independent estimates of σ^2, each with $r - 1 = f$, say, degrees of freedom. (See below for the case when the degrees of freedom differ.) We proceed to calculate a test statistic involving these estimates as follows: Let $s_1^2, s_2^2, \ldots, s_t^2$ be the t independent estimates of σ^2 obtained from the observations under the t treatments. Let

$$M = (\ln 10) f \left(t \ln \frac{\sum s_i^2}{t} - \sum \ln s_i^2 \right)$$

and

$$C = 1 + \frac{t+1}{3tf}$$

If we now take

$$B = \frac{M}{C}$$

we have a test statistic for testing the null hypothesis positing homoscedasticity.

The statistic B is approximately distributed as a chi square with $t - 1$ degrees of freedom.

If the degrees of freedom of the different estimates of σ^2 differ, the following formulas are to be used in calculating M and C:

$$M = (\ln 10) \left(\sum f_i \ln \frac{\sum f_i s_i^2}{\sum f_i} - \sum f_i \ln s_i^2 \right)$$

$$C = 1 + \frac{1}{3(t-1)} \left(\sum \frac{1}{f_i} - \frac{1}{\sum f_i} \right)$$

Table 5.10 illustrates the calculations involved.

It should be noted that the chi-square approximation for the sampling distribution of B becomes less satisfactory if many of the f_i values (degrees of freedom of the independent estimates of σ^2) are less than 5. In that case instead of using the chi-square tables, one should use special tables prepared for the purpose (Pearson and Hartley, 1958, table 31). Special tables are also available (Pearson and Hartley, 1958, table 32) for testing the homoscedasticity hypothesis employing the ratio of the maximum of the s_i^2's to the minimum of the s_i^2's. This latter test, although less sensitive than Bartlett's test, will often be enough.

Both the above tests are sensitive to nonnormality, particularly to positive kurtosis. When the normality assumption is violated seriously, the tests described above would lead to too many erroneous rejections of the null hypothesis (see Box, 1953).

5.10   INTRACLASS CORRELATION

In Sec. 5.8 we described how the interviewer effect in survey responses can be studied using model II. Another approach to studying problems such as interviewer effect is based on the concept of *intraclass correlation*. Although the origin of this concept owes nothing to the logic underlying the model II analysis of variance, the calculations involved in the intraclass correlation have much in common with those for the model II analysis of variance.

When put as an alternative model for the study of interviewer effect, the intraclass-correlation approach can be described as follows. While the responses obtained by the jth interviewer are all distributed about the same mean μ_j and variance σ^2, any two of those responses have a common correlation coefficient, say ρ_I, which may be called the *intraclass correlation*. (This is assumed to be the same for all interviewers.) Thus if we refer to one response as x and another as y and form all possible pairs of

Table 5.10   BARTLETT'S TEST FOR HOMOGENEITY OF VARIANCE

Treatment	SS	df f_i	MSS s_i^2	$\ln s_i^2$	$f_i(\ln s_i^2)$	$\dfrac{1}{f_i}$
1	256.0	20	12.80	1.10721	22.14420	.050000
2	170.0	13	13.08	1.11661	14.51593	.076923
3	121.0	15	8.07	.90687	13.60305	.066667
4	99.0	10	9.90	.99563	9.95630	.100000
Total	646.0	58			60.21948	.293590

Computations

$$s^2 = \frac{\sum f_i s_i^2}{\sum f_i} = \frac{646.0}{58} = 11.13793$$

$$\sum f_i \ln s^2 = 58 \times 1.04680 = 60.71440$$

$$M = 2.3026[\sum f_i \ln s^2 - \sum (f_i \ln s_i^2)]$$
$$= 2.3026(60.71440 - 60.21948) = 1.13960$$

$$C = 1 + \frac{1}{3 \times 3}\left(\sum \frac{1}{f_i} - \frac{1}{\sum f_i}\right) = 1 + \tfrac{1}{9}(.29359 - .01724)$$
$$= 1.03071$$

$$\frac{M}{C} = \frac{1.13960}{1.03071} = 1.106 \quad \text{with 3 df}$$

responses for each interviewer, we would have an x array and a y array, the product-moment correlation coefficient obtained therefrom being an estimate of the intraclass correlation.

For a simple example, consider that five interviewers brought in the following responses:

Interviewer 1	(1,2,4)
Interviewer 2	(2,3,4)
Interviewer 3	(3,4,4)
Interviewer 4	(0,2,2)
Interviewer 5	(3,4,5)

The different possible pairs of responses "within" each interviewer and the calculation of the intraclass correlation are shown in Table 5.11. Note that in forming all possible pairs of observations within each interviewer, each observation appears as many times in the x column as it appears in the y column. Thus a pair of observations, say, (1,2), would appear as $(x = 1, y = 2)$ and $(x = 2, y = 1)$. Hence the sum of the entries in the x column will be equal to that of those in the y column, the same being true with respect to the SS.

If the number of observations in each class is the same, the intraclass correlation can be more directly estimated in the following manner:

$$r_I = \frac{\text{(number of observations per class)(between SS)} - \text{total SS}}{\text{(number of observations per class} - 1)\text{(total SS)}}$$

Table 5.12 illustrates this method of calculating r_I.

Table 5.11   DIFFERENT POSSIBLE PAIRS OF RESPONSES FORMED FROM THE RESPONSES OBTAINED BY EACH INTERVIEWER AND THE STEPS IN THE CALCULATION OF INTRACLASS CORRELATION

	Interviewer										Total	
	1		2		3		4		5			
	x	y	x	y	x	y	x	y	x	y	x	y
	1	2	2	3	3	4	0	2	3	4		
	1	4	2	4	3	4	0	2	3	5		
	2	1	3	2	4	3	2	0	4	3		
	2	4	3	4	4	4	2	2	4	5		
	4	1	4	2	4	3	2	0	5	3		
	4	2	4	3	4	4	2	2	5	4		
Sum	14	14	18	18	22	22	8	8	24	24	86	86
SS	42	42	58	58	82	82	16	16	100	100	298	298
$\sum xy$	28		52		80		8		94		262	

$$r_I = \frac{262 - 86^2/30}{\sqrt{298 - 86^2/30} \ \sqrt{298 - 86^2/30}} = .30$$

It can be shown algebraically that under the model described above involving intraclass correlation

$$E \text{ (between MSS)} = \sigma^2[1 + (r - 1)\rho_I]$$
$$E \text{ (within MSS)} = \sigma^2(1 - \rho_I)$$

where r is the number of observations in each class and ρ_I is the population value of the intraclass correlation.

From the above relationships we note that

$$\text{Between MSS} - \text{within MSS}$$

may be taken as an estimate of $\sigma^2 r \rho_I$ and that

$$\text{Between MSS} + (r - 1) \text{ (within MSS)}$$

may be taken as an estimate of $r\sigma^2$. This suggests that

$$\hat{\rho}_I = \frac{\text{between MSS} - \text{within MSS}}{\text{between MSS} + (r - 1) \text{ (within MSS)}}$$

may be taken as an estimate of the intraclass correlation. The estimate thus obtained may not, however, be always the same as the one obtained in the manner described in

Table 5.12   AN ALTERNATIVE PROCEDURE FOR CALCULATION OF INTRACLASS CORRELATION

	Interviewer					
	1	2	3	4	5	Total
	1	2	3	0	3	
	2	3	4	2	4	
	4	4	4	2	5	
Sum	7	9	11	4	12	43
$\sum y^2$	21	29	41	8	50	149
Sum^2	49	81	121	16	144	411

$$\text{Total SS} = 149 - \frac{43^2}{15} = 25.7333$$

$$\text{Between SS} = \frac{411}{3} - \frac{43^2}{15} = 13.7333$$

$$\text{Within SS} = 149 - \frac{411}{3} = 12.0000$$

$$r_I = \frac{3 \times 13.7333 - 25.7333}{2 \times 25.7333} = \frac{15.46667}{51.46667} = .30$$

Tables 5.11 and 5.12. The reason can be seen by noting that according to the procedure described in Table 5.12,

$$r_I = \frac{r \text{ (between SS)} - \text{total SS}}{(r - 1) \text{ (total SS)}}$$

$$= \frac{r \text{ (between SS)} - \text{(between SS + within SS)}}{(r - 1) \text{ (between SS + within SS)}}$$

$$= \frac{(r - 1) \text{ (between SS)} - \text{within SS}}{(r - 1) \text{ (between SS)} + (r - 1) \text{ (within SS)}}$$

which simplifies to

$$r_I = \frac{(t - 1) \text{ (between MSS)} - t \text{ (within MSS)}}{(t - 1) \text{ (between MSS)} + t(r - 1) \text{ (within MSS)}}$$

since

$$\text{Between MSS} = \frac{\text{between SS}}{t - 1}$$

and

$$\text{Within MSS} = \frac{\text{within SS}}{t(r - 1)}$$

It is not difficult to see that unless t is very large,

$$r_I = \frac{(t - 1) \text{ (between MSS)} - t \text{ (within MSS)}}{(t - 1) \text{ (between MSS)} + t(r - 1) \text{ (within MSS)}}$$

may differ from

$$\hat{\rho}_I = \frac{\text{between MSS} - \text{within MSS}}{\text{between MSS} + (r - 1) \text{ (within MSS)}}$$

Many statisticians seem to adopt in practice the $\hat{\rho}_I$ formula more frequently than the r_I formula for calculating the intraclass correlation. For an application of the concept of intraclass correlation in the analysis of interviewer effect see Kish (1962). In the literature on survey sampling one finds the concept of intraclass correlation playing a crucial role in the study of the effect of clustering (see Kish, 1965).

5.11 CONCLUDING OBSERVATIONS

Although the focus of attention in this chapter has been one-factor experiments, several of the ideas introduced in this chapter are of relevance to what follows in subsequent chapters. Particularly worth mentioning in this connection are the following:

    A Concepts
        1 Components of variance (Sec. 5.8)
        2 Decomposition of total sum of squares (Sec. 5.2)
        3 Degrees of freedom (Sec. 5.3)
        4 Estimable functions (Sec. 5.7)

It is also worth emphasizing that the Gauss-Markoff theorem referred to in this chapter is of relevance to all inference procedures involving the application of least-squares estimation in linear models.

EXERCISES

5.1 Suppose that the following are the results of a study of gasoline consumption of cars of four different makes.

Make	Consumption, mi/gal				
A	22	20	21	20	25
B	23	24	25	22	
C	14	15	17	16	19	20
D	13	12	18	17	15	

(a) Do cars of these four makes differ in gas consumption?

(b) Which makes are similar in gas consumption, and which ones are dissimilar? Or are all makes different from each other in this respect?

(c) Compare the average gas consumption of cars of makes A and C with the average gas consumption of cars of makes B and D.

5.2 Suppose the following data indicate how family income varies by type of residence:

Residence	Income, $1,000							
Urban	8	12	11	10	15	17	21	20
Rural farm	5	8	9	6				
Rural nonfarm	9	9	8	11	13			

(a) Is there sufficient evidence in this data set to infer that family income varies by type of residence?

(b) Do rural farm residents differ in family income from rural nonfarm residents?

(c) Do rural nonfarm residents differ in family income from urban residents?

(d) Is it valid to assume that on the average the income level of rural nonfarm residents lies midway between the income levels of urban residents and rural farm residents?

5.3 Suppose that the following data represent variation in crime rate by type of community:

Community type	Crime rate					
Administrative centers	7	10	6	8	5	
Education centers	4	3	5	3		
Industrial centers	9	12	11	8	10	9

(a) You are asked to fit by the method of least squares a model of the type

$$y = \mu x_0 + \beta_1 x_1 + \beta_2 x_2 + \beta_3 x_3 + \varepsilon$$

where y = crime rate and

$$x_0 = 1$$

$$x_1 = \begin{cases} 1 & \text{for administrative centers} \\ 0 & \text{for all others} \end{cases}$$

$$x_2 = \begin{cases} 1 & \text{for education centers} \\ 0 & \text{for all others} \end{cases}$$

$$x_3 = \begin{cases} 1 & \text{for industrial centers} \\ 0 & \text{for all others} \end{cases}$$

Verify that the normal equations for fitting the above model to the data set at hand are

$$y_{..} = 15\hat{\mu} + 5\hat{\beta}_1 + 4\hat{\beta}_2 + 6\hat{\beta}_3$$
$$y_{.1} = 5\hat{\mu} + 5\hat{\beta}_1$$
$$y_{.2} = 4\hat{\mu} + 4\hat{\beta}_2$$
$$y_{.3} = 6\hat{\mu} + 6\hat{\beta}_3$$

Solve the normal equations with the help of the side condition

$$\hat{\beta}_1 + \hat{\beta}_2 + \hat{\beta}_3 = 0$$

(b) Obtain the SS due to regression and the SS due to deviation from regression.

(c) Verify that the SS due to regression is the same as the between SS you would obtain if you were to follow the procedure described in Sec. 5.2.

(d) Verify that the SS due to deviation from regression is the same as the within SS you would obtain if you were to follow the procedure described in Sec. 5.2.

5.4 Suppose that you are asked to fit by the method of least squares the following model for the data in Exercise 5.3:

$$y = \mu x_0 + \beta_1 x_1 + \beta_2 x_2 + 0 x_3 + \varepsilon$$

where x_0, x_1, x_2, and x_3 are defined as in Exercise 5.3. Note that the coefficient of x_3 is arbitrarily set equal to zero in this model.

(a) Verify that the normal equations for fitting this model are

$$y_{..} = 15\hat{\mu} + 5\hat{\beta}_1 + 4\hat{\beta}_2$$
$$y_{.1} = 5\hat{\mu} + 5\hat{\beta}_1$$
$$y_{.2} = 4\hat{\mu} \qquad + 4\hat{\beta}_2$$

(b) Verify that the solutions of these normal equations are exactly the same as the ones you would have obtained in Exercise 5.3 if you had used the side condition $\hat{\beta}_3 = 0$.

(c) Verify that the above procedure is equivalent to fitting the model

$$y = \mu x_0 + \beta_1 x_1 + \beta_2 x_2 + \varepsilon$$

where x_0 is defined as in Exercise 5.3 and x_1 and x_2 are defined as follows:

$x_1 = 1$	$x_2 = 0$	for all administrative centers
$x_1 = 0$	$x_2 = 1$	for all education centers
$x_1 = 0$	$x_2 = 0$	for all industrial centers

(Regressors defined in this fashion are referred to as *dummy variables* in the literature.)

5.5 Demonstrate that with reference to the data set in Exercise 5.3, setting $\hat{\beta}_1 = 0$ as the side condition is equivalent to working with the following dummy variables as regressors:

$x_2 = 0$	$x_3 = 0$	for administrative centers
$x_2 = 1$	$x_3 = 0$	for education centers
$x_2 = 0$	$x_3 = 1$	for industrial centers

5.6 The term likelihood was used without any elaboration in Sec. 5.7. To understand the meaning of this term let us imagine that n values $(y_1, y_2, \ldots, y_n)$ have been selected at random from a normal universe with mean μ and variance σ^2. We now ask: In repeated sampling in what proportion of repetitions may we expect the particular set of values $(y_1, y_2, \ldots, y_n)$ to be selected? It is not difficult to see that in the present case the answer to this question is a constant multiple of the product of the n ordinates at $y_1, y_2, \ldots, y_n$ of the normal curve with mean μ and variance σ^2, that is, a constant multiple of

$$L = \left(\frac{1}{\sqrt{2\pi}\sigma}\right)^n \exp\left[-\frac{1}{2\sigma^2}\sum_{i=1}^{n}(y_i - \mu)^2\right]$$

The quantity L thus defined is called the likelihood of the observed data set.

(a) What is the likelihood of the observations in Table 5.2 given that model (5.6) and the Ω assumptions hold?

(b) What is the likelihood of the observations in Table 5.2 given that model (5.5) and the Ω assumptions hold?

5.7 Any exponential function of the form $e^{-(2-\beta)^2}$ is maximum when the exponent $(2 - \beta)^2$ is minimum. In general, maximizing a function with a negative exponent involves minimizing the magnitude of the exponent. Using this principle, show that the values of μ and β_j's that maximize

$$L = \left(\frac{1}{\sqrt{2\pi\sigma}}\right)^n \exp\left[-\frac{1}{2\sigma^2}\sum_i \sum_j (y_{ij} - \mu - \beta_j)^2\right]$$

is the same as the values of μ and β_j's that minimize

$$\sum_i \sum_j (y_{ij} - \mu - \beta_j)^2$$

This illustrates the equivalence of the maximum-likelihood estimates and the least-squares estimates under the Ω assumptions.

5.8 Referring to the simple regression model $y = \alpha + \beta x + \varepsilon$ (see Chap. 4) in which x is an interval-scaled variable, show that the least-squares estimates of α and β are linear estimates. Can you extend the above result to the multiple-regression situation in which all regressors are interval-scaled variables?

5.9 Suppose that you have data on "success orientation" and "self-esteem" for a random sample of n Catholics and m Protestants. Outline the procedure you would employ to test the following hypotheses:

(a) The relation between self-esteem and success orientation is the same for both the religious groups, in the sense that "the expected change in success orientation per unit change in self-esteem is the same in both the religious groups."

(b) Catholics and Protestants differ in success orientation, but this difference cannot be explained by the corresponding differential in the relationship between self-esteem and success orientation.

(Before attempting this exercise you may want to review the material presented under covariance analysis in Chap. 4.)

6

PROCEDURES TO REDUCE RESIDUAL VARIATION: TWO-WAY LAYOUT

In the preceding chapter we saw that the statistical analysis of designed experiments is structured around the notion that the observed value of a study phenomenon can be expressed as the sum of two parts, the accounted for effects and the residual. The former are sometimes referred to as the effects under investigation and the latter as the uncontrolled or unaccounted for effects. From our discussion of one-factor designs it should be clear that the residual variation, i.e., the variance of the residuals, plays a crucial role in testing hypotheses concerning the effects under investigation. This is obvious from the fact that in all test statistics employed in the preceding chapter for the analysis of one-factor experiments we find the quantity within MSS appearing in the denominator. It may be recalled that within MSS is our best estimate (under the usual assumptions) of the residual variation in one-factor experiments. Obviously then for any given number of observations, i.e., for any given size of the experiment, the larger the residual variation the smaller the chance of detecting true differences, if any, between the effects under investigation. This motivates the experimenter to search for methods to reduce the residual variation. One method of doing so is to identify and keep constant throughout the experiment one or more of the uncontrolled factors. An alternative method is to transfer, if possible, some of the residual factors to the category of effects under investigation. This chapter deals with a number of designs in which this second strategy is employed.

6.1 BLOCK DESIGNS

An important class of experimental designs in which two or more causal factors are explicitly recognized and accounted for, or regarded as factors under investigation, is the *block designs*. All block designs are based on the principle of grouping the experimental units, e.g., automobile tires, rural communities, human subjects, in such a way that each block consists of more or less homogeneous units, i.e., units more or less alike with respect to the causal factors used for blocking. A few examples are given below.

1 If we are interested in the wearing qualities of automobile tires, it is natural to consider the four tires of the same automobile as a block. All the four tires of a given automobile can be taken to be exposed to more or less the same conditions at any given time, especially if the tires are rotated frequently. This cannot be said about a group of tires consisting of the front tires of one automobile and the rear ones of another, for different automobiles may traverse different terrain, be driven by different drivers, etc.

2 In an experiment aimed at comparing different methods of teaching, the subjects may be blocked according to their age, years of schooling, and background knowledge ascertained perhaps by pretesting.

3 In a family-planning action research project in which different methods of communication are to be compared, a block may consist of a number of areal units, e.g., rural communities, that are more or less alike with respect to demographic and socioeconomic characteristics.

Main Features

The above examples help us note the following points.

1 Blocking essentially involves the same strategy as "keeping constant" some of the causal factors of the study phenomenon.

2 A number of causal factors may be simultaneously "kept constant" in a particular blocking procedure.

3 Blocking does not necessarily involve "keeping constant" a given set of causal factors in all possible ways. In the experiment involving automobile tires, for example, all possible terrain and all possible driving conditions need not be considered. The findings about the comparative performance of four tires under the worst possible conditions are likely to hold true under better conditions.

4 Often the experimenter may consider one or more of the causal factors used for blocking as *nuisance* variables in the sense that their effects on the study phenomenon are of no intrinsic interest. The appropriate blocking strategy would differ according as a given block variable, i.e., variable used for blocking, is a nuisance variable or not. If, for example, sex is considered a nuisance variable, it is enough to confine the experiment to units of any one sex, but then the findings of the experiments may not necessarily be applicable to subjects of the other sex. If, however, the experimenter is interested in assessing the effect of sex difference on the study phenomenon, e.g., sex difference in the responsiveness to given teaching methods, one or more blocks should consist of males only and one or more blocks of females.

5 Sometimes the experimenter may be faced with the following situation. He may be interested in the comparative effects of the different categories of a block variable, but the different possible categories of the block variable may be so numerous that all of them cannot be simultaneously incorporated in one experiment. The variable labeled "demographic and socioeconomic characteristics" in one of the examples given above illustrates such a situation. The question then arises: how to select a few categories of the block variable to be incorporated in the experiment. The strategy usually followed is to select a random sample from among the different possible categories. If random sampling is employed, the findings can be generalized to cover all the possible categories. Needless to say, if probability sampling has not been employed, the findings are strictly applicable only to the categories actually used in the experiment.

6 In many situations the causal factors used for blocking may be factors that are not amenable to manipulation by the experimenter. Sex, age, race, and genetic composition are examples of such factors. There are, of course, situations in which the factor used for blocking is one that can be manipulated by the experimenter. An example is the level of information about a relevant topic of the experimental subjects. Whatever the nature of the factor chosen for blocking, a noteworthy feature is the way the factor in question is handled in the conduct of the experiment. No attempt is made by the experimenter to manipulate the factor; the experimental subjects are simply classified into various categories on the basis of the observed values (scores) on that factor. For example, if age is a factor chosen for blocking, all the experimenter does is to stratify the subjects into various age groups. Factors that are manipulated by the experimenter are therefore of a status distinct from blocking factors. (We refer to the former factors as treatments.) Because of this difference in status, inferences concerning the causal connection between the blocking variables and the response variable will be relatively more obscure than the inferences concerning the causal connection between the treatment variable and the response variable unless the treatment involves a very complex stimulus.

6.2 RANDOMIZED-BLOCKS DESIGN

In order to see the role played by the blocking procedure in reducing the standard errors of treatment contrasts, let us first examine the simplest block design, namely, the *randomized-blocks design*.

Example We shall start with an example. An investigator wanted to compare four teaching methods. He grouped 20 students who volunteered as subjects for the experiment into five groups, making each group as homogeneous as possible with respect to such factors as age, number of school years completed, score received in a pretest, etc. The students in each block were then assigned at random to the different teaching methods; each method had assigned to it one and only one student in each block. In order to maintain uniformity in the exposure of different subjects to any given teaching method, video-tape recording of the instructor giving the lecture or demonstration was used. Each student was exposed to the appropriate video tape independently of the others. At the end of the exposure, all students were given a common test. Suppose the figures in Table 6.1 represent the test results, i.e., the data to be analyzed. The aim is to find out whether the four teaching methods differ in their effectiveness as reflected in the scores in the test.

Table 6.1 AN EXPERIMENT ON TEACHING METHODS USING RANDOMIZED-BLOCKS DESIGN

A Data: Scores of Students in a Common Test at the End of the Experiment

Block	Teaching method (treatment)				Total
	1	2	3	4	
1	7	1	6	15	29
2	8	4	4	11	27
3	6	5	9	16	36
4	3	2	6	15	26
5	6	3	10	13	32
Total	30	15	35	70	150

B Analysis Procedure

Total number of observations $= n = 20$

$$\text{Correction for mean} = C = \frac{(\text{grand total})^2}{n} = \frac{150^2}{20} = 1{,}125.0$$

$$\text{Total SS} = 7^2 + 8^2 + \cdots + 15^2 + 13^2 - C$$
$$= 1{,}514.0 - 1{,}125.0 = 389.0$$
$$\text{Treatment SS} = \tfrac{1}{5}(30^2 + 15^2 + 35^2 + 70^2) - C$$
$$= 1{,}450.0 - 1{,}125.0 = 325.0$$
$$\text{Block SS} = \tfrac{1}{4}(29^2 + 27^2 + 36^2 + 26^2 + 32^2) - C$$
$$= 1{,}141.5 - 1{,}125.0 = 16.5$$
$$\text{Residual SS} = \text{total SS} - \text{block SS} - \text{treatment SS}$$
$$= 389.0 - 325.0 - 16.5 = 47.5$$

C Analysis-of-Variance Table

Source of variation	SS	df	MSS	*F* ratio
Between treatments	325.0	3	108.33	27.36
Between blocks	16.5	4	4.12	1.04
Residual	47.5	12	3.96	
Total	389.0	19		

The conventional method of analysis starts with calculating the treatment totals, the block totals, and the grand total. The correction for the mean is then obtained as $C = $ (grand total)$^2/n$, where n is the total number of observations. The total SS and the SS due to treatment differences are calculated in the same way as in the one-way classification analysis of variance. The extra SS to be calculated in the present case concerns the variation attributable to the block effects. The rule for calculating the block SS is the same as that for calculating the treatment SS: first find the SS of the block totals; then divide that sum by the number of observations per block; finally subtract from the result so obtained C, the correction for the mean.

The residual SS is obtained by subtraction:

$$\text{Residual SS} = \text{total SS} - \text{block SS} - \text{treatment SS}$$

These calculations are illustrated in Table 6.1B, and the complete analysis of variance is shown in Table 6.1C.

From the analysis-of-variance table, we note that the F ratio for the treatments is 27.36 with 3 and 12 degrees of freedom, which is highly significant, the corresponding tabulated values of F being 3.49 for the 5 percent level and 5.95 for the 1 percent level of significance. (The assumptions underlying the test will be discussed shortly.) The F ratio of blocks, however, is not significant, which incidentally tells us that the particular blocking procedure attempted in this experiment did not accomplish much. (A method of estimating the gain due to blocking will be discussed shortly.)

If the treatment SS are found to be significant, it may be desirable to make further investigation of the differences between the treatment effects. This can be done following the procedure described in Chap. 5 for multiple comparisons. The LSD method or any one of several other methods may be employed for the purpose. In adapting the formulas of Chap. 5 for application to the randomized-blocks designs one should replace within MSS by residual MSS. The remaining steps are exactly the same as those described for one-way classifications in Chap. 5.

We shall illustrate the application of one multiple-comparison procedure, the Newman-Keuls method, to the data in Table 6.1. First we arrange the treatment means in ascending order of magnitude:

Treatment	$y_{.J}$	$\bar{y}_{.J}$
2	15	3
1	30	6
3	35	7
4	70	14

We now calculate

$$N_v = Q_v \sqrt{\frac{\text{residual MSS}}{r}}$$

where $v = $ number of steps between ordered treatment means

$r = $ number of observations per treatment

$Q_v = $ tabulated value of Studentized range

For $v = 2$, 3, and 4 and for a residual MSS with 12 degrees of freedom, we have, corresponding to the 5 percent level of significance, from Appendix Table D.4, $Q_2 = $

3.08, $Q_3 = 3.77$, and $Q_4 = 4.20$. Hence $N_2 = 3.08\sqrt{3.96/5} = 2.74$, $N_3 = 3.77\sqrt{3.96/5} = 3.36$, and $N_4 = 4.20\sqrt{3.96/5} = 3.74$. Testing the treatment differences, we notice

$$\text{Treatment } 4 - \text{treatment } 2 = 14 - 3 = 11 > N_4$$
$$\text{Treatment } 4 - \text{treatment } 1 = 14 - 6 = 8 > N_3$$
$$\text{Treatment } 4 - \text{treatment } 3 = 14 - 7 = 7 > N_2$$
$$\text{Treatment } 3 - \text{treatment } 2 = 7 - 3 = 4 > N_3$$
$$\text{Treatment } 3 - \text{treatment } 1 = 7 - 6 = 1 < N_2$$
$$\text{Treatment } 1 - \text{treatment } 2 = 6 - 3 = 3 > N_2$$

We thus see that all pairwise comparisons, except the one between treatments 1 and 3, are significant.

Mathematical Model

The data obtained in an experiment using a randomized-blocks design with t treatments and r blocks illustrates a two-way, $r \times t$, classification with one observation in each cell. Table 6.2 gives an algebraic representation of data sets from such experiments. Each treatment occurs once in each block. The observation under treatment j in the ith block is usually represented as

$$y_{ij} = \mu + \alpha_i + \beta_j + \varepsilon_{ij} \qquad (6.1)$$

where μ = constant which may be interpreted as representing combined effect of all constant factors

α_i = effect due to characteristics peculiar to ith block

β_j = effect of treatment j

$\varepsilon_{ij} = y_{ij} - \mu - \alpha_i - \beta_j$ = the residual effect

The last represents the total of all variable but unaccounted for effects.

A minimal assumption usually made about the residuals is that they all have expectation zero, that is, $E(\varepsilon_{ij}) = 0$, for all i and j. It is interesting to note that if the residuals all have a common expected value $\mu^* \neq 0$, then we can absorb μ^* in μ of

Table 6.2   ALGEBRAIC REPRESENTATION OF DATA FROM A TWO-WAY CLASSIFICATION WITH ONE OBSERVATION IN EACH CELL

Row classification (block)	Column classification (treatment)				Total	Mean
	1	2	$\cdots$	t		
1	y_{11}	y_{12}	$\cdots$	y_{1t}	$y_1.$	$\bar{y}_1.$
2	y_{21}	y_{22}	$\cdots$	y_{2t}	$y_2.$	$\bar{y}_2.$
r	y_{r1}	y_{r2}	$\cdots$	y_{rt}	$y_r.$	$\bar{y}_r.$
Total	$y_{.1}$	$y_{.2}$	$\cdots$	$y_{.t}$	$y_{..}$	
Mean	$\bar{y}_{.1}$	$\bar{y}_{.2}$	$\cdots$	$\bar{y}_{.t}$		$\bar{y}_{..}$

(6.1) and treat as residuals the deviations of the original residuals from μ^*. These deviations will then have expectation zero; however, we shall not be able to obtain separate estimates for μ and μ^*. Fortunately this creates no problem. Whether the first term on the right-hand side of (6.1) represents μ or $\mu + \mu^*$ or some other constant does not matter as far as making inferences about comparisons between β_j's and α_i's is concerned. This will become clear in a moment. Suffice it to emphasize here that it makes no practical difference whether the residuals all have zero expectations or all have a *common* nonzero expectation. What really makes a difference is when the residuals differ between themselves in expected values. Before examining the implications of such a situation let us consider the implications of model (6.1) under the condition of zero expectation for all the residuals.

First we notice that we can rewrite (6.1) as a multiple-regression model in the form

$$y_{ij} = \mu x_0 + \alpha_1 x_1 + \cdots + \alpha_r x_r + \beta_1 z_1 + \cdots + \beta_t z_t + \varepsilon_{ij} \qquad (6.2)$$

where $x_1, x_2, \ldots, x_r$ and $z_1, z_2, \ldots, z_t$ are a set of $r + t$ regressors such that for the observation under treatment j in the ith block, i.e., for y_{ij}, $x_i = 1$ but $x_k = 0$ if $k \neq i$; and $z_j = 1$ but $z_l = 0$ if $l \neq j$; while x_0 is equal to 1 for all observations. To fix our ideas let us confine attention for the moment to a particular design with $t = 4$ treatments and $r = 3$ blocks. Table 6.3 presents the y values arranged in a standard order and the corresponding x and z values.

Table 6.3 y VALUES AND REGRESSOR VALUES FOR A RANDOMIZED-BLOCKS DESIGN WITH FOUR TREATMENTS AND THREE BLOCKS, EACH TREATMENT OCCURRING ONCE IN EACH BLOCK

y	x_0†	Block regressors			Treatment regressors			
		x_1	x_2	x_3	z_1	z_2	z_3	z_4
Block 1								
y_{11}	1	1	0	0	1	0	0	0
y_{12}	1	1	0	0	0	1	0	0
y_{13}	1	1	0	0	0	0	1	0
y_{14}	1	1	0	0	0	0	0	1
Block 2								
y_{21}	1	0	1	0	1	0	0	0
y_{22}	1	0	1	0	0	1	0	0
y_{23}	1	0	1	0	0	0	1	0
y_{24}	1	0	1	0	0	0	0	1
Block 3								
y_{31}	1	0	0	1	1	0	0	0
y_{32}	1	0	0	1	0	1	0	0
y_{33}	1	0	0	1	0	0	1	0
y_{34}	1	0	0	1	0	0	0	1

† Constant.

The normal equations are easily written as shown below using the familiar dot notation.

$$\sum \varepsilon x_0 = 0: \quad y_{..} = 12\hat{\mu} + 4\hat{\alpha}_1 + 4\hat{\alpha}_2 + 4\hat{\alpha}_3 + 3\hat{\beta}_1 + 3\hat{\beta}_2 + 3\hat{\beta}_3 + 3\hat{\beta}_4$$

$$\sum \varepsilon x_1 = 0: \quad y_{1.} = 4\hat{\mu} + 4\hat{\alpha}_1 \qquad\qquad\quad + \hat{\beta}_1 + \hat{\beta}_2 + \hat{\beta}_3 + \hat{\beta}_4$$

$$\sum \varepsilon x_2 = 0: \quad y_{2.} = 4\hat{\mu} \qquad\quad + 4\hat{\alpha}_2 \qquad\quad + \hat{\beta}_1 + \hat{\beta}_2 + \hat{\beta}_3 + \hat{\beta}_4$$

$$\sum \varepsilon x_3 = 0: \quad y_{3.} = 4\hat{\mu} \qquad\qquad\qquad + 4\hat{\alpha}_3 + \hat{\beta}_1 + \hat{\beta}_2 + \hat{\beta}_3 + \hat{\beta}_4$$

$$\sum \varepsilon z_1 = 0: \quad y_{.1} = 3\hat{\mu} + \hat{\alpha}_1 + \hat{\alpha}_2 + \hat{\alpha}_3 + 3\hat{\beta}_1$$

$$\sum \varepsilon z_2 = 0: \quad y_{.2} = 3\hat{\mu} + \hat{\alpha}_1 + \hat{\alpha}_2 + \hat{\alpha}_3 \qquad + 3\hat{\beta}_2$$

$$\sum \varepsilon z_3 = 0: \quad y_{.3} = 3\hat{\mu} + \hat{\alpha}_1 + \hat{\alpha}_2 + \hat{\alpha}_3 \qquad\qquad + 3\hat{\beta}_3$$

$$\sum \varepsilon z_4 = 0: \quad y_{.4} = 3\hat{\mu} + \hat{\alpha}_1 + \hat{\alpha}_2 + \hat{\alpha}_3 \qquad\qquad\qquad + 3\hat{\beta}_4$$

(6.3)

We have eight equations in eight unknowns, but not all eight equations are independent of each other. In fact, the sum of the second through fourth equations gives the first equation, as does the sum of the fifth through eighth equations. We thus have only six independent equations in eight unknowns. Thus unique solutions for the unknowns are not possible. It is easily verified that any one of a number of solution sets satisfies the normal equations. Some examples of such solution sets are given in Table 6.4. Solution set 1 uses as side conditions $\hat{\mu} = 0$ and $\sum \hat{\beta}_j = 0$, solution set 2 uses instead $\hat{\mu} = 0$ and $\sum \hat{\alpha}_i = 0$. Most textbooks make use of solution set 3, which is based on the side conditions $\sum \hat{\beta}_j = 0$ and $\sum \hat{\alpha}_i = 0$. It is easy to verify that irrespective of the solution set used the predicted value for the (i, j) cell will be

$$\hat{y}_{ij} = \bar{y}_{i.} + \bar{y}_{.j} - \bar{y}_{..}$$

that is, Treatment mean + block mean − grand mean

Table 6.4 SOLUTION SETS FOR THE NORMAL EQUATIONS (6.3)

Identification number	Solution set		Side conditions
1	$\hat{\mu} = 0$ $\hat{\alpha}_1 = \bar{y}_{1.}$ $\hat{\alpha}_2 = \bar{y}_{2.}$ $\hat{\alpha}_3 = \bar{y}_{3.}$	$\hat{\beta}_1 = \bar{y}_{.1} - \bar{y}_{..}$ $\hat{\beta}_2 = \bar{y}_{.2} - \bar{y}_{..}$ $\hat{\beta}_3 = \bar{y}_{.3} - \bar{y}_{..}$ $\hat{\beta}_4 = \bar{y}_{.4} - \bar{y}_{..}$	$\hat{\mu}$ and $\hat{\beta}_1 + \hat{\beta}_2 + \hat{\beta}_3 + \hat{\beta}_4$ both set equal to zero
2	$\hat{\mu} = 0$ $\hat{\alpha}_1 = \bar{y}_{1.} - \bar{y}_{..}$ $\hat{\alpha}_2 = \bar{y}_{2.} - \bar{y}_{..}$ $\hat{\alpha}_3 = \bar{y}_{3.} - \bar{y}_{..}$	$\hat{\beta}_1 = \bar{y}_{.1}$ $\hat{\beta}_2 = \bar{y}_{.2}$ $\hat{\beta}_3 = \bar{y}_{.3}$ $\hat{\beta}_4 = \bar{y}_{.4}$	$\hat{\mu}$ and $\hat{\alpha}_1 + \hat{\alpha}_2 + \hat{\alpha}_3$ both set equal to zero
3	$\hat{\mu} = \bar{y}_{..}$ $\hat{\alpha}_1 = \bar{y}_{1.} - \bar{y}_{..}$ $\hat{\alpha}_2 = \bar{y}_{2.} - \bar{y}_{..}$ $\hat{\alpha}_3 = \bar{y}_{3.} - \bar{y}_{..}$	$\hat{\beta}_1 = \bar{y}_{.1} - \bar{y}_{..}$ $\hat{\beta}_2 = \bar{y}_{.2} - \bar{y}_{..}$ $\hat{\beta}_3 = \bar{y}_{.3} - \bar{y}_{..}$ $\hat{\beta}_4 = \bar{y}_{.4} - \bar{y}_{..}$	$\hat{\alpha}_1 + \hat{\alpha}_2 + \hat{\alpha}_3$ and $\hat{\beta}_1 + \hat{\beta}_2 + \hat{\beta}_3 + \hat{\beta}_4$ both set equal to zero

Furthermore, the least-squares estimates of all treatment contrasts and all block contrasts are unique. As described in Chap. 5, any linear combination of treatment effects such as $\sum l_j \beta_j$ is called a treatment contrast if $\sum l_j = 0$. Block contrasts are similarly defined. By least-squares estimates of $\sum l_j \beta_j$ is meant $\sum l_j \hat{\beta}_j$, where $\hat{\beta}_j$'s are taken from any solution set satisfying the normal equations (and a set of side conditions). It is easily verified that irrespective of the solution set used, the least-squares estimate of $\sum l_j \beta_j$ for the design represented in Table 6.3 is $\sum l_j \bar{y}_{.j}$. Similarly, it can easily be verified that irrespective of the solution set used, the least-squares estimate of $\sum l_i^* \alpha_i$ for the design represented in Table 6.3 is $\sum l_i^* \bar{y}_{i.}$.

Another important point worth noting is that irrespective of the side conditions used, our estimates of the residuals are unique:

$$\hat{\varepsilon}_{ij} = \text{observed } y_{ij} - \text{predicted } y_{ij} = y_{ij} - \hat{y}_{ij} = y_{ij} - \bar{y}_{i.} - \bar{y}_{.j} + \bar{y}_{..}$$

Although we have made the above observations with specific reference to a randomized-blocks design with four treatments and three blocks, they apply to any randomized-blocks design.

We now turn our attention to the identity

$$y_{ij} - \bar{y}_{..} = (\bar{y}_{.j} - \bar{y}_{..}) + (\bar{y}_{i.} - \bar{y}_{..}) + (y_{ij} - \bar{y}_{.j} - \bar{y}_{i.} + \bar{y}_{..})$$

A little algebra will now lead us to the important result (for the case of r blocks and t treatments)

$$\sum_i \sum_j (y_{ij} - \bar{y}_{..})^2 = r \sum_j (\bar{y}_{.j} - \bar{y}_{..})^2 + t \sum_i (\bar{y}_{i.} - \bar{y}_{..})^2$$
$$+ \sum_i \sum_j (y_{ij} - \bar{y}_{.j} - \bar{y}_{i.} + \bar{y}_{..})^2$$

This equation states that

$$\text{Total SS} = \text{treatment SS} + \text{block SS} + \text{residual SS}$$

It is important to note that this equation holds whether or not our assumptions about the observations are true. The assumptions about the observations become relevant only when we consider the sampling distributions of the different SS. Under the following conditions it can be shown that treatment SS, block SS, and residual SS are independent of each other and that each is distributed as a chi square:

1 The treatment effects are equal to one another: $\beta_1 = \beta_2 = \cdots = \beta_t = \beta$. This reduces model (6.1) to

$$y_{ij} = \mu + \alpha_i + \beta + \varepsilon_{ij}$$

2 The block effects are equal to one another $\alpha_1 = \alpha_2 = \cdots = \alpha_r = \alpha$. This reduces (6.1) to

$$y_{ij} = \mu + \alpha + \beta_j + \varepsilon_{ij}$$

3 The residuals satisfy the Ω assumptions of Chap. 5; i.e., they all have expectation 0, are independent of each other, have common variance $= \sigma^2$, and are normally distributed.

If conditions 1 and 3 hold, it can be shown that (treatment SS)/σ^2, that is, $r \sum (\bar{y}_{.j} - \bar{y}_{..})^2/\sigma^2$, and (residual SS)/$\sigma^2$, that is, $\sum \sum (y_{ij} - \bar{y}_{i.} - \bar{y}_{.j} + \bar{y}_{..})^2/\sigma^2$, are distributed as chi squares independently of each other, the former with $t - 1$ degrees of freedom and the latter with $(r - 1)(t - 1)$ degrees of freedom. It follows that under these conditions the ratio (treatment MSS)/(residual MSS) is distributed as F with $t - 1$, $(r - 1)(t - 1)$ degrees of freedom.

If conditions 2 and 3 hold, it can be shown that (block SS)/σ^2, that is, $t \sum (\bar{y}_{i.} - \bar{y}_{..})^2/\sigma^2$, and (residual SS)/$\sigma^2$ are distributed as chi squares independently of each other, the former with $r - 1$ degrees of freedom and the latter with $(r - 1)(t - 1)$ degrees of freedom. Under these conditions the ratio (block MSS)/ (residual MSS) is distributed as F with $r - 1$, $(r - 1)(t - 1)$ degrees of freedom.

We use the ratio (treatment MSS)/(residual MSS) for testing the hypothesis that the treatment effects are equal to one another. Similarly for testing the hypothesis that the block effects are equal to one another we use the ratio (block MSS)/(residual MSS). The rationale for this can be seen if we look at the expected values of the MSS under model (6.1). These are shown in Table 6.5. From the expected values shown in Table 6.5 it is obvious that if the treatment effects are not equal to one another, the ratio (treatment MSS)/(residuals MSS) may be expected to be greater than 1 beyond what may be attributed to sampling fluctuations. If, on the other hand, the treatment effects are equal to one another, this ratio may be expected to be equal to 1, except for sampling fluctuations. This justifies the use of the particular ratio in question for testing the hypothesis that the treatment effects are equal to one another. The justification for using the ratio (block MSS)/(residual MSS) for testing the significance of the difference between block effects can be similarly seen from Table 6.5.

Before concluding this section, it may be of some interest to emphasize the following results.

RESULT 1   The SS due to regression if we fit model (6.2), that is,

$$y_{ij} = \mu x_0 + \alpha_1 x_1 + \cdots + \alpha_r x_r + \beta_1 z_1 + \cdots + \beta_t z_t + \varepsilon_{ij}$$

is exactly equal to the sum of treatment SS and block SS obtained above. In symbols

$$\text{Regression } SS\ (\mu, \alpha_i\text{'s}, \beta_j\text{'s}) = \text{treatment SS} + \text{block SS}$$

where regression SS $(\mu, \alpha_i\text{'s}, \beta_j\text{'s})$ is a shorthand notation for the expression "sum of

Table 6.5   EXPECTED VALUES OF MSS IN RANDOMIZED-BLOCKS DESIGNS WITH t TREATMENTS AND r BLOCKS, EACH TREATMENT OCCURRING ONCE IN EACH BLOCK

MSS	Expected value
Treatment MSS $= \dfrac{r \sum (\bar{y}_{.j} - \bar{y}_{..})^2}{t - 1}$	$\sigma^2 + \dfrac{r \sum (\beta_j - \bar{\beta}_.)^2}{t - 1}$
Block MSS $= \dfrac{t \sum (\bar{y}_{i.} - \bar{y}_{..})^2}{r - 1}$	$\sigma^2 + \dfrac{t \sum (\alpha_i - \bar{\alpha}_.)^2}{r - 1}$
Residual MSS $= \dfrac{\sum \sum (y_{ij} - \bar{y}_{.j} - \bar{y}_{i.} + \bar{y}_{..})^2}{(r - 1)(t - 1)}$	σ^2

squares due to regression when the regression model involves the parameters μ, $(\alpha_1, \alpha_2, \ldots, \alpha_r)$, and $(\beta_1, \ldots, \beta_t)$."

This result can be demonstrated easily. Recall that when we fit the above model to the data set in Table 6.2, we get $\bar{y}_{i.} + \bar{y}_{.j} - \bar{y}_{..}$ as the predicted value for the (i, j) cell. The mean of these values is seen to be $\bar{y}_{..}$. Now the SS due to regression is simply the SS of the deviations of the predicted values from their mean. Hence

$$\text{Regression SS } (\mu, \alpha_i\text{'s}, \beta_j\text{'s}) = \sum_i \sum_j (\bar{y}_{i.} + \bar{y}_{.j} - \bar{y}_{..} - \bar{y}_{..})^2$$

which simplifies to

$$t \sum_i (\bar{y}_{i.} - \bar{y}_{..})^2 + r \sum_j (\bar{y}_{.j} - \bar{y}_{..})^2$$

or   Block SS + treatment SS

RESULT 2   If we ignore the blocking in Table 6.2 and fit the model

$$y_{ij} = \mu x_0 + \beta_1 z_1 + \cdots + \beta_t z_t + \varepsilon_{ij}$$

the SS due to regression will be exactly the same as treatment SS. This follows from the results of Chap. 5. In symbols,

$$\text{Regression SS } (\mu, \beta_j\text{'s}) = \text{treatment SS}$$

RESULT 3   Similarly, if we confine our attention to the row classification in Table 6.2, i.e., if we ignore the treatments, we fit the model

$$y_{ij} = \mu x_0 + \alpha_1 x_1 + \cdots + \alpha_r x_r + \varepsilon_{ij}$$

and the resulting SS due to regression will be exactly equal to the block SS. Or in symbols,

$$\text{Regression SS } (\mu, \alpha_i\text{'s}) = \text{block SS}$$

RESULT 4   Putting together results 1 to 3 for a randomized-blocks design in which each treatment occurs once in each block, we have

Regression SS $(\mu, \alpha_i\text{'s}, \beta_j\text{'s})$ = regression SS $(\mu, \alpha_i\text{'s})$ + regression SS $(\mu, \beta_j\text{'s})$

Whenever this result holds, we say that there is *orthogonality* between treatments and blocks in the design. The advantage of orthogonality is that we can calculate treatment SS by regarding the data set as a one-way classification according to treatments and then calculate block SS by treating the same data set as a one-way classification according to blocks.

Gain By Blocking

We shall now describe a procedure that is generally used to assess the gain in precision furnished by blocking. Let $\hat{\sigma}_{RB}^2$ be the estimate of the residual variance, i.e., the variance of ε_{ij}'s, supplied by a randomized-blocks design, and let $\hat{\sigma}_{CR}^2$ be the corresponding estimate supplied by a completely randomized design. If r is the number of blocks used in the randomized-blocks design, the estimated variance of a treatment mean is $(1/r)\hat{\sigma}_{RB}^2$. Similarly if n is the number of observations under each treatment in a completely randomized design, the estimated variance of a treatment mean is $(1/n)\hat{\sigma}_{CR}^2$. For these two variances to be equal, we should have

$$\frac{1}{r}\hat{\sigma}_{RB}^2 = \frac{1}{n}\hat{\sigma}_{CR}^2 \quad \text{or} \quad \frac{n}{r} = \frac{\hat{\sigma}_{CR}^2}{\hat{\sigma}_{RB}^2}$$

If n/r is greater than 1, it means that the randomized-blocks design is able to produce the same precision as the completely randomized design by using fewer observations. It is therefore meaningful to use $\hat{\sigma}_{CR}^2/\hat{\sigma}_{RB}^2$ as a measure of the relative gain furnished by blocking. This ratio can be estimated from the figures available in the analysis-of-variance table associated with a given randomized-blocks experiment.

To see how this can be done, imagine conducting a new randomized-blocks experiment with the same r blocks of the given experiment and a common treatment applied t times in each block. The results of this experiment can be expected to validly reflect the variation due to block differences and the residual variation in the original randomized-blocks experiment. The total SS of the new experiment can be partitioned into between SS and within SS as in a one-factor experiment, the factor involved in this case being that represented by the blocks. Let B and W stand for the corresponding between MSS and residual MSS, respectively.

The estimated variance of any observed mean in this new experiment is proportional to W. A mean based on n observations has an estimated variance of W/n, and so on. W is the same as $\hat{\sigma}_{RB}^2$ mentioned above, since W represents the variation left over after removing treatment differences by design, i.e., by using a common treatment throughout, and block differences by analysis, i.e., by partitioning out block SS.

If the blocking had been ignored altogether, we would have had a completely randomized experiment with a common treatment used rt times. The variance of any observed mean would then be estimated as proportional to the total MSS, which algebraically is seen to be equal to

$$\frac{(r-1)B + r(t-1)W}{rt-1}$$

where B and W stand for the quantities defined earlier. This total MSS represents the variation left over after treatment differences have been removed by design, i.e., by using a common treatment throughout. Hence this is the same as $\hat{\sigma}_{CR}^2$ mentioned above. Thus we have

$$\frac{\hat{\sigma}_{CR}^2}{\hat{\sigma}_{RB}^2} = \frac{(r-1)B + r(t-1)W}{(rt-1)W}$$

Now recall that B represents the same thing as block MSS and W the same thing as residual MSS in the given randomized-blocks experiment. Thus we get the formula

$$\frac{\hat{\sigma}_{CR}^2}{\hat{\sigma}_{RB}^2} = \frac{(r-1)(\text{block MSS}) + r(t-1)(\text{residual MSS})}{(rt-1)(\text{residual MSS})}$$

for calculating the relative gain due to blocking.

Using the results of Table 6.1, we have

$$\frac{\hat{\sigma}_{CR}^2}{\hat{\sigma}_{RB}^2} = \frac{(4 \times 4.12) + (5 \times 3 \times 3.96)}{(5 \times 4 - 1)(3.96)} = 1.01$$

This means that very little has been gained by blocking in the experiment reported in Table 6.1.

Mixed and Random Models

In the example given in Table 6.1, the blocks were not considered to be a random sample from a universe of blocks; nor were the treatments considered a random sample from a universe of teaching methods. If the blocks or the treatments are considered as a random sample, the corresponding effects should be treated as random variables in the model representing the observations. If both block and treatment effects are random effects, the model is referred to as a *random* model, and if one of the two sets is considered as random variables while the other is considered as fixed, the model is referred to as a *mixed* one.

Let us consider first a mixed model in which the block effects are assumed to be a random sample from a corresponding universe while the treatment effects are assumed to be constants. The relevant model would then be

$$y_{ij} = \mu + a_i + \beta_j + \varepsilon_{ij}$$

Universe of a_i's has mean 0 and variance σ_a^2

ε_{ij}'s are $\text{NID}(0,\sigma^2)$ (6.4)

a_i's and ε_{ij}'s are independent

where $\text{NID}(0,\sigma^2)$ stands for normally and independently distributed with mean 0 and variance σ^2. The hypotheses we are interested in testing may be

H_T: There is no difference between the treatment effects

H_B: There is no difference between the block effects

The hypothesis H_T is equivalent to the one that states that $S_\beta^2 = 0$, where $S_\beta^2 = \sum (\beta_j - \bar{\beta}_.)^2/(t - 1)$. Similarly, the hypothesis H_B is equivalent to the one that posits that $\sigma_a^2 = 0$. It should be noted that S_β^2 represents the variation between the t particular values of β's actually involved in the experiment whereas σ_a^2 represents the variation between all possible a's of which the particular values actually involved in the experiment are only a sample. To test the hypotheses just described, the first step is to complete the analysis-of-variance table as described in Table 6.1. The next step is to examine the expected values of the MSS. In the model under consideration, the expected value of block MSS can be shown to be $\sigma^2 + t\sigma_a^2$, and that of the treatment MSS can be shown to be $\sigma^2 + r \sum (\beta_j - \bar{\beta}_.)^2/(t - 1)$. As in the fixed-effects model, the expected value of the residual MSS is σ^2. These results are summarized in Table 6.6.

The hypothesis that $\sum (\beta_j - \bar{\beta}_.)^2/(t - 1) = 0$ is tested on the basis of the F ratio (treatment MSS)/(residual MSS), and the hypothesis that $\sigma_a^2 = 0$ is tested on the basis of the F ratio (block MSS)/(residual MSS). The logic of this procedure can be seen by noting that the expected values of the treatment MSS and residual MSS

Table 6.6   EXPECTED VALUES OF MSS UNDER MIXED AND
RANDOM MODELS: RANDOMIZED-BLOCKS DESIGN

| MSS | Mixed model | | Random model |
	Blocks random, treatments fixed	Blocks fixed, treatment random	
Block	$\sigma^2 + t\sigma_a^2$	$\sigma^2 + tS_\alpha^2$	$\sigma^2 + t\sigma_a^2$
Treatment MSS	$\sigma^2 + rS_\beta^2$	$\sigma^2 + r\sigma_b^2$	$\sigma^2 + r\sigma_b^2$
Residual MSS	σ^2	σ^2	σ^2

$$S_\alpha^2 = \frac{\sum_i (\alpha_i - \bar{\alpha}.)^2}{r - 1} \qquad S_\beta^2 = \frac{\sum_j (\beta_j - \bar{\beta}.)^2}{t - 1}$$

σ_a^2 = variance of universe of block effects
σ_b^2 = variance of universe of treatment effects

will be equal if and only if the β_j's do not differ between themselves and similarly that the expected values of the block MSS and residual MSS will be equal if and only if $\sigma_a^2 = 0$.

To illustrate the above procedure, suppose an investigator selected 36 villages at random from the entire set of villages in a country and classified them into 12 categories of three each on the basis of relevant demographic and socioeconomic characteristics. Suppose he used the 12 categories thus obtained as blocks in a randomized-blocks experiment designed to compare three methods of family-planning action programs. Suppose the data in Table 6.7 represent the results of the experiment.

Table 6.7   DATA FROM A
RANDOMIZED-BLOCKS
EXPERIMENT COMPARING
THREE FAMILY-PLANNING
ACTION PROGRAMS

| | Treatment | | | |
Block	1	2	3	Total
1	8	6	9	23
2	2	1	7	10
3	4	4	9	17
4	5	4	11	20
5	8	5	9	22
6	5	2	9	16
7	6	9	12	27
8	8	8	10	26
9	5	10	16	31
10	7	10	9	26
11	4	2	10	16
12	8	9	11	28
Total	70	70	122	262

The analysis of variance completed as described in Table 6.1 is given in Table 6.8. To test the hypothesis that $\sigma_a^2 = 0$, we get the variance ratio 12.72/3.60 with 11 and 22 degrees of freedom, which is significant at the 1 percent level. To test the hypothesis that the treatment effects are not different, we use the ratio 75.11/3.60 with 2 and 22 degrees of freedom, which is also found to be highly significant. If one wishes, multiple-comparison procedures can be applied at this point, as was done earlier in connection with the data in Table 6.1.

The random model in which block effects as well as treatment effects are assumed to be random variables may be described as

$$y_{ij} = \mu + a_i + b_j + \varepsilon_{ij}$$

Universe of a_i's has mean 0 and variance σ_a^2

Universe of b_j's has mean 0 and variance σ_b^2 (6.5)

ε_{ij} are NID$(0,\sigma^2)$

a_i, b_j, and ε_{ij} are independent

The expected values of the MSS associated with the above model are shown in Table 6.6. As in the fixed and mixed models, the ratio of block MSS to residual MSS is used to test the hypothesis that $\sigma_a^2 = 0$ (which implies that the block effects are equal), and the ratio of the treatment MSS to residual MSS is used to test the hypothesis that $\sigma_b^2 = 0$ (which implies that the treatment effects are equal).

The following example illustrates a situation in which a random model like that just described is appropriate. Suppose an investigator selects 60 city blocks at random and classifies them into 10 categories of 6 each according to median family income. He assigns one block in each category at random to each of 6 interviewers (assumed to be selected at random from a population of interviewers). The treatment variable of interest is the interviewer effect on the response received to an attitudinal question, the block variable being the median income (socioeconomic status) of the city block. Here both the sets of effects may be regarded as random effects.

Consequences of Wrong Assumptions about Expectations of Residuals

So far we have confined our attention to the situation in which the residuals in the model

$$y_{ij} = \mu + \alpha_i + \beta_j + \varepsilon_{ij}   (6.1)$$

Table 6.8 ANALYSIS-OF-VARIANCE TABLE
FOR THE DATA IN TABLE 6.7

Source of variation	SS	df	MSS
Between treatments	150.22	2	75.11
Between blocks	139.89	11	12.72
Residual	79.11	22	3.60
Total	369.22	35	

all have a common expected value. In practice there is no assurance that ε_{ij}'s do not differ in expected value. Consequently we are often concerned with the validity of this assumption. We shall now examine the consequences of making the assumption that $E(\varepsilon_{ij}) = 0$ for all i and j when in fact some or all ε_{ij}'s differ in expected value.

Suppose each ε_{ij} has its own characteristic expected value, say γ_{ij}. We may then write ε_{ij} as the sum of two parts: $\gamma_{ij} + (\varepsilon_{ij} - \gamma_{ij})$, the first part, is $E(\varepsilon_{ij})$, while the second is the deviation of ε_{ij} from its expected value. For simplicity we use η_{ij} for the deviation $\varepsilon_{ij} - \gamma_{ij}$ and write

$$\varepsilon_{ij} = \gamma_{ij} + \eta_{ij} \qquad \text{for all } i \text{ and } j \qquad (6.6)$$

Since η_{ij} is the deviation of ε_{ij} from $E(\varepsilon_{ij})$, we have $E(\eta_{ij}) = 0$ for all i and j.

The model

$$y_{ij} = \mu + \alpha_i + \beta_j + \varepsilon_{ij} \qquad (6.1)$$

now becomes

$$y_{ij} = \mu + \alpha_i + \beta_j + \gamma_{ij} + \eta_{ij} \qquad (6.7)$$

where $E(\eta_{ij}) = 0$ for all i and j.

In most textbooks the following side conditions are associated with this model.

$$\begin{aligned}
\alpha_{.} &= \sum_i \alpha_i = 0 \\[4pt]
\beta_{.} &= \sum_j \beta_j = 0 \\[4pt]
\gamma_{i.} &= \sum_j \gamma_{ij} = 0 \qquad i = 1, 2, \ldots, r \\[4pt]
\gamma_{.j} &= \sum_i \gamma_{ij} = 0 \qquad j = 1, 2, \ldots, t
\end{aligned} \qquad (6.8)$$

Although these side conditions together are $r + t + 2$ in number, only $r + t + 1$ of them are independent. To see this note that if $\gamma_{i.} = 0$ for $i = 1, 2, \ldots, r$, then obviously $\gamma_{..} = 0$. It follows therefore that if any $t - 1$ of $\gamma_{.j}$'s are set equal to zero, the remaining $\gamma_{.j}$ will automatically be zero. Hence the $r + t$ side conditions involving γ_{ij}'s in (6.8) contain only $r + t - 1$ independent side conditions.

The reason for this number of side conditions is not difficult to see. If we write the normal equations corresponding to (6.7) or its equivalent multiple-regression equation, we notice that there are $rt + t + r + 1$ parameters to be estimated but only rt independent equations available. Consequently we need $r + t + 1$ side conditions to solve for the unknowns. Some of these must necessarily involve γ_{ij}'s. There is, of course, no need to make use of the particular set of side conditions given in (6.8). Any one of a number of other interesting sets of side conditions, will serve. Reference may be made in this connection to Table 6.4. We shall stick with the side conditions shown in (6.8), however, mainly because these are the ones most frequently used in the literature.

Before we proceed it may be useful to emphasize one point. We reached model (6.7) first by stipulating that the sum of the block effect α_i, the treatment effect β_j, a constant μ, and the residual ε_{ij} with expectation 0 produced y_{ij}, the observation in the (i,j) cell. Then we asked whether the assumption that each residual has zero expectation is tenable. Thus we were led to recognize the possibility that each residual may have its own characteristic mean. In other words, we started with a simple causal model, then raised questions about the validity of the model, and to accommodate our suspicions modified our simple model, making it more complex. One may of course start with a complex causal model and simplify it step by step. Thus, for example, one may start with model (6.7) and simplify it, leading to models like (6.1). This latter procedure amounts to taking the initial position that each observation has its own characteristic explanation and later modifying it by hypothesizing that there are subsets of observations with common explanations (see Chap. 3).

Turning to the consequences of making wrong assumptions concerning the expected values of the residuals, let us suppose that a data set at hand is best described by model (6.7) but through ignorance or indifference we decided to fit model (6.1) to it. Will our inferences be affected? If so, in what way?

Recall that when the side conditions $\alpha_. = 0$ and $\beta_. = 0$ are used, the least-squares estimates of α_i's and β_j's of model (6.1) are, according to solution sets 3 in Table 6.4,

$$\hat{\alpha}_i = \bar{y}_{i.} - \bar{y}_{..} \qquad i = 1, 2, \ldots, r$$
$$\hat{\beta}_j = \bar{y}_{.j} - \bar{y}_{..} \qquad j = 1, 2, \ldots, t$$

We now ask:

1   Are the least-squares estimates of treatment (block) contrasts arrived at from the above solutions unbiased?
2   Does the sensitivity of tests of significance of treatment (block) contrast become reduced by using a wrong model?

The answers to both these questions are yes.

Consider the treatment contrast $\psi = \sum l_j \beta_j$, where $\sum l_j = 0$. The least-squares estimate of ψ is

$$\hat{\psi} = \sum l_j \hat{\beta}_j = \sum l_j(\bar{y}_{.j} - \bar{y}_{..})$$

Now, according to model (6.7) and side conditions (6.8),

$$\bar{y}_{.j} - \bar{y}_{..} = \beta_j + \bar{\eta}_{.j} - \bar{\eta}_{..}$$

Also $E(\bar{\eta}_{.j}) = E(\bar{\eta}_{..}) = 0$ by virtue of the assumption that $E(\eta_{ij}) = 0$ for all i and j. Consequently $E(\bar{y}_{.j} - \bar{y}_{..}) = E(\beta_j) = \beta_j$, and hence

$$E(\hat{\psi}) = E\left[\sum l_j(\bar{y}_{.j} - \bar{y}_{..})\right] = \sum l_j E(\bar{y}_{.j} - \bar{y}_{..}) = \sum l_j E(\beta_j) = \sum l_j \beta_j = \psi$$

Thus $\hat{\psi}$ is an unbiased estimate of ψ. Similar results hold for block contrasts too. In other words, as far as providing unbiased estimates of treatment (block) contrasts is concerned, the use of model (6.1) instead of the more valid model (6.7) does not make any difference. But the situation changes when testing the significance of the estimated contrasts. Consider the problem of testing the significance of $\hat{\psi}$. If the investigator uses model (6.1), under the usual assumptions about the residuals, the variance of $\hat{\psi}$ is given by

$$\text{var } \hat{\psi} = \frac{1}{r}\left(\sum l_j^2\right)\sigma^2$$

and as an estimate of σ^2 in the above expression he would use

$$\frac{1}{(r-1)(t-1)} \sum_i \sum_j (y_{ij} - \bar{y}_{i.} - \bar{y}_{.j} + \bar{y}_{..})^2$$

The problem is that if (6.7) is the correct model, the use of the above quantity as an estimate of σ^2 amounts to using an overestimate of var $\hat{\psi}$ in the test of significance of $\hat{\psi}$ because under model (6.7) and side conditions (6.8)

$$E\left[\frac{1}{(r-1)(t-1)} \sum_i \sum_j (y_{ij} - \bar{y}_{i.} - \bar{y}_{.j} + \bar{y}_{..})^2\right] = \sigma^2 + \frac{\sum_i \sum_j \gamma_{ij}^2}{(t-1)(r-1)}$$

Obviously, if one uses an overestimate of the variance of $\hat{\psi}$ for tests of significance, the probability of declaring ψ to be zero when in fact it is different from zero will be greater than when the variance of $\hat{\psi}$ is not overestimated. Therefore, if one uses model (6.1) when in fact model (6.7) best describes the data, the usual test procedures are likely to be less efficient.

6.3   THE CONCEPT OF ADDITIVITY

At this point it is pertinent to introduce the concept of additivity between block (row) and treatment (column) effects. If the treatment effect β_j simply adds onto the block effect α_i to produce their combined effect, we say that the two effects are *additive*. If, on the other hand, the treatment effect β_j combines with the block effect α_i in a nonlinear fashion involving product terms such as $\alpha_i\beta_j$, we say that the two effects are *nonadditive*. Obviously there are an unlimited number of ways in which α_i and β_j can combine in a nonlinear fashion. Some examples are

$$\alpha_i + \beta_j + \theta\alpha_i\beta_j \qquad \text{where } \theta = \text{const}$$
$$\alpha_i + \beta_j + (\theta\alpha_i + \phi\beta_j)^2 \qquad \text{where } \theta, \phi = \text{const}$$
$$\alpha_i + \beta_j + \gamma_{ij} \qquad \begin{array}{l}\text{where } \gamma_{ij} = \text{nonlinear function of } \alpha_i \text{ and } \beta_j, \\ \text{involving product terms such as } \alpha_i\beta_j\end{array}$$

In the above sense the model

$$y_{ij} = \mu + \alpha_i + \beta_j + \varepsilon_{ij} \qquad (6.1)$$

where $E(\varepsilon_{ij}) = 0$ for all i and j, is an additive model, while

$$y_{ij} = \mu + \alpha_i + \beta_j + \gamma_{ij} + \eta_{ij} \qquad (6.7)$$

where $E(\eta_{ij}) = 0$ for all i and j and γ_{ij} is a nonlinear function of α_i and β_j involving product terms such as $\alpha_i\beta_j$, is a nonadditive model. Obviously

$$y_{ij} = \mu + \alpha_i + \beta_j + \theta\alpha_i\beta_j + \eta_{ij} \qquad (6.9)$$

where θ is a constant, is a particular case of (6.7).

The reader may notice that when (6.7) was introduced earlier, no mention was made of viewing γ_{ij} as a function of α_i and β_j. It deserves to be emphasized that viewing γ_{ij} as a function of α_i and β_j does not amount to putting any restriction on the nature of γ_{ij}, for given any three numbers (γ_{ij}, α_i, and β_j), one of them can always be expressed as a mathematical function of the other two. However, if we assume that α_i and β_j combine identically in the same fashion for all i and j to become γ_{ij}, it amounts to restricting the scope of γ_{ij}'s.

In many textbooks additive models are distinguished from nonadditive models in another way. The additive models in two-way classifications, e.g., randomized-blocks designs, imply that the difference between the expected values of observations under any two treatments (columns) is the same in all blocks (rows), while this may not be true of nonadditive models. To understand what this means let us consider the difference $E(y_{ij}) - E(y_{il})$ under models (6.1) and (6.9). Under (6.1) we notice that

$$E(y_{ij}) - E(y_{il}) = (\mu + \alpha_i + \beta_j) - (\mu + \alpha_i + \beta_l) = \beta_j - \beta_l$$

which remains the same in all blocks, i.e., for all i. Under (6.9), however, the corresponding difference varies from one block to another as the block effect varies from one block to another, for

$$E(y_{ij}) - E(y_{il}) = (\mu + \alpha_i + \beta_j + \theta\alpha_i\beta_j) - (\mu + \alpha_i + \beta_l + \theta\alpha_i\beta_l)$$
$$= (\beta_j - \beta_l) + \theta\alpha_i(\beta_j - \beta_l) = (\beta_j - \beta_l)(1 + \theta\alpha_i)$$

It may also be noted that many authors use the term *interaction* for nonadditivity. Let us illustrate the idea of nonadditivity (interaction) with some examples. Suppose that among United States whites the annual earning of college graduates exceeds on the average that of high school graduates (with no college) by $5,000. Now suppose that among nonwhites the corresponding difference is only $2,500. We would then say that there is nonadditivity between race and education as far as earning power is concerned. College education for some reason fails to produce the same amount of increase in earning power among nonwhites as among whites.

To give another example, it is often contended that the impact of family-planning action programs may be negligible in populations below a certain threshold of social and economic development but may be dramatic in populations above the threshold. Put in another way, the social and economic levels of development (blocks) and family-planning action-program inputs (treatments) are likely to be nonadditive as far as bringing about changes in contraceptive practice and the like is concerned.

For a third example, we turn to laboratory experiments in social psychology. Wiggins et al. (1965) found that high-status persons tend to be punished less severely than low-status persons when they interfere with the attainment of group goals; this is true, however, only when the interference is of a minor nature and not one that makes the attainment of group goals virtually impossible. In this case we say that there seems to be an interaction between status and degree of interference as far as their relationship to the severity of punishment is concerned.

When we analyze real data, there is no guarantee that block (row) and treatment (column) effects will be exactly additive. We make use of additive models because of their simplicity and also because the additive model is often a close approximation to more complex (nonadditive) models. This leads us to the question: How do we test whether an additive model adequately describes the patterns discernible in a given data set? Clearly, the test consists in determining whether the fit of model (6.1) to the data set at hand differs significantly from the fit of (6.7).

Tukey's Test of Additivity

When we are confronted with the task of comparing the fit of model (6.1) with that of model (6.7) to a given data set, the situation in which each treatment occurs only once in each block is to be distinguished from that in which each treatment occurs twice or more in each block. This section deals with the first situation, leaving the second to be dealt with in Sec. 6.4.

Situations in which each treatment occurs only once in each block are not infrequent in social engineering studies. If, for example, regions of a country serve as experimental units, it is likely that units similar in various respects chosen to serve as a block may be so few that each treatment can be applied only once in each block.

Now if we use the method of least squares to fit model (6.7) to a data set in which each treatment occurs only once in each block, we shall not be able to obtain any valid estimate of the residuals. For the least-squares estimates of the α's, β's, and γ's in (6.7) subject to the side conditions (6.8) are

$$\hat{\mu} = \bar{y}_{..}$$
$$\hat{\alpha}_i = \bar{y}_{i.} - \bar{y}_{..} \qquad\qquad i = 1,\dots,r$$
$$\hat{\beta}_j = \bar{y}_{.j} - \bar{y}_{..} \qquad\qquad j = 1,\dots,t$$

and

$$\hat{\gamma}_{ij} = y_{ij} - \bar{y}_{i.} - \bar{y}_{.j} + \bar{y}_{..} \qquad\qquad \begin{array}{l} i = 1,\dots,r \\ j = 1,\dots,t \end{array}$$

It follows immediately that the estimate of the residual ε_{ij} is automatically zero. For

$$\hat{\varepsilon}_{ij} = y_{ij} - \text{predicted } y_{ij} = y_{ij} - (\hat{\mu} + \hat{\alpha}_i + \hat{\beta}_j + \hat{\gamma}_{ij})$$
$$= y_{ij} - (\bar{y}_{..} + \bar{y}_{i.} - \bar{y}_{..} + \bar{y}_{.j} - \bar{y}_{..} + y_{ij} - \bar{y}_{i.} - \bar{y}_{.j} + \bar{y}_{..}) = 0$$

This means that irrespective of the residual variance underlying the data set at hand, our estimate of it will always be 0. In other words, model (6.7) does not permit us to get any valid estimate of the residual variance underlying a data set in which each treatment occurs only once in each block. In fitting (6.7) to such a data set we use up all the information to estimate the parameters in the model, leaving no information at all to gain any insight into the magnitude of the residual variation. Since all our tests are based upon what the data set tells us about the underlying residual variance, we have here a situation in which tests of hypotheses concerning treatment effects, block effects, or nonadditivity are impossible.

We have been led to this unfortunate situation because of our unwillingness to make any assumption regarding the parameters in (6.7). In particular, we can trace the source of our difficulty in part to our unwillingness to say anything specific about the parameters γ_{ij} in (6.7). Tukey (1949) looked into this matter and demonstrated that if we are willing to stipulate that γ_{ij} if present must be of the form $\theta\alpha_i\beta_j$, as shown in (6.9), we can perform a valid test of the hypothesis that nonadditivity of the stipulated form is negligible.

Before describing Tukey's test procedure, it is of interest to note that the assumption that $\gamma_{ij} = \theta\alpha_i\beta_j$ is equivalent to the assumption that γ_{ij} is a quadratic function of α_i and β_j of the form

$$\gamma_{ij} = C + (A_1\alpha_i + A_2\alpha_i^2) + (B_1\beta_j + B_2\beta_j^2) + \theta\alpha_i\beta_j \qquad (6.10)$$

This, as Scheffé (1959) has shown, is a consequence of the side conditions

$$\alpha_. = \beta_. = \gamma_{i.} = \gamma_{.j} = 0 \qquad (6.8)$$

Some readers may want to skip the proof given below. To prove this we observe that since $\bar{\alpha}_. = \bar{\beta}_. = \bar{\gamma}_{i.} = \bar{\gamma}_{.j} = 0$, (6.10) implies

$$\bar{\gamma}_{i.} = C + A_1\alpha_i + A_2\alpha_i^2 + \frac{B_2 \sum_j \beta_j^2}{t} = 0$$

$$\bar{\gamma}_{.j} = C + B_1\beta_j + B_2\beta_j^2 + \frac{A_2 \sum_i \alpha_i^2}{r} = 0$$

Hence

$$A_1\alpha_i + A_2\alpha_i^2 = -C - \frac{B_2 \sum_j \beta_j^2}{t}$$

$$B_1\beta_j + B_2\beta_j^2 = -C - \frac{A_2 \sum_i \alpha_i^2}{r}$$

Substituting these in (6.10) gives

$$\gamma_{ij} = -C - \frac{B_2 \sum_j \beta_j^2}{t} - \frac{A_2 \sum_i \alpha_i^2}{r} + \theta\alpha_i\beta_j \qquad (6.11)$$

Again we apply the side condition $\gamma_{i.} = 0$. From (6.11) we have

$$0 = -C - \frac{B_2 \sum_j \beta_j^2}{t} - \frac{A_2 \sum_i \alpha_i^2}{r}$$

Inserting this in (6.11) gives

$$\gamma_{ij} = \theta\alpha_i\beta_j$$

Tukey's nonadditive model is then

$$y_{ij} = \mu + \alpha_i + \beta_j + \theta\alpha_i\beta_j + \eta_{ij} \qquad (6.9)$$

Associated with this are the side conditions $\alpha_{i.} = \beta_{.j} = 0$. We shall also assume that η_{ij}'s are independently normally distributed with 0 mean and variance σ^2.

It can be shown that the least-squares estimate of the parameters in (6.9) is given by

$$\hat{\alpha}_i = \bar{y}_{i.} - \bar{y}_{..} \qquad i = 1, 2, \ldots, r$$

$$\hat{\beta}_j = \bar{y}_{.j} - \bar{y}_{..} \qquad j = 1, \ldots, t$$

$$\hat{\theta} = \frac{\sum_i \sum_j \hat{\alpha}_i\hat{\beta}_j y_{ij}}{\sum_i \hat{\alpha}_i^2 \sum_j \hat{\beta}_j^2} = \frac{\sum_i \sum_j (\bar{y}_{i.} - \bar{y}_{..})(\bar{y}_{.j} - \bar{y}_{..})y_{ij}}{\sum_i (\bar{y}_{i.} - \bar{y}_{..})^2 \sum_j (\bar{y}_{.j} - \bar{y}_{..})^2}$$

The SS due to nonadditivity with 1 degree of freedom is given by

$$\text{Nonadditivity SS} = \hat{\theta} \sum_i \sum_j (\bar{y}_{i.} - \bar{y}_{..})(\bar{y}_{.j} - \bar{y}_{..})y_{ij} \qquad (6.12)$$

Table 6.9   ANALYSIS OF VARIANCE ASSOCIATED WITH MODEL
(6.9)

Source of variation	SS†	df
Between treatments	$S_T = \dfrac{\sum y_{.j}^2}{r} - C$	$t - 1$
Between blocks	$S_B = \dfrac{\sum y_{i.}^2}{t} - C$	$r - 1$
Nonadditivity	$S_N = \dfrac{[\sum \sum (\bar{y}_{i.} - \bar{y}_{..})(\bar{y}_{.j} - \bar{y}_{..})y_{ij}]^2}{\sum (\bar{y}_{i.} - \bar{y}_{..})^2 \sum (\bar{y}_{.j} - \bar{y}_{..})^2}$	1
Residual	Total SS $- S_T - S_B - S_N$	$rt - r - t$
Total	$\sum_i \sum_j y_{ij}^2 - C$	$rt - 1$

† C in this column stands for the correction for the mean.

To test whether nonadditivity of the kind specified in (6.9) is significant, the analysis-of-variance table is first completed as described in Table 6.1 assuming additivity. The SS due to nonadditivity is calculated according to formula (6.12). The SS thus calculated is deducted from the residual SS obtained under the assumption of additivity; the remainder is used as the new residual SS with the appropriate degrees of freedom. Table 6.9 shows the structure of the analysis of variance associated with model (6.9). The nonadditivity SS is tested against the residual SS obtained by the formula

Residual SS = total SS − block SS − treatment SS − nonadditivity SS

Application of this procedure to the data in Table 6.7 yields the results shown in Table 6.10, which show that the nonadditivity SS was not found to be significant.

Table 6.10   TEST OF ADDITIVITY IN TABLE 6.7

Source of variation	SS	df	MSS	F ratio
Between treatments	150.22	2		
Between blocks	139.89	11		
Nonadditivity	3.12	1	3.12	.86
Residual	75.99	21	3.62	
Total	369.22	35		

It should be borne in mind that Tukey's test will not detect all kinds of nonadditivities. Table 6.11 illustrates this point. In the second block we find no difference between the treatments, which is in marked contrast to what we find in the other two blocks. If there were additivity, we would expect the comparisons between any two treatments to hold the same way in all the blocks. On commonsense considerations, then, one would suspect nonadditivity in the data in Table 6.11. But the SS due to nonadditivity obtained according to Tukey's procedure turns out to be equal to zero, as can be easily verified by taking the sum of the cross products of the figures in the last two columns.

The situation becomes complicated if on the basis of the test described above or on some other basis it is judged that additivity does not prevail. One main difficulty is that under such circumstances there is no way of unbiasedly estimating the error variance (see Scheffé, 1959, p. 134). The available estimate is an overestimate, as already mentioned, and therefore the use of the available estimate in the usual test procedures leads to considerable loss of efficiency. A second difficulty concerns giving practical interpretations to estimated treatment contrasts. Consider, for example, the treatment contrast $\beta_1 - \beta_2$. We have seen that irrespective of whether γ_{ij}'s are all zero, $\bar{y}_{.1} - \bar{y}_{.2}$ is an unbiased estimate of $\beta_1 - \beta_2$. Suppose that $\bar{y}_{.1} - \bar{y}_{.2}$ is significantly positive. If additivity prevails, this finding can be interpreted to mean that the effect of treatment 1 on the study phenomenon is greater than the corresponding effect of treatment 2 by an estimated amount equal to $\bar{y}_{.1} - \bar{y}_{.2}$ in all the blocks used in the experiment. If, on the other hand, additivity does not prevail, all

Table 6.11   A NONADDITIVE TWO-WAY CLASSIFICATION IN WHICH TUKEY'S TEST FAILS TO DETECT THE NONADDITIVITY

Block	Treatment 1	Treatment 2	Treatment 3	Block mean $\bar{y}_i.$	$\bar{y}_i. - \bar{y}_{..}$	$\sum_j (\bar{y}_{.j} - \bar{y}_{..})y_{ij}$†
1	5	8	5	6	1	4
2	5	5	5	5	0	0
3	3	6	3	4	−1	4
Treatment mean:						
$\bar{y}_{.j}$	$\frac{13}{3}$	$\frac{19}{3}$	$\frac{13}{3}$	$\bar{y}_{..} = 5$		
$\bar{y}_{.j} - \bar{y}_{..}$	$-\frac{2}{3}$	$\frac{4}{3}$	$-\frac{2}{3}$			

† The figures in the last column have been obtained as follows:

$$\begin{aligned}\text{Top figure: } 4 &= 5(-\tfrac{2}{3}) + 8(\tfrac{4}{3}) + 5(-\tfrac{2}{3}) && \text{from block 1 and last row}\\ \text{Middle figure: } 0 &= 5(-\tfrac{2}{3}) + 5(\tfrac{4}{3}) + 5(-\tfrac{2}{3}) && \text{from block 2 and last row}\\ \text{Bottom figure: } 4 &= 3(-\tfrac{2}{3}) + 6(\tfrac{4}{3}) + 3(-\tfrac{2}{3}) && \text{from block 3 and last row}\end{aligned}$$

In preparing the table, calculation of nonadditivity sum of squares is facilitated if we write formula (6.12) as

$$\frac{[\sum_i (\bar{y}_i. - \bar{y}_{..}) \sum_j (\bar{y}_{.j} - \bar{y}_{..})y_{ij}]^2}{\sum_i (\bar{y}_i. - \bar{y}_{..})^2 \sum_j (\bar{y}_{.j} - \bar{y}_{..})^2}$$

that we can say about the above finding that $\bar{y}_{.1} - \bar{y}_{.2}$ is positive is that when the effect of treatment 1 is averaged over all the blocks used in the experiment, the result is greater than the corresponding figure for treatment 2 by $\bar{y}_{.1} - \bar{y}_{.2}$; there is, however, the possibility that the effect of treatment 2 may be higher than that of treatment 1 in some of the blocks. When that is the case, the findings in each block should be separately reported and described; summary comparisons may not be of practical interest.

Because of the difficulties in testing and interpreting the findings when nonadditivity is present, it is sometimes wise to see whether a mathematical transformation of data will eliminate nonadditivity, or at least reduce it considerably, unless nonadditivity itself is of theoretical significance. Unfortunately there is no single procedure that can be recommended for all situations. One usually starts with the variance-stabilizing transformations of the kind described below, but unfortunately they are not always reconcilable with those which produce additivity; when a transformation is employed to produce additivity, variance stability may be destroyed, and vice versa.

Transformations of Observations

Transformations are sometimes used to reduce nonadditivity or to reduce nonnormality, but most often they are used to stabilize variance. Three common transformations are briefly described below.

The square-root transformation This involves using instead of the original observations, y_{ij}, their square roots, $\sqrt{y_{ij}}$, or if y_{ij}'s are small, say less than 5, $\sqrt{y_{ij} + \frac{1}{2}}$, $\sqrt{y_{ij} + 1}$, or $\sqrt{y_{ij}} + \sqrt{y_{ij} + 1}$ (Mosteller and Youtz, 1961). This transformation will be found useful if the variances of y_{ij} are proportional to the expected values $E(y_{ij})$. Such a situation is not unusual if the data are in the form of frequency counts, e.g., number of yes answers to an interview question.

It should be noted that if the treatment and block effects are additive in the original scale, they will not be additive in the square-root scale, and vice versa. However, if either the treatment or the block effects are small, effects that are additive on one scale may be so in the other scale too.

The logarithmic transformation This involves using instead of y_{ij}'s their logarithms. The most common situation in which this transformation is applied is when y_{ij} represents the frequency of occurrence of an event, e.g., number of deaths. If some y_{ij}'s are zero, then $\log (y_{ij} + 1)$ are used for all i and j. The logarithmic transformation is useful if variance of y_{ij} is proportional to $[E(y_{ij})]^2$. In some situations it may be found that the effect of one treatment is a given percentage higher than that of another instead of greater by a certain absolute number of units. For example, β_2 may be 20 percent higher than β_1 rather than greater by 10 units, β_3 may be 25 percent higher than β_2 rather than greater by 12 units, and so on. In such situations, if we employ the logarithmic transformation, we may achieve not only variance stability but also a reduction in nonadditivity.

The arcsine transformation This involves using instead of y_{ij} the angle whose sine is equal to the square root of y_{ij}. For example, if the original score y_{ij} equals 0.50, its square root is approximately equal to 0.707. From the tables for trigonometric functions, we find that the angle whose sine is equal to 0.707 is 45°. Hence we use in the analysis the transformed value 45 instead of the original value 0.50. This particular transformation is useful when y_{ij}'s are proportions. Note that when y_{ij}'s are proportions, each based on n observations, we have, if $E(y_{ij}) = P_{ij}$, the variance of y_{ij} given by $P_{ij}(1 - P_{ij})/n$, which differs little over P_{ij}'s lying between .3 and .7. Hence the transformation in question is unlikely to make a noticeable change in the conclusions unless the underlying proportions range from 0 to .3 or from .7 to 1.

A general formula[1] All three transformations can be seen to be particular cases of a general rule given below.

If σ_{ij}, the standard deviation of y_{ij}, is proportional to $f(g_{ij})$, where g_{ij} represents the expected value of y_{ij}, the appropriate transformation to achieve variance stability is given by

$$y_{ij}^* = \int \frac{1}{f(g_{ij})} \, dg_{ij}$$

Disadvantages It should be emphasized that there are certain substantive disadvantages in using transformations of the above types merely to satisfy the assumptions of one's model. For example, in a two-way layout, it is easier to think about the meaning and substantive significance of the row and column effects if the observations are proportions (p_{ij}'s), while it will be difficult to do so if we are dealing with arcsin $\sqrt{p_{ij}}$ instead.

6.4 THE TWO-WAY CLASSIFICATION WITH EQUAL NUMBERS (TWO OR MORE) OF OBSERVATIONS IN THE CELLS

Mathematical Model

Suppose in an experiment involving t treatments and r blocks, each cell gives n ($= 2$ or more) observations. We shall assume that nrt experimental units are divided into r blocks of nt units each in such a way that the units in each block are as homogeneous as possible with respect to the block variables. The nt units in each block are then randomly allocated to the t treatments so that each treatment gets n units. Randomization is done separately in each block. We shall further assume

$$y_{ijk} = \mu + \alpha_i + \beta_j + \gamma_{ij} + \varepsilon_{ijk} \qquad k = 1, 2, \ldots, n$$

$$\varepsilon_{ijk}\text{'s are independently } N(0, \sigma^2) \tag{6.13}$$

$$\alpha_. = \beta_. = \gamma_{i.} = \gamma_{.j} = 0$$

[1] This may be omitted on first reading.

The steps involved in the analysis of variance can be summarized as follows. First we calculate the grand total $y_{...} = \sum_i \sum_j \sum_k y_{ijk}$ of the nrt observations, from which the correction for mean is obtained as usual:

$$C = \frac{y_{...}^2}{nrt}$$

The different SS are then calculated:

Total SS = total of squares of all observations − C

Block SS = $\dfrac{\text{total of squares of block totals}}{\text{number of observations per block}}$ − C

Treatment SS = $\dfrac{\text{total of squares of treatment totals}}{\text{number of observations per treatment}}$ − C

Cell SS = $\dfrac{\text{total of squares of cell totals}}{\text{number of observations per cell}}$ − C

Nonadditivity SS = cell SS − block SS − treatment SS

Residual SS = total SS − cell SS

The analysis-of-variance table takes the form shown in Table 6.12.

Example As an illustration suppose the data shown in Table 6.13 came from an experiment in social engineering in which the following three methods of communication are compared: (1) film shows followed by group discussion, (2) group seminars, and (3) individual contacts by social engineers. The blocking was done according to the demographic and socioeconomic characteristics of the experimental units (villages). The data represent the change in fertility level that occurred during the period of the experiment.

Table 6.12 ANALYSIS OF VARIANCE: TWO-WAY CLASSIFICATION WITH n (= 2 OR MORE) OBSERVATIONS IN EACH CELL

Source of variation	SS	df
Between treatments	$\dfrac{\sum y_{.j.}^2}{nr} - C$	$t - 1$
Between blocks	$\dfrac{\sum y_{i..}^2}{nt} - C$	$r - 1$
Nonadditivity	Cell SS − block SS − treatment SS	$(t - 1)(r - 1)$
Between cells	$\dfrac{\sum \sum y_{ij.}^2}{n} - C$	$rt - 1$
Residual	Total SS − cell SS	$rt(n - 1)$
Total	$\sum_i \sum_j \sum_k y_{ijk}^2 - C$	$nrt - 1$

Table 6.13*B* gives the cell totals, block totals, treatment totals, and grand total.

Correction for mean $C = \dfrac{265^2}{18} = 3,901.389$

$$\text{Total SS} = 15^2 + 20^2 + \cdots + 13^2 + 15^2 - C = 4,565 - C = 663.611$$

$$\text{Block SS} = \tfrac{1}{6}(74^2 + 82^2 + 109^2) - C = \tfrac{1}{6}(24,081) - C = 112.111$$

$$\text{Treatment SS} = \tfrac{1}{6}(132^2 + 70^2 + 63^2) - C = \tfrac{1}{6}(26,293) - C = 480.778$$

$$\text{Cell SS} = \tfrac{1}{2}(35^2 + 43^2 + \cdots + 28^2) - C = \tfrac{1}{2}(9,049) - C = 623.111$$

$$\text{Nonadditivity SS} = \text{cell SS} - \text{block SS} - \text{treatment SS} = 30.222$$

$$\text{Error SS} = \text{total SS} - \text{cell SS} = 40.500$$

Table 6.13   RANDOMIZED-BLOCKS DESIGN WITH TWO OBSERVATIONS IN EACH CELL

A   Data

Block	Treatment		
	1	2	3
1	15,20	13,10	7, 9
2	20,23	11, 9	9,10
3	25,29	12,15	13,15

B   Cell, Block, Treatment, and Grand Total Calculated from Data in Part *A*

Block	Treatment			Total
	1	2	3	
1	35	23	16	74
2	43	20	19	82
3	54	27	28	109
Total	132	70	63	265

C   Analysis of Variance

Source of variation	SS	df	MSS	*F* ratio
Between treatments	480.78	2	240.39	53.42
Between blocks	112.11	2	56.05	12.46
Nonadditivity	30.22	4	7.55	1.68
Between cells	623.11	8		
Error	40.50	9	4.50	
Total	663.61	17		

The analysis of variance is presented in Table 6.13C. Note that if there are two or more observations per cell, it is possible to test whether nonadditivity is negligible without imposing any constraints on the form of nonadditivity. However, if non-additivity is found to be nonnegligible, there will be the usual difficulty of assigning practical meaning to the findings on treatment contrasts and/or contrasts involving block means.

One may analyze data sets like that in Table 6.13A by the multiple-regression approach. The kth observation in the (i,j) cell of Table 6.13A can be written

$$y_{ijk} = \mu x_0 + \sum_{i=1}^{3} \alpha_i x_i + \sum_{j=1}^{3} \beta_j z_j + \sum_{i=1}^{3} \sum_{j=1}^{3} \gamma_{ij} v_{ij} + \varepsilon_{ijk} \qquad (6.14)$$

where $x_0 = 1$, (x_1,x_2,x_3), (z_1,z_2,z_3), and $(v_{11},v_{12},\ldots,v_{33})$ are $3 + 3 + 9 = 15$ regressors such that for a given observation in the (i,j) cell, $x_i = 1$ but $x_k = 0$ if $k \neq i$; $z_j = 1$ but $z_l = 0$ if $l \neq j$; and $v_{ij} = 1$ but $v_{kl} = 0$ if either $k \neq i$ or $l \neq j$.

If we fit (6.14) by the method of least squares using as side conditions $\alpha_. = \beta_. = \gamma_{i.} = \gamma_{.j} = 0$, the last two considered as true for all i and j, we get the following estimates

$$\hat{\mu} = \bar{y}_{...}$$
$$\hat{\alpha}_i = \bar{y}_{i..} - \bar{y}_{...} \qquad i = 1, 2, 3$$
$$\hat{\beta}_j = \bar{y}_{.j.} - \bar{y}_{...} \qquad j = 1, 2, 3$$
$$\hat{\gamma}_{ij} = \bar{y}_{ij.} - \bar{y}_{i..} - \bar{y}_{.j.} + \bar{y}_{...} \qquad \begin{aligned} i &= 1, 2, 3 \\ j &= 1, 2, 3 \end{aligned}$$

According to these estimates, the predicted value corresponding to any given y value in the (i,j) cell is

$$\hat{y}_{ijk} = \hat{\mu} + \hat{\alpha}_i + \hat{\beta}_j + \hat{\gamma}_{ij} = \bar{y}_{ij.} \qquad \begin{aligned} i &= 1, 2, 3 \\ j &= 1, 2, 3 \end{aligned}$$

This leads to the following sum of squares due to regression:

$$\text{Regression SS } (\mu,\alpha_i\text{'s},\beta_j\text{'s},\gamma_{ij}\text{'s}) = \sum_i \sum_j \sum_k (\bar{y}_{ij.} - \bar{y}_{...})^2$$

with 8 degrees of freedom. This is easily verified to be the cell SS obtained in Table 6.13C. The sum of squares due to deviation from regression is

$$\text{Residual SS} = \sum_i \sum_j \sum_k (y_{ijk} - \bar{y}_{ij.})^2 \qquad (6.15)$$

with $17 - 8 = 9$ degrees of freedom.

If we set $\gamma_{ij} = 0$ for all i and j in (6.14), we have the multiple-regression equation corresponding to the hypothesis of no interaction. The equation is

$$y_{ijk} = \mu + \sum_{i=1}^{3} \alpha_i x_i + \sum_{j=1}^{3} \beta_j z_j + \varepsilon_{ijk} \qquad (6.16)$$

Fitting this equation by the method of least squares subject to the side conditions $\alpha_. = \beta_. = 0$ yields the estimates

$$\hat{\mu} = \bar{y}...$$
$$\hat{\alpha}_i = \bar{y}_{i..} - \bar{y}... \qquad i = 1, 2, 3$$
$$\hat{\beta}_j = \bar{y}_{.j.} - \bar{y}... \qquad j = 1, 2, 3$$

which lead to the SS due to regression

$$\text{Regression SS } (\mu,\alpha_i\text{'s},\beta_j\text{'s}) = \sum_i \sum_j \sum_k (\bar{y}_{i..} + \bar{y}_{.j.} - \bar{y}... - \bar{y}...)^2$$
$$= \sum_i \sum_j \sum_k (\bar{y}_{i..} - \bar{y}...)^2 + \sum_i \sum_j \sum_k (\bar{y}_{.j.} - \bar{y}...)^2$$

with 4 degrees of freedom. This SS is exactly equal to the sum of treatment SS and block SS obtained in Table 6.13C.

The SS due to interaction (nonadditivity) is given by

$$\text{Nonadditivity SS} = \text{regression SS } (\mu,\alpha_i\text{'s},\beta_j\text{'s},\gamma_{ij}\text{'s}) - \text{regression SS } (\mu,\alpha_i\text{'s},\beta_j\text{'s}) \quad (6.17)$$

with $8 - 4 = 4$ degrees of freedom. This is the same as cell SS − treatment SS − block SS obtained in Table 6.13C. To test the significance of nonadditivity we use the F ratio (nonadditivity MSS)/(residual MSS) with (4,9) degrees of freedom, which is exactly what we did in Table 6.13C.

It may be interesting to note that if we set $\beta_1 = \beta_2 = \beta_3 = 0$ in (6.16), we get the equation

$$y_{ijk} = \mu + \sum_{i=1}^{3} \alpha_i x_i + \varepsilon_{ijk} \qquad (6.18)$$

which is the multiple-regression equation that represents the one-way classification considering rows only in Table 6.13A.

Needless to say, the SS due to regression corresponding to (6.18) is exactly the same as the block SS obtained in Table 6.13C. In symbols, regression SS $(\mu,\alpha_i\text{'s})$ = block SS. Similarly by fitting

$$y_{ijk} = \mu + \sum_{j=1}^{3} \beta_j z_j + \varepsilon_{ijk}$$

we get an SS due to regression which is the same as the treatment SS obtained in Table 6.13C. Or, in symbols, regression SS $(\mu,\beta_j\text{'s})$ = treatment SS.

We have already remarked that regression SS$(\mu,\alpha_i\text{'s},\beta_j\text{'s})$ = block SS + treatment SS. Hence we have

$$\text{Regression SS } (\mu,\alpha_i\text{'s},\beta_j\text{'s}) = \text{regression SS } (\mu,\alpha_i\text{'s}) + \text{regression SS } (\mu,\beta_j\text{'s}) \quad (6.19)$$

Recall that when this relationship obtains in any design, we call the treatments and blocks orthogonal.

Random and Mixed Models

Suppose the blocks used in an experiment like that described in the above subsection are to be considered as a random sample from a universe of, say, R blocks. Similarly, let the treatments be considered as a random sample from a possible collection of T

treatments. We use the following model to represent the data from such an experiment

$$y_{ijk} = \mu + a_i + t_j + g_{ij} + \varepsilon_{ijk} \qquad i = 1, 2, \dots, r$$
$$j = 1, 2, \dots, t$$
$$k = 1, 2, \dots, n$$

Universe of a_i's has mean 0 and variance σ_a^2

Universe of t_j's has mean 0 and variance σ_t^2 \qquad (6.20)

Universe of g_{ij}'s has mean 0 and variance σ_g^2

ε_{ijk}'s are independently $N(0,\sigma^2)$

a_i, t_j, g_{ij}, and ε_{ijk} are mutually independent

The expected values of the different SS in the analysis of variance would then be as shown in Table 6.14.

Table 6.14 **EXPECTED VALUES OF MEAN SUMS OF SQUARES UNDER RANDOM- AND MIXED-MODEL RANDOMIZED-BLOCK DESIGNS WITH t TREATMENTS AND r BLOCKS EACH TREATMENT OCCURRING n TIMES IN EACH BLOCK**

Source of variation	df	Expected values of MSS
A Random Model		
Treatments	$t - 1$	$\sigma^2 + n\left(1 - \dfrac{r}{R}\right)\sigma_g^2 + nr\sigma_t^2$
Blocks	$r - 1$	$\sigma^2 + n\left(1 - \dfrac{t}{T}\right)\sigma_g^2 + nt\sigma_a^2$
Nonadditivity	$(r-1)(t-1)$	$\sigma^2 + n\sigma_g^2$
Residual	$rt(n-1)$	σ^2
B Mixed Model (Blocks Fixed)		
Treatments	$t - 1$	$\sigma^2 + nr\sigma_t^2$
Blocks	$r - 1$	$\sigma^2 + n\left(1 - \dfrac{t}{T}\right)\sigma_g^2 + \dfrac{nt\sum \alpha_i^2}{r-1}$
Nonadditivity	$(r-1)(t-1)$	$\sigma^2 + n\sigma_g^2$
Residual	$rt(n-1)$	σ^2
C Mixed Model (Treatments Fixed)		
Treatments	$t - 1$	$\sigma^2 + n\left(1 - \dfrac{r}{R}\right)\sigma_g^2 + \dfrac{nr\sum \beta_j^2}{t-1}$
Blocks	$r - 1$	$\sigma^2 + nt\sigma_a^2$
Nonadditivity	$(r-1)(t-1)$	$\sigma^2 + n\sigma_g^2$
Residual	$rt(n-1)$	σ^2

The corresponding results for mixed models can be obtained from Table 6.14A by setting $r = R$ or $t = T$ according as the block effects or the treatment effects are considered as constants. If the block effects are considered fixed, the term σ_a^2 in the expected values should be replaced by $\sum \alpha_i^2/(r - 1)$ (see Table 6.14B), and similarly if the treatment effects are considered fixed, the term σ_t^2 should be replaced by $\sum \beta_j^2/(t - 1)$ (see Table 6.14C). Note, however, that $\sum \alpha_i^2/(r - 1)$ represents the variance of the fixed block effects $\alpha_1, \alpha_2, \ldots, \alpha_r$ by virtue of the side condition $\sum \alpha_i = 0$ and similarly that $\sum \beta_j^2/(t - 1)$ represents the variance of the fixed treatment effects $\beta_1, \beta_2, \ldots, \beta_t$, by virtue of the side condition $\sum \beta_j = 0$.

Given the expected MSS as in Table 6.14, it should be easy for you to identify the ratios of MSS that are appropriate for testing given hypotheses such as

H_T: There is no difference between treatment effects $\sigma_b^2 = 0$

H_B: There is no difference between block effects $\sigma_a^2 = 0$

H_N: The nonadditivity is zero $\sigma_g^2 = 0$

To test a certain hypothesis H, *the appropriate variance ratio is obtained by taking those two MSS whose expectations are equal under the hypothesis H.* We note, for example, that in Table 6.14, the nonadditivity MSS and residual MSS have the same expected value if and only if $\sigma_g^2 = 0$; hence the appropriate ratio for testing the hypothesis H_N, described above, is (nonadditivity MSS)/(residual MSS).

Suppose we want to test the hypothesis H_T described above. Let us stipulate that it is the random model that best describes the data situation. Suppose further that the hypothesis H_N concerning nonadditivity has been rejected. Let us assume also that the ratio r/R is negligibly small. Under these circumstances we note that to test H_T the appropriate ratio is (treatment MSS)/(nonadditivity MSS). If, however, the model that best describes the data is the mixed model represented in Table 6.14B, the ratio to be used to test the hypothesis H_T would be (treatment MSS)/(residual MSS), which is different from the one used for the same purpose under the random model represented in Table 6.14A. Thus, we have one test when the block effects are considered as random and another when they are considered fixed if nonadditivity prevails. If nonadditivity is absent, however, the ratio (treatment MSS)/(residual MSS) is applicable whether block effects are fixed or random.

A More Inclusive Model

Suppose in a randomized-blocks experiment of the type described in Table 6.13A that the observations in each block are suspected of being subject to measurement errors. Note that we are dealing with measurement errors in the response variable only. Let us assume that the causes producing measurement errors operate differently in different cells. Let us assume also that the augmentation or depletion due to measurement error is the same for all the observations in a given cell. We have then the following model:

$$y_{ijk} = \mu + a_i + t_j + g_{ij} + e_{ij} + \varepsilon_{ijk} \qquad k = 1, 2, \ldots n$$

Universe of a_i's has mean 0 and variance σ_a^2

Universe of t_j's has mean 0 and variance σ_t^2

Universe of g_{ij}'s has mean 0 and variance σ_g^2 $\qquad\qquad$ (6.21)

Universe of measurement errors e_{ij}'s has mean 0 and variance σ_e^2

ε_{ijk}'s are independently $N(0, \sigma^2)$

$a_i, t_j, g_{ij}, e_{ij},$ and ε_{ijk} are mutually independent

Assuming that the sampling fractions of the blocks and treatments are negligibly small, the analysis of variance and the expected MSS associated with (6.21) are shown in Table 6.15. We have given the label error cum nonadditivity to the SS

$$n \sum \sum (\bar{y}_{ij.} - \bar{y}_{i..} - \bar{y}_{.j.} + \bar{y}_{...})^2$$

which is actually cell SS − block SS − treatment SS, since it covers variations due to nonadditivity as well as measurement errors.

It is clear from the expected values shown in the last column of Table 6.15 that we cannot separately test hypotheses concerning nonadditivity and measurement errors. In the random model (6.21) assumed in Table 6.15, we use the ratio (treatment MSS)/(error cum nonadditivity MSS) for testing the hypothesis $\sigma_t^2 = 0$ and the ratio (block MSS)/(error cum nonadditivity MSS) for testing the hypothesis $\sigma_a^2 = 0$.

6.5 THE TWO-WAY CLASSIFICATION WITH DISPROPORTIONAL NUMBERS OF OBSERVATIONS IN THE CELLS

Inapplicability of the Elementary Method

For one reason or another, the numbers of observations in the cells of a two-way classification may be unequal. This is frequently the case in nonexperimental studies in which the analyst classifies the units of analysis according to the independent

Table 6.15 ANALYSIS OF VARIANCE AND EXPECTED VALUES OF DIFFERENT MSS ASSOCIATED WITH MODEL (6.21)

Source of variation	SS	df	Expected value of MSS
Between treatments		$t - 1$	$\sigma^2 + n(\sigma_g^2 + \sigma_e^2) + nr\sigma_b^2$
Between blocks		$r - 1$	$\sigma^2 + n(\sigma_g^2 + \sigma_e^2) + nt\sigma_a^2$
Error cum nonadditivity	Cell SS − block SS − treatment SS	$(t - 1)(r - 1)$	$\sigma^2 + n(\sigma_g^2 + \sigma_e^2)$
Between cells		$rt - 1$	
Error		$rt(n - 1)$	σ^2
Total		$nrt - 1$	

variables of interest. In such situations the analyst may have no control over how the units of analysis distribute themselves among the cells. In experimental studies also such situations may arise for a variety of reasons: experimental units may not be available in equal numbers in all blocks; subjects who agreed to participate in the experiment may change their minds later on; observations available for some units may turn out to be useless in the end due to contamination by extraneous factors; and so on. When the numbers of observations differ between cells, the elementary method of analysis, in which treatment SS and block SS are calculated directly from the treatment and block totals and the sum treatment SS + block SS is subtracted from cell SS to obtain the nonadditivity SS, may not be applicable. This can be demonstrated by a simple example. An investigator wanted to examine whether the sex of the test administrator affects the performance of disadvantaged children as opposed to "normal" children in a specific test; 12 disadvantaged children and 12 "normal" children were chosen for the experiment. Each group was randomly subdivided into two groups of 6 each. A man administered the test to one subgroup of disadvantaged children and to one subgroup of "normal" children. A woman did the same to the other subgroups. Some children withdrew their cooperation during the course of the test. The final results were as shown in Table 6.16. From the row means we notice that disadvantaged children scored somewhat less than the "normal" children, 13.75 as opposed to 16.25, a difference of 2.5. From the column means we notice that the children's performance in the test was slightly better when the test was administered by a woman than when it was administered by a man (16.25 against 13.75, a difference of 2.5).

The subclass means per child, however, reveal a somewhat different picture: disadvantaged children scored 5 points less than the "normal" children under both male and female test administrators. Also among both groups of children, the test performance was better by 5 points under female test administrator than under male test administrator. The comparison between the row means thus in this case leads to

Table 6.16   DATA ON THE IMPACT OF THE SEX OF THE TEST ADMINISTRATOR ON THE TEST RESULTS

Children	Test administrator		Row means
	Female	Male	
Disadvantaged:			
Total score	90	20	
Number	6	2	
Mean score	15	10	13.75
"Normal":			
Total score	40	90	
Number	2	6	
Mean score	20	15	16.25
Column means	16.25	13.75	

underestimating the difference between disadvantaged and "normal" children. Similarly the comparison between the column means underestimates the "effect" of the sex of the test administrator. Why? The reason is to be found in the fact that in the final data set the better-performing children are disproportionately more represented among the disadvantaged children and disproportionately less among the "normal" children.

If we apply the elementary method of analysis and calculate nonadditivity SS by subtracting the column SS + row SS from cell SS, we may run into problems. The cell SS in this case is

$$\frac{90^2}{6} + \frac{20^2}{2} + \frac{40^2}{2} + \frac{90^2}{6} - \frac{240^2}{16} = 100$$

The SS due to children's background, from row totals, is

$$\frac{110^2}{8} + \frac{30^2}{8} - \frac{240^2}{16} = 25$$

and the SS due to the sex of the test administrator, from column totals, is

$$\frac{130^2}{8} + \frac{110^2}{8} - \frac{240^2}{16} = 25$$

This leaves, according to the elementary procedure, a nonadditivity SS of $100 - 25 - 25 = 50$. But from the cell means we notice that the differences between the means are the same in each row. Hence additivity prevails. In a correct analysis the nonadditivity SS should be zero.

The Method of Fitting Constants

The proper way to calculate the SS due to nonadditivity is to subtract the SS due to regression when nonadditivity is ignored from the corresponding quantity when nonadditivity is not ignored. To illustrate the procedure let us start with an algebraic representation as shown in Table 6.17 of the data set in Table 6.16.

We first fit the following multiple-regression equation, which ignores nonadditivity:

$$y_{ijk} = \mu x_0 + \alpha_1 x_1 + \alpha_2 x_2 + \beta_1 z_1 + \beta_2 z_2 + \varepsilon_{ijk} \qquad (6.22)$$

where $x_0 = 1$ for all observations
$\quad x_i = 1$ for all observations in the ith row and 0 for all others
$\quad z_j = 1$ for all observations in the jth column and 0 for all others
We shall refer to this equation later as the *additive model*.

The normal equations are

$$y_{...} = 16\hat{\mu} + 8\hat{\alpha}_1 + 8\hat{\alpha}_2 + 8\hat{\beta}_1 + 8\hat{\beta}_2$$
$$y_{1..} = 8\hat{\mu} + 8\hat{\alpha}_1 \qquad\qquad + 6\hat{\beta}_1 + 2\hat{\beta}_2$$
$$y_{2..} = 8\hat{\mu} \qquad\qquad + 8\hat{\alpha}_2 + 2\hat{\beta}_1 + 6\hat{\beta}_2 \qquad (6.23)$$
$$y_{.1.} = 8\hat{\mu} + 6\hat{\alpha}_1 + 2\hat{\alpha}_2 + 8\hat{\beta}_1$$
$$y_{.2.} = 8\hat{\mu} + 2\hat{\alpha}_1 + 6\hat{\alpha}_2 \qquad\qquad + 8\hat{\beta}_2$$

Although we have five equations here, only three of them are independent. One of the first three equations is redundant, since the sum of the second and third equals the first. Similarly, given the first equation, one of the last two is redundant, since the sum of the last two equals the first. There being only three independent equations, we need two side conditions in order to solve for the unknowns. Let us use the side conditions

$$\hat{\alpha}_1 + \hat{\alpha}_2 = 0 \qquad \hat{\beta}_1 + \hat{\beta}_2 = 0 \qquad (6.24)$$

which are equivalent to stipulating that we are satisfied by calculating $\hat{\alpha}$'s and $\hat{\beta}$'s expressed as deviations from their respective means. Inserting (6.24) into the first equation of (6.23) gives $\bar{y}_{...}$ as the solution for $\hat{\mu}$. Substituting $\bar{y}_{...}$ for $\hat{\mu}$ in the second equation of (6.23) and remembering that $2\hat{\beta}_2 = -2\hat{\beta}_1$ by virtue of the side condition $\hat{\beta}_1 + \hat{\beta}_2 = 0$, we get

$$\hat{\alpha}_1 + \tfrac{1}{2}\hat{\beta}_1 = \bar{y}_{1..} - \bar{y}_{...}$$

Similarly the fourth equation in (6.23) gives, after substituting $\bar{y}_{...}$ for $\hat{\mu}$ and $-2\hat{\alpha}_1$ for $2\hat{\alpha}_2$,

$$\tfrac{1}{2}\hat{\alpha}_1 + \hat{\beta}_1 = \bar{y}_{.1.} - \bar{y}_{...}$$

Table 6.17 ALGEBRAICAL REPRESENTATION OF THE DATA IN TABLE 6.16

Block (row)	Treatment (column) 1	Treatment (column) 2	Row totals and row means
1	y_{111}	y_{121}	
	y_{112}	y_{122}	
	y_{113}		
	y_{114}		
	y_{115}		
	y_{116}		
	$y_{11.} = 90$	$y_{12.} = 20$	$y_{1..} = 110$
	$n_{11} = 6$	$n_{12} = 2$	$n_{1.} = 8$
	$\bar{y}_{11.} = 15$	$\bar{y}_{12.} = 10$	$\bar{y}_{1..} = 13.75$
2	y_{211}	y_{221}	
	y_{212}	y_{222}	
		y_{223}	
		y_{224}	
		y_{225}	
		y_{226}	
	$y_{21.} = 40$	$y_{22.} = 90$	$y_{2..} = 130$
	$n_{21} = 2$	$n_{22} = 6$	$n_{2.} = 8$
	$\bar{y}_{21.} = 20$	$\bar{y}_{22.} = 15$	$\bar{y}_{2..} = 16.25$
Column totals and column means	$y_{.1.} = 130$	$y_{.2.} = 110$	$y_{...} = 240$
	$n_{.1} = 8$	$n_{.2} = 8$	$n_{..} = 16$
	$\bar{y}_{.1.} = 16.25$	$\bar{y}_{.2.} = 13.75$	$\bar{y}_{...} = 15$

If in these equations we replace $\bar{y}_{1..} - \bar{y}_{...}$ and $\bar{y}_{.1.} - \bar{y}_{...}$ by their numerical values from Table 6.17, we obtain

$$\hat{\alpha}_1 + \tfrac{1}{2}\hat{\beta}_1 = -1.25 \qquad \tfrac{1}{2}\hat{\alpha}_1 + \hat{\beta}_1 = 1.25$$

These are easy to solve. The solutions are $\hat{\alpha}_1 = -2.5$ and $\hat{\beta}_1 = 2.5$ and they with the side conditions $\hat{\alpha}_1 + \hat{\alpha}_2 = 0$, $\hat{\beta}_1 + \hat{\beta}_2 = 0$ lead to the complete solution set

$$\hat{\mu} = 15.0 \qquad \hat{\alpha}_1 = -2.5 \qquad \hat{\alpha}_2 = 2.5 \qquad \hat{\beta}_1 = 2.5 \qquad \hat{\beta}_2 = -2.5$$

The predicted values corresponding to each observed y value can now be written as shown in Table 6.18A.

The SS due to regression is calculated in the usual manner as the SS deviations of the predicted y values from their mean. Remembering that there are six values in the $(1,1)$ cell, two in the $(1,2)$ cell, two in the $(2,1)$ cell, and six in the $(2,2)$ cell, we obtain the mean of the predicted y values from Table 6.18A as

$$\frac{(6 \times 15.0) + (2 \times 10.0) + (2 \times 20.0) + (6 \times 15.0)}{6 + 2 + 2 + 6}$$

which is 15.0, the same as the grand mean of the observed y values in Table 6.17. Consequently,

Regression SS $(\mu, \alpha_i\text{'s}, \beta_j\text{'s}) = 6(15.0 - 15.0)^2 + 2(10.0 - 15.0)^2$
$$+ 2(20.0 - 15.0)^2 + 6(15.0 - 15.0)^2 = 100.0 \quad (6.25)$$

Table 6.18   PREDICTED y VALUES

Block (row)	Treatment (column)	
	$j = 1$	$j = 2$
A	Calculated from the Additive Model (6.22) Fitted to the Data in Table 6.16	
$i = 1$	$\hat{y}_{11k} = \hat{\mu} + \hat{\alpha}_1 + \hat{\beta}_1$ $= 15.0 - 2.5 + 2.5$ $= 15.0$	$\hat{y}_{12k} = \hat{\mu} + \hat{\alpha}_1 + \hat{\beta}_2$ $= 15.0 - 2.5 - 2.5$ $= 10.0$
$i = 2$	$\hat{y}_{21k} = \hat{\mu} + \hat{\alpha}_2 + \hat{\beta}_1$ $= 15.0 + 2.5 + 2.5$ $= 20.0$	$\hat{y}_{22k} = \hat{\mu} + \hat{\alpha}_2 + \hat{\beta}_2$ $= 15.0 + 2.5 - 2.5$ $= 15.0$
B	Calculated from the Nonadditive Model (6.26) Fitted to the Data in Table 6.16	
$i = 1$	$\hat{y}_{11k} = \hat{\mu} + \hat{\alpha}_1 + \hat{\beta}_1 + \hat{\gamma}_{11}$ $= 15.0 - 2.5 + 2.5 + 0.0$ $= 15.0$	$\hat{y}_{12k} = \hat{\mu} + \hat{\alpha}_1 + \hat{\beta}_2 + \hat{\gamma}_{12}$ $= 15.0 - 2.5 - 2.5 + 0.0$ $= 10.0$
$i = 2$	$\hat{y}_{21k} = \hat{\mu} + \hat{\alpha}_2 + \hat{\beta}_1 + \hat{\gamma}_{21}$ $= 15.0 + 2.5 + 2.5 + 0.0$ $= 20.0$	$\hat{y}_{22k} = \hat{\mu} + \hat{\alpha}_2 + \hat{\beta}_2 + \hat{\gamma}_{22}$ $= 15.0 + 2.5 - 2.5 + 0.0$ $= 15.0$

The next step is to fit the multiple regression which does not ignore nonadditivity:

$$y_{ijk} = \mu x_0 + \alpha_1 x_1 + \alpha_2 x_2 + \beta_1 z_1 + \beta_2 z_2 + \gamma_{11} v_{11}$$
$$+ \gamma_{12} v_{12} + \gamma_{21} v_{21} + \gamma_{22} v_{22} + \varepsilon_{ijk} \qquad (6.26)$$

where $x_0 = 1$ for all observations

$x_i = 1$ for all observations in ith row and 0 for all others

$z_j = 1$ for all observations in jth column and 0 for all others

$v_{ij} = 1$ for all observations in (i,j) cell and 0 for all others

We shall refer to this as the *nonadditive model*.

The normal equations are

$$
\begin{aligned}
y_{...} &= 16\hat{\mu} + 8\hat{\alpha}_1 + 8\hat{\alpha}_2 + 8\hat{\beta}_1 + 8\hat{\beta}_2 + 6\hat{\gamma}_{11} + 2\hat{\gamma}_{12} + 2\hat{\gamma}_{21} + 6\hat{\gamma}_{22} \\
y_{1..} &= 8\hat{\mu} + 8\hat{\alpha}_1 \qquad\quad + 6\hat{\beta}_1 + 2\hat{\beta}_2 + 6\hat{\gamma}_{11} + 2\hat{\gamma}_{12} \\
y_{2..} &= 8\hat{\mu} \qquad\quad + 8\hat{\alpha}_2 + 2\hat{\beta}_1 + 6\hat{\beta}_2 \qquad\qquad\quad + 2\hat{\gamma}_{21} + 6\hat{\gamma}_{22} \\
y_{.1.} &= 8\hat{\mu} + 6\hat{\alpha}_1 + 2\hat{\alpha}_2 + 8\hat{\beta}_1 \qquad + 6\hat{\gamma}_{11} \qquad\quad + 2\hat{\gamma}_{21} \\
y_{.2.} &= 8\hat{\mu} + 2\hat{\alpha}_1 + 6\hat{\alpha}_2 \qquad + 8\hat{\beta}_2 \qquad + 2\hat{\gamma}_{12} \qquad\qquad + 6\hat{\gamma}_{22} \qquad (6.27) \\
y_{11.} &= 6\hat{\mu} + 6\hat{\alpha}_1 \qquad\quad + 6\hat{\beta}_1 \qquad + 6\hat{\gamma}_{11} \\
y_{12.} &= 2\hat{\mu} + 2\hat{\alpha}_1 \qquad\qquad\quad + 2\hat{\beta}_2 \qquad + 2\hat{\gamma}_{12} \\
y_{21.} &= 2\hat{\mu} \qquad\quad + 2\hat{\alpha}_2 + 2\hat{\beta}_1 \qquad\qquad\qquad + 2\hat{\gamma}_{21} \\
y_{22.} &= 6\hat{\mu} \qquad\quad + 6\hat{\alpha}_2 \qquad + 6\hat{\beta}_2 \qquad\qquad\qquad\qquad + 6\hat{\gamma}_{22}
\end{aligned}
$$

Of these nine equations only four are independent. There being nine unknowns to solve for, we need five side conditions. Let us use

$$
\begin{aligned}
\hat{\alpha}_1 + \hat{\alpha}_2 &= 0 \\
\hat{\beta}_1 + \hat{\beta}_2 &= 0 \\
6\hat{\gamma}_{11} + 2\hat{\gamma}_{12} &= 0 \qquad (6.28) \\
2\hat{\gamma}_{21} + 6\hat{\gamma}_{22} &= 0 \\
2\hat{\gamma}_{12} + 6\hat{\gamma}_{22} &= 0
\end{aligned}
$$

As will become clear in a moment, this particular set of side conditions makes solving (6.27) very easy, given the solutions to (6.23). It is not difficult to see that the last three side conditions in (6.28) necessarily imply

$$6\hat{\gamma}_{11} + 2\hat{\gamma}_{21} = 0 \qquad (6.29)$$

We notice that the side conditions involving $\hat{\gamma}_{ij}$'s reduce the first five equations in (6.27) to the five equations in (6.23), which we have already solved with the help of the side conditions $\hat{\alpha}_1 + \hat{\alpha}_2 = \hat{\beta}_1 + \hat{\beta}_2 = 0$, these being the same as the first two in (6.28). Hence the solutions arrived at earlier for (6.23) apply to the equations in (6.27) too. If we substitute those solutions into the last four equations in (6.27), we get the following solutions for $\hat{\gamma}_{ij}$'s:

$$
\begin{aligned}
\hat{\gamma}_{11} &= \bar{y}_{11.} - \hat{\mu} - \hat{\alpha}_1 - \hat{\beta}_1 = 15.0 - 15.0 + 2.5 - 2.5 = 0 \\
\hat{\gamma}_{12} &= \bar{y}_{12.} - \hat{\mu} - \hat{\alpha}_1 - \hat{\beta}_2 = 10.0 - 15.0 + 2.5 + 2.5 = 0 \\
\hat{\gamma}_{21} &= \bar{y}_{21.} - \hat{\mu} - \hat{\alpha}_2 - \hat{\beta}_1 = 20.0 - 15.0 - 2.5 - 2.5 = 0 \\
\hat{\gamma}_{22} &= \bar{y}_{22.} - \hat{\mu} - \hat{\alpha}_2 - \hat{\beta}_2 = 15.0 - 15.0 - 2.5 + 2.5 = 0
\end{aligned}
$$

Since the nonadditivity components all turned out to be zero, it is obvious that there is additivity in the data set in question. Let us nonetheless carry the formal analysis forward. Having solved for all the unknowns, we can calculate the predicted values under the nonadditive model (6.26) as shown in Table 6.18B. Notice in this connection that the predicted value under the nonadditive model corresponding to any y_{ijk} is the cell mean $\bar{y}_{ij}$. This is a general result worth remembering. The regression SS under the nonadditive model is found from the figures in Table 6.18B:

$$\text{Regression SS } (\mu, \alpha_i\text{'s}, \beta_j\text{'s}, \gamma_{ij}\text{'s}) = 6(15.0 - 15.0)^2 + 2(10.0 - 15.0)^2$$
$$+ 2(20.0 - 15.0)^2 + 6(15.0 - 15.0)^2$$
$$= 100.0$$

The SS due to nonadditivity is obtained as

$$\text{Nonadditivity SS} = \text{regression SS } (\mu, \alpha_i\text{'s}, \beta_j\text{'s}, \gamma_{ij}\text{'s})$$
$$- \text{regression SS } (\mu, \alpha_i\text{'s}, \beta_j\text{'s}) = 100.0 - 100.0 = 0$$

which is as we expected from the patterns of cell means in the original data set (Table 6.16).

In the present case the nonadditivity SS turns out to be exactly zero, but in most data situations this will not be the case. The question therefore arises of how to test the significance of nonadditivity SS. For a general $r \times t$ table with n_{ij} observations in the (i,j) cell, nonadditivity SS has $(r - 1)(t - 1)$ degrees of freedom. The SS due to deviation from the nonadditive regression has

$$\sum_i \sum_j n_{ij} - (r - 1) - (t - 1) - (r - 1)(t - 1)$$

or $n_{..} - rt + 1$ degrees of freedom. If the residuals, ε_{ijk}'s, are independently distributed as normal with 0 mean and common variance σ^2, the ratio (nonadditivity MSS)/(residual MSS), where residual stands for deviations from the nonadditive regression, is an F with $[(r - 1)(t - 1), n_{..} - rt + 1]$ degrees of freedom. If nonadditivity is found to be negligible, the analyst proceeds to test the significance of treatment (column) SS and block (row) SS.

It should be emphasized that for the above purpose the treatment SS and the block SS to be used are not the ones directly calculated according to the elementary procedure from the treatment and block totals, respectively. To describe the proper procedure, let us denote the treatment SS calculated from the treatment totals according to the elementary procedure as regression SS $(\mu, \beta_j\text{'s})$ and the corresponding block SS as regression SS $(\mu, \alpha_i\text{'s})$. Thus

$$\text{Regression SS } (\mu, \beta_j\text{'s}) = \sum \frac{(j\text{th treatment total})^2}{\text{no. of observations under } j\text{th treatment}}$$
$$- \text{ correction for mean}$$

$$\text{Regression SS } (\mu, \alpha_i\text{'s}) = \sum \frac{(i\text{th block total})^2}{\text{no. of observations in } i\text{th block}}$$
$$- \text{ correction for mean}$$

When the numbers of observations are the same in all cells, regression SS (μ, β_j's) and regression SS (μ, α_i's) add up exactly to regression SS (μ, α_i's, β_j's), the SS due to regression according to the additive model. If the numbers of observations are disproportional between columns or rows, regression SS (μ, α_i's) + regression SS (μ, β_j's) may not necessarily equal regression SS (μ, α_i's, β_j's) because when the numbers of cases are disproportional between columns (rows), regression SS (μ, α_i's) overlaps somewhat with regression SS (μ, β_j's). In other words, differences between treatment (column) means partly reflect differences between block (row) effects and differences between block (row) means partly reflect differences between treatment (column) effects. This may become clear if we examine our simple illustrative example in Table 6.16. Recall that since the nonadditivity component has been judged to be zero, the observed mean in the (i, j) cell can be taken to represent $\mu + \alpha_i + \beta_j$, except for sampling fluctuations. Weighting with the numbers of cases in respective cells, we find that (except for sampling fluctuations) the first treatment (column) mean represents $[6(\mu + \alpha_1 + \beta_1) + 2(\mu + \alpha_2 + \beta_1)]/8$, or $\mu + \beta_1 + (6\alpha_1 + 2\alpha_2)/8$, and the second treatment (column) mean represents $[2(\mu + \alpha_1 + \beta_2) + 6(\mu + \alpha_2 + \beta_2)]/8$, or $\mu + \beta_2 + (2\alpha_1 + 6\alpha_2)/8$. Consequently the difference between the two treatment means represents $(\beta_1 - \beta_2) + (\alpha_1 - \alpha_2)/2$, which involves in part the difference between block effects, α_1 and α_2. Similarly, it can be seen that, except for sampling fluctuations, the difference between the two block means represents

$$\frac{6(\mu + \alpha_1 + \beta_1) + 2(\mu + \alpha_1 + \beta_2)}{8} - \frac{2(\mu + \alpha_2 + \beta_1) + 6(\mu + \alpha_2 + \beta_2)}{8}$$

or $(\alpha_1 - \alpha_2) + (\beta_1 - \beta_2)/2$, which involves in part the difference between the treatment effects, β_1 and β_2. Under such circumstances, differences between treatment (column) means are not exclusively attributable to treatment differences, and differences between block (row) means are not exclusively attributable to block differences. In other words, the proper SS to test the null hypothesis that treatment differences do not exist is not regression SS (μ, β_j's), and the proper SS for testing the null hypothesis that block differences do not exist is not regression SS (μ, α_i's).

The question then arises whether there is any quantity that represents all (and nothing but) treatment differences and a corresponding one that represents all (and nothing but) block differences. Unfortunately when the cell frequencies are disproportional, it is not possible to obtain such quantities. What is possible is to obtain a statistic that represents the variation in the study phenomenon attributable to treatment differences *over and beyond* what is attributable to differences between block means. This is given by the difference

$$\text{Regression SS } (\mu, \alpha_i\text{'s}, \beta_j\text{'s}) - \text{regression SS } (\mu, \alpha_i\text{'s})$$

We shall refer to this as treatment SS eliminating blocks (cf. direct effects, discussed in Part Three). Similarly, we have

$$\text{Regression SS } (\mu, \alpha_i\text{'s}, \beta_j\text{'s}) - \text{regression SS } (\mu, \beta_j\text{'s})$$

representing the variation in the study phenomenon attributable to block differences *over and beyond* what is attributable to differences between treatment means. This may be referred to as block SS eliminating treatments.

If treatment SS eliminating blocks is significant, we declare that the treatment effects are not equal to each other. Similarly, if block SS eliminating treatments is significant, we declare that the block effects are not equal to each other. The analogy between this procedure and the significance testing of partials in regression analysis should be obvious. Table 6.19 presents the structure of the analysis-of-variance table in which both treatment SS eliminating blocks and block SS eliminating treatments are shown. It is worth emphasizing that in preparing Table 6.19 all SS except regression SS (μ,α_i's,β_j's) can be calculated in the elementary fashion. We have already pointed out that regression SS (μ,α_i's), that is, block SS ignoring treatments, and regression SS (μ,β_j's), that is, treatment SS ignoring blocks can be calculated from the block (row) and treatment (column) marginals by following the procedure for the calculation of between SS in the one-way analysis of variance. The cell sum of squares or regression SS (μ,α_i's,β_j's,γ_{ij}'s) can be directly calculated from the cell totals and the cell frequencies in the elementary fashion:

$$\text{Cell SS} = \sum_i \sum_j \frac{(\text{cell total})^2}{\text{cell frequency}} - \text{correction for mean}$$

The calculation of the total SS does not involve any new procedure either, as we simply subtract the correction for the mean from the SS of individual y values.

Table 6.19   ANALYSIS OF VARIANCE OF $r \times t$ TABLE WITH n_{ij} OBSERVATIONS IN THE (i,j) CELL

Source of variation	SS	df
1 Treatments ignoring blocks	Regression SS (μ,β_j's)	$t - 1$
Blocks eliminating treatments	Regression SS (μ,α_i's,β_j's) − regression SS (μ,β_j's)	$r - 1$
Nonadditivity	Regression SS (μ,α_i's,β_j's,γ_{ij}'s) − regression SS (μ,α_i's,β_j's)	$(r-1)(t-1)$
Cells	Regression SS (μ,α_i's,β_j's,γ_{ij}'s)	$rt - 1$
Residual	Total SS − cell SS	$n_{..} - rt + 1$
Total	$\sum_i \sum_j \sum_k y_{ijk}^2 - C$	$n_{..} - 1$
2 Blocks ignoring treatments	Regression SS (μ,α_i's)	$r - 1$
Treatments eliminating blocks	Regression SS (μ,α_i's,β_j's) − regression SS (μ,α_i's)	$t - 1$
Nonadditivity	Regression SS (μ,α_i's,β_j's,γ_{ij}'s) − regression SS (μ,α_i's,β_j's)	$(r-1)(t-1)$
Cells	Regression SS (μ,α_i's,β_j's,γ_{ij}'s)	$rt - 1$
Residual	Total SS − cell SS	$n_{..} - rt + 1$
Total	$\sum_i \sum_j \sum_k y_{ijk}^2 - C$	$n_{..} - 1$

Nonadditivity SS and residual SS are obtained by subtraction. Thus the only SS that requires a special computation procedure is regression SS $(\mu, \alpha_i\text{'s}, \beta_j\text{'s})$, that is, the SS due to regression under the additive model. To obtain this SS we write the normal equations, define the side conditions, solve for the unknowns, obtain the predicted values in each cell, and calculate the SS of deviations of the predicted values from their mean.

It is particularly worth noting that all calculations except that of total SS can be carried out knowing only the cell totals and the cell frequencies. In some textbooks the method described above is referred to as that of fitting constants, following Yates (1934).

6.6   ESTIMATION OF TREATMENT (BLOCK) CONTRASTS AND THE ASSOCIATED STANDARD ERRORS

Let us consider a data set which is slightly more complex than the one considered in the preceding section. Subjects are divided into three groups according as they hold a favorable, neutral, or unfavorable attitude toward legalizing abortion, and are then exposed to a well-balanced discussion of the pros and cons of the abortion issue. Each group of subjects was divided randomly into three subgroups, and one subgroup was tested immediately after the discussion, another after 1 week's delay, and the third after 2 week's delay. The test scores represented the degree to which subjects remember the pros and cons of the abortion issue covered in the discussion. Table 6.20 contains the test scores.

The research questions were (1) whether the initial attitude toward an issue affects the power to remember various points about the issue and (2) whether the tendency to forget things is linearly related to the time interval between the discussion session and the test (see Edwards, 1941, for a report on a similar study).

For the purpose of the present exposition we shall refer to the initial attitude as the block variable and the time intervals between the discussion session and the administration of the test as treatments. We have thus three blocks and three treatments.

As mentioned in Sec. 6.5, much of the calculation involved in the analysis of variance of data sets like this can be carried out in the elementary manner. From Table 6.20 we calculate

$$\text{Correction for the mean} = C = \frac{105^2}{22} = 501.136$$

$$\text{Treatment (column) SS} = \text{regression SS } (\mu, \beta_j\text{'s})$$

$$= \frac{76^2}{8} + \frac{14^2}{7} + \frac{15^2}{7} - C = 281.007$$

$$\text{Block (row) SS} = \text{regression SS } (\mu, \alpha_i\text{'s})$$

$$= \frac{29^2}{7} + \frac{37^2}{7} + \frac{39^2}{8} - C = 4.703$$

$$\text{Total SS} = 9^2 + 8^2 + \cdots + 3^2 - C$$
$$= 799.000 - C$$
$$= 297.864$$
$$\text{Cell SS} = \text{regression SS } (\mu, \alpha_i\text{'s}, \beta_j\text{'s}, \gamma_{ij}\text{'s})$$
$$= \frac{17^2}{2} + \frac{7^2}{3} + \cdots + \frac{7^2}{3} - C$$
$$= 788.333 - C$$
$$= 287.197$$
$$\text{Residual SS} = \text{total SS} - \text{cell SS}$$
$$= 297.864 - 287.197$$
$$= 10.667$$

We now proceed to fit the additive model

$$y_{ijk} = \mu x_0 + \alpha_1 x_1 + \alpha_2 x_2 + \alpha_3 x_3 + \beta_1 z_1 + \beta_2 z_2 + \beta_3 z_3 + \varepsilon_{ijk}$$

where $x_0 = 1$ for all observations
 $x_i = 1$ for all observations in ith row and 0 for all others
 $z_j = 1$ for all observations in jth column and 0 for all others

Table 6.20   DATA† FROM A STUDY OF TWO FACTORS AFFECTING TENDENCY TO FORGET VARIOUS ASPECTS OF A SENSITIVE ISSUE

Initial attitude toward issue (blocks)	Time interval between discussion session and administration of test (treatments)			Block totals
	Immediately after	After 1-week delay	After 2-week delay	
Favorable	$y_{111} = 9$	$y_{121} = 2$	$y_{131} = 3$	
	$y_{112} = 8$	$y_{122} = 2$	$y_{132} = 2$	
		$y_{123} = 3$		
	$y_{11.} = 17$	$y_{12.} = 7$	$y_{13.} = 5$	$y_{1..} = 29$
	$n_{11} = 2$	$n_{12} = 3$	$n_{13} = 2$	$n_{1.} = 7$
Neutral	$y_{211} = 11$	$y_{221} = 1$	$y_{231} = 2$	
	$y_{212} = 10$	$y_{222} = 2$	$y_{232} = 1$	
	$y_{213} = 10$			
	$y_{21.} = 31$	$y_{22.} = 3$	$y_{23.} = 3$	$y_{2..} = 37$
	$n_{11} = 3$	$n_{22} = 2$	$n_{23} = 2$	$n_{2.} = 7$
Unfavorable	$y_{311} = 11$	$y_{321} = 2$	$y_{331} = 3$	
	$y_{312} = 8$	$y_{322} = 2$	$y_{332} = 1$	
	$y_{313} = 9$		$y_{333} = 3$	
	$y_{31.} = 28$	$y_{32.} = 4$	$y_{33.} = 7$	$y_{3..} = 39$
	$n_{31} = 3$	$n_{32} = 2$	$n_{33} = 3$	$n_{3.} = 8$
Treatment totals	$y_{.1.} = 76$	$y_{.2.} = 14$	$y_{.3.} = 15$	$y_{...} = 105$
	$n_{.1} = 8$	$n_{.2} = 7$	$n_{.3} = 7$	$n_{..} = 22$

† Fictitious.

The normal equations are

$$
\begin{aligned}
y_{\ldots} &= 22\hat{\mu} + 7\hat{\alpha}_1 + 7\hat{\alpha}_2 + 8\hat{\alpha}_3 + 8\hat{\beta}_1 + 7\hat{\beta}_2 + 7\hat{\beta}_3 \\
y_{1..} &= 7\hat{\mu} + 7\hat{\alpha}_1 \qquad\qquad\quad + 2\hat{\beta}_1 + 3\hat{\beta}_2 + 2\hat{\beta}_3 \\
y_{2..} &= 7\hat{\mu} \qquad\quad + 7\hat{\alpha}_2 \qquad\quad + 3\hat{\beta}_1 + 2\hat{\beta}_2 + 2\hat{\beta}_3 \\
y_{3..} &= 8\hat{\mu} \qquad\qquad\qquad + 8\hat{\alpha}_3 + 3\hat{\beta}_1 + 2\hat{\beta}_2 + 3\hat{\beta}_3 \\
y_{.1.} &= 8\hat{\mu} + 2\hat{\alpha}_1 + 3\hat{\alpha}_2 + 3\hat{\alpha}_3 + 8\hat{\beta}_1 \\
y_{.2.} &= 7\hat{\mu} + 3\hat{\alpha}_1 + 2\hat{\alpha}_2 + 2\hat{\alpha}_3 \qquad\quad + 7\hat{\beta}_2 \\
y_{.3.} &= 7\hat{\mu} + 2\hat{\alpha}_1 + 2\hat{\alpha}_2 + 3\hat{\alpha}_3 \qquad\qquad\qquad + 7\hat{\beta}_3
\end{aligned}
\tag{6.30}
$$

One of the equations among the first four is redundant, as is one among the first one and the last three. We may replace any one of the first four equations by a side condition and any one of the last three by another independent side condition. This gives us as many independent equations as there are unknowns. Theoretically, solving these equations is a straightforward matter. But with seven equations the computations may be somewhat complex. If one has access to a computer, the solutions can easily be obtained by matrix inversion. We shall indicate the nature of this procedure a little later. For the moment let us suppose that we have access only to a desk calculator. Under such circumstances it will be helpful if we can reduce the problem of solving seven equations to one of solving fewer equations. Let us see how this can be done in the present case.

From the last three equations in (6.30) we have

$$
\hat{\beta}_1 = \bar{y}_{.1.} - \hat{\mu} - \frac{2\hat{\alpha}_1 + 3\hat{\alpha}_2 + 3\hat{\alpha}_3}{8}
$$

$$
\hat{\beta}_2 = \bar{y}_{.2.} - \hat{\mu} - \frac{3\hat{\alpha}_1 + 2\hat{\alpha}_2 + 2\hat{\alpha}_3}{7}
\tag{6.31}
$$

$$
\hat{\beta}_3 = \bar{y}_{.3.} - \hat{\mu} - \frac{2\hat{\alpha}_1 + 2\hat{\alpha}_2 + 3\hat{\alpha}_3}{7}
$$

Inserting these values for $\hat{\beta}_j$'s into the second equation in (6.30), we get

$$
\begin{aligned}
y_{1..} &= 7\hat{\mu} + 7\hat{\alpha}_1 + 2\left(\bar{y}_{.1.} - \hat{\mu} - \frac{2\hat{\alpha}_1 + 3\hat{\alpha}_2 + 3\hat{\alpha}_3}{8}\right) \\
&\quad + 3\left(\bar{y}_{.2.} - \hat{\mu} - \frac{3\hat{\alpha}_1 + 2\hat{\alpha}_2 + 2\hat{\alpha}_3}{7}\right) \\
&\qquad\quad + 2\left(\bar{y}_{.3.} - \hat{\mu} - \frac{2\hat{\alpha}_1 + 2\hat{\alpha}_2 + 3\hat{\alpha}_3}{7}\right) \\
&= [7 - 2(\tfrac{2}{8}) - 3(\tfrac{3}{7}) - 2(\tfrac{2}{7})]\hat{\alpha}_1 \\
&\qquad\qquad\qquad - [2(\tfrac{3}{8}) + 3(\tfrac{2}{7}) + 2(\tfrac{2}{7})]\hat{\alpha}_2 \\
&\quad - [2(\tfrac{3}{8}) + 3(\tfrac{2}{7}) + 2(\tfrac{3}{7})]\hat{\alpha}_3 \\
&\qquad\qquad\qquad + (2\bar{y}_{.1.} + 3\bar{y}_{.2.} + 2\bar{y}_{.3.}) \\
&= \tfrac{260}{56}\hat{\alpha}_1 - \tfrac{122}{56}\hat{\alpha}_2 - \tfrac{138}{56}\hat{\alpha}_3 \\
&\qquad\qquad\qquad + (2\bar{y}_{.1.} + 3\bar{y}_{.2.} + 2\bar{y}_{.3.})
\end{aligned}
$$

This can be rewritten as

$$\tfrac{260}{56}\hat{\alpha}_1 - \tfrac{122}{56}\hat{\alpha}_2 - \tfrac{138}{56}\hat{\alpha}_3 = y_1.. - (2\bar{y}._1. + 3\bar{y}._2. + 2\bar{y}._3.)$$

The right-hand side of this equation is seen to be, from the marginals in Table 6.20,

$$29 - [2(\tfrac{76}{8}) + 3(\tfrac{14}{7}) + 2(\tfrac{15}{7})] = -\tfrac{16}{56}$$

Thus we have one equation involving $\hat{\alpha}_i$'s only:

$$\tfrac{260}{56}\hat{\alpha}_1 - \tfrac{122}{56}\hat{\alpha}_2 - \tfrac{138}{56}\hat{\alpha}_3 = -\tfrac{16}{56} \qquad (6.32)$$

Similarly we get another equation of this form by substituting for $\hat{\beta}_j$'s from (6.31) in the third equation of (6.30):

$$-\tfrac{122}{56}\hat{\alpha}_1 + \tfrac{265}{56}\hat{\alpha}_2 - \tfrac{143}{56}\hat{\alpha}_3 = \tfrac{12}{56} \qquad (6.33)$$

We now introduce as a side condition

$$\hat{\alpha}_3 = 0 \qquad (6.34)$$

This side condition is different from the type we have frequently used so far in this chapter. But as pointed out already, our inference concerning treatment (block) contrasts will not be affected by the particular side condition(s) we employ for solving the normal equations. The use of (6.34) as a side condition will reduce the computations involved considerably. Inserting (6.34) into (6.32) and (6.33), we get the following two equations in the two unknowns $\hat{\alpha}_1$ and $\hat{\alpha}_2$:

$$\begin{aligned} \tfrac{260}{56}\hat{\alpha}_1 - \tfrac{122}{56}\hat{\alpha}_2 &= -\tfrac{16}{56} \\ -\tfrac{122}{56}\hat{\alpha}_1 + \tfrac{265}{56}\hat{\alpha}_2 &= \tfrac{12}{56} \end{aligned} \qquad (6.35)$$

This is a two-equation system. These can be solved in the usual manner, but we shall solve them using the method of matrix inversion. The coefficients in the above two-equation system can be written as a 2×2 array:

$$\begin{bmatrix} \tfrac{260}{56} & -\tfrac{122}{56} \\ -\tfrac{122}{56} & \tfrac{265}{56} \end{bmatrix}$$

This is a matrix with 2 rows and 2 columns, called a 2×2 matrix. Mathematicians have extended the notion of reciprocal of a number to the *inverse* of a matrix (see Appendix C for details). To understand what is meant by the inverse of a matrix it is helpful to know what is meant by the identity matrix. Any matrix whose leading (top-left to bottom-right) diagonal is composed of 1s with all other elements 0 is called an identity matrix. Thus

$$\begin{bmatrix} 1 & 0 \\ 0 & 1 \end{bmatrix}$$

is a 2×2 identity matrix. The inverse of a matrix is another matrix such that the product of the two matrices is an identity matrix. The inverse of

$$\begin{bmatrix} \tfrac{260}{56} & -\tfrac{122}{56} \\ -\tfrac{122}{56} & \tfrac{265}{56} \end{bmatrix}$$

is

$$\begin{bmatrix} c_{11} & c_{12} \\ c_{21} & c_{22} \end{bmatrix}$$

if

$$\begin{bmatrix} c_{11} & c_{12} \\ c_{21} & c_{22} \end{bmatrix} \begin{bmatrix} \tfrac{260}{56} & -\tfrac{122}{56} \\ -\tfrac{122}{56} & \tfrac{265}{56} \end{bmatrix} = \begin{bmatrix} 1 & 0 \\ 0 & 1 \end{bmatrix}$$

This matrix equation simply means the following: the products of the first row vector of the inverse with the first and second column vectors of the given matrix must be 1 and 0, respectively. The products of the second row vector of the inverse with the first and second column vectors of the given matrix must be 0 and 1 respectively.

Once the inverse of the coefficient matrix is at hand, the solutions for the unknowns in (6.35) are obtained by the following rule:

$\hat{\alpha}_i$ = product of ith row of inverse with column vector on right-hand side of (6.35)

In the present simple case it is easy to verify that

$$\begin{bmatrix} \dfrac{265 \times 56}{54,016} & \dfrac{122 \times 56}{54,016} \\[2ex] \dfrac{122 \times 56}{54,016} & \dfrac{260 \times 56}{54,016} \end{bmatrix} \tag{6.36}$$

is the inverse of the coefficient matrix at hand. To obtain the inverse we interchanged the elements of the leading diagonal of the given matrix, retained the off-diagonal elements with their signs changed, and finally divided each element of the rearrangement by

D = product of elements of the leading diagonal *minus* product of off-diagonal elements

$$= \frac{260}{56}\frac{265}{56} - \frac{122}{56}\frac{122}{56} = \frac{54,016}{56 \times 56}$$

This method will work for all 2×2 matrices. To invert more complicated matrices, more complex procedures are necessary. Fortunately, computer programs are available for the purpose. (See Appendix C.)

Using the inverse obtained above, we obtain the solutions

$$\hat{\alpha}_1 = -\frac{16}{56}\frac{265 \times 56}{54,016} + \frac{12}{56}\frac{122 \times 56}{54,016} = -0.0514$$

$$\hat{\alpha}_2 = -\frac{16}{56}\frac{122 \times 56}{54,016} + \frac{12}{56}\frac{260 \times 56}{54,016} = +0.0216$$

We have already set as a side condition

$$\hat{\alpha}_3 = 0$$

This completes the estimation of the α parameters. Substituting these estimates into (6.31), we get

$$\hat{\beta}_1 + \hat{\mu} = \bar{y}_{.1.} - \frac{2\hat{\alpha}_1 + 3\hat{\alpha}_2 + 3\hat{\alpha}_3}{8} = \bar{y}_{.1.} - \frac{2\hat{\alpha}_1 + 3\hat{\alpha}_2}{8} = 9.505$$

$$\hat{\beta}_2 + \hat{\mu} = \bar{y}_{.2.} - \frac{3\hat{\alpha}_1 + 2\hat{\alpha}_2}{7} = 2.016$$

$$\hat{\beta}_3 + \hat{\mu} = \bar{y}_{.3.} - \frac{2\hat{\alpha}_1 + 2\hat{\alpha}_2}{7} = 2.151$$

The predicted values in each cell can now be obtained from the estimates of $\beta_j + \mu$ and α_i: the sum of $\hat{\alpha}_i$ and $\hat{\beta}_j + \hat{\mu}$ gives the predicted value for the (i,j) cell.

The sum of squares due to regression can be calculated directly from the predicted values thus obtained.

Since $\hat{\beta}_j + \hat{\mu}$ are expressed in terms of $\hat{\alpha}_i$'s, it follows that the predicted values in each cell can also be expressed solely in terms of $\hat{\alpha}_i$'s. It seems reasonable therefore to expect that the sum of squares due to regression can be calculated directly from $\hat{\alpha}_i$'s, without going through the intermediate step of estimating β_j's and μ. It can be shown algebraically that the block (row) SS eliminating treatments is equal to the sum of the products of $\hat{\alpha}_i$'s with the quantities on the right-hand side of (6.35). In symbols

$$\text{Regression SS } (\mu,\alpha_i\text{'s},\beta_j\text{'s}) - \text{regression SS } (\mu,\beta_j\text{'s}) = \sum \hat{\alpha}_i Q_i$$

where Q_i stand for the quantities on the right-hand side of (6.35). It is seen that $\sum \hat{\alpha}_i Q_i = -0.05139(-\frac{16}{56}) + 0.02162(\frac{12}{56}) = 0.01932$. We have already obtained regression SS $(\mu,\beta_j\text{'s})$ as 281.007. Hence

$$\text{Regression SS } (\mu,\alpha_i\text{'s},\beta_j\text{'s}) = 281.007 + 0.019 = 281.026$$

Subtracting this SS from the cell SS obtained earlier, we get $287.197 - 281.026 = 6.171$ as nonadditivity SS. The complete analysis-of-variance table is shown in Table 6.21.

We find that nonadditivity is negligible. The next step is to test the significance of block SS eliminating treatment and treatment SS eliminating blocks. Table 6.21 shows that the former is not significant while the latter is highly significant. We interpret this to mean that the initial attitude toward an issue does not have marked influence on the power to remember various aspects of the issue. Time, however, plays a crucial role; people tend to forget as time progresses things they heard in the past. But is this tendency linearly related to time? Does one forget as much in the second week as in the first week? In other words, according to our symbols, is it valid to assume that $\hat{\beta}_2 = (\hat{\beta}_1 + \hat{\beta}_3)/2$? This leads us to the problem of testing the significance of treatment contrasts.

Table 6.21   ANALYSIS OF VARIANCE OF THE DATA IN TABLE 6.20

Source of variation	SS	df	MSS	F ratio
Treatments ignoring blocks	281.007	2		
Blocks eliminating treatments	.019	2	.0095	
Nonadditivity	6.171	4	1.543	1.88
Cells	287.197	8		
Residual	10.667	13	.821	
Total	297.864	21		
Treatments eliminating blocks	276.323	2	138.162	168.28
Blocks ignoring treatments	4.703	2		
Nonadditivity	6.171	4		
Cells	287.197	8		
Residual	10.667	13	.821	
Total	297.864	21		

Significance Testing

It is important to remember that treatment (block) contrasts are not ordinarily tested for significance if additivity does not prevail. In the present case we have no reason to believe that additivity does not prevail. When this is so, it is easy to estimate β_j's in terms of $\hat{\alpha}_i$'s and $\hat{\mu}$ using the additive model, as already indicated. From the expressions for β_j's in terms of α_i's and μ presented earlier we notice

$$
\hat{\beta}_2 - \frac{\hat{\beta}_1 + \hat{\beta}_3}{2} = \bar{y}_{.2.} - \frac{3\hat{\alpha}_1 + 2\hat{\alpha}_2}{7} - \frac{\bar{y}_{.1.}}{2} + \frac{2\hat{\alpha}_1 + 3\hat{\alpha}_2}{16}
$$

$$
- \frac{\bar{y}_{.3.}}{2} + \frac{2\hat{\alpha}_1 + 2\hat{\alpha}_2}{14}
$$

$$
= \bar{y}_{.2.} - \frac{\bar{y}_{.1.}}{2} - \frac{\bar{y}_{.3.}}{2} - (\tfrac{3}{7} - \tfrac{2}{16} - \tfrac{2}{14})\hat{\alpha}_1
$$

$$
- (\tfrac{2}{7} - \tfrac{3}{16} - \tfrac{2}{14})\hat{\alpha}_2
$$

$$
= \bar{y}_{.2.} - \frac{\bar{y}_{.1.}}{2} - \frac{\bar{y}_{.3.}}{2} - \tfrac{18}{112}\hat{\alpha}_1 + \tfrac{5}{112}\hat{\alpha}_2
$$

In general, estimates of any treatment contrasts of the form

$$
\psi = \sum l_j \beta_j
$$

where $\sum l_j = 0$, can be expressed as

$$
\hat{\psi} = \sum l_j \bar{y}_{.j.} - \sum k_i \hat{\alpha}_i \qquad (6.37)
$$

where k_i's are constants. To test the significance of treatment contrasts we therefore need variances of statistics of the form (6.37). To obtain an estimate of the variance of (6.37) we multiply by residual MSS in the analysis of variance the expression

$$
\sum \frac{l_j^2}{n_{.j}} + \sum k_i^2 c_{ii} + \sum\sum_{i \neq j} k_i k_j c_{ij}
$$

where c_{ii} is the ith, from top left, diagonal element of the matrix (6.36) and c_{ij} is the jth element in the ith row of the same matrix.

Let us calculate the variance of the treatment contrast $\hat{\psi}$. We have

$$
\hat{\psi} = \hat{\beta}_2 - \frac{\hat{\beta}_1 + \hat{\beta}_3}{2}
$$

$$
= \bar{y}_{.2.} - \frac{\bar{y}_{.1.}}{2} - \frac{\bar{y}_{.3.}}{2} - \tfrac{18}{112}\hat{\alpha}_1 + \tfrac{5}{112}\hat{\alpha}_2
$$

$$
= -3.8122
$$

We estimate the variance of $\hat{\psi}$ as residual MSS times the following multiplier:

$$
\frac{1}{7} + \frac{1}{4 \times 8} + \frac{1}{4 \times 7} + \left(\frac{18}{12}\right)^2 c_{11} + \left(\frac{5}{112}\right)^2 c_{11} - 2\frac{18}{112}\frac{5}{112}c_{12}
$$

$$
= \frac{1}{7} + \frac{1}{4 \times 8} + \frac{1}{4 \times 7} + \left(\frac{18}{112}\right)^2 \frac{265 \times 56}{54{,}016} + \left(\frac{5}{112}\right)^2 \frac{260 \times 56}{54{,}016}
$$

$$
- 2\frac{18 \times 5}{112^2}\frac{122 \times 56}{54{,}016}
$$

$$
= .2156
$$

Thus the standard error of $\hat{\psi}$ is estimated to be

$$\sqrt{.2156 \times 10.6667} = 1.516$$

The ratio $3.8122/1.516 = 2.5146$ (which may be treated as a Student's t with 13 degrees of freedom) indicates clearly that $\hat{\psi}$ is significant. We infer therefore that the tendency to forget things is not linearly related to time.

It may be useful to summarize the procedure described in this section for obtaining the variances and covariances of the least-squares estimates of treatment and block effects. For the purpose of presenting this summary we shall assume that the reference is to a design involving t treatments and r blocks, the treatment effects being β_1, $\beta_2, \ldots, \beta_t$ and the block effects being $\alpha_1, \alpha_2, \ldots, \alpha_r$. It is also important to emphasize that the procedure described in this section assumes that additivity prevails in the data set at hand. The steps involved in obtaining the variances and covariances of block and treatment effects are the following:

1 Write the normal equations assuming additivity [see Eq. (6.30)].

2 Express each treatment effect $\hat{\beta}_j$ in terms of the block effects $\hat{\alpha}_i$'s and the constant term $\hat{\mu}$ [see Eqs. (6.31)].

3 Substituting the expressions for $\hat{\beta}_j$'s obtained in step 2 in the remaining normal equations and using a side condition involving $\hat{\alpha}_i$'s, for example, $\hat{\alpha}_r = 0$, derive a set of $r - 1$ equations containing $r - 1$ of the $\hat{\alpha}_i$'s, say $\hat{\alpha}_1, \hat{\alpha}_2, \ldots,$ $\hat{\alpha}_{r-1}$ [see Eqs. (6.35)].

4 Write the coefficient matrix of the equations derived in step 3.

5 Obtain the inverse of this coefficient matrix [see (6.36)]. For the purpose of this summary we shall denote by c_{ij} the jth element of the ith row of the inverse under reference.

6 Estimate the variances and covariances of $\hat{\alpha}_i$'s, $i = 1, 2, \ldots, r - 1$, according to the formulas

$$\text{Estimate of var } \hat{\alpha}_i = c_{ii} \text{ (residual MSS)}$$

$$\text{Estimate of covar } (\hat{\alpha}_i, \hat{\alpha}_j) = c_{ij} \text{ (residual MSS)}$$

7 Notice that each $\hat{\beta}_j$ can be expressed in terms of the observed treatment means and a linear combination of $\hat{\alpha}_i$'s. Apply the formula for the variances and covariance of linear combinations of random variables to obtain the variances and covariances of $\hat{\beta}_1, \ldots, \hat{\beta}_t$.

8 Once estimates of the variances and covariances of $\hat{\alpha}_i$'s and $\hat{\beta}_j$'s are computed, the calculation of standard errors of treatment (block) contrast is straightforward.

It is pertinent to make two remarks at this point. In step 2 the decision to express $\hat{\beta}_j$'s in terms of $\hat{\alpha}_i$'s is a wise strategy if the number of $\hat{\alpha}_i$'s, that is, the number of blocks, is fewer than the number of $\hat{\beta}_j$'s, that is, the number of treatments. Otherwise it is wise to express $\hat{\alpha}_i$'s in terms of $\hat{\beta}_j$'s and proceed accordingly in step 3 to obtain a set of equations in $\hat{\beta}_j$'s. This leads to fewer equations to solve than is otherwise the case.

Another remark that is pertinent in this connection concerns the reason for steps 2 and 3 in the above sequence. Notice that steps 2 and 3 aim at reducing the problem of solving the full set of normal equations to one of solving fewer equations.

This is worth aiming at if one has to complete the calculations without the help of the digital computer. If there is no access to the computer, the fewer the number of equations to solve the better off we shall be from the point of view of reducing the complexity of computations. But if one has access to the computer, steps 2 and 3 are unnecessary. We can solve the full set of normal equations directly after inserting the required number of side conditions. Such a procedure has the advantage that the variances and covariances of $\hat{\alpha}_i$'s and $\hat{\beta}_j$'s can be obtained more directly. The details of the procedure are shown in the next section.

6.7 THE DESIGN MATRIX AND ITS USES IN DATA ANALYSIS

The Preparation of the Design Matrix

Recall that the observations from any experiment involving r blocks and t treatments with treatment j occurring n_{ij} times in block i can be expressed, assuming additivity, as

$$y_{ijk} = \mu x_0 + \sum_{i=1}^{r} \alpha_i x_i + \sum_{j=1}^{t} \beta_j z_j + \varepsilon_{ijk}$$

where $x_0 = 1$ for all y's, $x_i = 1$ for all y's in block i and 0 for all others, and $z_j = 1$ for all y's under treatment j and 0 for all others. Suppose that we arrange the y values in some order and correspondingly arrange the values of $x_0, x_1, \ldots, x_r, z_1, z_2, \ldots, z_t$. The resulting row-by-column arrangement of the values of $x_0, x_1, \ldots$ may be called the design matrix associated with the data set at hand under the additive model. For a randomized-block design with three blocks and four treatments, we may regard the following as the associated design matrix under the additive model:

$$\mathbf{X} = \begin{bmatrix} 1 & 1 & 0 & 0 & 1 & 0 & 0 & 0 \\ 1 & 1 & 0 & 0 & 0 & 1 & 0 & 0 \\ 1 & 1 & 0 & 0 & 0 & 0 & 1 & 0 \\ 1 & 1 & 0 & 0 & 0 & 0 & 0 & 1 \\ 1 & 0 & 1 & 0 & 1 & 0 & 0 & 0 \\ 1 & 0 & 1 & 0 & 0 & 1 & 0 & 0 \\ 1 & 0 & 1 & 0 & 0 & 0 & 1 & 0 \\ 1 & 0 & 1 & 0 & 0 & 0 & 0 & 1 \\ 1 & 0 & 0 & 1 & 1 & 0 & 0 & 0 \\ 1 & 0 & 0 & 1 & 0 & 1 & 0 & 0 \\ 1 & 0 & 0 & 1 & 0 & 0 & 1 & 0 \\ 1 & 0 & 0 & 1 & 0 & 0 & 0 & 1 \end{bmatrix}$$

Reference to Table 6.3 will reveal that the arrangement of the values of x_0, x_1, x_2, x_3, z_1, z_2, z_3, and z_4 presented in that table taken as such and put into a row-by-column pattern gives the design matrix shown above. This indicates that no new principle is involved in preparing the design matrix associated with a data set under any given model. We just write down in a systematic manner the values of x_0 and the regressors associated with the observed y values. In designs like the one in Table 6.3 if we arrange the y values in a standard order and then exhibit the values of x_0 and the regressors corresponding to the y values so arranged, the design matrix will be highly patterned. The advantage of preparing the design matrix in a patterned manner is that by looking at the patterns one can readily identify redundancies (dependencies), if any, between the rows (columns) of the design matrix.

The Observational Equations

It is easy to verify that given the design under consideration, the equations representing the observations in terms of x_0, x_1, x_2,... can be written in matrix form as

$$\mathbf{Y} = \mathbf{XB} + \mathbf{\varepsilon} \qquad (6.38)$$

where $\mathbf{X}$ is the design matrix shown above, and $\mathbf{Y}$, $\mathbf{B}$, and $\mathbf{\varepsilon}$ are column vectors respectively of the observed y values, the regression weights, and the residuals, i.e.,

$$
\mathbf{Y} = \begin{bmatrix} y_{11} \\ y_{12} \\ y_{13} \\ y_{14} \\ y_{21} \\ y_{22} \\ y_{23} \\ y_{24} \\ y_{31} \\ y_{32} \\ y_{33} \\ y_{34} \end{bmatrix}
\qquad
\mathbf{B} = \begin{bmatrix} \mu \\ \alpha_1 \\ \alpha_2 \\ \alpha_3 \\ \beta_1 \\ \beta_2 \\ \beta_3 \\ \beta_4 \end{bmatrix}
\qquad
\mathbf{\varepsilon} = \begin{bmatrix} \varepsilon_{11} \\ \varepsilon_{12} \\ \varepsilon_{13} \\ \varepsilon_{14} \\ \varepsilon_{21} \\ \varepsilon_{22} \\ \varepsilon_{23} \\ \varepsilon_{24} \\ \varepsilon_{31} \\ \varepsilon_{32} \\ \varepsilon_{33} \\ \varepsilon_{34} \end{bmatrix}
$$

The g-Inverse Method of Solving the Normal Equations

The normal equations that give the least-squares estimates of the regression weights can be written as

$$(\mathbf{X'X})\hat{\mathbf{B}} = \mathbf{X'Y} \qquad (6.39)$$

where $\mathbf{X}'$ is the transpose of the design matrix $\mathbf{X}$. The reader may want to verify that (6.39) is simply a matrix representation of (6.3). In the present case $\mathbf{X}'\mathbf{X}$ is easily found to be

$$
\begin{bmatrix}
12 & 4 & 4 & 4 & 3 & 3 & 3 & 3 \\
4 & 4 & 0 & 0 & 1 & 1 & 1 & 1 \\
4 & 0 & 4 & 0 & 1 & 1 & 1 & 1 \\
4 & 0 & 0 & 4 & 1 & 1 & 1 & 1 \\
3 & 1 & 1 & 1 & 3 & 0 & 0 & 0 \\
3 & 1 & 1 & 1 & 0 & 3 & 0 & 0 \\
3 & 1 & 1 & 1 & 0 & 0 & 3 & 0 \\
3 & 1 & 1 & 1 & 0 & 0 & 0 & 3
\end{bmatrix}
\tag{6.40}
$$

If it were possible to invert $\mathbf{X}'\mathbf{X}$, we would have been able to solve (6.39) by pre-multiplying the right-hand side of (6.39) by the inverse of $\mathbf{X}'\mathbf{X}$. In the present case $\mathbf{X}'\mathbf{X}$ cannot be inverted because its rows (columns) are not all independent of each other. In fact it is easily seen that the sum of the second, third, and fourth rows (columns) is equal to the first row (column). Similarly, the sum of the last four rows (columns) also gives the first row (column). Matrices like this whose rows (columns) are interdependent cannot be inverted. We can, however, calculate what mathematicians call a generalized inverse (or g-inverse) of $\mathbf{X}'\mathbf{X}$ and use it instead of the true inverse (which in this case does not exist). There are a number of g-inverses for matrices such as $\mathbf{X}'\mathbf{X}$. One g-inverse of $\mathbf{X}'\mathbf{X}$ can be obtained in the following manner.

Omit redundant rows and the corresponding columns, thus reducing $\mathbf{X}'\mathbf{X}$ to a smaller matrix which has a true inverse. Find the true inverse of the reduced matrix and then augment the inverse by inserting rows and columns of zeros from where the redundant rows and columns were removed. The matrix so obtained is a g-inverse of $\mathbf{X}'\mathbf{X}$.

Notice that in applying the above procedure it is possible to remove redundant rows and columns in a number of ways. For example, from (6.40) we may remove any one of the first four rows and the corresponding column to remove the redundancy among the first four rows (columns). The same flexibility applies to removing the redundancy between the remaining rows and columns. Different ways of removing the redundancies in question may lead to different g-inverses, but this need not bother us for reasons which will become clear later. Let us for the moment remove the fourth and the last rows and the corresponding columns of (6.40). The resulting reduced matrix is

$$
\begin{bmatrix}
12 & 4 & 4 & 3 & 3 & 3 \\
4 & 4 & 0 & 1 & 1 & 1 \\
4 & 0 & 4 & 1 & 1 & 1 \\
3 & 1 & 1 & 3 & 0 & 0 \\
3 & 1 & 1 & 0 & 3 & 0 \\
3 & 1 & 1 & 0 & 0 & 3
\end{bmatrix}
$$

It can be verified that the inverse of this matrix is

$$\frac{1}{12}\begin{bmatrix} 6 & -3 & -3 & -4 & -4 & -4 \\ -3 & 6 & 3 & 0 & 0 & 0 \\ -3 & 3 & 6 & 0 & 0 & 0 \\ -4 & 0 & 0 & 8 & 4 & 4 \\ -4 & 0 & 0 & 4 & 8 & 4 \\ -4 & 0 & 0 & 4 & 4 & 8 \end{bmatrix}$$

To get a g-inverse of (6.40) we now insert in this inverse a row of zeros after the third row and another row of zeros after the last row. We also insert the corresponding columns of zeros, getting

$$\mathbf{C} = \frac{1}{12}\begin{bmatrix} 6 & -3 & -3 & 0 & -4 & -4 & -4 & 0 \\ -3 & 6 & 3 & 0 & 0 & 0 & 0 & 0 \\ -3 & 3 & 6 & 0 & 0 & 0 & 0 & 0 \\ 0 & 0 & 0 & 0 & 0 & 0 & 0 & 0 \\ -4 & 0 & 0 & 0 & 8 & 4 & 4 & 0 \\ -4 & 0 & 0 & 0 & 4 & 8 & 4 & 0 \\ -4 & 0 & 0 & 0 & 4 & 4 & 8 & 0 \\ 0 & 0 & 0 & 0 & 0 & 0 & 0 & 0 \end{bmatrix}$$

as a g-inverse of $\mathbf{X}'\mathbf{X}$. We can now use $\mathbf{C}$ for two purposes: (1) to solve the normal equations (6.39) and (2) to obtain the variances and covariances of treatment and block contrasts. To solve the normal equations (6.39) we simply premultiply the right-hand side of (6.39) with $\mathbf{C}$ and write

$$\hat{\mathbf{B}} = \mathbf{C}\mathbf{X}'\mathbf{Y} \qquad (6.41)$$

as the solutions for the regression weights in $\hat{\mathbf{B}}$. Notice in this connection that

$$\mathbf{X}'\mathbf{Y} = \begin{bmatrix} y_{..} \\ y_{1.} \\ y_{2.} \\ y_{3.} \\ y_{.1} \\ y_{.2} \\ y_{.3} \\ y_{.4} \end{bmatrix}$$

Hence (6.41) means

$$
\begin{bmatrix}
\hat{\mu} \\
\hat{\alpha}_1 \\
\hat{\alpha}_2 \\
\hat{\alpha}_3 \\
\hat{\beta}_1 \\
\hat{\beta}_2 \\
\hat{\beta}_3 \\
\hat{\beta}_4
\end{bmatrix}
= \frac{1}{12}
\begin{bmatrix}
6 & -3 & -3 & 0 & -4 & -4 & -4 & 0 \\
-3 & 6 & 3 & 0 & 0 & 0 & 0 & 0 \\
-3 & 3 & 6 & 0 & 0 & 0 & 0 & 0 \\
0 & 0 & 0 & 0 & 0 & 0 & 0 & 0 \\
-4 & 0 & 0 & 0 & 8 & 4 & 4 & 0 \\
-4 & 0 & 0 & 0 & 4 & 8 & 4 & 0 \\
-4 & 0 & 0 & 0 & 4 & 4 & 8 & 0 \\
0 & 0 & 0 & 0 & 0 & 0 & 0 & 0
\end{bmatrix}
\begin{bmatrix}
y_{..} \\
y_{1.} \\
y_{2.} \\
y_{3.} \\
y_{.1} \\
y_{.2} \\
y_{.3} \\
y_{.4}
\end{bmatrix}
$$

We at once get

$$\hat{\mu} = \tfrac{1}{12}(6y_{..} - 3y_{1.} - 3y_{2.} - 4y_{.1} - 4y_{.2} - 4y_{.3})$$

But $y_{.1} + y_{.2} + y_{.3} + y_{.4} = y_{..}$, and $y_{1.} + y_{2.} + y_{3.} = y_{..}$ Hence

$$-3y_{1.} - 3y_{2.} = -3y_{..} + 3y_{3.}$$

and

$$-4y_{.1} - 4y_{.2} - 4y_{.3} = -4y_{..} + 4y_{.4}$$

Thus the solution for $\hat{\mu}$ obtained above simplifies to

$$
\begin{aligned}
\hat{\mu} &= \tfrac{1}{12}(6y_{..} - 3y_{..} + 3y_{3.} - 4y_{..} + 4y_{.4}) \\
&= \tfrac{1}{12}(3y_{3.} + 4y_{.4} - y_{..}) \\
&= \frac{y_{3.}}{4} + \frac{y_{.4}}{3} - \frac{y_{..}}{12} \\
&= \bar{y}_{3.} + \bar{y}_{.4} - \bar{y}_{..}.
\end{aligned}
$$

Similarly,

$$
\begin{aligned}
\hat{\alpha}_1 &= \tfrac{1}{12}(-3y_{..} + 6y_{1.} + 3y_{2.}) = \bar{y}_{1.} - \bar{y}_{3.}. \\
\hat{\alpha}_2 &= \tfrac{1}{12}(-3y_{..} + 3y_{1.} + 6y_{2.}) = \bar{y}_{2.} - \bar{y}_{3.}. \\
\hat{\alpha}_3 &= 0 \\
\hat{\beta}_1 &= \tfrac{1}{12}(-4y_{..} + 8y_{.1} + 4y_{.2} + 4y_{.3}) = \bar{y}_{.1} - \bar{y}_{.4} \\
\hat{\beta}_2 &= \tfrac{1}{12}(-4y_{..} + 4y_{.1} + 8y_{.2} + 4y_{.3}) = \bar{y}_{.2} - \bar{y}_{.4} \\
\hat{\beta}_3 &= \tfrac{1}{12}(-4y_{..} + 4y_{.1} + 4y_{.2} + 8y_{.3}) = \bar{y}_{.3} - \bar{y}_{.4} \\
\hat{\beta}_4 &= 0
\end{aligned}
$$

These are exactly the same solutions we would have arrived at for the normal equations (6.3) had we employed as side conditions $\hat{\alpha}_3 = 0$ and $\hat{\beta}_4 = 0$. Then why do we bother to adopt the g-inverse approach? This brings us to the second use of g-inverse.

Estimation of Variances of Treatment (Block) Contrasts

The elements of any g-inverse of $\mathbf{X'X}$ can readily be used to compute the variances and covariances of the least-squares estimates of treatment or block contrasts. To describe how this can be done, let us denote the elements of $\mathbf{C}$ by C_{ij}, where the first subscript denotes the row and the second subscript denotes the column under reference. Thus C_{23} is the third element in the second row, C_{32} the second element in the third row, and so on. Also let us refer to the estimated regression weights by their order in the column vector $\hat{\mathbf{B}}$. Thus $\hat{\mu}$ is the first element, $\hat{\alpha}_1$ the second element, and so on.

For calculating the variances and covariances of treatment or block contrasts, we formally regard $C_{ii}\sigma^2$, where σ^2 is the residual variance, as the variance of the ith element in $\hat{\mathbf{B}}$ and $C_{ij}\sigma^2$ as the covariance between the ith and the jth element in $\hat{\mathbf{B}}$. Noticing, for example, that $\hat{\beta}_1$, $\hat{\beta}_2$, and $\hat{\beta}_3$ are respectively the fifth, sixth, and seventh elements in $\hat{\mathbf{B}}$, we formally regard $C_{55}\sigma^2$, $C_{66}\sigma^2$, and $C_{77}\sigma^2$, each of which is equal to $\frac{8}{12}\sigma^2$ in the present case, as the variances of $\hat{\beta}_1$, $\hat{\beta}_2$, and $\hat{\beta}_3$, respectively, and $C_{56}\sigma^2$, $C_{57}\sigma^2$, $C_{67}\sigma^2$, as the covariance between $\hat{\beta}_1$ and $\hat{\beta}_2$, $\hat{\beta}_1$ and $\hat{\beta}_3$, and $\hat{\beta}_2$ and $\hat{\beta}_3$, respectively. Now if, for example, the treatment contrast we are interested in is $\psi = \beta_1 - 2\beta_2 + \beta_3$, we have $\hat{\psi} = \hat{\beta}_1 - 2\hat{\beta}_2 + \hat{\beta}_3$ as the least-squares estimate of ψ, and its variance is obtained as follows:

$$
\begin{aligned}
\operatorname{var} \hat{\psi} &= \operatorname{var} (\hat{\beta}_1 - 2\hat{\beta}_2 + \hat{\beta}_3) \\
&= \operatorname{var} \hat{\beta}_1 + 4 \operatorname{var} \hat{\beta}_2 + \operatorname{var} \hat{\beta}_3 \\
&\quad - 4 \operatorname{covar} (\hat{\beta}_1, \hat{\beta}_2) + 2 \operatorname{covar} (\hat{\beta}_1, \hat{\beta}_3) - 4 \operatorname{covar} (\hat{\beta}_2, \hat{\beta}_3) \\
&= \sigma^2 \left[\tfrac{8}{12} + 4(\tfrac{8}{12}) + (\tfrac{8}{12}) - 4(\tfrac{4}{12}) + 2(\tfrac{4}{12}) - 4(\tfrac{4}{12}) \right] = 2\sigma^2
\end{aligned}
$$

Using residual MSS as our estimate of σ^2, we take 2(residual MSS) as our estimate of var $\hat{\psi}$.

It should be emphasized that no matter which g-inverse of $\mathbf{X'X}$ one employs in the above calculations, the least-squares estimate of $\hat{\psi}$ as well as the estimate of var $\hat{\psi}$ are unique. The verification of this result is left as an exercise.

An examination of the elements of $\mathbf{C}$ clearly reveals that $\hat{\alpha}_i$'s and $\hat{\beta}_j$'s are uncorrelated. Notice that $C_{25}, \ldots, C_{28}$; $C_{35}, \ldots, C_{38}$; and $C_{45}, \ldots, C_{48}$ are all zero, indicating that the second, third, and fourth elements of $\hat{\mathbf{B}}$, that is, $\hat{\alpha}_1, \hat{\alpha}_2, \hat{\alpha}_3$, are uncorrelated with the fifth, $\ldots$, eighth elements of $\hat{\mathbf{B}}$, that is, $\hat{\beta}_1, \hat{\beta}_2, \hat{\beta}_3, \hat{\beta}_4$. This will be the case in all designs in which treatments and blocks are orthogonal.

To illustrate the application of the above method to designs in which treatments and blocks are not orthogonal let us consider the design presented in the preceding section. To construct the design matrix we prepare the data set in Table 6.20 as shown in Table 6.22. Notice that in Table 6.22 we have arranged the y values in an obvious order: block 1 followed by blocks 2 and 3 in that order and within each block treatment 1 followed by treatments 2 and 3 in that order. The design matrix $\mathbf{X}$ associated with this data set under the additive model is simply the row-by-column arrangement of values of x_0, x_1, x_2, x_3, z_1, z_2, and z_3 in Table 6.22. It is not difficult to verify that given this design matrix, we have

$$\mathbf{X'X} = \begin{bmatrix} 22 & 7 & 7 & 8 & 8 & 7 & 7 \\ 7 & 7 & 0 & 0 & 2 & 3 & 2 \\ 7 & 0 & 7 & 0 & 3 & 2 & 2 \\ 8 & 0 & 0 & 8 & 3 & 2 & 3 \\ 8 & 2 & 3 & 3 & 8 & 0 & 0 \\ 7 & 3 & 2 & 2 & 0 & 7 & 0 \\ 7 & 2 & 2 & 3 & 0 & 0 & 7 \end{bmatrix}$$

To remove the redundancies in rows and columns of $\mathbf{X'X}$ let us delete the fourth and seventh rows and the corresponding columns. The resulting matrix is

$$\begin{bmatrix} 22 & 7 & 7 & 8 & 7 \\ 7 & 7 & 0 & 2 & 3 \\ 7 & 0 & 7 & 3 & 2 \\ 8 & 2 & 3 & 8 & 0 \\ 7 & 3 & 2 & 0 & 7 \end{bmatrix}$$

Table 6.22 PREPARATION OF THE DATA SET IN TABLE 6.20 FOR
CONSTRUCTION OF THE DESIGN MATRIX ASSOCIATED
WITH THE DATA SET UNDER THE ADDITIVE MODEL

y	x_0	x_1	x_2	x_3	z_1	z_2	z_3
y_{111}	1	1	0	0	1	0	0
y_{112}	1	1	0	0	1	0	0
y_{121}	1	1	0	0	0	1	0
y_{122}	1	1	0	0	0	1	0
y_{123}	1	1	0	0	0	1	0
y_{131}	1	1	0	0	0	0	1
y_{132}	1	1	0	0	0	0	1
y_{211}	1	0	1	0	1	0	0
y_{212}	1	0	1	0	1	0	0
y_{213}	1	0	1	0	1	0	0
y_{221}	1	0	1	0	0	1	0
y_{222}	1	0	1	0	0	1	0
y_{231}	1	0	1	0	0	0	1
y_{232}	1	0	1	0	0	0	1
y_{311}	1	0	0	1	1	0	0
y_{312}	1	0	0	1	1	0	0
y_{313}	1	0	0	1	1	0	0
y_{321}	1	0	0	1	0	1	0
y_{322}	1	0	0	1	0	1	0
y_{331}	1	0	0	1	0	0	1
y_{332}	1	0	0	1	0	0	1
y_{333}	1	0	0	1	0	0	1

and its inverse corrected to four decimal places is

$$
\begin{bmatrix}
.2079 & -.1146 & -.1132 & -.1368 & -.1265 \\
-.1146 & .2747 & .1265 & -.0015 & -.0392 \\
-.1132 & .1265 & .2695 & -.0195 & -.0181 \\
-.1368 & -.0015 & -.0195 & .2695 & .1431 \\
-.1265 & -.0392 & -.0181 & .1431 & .2913
\end{bmatrix}
$$

Recalling that we deleted the fourth and seventh rows and the corresponding columns of $X'X$ to get a matrix which has a true inverse, we now insert rows and columns of zeros in the inverse just obtained after the third and after the last rows (columns) and thus get as a g-inverse of $X'X$

$$
C =
\begin{bmatrix}
.2079 & -.1146 & -.1132 & .0 & -.1368 & -.1265 & .0 \\
-.1146 & .2747 & .1265 & .0 & -.0015 & -.0392 & .0 \\
-.1132 & .1265 & .2695 & .0 & -.0195 & -.0181 & .0 \\
.0 & .0 & .0 & .0 & .0 & .0 & .0 \\
-.1368 & -.0015 & -.0195 & .0 & .2695 & .1431 & .0 \\
-.1265 & -.0392 & -.0181 & .0 & .1431 & .2913 & .0 \\
.0 & .0 & .0 & .0 & .0 & .0 & .0
\end{bmatrix}
$$

It is easily verified that the normal equations

$$X'X\hat{B} = X'Y$$

are satisfied by the following solutions set (which, it may be noted, have been calculated using more than four decimal places for the elements of C).

$$
\hat{B} =
\begin{bmatrix}
\hat{\mu} \\
\hat{\alpha}_1 \\
\hat{\alpha}_2 \\
\hat{\alpha}_3 \\
\hat{\beta}_1 \\
\hat{\beta}_2 \\
\hat{\beta}_3
\end{bmatrix}
= C
\begin{bmatrix}
y_{...} \\
y_{1..} \\
y_{2..} \\
y_{3..} \\
y_{.1.} \\
y_{.2.} \\
y_{.3.}
\end{bmatrix}
= C
\begin{bmatrix}
105 \\
29 \\
37 \\
39 \\
76 \\
14 \\
15
\end{bmatrix}
=
\begin{bmatrix}
2.1514 \\
-.0514 \\
.0216 \\
.0000 \\
7.3534 \\
-.1355 \\
.0000
\end{bmatrix}
$$

The estimates of treatment and block contrasts obtained from these solutions are identical, except for rounding errors, to the corresponding estimates obtained by the method described in the preceding section. The advantage of the present method is that the elements of C can be formally regarded as σ^2 times the variances and covariances of $\hat{\mu}$, $\hat{\alpha}_i$'s, and $\hat{\beta}_j$'s. Consequently calculations of standard errors of treatment and block contrasts are made considerably easier if we follow the g-inverse method.

The Design Matrix in Multiple-Regression Analysis

Before concluding this section it is pertinent to remark that the concept of the design matrix is applicable to the multiple-regression analysis in which all regressors are interval-scaled variables. In this case the design matrix is simply the values of the regressors including the constant term arranged in a row-by-column fashion, columns corresponding to the regressors and the constant term and rows to the observed y values. If X is the design matrix associated with a given data set under a given model,

$$Y = XB + \varepsilon$$

represent the observational equations, i.e., equations representing the observed y values in terms of the regressors and the residual. The normal equations are

$$(X'X)\hat{B} = X'Y$$

When the regressors are all interval-scaled variables, there will be no dependencies between rows or columns of $X'X$. Consequently $X'X$ can be inverted. Let C be the inverse of $X'X$. Then the solutions to the normal equations are

$$\hat{B} = CX'Y$$

which may be written

$$\hat{B} = (X'X)^{-1}X'Y$$

using the conventional notation $(X'X)^{-1}$ for the true inverse of $(X'X)$.

The elements of C multiplied by σ^2 the residual variance are the variances and covariances of the estimated regression weights. Another way of saying the same thing is that $\sigma^2 C$ is the *dispersion matrix* of the estimated regression weights. Also it is important to emphasize that all functions of regression weights are estimable when all regressors are interval-scaled variables.

EXERCISES

6.1 The following data are from a randomized-blocks experiment.

	Treatment			
Block	A	B	C	D
1	.0	.9	1.0	.7
2	.6	1.1	1.4	.8
3	.0	.8	1.5	1.0

(a) Analyze the data and state your conclusions.

(b) Estimate the standard error of the treatment constrast $(A + B) - (C + D)$.

6.2 Compare the following two experiments in terms of the underlying mathematical model and the appropriate testing procedure for testing treatment differences.

(a) There are bt experimental units divided into b blocks of t units each in such a manner that each block contains more or less homogeneous units. The t treatments are randomly assigned to the t experimental units in each block, a different random allocation being made in each block. Each experimental unit contains several observational units; e.g., villages may be the experimental units and families within the villages the observational units. Or classes may be the experimental units and students within the classes the observational units. Assume that in each experimental unit the number of observational units is equal to n and that the n observational units are selected at random from the corresponding universe of observational units in each experimental unit.

(*b*) There are *btn* experimental units. These are divided into *b* blocks of *tn* experimental units in such a way that the experimental units belonging to each block are as homogeneous as possible. The *t* treatments are allocated at random to the *tn* units in each block so that each treatment falls on exactly *n* units. Randomization is done separately in each block. Each experimental unit yields one observation on the response variable.

6.3 Imagine that the following data were obtained in an interview survey.

"DESIRED FAMILY SIZE" BY RELIGIOUS BACKGROUND AND WORK EXPERIENCE OF THE WIFE

Work experience	Religious affiliation		
	Catholic	Protestant	Other
None	4,3,3,5	3,2,4,1,2	3,2,1
Some	3,4,1,3,2,3	3,2,1,2,2,2	1,1,2,2,2

Complete the analysis of variance.

6.4 The following data are from the Social Survey in England and Wales in 1949, in which a nationwide sample of approximately 10,000 cases was employed. The figures reported below pertain, however, only to couples who had been married 20 years or longer by the time of the survey. For each origin-destination category the average number of live births and the number of couples are shown. Origin = husband's father's occupational class, and destination = husband's occupational class at the time of the survey. The total SS of numbers of children of couples in the sample was 10,074.3. (This is the corrected SS.) Complete the two-way classification analysis of variance. Test for nonadditivity. Duncan (1966*a*) argued that if there is additivity in the data set, it should be inferred that there is no mobility effect on fertility. "One is not entitled to discuss

AVERAGE NUMBER OF LIVE BIRTHS PER COUPLE $\bar{y}$ AND THE NUMBER OF COUPLES IN THE SAMPLE n IN EACH ORIGIN-DESTINATION CATEGORY

Origin (husband's father's social, i.e., occupational class)	Destination (husband's social, i.e., occupational, class at time of survey)									
	1 (high)		2		3		4 (low)		All	
	$\bar{y}$	n	$\bar{y}$	n	$\bar{y}$	n	$\bar{y}$	n	$\bar{y}$	n
1 (high)	1.74	65	1.79	43	1.96	23	2.00	11	1.81	142
2	2.05	38	2.14	197	2.51	150	2.97	68	2.38	453
3	1.87	37	2.01	154	2.67	431	3.69	244	2.81	866
4 (low)	2.40	5	3.20	45	3.22	162	3.68	220	3.44	432
All	1.88	145	2.17	439	2.73	766	3.56	543	2.77	1,893

'effects' of mobility (or other status discrepancy measures) until he has established that the apparent effect cannot be due merely to a simple combination of effects of the variables used to define mobility" (Duncan, 1966a, p. 91). Describe the abstract experimental model Duncan superimposed on the data set in question and discuss the validity of his assumptions.

6.5 Consider a two-way layout with 2 columns and r rows. Let the column (treatment) effects be denoted by β_1 and β_2, respectively. Let there be n_{ij} observations in the (i,j) cell, i.e., in the ith row of the jth column. Assuming additivity and homoscedasticity, show that the normal equations giving the least-squares estimate of β_1 and β_2 can be reduced to the following, using the side condition $\hat{\beta}_2 = 0$:

$$C\hat{\beta}_1 = Q$$

where

$$C = n_{.1} - \left(\frac{n_{11}^2}{n_{1.}} + \frac{n_{21}^2}{n_{2.}} + \cdots + \frac{n_{r1}^2}{n_{r.}} \right)$$

and

$$Q = y_{.1.} - \left(\frac{n_{11} y_{1..}}{n_{1.}} + \frac{n_{21} y_{2..}}{n_{2.}} + \cdots + \frac{n_{r1} y_{r..}}{n_{r.}} \right)$$

Hence show that

$$\hat{\beta}_1 = \frac{\sum w_i (\bar{y}_{i1.} - \bar{y}_{i2.})}{\sum w_i}$$

where w_i is the harmonic mean of the cell frequencies n_{i1} and n_{i2}, that is $w_i = 2n_{i1}n_{i2}/n_{i.}$. This demonstrates that the least-squares estimate of the treatment contrast $\beta_1 - \beta_2$ in the case of $2 \times r$ table, i.e., a design with 2 treatments and r replications, is simply a weighted average of the corresponding contrasts of the cell means, the weights being the harmonic means of the cell frequencies involved.

6.6 A quick preliminary analysis of a two-way layout with disproportionate cell frequencies is the following:

Step 1 Compute the residual SS in the usual manner by subtracting the cell SS from the total SS. Let its degrees of freedom be f. Divide the residual SS by f and multiply the quotient by the arithmetic mean of the reciprocals of the cell frequencies. Let the symbol $\hat{\sigma}^2$ represent the resulting quantity.

Step 2 Consider the table of cell means as a table with one "observation" in each cell, and complete the analysis of variance table in the usual elementary manner. The difference total SS − row SS − column SS is to be taken as representing the row-by-column interaction SS with $(r - 1)(t - 1)$ degrees of freedom, where r is the number of rows and t the number of columns.

To test the significance of the row effects, the column effects, and the interaction, we use $\hat{\sigma}^2$ obtained in Step 1 as our estimate of the residual variance. Thus (row MSS)/$\hat{\sigma}^2$ is regarded as an F with $r - 1$ and f degrees of freedom, and so on. Provide a justification for this procedure. *Hint:* Suppose the original observations are independently normal with a common population variance σ^2. Let there be n_{ij} observations in the (i,j) cell and let the cell mean in the (i,j) cell be $\bar{y}_{ij.}$. Then the variance of the (i,j)-cell mean will be σ^2/n_{ij}. Hence the table of cell means when viewed as a table with one observation in each cell represents a data situation characterized by heteroscedasticity to the degree that n_{ij}'s differ between cells. But with unequal variances and equal cell numbers ($= 1$ in the present case), the F test behaves approximately the same as though the variances were equal to the average of the cell variances (see Scheffé, 1959, pp. 351–358), which in the present case is equal to

$$\frac{1}{rt} \sum \sum \frac{\sigma^2}{n_{ij}}$$

or σ^2 times the arithmetic mean of the reciprocal of the cell frequencies. *Note:* This method is usually recommended as a substitute for the exact least-squares method in situations where heteroscedasticity is relatively serious, i.e., where σ^2 varies between cells.

6.7 Apply the above approximate procedure to the two-way layout in Table 6.20. The main advantage of the approximate procedure described above is that we can get a quick test for nonadditivity.

7

MORE ON PROCEDURES TO REDUCE RESIDUAL VARIATION

This chapter is a continuation of Chap. 6. We continue to examine procedures for reducing residual variation. Among the designs presented in this chapter are hierarchical (nested) designs, Latin squares, and extensions of Latin squares (Greco-Latin squares, and hyper-Greco-Latin squares). Certain aspects of the covariance analysis not explicitly brought out in Part One are briefly reviewed. Finally, some attention is given to data situations in which the response (dependent) variable is qualitative rather than quantitative.

7.1 HIERARCHICAL (NESTED) DESIGNS

Let us consider a model which superficially resembles the ones we have seen in Chap. 6 but which essentially has a different basis. We shall first describe a real situation in which an experimental study of the kind we are concerned with in this section is appropriate. In a low-income country, a loan program for farmers was in operation for several years, but very few of the small farmers derived any benefit from it. A study commission appointed by the government to look into the problem found that the small farmers were not able to take advantage of the program because they could not provide the collateral demanded by the existing rules. The commission recommended that the government should on an experimental basis allow small farmers to borrow money on the basis of simple promissory notes. Suppose that the government decided to try out the commission's suggestion. Let us assume that the following strategy was adopted in designing an experimental study. It was decided that the relaxed procedures be tried out only in a few selected places. The experimental places were selected in accordance with the hierarchy of the existing administrative organization. First a random sample of four revenue districts was selected. Then within each district, a random sample of three subdistricts was selected, and within each subdistrict, a random sample of two places (villages) was selected. All farmers in the experimental villages were contacted and were asked to apply for loans if interested. It was explained to each farmer that there was an upper limit on the money each farmer would be allowed to borrow. In each experimental village a fixed amount of money was distributed. To be fair, the recipients were selected from among the applicants at random. The recipients were then asked to sign a simple promissory note pledging that the money would be paid back before the beginning of the next season, i.e., within 1 year from the date of borrowing. Suppose the data given in Table 7.1 show the unpaid total balance in each of the experimental villages as of the beginning of the next agricultural season. (Here we shall take this unpaid balance as the price paid for relaxing the rules.)

Mathematical Model and Analysis Procedure

The village data shown in Table 7.1 can be viewed as being classified at two levels, first according to revenue districts and then according to subdistricts within the revenue districts. Since random sampling has been employed, the appropriate model to represent the data in Table 7.1 is a random model. The usual model for representing data like those in Table 7.1 is

$$y_{ijk} = \mu + u_i + v_{ij} + \varepsilon_{ijk}$$

$$u_i\text{'s are NID}(0, \sigma_u^2)$$

$$v_{ij}\text{'s are NID}(0, \sigma_v^2)$$

$$\varepsilon_{ijk}\text{'s are NID}(0, \sigma^2)$$

$$u_i, v_{ij}, \text{ and } \varepsilon_{ijk} \text{ mutually independent}$$

Here u's refer to revenue districts and v's to subdistricts. Note that in the above model we do not have a term with subscript j alone, which is in marked contrast to the model we have employed for the general two-way classification. The reason for this is that the effect of the characteristics of a subdistrict cannot be thought of in the same way as the effect of a given treatment. While in a two-way classification we think of a given treatment being applied in each block, in the case of the experiment reported in Table 7.1, we cannot think of the characteristic properties of a given subdistrict being applied in each revenue district. Therefore, there is no conceptual correspondence between the effect of any of the subdistricts under one revenue district and that of any of the subdistricts under another revenue district. We should note also that since there is no correspondence between v_{ij} and v_{lj}, $i \neq l$, the question of nonadditivity does not arise here. Recall that in the two-way classification, the question of nonadditivity arose because we were concerned whether the comparison of two treatment effects remains the same from one block to another. Since no such comparison is possible in the hierarchical classification, the question of nonadditivity does not arise at all.

The null hypotheses of interest in the case of the experiment reported in Table 7.1 are

H_u: There is no difference between the impacts of the district characteristics, e.g., district administration, on the study phenomenon; this is equivalent to positing that $\sigma_u^2 = 0$

$H_{v/u}$: There is no difference between the impacts of the subdistrict characteristics, within any district, on the study phenomenon; this is equivalent to positing that $\sigma_v^2 = 0$

Table 7.1 HIERARCHICAL CLASSIFICATION: THE UNPAID BALANCE OF LOANS IN DIFFERENT EXPERIMENTAL UNITS (VILLAGES), THE FIRST-LEVEL UNITS BEING REVENUE DISTRICTS AND THE SECOND-LEVEL UNITS BEING SUBDISTRICTS WITHIN REVENUE DISTRICTS

Revenue district i	Sub-district j	Village k	Unpaid balance y_{ijk}	$y_{ij.}$	$y_{i..}$	$y_{...}$
1	1	1	32.8			
		2	30.9	63.7		
	2	1	35.4			
		2	32.4	67.8		
	3	1	28.3			
		2	27.9	56.2	187.7	
2	1	1	24.5			
		2	24.3	48.8		
	2	1	17.8			
		2	18.9	36.7		
	3	1	20.8			
		2	21.2	42.0	127.5	
3	1	1	26.9			
		2	27.6	54.5		
	2	1	34.0			
		2	37.5	71.5		
	3	1	25.4			
		2	23.9	49.3	175.3	
4	1	1	35.9			
		2	35.3	71.2		
	2	1	41.6			
		2	40.2	81.8		
	3	1	33.2			
		2	32.9	66.1	219.1	709.6

To test these hypotheses, we first calculate the sum of squares (SS) due to different sources of variation in the usual manner:

Correction for the mean:
$$C = \frac{709.6^2}{24} = 20{,}980.5067$$

$$\text{Total SS} = 32.8^2 + \cdots + 32.9^2 - C = 1{,}009.3733$$

$$\text{Revenue districts SS} = \tfrac{1}{6}(187.7^2 + \cdots + 219.1^2) - C = 723.2333$$

$$\text{Subdistricts SS} = \tfrac{1}{2}(63.7^2 + \cdots + 66.1^2) - C = 993.5833$$

The SS due to *subdistricts within revenue districts* is obtained by subtraction:

SS due to subdistricts within revenue districts = subdistricts SS − revenue districts SS

Note that this amounts to pooling the subdistricts variations within different districts. From the total SS we subtract subdistrict SS to get residual SS. The analysis-of-variance table is completed as shown in Table 7.2. From the expected values of the MSS it can be seen that the appropriate ratio to test H_u is

$$\frac{\text{Revenue districts MSS}}{\text{Subdistricts-within-revenue-districts MSS}}$$

Table 7.2   ANALYSIS OF VARIANCE OF DATA IN TABLE 7.1

A   Analysis of Variance

Source of variation	SS	df	MSS
Between revenue districts	723.2333	3	241.0777
Between subdistricts within revenue districts	270.3500	8	33.7938
Residual	15.7900	12	1.3158
Total	1,009.3733	23	

B   Expected Values of MSS†

Source of variation	MSS	Expected value of MSS
Between revenue districts	241.0777	$\sigma^2 + n\sigma_v^2 + bn\sigma_u^2$
Between subdistricts within revenue districts	33.7938	$\sigma^2 + n\sigma_v^2$
Residual	1.3158	σ^2

C   Estimates of Components of Variance

$$\hat{\sigma}_u^2 = \tfrac{1}{3} \times \tfrac{1}{2}(241.0777 - 33.7938) = 34.5473$$
$$\hat{\sigma}_v^2 = \tfrac{1}{2}(33.7938 - 1.3158) = 16.2390$$

† b = number of subdistricts within each revenue district = 3; n = number of villages within each subdistrict = 2.

since if H_u is true, the expected values of the two MSS will be the same. Similarly, to test $H_{v/u}$, we use the ratio

$$\frac{\text{Subdistricts-within-revenue-districts MSS}}{\text{Residual MSS}}$$

For the first hypothesis, the ratio with 3 and 8 degrees of freedom is significant at the 1 percent level. So is the ratio for the second hypothesis, with 8 and 12 degrees of freedom. There is thus considerable variation in the unpaid balance from one district to another and from one subdistrict to another within each district. The analyst would therefore recommend that the causes for these variations be carefully studied.

It can be shown that the estimated variance of the observed average value of the study variate per village is the revenue district MSS divided by the number of villages used in obtaining the average. Thus for the overall average $\bar{y}_{\dots}$ $(= 29.6)$ in Table 7.1, the estimated variance is $241.078/24 = 10.045$. This estimated variance can be expressed in terms of the estimates of σ^2, $\sigma_u{}^2$, and $\sigma_v{}^2$ obtained from Table 7.2 as:

$$10.045 = \tfrac{1}{24}(\hat{\sigma}^2 + n\hat{\sigma}_v{}^2 + nb\hat{\sigma}_u{}^2)$$

$$= \tfrac{1}{24}(1.3158 + 2 \times 16.2390 + 6 \times 34.5473)$$

$$= \frac{1.3158}{24} + \frac{16.2390}{12} + \frac{34.5473}{4}$$

The advantage of expressing it this way is that one can judge which source of variation contributes more and which contributes less to the variance of the mean. In the present case, for example, most of the variance is attributable to between-districts variation.

A More General Model

Let us now consider the general two-level hierarchical classification (nested design), in which the number of subclasses differs from one first-level class to another and the number of observations differs from one subclass to another. Let there be a first-level classes. Let the number of subclasses within the ith first level class be s_i. Also let the number of observations in the jth subclass of the ith first-level class be n_{ij}. The observations may now be denoted by y_{ijk}, where the first subscript is for the first-level class, the second for the subclass within the ith first-level class, and the third for the individual observation within the jth subclass within the ith first-level class.

The easiest way to get the SS in the analysis of variance associated with such a data set is as shown below. First we calculate

$$S_1 = \sum_i \sum_j \sum_k y_{ijk}^2 \qquad S_2 = \sum_i \sum_j \frac{y_{ij.}^2}{n_{ij}}$$

$$S_3 = \sum_i \frac{y_{i..}^2}{n_{i.}} \qquad S_4 = \frac{y_{...}^2}{n_{..}}$$

Then the sum of squares and the degrees of freedom in the analysis of variance can be obtained as shown in Table 7.3. Let us now assume that the observations in the data set under reference can be represented as

$$y_{ijk} = \mu + u_i + v_{ij} + \varepsilon_{ijk}$$

We shall assume that the first-level classes are a random sample (with equal probability) of all possible such classes, that the second-level classes within each first-level class are a random sample (with equal probability) of all possible such subclasses, and finally that the individual observations within each second-level class are a random sample with equal probability from all possible such individuals. We shall also assume that the variability at each level of the classification is constant over all the possible classes at that level, i.e., that the variance of u_i is σ_u^2, the variance of v_{ij} within each i class is σ_v^2, and the variance of ε_{ijk} within each (i,j) class is σ^2. Note that we have not yet made any normality assumption.

The analysis of the data involves estimating σ_u^2, σ_v^2, and σ^2. It can be shown that

$$E(S_1 - S_2) = \left(n_{..} - \sum_i \sum_j 1\right)\sigma^2 = \left(\sum_i \sum_j (n_{ij} - 1)\right)\sigma^2$$

$$E(S_2 - S_3) = \left(n_{..} - \sum_i \sum_j \frac{n_{ij}^2}{n_{i.}}\right)\sigma_v^2 + \left(\sum_i \sum_j 1 - \sum_i 1\right)\sigma^2$$

$$E(S_3 - S_4) = \left(n_{..} - \sum_i \frac{n_{i.}^2}{n_{..}}\right)\sigma_u^2 + \left(\sum_i \sum_j \frac{n_{ij}^2}{n_{i.}} - \sum_i \sum_j \frac{n_{ij}^2}{n_{..}}\right)\sigma_v^2$$
$$+ \left(\sum_i 1 - 1\right)\sigma^2$$

Note that the coefficient of σ^2 in each of the expectations is equal to the degrees of freedom of the SS involved. For example, for the SS due to between-individuals variation, that is, $S_1 - S_2$, the coefficient of σ^2 is

$$\sum_i \sum_j (n_{ij} - 1)$$

Table 7.3 ANALYSIS OF VARIANCE OF THE GENERAL HIERARCHICAL CLASSIFICATION INVOLVING TWO LEVELS

Source of variation	SS	df
Between first-level classes or units	$S_3 - S_4 = \sum_i \frac{y_{i..}^2}{n_{i.}} - \frac{y_{...}^2}{n_{..}}$	$(n_{..} - 1) - \sum_i (n_{i.} - 1)$
Between second-level classes within first-level classes	$S_2 - S_3 = \sum_i \sum_j \frac{y_{ij.}^2}{n_{ij}} - \sum_i \frac{y_{i..}^2}{n_{i.}}$	$\sum_i (n_{i.} - 1) - \sum_i \sum_j (n_{ij} - 1)$
Between individuals within second-level classes	$S_1 - S_2 = \sum_i \sum_j \sum_k y_{ijk}^2 - \sum_i \sum_j \frac{y_{ij.}^2}{n_{ij}}$	$\sum_i \sum_j (n_{ij} - 1)$
Total	$S_1 - S_4 = \sum_i \sum_j \sum_k y_{ijk}^2 - \frac{y_{...}^2}{n_{..}}$	$n_{..} - 1$

which is the same as the degrees of freedom of $S_1 - S_2$. Similarly, it can be easily seen that for $S_2 - S_3$

$$\sum_i (n_{i.} - 1) - \sum_i \sum_j (n_{ij} -) = \sum_i \left(\left(\sum_j n_{ij} \right) - 1 \right) - \sum_i \sum_j (n_{ij} - 1)$$

$$= \sum_i \sum_j 1 - \sum_i 1$$

and that for $S_3 - S_4$,

$$(n_{..} - 1) - \sum_i (n_{i.} - 1) = \sum_i n_{i.} - 1 - \sum_i n_{i.} + \sum_i 1$$

$$= \sum_i 1 - 1$$

Given the above expressions for the expected values of the different SS, we can easily get the expected values of the MSS. We find them to be

SS	Expectation of MSS
$S_3 - S_4$	$\sigma^2 + k_1 \sigma_v^2 + k_2 \sigma_u^2$
$S_2 - S_3$	$\sigma^2 + k_3 \sigma_v^2$
$S_1 - S_2$	σ^2

$$k_1 = \frac{\sum_i \sum_j \dfrac{n_{ij}^2}{n_{i.}} - \sum_i \sum_j \dfrac{n_{ij}^2}{n_{..}}}{\sum_i 1 - 1}$$

$$k_2 = \frac{n_{..} - \sum_i \dfrac{n_{i.}^2}{n_{..}}}{\sum_i 1 - 1}$$

$$k_3 = \frac{n_{..} - \sum_i \sum_j \dfrac{n_{ij}^2}{n_{i.}}}{\sum_i \sum_j 1 - \sum_i 1}$$

In the balanced case, i.e., when the numbers of individual observations are the same (say, n) from one class level to another, and when the numbers of second-level classes are the same (say, b) from one first-level class to another, if there are a first-level classes, we have

$$n_{ij} = n$$

$$n_{i.} = bn$$

$$n_{..} = abn$$

$$\sum_i 1 = a \qquad \sum_i \sum_j 1 = ab$$

$$\sum_i \sum_j \frac{n_{ij}^2}{n_{..}} = \sum_i \sum_j \frac{n^2}{abn} = n$$

$$\sum_i \sum_j \frac{n_{ij}^2}{n_{i.}} = \sum_i \sum_j \frac{n^2}{bn} = an$$

and

$$\sum_i \frac{n_{i.}^2}{n_{..}} = \sum_i \frac{b^2 n^2}{abn} = bn$$

Hence

$$k_1 = \frac{an - n}{a - 1} = n$$

$$k_2 = \frac{abn - bn}{a - 1} = bn$$

and

$$k_3 = \frac{abn - an}{ab - a} = n$$

The expectations of the MSS in the balanced case thus are:

SS	Expected value of MSS
Between the first-level classes	$\sigma^2 + n\sigma_v^2 + bn\sigma_u^2$
Between the second-level classes within first-level classes	$\sigma^2 + n\sigma_v^2$
Between individuals within second-level classes	σ^2

The estimation of the components of variance, σ^2, σ_u^2, and σ_v^2, is now easy. If SS_a, SS_b and SS_c denote the SS's listed above in the order listed, and if MSS_a, MSS_b, and MSS_c, denote the corresponding MSS's, then

$$\hat{\sigma}_u^2 = \frac{MSS_a - MSS_b}{bn} \qquad \hat{\sigma}_v^2 = \frac{MSS_b - MSS_c}{n} \qquad \hat{\sigma}^2 = MSS_c$$

If we now make the additional assumption that all the class variables in the model are normally distributed, tests of significance of each component can be made by using the appropriate variance ratio. To test the significance of $\hat{\sigma}_u^2$ we use the ratio MSS_a/MSS_b and to test the significance of $\hat{\sigma}_v^2$ we use MSS_b/MSS_c. (This was what we did in Table 7.2.)

Remarks

Nested and crossed designs   Suppose we are interested in comparing three methods of teaching introductory sociology (see Davies, Gross, and Short, 1958, for a complex experiment on this subject). Let the methods taken for comparison be A, B, and C. A nested design which may be suitable for the study is the following. Select nine teachers and number them randomly from 1 to 9. Assign them to the three methods as shown below:

Method	A	B	C
Teacher	1–3	4–6	7–9

This design may be described as a nested design with one treatment (method of teaching) and one nested factor (teacher).

An alternative to the above design is one in which only three teachers are involved, each teacher using on different groups of students each of the three methods. This design looks like the following:

	Method		
Teacher	A	B	C
1			
2			
3			

In this design, we cross the teacher factor with the treatment (method of teaching); hence it may be described as a *crossed* design.

Notice that in the nested design we use nine teachers and nine different groups of students, while in the crossed design we use only three teachers and nine different groups of students. It may be worth pointing out that when the same teacher uses different methods of teaching in succession, there may be certain problems in the interpretation of the results, as will become clear when we discuss (Chap. 9) experimental designs in which each subject (experimental unit) is used repeatedly.

Fixed and random effects Nested designs are sometimes referred to as samples within samples (see, for example, Snedecor and Cochran, 1967, pp. 285–294). In fact, the nested model has found more frequent use in sampling studies than in experimental investigations. Sampling statisticians are often interested in examining the components of variance of estimates obtained on the basis of multistage sampling. In multistage sampling, one selects several first-stage units, from each of which a number of second-stage units are then selected, the process being then continued stage by stage until the final-stage unit, i.e., the unit of observation, is reached. In a particular study, the first-stage units may be counties, the second-stage units may be urban places within counties, the third-stage units blocks, and the final-stage units households.

If all the first-stage units in the universe have been included in the sample, the characteristics of those units are regarded as having fixed effects. If, on the other hand, the first-stage units have been sampled, the corresponding characteristics are regarded as having random effects. Thus it is possible to have in a nested design fixed as well as random effects. If no sampling has been done at any stage, we have fixed effects at all levels. If sampling has been confined to second- and higher-stage units, we have fixed effects at the first stage and random effects at later stages, and so on. For an exposition of multistage sampling see Kish (1965) or Schuessler (1971). Sampling texts give special attention to situations in which the numbers of population units at different stages of sampling are finite rather than infinite. In our exposition we have regarded all universes as infinite in size.

Higher-level nesting Nesting can be carried to any number of stages, as implied in the above remarks. To give an example, in the study of components of variance of distance traveled for shopping by urban residents, we may select towns as the first-stage units, blocks within towns as the second-stage units, households within blocks as the third-stage units, persons within households as the fourth-stage units, and finally days within weeks for persons as the last-stage units. The model and the analysis procedure presented earlier can be readily extended to data situations of this type (see, for example, Kempthorne, 1969).

Partial nesting   In certain experiments, a given factor may be nested within a treatment but crossed with a block variable. An example adapted from Herman, Potterfield, Dayton, and Amershek (1969), is the following:

		Treatment					
		Teacher-centered instruction			Pupil-centered instruction		
	Nested factor (teacher)	1	2	3	4	5	6
Block (sex of students) Male Female							

Notice that in this design the teacher factor is nested within the treatment (method of instruction) but crossed with the block variable (the sex of the students). Such designs, called *partially nested* designs, are to be distinguished from completely nested designs, in which the nested factor(s) is (are) not crossed with any factor. See Dayton (1970) for further details about partially nested designs.

7.2   LATIN SQUARES

The randomized-blocks design discussed in Chap. 6 may be considered as a one-dimensional blocking method to control heterogeneity of the experimental units. Recall, however, that simply because the blocking method is one-dimensional it does not mean that more than one causal factor cannot be taken into account simultaneously in randomized-blocks designs. In classifying human subjects into blocks in learning experiments, for example, one may consider simultaneously such factors as sex, race, age, intelligence, etc. But when several factors are considered simultaneously for blocking, it may be necessary to use a large number of blocks in the experiment in order to guarantee generalizability of the findings to the whole range of each of the block factors. The question therefore arises whether the purpose behind the simultaneous introduction of several block factors can be achieved without making the experiment excessively large. If all block factors have the same number of categories, say r, and if the number of treatments is also the same, i.e., it equals r, then the use of Latin squares, Greco-Latin squares, or hyper-Greco-Latin squares would enable the experimenter to introduce several block factors simultaneously using only r

blocks. We shall consider Latin squares in some detail and only briefly mention the properties of Greco-Latin squares and hyper-Greco-Latin squares.

Suppose that we wish to compare the following social-engineering methods in a family-planning action-research program:

A   Clinical approach, which consists of making family-planning services and supplies available to the public through clinics.

B   Clinical approach integrated with a comprehensive health program, in which the services and supplies made available through clinics, as described in *A*, are integrated with a comprehensive health program covering sanitation, protected water supply, immunization against infectious diseases, etc.

C   Adult-education program integrated with the comprehensive health program included in B with the provision that family-planning clinics would be set up and the associated services rendered in accordance with the demand that may be generated. In this approach, the adult-education program would cover not only matters of general knowledge (including reading, writing, etc.) but ways to limit family size and the reasons why family size should be limited.

Now suppose that we wish to introduce as block factors the *sociocultural milieu* (level of economic development, educational facilities and services, medical facilities and services, organizational development for community action, etc.) and *subcultures*, e.g., ethnic and racial groupings, language regions.

Let us suppose that *sociocultural milieu* has been scaled and that "low," "middle," and "high" values on the scale have been defined. Let us also suppose that from among the different subcultures, three have been selected, e.g., three language regions, for the purpose of this experiment.

We have then three treatments, *A*, *B*, *C*, and a two-dimensional blocking, each dimension having three categories. We may use a Latin-square arrangement for the experiment under the circumstances. The layout may look like the following:

Sociocultural milieu	Subculture		
	1	2	3
Low	A	B	C
Middle	B	C	A
High	C	A	B

In this layout, each treatment is applied once and only once to each sociocultural milieu and once and only once to each subculture. From now on we shall refer to the two-dimensional blocking as one involving rows and columns. In the above layout, rows refer to the sociocultural-milieu dimension and columns to the subcultural dimension.

Following the above convention, *we may define a Latin-square layout as one in which each treatment appears once and only once in each row and once and only once in each column.*

It is important to remember that in Latin squares the numbers of rows, columns, and treatments must be equal. This requirement obviously restricts the usefulness of

Latin-square designs somewhat. For example, in the experiment described above, if the experimenter wanted to use more than three subcultures in the blocking procedure, a Latin-square layout would not be applicable, unless the number of treatments and the number of categories of the other block variable could be similarly increased.

For an extensive list of Latin squares and a discussion of their construction, refer to Fisher and Yates (1973). Given below are a few of the designs shown in that publication:

A B	A B C	A B C D	A B C D E	A B C D E F
B A	B C A	B A D C	B A E C D	B C F A D E
	C A B	C D B A	C D A E B	C F B E A D
		D C A B	D E B A C	D E A B F C
			E C D B A	E A D F C B
				F D E C B A

Obviously we can describe each of the above layouts using numbers instead of letters. Thus if we replace A by 1, B by 2, etc., in the above layout of a 3×3 square, we would have the following description:

$$
\begin{array}{ccc}
1 & 2 & 3 \\
2 & 3 & 1 \\
3 & 1 & 2
\end{array}
$$

Whether described in terms of letters or numbers, the layouts presented above show a common characteristic. In each case, the letters (numbers) in the first row and in the first column appear in the alphabetical (numerical) order. Such Latin squares (or, simply, squares) are called *standard squares*. From each standard square, additional squares can be generated by rearrangement of letters (numbers). Reference may be made to Fisher and Yates (1973) for rules to be followed in selecting a Latin square from among the possible Latin squares of a given dimension.

Analysis

Let us turn our attention to the analysis of data from a Latin-square design. Consider a Latin square of dimension r, that is, one in which there are r rows, r columns, and r treatments. In the fixed-effects case, assuming additivity, normality, and homoscedasticity, we represent the observation in the ith row, jth column, and kth treatment as $y_{ijk} = \mu + \alpha_i + \beta_j + \gamma_k + \varepsilon_{ijk}$ with the side conditions $\sum \alpha_i = \sum \beta_j = \sum \gamma_k = 0$ and the assumption that ε_{ijk}'s are NID$(0,\sigma^2)$. Note particularly here that there are only r^2 observations. The SS's attributable to different sources of variation and the associated degrees of freedom together with expectations of MSS's are shown in Table 7.4.

The hypotheses

$$
\begin{aligned}
H_\alpha: &\quad \alpha_1 = \alpha_2 = \cdots = \alpha_r = 0 \\
H_\beta: &\quad \beta_1 = \beta_2 = \cdots = \beta_r = 0 \\
H_\gamma: &\quad \gamma_1 = \gamma_2 = \cdots = \gamma_r = 0
\end{aligned}
$$

are tested by using the variance ratios (row MSS)/(residual MSS), (column MSS)/(residual MSS), and (treatment MSS)/(residual MSS), respectively. If the additive model presented above adequately represents the data at hand, the above variance ratios will each be distributed under the respective null hypotheses as an F with $r - 1$ and $(r - 1)(r - 2)$ degrees of freedom.

It can be shown that the least-squares estimates of α_i, β_j, and γ_k, subject to the side conditions $\sum \hat{\alpha}_i = \sum \hat{\beta}_j = \sum \hat{\gamma}_k = 0$, are given by

$$\hat{\alpha} = \bar{y}_{i..} - \bar{y}_{...} \qquad \hat{\beta} = \bar{y}_{.j.} - \bar{y}_{...} \qquad \hat{\gamma} = \bar{y}_{..k} - \bar{y}_{...}$$

and consequently, contrasts involving α_i's, β_j's, or γ_k's can be estimated and the variance of those estimates computed as shown:

Contrast	Unbiased estimate	Variance of the estimate
$\sum_i l_i \alpha_i$	$\sum_i l_i \bar{y}_{i..}$	$\dfrac{\sigma^2}{r} \sum_i l_i^2$
$\sum_j l_j^* \beta_j$	$\sum_j l_j^* \bar{y}_{.j.}$	$\dfrac{\sigma^2}{r} \sum_j l_j^{*2}$
$\sum_k l_k^{**} \gamma_k$	$\sum_k l_k^{**} \bar{y}_{..k}$	$\dfrac{\sigma^2}{r} \sum_k l_k^{**2}$

Example To illustrate the procedure described above consider the data set shown in Table 7.5. Suppose that we wish to test the following null hypotheses:

1 There is no difference between the row effects. This corresponds to the statement that the effects of sociocultural milieu do not differ between the low, medium, and high categories.

Table 7.4 SUMS OF SQUARES AND EXPECTED VALUES OF MEAN SUMS OF SQUARES ASSOCIATED WITH $r \times r$ LATIN SQUARES

Source of variation	SS	df	Expected value of MSS
Between rows	$\dfrac{1}{r}\sum_i y_{i..}^2 - \dfrac{y_{...}^2}{r^2}$	$r - 1$	$\sigma^2 + \dfrac{r}{r-1}\sum \alpha_i^2$
Between columns	$\dfrac{1}{r}\sum_j y_{.j.}^2 - \dfrac{y_{...}^2}{r^2}$	$r - 1$	$\sigma^2 + \dfrac{r}{r-1}\sum \beta_j^2$
Between treatments	$\dfrac{1}{r}\sum_k y_{..k}^2 - \dfrac{y_{...}^2}{r^2}$	$r - 1$	$\sigma^2 + \dfrac{r}{r-1}\sum \gamma_k^2$
Residual (error)	Total SS − row SS − column SS − treatment SS	$(r-1)(r-2)$	σ^2
Total	$\sum_i \sum_j \sum_k y_{ijk}^2 - \dfrac{y_{...}^2}{r^2}$	$r^2 - 1$	

2   There is no difference between the column effects. This corresponds to the statement that the effects of the three chosen subcultures do not differ between themselves.

3   There is no difference between the treatment effects.

The analysis of variance shown in Table 7.5 reveals that on the basis of the evidence at hand, it is not possible to reject any of the null hypotheses described above. This brings us to one of the major problems in using Latin-square designs when the number of treatments is as few as 3. As can be seen from Table 7.5, in a 3 × 3 Latin-square experiment, the residual SS has only 2 degrees of freedom. When the residual SS has few degrees of freedom, the length of the confidence interval for any given true difference is likely to be large, and consequently the probability of obtaining a significant result is likely to be small. Because of this problem it is often suggested that Latin-square designs not be used if the number of treatments is less than 5 unless each cell contains two or more observations.

To illustrate what may be expected if the number of observations is more than 1 in a Latin-square experiment with three treatments, let us analyze the data shown in Table 7.6 which, as can be easily seen, are very similar to the data shown in Table 7.5 except that each cell has two observations. The results of the analysis of the data in Table 7.6 are presented in that table. The observed treatment means in Tables 7.5 and 7.6 are shown below for easy comparison:

Treatment	Table 7.5	Table 7.6
A	6.67	7.00
B	12.00	12.05
C	4.33	3.62

Table 7.5   ANALYSIS OF A LATIN-SQUARE EXPERIMENT

A   Data

Row (sociocultural milieu)	Column (subculture)			Total
	1	2	3	
1	*A* 9.5	*B* 11.8	*C* 3.3	24.6
2	*B* 9.2	*C* 5.1	*A* 4.0	18.3
3	*C* 4.6	*A* 6.5	*B* 15.0	26.1
Total	23.3	23.4	22.3	69.0

B   Analysis of Variance

Source of variation	SS	df	MSS	*F* ratio
Between rows	11.42	2	5.710	
Between columns	.25	2	.125	
Between treatments	92.67	2	46.335	4.19
Residual (error)	22.10	2	11.050	
Total	126.44	8		

Notice the close resemblance between the two arrays of means. Yet one experiment (Table 7.5) failed in declaring the treatment differences significant while the other (Table 7.6) did not.

Although the above illustration oversimplifies the difference between two experiments—one in which there is only one observation per cell and the other in which there are two observations per cell—it is enough to emphasize that in experiments using Latin-square designs, it is better to have more than one observation per cell if the number of treatments is as small as 3. Some authors recommend that an acceptable minimum for the degrees of freedom of the error SS is 12. If that recommendation is accepted, all Latin-square designs involving fewer than five treatments should be considered impractical unless two or more observations are available in each cell.

Table 7.6 **A LATIN-SQUARE EXPERIMENT WITH TWO OBSERVATIONS IN EACH CELL**

A Data

Row (sociocultural milieu)	Column (subculture)						Total
		1		2		3	
1	A	9.5	B	11.8	C	3.3	
		7.1		13.5		1.9	
			16.6		25.3		5.2	47.1
2	B	9.2	C	5.1	A	4.0	
		10.8		3.8		6.5	
			20.0		8.9		10.5	39.4
3	C	4.6	A	6.5	B	15.0	
		3.0		8.4		12.0	
			7.6		14.9		27.0	49.5
Total			44.2		49.1		42.7	136.0

B Analysis of Variance

Source of variation	SS	df	MSS	F ratio
Between rows	9.28	2	4.640	2.30
Between columns	3.73	2	1.865	
Between treatments	216.14	2	108.070	53.61
Nonadditivity	13.75	2	6.875	3.41
Between cells	242.90	8	30.363	15.06
Error	18.14	9	2.016	
Total	261.04	17		

Remarks

Before concluding this section, let us briefly note the following points pertaining to $r \times r$ Latin-square designs with one observation per cell.

Blocking effects The effectiveness of either the row or the column blocking can be estimated from the results of the analysis of variance of a Latin square. The expression

$$\frac{\text{Row MSS} + (r - 1)(\text{error MSS})}{r}$$

is an estimate of the error MSS which would have been obtained if the row blocking had not been used, i.e., if a randomized blocks design with the columns of the Latin-square design as the blocks had been used. The ratio

$$\frac{\text{Row MSS} + (r - 1)(\text{error MSS})}{r\,(\text{error MSS})}$$

obtained from the analysis of variance of an $r \times r$ Latin-square experiment, therefore, gives an estimate of the relative efficiency of the row blocking in the Latin-square design (as compared with a randomized-blocks design that uses the columns of the Latin-square design as the blocks). Similar results can be obtained for the column blocking.

Comparison with a completely randomized design The efficiency of a Latin-square design compared with complete randomization can be estimated by the expression

$$\frac{\text{Column MSS} + \text{row MSS} + (r - 1)(\text{error MSS})}{(r + 1)(\text{error MSS})}$$

Tukey's test of nonadditivity When each cell in a Latin-square experiment contains only one observation, nonadditivity cannot be tested in a general fashion, but the following method, from Tukey (1955), may be useful. Obtain the predicted $\hat{y}_{ijk}$ values under the additive model ($\hat{y}_{ijk} = \hat{\mu} + \hat{\alpha}_i + \hat{\beta}_j + \hat{\gamma}_k$) and calculate the residuals $d_{ijk} = y_{ijk} - \hat{y}_{ijk}$. Test for significance the coefficient of regression of d_{ijk} on the residuals of any arbitrary function of $\hat{y}_{ijk}$, for example, $A(\hat{y}_{ijk} - B)^2$, where A and B are arbitrarily chosen constants. If the regression is significant, nonadditivity of this particular form is assumed to be present. Obviously, instead of a quadratic function one may employ any other nonlinear function of $\hat{y}$.

Test of nonadditivity when each cell has two or more cases When there are two or more observations per cell, we can estimate the SS due to nonadditivity separately by subtracting from the cell SS the total of the SS due to rows, columns, and treatments.

Greco-Latin square A Greco-Latin square is equivalent to a pair of *orthogonal Latin squares*. Two Latin squares are said to be *orthogonal* if when one is superimposed on the other, every letter of one square appears once and only once with every letter of the other. Consider, for example, the following two Latin squares:

$$
\begin{array}{cccc}
A & B & C & D \\
B & A & D & C \\
C & D & A & B \\
D & C & B & A
\end{array}
\qquad
\begin{array}{cccc}
1 & 2 & 3 & 4 \\
4 & 3 & 2 & 1 \\
2 & 1 & 4 & 3 \\
3 & 4 & 1 & 2
\end{array}
$$

We have used letters to describe one square and numbers to describe the other to make it easier to distinguish one from the other. An alternative procedure is to use Latin letters for one square and Greek letters for the other, as is conventionally done (it is this convention which is responsible for the name Greco-Latin square). If we now superimpose the number square on the other and write the numbers as subscripts, we get

$$
\begin{array}{cccc}
A_1 & B_2 & C_3 & D_4 \\
B_4 & A_3 & D_2 & C_1 \\
C_2 & D_1 & A_4 & B_3 \\
D_3 & C_4 & B_1 & A_2
\end{array}
$$

This arrangement, as can be easily verified, exhibits the property mentioned above, namely, that every letter appears once and only once with every number. This square is therefore a Greco-Latin square.

As pointed out in the introduction to this section, Greco-Latin squares can be used for triple blocking. In the 4×4 square shown above, for example, the rows can be used to represent one block factor, the columns to represent another, and the letters a third. The numbers would then stand for the treatments.

Hyper-Greco-Latin square   The idea of using a pair of orthogonal Latin squares to obtain a Greco-Latin square can be extended to obtaining *hyper-Greco-Latin* squares. For a square of size $r \times r$, a hypersquare can be constructed by superimposing 3 or up to $r - 1$ orthogonal Latin squares, if available. The following is an example, in which three orthogonal Latin squares (one described in terms of capital letters, another in terms of lowercase letters, and the third in terms of numbers) are superimposed:

$$
\begin{array}{cccc}
A_{a_1} & B_{b_2} & C_{c_3} & D_{d_4} \\
B_{c_4} & A_{d_3} & D_{a_2} & C_{b_1} \\
C_{d_2} & D_{c_1} & A_{b_4} & B_{a_3} \\
D_{b_3} & C_{a_4} & B_{d_1} & A_{c_2}
\end{array}
$$

In this case the degrees of freedom will be divided among different sums of squares as follows, assuming one observation per cell:

Source of variation	df
Between rows	3
Between columns	3
Between A, B, C, D	3
Between a, b, c, d	3
Between 1, 2, 3, 4	3
Total	15

As can be easily seen from the above table, there is no degree of freedom left for residual SS. The total SS has been completely accounted for by the four block factors (rows, columns, A, B, C, D and a, b, c, d) together with the treatments (1, 2, 3, 4). In order to perform tests of significance in such a case, a separate estimate of the error variance (perhaps based on a priori knowledge) must be obtained. In our judgment, the use of hyper-Greco-Latin squares in social research would be confined to artificial experiments simulated in the computer. For a detailed discussion of hyper-Greco-Latin squares, reference may be made to Federer (1955).

7.3 CONCOMITANT VARIABLES AND THE COVARIANCE ANALYSIS

As indicated in Sec. 6.1, blocking is one method that can be employed to reduce the magnitude of the residual variation in the experiment. We have seen how this strategy works in randomized-blocks, Latin-square, and other designs. It may be recalled that the principle behind blocking is that when experimental units blocked together are similar in all essential respects, differences (if any) in their responses observed in the experiment are attributable to treatment differences. Experimental units are said to be similar in all essential respects only if when exposed to a common treatment their responses do not differ appreciably. Such units may also be described as homogeneous units. Ideally, then, only completely homogeneous units should be put in the same block. In practice it is impossible to find completely homogeneous units because it is impossible to control physically all factors that affect the response variable. Suppose information is available for all experimental units on certain factors that affect the response variable but were not physically controlled. Such variables are termed *concomitant variables*, and measurements on them are referred to as *concomitant information*. The score received by a subject in a preliminary test in a psychological experiment and observations on family growth patterns of populations prior to the introduction of a family-planning action program are examples of concomitant observations. The question we are concerned with here is whether concomitant information can be made use of in reducing the residual variation, and if so, how.

The method of analysis that employs data on concomitant variables to reduce the residual variation is known as *the analysis of covariance*. Chapter 4 has discussed some aspects of this topic. Certain ideas not explicitly brought out there are discussed in this section.

It is useful to distinguish between two contexts in which the analysis of covariance is usually employed. One context may be described in terms of the causal diagram shown in Fig. 7.1. The essential feature of this situation is that the treatments do not affect the concomitant variables. In experiments this is often ensured by taking measurements on the concomitant variables before making the assignment of treatments to units or at the latest before the effect of treatments has had time to develop. In certain experiments even if the concomitant variables are measured while the experiment is in progress or at the end of the experiment, there may be sufficient knowledge or reason to believe that the concomitant variables were unaffected by the treatments. To be distinguished from such a situation is the one in which it cannot be

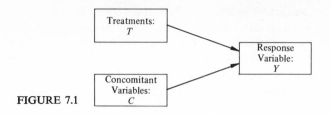

FIGURE 7.1

assumed that the treatments do not affect the concomitant variables. Either of the causal diagrams in Figs. 7.2 and 7.3 may represent this second situation.

Many books on experimental design and analysis emphasize that the covariance analysis is appropriate only in situations represented by Fig. 7.1. We shall comment on this proscription after we review the formal theory of the covariance analysis.

For the moment we shall broadly define the analysis-of-covariance model as a regression model in which some of the regressors are qualitative variables while others are quantitative variables. Qualitative variables are incorporated into the regression by defining them in terms of variables that take only values 0 or 1. Quantitative variables, on the other hand, are incorporated into the regression by means of their observed numerical values.

In the covariance analysis all concomitant variables are assumed to be quantitative variables. Suppose there are under consideration m qualitative variables, $x_1, x_2, \ldots, x_m$ (including a constant term), and k quantitative variables $q_1, q_2, \ldots, q_k$. Assuming the regression of y on $q_1, q_2, \ldots, q_k$ to be linear, we can represent the ith observation as

$$y_i = \sum_{j=1}^{m} \beta_j x_{ij} + \sum_{l=1}^{k} \gamma_l q_{il} + \varepsilon_i \qquad i = 1, 2, \ldots, n$$

It deserves to be emphasized that there is no loss of generality in assuming the regression of y on $q_1, q_2, \ldots, q_k$ to be linear. For the k variables $q_1, q_2, \ldots, q_k$ may collectively represent any one of the following:

1   k separate concomitant variables
2   Fewer than k concomitant variables and several products of the type $x_j q_l$
3   Fewer than k concomitant variables and several nonlinear functions such as q_l^2, $\sqrt{q_l}$, etc., and several products of the types $x_j q_l$, $x_j q_l^2$, $x_j \sqrt{q_l}$, etc.

The model presented above can therefore be regarded as a very general model indeed. It covers situations in which interactions, in the sense of Chap. 4, are suspect-

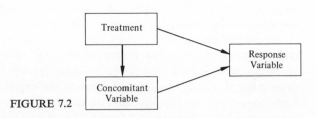

FIGURE 7.2

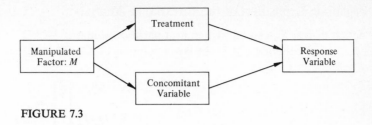

FIGURE 7.3

ed, since some of the q's may stand for products of the type $x_j q_l$. The model also covers nonlinear regressions of y on q's, since some of the q's may stand for nonlinear functions of the observed concomitant variables.

If we consider the observations on the concomitant variables as fixed (see Chap. 5 for a discussion of the difference between fixed and random effects), the situation represented by the model described above can be viewed as an ordinary regression situation in which there are $m + k$ parameters, $\beta_1, \beta_2, \ldots, \beta_m; \gamma_1, \gamma_2, \ldots, \gamma_k$, an observation vector of y values, $y_1, y_2, \ldots, y_n$, and a new design matrix

$$\mathbf{X}_* = \begin{bmatrix} x_{11} & x_{12} & \cdots & x_{1m} & q_{11} & q_{12} & \cdots & q_{1k} \\ x_{21} & x_{22} & \cdots & x_{2m} & q_{21} & q_{22} & \cdots & q_{2k} \\ \multicolumn{8}{c}{\cdots\cdots\cdots\cdots\cdots\cdots\cdots\cdots\cdots\cdots\cdots} \\ x_{n1} & x_{n2} & \cdots & x_{nm} & q_{n1} & q_{n2} & \cdots & q_{nk} \end{bmatrix}$$

When it is viewed in this way, it becomes clear that there are no new problems in applying the least-squares theory either in the estimation of the parameters or in testing hypotheses about them. Let $(\mathbf{X}'_* \mathbf{X}_*)_g^{-1}$ be a g-inverse of $(\mathbf{X}'_* \mathbf{X}_*)$, the subscript g denoting that the reference is to a g-inverse.

A solution set for the $m + k$ parameters $\beta_1, \beta_2, \ldots, \beta_m; \gamma_1, \gamma_2, \ldots, \gamma_k$ is obtained as

$$\begin{bmatrix} \hat{\beta}_1 \\ \hat{\beta}_2 \\ \vdots \\ \hat{\beta}_m \\ \hat{\gamma}_1 \\ \hat{\gamma}_2 \\ \vdots \\ \hat{\gamma}_k \end{bmatrix} = (\mathbf{X}'_* \mathbf{X}_*)_g^{-1} \mathbf{X}'_* \begin{bmatrix} y_1 \\ y_2 \\ \vdots \\ y_n \end{bmatrix}$$

Also the elements of $\sigma^2 (\mathbf{X}'_* \mathbf{X}_*)_g^{-1}$, where σ^2 is the residual variance, can be formally treated as the variances and covariances of $\hat{\beta}_1, \hat{\beta}_2, \ldots, \hat{\beta}_m; \hat{\gamma}_1, \hat{\gamma}_2, \ldots, \hat{\gamma}_k$. With these at hand we can proceed to test the significance of any contrasts involving $\hat{\beta}$'s or $\hat{\gamma}$'s in the usual manner. Note that we use residual MSS as our estimate of σ^2 and that with respect to any contrast such as $\psi = \sum l_j \beta_j$, where $\sum l_j = 0$, the least-squares estimate of it will be the corresponding estimate ignoring the concomitant information *minus* an adjustment for the effects of the concomitant variables. An algebraic demonstration of this result is given below.

An Algebraic Exposition of the Covariance Analysis[1]

The design matrix X_* can be written as a partitioned matrix $[X \mid Q]$, where X represents the first m columns and Q the last k columns of X_*. The normal equations can be written

$$\begin{bmatrix} X'X & \vdots & X'Q \\ \cdots & \vdots & \cdots \\ Q'X & \vdots & Q'Q \end{bmatrix} \begin{bmatrix} \hat{\beta} \\ - \\ \hat{\gamma} \end{bmatrix} = \begin{bmatrix} X' \\ -- \\ Q' \end{bmatrix} Y$$

where $\hat{\beta}$ = column vector consisting of $\hat{\beta}_1, \ldots, \hat{\beta}_m$

$\hat{\gamma}$ = column vector consisting of $\hat{\gamma}_1, \ldots, \hat{\gamma}_k$

Y = column vector consisting of $y_1, y_2, \ldots, y_n$

This equation can be written as two matrix equations:

$$(X'X)\hat{\beta} + (X'Q)\hat{\gamma} = X'Y \qquad (7.1)$$

$$(Q'X)\hat{\beta} + (Q'Q)\hat{\gamma} = Q'Y \qquad (7.2)$$

At this point we imagine regressing separately y as well as each q on the m regressors $x_1, x_2, \ldots, x_m$. In obvious notation the normal equations associated with these regressions are seen to be

$$(X'X)\hat{\delta} = X'Y \qquad (7.3)$$

$$(X'X)\hat{\alpha}_i = X'Q_i \qquad i = 1, 2, \ldots, k \qquad (7.4)$$

where Q_i is the ith column of Q. Putting together the k equations (7.4), we may write

$$(X'X)\hat{\alpha} = X'Q \qquad (7.5)$$

where $\hat{\alpha}$ is a matrix with $\hat{\alpha}_i$ of (7.4) as the ith column. Let $\hat{\alpha}$ be any solution set satisfying (7.5). Premultiply (7.1) by $\hat{\alpha}'$. Notice that

$$\hat{\alpha}'(X'X) = (X'X\hat{\alpha})'$$

But since $\hat{\alpha}$ satisfies (7.5), $(X'X)\hat{\alpha} = X'Q$. Consequently $(X'X\hat{\alpha})' = Q'X$. It follows therefore that the result obtained by premultiplying (7.1) with $\hat{\alpha}'$ is

$$Q'X\hat{\beta} + \hat{\alpha}'X'Q\hat{\gamma} = \hat{\alpha}'X'Y \qquad (7.6)$$

Subtract (7.6) from (7.2):

$$(Q'Q - \hat{\alpha}'X'Q)\hat{\gamma} = (Q' - \hat{\alpha}'X')Y \qquad (7.7)$$

This is an equation system containing exclusively the unknowns $\hat{\gamma}_1, \hat{\gamma}_2, \ldots, \hat{\gamma}_k$, since we know Q, X, Y, and $\hat{\alpha}$. Let $\hat{\gamma}$ be any solution set satisfying (7.7) and $\hat{\delta}$ be any solution set satisfying (7.3). Now obtain $\hat{\beta}$ as

$$\hat{\hat{\beta}} = \hat{\hat{\delta}} - \hat{\alpha}\hat{\hat{\gamma}}$$

---

[1] The balance of Sec. 7.3 may be skimmed or omitted on the first reading.

We can show that $\hat{\hat{\beta}}$ and $\hat{\hat{\gamma}}$ obtained in the above fashion will satisfy (7.1) and (7.2). Substituting $\hat{\hat{\beta}}$ for $\hat{\beta}$ and $\hat{\hat{\gamma}}$ for $\hat{\gamma}$ on the left-hand side of (7.1), we get

$$(\mathbf{X'X})(\hat{\hat{\delta}} - \hat{\alpha}\hat{\hat{\gamma}}) + (\mathbf{X'Q})\hat{\hat{\gamma}}$$

which when expanded becomes

$$(\mathbf{X'X})\hat{\hat{\delta}} - (\mathbf{X'X}\hat{\alpha} - \mathbf{X'Q})\hat{\hat{\gamma}} \qquad (7.8)$$

But since $\hat{\alpha}$ satisfies (7.5), $\mathbf{X'X}\hat{\alpha} = \mathbf{X'Q}$. Consequently $(\mathbf{X'X}\hat{\alpha} - \mathbf{X'Q})\hat{\hat{\gamma}}$ is a zero matrix. Also since $\hat{\delta}$ satisfies (7.3), we have

$$(\mathbf{X'X})\hat{\hat{\delta}} = \mathbf{X'Y}$$

which is the right-hand side of (7.1). This proves that $\hat{\hat{\beta}}$ and $\hat{\hat{\gamma}}$ arrived at in the manner described above satisfy the normal equations (7.1). Similarly it can be shown that they satisfy (7.2) too. Hence $\hat{\hat{\beta}}$ and $\hat{\hat{\gamma}}$ may be viewed as least-squares estimates of β and γ.

Notice that the formula $\hat{\hat{\beta}} = \hat{\delta} - \hat{\alpha}\hat{\hat{\gamma}}$ implies the following. The least-squares estimates of the effects of x's, that is, $\hat{\hat{\beta}}_j$'s, are the corresponding estimates obtained by ignoring the concomitant information, in other words, $\hat{\delta}_j$'s, *minus* a weighted sum of the regression weights of the concomitant variables. If β_j's represent treatment effects, the estimates of treatment contrasts arrived at in the above fashion are adjusted estimates in the sense that they are what emerge after the effects of the concomitant variables on the response variable have been removed.

To make the above point clearer it may be helpful to consider a simple experiment. Suppose there is only one concomitant variable q, and suppose further that equal numbers of cases have been observed in each cell. If β_j and β_l are the treatment effects of the jth and lth treatments, respectively, the least-squares estimate of the contrast $\beta_j - \beta_l$ that the covariance analysis yields is seen to be of the form

jth treatment mean $-$ lth treatment mean

$-$ const $\times$ (mean of q under treatment j $-$ mean of q under treatment l)

Clearly, then, if the treatments affected the concomitant variable, the difference between the means of q under the jth and lth treatments would contain part of the treatment effect on the response variable. Consequently the interpretation of the estimate of the treatment contrast obtained in the above manner is not the same as what it would have been if q had not been affected by the treatments. In the former situation the least-squares estimates of treatment contrasts, i.e., the *adjusted treatment contrasts*, do not tell the full story concerning the roles of the treatments contrasted. Only when the concomitant variables are unaffected by the treatments do the adjusted treatment contrasts validly reflect the comparative impacts on the response variable of the treatments compared (see Cochran and Cox, 1957; Cox, 1958).

When the treatments do affect the concomitant variables, the investigator may be interested in shedding light on how the treatments produced their effects. One procedure that may be adopted for such investigation is the simultaneous-equations method covered in Part Three. Reference may also be made to Mallios (1970). The main point to remember in this connection is that when the treatment variable T is causally connected to C, that is, $T \rightarrow C$, the arrow indicating the causal direction, the effect of C on the response variable is also attributable to T. But this cannot be called a *direct effect* of T (cf. Chap. 10). The effect of T that operates through C may be called instead an *indirect effect* of T.

7.4   QUALITATIVE RESPONSE VARIABLES

So far our attention has been confined to data situations involving quantitative response (dependent) variables, but often the analyst may be confronted with experiments in which the response is coded as success or failure; adopter (of a new technique) or nonadopter; survivor or nonsurvivor; etc. In nonexperimental research, too, such situations are frequent. For example, the dependent variable in an election study may be the stated intention of respondents to vote Democratic or Republican. In this section we shall briefly indicate a few common methods of analyzing data sets of this kind.

Least-Squares Method Applied to Proportions

In data situations where the response (dependent) variable is dichotomous, interest may center on the analysis of the proportions of units falling into one category of the response (dependent) variable. Thus we may be interested in analyzing factors affecting individuals' propensity to succeed, or farmers' tendency to adopt a new farming technique, and so on. Insofar as such propensities or tendencies are reflected in the observed proportion of successes or proportion of adopters, etc., the analytical problem becomes one of estimating the effects of various factors on these proportions. There is no loss of generality if we refer below to the proportion of successes as our dependent variable.

Let the independent variables be one with r categories and one with c categories. The former we shall refer as the row variable and the latter as the column variable. Let there be n_{ij} observations in the (i,j) cell, and let the number of successes be g_{ij}. Then the observed proportion of successes in the (i,j) cell is $p_{ij} = g_{ij}/n_{ij}$, and the data set at hand may be viewed as rc proportions arranged in r rows and c columns (see Table 7.7). Our interest is in examining whether the row effects and column effects are additive and, if so, to make comparisons between rows and columns. If nonadditivity is detected, we may be interested in investigating its nature.

Table 7.7   A TWO-WAY TABLE OF PROPORTIONS

Category of row variable	Item of information	Category of column variable			
		1	2	$\cdots$	c
1	Number of cases	n_{11}	n_{12}	$\cdots$	n_{1c}
	Number of successes	g_{11}	g_{12}	$\cdots$	g_{1c}
	Proportion of successes	p_{11}	p_{12}	$\cdots$	p_{1c}
2	Number of cases	n_{21}	n_{22}	$\cdots$	n_{2c}
	Number of successes	g_{21}	g_{22}	$\cdots$	g_{2c}
	Proportion of successes	p_{21}	p_{22}	$\cdots$	p_{2c}
$\cdots$					
r	Number of cases	n_{r1}	n_{r2}	$\cdots$	n_{rc}
	Number of successes	g_{r1}	g_{r2}	$\cdots$	g_{rc}
	Proportion of successes	p_{r1}	p_{r2}	$\cdots$	p_{rc}

One method of analyzing proportions arranged in the form shown in Table 7.7 is to apply the ordinary least-squares procedure to the observed p_{ij}'s. The question immediately arises whether this is a valid procedure. In examining this question it may be useful to distinguish between the situation in which all p_{ij}'s are based on equal numbers of cases and that in which the base numbers differ between cells. The former situation may be characteristic of controlled experiments, while the latter is character- istic of most survey-data situations.

Equal n's   If the observed p_{ij}'s in Table 7.7 are all based on n cases each, and if the population proportion corresponding to p_{ij} is π_{ij}, then under random sampling p_{ij} has expectation π_{ij} and variance $\pi_{ij}(1 - \pi_{ij})/n$. We may regard p_{ij} as the mean of n observations each of which is coded 0 or 1 according as it is a failure or a success. Thus the data situation under discussion may be described as a two-way table of cell means, each mean being based on n observations. With equal numbers of cases in each cell, inferences about column and row effects drawn from the least-squares analysis are robust to departures from normality and homoscedasticity. Hence in this case we may apply the ordinary least-squares procedure to analyze the column and row effects on proportions arranged in the form of Table 7.7. One problem, however, deserves to be mentioned. The ordinary least-squares procedure does not impose any constraints on the observed proportions, e.g., that their expectations must lie between 0 and 1. Consequently, there is no guarantee that when we estimate the expected values in each cell from the fitted least-squares equation we shall always get numbers between 0 and 1. If any of the estimated expected values fall outside the range (0,1), we may be faced with an anomaly in that what is purported to be a propor- tion takes on a value below 0 or above 1. In practice, if n is large, say greater than 20, and if none of the population proportions (π_{ij}'s) lies below 0.25 or above 0.75, the anomalous situation referred to above may be unlikely, although there is no guarantee that it will not arise. Unfortunately the analyst seldom knows the true range of the population proportions. A working rule therefore may be necessary to help decide when not to apply the ordinary least-squares procedure to the analysis of proportions. The following rule may be used for the purpose:

If any of the observed proportions is less than 0.25 or greater than 0.75, do not apply the ordinary least-squares procedure for analyzing proportions arranged in the form of Table 7.7.

Unequal n's   What if n_{ij}'s in Table 7.7 differ between themselves? In this case the observed proportion p_{ij} has expectation π_{ij} and variance $\pi_{ij}(1 - \pi_{ij})/n_{ij}$. Hetero- scedasticity may then be extreme. Since the cell means (p_{ij}'s) are based on unequal n's, the inferences about the column and row effects may be seriously affected by heteroscedasticity. In such situations we can transform the heteroscedastic residuals to homoscedastic ones by dividing each residual by its standard deviation. This strategy is based on the well-known result that if a random variable ε_i has variance σ_i^2, then a multiple k of ε_i has variance $k^2\sigma_i^2$; hence if we choose $k = 1/\sigma_i$, we get $k\varepsilon_i$ as a random variable with unit variance. Obviously to apply this procedure to the problem at hand we should know the variances of the residuals associated with each cell, which is the same as the variances of each p_{ij}. If we know the population values π_{ij}'s, we can immediately calculate $\pi_{ij}(1 - \pi_{ij})/n_{ij}$, which is the variance of p_{ij}. But seldom do we know the π_{ij} values associated with tables like Table 7.7. Hence in practice what one does is to make use of a consistent estimate of $\pi_{ij}(1 - \pi_{ij})/n_{ij}$ instead of $\pi_{ij}(1 - \pi_{ij})/n_{ij}$ itself. It is well known that under random sampling $p_{ij}(1 - p_{ij})/n_{ij}$ is a consistent estimate of $\pi_{ij}(1 - \pi_{ij})/n_{ij}$. Accordingly, in order to reduce the heteroscedasticity in Table 7.7 we divide each residual in the linear model for p_{ij} by $\sqrt{p_{ij}(1 - p_{ij})/n_{ij}}$. We then fit the linear model in question to the observed p_{ij}'s by minimizing the SS of the transformed residuals. Thus if the linear model we propose to fit to the observed proportions in Table 7.7 is

$$p_{ij} = \mu x_0 + \sum_{i=1}^{r} x_i \alpha_i + \sum_{j=1}^{c} z_j \beta_j + \varepsilon_{ij} \qquad (7.9)$$

where $x_0 = 1$ for all p's

$x_i = 1$ for all p's in ith row and 0 for all others

$z_j = 1$ for all p's in jth column and 0 for all others

we would divide ε_{ij}, that is, $p_{ij} - \mu x_0 - x_i \alpha_i - z_j \beta_j$, by $\sqrt{p_{ij}(1 - p_{ij})/n_{ij}}$ and then minimize the SS of the resulting quantities. The normal equations are $\sum \sum w_{ij} \varepsilon_{ij} = 0$; $\sum \sum x_k w_{ij} \varepsilon_{ij} = 0$, $k = 1, 2, \ldots, r$; $\sum \sum z_l w_{ij} \varepsilon_{ij} = 0$, $l = 1, 2, \ldots, c$; where $1/w_{ij} = p_{ij}(1 - p_{ij})/n_{ij}$, for all i and j. To solve these equations we use appropriate side conditions.

As mentioned in connection with the case of equal n's, here also there is no guarantee that the expected values of p's estimated from the fitted equations will always lie between 0 and 1. The working rule given above for equal n's may be followed here too. That is, avoid applying the above procedure if any of the observed p's is close to 0 or 1. Needless to add, if any of the p's is exactly 0 or 1, the corresponding w_{ij} becomes undefined and hence the method is inapplicable.

The weighted least squares   The method just described, in which the residuals in a linear model are divided by their respective standard deviations and the SS of the resulting transformed residuals is minimized, is known as the *method of weighted least squares*. To see a justification of this procedure, let us revert for a moment to the problem of fitting a simple linear regression of y on x, given n pairs (y_i, x_i). We start with the observational equations

$$y_i = \alpha + \beta x_i + \varepsilon_i$$

Suppose that

$$E(\varepsilon_i) = 0 \qquad \text{for all } i$$

$$E(\varepsilon_i, \varepsilon_j) = 0 \qquad \text{for all } i \neq j$$

$$\text{var } \varepsilon_i = \sigma_i^2$$

If in addition to the above we also assume normality for the residuals, we have for any α and β the likelihood of the observations $y_1, y_2, \ldots, y_n$ given by

$$L = \frac{1}{(2\pi)^{n/2} \sigma_1 \sigma_2 \cdots \sigma_n} \exp\left[-\frac{1}{2} \sum \frac{(y_i - \alpha - \beta x_i)^2}{\sigma_i^2} \right]$$

This likelihood is maximized when the negative exponent is minimized. In other words, the maximum-likelihood estimates of α and β in the above situation are those values which minimize the weighted SS

$$\sum \frac{1}{\sigma_i^2} (y_i - \alpha - \beta x_i)^2$$

the weights being inversely proportional to the variance of ε_i, that is, of $y_i - \alpha - \beta x_i$, $i = 1, 2, \ldots, n$. This demonstrates that just as the ordinary least-squares procedure yields the same estimates as the maximum-likelihood method under normality, independence, and homoscedasticity, the weighted least-squares procedure yields the same estimates as the maximum-likelihood procedure under normality, independence, and heteroscedasticity.

Testing goodness of fit Once a given model has been fitted, the researcher may want to know whether the fit is satisfactory. If the additive model (7.9) provides a satisfactory fit to the data in the form of Table 7.7, we declare that nonadditivity between rows and columns in the data is negligible. If model (7.9) with all β's set equal to zero provides a satisfactory fit to the data, we declare that the column effects do not differ between themselves. Similarly if model (7.9) with all α's set equal to zero provides a satisfactory fit, we declare that the row effects do not differ between themselves. And so on. To test the goodness of fit to any given data set of any given model, we proceed as described below.

First we estimate the expected values of the observed proportions in each cell. In Table 7.7 the expected value of the proportion in the (i,j) cell, according to model (7.9), is given by

$$\hat{\pi}_{ij} = \hat{\mu} + \hat{\alpha}_i + \hat{\beta}_j \qquad (7.10)$$

where $\hat{\mu}$, $\hat{\alpha}_i$, $\hat{\beta}_j$ are estimates of μ, α_i, and β_j, respectively, obtained by fitting (7.9) as described above.

The next step in testing the goodness of fit of (7.9) is to calculate the following quantity to be used as a goodness-of-fit statistic:

$$\sum_i \sum_j w_{ij}(p_{ij} - \hat{\pi}_{ij})^2 \qquad (7.11)$$

where $1/w_{ij} = p_{ij}(1 - p_{ij})/n_{ij}$. To use (7.11) as a goodness-of-fit statistic we need to know its sampling distribution.

In large samples, as already mentioned, p_{ij} may be regarded as a normal variate with mean π_{ij} and variance $\pi_{ij}(1 - \pi_{ij})/n_{ij}$. Therefore, for large n_{ij}, say greater than 20,

$$\frac{(p_{ij} - \pi_{ij})^2}{\pi_{ij}(1 - \pi_{ij})/n_{ij}} \qquad (7.12)$$

behaves approximately as a squared standard normal deviate and hence has approximately a chi-square distribution with 1 degree of freedom. It follows from this that if the observations in all cells in Table 7.7 have been independently drawn at random,

$$\sum_i \sum_j \frac{(p_{ij} - \pi_{ij})^2}{\pi_{ij}(1 - \pi_{ij})/n_{ij}} \qquad (7.13)$$

will be approximately distributed as a chi square with rc degrees of freedom.

The sampling distribution of the statistic (7.11) can be derived from that of (7.13). We may view (7.11) as a quantity to which (7.13) converts itself when the terms of the latter are subjected to certain constraints. The constraints involved are the relationships reflected in the normal equations used in fitting (7.9). Altogether $1 + (r - 1) + (c - 1)$, or $r + c - 1$, independent relationships are involved in fitting (7.9). We may therefore view (7.11) as the SS of rc terms, each of which is approximately a standard normal deviate but between which $r + c - 1$ relationships prevail. It is thus seen that (7.11) has approximately a chi-square distribution with $rc - r - c + 1$ or $(r - 1)(c - 1)$ degrees of freedom.

If we fit (7.9) setting all α's equal to zero, the resulting quantity analogous to (7.11) will be approximately distributed as a chi square with $rc - c$ degrees of freedom. Similarly if (7.9) is fitted with all β's set equal to zero, the resulting quantity analogous to (7.11) will have a chi-square distribution with $rc - r$ degrees of freedom.

A few words of caution are pertinent here. When we say that the quantity (7.13) has approximately a chi-square distribution if n_{ij}'s are greater than 20, we are assuming that in any given cell the observations are drawn at random from a single *binomial distribution* and that the sampling in any one cell is independent of that in any other. Sometimes the data in a cell may actually belong to several binomials with different p's. Also the sampling in a given cell often is not independent of that in others. To illustrate the former situation, consider a study of high school dropouts. The basic measurement of interest might be the proportion of dropouts during a given year among students enrolled at the beginning of that year. Suppose the proportions are arranged in a two-way format, say, by age and sex. It may very well be that the data in a cell, e.g., boys aged 15 years, might come from several different binomials depending upon the degree to which the propensity to drop out varies with race, socio-economic status, familial solidarity, etc. In such situations it would be advisable to enter into the cross tabulations as many pertinent variables as possible (see Chap. 1 for a few comments on this issue). The main problem an analyst may be faced with in this connection is that there may not be enough cases in the sample to take into account more than a few pertinent variables. Hence the possibility that what we regard as observations from a single binomial may in fact be a collection of observations from several binomials with different p's should always be recognized as a possible threat to the validity of the test procedure described above.

As for the possibility of nonindependence of the samples in different cells, the common situation in which this is a very serious threat is when the data are from surveys based on cluster-sampling procedures. Such sample-design effects have not been systematically investigated so far.

Logit Transformation

Instead of dealing with the observed p's, it is often recommended that the quantity

$$y_{ij} = \ln \frac{p_{ij}}{1 - p_{ij}} \qquad (7.14)$$

be used as the dependent variable. The ratio $p_{ij}/(1 - p_{ij})$ is referred to as the odds in favor of success if p_{ij} is the proportion of successes. The notion of odds is frequently employed in common parlance. For example, we say "the odds are 4 to 1 in favor" of a particular event if the probability of occurrence of that event is 0.8. The quantity y_{ij} defined in (7.14) may be called *log odds*. Some textbooks refer to y_{ij} as the *logit*.

The merit of working with the odds is that whereas p's range from 0 to 1, the corresponding odds range from 0 to ∞ and the log odds range from $-\infty$ to $+\infty$. This latter, it may be noticed, is the range of a normal variate. By stretching the range in the above manner, the logit scale makes it unnecessary for the analyst to worry about the possibility that estimated expected values in any cell may go outside plausible limits. A substantive disadvantage of working with the logit scale is that it may be difficult to think about the meaning and practical significance of estimated row and column effects. But certain results of the logit-scale analysis can be translated into results applicable to odds by taking antilogarithms. For example, suppose $\hat{\beta}_j - \hat{\beta}_1$ is an estimated contrast of, say, two column effects in the logit scale. By finding the antilogarithm of $\hat{\beta}_j - \hat{\beta}_1$ we get a figure which can be interpreted as the ratio of the effect of row j to that of row 1 on the odds in question.

Turning to the mechanics of the analysis in the logit scale, in large samples the variance of y_{ij} defined in (7.14) can be approximately taken as $1/n_{ij}p_{ij}(1 - p_{ij})$, where p_{ij} is the observed proportion of successes in the (i,j) cell and n_{ij} is the base number in that cell. In the weighted least-squares analysis we use as weights $w_{ij} = n_{ij}p_{ij}(1 - p_{ij})$.

Various small-sample adjustments have been proposed in the literature to extend the validity of the above method. One of the most commonly accepted procedures may be described as follows. If p is a binomial proportion obtained from g successes in n independent trials in a typical two-way table, calculate the log odds as

$$y = \ln \frac{g + \frac{1}{2}}{n - g + \frac{1}{2}} \qquad (7.15)$$

Notice that this adjustment amounts to adding $\frac{1}{2}$ to the numbers of successes and failures. The weight to be assigned to y in the weighted least-squares analysis is calculated as

$$w = \frac{(g + \frac{1}{2})(n - g + \frac{1}{2})}{n + 1} \qquad (7.16)$$

which is the same as

$$(n + 1) \frac{g + \frac{1}{2}}{n + 1} \frac{n - g + \frac{1}{2}}{n + 1}$$

or an adjusted form of $np(1 - p)$. These approximations extend the validity of the analysis to tables with a number of cells containing as few as five cases (see, for example, Gart and Zweifel, 1967).

The goodness-of-fit statistic to be used in logit analysis is

$$\sum_i \sum_j w_{ij}(y_{ij} - \hat{y}_{ij})^2 \qquad (7.17)$$

where $\hat{y}_{ij}$ is the cell value, corresponding to the observed y_{ij}, estimated according to the given model, for example, $y_{ij} = \mu x_0 + \sum \alpha_i x_i + \sum \beta_j z_j + \varepsilon_{ij}$ as in (7.9), and w_{ij} is the weight attached to y_{ij} in the weighted least-squares analysis. This statistic is approximately distributed as a chi square whose degrees of freedom are to be calculated in the manner described earlier for the p-scale analysis.

Analysis of Polychotomous Dependent Variables

Often in social research we may be confronted with polychotomous response (dependent) variables. Two types of such variables may be distinguished, the ordinal type and the nominal type. Examples of the ordinal type are (1) opinions coded agree, ambiguous, and disagree; (2) illness coded critical, serious, mild, and negligible. Examples of the nominal type are (1) voting behavior coded Democratic, Republican, and Independent; (2) reasons for holding a given view coded personal, familial, and extrafamilial.

When dealing with polychotomous dependent variables, we imagine a two-way (or multiway, as the case may be) table of multinomial universes from each of which a random sample has been independently drawn. An example is presented in Table 7.8. A general data set of the form shown in Table 7.8 may be described as one with r rows and c columns, the (i,j) cell containing a random sample of size n_{ij} from a multinomial universe of m categories. Let us denote the observed frequencies in the m categories by $n_{ij1}, n_{ij2}, \ldots, n_{ijm}$. The observed proportion may be denoted by $p_{ij1}, p_{ij2}, \ldots, p_{ijm}$. Let the corresponding population proportions be $\pi_{ij1}, \pi_{ij2}, \ldots, \pi_{ijm}$. Obviously,

$$\sum_{k=1}^{m} \pi_{ijk} = 1 \qquad \sum_{k=1}^{m} p_{ijk} = 1$$

From the sampling theory associated with the multinomial distribution we have

$$E(p_{ijk}) = \pi_{ijk} \qquad k = 1, 2, \ldots, m$$

$$\text{var } p_{ijk} = \frac{\pi_{ijk}(1 - \pi_{ijk})}{n_{ij}} \qquad k = 1, 2, \ldots, m$$

$$\text{covar}(p_{ijk}, p_{ijk'}) = - \frac{\pi_{ijk}\pi_{ijk'}}{n_{ij}} \qquad k \neq k'$$

Table 7.8 DISPOSAL OF CASES FILED IN U.S. DISTRICT COURTS (FICTITIOUS DATA)

	Disposal of case					
	Defendant brought to trial and found guilty		Defendant brought to trial and found not guilty		Defendant not brought to trial or case settled out of court	
Region	White	Black	White	Black	White	Black
Northeast	50	30	7	8	8	3
North Central	14	10	3	5	1	2
South	19	25	12	20	6	3
West	25	30	18	19	4	6

The case of ordered multinomial categories   If the qualitative categories represented by the multinomial are ordered, we may think of computing a combined score of the following type for the (i, j) cell:

$$y_{ij} = \sum_{k=1}^{m} l_k p_{ijk} \qquad (7.18)$$

where l_k, $k = 1, 2, \ldots, m$, are m constants. For example, suppose there are five categories to the multinomial. Let the categories represent ordered responses to an opinion question, the responses being strongly agree, agree, ambivalent, disagree, and strongly disagree. These responses may be given weights 2, 1, 0, -1, and -2, respectively. The chief objection to assigning scores to the categories of the multinomial is that the method is quite arbitrary. Textbooks suggest the following procedure to avoid arbitrariness in assigning scores to the categories involved. If there are m ordered traits obtain the expected values of the ordered elements of a random sample of size m from an appropriate standard population, e.g., a normal universe with mean 0 and variance 1, and use these values as numerical scores for the ordered qualitative traits in question (see Bross, 1958; Snedecor and Cochran, 1967, pp. 243–246; Fisher and Yates, 1973, table XXI).

Once combined scores of the above type are computed for each cell, the analysis procedure is straightforward. For the application of the weighted least-squares method we require estimates, if not the actual values, of the variances of the combined scores in different cells. We notice that given (7.18),

$$\text{var } y_{ij} = \sum_{k=1}^{m} l_k^2 \text{ var } p_{ijk} + \sum_{k=1}^{m} \sum_{\substack{k'=1 \\ k \neq k'}}^{m} l_k l_{k'} \text{ covar } (p_{ijk}, p_{ijk'})$$

Estimate of var y_{ijk} is obtained by substituting in the above formula estimates of var p_{ijk} and of covar $(p_{ijk}, p_{ijk'})$. No new principle is involved in the remaining part of the analysis.

Unordered multinomial categories   If the multinomial populations involved represent nominal categories, there would appear to be little sense in assigning scores. A few remarks about this situation are given in connection with a discussion of measurement errors in Chap. 10. Instead of repeating those comments here, we shall simply indicate what is commonly done in the case of nominal polychotomous dependent variables. We decide that some particular category of the dependent variable is of interest and pool all the other categories, thus converting the multinomial case to one involving the binomial. For example, suppose our dependent variable is how the U.S. District Courts dispose cases filed with them. Let the categories of interest be

  G Defendant brought to trial and found guilty
  A Defendant brought to trial and acquitted
  S Defendant not brought to trial or case settled out of court

The analyst may be interested in cases of type G as a proportion of all cases, or of type S as a proportion of all cases. In either case the problem reverts to one of analyzing binomial proportions.

An alternative approach is to view the given data set as a multidimensional contingency table and to translate hypotheses of interest into pairwise independence, conditional independence, multiple independence, and multifactor interactions. Before outlining this strategy we must be familiar with certain of the concepts introduced in the next chapter, and so we shall revert to this issue at the end of Chap 8.

7.5 CONCLUDING REMARKS

This chapter and the preceding one have been concerned with procedures to reduce residual variation in experiments. We discussed two procedures, blocking and the use of concomitant information. Among designs that use the former procedure are randomized blocks, Latin squares, Greco-Latin squares, and hyper-Greco-Latin squares. There are of course many other designs that make use of blocking in one form or other. In all these designs, it should be noted, one may make use of concomitant information, if available, to further reduce the residual variation.

Several new concepts were introduced in these chapters. Among them are the design matrix, generalized inverses of a matrix that has no true inverse, dispersion matrix of estimated regression weights, nonadditivity (interaction), orthogonality between treatments and blocks, and weighted least squares.

The distinction made in Chap. 5 between fixed effects and random effects was carried forward to these chapters. Designs in which all effects are random effects have been called random designs, and those in which all effects are fixed effects have been called fixed designs, leaving the term mixed designs for those in which some effects are random and some fixed.

The structure of hierarchical designs and the rationale behind them were discussed and illustrated briefly. Brief attention was also given to the analysis of proportions.

It is perhaps worth emphasizing that the concept of the design matrix and its generalized inverse make it possible to approach in a unified manner the least-squares theory covering the analysis of variance, the analysis of covariance, and the multiple-regression analysis (involving quantitative regressors).

EXERCISES

7.1 The following data are from an experiment run on a Latin-square design.

	Column			
Row	1	2	3	4
1	$C = 9$	$D = 8$	$A = 10$	$B = 5$
2	$B = 11$	$C = 9$	$D = 8$	$A = 10$
3	$A = 14$	$B = 10$	$C = 7$	$D = 6$
4	$D = 10$	$A = 11$	$B = 9$	$C = 5$

(a) Analyze the data and prepare a short report on your findings.

(b) If one of the sources of variation, e.g., row, turns out to be not significant, is it advisable to pool the SS due to that source with the residual SS and retest the remaining SS? Justify your answer.

7.2 From a randomized-blocks experiment the following data were obtained:

	Treatment							
	A		B		C		D	
Block	y	x	y	x	y	x	y	x
1	5.0	2.0	6.0	2.0	9.5	2.5	13.5	2.5
2	5.5	3.0	7.0	3.5	11.5	6.5	13.5	3.0
3	5.0	3.5	7.0	3.5	11.0	6.0	14.0	4.0
4	7.0	6.0	7.0	4.0	12.0	8.5	14.5	5.0
5	8.5	8.5	6.5	4.0	13.0	10.0	15.5	6.0

Notice that x represents a concomitant variate.

(a) Apply the g-inverse method to carry out the covariance analysis.

(b) Test the hypothesis that there is no difference between the adjusted means of treatments A and B. (Adjusted mean denotes treatment mean adjusted for the influence of the concomitant variate.)

7.3 You are given several data situations below, and a hypothesis, a research question, or estimation problem associated with each. Indicate briefly what statistical technique you would employ to deal with the inference problem posed. If the given data set is deficient or inadequate to deal with the problem posed, indicate how.

(a) Data are available on whether respondents in an interview survey expect to buy a new car within the next 12 months. Data are also available on such background characteristics as region of residence and family income. It is required to investigate whether region of residence and family income are jointly sufficient to predict reasonably well the propensity to buy a new car within the next 12 months.

(b) You are given data on y, x_1, x_2, z, where y stands for yearly salary, x_1 for years of work experience, x_2 for professional standing (measured by number of publications such as books and articles), and z for sex. The data are from a survey of a random sample of male and female members on the faculty of 4-year colleges and universities. It is required to test whether there is sex discrimination in salary level.

(c) The ratio of 1970 population to 1960 population of each county in the United States together with County Data Book and Census Reports are given to you. You are asked to investigate factors associated with population change during the period 1960–1970.

(d) A sample of high school graduates of 1972 in the United States is proposed to be followed up in 1992. In the follow-up survey data are proposed to be collected on the social-mobility experience of respondents. Assume that you will have data on:
(1) Father's occupational class: 1972
(2) Respondent's occupational class: 1992
(3) Respondent's educational level
(4) Respondent's race
It is required to test whether there is race difference in mobility chances.

7.4 You are given an $I \times J$ table of proportions of "successes." A multiplicative model seems to be appropriate. Accordingly you write

$$\pi_{ij} = \text{probability of success in } (i,j) \text{ cell} = E(p_{ij}) = \mu \alpha_i \beta_j$$

Since you are dealing with probabilities of successes, it occurs to you that the model just written must have a companion to represent the corresponding probability of "failure," and hence you write

$$1 - \pi_{ij} = \text{probability of failure in } (i,j) \text{ cell} = E(1 - p_{ij}) = \mu^* \alpha_i^* \beta_j^*$$

Automatically you have

$$\mu \alpha_i \beta_j + \mu^* \alpha_i^* \beta_j^* = 1$$

Show that you can now express π_{ij} as

$$\pi_{ij} = \frac{e^{\theta + \phi_i + \eta_j}}{1 + e^{\theta + \phi_i + \eta_j}}$$

where

$$e^\theta = \frac{\mu}{\mu^*} \qquad e^{\phi_i} = \frac{\alpha_i}{\alpha_i^*} \qquad e^{\eta_j} = \frac{\beta_j}{\beta_j^*}$$

Show further that according to the above representation,

$$\ln \frac{\pi_{ij}}{1 - \pi_{ij}} = \theta + \phi_i + \eta_j$$

This exercise demonstrates that the multiplicative model you started with implies a linear additive model in the logit scale.

7.5 Suppose that there are two groups of individuals, group 1 and group 2, of sizes n_1 and n_2, respectively. On each individual a binary response, e.g., success or failure, is obtained. The groups may be thought of as corresponding to two treatments (or a treatment and a control). Let the observed proportions of successes be p_1 and p_2, respectively. Let y_1 and y_2 be the logit transformations of p_1 and p_2, respectively. If you are willing to assume that

$$E(y_j) = \mu + \beta_j$$

you have a linear logit model for representing the relationship between the treatments and the corresponding responses. How would you extend this model if there are k independent 2×2 tables comprising the binary responses under k different conditions? *Hint:* Your model would be $E(y_{ij}) = \mu + \alpha_i + \beta_j$. Modify your model to conform to each of the following:
(*a*) No treatment differences
(*b*) No block (condition) differences
(*c*) No treatment differences *and* no block differences

7.6 Suppose for each individual there are m explanatory variables, $x_1, x_2, \ldots, x_m$, regarded as nonrandom (fixed), and a binary response. It is required to assess the relationship between the probability of success and the variables $x_1, x_2, \ldots, x_m$. Outline the analysis procedure you would employ. Notice that in this case the dependent variable is not a proportion but a binary response, e.g., success or failure of a particular medical treatment; death or survival over the first year of life; etc. The independent variables $x_1, x_2, \ldots, x_m$ may be quantitative or qualitative properties of individuals presumed to influence the response. For specific illustrations see Walker and Duncan (1967) or Feldstein (1966). Also see Morgan and Messenger (1973) and Andrews and Messenger (1973).

7.7 Suppose we want to study the relationship between factor A, for example, smoking, and a response variable which is binary, e.g., suffering from lung cancer or not. Let factor A have two levels, and let the population probabilities be as shown below (with $\pi_{11} + \pi_{12} + \pi_{21} + \pi_{22} = 1$):

	Factor A	
	Smoker	Nonsmoker
Sufferer	π_{11}	π_{12}
Nonsufferer	π_{21}	π_{22}

Suppose that we are interested in comparing the probabilities of not suffering from lung cancer between smokers and nonsmokers. Consider the following sampling schemes and indicate which ones permit estimating the comparison.

(*a*) A random sample from the whole population

(*b*) Two random samples (may be of equal size) from smokers and nonsmokers

(*c*) Two random samples (may be of equal size) from people suffering from lung cancer and those free from lung cancer

Show that the logistic difference of the probabilities of not suffering from lung cancer between smokers and nonsmokers can be estimated from each of the above three samples (see Cox, 1970, pp. 22–23).

8

FACTORIAL EXPERIMENTS
AND AN APPLICATION OF THEIR STRUCTURE
TO THE ANALYSIS OF CATEGORICAL DATA

It is a common problem in experimental research that the investigator becomes interested in studying the effects of each of a number of causal factors on a given response variable. For example, in a Puerto Rican experiment on social change, Hill, Stycos, and Back (1959) wanted to assess the effects of each of the following two factors on the family-planning behavior of couples: (1) ways of delivering a message and (2) content of the message. Three ways of delivering messages were taken for trial: (1) through small group meetings (12 families per meeting), (2) through large group meetings (28 families per meeting), and (3) through pamphlets (to 40 families), in the experimental community. For the second factor, content of the message, the following levels were chosen for trial: (1) emphasis on values, (2) emphasis on communication between husband and wife in the family, and (3) emphasis on both.

In earlier times, the advice researchers used to be given when they wanted to study the effects of two or more factors on a response variable was to study each factor separately, i.e., to devote a separate experiment to the investigation of each factor. Fisher pointed out for the first time, in his *Design of Experiments*, that there are certain definite advantages to be gained by combining the study of several factors in the same experiment. An experiment in which the effects of two or more factors on a response variable are evaluated simultaneously is called a *factorial experiment*.

Factorial experimentation is extensively used in research programs, particularly in industry. In behavioral sciences, too, according to one author, factorial design is "the most widely used design, as an examination of recent volumes of behavioral science journals will verify" (Kirk, 1968, p. 171). If factorial experimentation has been extensively used in research programs, it is because the factorial approach is highly efficient and comprehensive. This chapter deals with various aspects of factorial experiments, and as the title indicates, some attention is also paid to an application of their structure to the analysis of categorical data.

8.1   THE ONE-FACTOR-AT-A-TIME APPROACH VERSUS THE FACTORIAL APPROACH

Why the factorial approach is considered highly efficient and comprehensive can be seen by comparing it with the one-factor-at-a-time approach. For simplicity, suppose that there are only two factors to be studied and that each factor has only two *levels* (doses, categories). As a concrete example, let us consider the following problem.

A weighing problem   The physical weights of two objects, A and B, are to be determined. There is only one balance available, but unfortunately it is one with a bias in that it gives nonzero readings when no object is placed on it. It has been decided that in order to have the required accuracy for each object, four independent determinations of the weights are to be averaged.

The one-at-a-time approach to the above problem involves *replicating* (repeating) the following three steps for each object: (1) record the reading on the balance when no object rests on it; (2) place the given object on the balance and record the resulting reading; (3) subtract the reading obtained in step 1 from that obtained in step 2. The difference thus obtained in any *replication* (repetition) represents one determination of the weight of the given object. If the replications are done independently of each other, the average of the determinations of the weight in the different replications yields the required estimated weight for each object. If we follow the above approach, we need 16 readings in all, 8 for A and 8 for B.

Let us now consider the factorial approach to the above problem. Suppose that the eight readings shown in Table 8.1 have been taken. Imagine that the two factors in the experiment are A and B, each with two levels (presence and absence on the balance). The responses in this case are represented by the readings on the balance. The four treatment combinations applied are enumerated in the first column of

Table 8.1   SYMBOLIC REPRESENTATION OF DATA FROM A WEIGHING EXPERIMENT INVOLVING A 2^2 FACTORIAL ARRANGEMENT EXECUTED OVER A COMPLETELY RANDOMIZED DESIGN

Treatment combination	Individual observation	Treatment total	Treatment mean
(a_0b_0): A and B absent	$y_1(a_0b_0), y_2(a_0b_0)$	$y(a_0b_0)$	$\bar{y}(a_0b_0)$
(a_1b_0): A present	$y_1(a_1b_0), y_2(a_1b_0)$	$y(a_1b_0)$	$\bar{y}(a_1b_0)$
(a_0b_1): B present	$y_1(a_0b_1), y_2(a_0b_1)$	$y(a_0b_1)$	$\bar{y}(a_0b_1)$
(a_1b_1): A and B present	$y_1(a_1b_1), y_2(a_1b_1)$	$y(a_1b_1)$	$\bar{y}(a_1b_1)$

Table 8.1. The individual readings (responses) are represented by symbols such as $y_1(a_0b_0)$, etc. The subscripts to a and b in these symbols indicate whether the objects denoted by the corresponding capital letters are present or absent on the balance when the reading is taken. Thus, if both a and b have subscript 0, that means that the corresponding reading was taken when A and B were absent on the balance, if a has subscript 1 and b 0, the corresponding reading was taken when A was present but B absent, and so on. The totals of repeated readings are represented by symbols such as $y(a_0b_0)$, $y(a_1b_0)$, etc., and the corresponding means, by symbols such as $\bar{y}(a_0b_0)$, $\bar{y}(a_1b_0)$, etc. It is easy to see that each of the following differences provides an estimate of the weight of object A:

$$\bar{y}(a_1b_0) - \bar{y}(a_0b_0) \qquad (8.1a)$$

$$\bar{y}(a_1b_1) - \bar{y}(a_0b_1) \qquad (8.1b)$$

These differences represent the average increase in the readings on the balance, i.e., response, due to the weight of A in two different circumstances: (8.1a) when B is absent on the balance and (8.1b) when B is present.

In this experiment, there is no reason to suspect that the presence or absence of B on the balance would affect the average increase in the readings due to the weight of A. Hence it is meaningful to take the average of the two estimates (8.1a) and (8.1b) as an estimate of the weight of A. Thus we have, denoting the estimated weight of A by (A),

$$(A) = \tfrac{1}{2}[\bar{y}(a_1b_0) - \bar{y}(a_0b_0) + \bar{y}(a_1b_1) - \bar{y}(a_0b_1)] \qquad (8.2)$$

Let us now assume that the readings on the balance can be represented by the following model:

Reading = bias + weight of object on balance + combined effect due
          to unaccounted for factors including measurement errors   (8.3)

where the last component has expectation zero and variance σ^2. Since each term within the brackets in (8.2), being a mean of two independent readings, has a variance of $\sigma^2/2$, it follows at once that the estimate (A) given above has a variance of $(\tfrac{1}{2})^2[4(\sigma^2/2)] = \tfrac{1}{2}\sigma^2$. Applying the same logic used above in arriving at the estimate (A), we get

$$(B) = \tfrac{1}{2}[\bar{y}(a_0b_1) - \bar{y}(a_0b_0) + \bar{y}(a_1b_1) - \bar{y}(a_1b_0)] \qquad (8.4)$$

for an estimate of the weight of B. As with (A), the variance of (B) is also $\tfrac{1}{2}\sigma^2$.

Using symbols analogous to those used above, we can denote the estimates obtained for the weights of A and B by the one-at-a-time approach as

$$(A)^* = \bar{y}(a_1) - \bar{y}(a_0) \qquad (8.5)$$

and

$$(B)^* = \bar{y}(b_1) - \bar{y}(b_0) \qquad (8.6)$$

Remembering that each of the mean values on the right-hand side of (8.5) and (8.6) represents an average of four readings, and applying model (8.3), we find that $(A)^*$ has a variance of $2(\tfrac{1}{4})\sigma^2 = \tfrac{1}{2}\sigma^2$ and that the variance of $(B)^*$ is also $\tfrac{1}{2}\sigma^2$.

Thus the factorial approach requires only 8 observations, as opposed to 16 by the one-at-a-time approach, to estimate the weights of A and B with the same variance

$\frac{1}{2}\sigma^2$. This clearly demonstrates the greater efficiency of the factorial approach compared with the single-factor approach. This higher efficiency results because in the factorial approach each observation supplies information on the effect of every factor, whereas in the single-factor approach no observation supplies information on more than one factor. Note particularly that in the factorial approach the observations at the various levels of one factor constitute replications of observations on the other factor(s).

Another advantage of factorial experimentation is that the observations yielded by a factorial experiment reveal the relationship, if any, between the factors included in the experiment with respect to their impact on the response variable. We saw earlier that the factorial experiment provides separate estimates of the weight of A at each level of B, that is, in the presence as well as in the absence of B [see (8.1a) and (8.1b)]. These estimates help answer the question: Is the effect of factor A (in the present case, the increase in the reading caused by the weight of A) the same at both levels of B? The following comparison provides the answer:

$$[\bar{y}(a_1b_1) - \bar{y}(a_0b_1)] - [\bar{y}(a_1b_0) - \bar{y}(a_0b_0)] \qquad (8.7)$$

This expression measures the difference between the effect of A when B is at one level and the effect of A when B is at the other level. Similarly, the question whether the effect of B differs according as A is present or absent is answered by the comparison

$$[\bar{y}(a_1b_1) - \bar{y}(a_1b_0)] - [\bar{y}(a_0b_1) - \bar{y}(a_0b_0)] \qquad (8.8)$$

Note that (8.7) and (8.8) are identical. If the comparison (8.7) or (8.8) is zero, we say that there is no *interaction* between the factors A and B, or simply that there is no *AB interaction*, or that A and B are additive (cf. Chap. 6, pages 279 to 280 and also the discussion that follows on pages 365 to 367). In the weighing experiment this statement means the following: The estimate of the weight of A obtained when B rests on the balance is not different from the estimate of the weight of A obtained when B is not resting on the balance. Obviously the single-factor approach is incapable of supplying any information on the *AB interaction*.

In the terminology of factorial experiments, the quantities represented by (8.2) and (8.4) are referred to as the *main effects* of A and B, respectively, the adjective *main* being used as a reminder that the reference is, in each case, to an average of the effects taken over the levels of the other factor. Conventionally, the AB interaction is measured by the product of (8.7) or (8.8) with the same external multiplier used in obtaining the main effects. Using the notation (AB) for the AB interaction, we have thus

$$(AB) \stackrel{\vee}{=} \frac{1}{2}[\bar{y}(a_1b_1) - \bar{y}(a_0b_1) - \bar{y}(a_1b_0) + \bar{y}(a_0b_0)] \qquad (8.9)$$

The formulas given above for the main effects and interactions are summarized in Table 8.2. The systematic nature of the coefficients shown in Table 8.2 is particularly noteworthy. For the main effect (A) we find that each treatment total containing a_0 has the coefficient -1 while every treatment total containing a_1 has the coefficient $+1$. Similarly, for the main effect (B), every treatment total containing b_0 has the coefficient -1 while every treatment total containing b_1 has the coefficient $+1$. The coefficients pertaining to (AB) are simply the products of the corresponding coefficients of (A)

Table 8.2 CALCULATION OF MAIN EFFECTS AND INTERACTION IN A 2^2 FACTORIAL EXPERIMENT

Factorial effect	Coefficient for the treatment totals				Factorial total[†]	Estimate of factorial effect	SS due to factorial effects
	$y(a_0b_0)$	$y(a_1b_0)$	$y(a_0b_1)$	$y(a_1b_1)$			
Main effect (A)	-1	$+1$	-1	$+1$	L_A	$\dfrac{L_A}{2r}$	$\dfrac{L_A^2}{4r}$
Main effect (B)	-1	-1	$+1$	$+1$	L_B	$\dfrac{L_B}{2r}$	$\dfrac{L_B^2}{4r}$
Interaction (AB)	$+1$	-1	-1	$+1$	L_{AB}	$\dfrac{L_{AB}}{2r}$	$\dfrac{L_{AB}^2}{4r}$

† L_A, L_B, and L_{AB} stand for linear combinations of the treatment totals obtained by applying the coefficients shown in the rows for (A), (B), and (AB), respectively. Thus, for example, $L_A = -y(a_0b_0) + y(a_1b_0) - y(a_0b_1) + y(a_1b_1)$.

and (B). These features are easy to remember and can easily be generalized to 2^3 (three factors each at two *levels*), 2^4 (four factors each at two *levels*), etc., factorial designs. The term levels, it may be recalled, is used for the number of categories recognized for any given factor.

8.2 ANALYSIS OF THE 2^n FACTORIAL EXPERIMENT

The 2^2 Factorial

In order to illustrate the analysis of the 2^2 factorial experiment, let us consider the following adaptation of the experiment on social change (Hill, Stycos, and Back, 1959) referred to earlier. (We shall be using hypothetical figures for our numerical illustration.)

Two factors, each having two levels, were selected for study. The factors and their respective levels were

A Method of delivering message
  a_0 Through small group meetings
  a_1 Through large group meetings
B Content of message
  b_0 Emphasis on values
  b_1 Emphasis on interspouse communication

Twelve villages, more or less similar in demographic and socioeconomic characteristics, were selected to be used as experimental units. The other main consideration in their selection was simply administrative convenience. The treatment combinations were assigned to the experimental units at random in such a manner that each combination was assigned to three units. In each village, all eligible couples, e.g., wives below the end of reproductive period and their husbands, were exposed to the treatment combination assigned to the village. The response variable consisted of change in contraceptive practices of the population. Before starting the experiment and after completing administration of the treatments, data were collected on contraceptive practices of the couples in the villages. From the before and after data thus collected, aggregate measures of changes in contraceptive practices were obtained for each village. Let us suppose that the data shown in Table 8.3 pertain to the above experiment.

We shall see later on (cf. Sec. 9.7) that there are certain problems in dealing with change data obtained from before and after observations. At this moment, however, we shall proceed to analyze the data in Table 8.3 ignoring problems, if any, peculiar to change data obtained in the manner described above.

The first point to note here is that the data in Table 8.3 correspond to those from a completely randomized design with equal numbers of observation in each class. Naturally, therefore, the first step in the analysis is to decompose total SS into between (treatment) SS and within (residual) SS in the usual manner. We find

$$\text{Total SS} = 3.4^2 + 3.6^2 + \cdots + 2.2^2 + 2.9^2 - \text{correction for mean}$$
$$= 137.72 - 104.43 = 33.29$$
$$\text{Treatment SS} = \tfrac{1}{3}(11.0^2 + 1.6^2 + 15.1^2 + 7.7^2) - \text{correction for mean}$$
$$= 136.95 - 104.43 = 32.52$$
$$\text{Residual SS} = \text{total SS} - \text{treatment SS} = 33.29 - 32.52 = .77$$

The next step is to decompose treatment SS into orthogonal components attributable to the factorial effects. This is done by obtaining the factorial-effect totals according

Table 8.3 INCREASES IN BIRTHS AVERTED PER 1,000 POPULATION IN A 2^2 FACTORIAL EXPERIMENT

	Treatment combination†					
	(a_0b_0)	(a_1b_0)	(a_0b_1)	(a_1b_1)		
	3.4	0.4	5.0	2.6		
	3.6	0.9	4.8	2.2		
	4.0	0.3	5.3	2.9		
Total	11.0	1.6	15.1	7.7		
					Factorial total	Estimate of factorial effect
Main effect (A)	-1	$+1$	-1	$+1$	-16.8	-2.80
Main effect (B)	-1	-1	$+1$	$+1$	10.2	1.70
Interaction (AB)	$+1$	-1	-1	$+1$	2.0	.33

† In the treatment combinations, a_0 stands for delivering message through small group meetings and a_1 for delivering message through large group meetings, b_0 for emphasizing values and b_1 for emphasizing interspouse communication, in the message delivered.

to the formulas given in Table 8.2. The calculations of the factorial-effect totals are shown in Table 8.3. Since each treatment total is based on three observations, the SS due to the factorial effects are obtained by dividing the squares of the factorial totals by $2^2 \times 3$. We thus get

$$\text{SS due to main effect } (A) = \frac{(-16.8)^2}{12} = 23.52$$

$$\text{SS due to main effect } (B) = \frac{10.2^2}{12} = 8.67$$

$$\text{SS due to interaction } (AB) = \frac{2.0^2}{12} = .33$$

It can now be verified that the SS attributable to the factorial effects add up to the treatment SS obtained above: $23.52 + 8.67 + .33 = 32.52$. The complete analysis can now be set up as in Table 8.4.

The F ratio for interaction is 3.44. The probability of getting an F ratio higher than 3.44 under the null hypothesis of no interaction is about .10. We shall infer that interaction is absent here. (Sometimes researchers use higher significance levels when deciding whether interaction is to be ignored or not. This is a matter to be handled according to the judgment of the researcher.) If we assume that interaction is negligible, we can concentrate our attention on the main effects. The F ratios for both main effects are found to be highly significant. The interpretation of this finding will be easier if we examine the factorial-effect means. Since each treatment total is based on three observations, the factorial-effect means can be obtained by dividing the corresponding factorial-effect totals by $2 \times 3(= 6)$. The results are shown in Table 8.3. We find that the main effect of factor A (method of delivering message) is estimated as -2.80. In accordance with formula (8.2), this means that the effect of level a_0 of A is greater than that of a_1. Delivering messages through small group meetings is thus found to be more effective than using large group meetings for that purpose. Now looking at the main effect of B, we find that, in accordance with formula (8.4), level b_1 is more effective than b_0; that is, emphasizing interspouse communication produces larger impact than emphasizing values. It is interesting in this connection to note that the statistical significance of the main effects and interaction can be tested using Student's t. The standard error of each factorial-effect mean is $\sqrt{\text{residual MSS}/r} = \sqrt{.096/3} = .18$.

Table 8.4   ANALYSIS OF VARIANCE

Source of variation	SS	df	MSS	F ratio
Main effect (A)	23.52	1	23.52	245.00
Main effect (B)	8.67	1	8.67	90.31
Interaction (AB)	.33	1	.33	3.44
Error	.77	8	.096	
Total	33.29	11		

Perhaps it should be emphasized here that statistical significance and practical significance are not the same. For example, whether a difference of 2.8 births per 1,000 population is large enough to outweigh the additional cost involved in administering the program through small group meetings is something that cannot be judged in terms of the analysis given above.

REMARK 1   From the formulas (8.2), (8.4), and (8.9) giving (A), (B), and (AB) in terms of the individual means of the treatment combinations, it is possible to express the individual means of the treatment combinations in terms of (A), (B), (AB), and a fourth quantity M, which represents the general mean of the whole set of observations in the experiment. This can be seen by writing the coefficients of $\bar{y}(a_0 b_0)$, $\bar{y}(a_1 b_0)$, $\bar{y}(a_0 b_1)$, and $\bar{y}(a_1 b_1)$ in the linear combinations giving $2M$, (A), (B), and (AB), as shown below:

	$\bar{y}(a_0 b_0)$	$\bar{y}(a_1 b_0)$	$\bar{y}(a_0 b_1)$	$\bar{y}(a_1 b_1)$
$2M$	$\frac{1}{2}$	$\frac{1}{2}$	$\frac{1}{2}$	$\frac{1}{2}$
(A)	$-\frac{1}{2}$	$\frac{1}{2}$	$-\frac{1}{2}$	$\frac{1}{2}$
(B)	$-\frac{1}{2}$	$-\frac{1}{2}$	$\frac{1}{2}$	$\frac{1}{2}$
(AB)	$\frac{1}{2}$	$-\frac{1}{2}$	$-\frac{1}{2}$	$\frac{1}{2}$

It is easily verified that

$$2\bar{y}(a_1 b_1) = 2M + (A) + (B) + (AB)$$
$$2\bar{y}(a_0 b_1) = 2M - (A) + (B) - (AB)$$
$$2\bar{y}(a_1 b_0) = 2M + (A) - (B) - (AB)$$
$$2\bar{y}(a_0 b_0) = 2M - (A) - (B) + (AB)$$

from which

$$
\begin{aligned}
\bar{y}(a_1 b_1) &= M + \tfrac{1}{2}[(A) + (B) + (AB)] \\
\bar{y}(a_0 b_1) &= M + \tfrac{1}{2}[-(A) + (B) - (AB)] \\
\bar{y}(a_1 b_0) &= M + \tfrac{1}{2}[(A) - (B) - (AB)] \\
\bar{y}(a_0 b_0) &= M + \tfrac{1}{2}[-(A) - (B) + (AB)]
\end{aligned}
\qquad (8.10)
$$

REMARK 2   From (8.10) it follows that

$$[\bar{y}(a_1 b_1) - M]^2 + \cdots + [\bar{y}(a_0 b_0) - M]^2 = (A)^2 + (B)^2 + (AB)^2 \qquad (8.11)$$

Multiplying both sides by r, we get from (8.11) the decomposition of treatment SS:

$$\text{Treatment SS} = \text{SS due to } (A) + \text{SS due to } (B) + \text{SS due to } (AB)$$

REMARK 3   In the social-change experiment mentioned at the beginning of this section, the format of a completely randomized design was used. The experimental units in that example were randomly assigned to different treatment combinations, subject to the restriction that each treatment combination was to be assigned to the same number of units as each of the others. If the experimental units are heterogeneous, we may employ one-way, two-way, or, in general, multiway, blocking with a

Table 8.5   OUTLINE OF ANALYSIS:
            2^2 FACTORIAL IN RANDOMIZED
            BLOCKS

Source of variation	df
Main effect (A)	1
Main effect (B)	1
Interaction (AB)	1
	—
Treatments	3
Blocks	$r - 1$
Residual (error)	$(4r - 1) - (r - 1) - 3$
Total	$4r - 1$

view to improving the efficiency of the design.   To employ one-way blocking, a randomized-blocks arrangement can be used, applying in each block each of the treatment combinations.   The analysis would then proceed as usual except that the treatment SS should be decomposed into the SS attributable to each of the factorial effects.   An outline of the analysis of a 2^2 factorial experiment in a randomized-blocks design with r blocks will be as shown in Table 8.5.

If a Latin-square design is used for the purpose of double blocking, the treatments in the square should represent the different treatment combinations.   This means that for a 2^2 factorial experiment, the appropriate Latin square is the 4×4 square.   In this case, the analysis will follow the outline shown in Table 8.6.

In the above outlines, we have made the assumption that the factorial effects remain the same from one block to another in the randomized-blocks design and from one row (column) to another in the Latin-square design.   This amounts to the additivity assumption discussed in the preceding chapter.   It may be useful to review those discussions at this point.

Table 8.6   OUTLINE OF ANALYSIS OF
            VARIANCE: 2^2 FACTORIAL IN A
            LATIN SQUARE

Source of variation	df
Main effect (A)	1
Main effect (B)	1
Interaction (AB)	1
	—
Treatments	3
Rows	3
Columns	3
Residual (error)	6
Total	15

The 2^2 Factorial with Nonnegligible Interaction

Irrespective of whether a completely randomized design, a randomized-blocks design, or a Latin square is used in the 2^2 factorial experiment, if the AB interaction is found to be statistically significant or judged to be nonnegligible on some other basis, the proper procedure is to direct attention to the four treatment means (totals). A close examination of the individual treatment means may shed light on the nature of the interaction and perhaps help find an explanation for its presence. If the interaction is nonnegligible, there is usually little practical interest in the estimated main effects even if they are found to be statistically significant.

The no-interaction situation can be described as one involving factorial effects that are *additive*. To understand what this means, consider the situation in which a_0 stands for the absence and a_1 for the presence of factor A, and similarly b_0 stands for the absence and b_1 for the presence of factor B. Suppose that

$$E[\bar{y}(a_0 b_0)] = \mu$$
$$E[\bar{y}(a_1 b_0)] = \mu + \alpha$$
$$E[\bar{y}(a_0 b_1)] = \mu + \beta$$

The second equation means that when A is present without B, the result will be to augment the expected value of the treatment mean by an amount α. Similarly, the third equation means that if B is present without A, the result will be to augment the expected value of the treatment mean by an amount equal to β. If both A and B are present, and if their effects are additive, the expected value of the treatment mean will be given by

$$E[\bar{y}(a_1 b_1)] = \mu + \alpha + \beta$$

If this (additive) model holds, it can be easily verified that the expected value of the AB interaction is zero:

$$E(AB) = \tfrac{1}{2}E[\bar{y}(a_0 b_0) - \bar{y}(a_1 b_0) - \bar{y}(a_0 b_1) + \bar{y}(a_1 b_1)]$$
$$= \tfrac{1}{2}[\mu - (\mu + \alpha) - (\mu + \beta) + (\mu + \alpha + \beta)] = 0$$

Note that $E(AB) = 0$ does not mean that in any particular experiment the observed interaction will be zero. It means only that if the experiment is repeated under constant conditions an indefinite number of times and the interactions obtained in those experiments averaged, the result will be zero. Nearly all experiments of the type under discussion here will show some apparent interaction. The test of significance of interaction discussed earlier (see Table 8.4) was aimed at answering whether an observed interaction is consistent with the claim that $E(AB) = 0$.

Given the results of any particular experiment, it is usually instructive to represent the treatment means graphically to see whether the interaction is conspicuous. The treatment means calculated from Table 8.3 are shown in Fig. 8.1a.

The bottom line shows the drop in the treatment mean that is attributable to the change of level of A from a_0 to a_1 when B is at its level b_0, and the upper line shows the corresponding drop in the treatment mean when B is at its level b_1. In this particular case, the two lines are more or less parallel, indicating that the interaction is negligible. If the lines are far from being parallel, that should be taken as a warning that

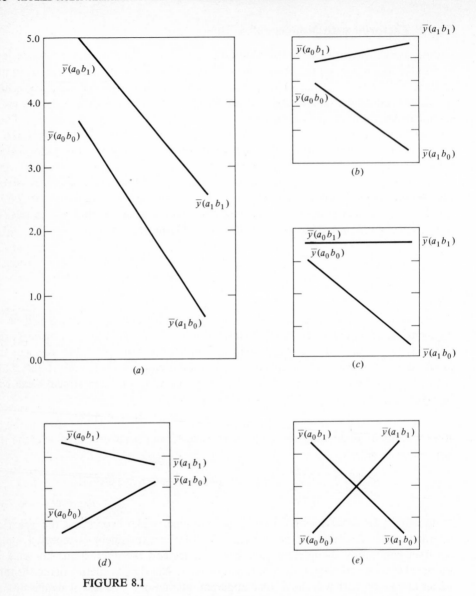

FIGURE 8.1

the interaction is not negligible. Figure 8.1*b* to *e* illustrates several such situations. See also Fig. 4.3*a* and *b*.

The above account of interaction is tied to the levels of factors actually used in the experiment. With quantitative factors, e.g., different strengths of stimuli or different lengths of time a treatment is applied, interest may center on generalizing the results about interaction obtained from an experiment to cover a continuous range of levels of the factors involved. To facilitate this, the concept of *response surface* or *response* function has been evolved. With two quantitative factors, the simplest response function is

$$y = \alpha_0 + \alpha_1 x_1 + \alpha_2 x_2 + \varepsilon \qquad (8.12)$$

where α_0, α_1, and α_2 are constants to be determined so as to fit the model to the data at hand, x_1 and x_2 represent the amounts of the two factors involved in a treatment combination, and y is the corresponding response. The last term on the right-hand side of (8.12) represents the combined effect of uncontrolled factors. If the above response surface is judged to represent a given set of data, one may proceed to determine the α's by the usual method of least squares which minimizes the SS of the error terms ε's. But a significant interaction is a warning that (8.12) is not an adequate representation of the observations at hand. Naturally, as the complexity of the response surface increases, more coefficients must be estimated, and the number of experimental points (treatment combinations actually used in the experiment) must be increased. Several experimental designs have been developed that minimize the amount of work involved in estimating the coefficients in given response surfaces. Davies (1956), Myers (1971), and others give excellent treatment of response-surface experimentation.

Unless interaction is of theoretical interest, it may be worthwhile to investigate whether additivity can be attained on a transformed scale. The simplest example is that of multiplicative effects. A logarithmic transformation of the data before analysis would then make effects additive.

The 2^3 Factorial

The results of the preceding subsections for the 2^2 factorial can be easily extended to the 2^3 factorial, which involves three factors each at two levels. Let A, B, and C stand for the three factors, a_0 and a_1 for the two levels of A, b_0 and b_1 for those of B, and c_0 and c_1 for those of C. The different treatment combinations and the conventional notations for them are shown below.

Treatment combination	Conventional symbol
$a_0 b_0 c_0$	(1)
$a_1 b_0 c_0$	a
$a_0 b_1 c_0$	b
$a_1 b_1 c_0$	ab
$a_0 b_0 c_1$	c
$a_1 b_0 c_1$	ac
$a_0 b_1 c_1$	bc
$a_1 b_1 c_1$	abc

As can be easily seen, in the conventional notation, the presence of any lower-case letter in the treatment combination indicates that the factor represented by the corresponding capital letter occurs at the level denoted by subscript 1, and the absence of a, b, or c in the treatment combination indicates that the factor represented by the corresponding capital letter occurs at the level denoted by subscript 0. The symbol (1) is used to denote the situation in which all the three treatments occur at levels denoted by subscript 0. Obviously, there is no loss of generality if we stipulate that for convenience of reference the subscript 0 shall be taken to represent the absence of the corresponding factor and the subscript 1, to represent its presence.

The 2^3 factorial experiment can be executed as a completely randomized design, as a randomized-blocks design, or as a Latin-square design. We need discuss here only the decomposition of the treatment SS. Denoting the treatment means by the symbols $\bar{y}(1)$, $\bar{y}(a)$, $\bar{y}(b)$, $\bar{y}(ab)$, etc., we note that each of the following differences gives an estimate of the effect of factor A:

$$\bar{y}(a) - \bar{y}(1) \qquad (8.13a)$$

$$\bar{y}(ab) - \bar{y}(b) \qquad (8.13b)$$

$$\bar{y}(ac) - \bar{y}(c) \qquad (8.13c)$$

$$\bar{y}(abc) - \bar{y}(bc) \qquad (8.13d)$$

The average of the above four quantities gives (A), the main effect of factor A. Similarly, the average of the differences $\bar{y}(b) - \bar{y}(1)$, $\bar{y}(ab) - \bar{y}(a)$, $\bar{y}(bc) - \bar{y}(c)$, and $\bar{y}(abc) - \bar{y}(ac)$, gives (B), the main effect of factor B, and the average of $\bar{y}(c) - \bar{y}(1)$, $\bar{y}(ac) - \bar{y}(a)$, $\bar{y}(bc) - \bar{y}(b)$, and $\bar{y}(abc) - \bar{y}(ab)$ gives (C), the main effect of C.

We now notice that the comparisons

$$\tfrac{1}{2}\{[\bar{y}(ab) - \bar{y}(b)] - [\bar{y}(a) - \bar{y}(1)]\} \qquad (8.14a)$$

and
$$\tfrac{1}{2}\{[\bar{y}(abc) - \bar{y}(bc)] - [\bar{y}(ac) - \bar{y}(c)]\} \qquad (8.14b)$$

both provide estimates of the AB interaction, the former pertaining to the situation when the third factor C is at one level and the latter to the situation when C is at the other level. The average of these estimates can be taken as a measure of the AB interaction. Similarly, the average of

$$\tfrac{1}{2}\{[\bar{y}(ac) - \bar{y}(a)] - [\bar{y}(c) - \bar{y}(1)]\} \qquad (8.15a)$$

and
$$\tfrac{1}{2}\{[\bar{y}(abc) - \bar{y}(ab)] - [\bar{y}(bc) - \bar{y}(b)]\} \qquad (8.15b)$$

gives the AC interaction, and the average of

$$\tfrac{1}{2}\{[\bar{y}(bc) - \bar{y}(b)] - [\bar{y}(c) - \bar{y}(1)]\} \qquad (8.16a)$$

and
$$\tfrac{1}{2}\{[\bar{y}(abc) - \bar{y}(ab)] - [\bar{y}(ac) - \bar{y}(a)]\} \qquad (8.16b)$$

the BC interaction.

Thus in the 2^3 factorial experiment there are three two-factor interactions, (AB), (AC), and (BC). To measure (AB), we imagine that the 2^2 factorial involving A and B is repeated at both the levels of C and the AB interactions estimated separately at those repetitions are averaged. Similarly, we get (AC) and (BC). Since each of the two-factor interactions in the 2^3 factorial represents an average of two quantities, there is some interest in knowing whether the constituent quantities differ between themselves. This can be seen by taking the difference

$$\text{Expression (8.14}b) - \text{expression (8.14}a) \qquad (8.17)$$

which incidentally is identical to

$$\text{Expression } (8.15b) - \text{expression } (8.15a)$$

and
$$\text{Expression } (8.16b) - \text{expression } (8.16a)$$

The comparison (8.17) multiplied by $\frac{1}{2}$ is termed the three-factor interaction involving A, B, and C, or simply, the ABC interaction, denoted by (ABC). The formulas obtained above for the main effects and interactions are summarized in Table 8.7. As with the 2^2 factorial (see Table 8.2), it is instructive to take note of the systematic pattern of the coefficients in Table 8.7. For (A), we have a minus sign whenever the treatment combination does not contain a and a plus sign whenever the treatment combination contains a. Similarly, for (B), all treatment combinations containing b have a plus sign and all the others, a minus sign, and for (C), all combinations containing c have a plus sign, and all the others a minus sign. As for the two-factor interactions, the signs under each treatment combination can be easily obtained from those for the main effects: the signs for (AB) can be obtained by taking the products of the corresponding signs for (A) and (B), those for (AC) can be similarly obtained from those of (A) and (C), and those for (BC), from those of (B) and (C). The signs for (ABC) can be obtained in a number of ways: by taking the products of the corresponding signs for (A), (B), and (C), or for (AB) and (C), or for (A) and (BC), or for (AC) and (B). The above procedure of writing down the coefficients for the different factorial effects can be easily extended to 2^4, 2^5, etc., factorials. Once the coefficients are written down, as in Table 8.7, we can write the treatment totals at the top of the table and obtain the factorial-effect totals by combining the treatment totals according to the signs shown for each factorial effect. Let the factorial totals thus obtained be L_A, L_B, L_C, L_{AB}, etc. Then if each treatment total is based on r observations, the SS due to the different factorial effects are given by $(1/2^3)L_A{}^2/r$, $(1/2^3)L_B{}^2/r$, etc., each with 1 degree of freedom. As in the case of the 2^2 factorial, it is easy to verify that

$$\text{Treatment SS} = \frac{L_A{}^2 + L_B{}^2 + \cdots + L_{ABC}^2}{2^3 r}$$

Table 8.7   COEFFICIENTS (EXCEPTING THE EXTERNAL MULTIPLIER $\frac{1}{4}$) FOR FACTORIAL EFFECTS IN A 2^3 FACTORIAL EXPERIMENT

Factorial effect	Treatment combination							
	(1)	a	b	ab	c	ac	bc	abc
(A)	−	+	−	+	−	+	−	+
(B)	−	−	+	+	−	−	+	+
(C)	−	−	−	−	+	+	+	+
(AB)	+	−	−	+	+	−	−	+
(AC)	+	−	+	−	−	+	−	+
(BC)	+	+	−	−	−	−	+	+
(ABC)	−	+	+	−	+	−	−	+

Table 8.8 OUTLINE OF ANALYSIS OF VARIANCE: 2^3 FACTORIAL IN A COMPLETELY RANDOMIZED DESIGN

Source of variation	df
Main effects:	
(A)	1
(B)	1
(C)	1
Two-factor interactions:	
(AB)	1
(AC)	1
(BC)	1
Three-factor interaction (ABC)	1
Residual	$(8r-1)-7$
Total	$8r-1$

The 7 degrees of freedom of the treatment SS are thus partitioned into

Source of variation	df
(A)	1
(B)	1
(C)	1
(AB)	1
(AC)	1
(BC)	1
(ABC)	1

If a completely randomized design has been used for the experiment with r observations under each treatment combination, the analysis of variance will be of the form shown in Table 8.8. If a randomized-blocks design has been used for the experiment with r blocks, the outline of the analysis of variance will be as shown in Table 8.9. Of course in the detailed analysis-of-variance table, the SS for each of the main effects and each of the two-factor interactions should be shown separately. With a 8×8 Latin-square design, the outline of the analysis of variance will be as shown in Table 8.10.

Table 8.9 OUTLINE OF ANALYSIS OF VARIANCE: 2^3 FACTORIAL IN RANDOMIZED BLOCKS

Source of variation	df
Main effects each with 1 df	3
Two-factor interactions each with 1 df	3
Three-factor interaction	1
Blocks	$r-1$
Residual (error)	$(8r-1)-(r-1)-7$
Total	$8r-1$

Table 8.10   OUTLINE OF ANALYSIS OF
VARIANCE: 2^3 FACTORIAL IN A
LATIN SQUARE

Source of variation	df
Main effects	3
Two-factor interactions	3
Three-factor interaction	1
Rows	7
Columns	7
Residual (error)	$(8^2 - 1) - 7 - 7 - 7$
Total	$8^2 - 1$

Example   To illustrate this analysis we shall use the data in Table 8.11, which have
been taken from a nonexperimental study (Keyfitz, 1953) examining the relationship
of family size to several "causal factors," such as distance of residence from city,
income level of the county of residence, ethnicity, etc. Altogether six causal factors

Table 8.11   FAMILY SIZE DATA PUT INTO A 2^3 FACTORIAL FORMAT†

Age at marriage years	15–19				20–24				
Income level of county of residence	Low		High		Low		High		
Distance from city	Far	Near	Far	Near	Far	Near	Far	Near	
Block	(1)	a	b	ab	c	ac	bc	abc	Block total
1	9.4	7.4	10.9	8.3	10.3	8.3	10.6	7.1	72.3
2	12.9	9.7	12.8	10.5	8.9	9.4	9.4	7.6	81.2
3	10.7	12.9	12.9	8.7	9.8	6.7	9.8	10.3	81.8
4	10.9	11.3	14.3	12.2	9.8	7.1	11.2	8.8	85.6
5	10.1	10.0	12.1	10.8	10.4	7.6	9.0	10.9	80.9
6	8.3	9.0	10.6	11.0	8.4	8.6	9.9	8.6	74.4
7	14.5	11.0	12.5	13.2	9.8	8.6	11.3	9.9	90.8
8	12.8	9.9	12.0	11.0	9.6	8.6	9.0	8.4	81.3
Treatment total	89.6	81.2	98.1	85.7	77.0	64.9	80.2	71.6	648.3

Factorial effect	Multipliers for factorial effect total								Factorial effect total
(A)	−	+	−	+	−	+	−	+	−41.5
(B)	−	−	+	+	−	−	+	+	22.9
(C)	−	−	−	−	+	+	+	+	−60.9
(AB)	+	−	−	+	+	−	−	+	−.5
(AC)	+	−	+	−	−	+	−	+	.1
(BC)	+	+	−	−	−	−	+	+	−3.1
(ABC)	−	+	+	−	+	−	−	+	7.5

† Adapted from Keyfitz (1953).

were taken for examination, and a 2^6 factorial arrangement was employed for the investigation. The six factors taken for examination and their levels were:

Current age (45 to 54 years and 55 to 74 years)
Age at marriage (15 to 19 years and 20 to 24 years)
Years of schooling (0 to 6 years and 7 or more years)
Income level of county of residence (low and high)
Ethnicity (French and mixed)
Distance from city (far and near)

The average family sizes observed by Keyfitz in his 2^6 classification are taken as the observations for the present analysis (involving one observation per cell). The variables current age, years of schooling, and ethnicity have been used as block variables to provide eight blocks in a randomized-blocks design. (Needless to say, we are treating the data as though they had come from an experiment.) The treatment combinations chosen for examination are

	Distance from city	Income level of county	Age at marriage, years
(1)	Far	Low	15–19
a	Near	Low	15–19
b	Far	High	15–19
ab	Near	High	15–19
c	Far	Low	20–24
ac	Near	Low	20–24
bc	Far	High	20–24
abc	Near	High	20–24

From the treatment totals shown in Table 8.11 the factorial-effect totals can be obtained by straightforward application of the appropriate multipliers ($+1$ or -1) to the treatment totals. The results obtained in this manner are shown at the right-hand side of the bottom half of Table 8.11.

Squaring the factorial-effect total and dividing the square by $2^3 r$ ($= 2^3 \times 8$) gives the corresponding SS.

We can now verify that the treatment SS obtained in the usual manner is equal to the sum of the factorial-effect SS:

$$\text{Treatment SS} = \tfrac{1}{8}(89.6^2 + 81.2^2 + \cdots + 71.6^2) - \frac{648.3^2}{64} = 94.09$$

which is equal to

$$\tfrac{1}{64}[(-41.5)^2 + 22.9^2 + (-.5)^2 + (-60.9)^2 + .1^2 + (-3.1)^2 + 7.5^2]$$

as can be verified.

The SS due to blocks is obtained as

$$\tfrac{1}{8}(72.3^2 + \cdots + 81.3^2) - \frac{648.3^2}{64} = 29.65$$

Table 8.12   ANALYSIS-OF-VARIANCE TABLE OF DATA IN TABLE 8.11

Source of variation	SS	df	MSS	F ratio
Main effect:				
(A)	26.9102	1	26.9102	17.88
(B)	8.1939	1	8.1939	5.44
(C)	57.9502	1	57.9502	38.50
Two-factor interaction:				
(AB)	.0039	1	.0039	
(AC)	.0002	1	.0002	
(BC)	.1502	1	.1502	
Three-factor interaction:				
(ABC)	.8789	1	.8789	
Blocks	29.6523	7	4.2360	2.81
Error	73.7538	49	1.5052	
Total	197.4936	63		

The analysis-of-variance table can now be set up as shown in Table 8.12. The only significant effects are the main effects (A), (B), and (C). The absence of any significant interaction gives some assurance that the effect of any one factor remains the same at different levels of the other factors. We need therefore report only the main effects. The main effects are shown in Table 8.13.

8.3   FACTORIAL EXPERIMENTS WITH MORE THAN TWO LEVELS FOR SOME FACTORS

Leaving the special case of two levels per factor, let us consider the general arrangement in which some of the factors have more than two levels.

A factorial experiment with two factors when one has three levels and the other has two levels can be illustrated by an adaptation of the Hill-Stycos-Back experiment

Table 8.13   MAIN EFFECTS OF FACTORS AFFECTING FAMILY SIZE

Factor	Direction of effect†	Estimate
Distance from city	Near − far	$\dfrac{L_A}{32} = -1.30$
Income level of county	High − low	$\dfrac{L_B}{32} = .72$
Age at marriage	(20 to 24 years) − (15 to 19 years)	$\dfrac{L_C}{32} = -1.90$

† The effect of distance from city has been estimated to display the excess of the effect of being near over that of being far, and so on.

Table 8.14 TREATMENT COMBINATIONS IN A
3 × 2 FACTORIAL

A Method of delivering message	*B* Content of message	
	b_1 Emphasis on values	b_2 Emphasis on communication
a_1 Small meeting	(a_1b_1)	(a_1b_2)
a_2 Large meeting	(a_2b_1)	(a_2b_2)
a_3 Pamphlets	(a_3b_1)	(a_3b_2)

on social change, to which reference has already been made. We may consider factor A (method of delivering message) as having three levels (a_1, through small meeting; a_2, through large meeting; and a_3, through pamphlets) and factor B (content of message) as having two levels (b_1, emphasis on values; and b_2, emphasis on inter-spouse communication). There will be 3 × 2 (= 6) treatment combinations, and we may call the experiment a 3 × 2 factorial. The treatment combinations can be enumerated as shown in Table 8.14.

We can organize the 3 × 2 factorial experiment on any standard layout. Let us suppose that the experiment has been organized on a randomized-blocks layout with r blocks (replications). The first step in the analysis of the results would be to subdivide the total SS as shown in Table 8.15. The next step in the analysis is to subdivide the treatment SS obtained above into its components, the SS attributable to the main effects and interaction. To see how this is to be done, let us put the treatment totals into a two-way format, as in Table 8.16, using the notation $y(a_ib_j)$ for the treatment total corresponding to the combination (a_ib_j), $y(a_i)$ for the total of all observations involving the a_i level of A, and $y(b_j)$ for the total of all observations involving the b_j level of B. Our knowledge of the analysis of variance of two-way tables immediately leads to the partitioning shown in Table 8.17. To illustrate these computations, let us consider the data in Table 8.18. The details of the calculations involved in the partitioning of total SS and the decomposition of treatment SS are shown in Table 8.18 itself. The analysis of variance is set up in Table 8.19. Among the factorial effects, only main effect (A) and interaction (AB) are significant. As the interaction is significant, the next logical step is to investigate the effect of each factor separately at each

Table 8.15 OUTLINE OF ANALYSIS OF VARIANCE: 3 × 2
FACTORIAL IN RANDOMIZED BLOCKS

Source of variation	df
Treatment combinations	3 × 2 − 1
Blocks (replications)	$r − 1$
Residual (error)	$(r × 3 × 2 − 1) − (r − 1) − (3 × 2 − 1)$
Total	$r × 3 × 2 − 1$

Table 8.16 TREATMENT TOTALS IN 3 × 2 FACTORIAL

A Method of delivering message	B Content of message		Total
	b_1 Value	b_2 Communication	
a_1 Small group	$y(a_1b_1)$	$y(a_1b_2)$	$y(a_1)$
a_2 Large group	$y(a_2b_1)$	$y(a_2b_2)$	$y(a_2)$
a_3 Pamphlets	$y(a_3b_1)$	$y(a_3b_2)$	$y(a_3)$
Total	$y(b_1)$	$y(b_2)$	Grand total $= y(b_1) + y(b_2)$

level of the other factor. We may also want to examine closely the interaction SS with a view to finding how it can be accounted for.

Let us consider the meaning of the interaction SS first. As you may recall, in the 2^2 and 2^3 factorial experiments each interaction has only 1 degree of freedom, and it was easy for us to associate each degree of freedom with a specific comparison between the treatment means (totals). In the present case, however, the SS due to the AB interaction has 2 degrees of freedom. In order to understand the meaning of the observed interaction it would be helpful if we partitioned the SS with 2 degrees of freedom into two parts, each with 1 degree of freedom, in such a way that each part can be associated with a specific comparison between the treatment means (totals). Now, a given SS with 2 or more degrees of freedom can be partitioned into independent parts each with 1 degree of freedom in more than one way. Which mode of partitioning gives meaningful results depends upon the objectives the investigators had in mind when planning the experiment. Let us suppose, for purposes of illustration, that the investigators argued as follows when planning the above experiment.

Any effort at educating a group of subjects on certain topics or at persuading them to change a behavioral pattern is likely to be more effective if it is undertaken in a face-to-face situation. On this ground, it is to be expected that the group-meeting approach is likely to be more effective than the pamphlet approach. But even when

Table 8.17 PARTITIONING OF TREATMENT SUM OF SQUARES IN 3 × 2 FACTORIAL

Source of variation	SS	df
Main effect (A)	$\dfrac{1}{2r} \sum [y(a_i)]^2 - \dfrac{(\text{grand total})^2}{6r}$	2
Main effect (B)	$\dfrac{1}{3r} \sum [y(b_j)]^2 - \dfrac{(\text{grand total})^2}{6r}$	1
Interaction (AB)	Treatment SS $-$ SS due to (A) $-$ SS due to (B)	2
Treatments	$\dfrac{1}{r} \sum \sum [y(a_ib_j)]^2 - \dfrac{(\text{grand total})^2}{6r}$	5

there is face-to-face communication, the effectiveness of the effort expended will be greater if the group is small because if there are only a few people in the group, the likelihood is greater that the communicative interaction between the educator and the participants and/or between the participants themselves will be more effective. On this ground, it is to be expected that the small-group approach is likely to be more effective than the large-group approach.

Table 8.18   RESULTS (FICTITIOUS) OF A 3 × 2 FACTORIAL EXPERIMENT ON SOCIAL CHANGE

| Block | b_1 Emphasis on values | | | b_2 Emphasis on interspouse communication | | | Total |
	a_1 Small meeting	a_2 Large meeting	a_3 Pamphlet	a_1 Small meeting	a_2 Large meeting	a_3 Pamphlet	
1	7	6	6	7	5	7	38
2	7	6	4	6	5	5	33
3	6	5	3	4	3	5	26
4	4	2	1	3	2	2	14
Total	24	19	14	20	15	19	111
	$y(a_1b_1)$	$y(a_2b_1)$	$y(a_3b_1)$	$y(a_1b_2)$	$y(a_2b_2)$	$y(a_3b_2)$	

Partition of Total SS

$$\text{Correction for mean} = \frac{111^2}{24} = 513.375$$

$$\text{Total SS} = 7^2 + 7^2 + \cdots + 5^2 + 2^2 - 513.375 = 75.625$$

$$\text{Block SS} = \tfrac{1}{6}(38^2 + 33^2 + 26^2 + 14^2) - 513.375 = 54.125$$

$$\text{Treatment SS} = \tfrac{1}{4}(24^2 + 19^2 + \cdots + 15^2 + 19^2) - 513.375 = 16.375$$

$$\text{Residual SS} = \text{total SS} - \text{block SS} - \text{treatment SS}$$
$$= 75.625 - 54.125 - 16.375 = 5.125$$

Partition of Treatment SS

| a level of A | b level of B | | Total |
	b_1	b_2	
a_1	24	20	$y(a_1) = 44$
a_2	19	15	$y(a_2) = 34$
a_3	14	19	$y(a_3) = 33$
Total	$y(b_1) = 57$	$y(b_2) = 54$	Grand total = 111

$$\text{Main effect } (A) \text{ SS} = \tfrac{1}{8}(44^2 + 34^2 + 33^2) - 513.375 = 9.250$$

$$\text{Main effect } (B) \text{ SS} = \tfrac{1}{12}(57^2 + 54^2) - 513.375 = .375$$

$$\text{Interaction SS} = \text{treatment SS} - \text{main effect } (A) \text{ SS} - \text{main effect } (B) \text{ SS}$$
$$= 16.375 - 9.250 - .375 = 6.750$$

If the investigators had argued as above, it means that the following comparisons were of interest:

Effect of small-group approach − effect of large-group approach   (8.18)

Effect of small-group approach + effect of large-group approach
− 2 (effect of pamphlet approach)   (8.19)

Obviously any comparison like (8.18) and (8.19) can be estimated separately at each level of B. Thus

$$\bar{y}(a_1 b_1) - \bar{y}(a_2 b_1) \qquad (8.20a)$$

gives the estimate of (8.18) at the b_1 level of B, and

$$\bar{y}(a_1 b_2) - \bar{y}(a_2 b_2) \qquad (8.20b)$$

gives the corresponding estimate at the b_2 level of B. Similarly

$$\bar{y}(a_1 b_1) + \bar{y}(a_2 b_1) - 2\bar{y}(a_3 b_1) \qquad (8.21a)$$

and
$$\bar{y}(a_1 b_2) + \bar{y}(a_2 b_2) - 2\bar{y}(a_3 b_2) \qquad (8.21b)$$

are separate estimates of (8.19) at the two levels of B.

At once we notice that the difference between (8.20a) and (8.20b), i.e.,

$$[\bar{y}(a_1 b_2) - \bar{y}(a_2 b_2)] - [\bar{y}(a_1 b_1) - \bar{y}(a_2 b_1)] \qquad (8.22)$$

represents the influence of the level of B on a comparison between the levels of A and hence is a measure of the AB interaction. Similarly, the difference between (8.21a) and (8.21b), i.e.,

$$[\bar{y}(a_1 b_2) + \bar{y}(a_2 b_2) - 2\bar{y}(a_3 b_2)] - [\bar{y}(a_1 b_1) + \bar{y}(a_2 b_1) - 2\bar{y}(a_3 b_1)] \qquad (8.23)$$

is also a measure of the AB interaction. We have thus two measures of the AB interaction. Do these two measures represent independent parts of the overall AB interaction, with 2 degrees of freedom, observed in Table 8.19? The answer to this question depends upon whether the two expressions (8.22) and (8.23) are *orthogonal* contrasts. The following general rules apply here.

Table 8.19   ANALYSIS OF VARIANCE OF THE DATA IN TABLE 8.18

Source of variation	SS	df	MSS	F ratio
Main effect (A)	9.250	2	4.625	13.52†
Main effect (B)	.375	1	0.375	
Interaction (AB)	6.750	2	3.375	9.87†
Treatments	16.375	5	3.275	
Blocks	54.125	3	18.042	52.75†
Residual (Error)	5.125	15	.342	
Total	75.625	23		

† Significant at 1% level.

RULE 1   Let $T_1, T_2, \ldots, T_t$ be the treatment totals of the t treatment combinations in a factorial experiment, and let

$$L = h_1 T_1 + h_2 T_2 + \cdots + h_t T_t$$

where $h_1 + h_2 + \cdots + h_t = 0$, be a treatment contrast. Note that this may represent treatment comparisons such as (8.18) and (8.19) or expressions for interactions such as (8.22) and (8.23). Since all of them are linear combinations of treatment totals, we refer to them by the general term *treatment contrast*. Let $L^* = k_1 T_1 + k_2 T_2 + \cdots + k_t T_t$, where $k_1 + k_2 + \cdots + k_t = 0$, be another treatment contrast. Then L and L^* are said to be *orthogonal* if $h_1 k_1 + h_2 k_2 + \cdots + h_t k_t = 0$, that is, if the sum of the products of the coefficients of the corresponding treatment totals in the two expressions L and L^* is equal to zero. When this rule is applied, it is easy to verify that the expressions (8.22) and (8.23) represent orthogonal treatment contrasts:

$$(+1)(+1) + (-1)(+1) + 0(-2) + (-1)(-1) + (+1)(-1) + 0(+2) = 0$$

RULE 2   If the treatment contrasts L and L^* as defined above are orthogonal, the SS attributable to them are independent. The SS attributable to L is

$$\frac{L^2}{n \sum h_i^2}$$

where n is the number of observations in any treatment total (every treatment total being assumed to be based on n observations). This SS has 1 degree of freedom. Similarly the SS attributable to L^* is

$$\frac{L^{*2}}{n \sum k_i^2}$$

which also has 1 degree of freedom.

Applying this rule to (8.22) and (8.23), we find, using the treatment totals shown in Table 8.18, that the SS attributable to (8.22) is

$$\frac{(-24 + 19 + 20 - 15)^2}{4[(-1)^2 + (+1)^2 + (+1)^2 + (-1)^2]} = 0 \qquad (8.24)$$

and similarly that the SS attributable to (8.23) is

$$\frac{(-24 - 19 + 2 \times 14 + 20 + 15 - 2 \times 19)^2}{4[(-1)^2 + (-1)^2 + (+2)^2 + (+1)^2 + (+1)^2 + (-2)^2]} = 6.750 \qquad (8.25)$$

Table 8.20   SMALL-GROUP VERSUS LARGE-GROUP COMPARISON

Size of group	Mean response
Small	$\frac{1}{8}(24 + 20) = 5.50$
Large	$\frac{1}{8}(19 + 15) = 4.25$
Difference	$5.50 - 4.25 = 1.25$
Standard error	$\sqrt{\frac{2}{8}}(.816) = .45$

Table 8.21   COMPARISON BETWEEN THE MEETING APPROACH
AND THE PAMPHLET APPROACH

| | Mean response when: | | | |
	Values are emphasized	Interspouse communication is emphasized	Difference	SE†
Meetings	$\frac{43}{8} = 5.38$	$\frac{35}{8} = 4.38$	$5.38 - 4.38 = 1.00$	.45
Pamphlets	$\frac{14}{4} = 3.50$	$\frac{19}{4} = 4.75$	$3.50 - 4.75 = -1.25$	.64
Difference	1.88	−.37		
SE†	.55	.55		

† SE = standard error

These two SS's add up to the AB interaction SS obtained in the analysis of variance in Table 8.19. The SS's (8.24) and (8.25) have 1 degree of freedom each, and when they are tested against the mean error SS in Table 8.19, we find that (8.24) is not significant while (8.25) is significant at the 5 percent level.

Since (8.24) is not significant, we need present only the results of the comparison between the small-group approach and the large-group approach after averaging over both the categories of the other factor. Thus we have the summary shown in Table 8.20, which clearly brings out that the small-group approach is more effective than the large-group approach.

As far as the comparative effects of the meeting approach and the pamphlet approach are concerned, we need to present the results at both the levels of the other factor, since the interaction SS (8.25) was found to be significant. The results can be summarized as in Table 8.21, which clearly shows that the responses differ significantly between the meeting and pamphlet approaches only when values are emphasized. The interaction effect is clearly brought out. It may be interesting to point out also that emphasis on values seems to produce greater response than emphasis on interspouse communication when the meetings approach is adopted but not when the pamphlet approach is adopted.

The $a \times b$ Factorial

Let us consider the general case in which factor A has a levels and factor B has b levels. Let us assume that the experiment is carried out on the randomized-blocks design with r blocks. The total SS would then be partitioned as follows:

Source of variation	df
Treatment combinations	$ab - 1$
Blocks	$r - 1$
Residual (error)	$(abr - 1) - (ab - 1) - (r - 1)$
Total	$abr - 1$

The treatment SS can be partitioned, in the manner of a two-way classification analysis of variance, as

Source of variation	df
Main effect (A)	$a - 1$
Main effect (B)	$b - 1$
Interaction (AB)	$(ab - 1) - (a - 1) - (b - 1) = (a - 1)(b - 1)$
Treatments	$ab - 1$

After partitioning the total SS and treatment SS in the above manner, the investigator can concentrate on specific comparisons of interest to him between the levels of A or between the levels of B. There can be $a - 1$ orthogonal comparisons between the levels of A and $b - 1$ orthogonal comparisons between the levels of B. Consequently the SS due to the main effect of A can be partitioned into $a - 1$ SS's each with 1 degree of freedom, and similarly the SS due to the main effect of B can be partitioned into $b - 1$ SS's each with 1 degree of freedom. The interaction SS with $(a - 1)(b - 1)$ degrees of freedom can also be partitioned into SS's each with 1 degree of freedom. To illustrate how this can be done let us consider a 3×3 factorial organized over a completely randomized design with r observations under each treatment combination. Let A and B be the factors involved, and let the levels of A and B be respectively a_0, a_1, a_2 and b_0, b_1, b_2. The treatment combinations are then (a_0b_0), (a_0b_1), (a_0b_2), (a_1b_0), (a_1b_1), (a_1b_2), (a_2b_0), (a_2b_1), and (a_2b_2), which can be denoted without confusion by the subscripts of a and b, always writing the subscript of a first and that of b second. Thus (01) stands for the treatment combination a_0b_1, (11) for a_1b_1, and so on. The corresponding treatment totals can be represented by T_{00}, $T_{01}, \ldots, T_{22}$ where T_{ij} represents the total of observations under the treatment combination (a_ib_j), $i = 1, 2, 3; j = 1, 2, 3$. Each treatment total in the present case will be based on r observations.

Let us write the treatment totals at the top of a table and associate with each a coefficient so as to provide a contrast representing a main effect or an interaction, as shown in Table 8.22. There are two orthogonal contrasts each representing main effect (A), two representing main effect (B), and four representing interaction (AB). We refer to the two orthogonal contrasts representing main effect (A) as two components of main effect (A) and call them (A_1) and (A_2). Similarly we have two compo-

Table 8.22 COEFFICIENTS FOR CALCULATING SS's DUE TO MAIN EFFECTS AND INTERACTIONS EACH WITH 1 DEGREE OF FREEDOM IN A 3×3 FACTORIAL

Effect	T_{00}	T_{01}	T_{02}	T_{10}	T_{11}	T_{12}	T_{20}	T_{21}	T_{22}	SS of coefficients
A_1	-1	-1	-1	0	0	0	$+1$	$+1$	$+1$	6
A_2	$+1$	$+1$	$+1$	-2	-2	-2	$+1$	$+1$	$+1$	18
B_1	-1	0	$+1$	-1	0	$+1$	-1	0	$+1$	6
B_2	$+1$	-2	$+1$	$+1$	-2	$+1$	$+1$	-2	$+1$	18
A_1B_1	$+1$	0	-1	0	0	0	-1	0	$+1$	4
A_1B_2	-1	$+2$	-1	0	0	0	$+1$	-2	$+1$	12
A_2B_1	-1	0	$+1$	$+2$	0	-2	-1	0	$+1$	12
A_2B_2	$+1$	-2	$+1$	-2	$+4$	-2	$+1$	-2	$+1$	36

nents (B_1) and (B_2) for main effect (B), and four components (A_1B_1), (A_1B_2), (A_2B_1), and (A_2B_2) for interaction (AB).

The component A_1 represents a comparison between the a_0 and a_2 levels of A, while A_2 represents a comparison of the a_1 level with a_0 and a_2. In defining the coefficients representing A_1 and A_2 we have followed the principle developed earlier [see Eqs. (8.18) and (8.19)]. The coefficients representing B_1 and B_2 are similarly obtained. Once the orthogonal contrasts A_1, A_2, B_1, B_2 are defined as described above, the definitions of the interactions (A_1B_1), (A_1B_2), etc., are straightforward. The coefficients defining the contrast that represents interaction (A_iB_j) can be obtained by multiplying the corresponding coefficients of (A_i) and (B_j). Thus the fifth row of Table 8.22 is obtained by multiplying the corresponding entries in rows 1 and 3, the sixth row is obtained by multiplying the corresponding entries in rows 1 and 4, and so on.

Once the contrasts have been defined, the sum of squares associated with each is calculated according to the rules enumerated earlier in this section (see page 378). Thus the SS associated with $\sum \sum C_{ij}T_{ij}$ is

$$\frac{(\sum \sum C_{ij}T_{ij})^2}{r \sum \sum C_{ij}^2}$$

where r, it may be recalled, is the number of observations under each treatment combination. This SS has 1 degree of freedom.

The SS's due to A_1 and A_2 together make up the SS due to main effect (A) obtained in the manner described earlier in Tables 8.16 and 8.17. Similarly the SS's due to B_1 and B_2 together make up the SS due to main effect (B) with 2 degrees of freedom, and the SS's due to A_1B_1, A_1B_2, A_2B_1, and A_2B_2 together make up the SS due to interaction (AB) with 4 degrees of freedom. If interaction between the block variable and one or more of the treatments is suspected, it is desirable to have two or more observations under each block-treatment combination. This would permit analyzing block-treatment interactions by algebraically viewing the block variable as an extra factor (see below).

The $a \times b \times c$ Factorial

We shall now give the outline of analysis for a general factorial experiment with three factors A, B, and C with a, b, and c levels respectively. Let the experiment be organized on a randomized-blocks design with r blocks (replications).

1 First obtain the totals for each treatment combination and also obtain the block totals. Compute total SS, treatment SS, and block SS, as usual. Residual SS can be obtained by subtraction: total SS − treatment SS − block SS. The degrees of freedom associated with these SS's are obtained as follows:

	df
Treatment SS	$abc - 1$
Block SS	$r - 1$
Error SS	$(abcr - 1) - (abc - 1) - (r - 1)$
Total	$abcr - 1$

Table 8.23   OUTLINE OF ANALYSIS OF VARIANCE:
$a \times b \times c$ FACTORIAL IN RANDOMIZED
BLOCKS

Source of variation	df
Main effect:	
(A)	$a - 1$
(B)	$b - 1$
(C)	$c - 1$
Interaction:	
(AB)	$(a - 1)(b - 1)$
(AC)	$(a - 1)(c - 1)$
(BC)	$(b - 1)(c - 1)$
(ABC)	$(a - 1)(b - 1)(c - 1)$
Treatment combinations	$abc - 1$
Blocks	$r - 1$
Residual (error)	$(abcr - 1) - (abc - 1) - (r - 1)$
Total	$abcr - 1$

2 For each pair of factors, form a two-way table. The entries in the tables should be the sums of observations over the levels of the third factor and over the replications. Apply to the $A \times B$ table the procedure for the analysis of variance of two-way tables with rc observations per cell. The SS's obtained from the marginal totals give the SS due to the main effect of A and the SS due to the main effect of B. Subtract these two SS's from the cell SS associated with the $A \times B$ table. The remainder gives the SS due to the AB interaction. Similarly from the $A \times C$ table we calculate the SS due to the main effect of A, the SS due to the main effect of C, and the SS due to the interaction AC, and the $B \times C$ table gives the SS's due to the main effects of B and C and to the BC interaction.
3 From the treatment SS obtained in step 1 subtract the SS due to the three main effects, (A), (B), and (C), and the SS due to the three two-factor interactions, (AB), (AC), and (BC). The remainder is the SS due to the ABC interaction.
4 The analysis of variance can now be set up according to the outline in Table 8.23.

8.4   EXPECTED VALUES OF MEAN SUMS OF SQUARES

In Chap. 7 we saw the uses to which the expected values of the MSS in the analysis of variance can be put. They help estimate the components of total variation in the response variable, each component being attributable to one causal factor. The expected values of different MSS's also help decide on the appropriate test statistic, e.g., the appropriate variance ratio, for a given hypothesis. In this section we shall extend the results given in Chap. 7 to the factorial experiments.

Consider a two-factor experiment in which the two factors are A and B with a and b levels, respectively. Let the experiment be organized on a completely randomized design with r observations per treatment combination. Let the kth observed value under the (i,j)th treatment combination (ith level of A combined with the jth level of B) be represented as

$$y_{ijk} = \mu + A_i + B_j + (AB)_{ij} + \varepsilon_{ijk} \qquad (8.26)$$

where μ = general mean, or combined effect of all factors affecting all responses alike

A_i = effect of ith level of A

B_j = effect of jth level of B

$(AB)_{ij}$ = effect due to nonadditivity (interaction) associated with combination of ith level of A with jth level of B

ε_{ijk} = residual effect

It may be noted in passing that (8.26) can be rewritten in the multiple-regression form

$$y_{ijk} = \mu x_0 + \sum A_i x_i + \sum B_j z_j + \sum\sum (AB)_{ij} v_{ij} + \varepsilon_{ijk}$$

where $x_0 = 1$ for all y's

$x_i = 1$ for all y's under ith level of A and 0 for all others

$z_j = 1$ for all y's under jth level of B and 0 for all others

$v_{ij} = 1$ for all y's under (i,j) combination, i.e., ith level of A combined with jth level of B, and 0 for all others

If the underlying design is different from the completely randomized design, (8.26) should be modified to include block effects, etc.

In (8.26) A_i's and B_j's may be fixed or random. There should be no confusion if we use the same symbol for fixed as well as random effects here. Treating A_i's as random effects amounts to assuming that they form a random sample drawn from a corresponding universe. This is equivalent to assuming that the levels of A actually used in the experiment have been randomly selected from a corresponding universe of levels. This universe may be an actual one or a hypothetical one. We shall assume that the universe to which A_i's are assumed to belong is infinite in size and that it has mean 0 and variance σ_A^2. Similarly, when treating B_j's as random effects, we shall assume that they are a random sample from an infinite universe with mean 0 and variance σ_B^2. If A_i's or B_j's are treated as random effects, the interaction effects $(AB)_{ij}$'s are to be treated as random effects. When $(AB)_{ij}$'s are treated as random, we shall assume them to be a random sample from an infinite universe with mean 0 and variance σ_{AB}^2.

Recall that the calculation of the SS and the MSS is the same whether the effects are taken to be random or fixed. The expected values of the MSS are, however, different in fixed, random, and mixed models, as shown in Table 8.24. In showing the expected values of the MSS in Table 8.24, we have used the symbol S_A^2 for the variance

Table 8.24 EXPECTED VALUES OF MSS IN A TWO-FACTOR EXPERIMENT

MSS due to:	Fixed effects	Random effects	Mixed effects (A fixed, B random)
Main effect A	$\sigma^2 + rbS_A^2$	$\sigma^2 + r\sigma_{AB}^2 + rb\sigma_A^2$	$\sigma^2 + r\sigma_{AB}^2 + rb\sigma_A^2$
Main effect B	$\sigma^2 + raS_B^2$	$\sigma^2 + r\sigma_{AB}^2 + ra\sigma_B^2$	$\sigma^2 + raS_B^2$
AB interaction	$\sigma^2 + rS_{AB}^2$	$\sigma^2 + r\sigma_{AB}^2$	$\sigma^2 + r\sigma_{AB}^2$
Error	σ^2	σ^2	σ^2

of A_i's, S_B^2 for the variance of B_j's, and S_{AB}^2 for that of $(AB)_{ij}$'s actually involved in the experiment under discussion. In calculating these quantities, it is assumed that the divisors $a - 1$, $b - 1$, and $(a - 1)(b - 1)$, respectively, have been used. Thus $S_A^2 = \sum (A_i - \bar{A})^2/(a - 1)$, etc.

The expected values of the MSS's in any factorial experiment can be written by following the general rules given below.

Let the factors involved in the experiment be $A, B, C, D, \ldots$ with $a, b, c, d, \ldots$ levels, respectively. Any of these factors may be random or fixed. Suppose that we want to find the expected value of the MSS associated with a factorial effect U, where U stands for any one of the main effects, any one of the two-factor interactions, any one of the three-factor interactions, ... Thus U may stand for (A), (BC), $(CDEF)$, or any one such factorial effect.

RULE 1   The expected value of the MSS for U contains:

  1 A term in σ^2
  2 A term in σ_U^2
  3 Terms in σ_{UV}^2, where UV stands for letter combinations containing U and V, *V standing for one or more random effects but no fixed effects*

RULE 2   The coefficient of σ^2 is 1. The coefficient of any other term is r (the number of observations per treatment combination) times the product of letters $a, b, c, \ldots$ from which all letters corresponding to those appearing in the suffix of the variance term are canceled out.

To understand how to apply these rules, let us suppose we want the expected value of the MSS for the main effect (C) in a factorial experiment with A, B, C as the factors, their levels being $a, b,$ and c, respectively. Let us suppose that factor C is fixed while the other two factors are random. Then applying the first rule above, we have

$$E(\text{MSS}:C) = (?)\sigma^2 + (?)\sigma_C^2 + (?)\sigma_{AC}^2 + (?)\sigma_{BC}^2 + (?)\sigma_{ABC}^2 \qquad (8.27)$$

A systematic way of determining which variance symbols should be included in the expected value is to write down in a standard order all the possible combinations of the letters representing the factors. In the three-factor case under consideration, the standard order will be

$$A, \quad B, \quad AB, \quad C, \quad AC, \quad BC, \quad ABC$$

which is analogous to the order introduced in the case of 2^3 factorial, earlier. After writing down the letter combinations in the above order (or any other order), all we have to do is to pick all combinations containing the letter(s) corresponding to the factorial effect under consideration, i.e., whose expected value is required. In the present case we should look for all letter combinations containing C. We should then *exclude* from those combinations all those containing any fixed effects other than C. In the present case we need not exclude any combination, since both A and B are random. We have thus four letter combinations whose corresponding variance terms should be included in the expected value; these are $C, AC, BC,$ and ABC; the expected value of the MSS due to C should contain variance symbols with each of these letter combinations as subscripts. Thus we get the terms on the right-hand side of (8.27). Now to get the coefficient of different variance terms we apply rule 2. The coefficient of σ^2_{ABC} is obtained as

$$\frac{rabc}{\text{Product of small letters corresponding to } A, B, \text{ and } C} = \frac{rabc}{abc} = r$$

The coefficient of $\sigma_{AC}{}^2$ is

$$\frac{rabc}{\text{Product of small letters corresponding to } A \text{ and } C} = \frac{rabc}{ac} = rb$$

and so on.

If A and C are fixed effects but B is random, the terms containing variance symbols with A in the subscript will drop out according to rule 1, and thus we would have

$$E(\text{MSS}:C) = \sigma^2 + rab\sigma_C{}^2 + ra\sigma_{BC}{}^2$$

and so on.

8.5 ANALYSIS OF CATEGORICAL DATA

In this section we shall indicate how to apply to the analysis of categorical data some of the notions introduced in the preceding sections of this chapter. We shall first confine our attention to 2×2 tables. But our exposition will be organized in such a way that we shall be able to generalize the ideas presented for the 2×2 table to higher-dimensional contingency tables.

Let us start with a 2×2 frequency table representing a whole population, *not a sample* (see Table 8.25). A total of $N_{..}$ individuals in the population are classified into four mutually exclusive and exhaustive categories. Those in the (i,j) cell possess the ith level of A, the row attribute, and the jth level of B, the column attribute. It is easily seen that the probability that an individual selected at random belongs to the (i,j) cell is given by

$$P_{ij} = \frac{N_{ij}}{N_{..}} \qquad (8.28)$$

Table 8.25 A 2×2 POPULATION TABLE

Level of attribute A	Level of attribute B		Total
	b_1	b_2	
a_1	N_{11}	N_{12}	$N_{1.}$
a_2	N_{21}	N_{22}	$N_{2.}$
Total	$N_{.1}$	$N_{.2}$	$N_{..}$

Similarly, the probability that an individual belongs to the ith level of the row attribute is given by

$$P_{i.} = \frac{N_{i.}}{N_{..}} \qquad (8.29)$$

and the corresponding probability of membership in the jth category of the column attribute is given by

$$P_{.j} = \frac{N_{.j}}{N_{..}} \qquad (8.30)$$

Now from probability theory it is known that the row attribute and the column attribute in Table 8.25 are independent if and only if

$$P_{ij} = P_{i.}P_{.j} \qquad (8.31)$$

which translates into

$$\frac{N_{ij}}{N_{..}} = \frac{N_{i.}}{N_{..}} \frac{N_{.j}}{N_{..}}$$

or

$$N_{ij}N_{..} = N_{i.}N_{.j} \qquad (8.32)$$

Now if we set $i = 1$ and $j = 1$ in (8.32), we get

$$N_{11}(N_{11} + N_{12} + N_{21} + N_{22}) = (N_{11} + N_{12})(N_{11} + N_{21})$$

which simplifies to

$$N_{11}N_{22} = N_{12}N_{21} \qquad (8.33)$$

It is not difficult to verify that if we set $i = 1$ and $j = 2$, or $i = 2$, and $j = 1$, or $i = 2$ and $j = 2$, (8.32) invariably simplifies to (8.33). Consequently, in the present case, we may state that the hypothesis of independence between A and B in Table 8.25 is the same as the hypothesis expressed in Eq. (8.33). Now taking natural logarithms on both sides of (8.33) and bringing all terms to one side, we get

$$\ln N_{22} - \ln N_{21} - \ln N_{12} + \ln N_{11} = 0 \qquad (8.34)$$

The left-hand side of (8.34) can be written

$$U_{22} - U_{21} - U_{12} + U_{11} \qquad (8.35)$$

where $U_{ij} = \ln N_{ij}$. The expression (8.35) resembles the formula for calculating the AB interaction, except for the external multiplier, in the 2×2 factorial with U_{11}, U_{12}, U_{21}, and U_{22} as the responses under the various treatment combinations. This suggests the following. Suppose that we replace N_{ij}, for each i and j, in Table 8.25 by its

Table 8.26   NATURAL LOGARITHMS OF THE CELL FREQUENCIES IN TABLE 8.25

Level of attribute A	Level of attribute B	
	b_1	b_2
a_1	U_{11}	U_{12}
a_2	U_{21}	U_{22}

natural logarithm and obtain, say, Table 8.26. If we now regard Table 8.26 as representing the results of a 2×2 factorial experiment, the hypothesis that the AB interaction is nonexistent in the data situation represented by Table 8.26 is equivalent to the hypothesis of row-by-column independence in Table 8.25.

We shall now show that the hypothesis that the categories of A are equiprobable under each category of B is formally equivalent to the following joint hypothesis with reference to Table 8.26: the AB interaction and the main effect of A are both equal to zero.

If $(AB) = 0$ in Table 8.26, we have

$$U_{22} + U_{11} = U_{21} + U_{12} \qquad (8.36)$$

and if $(A) = 0$, we have

$$U_{22} - U_{12} + U_{21} - U_{11} = 0$$

or

$$U_{22} - U_{11} = U_{12} - U_{21} \qquad (8.37)$$

Equating the sum of the left-hand sides to the sum of the right-hand sides of (8.36) and (8.37), we get

$$U_{22} = U_{12} \qquad (8.38a)$$

and consequently from (8.37)

$$U_{11} = U_{21} \qquad (8.38b)$$

Now since $U_{ij} = \ln N_{ij}$, (8.38a) is equivalent to

$$N_{22} = N_{12} = \tfrac{1}{2} N_{.2}$$

and (8.38b) is equivalent to

$$N_{11} = N_{21} = \tfrac{1}{2} N_{.1}$$

This proves that the joint hypothesis $(AB) = 0$, $(A) = 0$ in Table 8.26 amounts to the hypothesis that the categories of A are equiprobable under each category of B.

Proceeding in the above fashion, it is possible to demonstrate that the hypothesis of equiprobability of all cells in Table 8.25 is formally equivalent to the following joint hypothesis:

$$(AB) = 0 \qquad (A) = 0 \qquad (B) = 0$$

The three hypotheses described above are summarized in Table 8.27. Obviously hypothesis 2 in Table 8.27 implies (but is not implied by) hypothesis 1. Similarly, hypothesis 3 implies (but is not implied by any of) the others. In this sense, the order in which the hypotheses are listed in Table 8.27 may be called hierarchical. As we go

Table 8.27 HYPOTHESES PERTAINING TO THE 2×2 TABLE

Hypothesis	Meaning in conventional terms
1 $(AB) = 0$	Attributes A and B are independent
2 $(AB) = 0$ $(A) = 0$	Conditional equiprobability: categories of attribute A are equiprobable in each category of attribute B
3 $(AB) = 0$ $(A) = 0$ $(B) = 0$	Unconditional equiprobability: all cells are equiprobable

from the top (low number) to bottom (high number), we go from less inclusive to more inclusive hypotheses.

We now turn to the question of testing these hypotheses, given a sample of observations. Suppose Table 8.28 represents a sample of $n_{..}$ observations classified according to the levels of attributes A and B. To test any of the hypotheses listed in Table 8.27, we first estimate the expected frequencies in each cell under the given hypothesis and then examine whether the estimated expected frequencies are satisfactorily close to the observed frequencies. In this latter step we make use of a test statistic. Either of the following statistics may be used for this purpose:

$$\sum_i \sum_j \frac{(\text{obs}_{ij} - \text{exp}_{ij})^2}{\text{exp}_{ij}} \qquad (8.39)$$

$$\sum_i \sum_j 2\,\text{obs}_{ij}\,\ln \frac{\text{obs}_{ij}}{\text{exp}_{ij}} \qquad (8.40)$$

where obs_{ij} stands for the observed frequency in the (i,j) cell and exp_{ij} stands for the corresponding estimated expected frequency. The statistic (8.39) is known as the goodness-of-fit chi square, while (8.40) is known as the likelihood-ratio statistic. Both follow the chi-square distribution whose degrees of freedom are to be calculated according to a rule which will be described later in this section. We shall not discuss the relative merits of one statistic over the other, except to point out that in certain contexts the likelihood-ratio statistic has certain attractive properties the other does not have (see Goodman, 1968, 1970).

The following formulas give maximum-likelihood estimates of the expected frequency in the (i,j) cell for hypotheses 1, 2, and 3, respectively, in Table 8.27:

$$H_1: \frac{n_i.n._j}{n_{..}} \qquad H_2: \frac{n._j}{2} \qquad H_3: \frac{n_{..}}{4}$$

We notice that in estimating the expected frequency in the (i,j) cell under hypothesis 1, we use only the marginal frequencies $n_i.$ and $n._j$ and, of course, the sum of the

Table 8.28 A SAMPLE OF $n_{..}$ OBSERVATIONS ARRANGED IN THE FORM OF A 2×2 TABLE

Level of attribute A	Level of attribute B		Total
	b_1	b_2	
a_1	n_{11}	n_{12}	$n_1.$
a_2	n_{21}	n_{22}	$n_2.$
Total	$n._1$	$n._2$	$n_{..}$

marginal frequencies, $n_{..}$ *but not any of the individual cell frequencies.* In this sense we may describe hypothesis 1 as one that fits the row and column marginal frequencies. It may be useful to introduce the notation $\{A\}$ for the marginals for attribute A, and $\{B\}$ for the marginals for attribute B. According to this notation, hypothesis 1 involves fitting $\{A\}$ and $\{B\}$. Similarly, hypothesis 2 of Table 8.27 involves fitting $\{B\}$, while hypothesis 3 involves fitting just the overall sample size $n_{..}$. It should be noted that hypothesis 2 of Table 8.27 has a companion which involves fitting only $\{A\}$. This hypothesis translated in terms of the AB interaction and the main effects amounts to setting (AB) and (B) equal to zero. In conventional terms this means that the levels of attribute B are equiprobable under each category of attribute A.

As stated earlier, once estimates of the expected frequencies under any given hypothesis are at hand, the next step in testing the hypothesis in question is to calculate the value of the test statistic (8.39) or (8.40). The value thus calculated is then compared with the tabulated values of the chi-square distribution. To do this we must know the degrees of freedom of the chi-square distribution. This is calculated by applying the following rule:

If k is the number of constraints imposed by the hypothesis in question on the expected frequencies, the degrees of freedom of the chi square should be the number of cells minus k.

An examination of the estimates of the expected frequencies given above under hypothesis 1 reveals that their row totals are constrained to equal the respective observed row marginals $(n_{1.}, n_{2.}, \dots)$ and their column totals are similarly constrained to equal the respective observed column marginals $(n_{.1}, n_{.2}, \dots)$. In an $I \times J$ table, there are thus $I + J$ constraints imposed by the above estimation procedure on the expected frequencies, but only $I + J - 1$ of these are independent constraints, since given any $I + J - 1$ of the $I + J$ constraints, the remaining one will automatically be satisfied. This means, applying the rule given above, that the degrees of freedom associated with hypothesis 1 in an $I \times J$ table will be $IJ - I - J + 1$ or $(I - 1)(J - 1)$. Similarly the degrees of freedom associated with hypotheses 2 and 3 are respectively $IJ - I$ or $I(J - 1)$ and $IJ - 1$. In the 2×2 table these become 1, 2, and 3, respectively, for hypotheses 1, 2, and 3 of Table 8.27.

For easy reference these results are summarized in Table 8.29. Before proceeding to examining the corresponding results for three-way contingency tables, two points may be noted, both applying to contingency tables of any dimension. It should be recognized that we are not necessarily dealing with results of experiments here. Clearly we have not made any distinction between dependent and independent variables in the above discussion. In such circumstances, concepts such as main effects and interactions introduced earlier in this book do not strictly apply. They are used in the present context for heuristic purposes, as they help identify various hypotheses of interest and make it easier to go from less inclusive hypotheses to more inclusive ones systematically.

A second point concerns a method aimed at improving the validity of the assumption that statistic (8.39) and (8.40) follow the chi-square distribution. The

Table 8.29 HIERARCHICAL HYPOTHESES FOR THE $I \times J$ CONTINGENCY TABLE

Hypothesis (interaction and/or main effect set equal to zero)	Meaning in conventional terms	Number of hypotheses of this kind	Maximum-likelihood estimate in (i,j) cell	Marginals fitted	df
1 (AB)	$(A \text{ Ind } B)$	1	$\dfrac{(n_{i.})(n_{.j})}{(n_{..})}$	$\{A\}, \{B\}$	$(I-1)(J-1)$
2 $(AB), (A)$	$(A \text{ Eq Pr}) \mid B$	2	$\dfrac{n_{.j}}{I}$	$\{B\}$	$J(I-1)$
3 $(AB), (A), (B)$	$(A \times B \text{ Eq Pr})$	1	$\dfrac{n_{..}}{IJ}$	$n_{..}$	$IJ - 1$

method involves replacing the observed frequency in each cell by the observed frequency plus $\frac{1}{2}$, for example, replacing n_{ij} by $n_{ij} + \frac{1}{2}$, for all i and j in two-way tables. See in this connection Gart and Zweifel (1967).

The $I \times J \times K$ Table

Suppose we have a three-way table of frequencies, with rows representing the levels of one attribute, say A, columns representing the levels of another, say B, and layers representing those of a third, say C. Let the attributes A, B, and C have levels I, J, and K, respectively. To fix our ideas it may be helpful to keep in mind a concrete data situation. For this purpose imagine a three-way table in which A stands for social-mobility status (up-mobile, nonmobile, down-mobile), B for socioeconomic status (high, medium, low), and C for self-esteem (high, medium, low).

Table 8.30 summarizes the results for the $I \times J \times K$ table following the format of Table 8.29. When we go from two-way to three-way tables, new conceptual problems arise, but the extension of three-way to higher-dimensional tables is straightforward.

The three-factor interaction   One of the hypotheses in three-way tables for which there is no analog in two-way tables is the one that sets the three-factor interaction (ABC) equal to zero (see hypothesis 1 in Table 8.30). The meaning of this hypothesis can best be understood if we consider data situations in which (ABC) will be different from zero. Suppose there is independence between attributes A and B when C is at one level but not when C is at another level. To give an example, suppose socioeconomic status and self-esteem are strongly associated among up-mobiles but are independent among nonmobiles. This situation makes the interaction (ABC) differ from zero. Such situations frequently occur in practice. Several examples can be found in Rosenberg (1968).

As indicated in Table 8.30, explicit formulas are not available for computing maximum-likelihood estimates of the expected frequencies in the cells under hypothesis 1. To compute these estimates one uses an iterative procedure, which may be briefly described as follows.

Table 8.30 HIERARCHICAL HYPOTHESES FOR THE $I \times J \times K$ CONTINGENCY TABLE

Hypothesis: interaction(s) and/or main effect(s) set equal to zero	Meaning in conventional terms	Number of hypotheses of this kind	Maximum-likelihood estimate of frequency in (i,j,k) cell	Marginals fitted	df
1 (ABC)	None	1	No explicit formula	$\{AB\}$, $\{BC\}$, $\{AC\}$	$(I-1)(J-1)(K-1)$
2 (ABC), (BC)	$(B \text{ Ind } C)\,\vert\,A$	3	$\dfrac{(n_{ij.})(n_{i.k})}{n_{i..}}$	$\{AB\}$, $\{AC\}$	$I(J-1)(K-1)$
3 (AB), (BC), (AC)	$(A \times B \text{ Ind } C)$	3	$\dfrac{(n_{ij.})(n_{..k})}{n_{...}}$	$\{AB\}$, $\{C\}$	$(IJ-1)(K-1)$
4 (ABC), (BC), (AC), (C)	$(C \text{ Eq Pr})\,\vert\,A \times B$	3	$\dfrac{n_{ij.}}{K}$	$\{AB\}$	$IJ(K-1)$
5 (ABC), (BC), (AC), (AB)	$(A \text{ Ind } B \text{ Ind } C)$	1	$\dfrac{(n_{i..})(n_{.j.})(n_{..k})}{(n_{...})(n_{...})}$	$\{A\}$, $\{B\}$, $\{C\}$	$IJK-I-J-K+2$
6 (ABC), (BC), (AC), (AB), (C)	$(A \text{ Ind } B)$ and $(C \text{ Eq Pr})\,\vert\,A \times B$	3	$\dfrac{(n_{i..})(n_{.j.})}{K(n_{...})}$	$\{A\}$, $\{B\}$	$IJK-I-J+1$
7 (ABC), (BC), (AC), (AB), (C), (B)	$(B \times C \text{ Eq Pr})\,\vert\,A$	3	$\dfrac{n_{i..}}{JK}$	$\{A\}$	$I(JK-1)$
8 (ABC), (BC), (AC), (AB), (C), (B), (A)	$(A \times B \times C \text{ Eq Pr})$	1	$\dfrac{n_{...}}{IJK}$	$n_{...}$	$IJK-1$

Let $\{AB\}$, $\{BC\}$, and $\{AC\}$ be the sets of marginals obtained by summing the observed (i,j,k) cell frequencies over the levels of C, A, and B, respectively. That is, $\{AB\}$ consists of n_{ij}., $i = 1, 2, \ldots, I$; $j = 1, 2, \ldots, J$; and so on. The iterative procedure mentioned above involves repeating a cycle of operations a sufficient number of times until a set of estimates of the (i,j,k) cell frequencies are obtained, which sum to each of the sets of marginals $\{AB\}$, $\{BC\}$, and $\{AC\}$. To describe a typical cycle of operations, let $n_{ijk}^{(0)}$ $(i = 1, 2, \ldots, I; j = 1, 2, \ldots, J; k = 1, 2, \ldots, K)$ be the set of (i,j,k) cell frequencies yielded by the preceding cycle of operations. The first cycle of operations may start with any arbitrary set of (i,j,k) cell frequencies, for example, 1 for all i, j, and k.

STEP 1 From $n_{ijk}^{(0)}$'s calculate a set of cell frequencies which fit the observed marginals $\{AB\}$. This is done by distributing $\{AB\}$ over the levels of C according to the patterns exhibited by $n_{ijk}^{(0)}$'s. In other words, calculate for each i, j, and k.

$$n_{ijk}^{(1)} = n_{ij}. \frac{n_{ijk}^{(0)}}{n_{ij}^{(0)}}$$

STEP 2 From $n_{ijk}^{(1)}$'s calculate another set of cell frequencies which fit the observed marginals $\{BC\}$. This is done by distributing $\{BC\}$ over the levels of A according to the patterns exhibited by $n_{ijk}^{(1)}$'s. In other words, calculate for each i, j, k

$$n_{ijk}^{(2)} = n_{.jk} \frac{n_{ijk}^{(1)}}{n_{.jk}^{(1)}}$$

STEP 3 From $n_{ijk}^{(2)}$'s calculate a new set of cell frequencies which fit the observed marginals $\{AC\}$. This is done by distributing $\{AC\}$ over the levels of B according to the patterns exhibited by $n_{ijk}^{(2)}$'s. In other words, calculate

$$n_{ijk}^{(3)} = n_{i.k} \frac{n_{ijk}^{(2)}}{n_{i.k}^{(2)}}$$

This completes one cycle of operations. If $n_{ijk}^{(3)}$'s fit all the three observed marginals $\{AB\}$, $\{BC\}$, $\{AC\}$, this is the end of the iterative process. Obviously, $n_{ijk}^{(3)}$'s fit $\{AC\}$. The check needed is whether they fit $\{AB\}$ and $\{BC\}$ also. If the fit is not satisfactory, go back to step 1, using $n_{ijk}^{(3)}$ instead of $n_{ijk}^{(0)}$, and continue the process. Only a few repetitions of the cycle will be necessary to obtain the desired degree of closeness of fit with the marginals (see, for example, Ireland and Kullback, 1968).

Conditional independence Another hypothesis in the three-way table which has no analog in the two-way table is one that posits conditional independence or partial independence between the attributes considered. Hypothesis 2 in Table 8.30 is of this kind. It holds that there is independence between attributes B and C at each level of A. This hypothesis is analogous to the hypothesis of zero partial correlation between two variables given the third in a three-variate normal universe. The companions to hypothesis 2 in Table 8.30 are $(A$ Ind $B) \mid C$ and $(A$ Ind $C) \mid B$.

Multiple independence   Yet another hypothesis which has no analog in two-way tables is hypothesis 3 shown in Table 8.30. This hypothesis is analogous to the hypothesis of zero multiple correlation in a three-variate normal population. We may describe hypothesis 3 in Table 8.30 in the following words: "A and B are jointly independent of C." Its companions are: (1) "A and C are jointly independent of B," and (2) "B and C are jointly independent of A."

Mutual independence   Hypothesis 5 in Table 8.30 also has no direct analog in the two-way table. If hypothesis 5 holds, the entire three-way table can be satisfactorily reproduced when only the one-way marginals $\{A\}$, $\{B\}$, and $\{C\}$ are known.

Needless to say, all other hypotheses in Table 8.30 have direct analogs in the two-way table. As indicated already, when we go from the three-way table to higher-way tables, no new conceptual problems arise. Hence a good grasp of the analysis procedure associated with the three-way table enables one to handle with confidence data situations involving more than three attributes. Reference may nonetheless be made to explicit treatments of higher-way tables in Goodman (1970, 1972a, b).

8.6   CONCLUDING REMARKS

This chapter has expounded the structure and analysis of factorial experiments. The use of the design matrix in the analysis of factorial experiments is indicated in the exercises that follow.

The idea of partitioning the treatment SS into components, each with 1 degree of freedom, provides a very helpful tool in gaining a practically useful interpretation of treatment effects on the response variable. This strategy has been exploited quite well by a number of writers in connection with the analysis of categorical data; see for example, Grizzle, Starmer, and Koch (1969), Koch, Johnson, and Tolley (1972), and Koch and Lemeshow (1972). An alternative way of analyzing categorical data, of course, is to employ the method outlined in the preceding section.

EXERCISES

8.1   Consider a 2^2 factorial experiment executed on a completely randomized design with two observations per treatment combination. Let A and B be the two factors. Let $x_0 = 1$ for all observations, $x_A = +1$ or -1 according as the observation belongs to the "high" or the "low" level of A, $x_B = +1$ or -1 according as the observation belongs to the high or low level of B, and $x_{AB} = x_A x_B$. Suppose you are asked to fit by the least-squares method the model

$$y = \beta_0 x_0 + \beta_A x_A + \beta_B x_B + \beta_{AB} x_{AB} + \varepsilon$$

(a) Verify that the observational equations can be expressed in matrix notation as

$$Y = \beta X + \varepsilon$$

where Y is a column vector of the observations arranged in such a manner that those under the treatment combination (low A, low B) appear at the top, followed by those under the combination (high A, low B), (low A, high B), and (high A, high B), in that order,

$$\mathbf{X} = \begin{bmatrix} 1 & -1 & -1 & 1 \\ 1 & -1 & -1 & 1 \\ 1 & 1 & -1 & -1 \\ 1 & 1 & -1 & -1 \\ 1 & -1 & 1 & -1 \\ 1 & -1 & 1 & -1 \\ 1 & 1 & 1 & 1 \\ 1 & 1 & 1 & 1 \end{bmatrix}$$

$$\boldsymbol{\beta} = \begin{bmatrix} \beta_0 \\ \beta_A \\ \beta_B \\ \beta_{AB} \end{bmatrix}$$

and ε is a column vector of residuals.

(b) Verify that

$$\mathbf{X'X} = \begin{bmatrix} 8 & 0 & 0 & 0 \\ 0 & 8 & 0 & 0 \\ 0 & 0 & 8 & 0 \\ 0 & 0 & 0 & 8 \end{bmatrix}$$

and that the inverse of $\mathbf{X'X}$ is

$$(\mathbf{X'X})^{-1} = \begin{bmatrix} \frac{1}{8} & 0 & 0 & 0 \\ 0 & \frac{1}{8} & 0 & 0 \\ 0 & 0 & \frac{1}{8} & 0 \\ 0 & 0 & 0 & \frac{1}{8} \end{bmatrix}$$

(c) If $\hat{\beta}_A, \hat{\beta}_B$, and $\hat{\beta}_{AB}$ are the least-squares estimates of β_A, β_B, and β_{AB}, respectively, verify that $\hat{\beta}_A = (A)/2$, $\hat{\beta}_B = (B)/2$, and $\hat{\beta}_{AB} = (AB)/2$, where (A), (B), and (AB) are respectively the main effects of A and B and the AB interaction defined according to the principle described in Sec. 8.1.

(d) The X matrix written above may be called the *design matrix*. Notice that the design matrix in the present case corresponds to the particular coding procedure carried out: $x_A = \pm 1$ according as the observation in question occurs under the high or the low level of A, and so on. If the coding is done in some other way, the design matrix takes a different form. In the present case $(\mathbf{X'X})$ is seen to have a true inverse. Since $\mathbf{X'X}$ is a diagonal matrix, i.e., one with nonzero elements only in the principal diagonal, the inversion of $\mathbf{X'X}$ is an easy matter. Recalling that $(\mathbf{X'X})^{-1}\sigma^2$, where σ^2 is the residual variance, is the dispersion matrix of the estimated regression weights, verify that in the present case the least-squares estimates of the regression weights are uncorrelated.

8.2 Repeat parts (a) and (b) of Exercise 8.1 for a 2^3 factorial with four observations per treatment combination. The model to be fitted by the least-squares method is

$$y = \beta_0 x_0 + \beta_A x_A + \beta_B x_B + \beta_C x_C + \beta_{AB} x_{AB} + \beta_{AC} x_{AC} + \beta_{BC} x_{BC} + \beta_{ABC} x_{ABC} + \varepsilon$$

where $x_0 = 1$ for all y values

$\qquad x_A = 1$ for all y values under high level of A and -1 for all those under low level of A

$\qquad x_B = 1$ for all y's under high level of B and -1 for all y's under low level of B

$\qquad x_C = 1$ for all y's under high level of C and -1 for all y's under low level of C

and $\qquad x_{AB} = x_A x_B \qquad x_{AC} = x_A x_C \qquad x_{BC} = x_B x_C \qquad x_{ABC} = x_A x_B x_C$

Verify that $\hat{\beta}_A = (A)/2$, $\hat{\beta}_B = (B)/2$, $\hat{\beta}_C = (C)/2$, $\hat{\beta}_{AB} = (AB)/2$, $\hat{\beta}_{AC} = (AC)/2$, $\hat{\beta}_{BC} = (BC)/2$, and $\hat{\beta}_{ABC} = (ABC)/2$, where (A), (B), ... are factorial effects defined as in Sec. 8.2.

8.3   In a 2^3 factorial experiment, the experimenter has evidence to believe that all the interactions are negligible. What model would then be appropriate for the data set at hand? Write the design matrix X assuming that there is only one observation per treatment combination.

8.4   Consider a 3×2 factorial experiment with one observation per treatment combination. Let A be the factor with three levels (low, medium, and high) and B be the one with two levels (low, high). Let

$$x_{A1} = \begin{cases} 1 & \text{if } A \text{ is at high level} \\ -1 & \text{if } A \text{ is at low level} \\ 0 & \text{if } A \text{ is at medium level} \end{cases}$$

$$x_{A2} = \begin{cases} 1 & \text{if } A \text{ is at high level} \\ 1 & \text{if } A \text{ is at low level} \\ -2 & \text{if } A \text{ is at medium level} \end{cases}$$

$$x_B = \begin{cases} 1 & \text{if } B \text{ is at high level} \\ -1 & \text{if } B \text{ is at low level} \end{cases}$$

$$x_{A1B} = x_{A1}x_B \qquad x_{A2B} = x_{A2}x_B$$

Also let $x_0 = 1$ for all observations. Write the design matrix and the normal equations for fitting by the method of least squares the following model

$$y = \beta_0 x_0 + \beta_{A1} x_{A1} + \beta_{A2} x_{A2} + \beta_B x_B + \beta_{A1B} x_{A1B} + \beta_{A2B} x_{A2B} + \varepsilon$$

8.5   Extend the procedure in Exercise 8.4 to
(a) A 3×3 factorial with one observation per treatment combination
(b) A $3 \times 2 \times 3$ factorial with one observation per treatment combination

8.6   The following data are from a survey on the influence of various possible sources of information on the knowledge of cancer. Respondents were classified according to whether they read newspapers, listened to radio, did solid reading, or attended lectures and also according to whether their knowledge of cancer was good or poor. Let us denote the four factors as

A   Newspaper reading
B   Listening to radio
C   Solid reading
D   Attending lectures

The observed proportions of respondents in each class with good knowledge of cancer are shown below, using the conventional notation for treatment combinations:

Treatment combination	Proportion with good knowledge of cancer	Treatment combination	Proportion with good knowledge of cancer
(1)	.176	d	.167
a	.325	ad	.538
b	.206	bd	.571
ab	.372	abd	.667
c	.447	cd	.273
ac	.532	acd	.600
bc	.500	bcd	.250
abc	.604	abcd	.742

Analyze the data considering the observed proportions as the y values. Compare your inferences with those resulting from a weighted-least-squares analysis using the logistic scale reported in Yates (1960, pp. 314–317).

9

CONFOUNDING, FRACTIONAL REPLICATION, INCOMPLETE NONFACTORIALS, AND CROSSOVER DESIGNS

This chapter has three main objectives: (1) to describe a number of ways in which factorial arrangements can be manipulated to suit the exigencies of different practical situations, (2) to mention briefly certain possible ways of handling design situations in which treatments are not necessarily arranged in factorial form but the size of the available blocks would not permit a full replication in each block, and (3) to review experiments in which the same subject (experimental unit) is used repeatedly.

9.1 SPLIT-PLOT DESIGN

First we deal with a design that is useful when, for practical reasons or otherwise, certain restrictions are to be introduced with respect to randomization of treatment combinations in factorial experiments. Let us consider an example from the field of industrial engineering.[1]

An experimenter wants to examine the effect of oven temperature (factor A) and baking time (factor B) on the length of life (response variable) of an electric component. Four levels are chosen for factor A, say 600, 620, 640, and 660°C, and three levels for factor B, say, 5, 10, and 15 min. A completely randomized layout would require the following procedure. Choose one of the baking times at random. Preheat

---

[1] This example has been adapted from a problem given in Wortham and Smith (1960, p. 112).

the oven to a temperature randomly chosen out of the four levels of factor A and bake a sample piece of the electric component in question for the duration of time selected. Repeat the whole procedure until all treatment combinations have been tried or (if more than one observations are to be taken for each treatment combination) until all the data have been obtained. For practical reasons, however, this is not the way one would run the experiment. Once an oven has been preheated to a selected temperature, it is not only economical but convenient to insert all the pieces to be baked at that temperature and then remove after 5, 10, or 15 min, as the case may be, the pieces to be baked that long. Complete randomization is too impractical and too expensive in such circumstances.

Examples of this kind can be found in many other fields of research. In agricultural experiments, cultural practices, e.g., irrigation, may have to be tried on whole groups of plots while varieties of seeds can be grown on individual plots separately. The following example comes from the area of education research. For the past several years, the CBS television network has been broadcasting, in collaboration with various universities, a series of courses under the title Sunrise Semester. Suppose the program administrators became interested in the following questions:

1 Is it advisable to follow each formal presentation by a question-answer period, as in classrooms?

2 Is it fruitful to have a systematic follow-up procedure organized through mail after the completion of each unit (one lecture or two)?

Answers to questions like these can be useful in setting norms to guide the organization and execution of university courses through national television networks, an undertaking which is still in its infancy but which may grow before long. Suppose it is decided that in order to find answers to these questions the following 2^2 factorial experiment is to be conducted.

A   Method of presentation

    a_1 One-half hour of formal presentation to be followed by one-half hour of question-answer period.

    a_2 One full hour of formal presentation.

B   Follow-up procedure

    b_1 Systematically organized follow-up by mail. Assignments, quizzes, etc., are mailed to the students after the completion of each unit (one lecture or two). The students return the work after completion; it is gone through by the instructor and returned with comments.

    b_2 No systematic follow-up; however, the students are welcome to raise questions by mail, which would be answered by mail.

It can be easily seen that by their very nature the levels of factor A cannot be randomized over individual students; they must be randomized over local television stations carrying the program. The levels of factor B, however, can be randomized over individual students registered in the course.

Similar examples can be imagined in social engineering experiments in which mass media such as television or radio are employed together with face-to-face contacts between social workers (social educators) and individual subjects.

Experiments of the type described above are called *split plot*, the whole group (*main plot*) being split into smaller groups (*subplots*) or subjects. The essential feature of the split-plot experiment is that the levels of one factor are randomized over the main plots (whole groups) whereas the levels of the other factor are randomized over the subplots within each main plot. As a consequence of this difference in randomization, the experimental error for the comparison between the levels of one treatment is not the same as that for the comparison between the levels of the other treatment.

In describing split-plot experiments, the term main treatments is usually employed to refer to the treatments, or levels of a factor, that are randomized over the main plots, and the term subtreatments refers to the treatments, or levels of a factor, that are randomized over the subplots within the main plots.

To illustrate how to analyze the data from a split-plot experiment, let us suppose that the program administrator of the Sunrise Semester series decides to conduct the experiment described above and that the data shown in Table 9.1 are obtained when the experiment is actually conducted.

For simplicity, we shall treat the entries in the cells of Table 9.1 as our observations, although they actually stand for means of satisfaction scores of several individual subjects.

The first step in the analysis of the data in Table 9.1A is to prepare two subtables from it. Table 9.1B shows the main-plot totals in each of the replications. Table 9.1C shows totals of observations in a two-way format, main treatments by subtreatments. From Table 9.1A, we calculate

$$C = \text{correction for mean} = \frac{61.0^2}{24} = 155.04$$

and

$$\text{Total SS} = 3.5^2 + 2.0^2 + \cdots + 2.0^2 - C = 167.50 - 155.04 = 12.46$$

From Table 9.1B we then calculate

$$\text{Cell (main plot) SS} = \tfrac{1}{2}(5.5^2 + 4.5^2 + \cdots + 5.0^2) - 155.04 = 5.21$$
$$\text{Main treatments SS} = \tfrac{1}{12}(32.0^2 + 29.0^2) - 155.04 = .38$$
$$\text{Replications (blocks) SS} = \tfrac{1}{4}(10.0^2 + \cdots + 10.5^2) - 155.04 = 3.46$$

and

$$\text{Main plot error SS} = \text{cell (main plots) SS} - \text{main treatments SS}$$
$$\qquad - \text{replications (blocks) SS}$$
$$= 5.21 - 0.38 - 3.46 = 1.37$$

Note that in calculating each of the above SS's, the following rule is observed: divide the square of each total by the number of observations under that total. It is also important to note that the same correction for the mean is used in obtaining each of the SS's. These rules apply to the calculations from Table 9.1C also.

Now, from Table 9.1C we calculate

Cell (main treatment by subtreatment) SS
$$= \tfrac{1}{6}(19.5^2 + 12.5^2 + 17.0^2 + 12.0^2) - 155.04$$
$$= 6.54$$

$$\text{Subtreatments SS} = \tfrac{1}{12}(36.5^2 + 24.5^2) - 155.04 = 6.00$$

Interaction (main treatment $\times$ subtreatment) SS

$$= \text{cell SS obtained from Table 9.1}C$$
$$- \text{main treatment SS obtained from Table 9.1}C$$
$$- \text{subtreatment SS obtained from Table 9.1}C$$
$$= 6.54 - 0.38 - 6.00 = 0.16$$

Table 9.1 EDUCATION RESEARCH STUDY IN WHICH A SPLIT-PLOT DESIGN
IS USED (FICTITIOUS DATA)

A Mean Satisfaction Score

Main treatment	Subtreatment	Replication (block)						Total
		1	2	3	4	5	6	
a_1 Question-answer period used	b_1 Regular follow-up	3.5	4.0	3.0	2.5	3.5	3.0	19.5
	b_2 No regular follow-up	2.0	2.5	1.5	1.5	2.5	2.5	12.5
a_2 Question-answer period not used	b_1 Regular follow-up	3.0	3.5	2.0	3.0	2.5	3.0	17.0
	b_2 No regular follow-up	1.5	2.5	1.0	2.5	2.5	2.0	12.0
Total		10.0	12.5	7.5	9.5	11.0	10.5	61.0

B Main-Plot Totals Calculated from Table 9.1*A*

Main treatment	Replication (block)						Total
	1	2	3	4	5	6	
a_1 Question-answer period used	5.5	6.5	4.5	4.0	6.0	5.5	32.0
a_2 Question-answer period not used	4.5	6.0	3.0	5.5	5.0	5.0	29.0
Total	10.0	12.5	7.5	9.5	11.0	10.5	61.0

C Totals of Observations for Main Treatment by Subtreatment Combinations
Obtained from Table 9.1*A*

Main treatment	Subtreatment		Total
	b_1 Regular follow-up	b_2 No regular follow-up	
a_1 Question-answer period used	19.5	12.5	32.0
a_2 Question-answer period not used	17.0	12.0	29.0
Total	36.5	24.5	61.0

The analysis of variance can now be set up as shown in Table 9.2. According to the results of the analysis of variance, only the subtreatment effects show any significant difference. In the present case this may be interpreted as meaning that, as far as satisfying the students in the particular program is concerned, what matters is whether there is regular follow-up organized after the completion of each unit (one lecture or two). The results may be summarized as shown in Table 9.3. The standard error for a main-treatment mean is to be calculated using the main-plot error MSS, and the standard error for a subtreatment-answer mean should be calculated using the subtreatment error MSS. If E_M stands for the main-plot error MSS and E_S for the subplot error MSS, the standard error for a main-treatment mean is $\sqrt{E_M/n_M}$ and that for a subplot mean is $\sqrt{E_S/n_S}$, where n_M is the number of observations on which the main treatment mean is based and n_S the number of observations on which the subtreatment mean is based. Note also that the standard error for any comparison between the subtreatment means or the main-treatment means can easily be obtained from the results given in Table 9.3. Thus, for example, the standard error of any linear combination of the subtreatment means, such as $h_1 \times 3.04 + h_2 \times 2.04$, is $\pm .095\sqrt{h_1^2 + h_2^2}$, and, similarly, the standard error of a comparison such as $k_1 \times 2.67 + k_2 \times 2.42$ between the main-treatment means is $\pm .151\sqrt{k_1^2 + k_2^2}$. In particular, the standard error of the difference $3.04 - 2.04$ is ± 0.13, and that of the difference $2.67 - 2.42$ is ± 0.21.

The Model for a Split-Plot Experiment

In order to see the justification of the mode of analysis described above, let us examine the model for the split-plot experiment. If a randomized-blocks layout has been used, the model will be

$$y_{ijk} = \mu + A_i + B_j + (AB)_{ij} + C_k + (AC)_{ik} + (BC)_{jk} + (ABC)_{ijk} + \varepsilon_{ijk}$$

Table 9.2 ANALYSIS OF VARIANCE OF THE SPLIT-PLOT
EXPERIMENT SHOWN IN TABLE 9.1

Source of variation	SS	df	MSS	*F* ratio
Main plots:				
Main treatments†	.38	1	.380	
Replications (blocks)	3.46	5	.692	
Main-plot error	1.37	5	.274	
Subplots:				
Subtreatments‡	6.00	1	6.00	55.05§
Interaction (main				
treatment × subtreatment)	.16	1	.16	
Subplot error¶	1.09	10	.109	
Total	12.46	23		

† Question-answer period used versus not used.
‡ Regular follow-up organized versus not organized.
§ Significant at 1% level.
¶ $1.09 = 12.46 - (0.38 + 3.46 + 1.37 + 6.00 + 0.16)$.

where μ = grand mean, or combined effect of all factors, each affecting all observations in same way

A_i = effect of ith (level of) main treatment

B_j = effect of jth block (replication)

$(AB)_{ij}$ = interaction effect of A_iB_j combination

C_k = effect of kth (level of) subtreatment

$(AC)_{ik}$ = interaction effect of A_iC_k combination

$(BC)_{jk}$ = interaction effect of B_jC_k combination

$(ABC)_{ijk}$ = interaction effect of $A_iB_jC_k$ combination

ε_{ijk} = effect of unaccounted for factors assumed to be NID$(0,\sigma^2)$

Let us assume that A, B, and C have a, b, and c levels, respectively. This is the same as assuming that there are a main treatments, b blocks or replications, and c subtreatments.

Several different data situations can be distinguished, according as the main treatments, blocks, and/or subtreatments are to be regarded as random samples from corresponding universes.

1 Levels of all factors, i.e., main treatments, blocks, and subtreatments, are assumed to be random samples from corresponding universes.

2 Blocks and main treatments are assumed to be random samples from corresponding universes; subtreatments are fixed.

3 Only blocks and subtreatments are assumed to be random samples from corresponding universes; main treatments are regarded as fixed.

4 Only blocks are regarded as random samples from a corresponding universe; main treatments and subtreatments are regarded as fixed.

Whenever the levels of any factor are considered as a random sample from a corresponding universe, we shall treat the universe in question as infinite. We shall also assume that the universe of the corresponding effects has mean zero and a finite variance. Whenever the levels of a factor are regarded as fixed, we shall assume the

Table 9.3 TREATMENT MEANS AND THEIR STANDARD ERROR

Main treatment	Subtreatment		Means	SE
	b_1 Regular follow-up	b_2 No regular follow-up		
a_1 Question-answer period used	3.25	2.08	2.67	
				$\pm\sqrt{\dfrac{.274}{12}}$ $= \pm.151$
a_2 Question-answer period not used	2.83	2.00	2.42	
Means	3.04	2.04		
SE	$\pm\sqrt{\dfrac{.109}{12}} = \pm 0.095$			

mean of the corresponding effects to be zero. Thus if the levels of A are regarded as fixed, we shall assume that $\sum A_i = 0$, where A_i stands for the effect of the ith level of A. Similarly, if the levels of C are regarded as fixed, we shall assume that $\sum C_k = 0$, where C_k stands for the effect of the kth level of C.

For a design with one observation under each combination, we have the expected values of different MSS's as shown in Table 9.4 under the random model.

From Table 9.4, we at once notice that when all three factors are considered random, no exact test is possible for (A), (B), or (C). Recall that to test (A) we need a ratio of the type

$$\frac{\text{MSS due to } (A)}{\text{MSS due to factorial effect other than } (A) \text{ or one due to error}}$$

such that when A has no effect, i.e., when the effects of all levels of A are equal, the expected value of the numerator is equal to that of the denominator. We are not able to find any ratio of the above type for testing (A), (B), or (C). Naturally, therefore, we are not interested in the situation in which A, B, and C are all to be considered random.

The case most frequent in research is the mixed one, in which A and C are considered fixed and B is considered random. Note that when B is considered random, the underlying assumption is that the blocks actually used in the experiment have been selected at random from a corresponding universe, hypothetical if not real. In particular cases the blocks used in the experiment may not have been strictly drawn at random from a specific universe; then the universe of blocks from which the blocks actually used are considered to have been drawn at random will be a hypothetical one. Since the mixed case just described is the most common in practice, we give in Table 9.5 for easy reference the expected values of some of the MSS resulting from experiments conducted under this model. From these results it follows immediately that the appropriate test statistic for testing (A) is

$$\frac{\text{MSS due to } (A)}{\text{MSS due to } (AB) \ (= \text{MSS due to main-plot error})}$$

Table 9.4   EXPECTED VALUES OF MEAN SUMS OF SQUARES: SPLIT-PLOT DESIGN, RANDOM MODEL

MSS due to:	Expected value
(A)	$\sigma^2 + bc\sigma_A{}^2 + c\sigma_{AB}{}^2 + b\sigma_{AC}{}^2 + \sigma_{ABC}^2$
(B)	$\sigma^2 + ac\sigma_B{}^2 + c\sigma_{AB}{}^2 + a\sigma_{BC}{}^2 + \sigma_{ABC}^2$
(AB)	$\sigma^2 + c\sigma_{AB}{}^2 + \sigma_{ABC}^2$
(C)	$\sigma^2 + ab\sigma_C{}^2 + b\sigma_{AC}{}^2 + a\sigma_{BC}{}^2 + \sigma_{ABC}^2$
(AC)	$\sigma^2 + b\sigma_{AC}{}^2 + \sigma_{ABC}^2$
(BC)	$\sigma^2 + a\sigma_{BC}{}^2 + \sigma_{ABC}^2$
(ABC)	$\sigma^2 + \sigma_{ABC}^2$
Error	σ^2 (not retrievable if there is only one observation per subplot)

and that the appropriate statistic to test C is

$$\frac{\text{MSS due to } (C)}{\text{MSS due to } (BC)}$$

If there is reason to believe that $\sigma_{BC}{}^2$ is not different from σ_{ABC}^2, we can pool the SS's due to (BC) and (ABC) and use the corresponding MSS in the denominator for testing (C). This is what we have done in the analysis of the data in Table 9.1.

It can also easily be seen that the appropriate test statistic to test the AC interaction (interactions between the main treatments and the subtreatments) is

$$\frac{\text{MSS due to } (AC)}{\text{MSS due to } (ABC)}$$

Some Comments on Split-Plot Experiments

Since in split-plot experiments, the main treatments (or the levels of the main treatment) are randomized over the main plots, the SS due to the main treatments contains variation in the study phenomenon attributable to differences between the main plots. It is not possible to separate the two variations. The effects of the main treatments and the effects of the main plots are inextricably *confounded* with each other. This means that the information available from split-plot experiments about the main treatments is not valid. The question immediately arises: Why use the main treatments at all in the experiment? Why not confine attention to the subtreatments? The answer is that while the split-plot experiment does not provide valid information about the main-treatment effects, it does provide valid information about the interactions between the main treatments and the subtreatments and also about the effects of the subtreatments. As already mentioned, the ratio

$$\frac{\text{MSS due to } (AC)}{\text{MSS due to } (ABC)}$$

Table 9.5   EXPECTED VALUES OF MSS: SPLIT-PLOT DESIGN: MIXED MODEL WITH ONLY BLOCK EFFECTS RANDOM†

MSS due to:	Expected value
(A)	$\sigma^2 + bcS_A{}^2 + c\sigma_{AB}{}^2$
(AB)	$\sigma^2 + c\sigma_{AB}{}^2$
(C)	$\sigma^2 + abS_C{}^2 + a\sigma_{BC}{}^2$
(AC)	$\sigma^2 + bS_{AC}{}^2 + \sigma_{ABC}^2$
(BC)	$\sigma^2 + a\sigma_{BC}{}^2$
(ABC)	$\sigma^2 + \sigma_{ABC}^2$

† $S_A{}^2 = \sum A_i{}^2/(a - 1)$, $S_C{}^2 = \sum C_k{}^2/(c - 1)$, and $S_{AC}{}^2 = \sum \sum (AC)_{ik}{}^2/(ac - 1)$.

provides an exact test for (AC) under the mixed model in which A and C are fixed and B is random. If, therefore, the interactions between the main treatments and the sub-treatments are of interest but the effects of the main treatments per se are not, the split-plot design may be an economic way of getting the desired information.

Split-plot experiments can thus be used under two different conditions: (1) when, for practical reasons, it is not possible to randomize the main treatments over the subplots and (2) when the interactions between the main treatments and the sub-treatments are of intrinsic interest but the effects of the main treatments per se are not.

If the interactions between the main treatments and the subtreatments are found to be significant, a test of significance of the effects of the subtreatments is of no interest. The subtreatment effects should then be examined under each level of the main treatment.

The split-plot design lends itself readily to various modifications. One obvious extension is to the split-split-plot design, in which each subplot is further split into sub-subplots. The computations in the analysis of variance in split-split-plot experiments are straightforward extensions of those for the simple split-plot experiment. For details of the computations and other modifications of split-plot designs, reference may be made to Cochran and Cox (1957). For a variety of applications of split-plot designs, see Hetherington and Carlson (1964), Lerea and Ward (1966), and Williams, Wark, and Minifie (1963).

9.2 CONFOUNDING IN FACTORIAL EXPERIMENTS

In the split-plot experiments we saw an instance of the principle that restricting randomization of treatment combinations in certain ways makes for the confounding of some comparisons with other comparisons. In this section we shall describe systematic procedures for confounding one or more treatment comparisons with other comparisons involving treatment effects or block effects in factorial experiments. The motivation for using the principle of confounding in factorial experiments usually stems from nonavailability of a sufficiently large number of homogeneous experimental units to permit unrestricted randomization of all the treatment combinations. To illustrate, suppose that there are two factors A and B to be compared, the levels of A, in the conventional notation, being 1 and a and those of B being 1 and b. The different treatment combinations are then (1), a, b, and ab. If a randomized-blocks design is employed, the unrestricted randomization of the treatment combinations would require blocks of size 4, that is, blocks containing four experimental units each. Suppose that the available experimental units do not permit blocking in such a way that each block contains four homogeneous units. Let us say that the most that can be done is to have blocks of size 2 if each block is to have only homogeneous units within it. Under such circumstances, the experimenter is forced to choose between unrestricted randomization, which involves using blocks containing heterogeneous units, and restricted randomization, which permits making use of blocks that are more or less homogeneous. The unrestricted-randomization strategy, under the circumstances, would result in larger experimental error (residual variation) and consequently would yield estimates with relatively lower precision. The restricted-randomization procedure, insofar as it involves randomizing over homogeneous units, on the other hand, would yield estimates with relatively greater precision. A drawback of the restricted-randomization procedure is that certain comparisons become confounded with block differences. We have seen an instance of this in the split-plot design. Let us consider another instance. Suppose the experimenter randomizes the treatment combinations (1) and a in one block while the other treatment combinations, b and ab, are randomized in another block. The design, except for randomization within blocks, would then be

Block 1   Block 2

(1)	a		b	ab

Of course, the experimenter may take several pairs of blocks, each of size 2, and in each pair allocate (1) and a to one block and b and ab to the other. There would then be more than one replication in the experiment. We use the term replication here to denote one application each of all treatment combinations. For our present purpose, we shall consider only designs with one replication. Incidentally, we need to distinguish from now on between blocks and replications. So far we have used these two terms interchangeably. But here, as can be seen from the design given above, two blocks constitute one replication.

Let us suppose, without loss of generality, that the level 1 of A stands for the absence of A and the other level, a, for its presence. Similarly, let the level 1 of B represent the absence of B and the level b its presence. Let us also suppose, for the purpose of illustration, that A gives an increase of 10 units in the response value irrespective of whether B is present or not, while B gives an increase of 15 units whether or not A is present. Let the characteristics of block 1 be such that they increase each response by 20 units and those of block 2 be such that they increase each response by 25 units. Assuming no experimental error, i.e., assuming that there are no varying but unaccounted for factors at work, and taking 100 as the basal value (which represents the common impact of all neglected but constant factors at work) we have the following responses for the different treatment combinations:

$$y(1) = 100 + 20 = 120$$
$$y(a) = 100 + 10 + 20 = 130$$
$$y(b) = 100 + 15 + 25 = 140$$

and
$$y(ab) = 100 + 10 + 15 + 25 = 150$$

Because there is no experimental error, we should expect the experiment to reproduce accurately the main effects and the interaction. Bear in mind that according to our construction of the data, the main effect of A is 10 and that of B is 15 and the AB interaction is 0. Let us now see whether the experiment given above actually reproduces these values. According to the familiar formulas

$$A = \tfrac{1}{2}[y(a) - y(1) + y(ab) - y(b)]$$
$$= \tfrac{1}{2}(130 - 120 + 150 - 140) = 10$$
$$B = \tfrac{1}{2}[y(b) - y(1) + y(ab) - y(a)]$$
$$= \tfrac{1}{2}(140 - 120 + 150 - 130) = 20$$

and
$$(AB) = \tfrac{1}{2}\{[y(ab) - y(b)] - [y(a) - y(1)]\}$$
$$= \tfrac{1}{2}[(150 - 140) - (130 - 120)] = 0$$

Looking at the above results, we note that (A) and (AB) have been correctly reproduced, while (B) has been overestimated by 5 units. It is interesting that the amount of overestimate in (B) obtained above is equal to the difference between the block effects, $25 - 20$. In actual practice we may not be able to isolate the difference between the block effects from the estimate obtained for (B) in the above fashion. Thus in practical situations we have the block effects inextricably confounded with the effects of the levels of B whenever, in a 2^2 factorial experiment, the treatment combinations (1) and a are allocated to one block and b and ab are allocated to another in a given replication.

Proceeding in the above fashion, it should be easy to verify that if the treatment combinations are allocated to the complementary blocks as

Block 1

(1)	b

Block 2

a	ab

the effect confounded with the block comparison is (A), and that if the allocations are

Block 1

a	b

Block 2

(1)	ab

the effect confounded is (AB).

Clearly then if in a 2^2 factorial blocks of size 2 are employed, one of the main effects or the interaction will have to be confounded with the block comparison. Which particular factorial effect is to be confounded with the block comparison in a given particular situation is a matter for the experimenter to decide. Once the factorial effect to be confounded has been chosen, the corresponding layout can be arrived at in a systematic fashion as described below.

A Systematic Way of Arriving at Confounded Arrangements

We shall describe two ways of arriving at the confounded arrangements. Both are applicable only to experiments in which all factors are at two levels each.

METHOD 1 One method makes use of the signs attached to the treatment means (totals) in the formulas giving the main effect or interaction chosen to be confounded. Thus in a 2^2 factorial experiment, if (A) is to be confounded, the formula giving (A) is consulted. We recall that (A) is given by

$$(A) = \tfrac{1}{2}[-\bar{y}(1) + \bar{y}(a) - \bar{y}(b) + \bar{y}(ab)]$$

We note that treatment means corresponding to the combinations (1) and b bear the minus sign and the others the plus sign. The arrangement in which (A) is confounded is therefore obtained by allocating treatment combinations (1) and b, that is, those with the minus sign in the above formula, to one block and the others to the other block.

This procedure can be generalized to 2^3, 2^4, etc., factorials. In a 2^3 factorial experiment, for example, if we decide to confound (A) with the blocks, we allocate (1), b, c, and bc to one block and a, ab, ac, and abc to the other.

METHOD 2   The expression *defining contrast* is used for the factorial effect that is confounded with the block comparison in any given replication. Given the defining contrast in any replication, the allocations of the treatment combinations to different blocks within a replication can be arrived at in another manner. The rule to be followed is simply this: put in one block all treatment combinations that have odd numbers of letters common with the small letters corresponding to the defining contrast and put the remaining treatment combinations in a second block. To illustrate, suppose we want to use (AB) as the defining contrast in a 2^3 factorial experiment. The treatment combinations having odd numbers of letters common with ab, the small-letter combination corresponding to (AB), are a, b, ac, and bc. These are to be allocated to one block and the rest to another. Similarly, if in a 2^4 factorial experiment, (A) is to be used as the defining contrast, the treatment combinations that should be allocated to one block are arrived at as a, ab, ac, ad, abc, abd, acd, and $abcd$, all of which have only one letter in common with a, the small-letter combination corresponding to (A). To give another example, if $(ABCD)$ is the defining contrast in a 2^4 experiment, the treatment combinations that are to be allocated to one block are

$$\underbrace{a \quad b \quad c \quad d}_{\text{1 common}} \qquad \underbrace{abc \quad abd \quad acd \quad bcd}_{\text{3 common}}$$

and those that are to be allocated to the second block are

$$\underbrace{(1)}_{\text{0 common}} \quad \underbrace{ab \quad ac \quad ad \quad bc \quad bd \quad cd}_{\text{2 common}} \qquad \underbrace{abcd}_{\text{4 common}}$$

As pointed out above, these methods have the disadvantage of being applicable only to 2^n factorials. For a general method applicable to all factorial experiments, see Kempthorne (1952) or Kirk (1968).

When there are several replications in a factorial experiment, it is possible to use one factorial effect as the defining contrast in some of the replications and not for all replications. If the same factorial effect is used as the defining contrast in all replications, the case is referred to as one of *complete confounding* of the factorial effect that serves as the defining contrast. If one defining contrast is used only in a subset of replications, the case is referred to as one of *partial confounding*. An example of partial confounding is

Replication	Complementary blocks				Defining contrast
1	(1)	a	b	ab	(B)
2	(1)	ab	a	b	(AB)

In replication 1, the main effect (B) has been used as the defining contrast, while in replication 2, it is the interaction (AB) that serves as the defining contrast. The advantage of partial confounding is that information on a given main effect or interaction can be had from the replications in which it is not serving as the defining contrast.

9.3 ANALYSIS OF THE 2^n FACTORIAL WITH CONFOUNDING

Suppose that the main effect (A) has been completely confounded in a 2^n factorial experiment. Let there be r replications. To analyze the data from this experiment, we first obtain in the usual manner, ignoring the fact that (A) has been confounded completely, the following sums of squares: total SS, SS's due to (A), (B), (AB), (C), (AC), ... The next step is to write down the block totals in a two-way format as shown in Table 9.6.

The row totals in Table 9.6 give the totals in different replications, and the totals in the cells are the block totals. The first column total gives the total under treatment combinations in which A occurs at level 1, and the second column total gives the total under treatment combinations in which A occurs at level a. The cell SS obtained from the above table represent the block SS, the row SS represent the replication SS, and the column SS represent the SS due to the main effect (A). The difference cell SS − replication SS − SS due to (A) represents the SS due to nonadditivity between (A) and the replications. It is not always necessary to partition the block SS into

> Replication SS
> SS due to (A)
> SS due to nonadditivity between (A) and replications

We have described this partitioning here in order to bring out clearly the constituents of the block SS. The residual (error) SS is obtained by subtraction:

Error SS = total SS − block SS − total of SSs due to different
factorial effects other than (A)

The outline of the analysis of variance is shown in Table 9.7, using for illustration the 2^2 factorial with r replications, (A) being confounded completely.

Table 9.6 ARRANGEMENT OF BLOCK TOTALS IN EACH REPLICATION BY LEVELS OF FACTOR A

| Replication | Level of A | | Total |
	1	a	
1			
2			
⋮			
r			
Total			

Table 9.7   OUTLINE OF ANALYSIS OF VARIANCE:
2^2 FACTORIAL WITH MAIN EFFECT OF A
CONFOUNDED

Source of variation		df
Main effect (B)		1
Interaction (AB)		1
Blocks:		
Main effect (A)	1	
Replications	$r - 1$	
Nonadditivity:		
(A) × replications	$r - 1$	
		$2r - 1$
Error		$(4r - 1) - (2r - 1) - 2$
Total		$4r - 1$

Example To illustrate the analysis described above and to bring out clearly the changes in degrees of freedom when confounding is employed we shall analyze a given set of data in two ways, first assuming no confounding and then assuming that the main effect of A has been completely confounded. For this purpose we shall reproduce the data reported in Table 8.3, stipulating that the figures given in different rows in that table represent different replications. Table 9.8A shows the data assum-

Table 9.8   DATA IN TABLE 8.3 REPRODUCED UNDER VARIOUS STIPULATIONS

A That Each Row in Table 8.3 Represents a Replication of a 2^2 Factorial Experiment
with a Randomized-Blocks Layout

	Treatment combination				
Replication	(1)	a	b	ab	Total
1	3.4	0.4	5.0	2.6	11.4
2	3.6	0.9	4.8	2.2	11.5
3	4.0	0.3	5.3	2.9	12.5
Total	11.0	1.6	15.1	7.7	35.4

Analysis

$$\text{Correction for mean} = \frac{35.4^2}{12} = 104.43$$

$$\text{Total SS} = 3.4^2 + \cdots + 2.9^2 - 104.43 = 33.29$$
$$\text{Treatment SS} = \tfrac{1}{3}(11.0^2 + \cdots + 7.7^2) - 104.43 = 32.52$$
$$\text{Replication SS} = \tfrac{1}{4}(11.4^2 + 11.5^2 + 12.5^2) - 104.43 = .18$$
$$\text{Error SS} = 33.29 - 32.52 - .18 = .59$$
$$\text{SS due to } (A) = \tfrac{1}{12}(-11.0 + 1.6 - 15.1 + 7.7)^2 = 23.52$$
$$\text{SS due to } (B) = \tfrac{1}{12}(-11.0 - 1.6 + 15.1 + 7.7)^2 = 8.67$$
$$\text{SS due to } (AB) = \tfrac{1}{12}(11.0 - 1.6 - 15.1 + 7.7)^2 = .33$$
$$\text{CHECK: } 23.52 + 8.67 + .33 = \text{treatment SS } (= 32.52)$$

(continued overleaf)

ing that no confounding has been employed, and Table 9.8B shows the same set of data assuming that (A) has been confounded in all replications. [In reproducing the data, the symbols a_0b_0, a_1b_0, a_0b_1, and a_1b_1 have been replaced by the conventional symbols (1), a, b, and ab, respectively.]

The steps involved in the analysis are described at the bottom of each table. The analysis of variance for the data in Table 9.8A is given in Table 9.9A, and that

Table 9.8 (*continued*)

B That Each Row in Table 8.3 Represents a Replication in a 2^2 Factorial and That Main Effect (A) Has Been Completely Confounded

Replication	First block (1)	b	Block total	Second block a	ab	Block total	Replication total
1	3.4	5.0	8.4	.4	2.6	3.0	11.4
2	3.6	4.8	8.4	.9	2.2	3.1	11.5
3	4.0	5.3	9.3	.3	2.9	3.2	12.5
Treatment total	11.0	15.1		1.6	7.7		35.4

Analysis

$$\text{Correction for mean} = \frac{35.4^2}{12} = 104.43$$

$$\text{Total SS} = 3.4^2 + \cdots + 2.9^2 - 104.43 = 33.29$$
$$\text{Block SS} = \tfrac{1}{2}(8.4^2 + 8.4^2 + 9.3^2 + 3.0^2 + 3.1^2 + 3.2^2) - 104.43 = 23.80$$
$$\text{Treatment SS}\dagger = \tfrac{1}{3}(11.0^2 + 15.1^2 + 1.6^2 + 7.7^2) - 104.43 = 32.52$$
$$\text{SS due to } (A)\dagger = \tfrac{1}{12}(-11.0 - 15.1 + 1.6 + 7.7)^2 = 23.52$$
$$\text{SS due to } (B) = \tfrac{1}{12}(-11.0 + 15.1 - 1.6 + 7.7)^2 = 8.67$$
$$\text{SS due to } (AB) = \tfrac{1}{12}(11.0 - 15.1 - 1.6 + 7.7)^2 = .33$$
CHECK: $23.52 + 8.67 + 0.33 = \text{treatment SS } (= 32.52)$

Block Totals Written in a Two-Way Table

Replication	Complementary blocks Block with the 1 level of A	Block with the a level of A	Total
1	8.4	3.0	11.4
2	8.4	3.1	11.5
3	9.3	3.2	12.5
Total	26.1	9.3	35.4

Partitioning of Block SS

From the two-way table in which the block totals form the cell totals:
$$\text{Cell SS} = \tfrac{1}{2}(8.4^2 + 8.4^2 + 9.3^2 + 3.0^2 + 3.1^2 + 3.2^2) - 104.43 = 23.80 \,(= \text{block SS})$$
$$\text{Row SS} = \tfrac{1}{4}(11.4^2 + 11.5^2 + 12.5^2) - 104.43 = .18 \,(= \text{replication SS})$$
$$\text{Column SS} = \tfrac{1}{6}(26.1^2 + 9.3^2) - 104.43 = 23.52 \,[= \text{SS due to } (A)]$$
$$\text{SS due to nonadditivity between replications and } (A) = 23.80 - .18 - 23.52 = .10$$

† Calculated for checking purposes only.

for the data in Table 9.8*B* in Table 9.9*B*. The main point that should be emphasized in the analysis shown in Table 9.8*B* and carried over to Table 9.9*B* concerns the partitioning of the block SS. As pointed out above, this partitioning is not a necessary step. However, if the block SS is partitioned as shown in Table 9.9*B*, it is legitimate to test the significance of the variation due to (*A*) in terms of the MSS associated with the nonadditivity between (*A*) and replications. This test, however, usually is of low sensitivity. It is also important to remember that even if the above test reveals that the variation due to (*A*) is significant, there is no way of estimating the effect due to (*A*), since the comparison that reveals (*A*) contains the block differences too.

Analysis in the Case of Partial Confounding

The analysis of a partially confounded experiment is usually more complicated than that of one with complete confounding. Let us consider, for illustrative purposes, a 2^3 factorial with four replications, all the interactions being confounded in one replication each. The design, except for randomization, is shown in Table 9.10.

It should be emphasized here that in practice one need not always confound all the interactions the same number of times. Nor should there be any restriction on the number of replications. We have used four replications in Table 9.10 for illustrative purposes only. One may decide to have more than four replications or less than four. One may confound (*AB*) in any three replications and (*ABC*) in the other or

Table 9.9 ANALYSIS OF VARIANCE OF DATA IN TABLE 9.8

A Data of Table 9.8*A*

Source of variation	SS	df	MSS	F ratio
Main effect (*A*)	23.52	1	23.52	235.2
Main effect (*B*)	8.67	1	8.67	86.7
Interaction (*AB*)	.33	1	.33	3.3
Replications	.18	2	.09	.9
Error	.59	6	.10	
Total	33.29	11		

B Data of Table 9.8*B*

Source of variation		SS	df	MSS	F ratio
Main effect (*B*)		8.67	1	8.67	72.25
Interaction (*AB*)		.33	1	.33	2.75
Blocks:					
Main effect (*A*)	23.52		1 ⎫		
Replications	.18		2 ⎬		
Nonadditivity	.10		2 ⎭		
		23.80	5	4.76	39.67
Error		0.49	4	.12	
Total		33.29	11		

Table 9.10 A PARTIALLY CONFOUNDED 2^3 FACTORIAL EXPERIMENT WITH FOUR REPLICATIONS EACH OF WHICH CONFOUNDS ONE OF THE INTERACTIONS

Replication	Complementary blocks								Confounded factorial effect
1	(1)	c	ab	abc	a	b	ac	bc	(AB)
2	(1)	b	ac	abc	a	c	ab	bc	(AC)
3	(1)	a	bc	abc	b	c	ab	ac	(BC)
4	(1)	ab	ac	bc	a	b	c	abc	(ABC)

(ABC) in any two replications and (BC) in the rest, and so on. The only important consideration to be taken into account when deciding about the effects to be confounded in partial confounding is that information on any given factorial effect will be extractable only from those replications in which that effect has not been confounded. So if the number of replications in which a given factorial effect is not confounded is not the same as the number of replications in which another factorial effect is not confounded, the information available from the experiment on these two factorial effects will differ in precision.

We now turn to the analysis of the experiment shown in Table 9.10. The calculation of total SS and of SS's due to (A), (B), and (C) is done in the usual manner. To calculate the SS due to (AB), we omit replication 1 (in which it has been confounded) and then proceed as usual. Similarly, for calculating the SS due to (AC) we omit replication 2, and so on. Thus the SS due to each of the interactions is estimated on the basis of three replications. Since all the main effects are estimated on the basis of four replications while each of the interaction effects is estimated on the basis of three

Table 9.11 OUTLINE OF ANALYSIS OF VARIANCE: 2^3 FACTORIAL WITH EACH INTERACTION PARTIALLY CONFOUNDED

Source of variation	df
(A)	1
(B)	1
(C)	1
(AB)	1'
(AC)	1'
(BC)	1'
(ABC)	1'
Blocks	7
Residual (error)	17
Total	31

replications, the estimates of the interaction effects are obviously less precise than the estimates of the main effects.

The residual (error) SS is obtained by subtraction. The analysis of variance is shown in the form outlined in Table 9.11. The degrees of freedom for the SS due to any partially confounded effect is conventionally indicated by a prime, to remind us that the corresponding information in the analysis of variance is partial, because of the partial confounding.

We shall not discuss complete or partial confounding in 3^n and other factorial experiments. Reference may be made to Kempthorne (1952) or to Kirk (1968) for a detailed discussion of the topic.

9.4   CONFOUNDING MORE THAN ONE EFFECT IN THE SAME REPLICATION

So far we have seen only cases in which at most one factorial effect is confounded in any replication. But situations may arise in practice where two or more effects have to be confounded simultaneously in the same replication. Consider, for example, a 2^4 factorial experiment. To try out all the treatment combinations in one replication without any restriction on randomization, we would need a block containing 16 homogeneous units. If simple confounding were to be employed, i.e., if the confounding in a given replication were to be restricted to one factorial effect, blocks of size 8 would be required. Suppose that blocks of 8 units are still too large to achieve homogeneity. Under such circumstances, if blocks of 4 homogeneous units can be found, we would prefer to conduct the experiment using blocks of size 4. But if blocks of 4 units each are used in each replication in a 2^4 factorial experiment, three factorial effects will be confounded with blocks in each replication. To illustrate this, let us consider the following 2^3 factorial experiment in which there is only one replication, but which employs four blocks, of 2 units each:

Block 1		Block 2		Block 3		Block 4	
a	ac	ab	abc	(1)	c	b	bc

Let us stipulate that the effects of different factors operating in the experiment have been as shown in Table 9.12. Taking 100 as the basal value (which represents the common impact of all neglected but constant factors at work), we find that the different treatment combinations have the responses shown in Table 9.13. From these responses we get the following estimates for the different factorial-effect means:

$$(A) = \quad 3.5 \qquad (AC) = 0$$
$$(B) = \quad 18.5 \qquad (BC) = 0$$
$$(AB) = -1.5 \qquad (ABC) = 0$$
$$(C) = \quad 20.0$$

We note that (C), (AC), (BC), and (ABC) have been correctly reproduced, whereas the same cannot be said about (A), (B), or (AB). It can be verified that the estimate obtained for (A) is in fact the sum of the main effect of A and the following block comparison:

$$\tfrac{1}{2}(\text{block } 1 + \text{block } 2 - \text{block } 3 - \text{block } 4) = \tfrac{1}{2}(8 + 10 - 13 - 18) = -6.5$$

Similarly, it can easily be verified that the estimate obtained for (B) above is in fact the sum of the main effect of B and the following comparison between the block effects:

$$\tfrac{1}{2}(-\text{block } 1 + \text{block } 2 - \text{block } 3 + \text{block } 4) = 3.5$$

and that the estimate obtained above for (AB) is actually the sum of the AB interaction effect and the following block comparison:

$$\tfrac{1}{2}(-\text{block } 1 + \text{block } 2 + \text{block } 3 - \text{block } 4) = -1.5$$

Here then we have an illustration of the principle that if a 2^n factorial experiment is run on four blocks of 2^{n-2} units each, three factorial effects will be confounded with blocks. This principle can be used to determine in a systematic manner the allocations of the treatment combinations to different blocks in any replication. The strategy is to choose any two factorial effects that, in the opinion of the experimenter, can be sacrificed, i.e., confounded with blocks, and use them to determine the allocations. The third effect confounded with the blocks is automatically determined according to the following combinatorial rule: given two factorial effects, for example, (AB) and (AC), to be confounded, the effect represented by the symbolic product of these factorial effects from which squared letters are dropped, for example, $(AB)(AC) = (A^2BC) = (BC)$, will automatically be confounded.

The procedure to determine, in 2^n factorial, the allocations of the treatment combinations to the different blocks in any replication, given two effects to be confounded, can be described in terms of the following steps.

STEP 1 Put in one group all treatment combinations that have odd numbers of letters common with the letters in one of the factorial effects chosen to be confounded and put the remaining treatment combinations in another group.

Table 9.12 SIMULATED DATA SITUATION:
2^3 FACTORIAL IN FOUR BLOCKS

Factor	Level	Effect	
A	1	0 ⎫	At all levels of B and C
	a	10 ⎭	
B	1	0 ⎫	At all levels of A and C
	b	15 ⎭	
C	1	0 ⎫	At all levels of A and B
	c	20 ⎭	
Block 1	—	8	
Block 2	—	10	
Block 3	—	13	
Block 4	—	18	
Error	—	0	

STEP 2 Apply the procedure described in step 1 to each of the groups obtained in step 1 but using the second factorial effect chosen to be confounded.

For example, suppose in a 2^3 factorial experiment, (AB) is one of the effects chosen, at will, to be confounded. Then the treatment combinations are to be put into two groups as shown below:

Group 1	a	b	ac	bc
Group 2	(1)	c	ab	abc

Note that group 1 contains all treatment combinations having one and only one letter corresponding to those in (AB). Group 2 contains all the remaining treatment combinations.

Now suppose (AC) is the second effect chosen to be confounded. Then the treatment combinations belonging to group 1 shown above are to be put into two groups as shown below:

Group 1(i)	a	bc
Group 1(ii)	b	ac

Note that group 1(i) contains all treatment combinations in group 1 that have one and only one letter common to the letters in (AC). Group 1(ii) contains all the remaining treatment combinations in group 1. Group 2 is to be subdivided into two groups in the same manner:

Group 2(i)	c	ab
Group 2(ii)	(1)	abc

The analysis described in Sec. 9.3 can be extended to the cases in which two or more effects are confounded with blocks in any replication. For some applications of confounded factorials see Hunt (1961) and Staats, Staats, and Heard (1959).

Table 9.13 OBSERVATIONS UNDER EACH TREATMENT
COMBINATION CALCULATED FROM TABLE 9.12

Treatment combination	Basal value	Treatment effect	Block effect	Error	Response
(1)	100	0	13	0	$y(1) = 113$
a	100	10	8	0	$y(a) = 118$
b	100	15	18	0	$y(b) = 133$
ab	100	$10 + 15$	10	0	$y(ab) = 135$
c	100	20	13	0	$y(c) = 133$
ac	100	$10 + 20$	8	0	$y(ac) = 138$
bc	100	$15 + 20$	18	0	$y(bc) = 153$
abc	100	$10 + 15 + 20$	10	0	$y(abc) = 155$

9.5 FRACTIONAL REPLICATIONS

So far we have been concerned with experiments in which the entire set of treatment combinations appears in each replication. Factorial experiments in which all treatment combinations appear in each replication are called *full factorials*, which are to be distinguished from *fractional factorials*, in which only a fraction of the full set of treatment combinations is run in each replication. The reason for running only a fraction of the full replicate in each replication may often be economic. Consider, for example, an investigation in which seven factors each at two levels are to be simultaneously studied. A full replicate would require the use of 128 experimental units in all, whether confounding is to be used or not. Often such large experiments are too costly. In such circumstances, the question arises whether the required information, e.g., estimates of given main effects and/or interactions, with a given precision, can be obtained by running a fraction, e.g., one-eighth, of the full replicate. The theory of fractional replication deals with this question.

Sometimes the reason for employing a fractional replicate, rather than a full replicate, may be a practical or technical one. An example of such a case can be found in the classical before-after experimental design. The following is a typical application of this design in social-engineering experiments.

A community is chosen for a social-reform experiment, e.g., one in which severe crackdown of drug-related crimes is organized along with an elaborate effort to treat cases of drug addiction. Before introducing the "reform" the "social health" of the community is measured in terms of a suitable index, e.g., death rate due to drug overdose. After the reform has been in operation for a reasonably long period, the social health is again measured. The experiment thus provides measurements of a study variate (social health) for an experimental unit (the community in question) before as well as after the introduction of the experimental stimulus (the social reform). Hence the experiment is said to have a before-after design.

In recent expositions (e.g., Ross and Smith, 1968), it has been held that a proper specification of the structure of the above design demands the recognition of a factor, called the *uncontrolled events*, presumed to be at work as the experiment is in progress and consequently influencing the magnitude of the after measurements. An example of such an uncontrolled event in a social-reform experiment may be the activities of an external agency, other than the experimenter, aiming to promote the very thing the social reform was aimed to bring about. The influence of the uncontrolled events may be slight in some cases and substantial in others. Since the experimenter is presumed to have no control over such events, it is considered safer to assume that such events are always present. Thus there are three factors assumed to be at work: A the pretest,[1] B the experimental stimulus, and C the uncontrolled events. Theoretically, any one of these factors may be present or absent; in other words, each factor has two levels. Hence we have a 2^3 factorial situation. However, instead of eight observations constituting the full replicate, only two observations are available, those represented, respectively, by the pretest and posttest mean scores. (We shall ignore for the moment the fact that the pretest scores and the posttest scores are obtained from the same

---

[1] Pretest here represents the before measurement.

subjects.) The pretest mean score is like the bias (error) of the balance in the weighing experiment we have described at the beginning of Chap. 8. The posttest mean score, on the other hand, represents the cumulative impact of the pretest, the experimental stimulus, and the set of uncontrolled events. These two mean scores may therefore be represented by $\bar{y}(1)$ and $\bar{y}(abc)$, respectively, using the conventional notation. If the experiment is confined to one community, it is not possible to get observations on any of the other treatment combinations: a, b, ab, c, ac, or bc. Thus we have in this case a situation in which it is impractical to obtain more than a fraction (one-fourth) of the full replicate.

To understand how inferences are to be drawn from fractional factorials, let us consider different half replicates that can be derived from a 2^2 factorial. (We take for consideration 2^2 factorial because of its simplicity. The results are directly generalizable to 2^3, 2^4, 2^5, etc., factorials. The fact that with only two factors each having two levels, one may not usually employ half replicates need not detain us here.)

Since by definition any half replicate of 2^2 factorial must contain 2 treatment combinations, it is easily seen that the following list exhausts the different possible half replicates from a 2^2 factorial:

Block

(i) | (1) | a

(ii) | (1) | b

(iii) | (1) | ab

(iv) | a | b

(v) | a | ab

(vi) | b | ab

Of these half-replicates, (i) and (vi), (ii) and (v), and (iii) and (iv) are complementary halves. We notice that if we put the complementary halves (i) and (vi) into two complementary blocks, we have a full replicate in which the main effect (B) is confounded. We can therefore say that the half replicates (i) and (vi) are uniquely defined by the main effect (B). In other words, (B) is the *defining contrast* of these two half replicates. Similarly, we notice that (A) is the defining contrast of the half replicates (ii) and (v) and that (AB) is the defining contrast of the half replicates (iii) and (iv). This leads us to the following rule for deriving half replicates.

RULE 1 Choose one of the factorial effects as the defining contrast. Subdivide the treatment combinations into two complementary blocks confounding the defining contrast with the blocks. Choose one of the blocks thus obtained. That gives the required half replicate.

The question immediately arises which factorial effect one should use as the defining contrast in a given situation. To find an answer to this question, recall our earlier discussion on confounding. It was pointed out in that connection that no replication can provide any estimate of the factorial effect confounded in that replication. Needless to say, if this is true of full replicates, it is true of any of the complementary half replicates too. Hence we have the following rule.

RULE 2   In order to get a fractional replicate, do not use as the defining contrast any factorial effect that must be estimated from the experiment run on that fractional replicate.

To get a quarter replicate, two defining contrasts are to be used in the fashion described in Sec. 9.4. Divide the full replicate into four complementary blocks; then any one of the four blocks will be a quarter replicate. To get a one-eighth replicate, we use three defining contrasts, divide a full replicate into eight complementary blocks, and then take any one block. It should be remembered in this connection that *any* three factorial effects cannot serve as three defining contrasts for the purpose of dividing a full replicate into eight complementary blocks. Thus, for example, if (AB) and (AC) have been used to divide the full replicate into four complementary blocks, (BC) will automatically be confounded with the blocks and therefore (BC) cannot serve as the third defining contrast to be used in subdividing each of the four blocks obtained by using (AB) and (AC) (see Sec. 9.4).

Suppose that we have obtained a fractional replicate and that the experiment has been carried out. How are we to interpret the results? Consider again the simple case of a half replicate in a 2^2 factorial. Suppose that the half replicate chosen for the experiment was

(1)	ab

Let us further suppose that a completely randomized design was employed and that the treatment means $\bar{y}(ab)$ and $\bar{y}(1)$ were obtained. The only comparison possible between the treatment means is

$$\bar{y}(ab) - \bar{y}(1) \qquad (9.1)$$

What does this comparison represent? To find out, recall Eqs. (8.10), obtained in Sec. 8.2, which express each treatment mean in terms of the factorial effects and the general mean M. From those equations we have

$$\bar{y}(ab) = \quad M + \tfrac{1}{2}[(A) + (B) + (AB)] \qquad (9.2)$$

and

$$\bar{y}(1) = M + \tfrac{1}{2}[-(A) - (B) + (AB)] \qquad (9.3)$$

from which by subtraction

$$\bar{y}(ab) - \bar{y}(1) = (A) + (B) \qquad (9.4)$$

In other words, the comparison furnished by the half replicate

(1)	ab

estimates the sum of the main effects of A and B. Unfortunately, we cannot tell (A) and (B) apart. They are inextricably confounded with each other, as it were. Such factorial effects are called *aliases*. From (9.4), it is clear that *if and only if* it is reasonably certain that (B) is negligible (and this is to be known from sources external to the experiment in question) can the comparison $\bar{y}(ab) - \bar{y}(1)$ be interpreted as representing the main effect (A). If (B) is not known to be negligible or cannot be judged to be so, the comparison $\bar{y}(ab) - \bar{y}(1)$ is of little use (see, however, Namboodiri, 1970). This leads us to the following rule.

RULE 3   Examine carefully for each factorial effect its aliases implicated by the fractional replicate taken for consideration. If a factorial effect of interest has non-negligible aliases, reject the fractional replicate in question.

Even if (B) is judged to be negligible, there may be difficulty in giving a practically useful interpretation to the comparison $\bar{y}(ab) - \bar{y}(1)$ in (9.4). The difficulty stems from the presence of the AB interaction. To understand the difficulty, let us consider the following simple experiment, with one observation per treatment combination.

The factor A may be either present or absent, B also may be either present or absent. A increases the response by 2 units when B is absent, that is, $y(a) - y(1) = 2$, and by 10 units when B is present that is, $y(ab) - y(b) = 10$. B, on the other hand, decreases the response by 4 units when A is absent, that is, $y(b) - y(1) = -4$, but increases it by 4 units when A is present that is, $y(ab) - y(a) = 4$. Assuming that the experimental error is negligible and that the basal value is 10, we have in the full replicate, taking one observation per treatment combination, $y(1) = 10$, $y(a) = 12$, $y(b) = 6$, $y(ab) = 16$. We at once get

$$(A) = \tfrac{1}{2}(-10 + 12 - 6 + 16) = 6$$

$$(B) = \tfrac{1}{2}(-10 - 12 + 6 + 16) = 0$$

$$(AB) = \tfrac{1}{2}(+10 - 12 - 6 + 16) = 4$$

Now suppose that we (the experimenter) have access only to the half replicate consisting of $y(ab) = 16$ and $y(1) = 10$. Using the contrast available from this half replicate, we would write down the equation

$$y(ab) - y(1) = 16 - 10 = (A) + (B) \qquad (9.5)$$

Let us suppose that we are willing to assume that (B) is negligible. On that basis, we would write from (9.5)

$$(A) = 6 \qquad (9.6)$$

Can we interpret the above result to mean that the factor A has a positive effect of 6 units? Can we say that this estimate holds at both the levels of B? If these questions are answered in the affirmative, it amounts to assuming that there is no interaction between A and B when, in fact, there is a positive interaction between them. Does the problem of interpretation become simplified if we assume that there is a positive interaction between A and B? The answer is negative because the half replicate under discussion does not provide enough information for interpreting the effect of A when the AB interaction is nonnegligible. If the AB interaction is assumed to be present, we should be able to say what the effect of A will be when B is present and/or what the effect will be when B is absent. But the half replicate consisting of $y(ab)$ and $y(1)$ does not provide enough information for such a description. In fact, we are not even able to say whether the estimate 6 obtained in (9.6) represents the effect of A when B is absent or whether it represents the effect of A when B is present; all that we can say is that it represents an average of the effects of A under the two levels of B. The last type of interpretation will be of little practical use, since those who want to use the experimental results may want to know the effect of A in specific contexts such as when B is present or when B is absent.

If the AB interaction is actually absent, the above difficulty does not arise, as can be seen from the following example. A may be either present or absent; B also may be either present or absent. The effect of A is to increase the response by 2 units at both levels of B, while B has no effect at either level of A: $y(a) - y(1) = 2$; $y(ab) - y(b) = 2$; $y(b) - y(1) = 0$; $y(ab) - y(a) = 0$. Basal value $= 10$. Experimental error is assumed to be zero. We have then in the full replicate taking one observation per treatment combination,

$$y(1) = 10 \qquad y(a) = 12 \qquad y(b) = 10 \qquad y(ab) = 12$$

from which

$$(A) = 2 \qquad (B) = 0 \qquad (AB) = 0$$

If we (the experimenter) have access only to the half replicate $y(ab)$, $y(1)$, and if we know that (B) is negligible we may, without any ambiguity, interpret the comparison $y(ab) - y(1) = 2$ as representing the effect of A, which we may assume to hold at both levels of B, insofar as we judge the AB interaction to be negligible, which in this case will be a valid judgment.

A Systematic Way to Identify the Aliases Estimated by a Given Contrast Between Treatment Means

Let us now go back to Eq. (9.4), which gives the aliases estimated by the contrast $y(ab) - y(1)$ obtained from the half replicate

ab	(1)

of the 2^2 factorial. In arriving at (9.4) we used the equations derived earlier that gave each treatment mean in the 2^2 factorial system in terms of the general mean and the

Table 9.14   DEFINITION OF CONTRASTS IN THE ONE-HALF REPLICATE [ab (1)] OF THE 2^2 FACTORIAL SYSTEM

	Treatment combinations	
Factorial effects	ab	(1)
(A)	$+1$	-1
(B)	$+1$	-1
(AB)	$+1$	$+1$

factorial effects. This procedure is often likely to be cumbersome. It would therefore be of interest to know easier ways to write down equations like (9.4). We start with a table of multipliers giving each factorial effect (see, for example, Table 8.2). From the table it is easy to write down equations like (9.4). For the half replicate consisting of (*ab*) and (1) from the 2^2 factorial system, we confine our attention to the multipliers under treatment combinations (*ab*) and (1), reproduced in Table 9.14. Note that we can prepare Table 9.14 without referring to Table 8.2. We simply apply the rules we applied in preparing Table 8.2. To estimate (*A*), we apply the multiplier $+1$ to all treatment combinations containing *a* and -1 to all treatment combinations that do not contain *a*. Similarly, to estimate (*B*), we apply $+1$ to all treatment combinations containing *b* and -1 to those that do not contain *b*. The multipliers for the *AB* inter-action are obtained by taking the products of the corresponding multipliers for (*A*) and (*B*). After preparing the tables of multipliers we compare each row with all other rows. In the present case we notice that the row for (*A*) is identical to that for (*B*), while the row for (*AB*) is different from the other two, and so we have (*A*) and (*B*) as aliases. Consequently the comparison $y(ab) - y(1)$ represents the sum (*A*) + (*B*). Let us apply this procedure to the following half replicate of the 2^3 factorial system:

a	*b*	*c*	*abc*

We first prepare the table of multipliers identifying the different factorial effects, shown in Table 9.15. From Table 9.15 it is seen that the multipliers for (*A*) are the same as those for (*BC*), that the multipliers for (*B*) are the same as those for (*AC*), and that the multipliers for (*C*) are the same as those for (*AB*). The estimating equations for the different factorial effects are therefore

$$\tfrac{1}{2}[\bar{y}(a) - \bar{y}(b) - \bar{y}(c) + \bar{y}(abc)] = (A) + (BC) \qquad (9.7)$$

$$\tfrac{1}{2}[-\bar{y}(a) + \bar{y}(b) - \bar{y}(c) + \bar{y}(abc)] = (B) + (AC) \qquad (9.8)$$

and
$$\tfrac{1}{2}[-\bar{y}(a) - \bar{y}(b) + \bar{y}(c) + \bar{y}(abc)] = (C) + (AB) \qquad (9.9)$$

The external multiplier $\tfrac{1}{2}$ is used in accordance with the principle that each comparison on the left of the above equations is an average of two comparisons. For example, $\tfrac{1}{2}[\bar{y}(a) - \bar{y}(b) - \bar{y}(c) + \bar{y}(abc)]$ is the average of $\bar{y}(a) - \bar{y}(c)$ and $\bar{y}(abc) - \bar{y}(b)$, each of which contains information on (*A*) + (*BC*).

Table 9.15 DEFINITION OF CONTRASTS IN THE ONE-HALF REPLICATE [*a b c abc*] OF THE 2^3 FACTORIAL SYSTEM

	Treatment combination			
Factorial effect	*a*	*b*	*c*	*abc*
(*A*)	$+1$	-1	-1	$+1$
(*B*)	-1	$+1$	-1	$+1$
(*C*)	-1	-1	$+1$	$+1$
(*AB*)	-1	-1	$+1$	$+1$
(*AC*)	-1	$+1$	-1	$+1$
(*BC*)	$+1$	-1	-1	$+1$
(*ABC*)	$+1$	$+1$	$+1$	$+1$

From (9.7) we notice that (A) and (BC) are aliases. Similarly, from (9.8), (B) and (AC) are found to be aliases, and from (9.9), (C) and (AB) are found to be aliases. Had we noticed to begin with that the half replicate under discussion has (ABC) as the defining contrast, it would have been possible for us to identify the aliases by applying the following simple rule.

RULE 4 To get the aliases in half replicates of a given factorial effect, obtain the symbolic product of the given factorial effect with the defining contrast of the half replicate and cancel out all the squared letters.

For example, in the half replicate under discussion, the alias of (A) is given by $(A) \times (ABC) = [A^2BC] = (BC)$, and the alias of (BC) is given by $(BC) \times (ABC) = [AB^2C^2] = (A)$.

The above rule can be extended to one-fourth, one-eighth, etc., replicates. In each one-fourth replicate, there are actually three defining contrasts involved. Two of these can be chosen at will when getting the one-fourth replicate. (These are to be used in the manner described earlier to subdivide the full replicate into complementary blocks.) The third one is represented by the symbolic product of the first two, from which all squared letters are canceled out. We have then the following rule for obtaining the aliases of any given effect in quarter replicates of 2^n systems.

RULE 5 Each effect in a quarter replicate in 2^n factorial systems has three aliases, and these can be obtained by multiplying the given effect with the defining contrasts and canceling out all the squared letters.

For example, consider the quarter replicate

(1)	abc

of the 2^3 factorial system. This quarter replicate, we have seen earlier, represents the classical before-after experimental design. It can be verified that the defining contrasts of this quarter replicate are (AB), (AC), and (BC). Hence by applying rule 5 we have the aliases of (A) given by

$$(A) \times (AB) = [A^2B] = (B)$$

$$(A) \times (AC) = [A^2C] = (C)$$

and

$$(A) \times (BC) = (ABC)$$

Clearly then the contrast available from the before-after experiment estimates the sum of four factorial effects: $(A) + (B) + (C) + (ABC)$. All the four factorial effects listed above are confounded with each other.

9.6 INCOMPLETE NONFACTORIAL DESIGNS

We have seen above that when treatments are arranged in factorial form, we can use the technique of confounding or arrangements involving fractional factorials if the available blocks are smaller than a complete replication would require. Sometimes, however, the treatments may not be arranged in factorial form, e.g., varieties of rice in an agricultural experiment, or the factorial structure may not be of importance. In such situations, if the available blocks are smaller than a full replication, we can carry out the experiment using incomplete nonfactorial designs.

There are four major classes of incomplete nonfactorial designs. Designs arranged in randomized incomplete blocks or in quasi-Latin squares, each either (completely) *balanced* or *partially balanced*. The main features of these four classes of designs are described below (see also Exercises 9.5 to 9.7).

The following is an illustration of a randomized balanced incomplete-block design with four treatments (*A*, *B*, *C*, and *D*) and blocks of size 2.

Block

1	A	B

2	C	D

3	A	C

4	B	D

5	A	D

6	B	C

Note that every pair of treatments appears together the same number of times (once in the present case) in the same block; for example, *A* and *B* appear together in block 1, *A* and *D* in block 5, *C* and *D* in block 2, and so on. This is why the design is called balanced. This property permits the use of the same standard error for comparing every pair of treatments. It is also interesting to note that blocks 1 and 2 together make a complete replication. Similarly blocks 3 and 4 make another replication and blocks 5 and 6 a third. Thus in this case three replications are needed to ensure balance. It is possible to construct such designs for any number of treatments *t* and any size of block *k*. But the minimum number of replications required for complete balance is determined by *t* and *k*. Some of the designs require too many replications and hence may not be of practical interest. Cochran and Cox (1957) provide a list of useful designs. The statistical analysis is also described in detail in that book.

The following is an illustration of a balanced-lattice design (balanced incomplete blocks arranged in quasi-Latin squares) (see Table 9.16). In this design every pair of treatments appears once in the same row (block) and once in the same column. All pairs of treatments are compared with the same precision. The number of treatments must always be a perfect square in order to be able to construct a balanced-lattice design. Balanced lattices do not exist for all treatment numbers that are perfect squares, for example, 36. A number of useful balanced-lattice designs are listed in Cochran and Cox (1957). The same book may be consulted for the method of statistical analysis applied to this type of design.

Table 9.16 BALANCED-LATTICE DESIGN

Replication	Row (block)	Column		
		1	2	3
1	1	A	B	C
	2	D	E	F
	3	G	H	I
		4	5	6
2	4	A	D	G
	5	B	E	H
	6	C	F	I
		7	8	9
3	7	A	F	H
	8	I	B	D
	9	E	G	C
		10	11	12
4	10	A	I	E
	11	F	B	G
	12	H	D	C

If we exclude from the balanced-lattice design the blocks constituting the last two replications, the remainder is a partially balanced incomplete-block design. Partially balanced designs are generally less useful than balanced designs and involve more complex statistical analysis. For a catalog of partially balanced designs for which the analysis is not unduly difficult refer to Cochran and Cox (1957) or Bose, Clatworthy, and Shrikhande (1954). Both these sources give plans and instructions for analysis with illustrations.

9.7 DESIGNS IN WHICH EACH SUBJECT IS USED REPEATEDLY[1]

In such diverse fields as agriculture, animal husbandry, education, psychology, market research, and social engineering, researchers often use experiments designed in such a way that each experimental unit (subject) is used repeatedly by exposing the unit to a sequence of different or identical treatments. Such designs are known by different names in the literature: *crossover* or *changeover* designs, (multiple) time-series designs, designs involving *repeated measurements* on the same subject, etc. A brief review of the recent works on such designs is given below.

[1] This section is taken with some modification from N. K. Namboodiri, "Experimental Designs in Which Each Subject Is Used Repeatedly," *Psychological Bulletin*, vol. 77, pp. 54–64, 1972, by permission of the American Psychological Association.

The following is a simple example of the kind of designs referred to above. In an effort to find out whether the race of the test administrator affects the performance of the subject in intelligence tests, each of a group of subjects of a given race was administered tests in succession by persons of different races, all persons using equivalent, if not identical, tests, and each test being separated far enough in time from the others to ensure that the results of one test did not affect those of the succeeding tests. Winer (1962, pp. 538–577) contains several examples.

The main attraction of crossover designs is that by properly choosing the design it is possible to make the precision of the estimate of the treatment differences dependent on the within-subject variance only, and since the within-subject variance, under certain conditions, will be much less than the between-subject variance, the aforementioned property of crossover designs makes for greater precision for a given cost, compared with designs in which each subject is used only once.

However, crossover designs raise certain special problems. These have to do with the possibility that the experimental error components of successive observations on the same subject may not be statistically independent. Nonindependence of error components may arise in one or the other of the following forms.

FORM 1   The effect of a treatment may extend into periods following the ones in which the treatment has been applied. Sheehe and Bross (1961) described an experiment that illustrates this situation. A study was planned in which five drugs were to be compared using a crossover design. Each patient was to serve as his own control. The control treatment consisted of giving the patient a neutral drug. A decision had to be made about the time interval at which the administration of one drug was to follow that of another. If the time interval between the administration of two drugs was to be as long as a week, it was feared that dropouts and the incompleteness of the sequences of treatments thereby resulting would create serious problems. It was therefore desirable to use periods of shorter duration, such as 3 days. There was no reason to suspect that by shortening the periods the chemical effects of one treatment would be influenced by that of the treatment given previously, since the chemical effect of each drug lasted only a few hours. But there seemed to be some kind of a psychological carryover effect at work. When an effective drug was given after a neutral one was administered, it tended to fail to produce any effect. It seemed as though the patient had lost confidence in the analgesics when a neutral drug was given: time (longer duration) alone seemed to be able to restore the patient's confidence.

FORM 2   The error components of successive observations on the same subject may behave like random variables with some definite, though not necessarily known, law of autocorrelation. For example, Williams (1949, 1950) found that in a uniformity trial involving one agricultural plot, the following discrete stationary random function fitted the yields in successive years:

$$E(y_i) = g$$
$$\text{var } y_i = \sigma^2$$
$$\text{cov } (y_i, y_j) = \rho^{(j-i)}\sigma^2 \qquad j > i$$

where $y_1, y_2, \ldots$ stand for the observations in successive years.

FORM 3   The error component of one observation may be related in a specific manner to the corresponding components of the preceding observations. Glass (1968) provided an illustration of this situation. Monthly data on traffic fatalities in Connecticut before and after 1956 (the year in which an unprecedented severe crackdown on speeding was instituted by the state government) were analyzed by using an exponentially weighted moving-average model that takes into account the nonstationary nature (caused by the 1956 reform) of the time series.

The particular form taken by the nonindependence of the error components of successive observations on each subject determines the specific nature of the problems in designing the experiment and in extracting full information from the data. Several different kinds of situations that have been recently considered in the literature are reviewed below.

Perhaps we should begin with the situation in which all observations can be taken to be statistically independent of each other. Finney (1964), Cochran and Cox (1957), and various other textbooks describe designs in which the same subject is used more than once but the successive observations can be assumed to be statistically independent. No new problems arise in the design or analysis of such experiments: the conventional procedures, applicable to situations in which each subject is used only once, apply also to situations in which each subject is used repeatedly, provided that the successive observations can be assumed to be statistically independent. Depending upon the number of treatments to be compared, the number of subjects available, and the number of occasions on which each subject can be used, etc., the experimenter may employ one or the other of the standard designs available in textbooks. Reference may be made to Finney (1964, pp. 273–286) for a catalog of designs likely to be of practical use in different circumstances of the above kind.

It may be of interest to note that an extreme form of crossover designs of the kind described above is the one in which the entire experiment is planned on one subject. Finney and Outhwaite (1956) present a statistical discussion of such experiments. An obvious disadvantage of such designs is that if the total time during which a given subject can be under observation is fixed, the number of treatments that can be compared may be severely limited.

If the effect of a treatment applied to a given subject in one period is likely to be carried over to the successive periods during which other treatments are applied to the same subject, then, in the analysis of the results, the methods applicable to the conventional designs (in which each subject is used only once) cannot be made use of as such. One method of preventing the carryover effects of treatments applied previously from influencing the effect of a treatment applied in a given period is to provide a sufficiently long rest period between the treatment periods. But, as many researchers have pointed out (e.g., Federer, 1955, p. 444), it is not always possible or desirable to use such a procedure. The experiment described by Sheehe and Bross (1961), to which reference has already been made, illustrates the point. The alternative is to design the experiment in such a way that the required estimates can be obtained, if possible, without going through complicated calculations. The estimates one is likely to be interested in may be the direct effects, i.e., effects during the period of application, free of carryover effects, carryover effects free of direct effects, or the sum of the direct and carryover effects in a given period. To see what is involved in designing an experiment that gives estimates of the above effects, let us consider the simple design shown in Table 9.17, which is discussed in detail in Cochran and Cox (1957, pp. 133–134).

Writing t_a, t_b, and t_c for the direct effects of treatments A, B, C, respectively, and r_a, r_b, and r_c for the carryover effects of those treatments, and assuming that the carryover effect of a treatment will last only during the period immediately following the one in which the treatment has been applied, we note that the observation for subject 1 in the third period (see Table 9.17) has the total treatment effect given by (assuming an additive model) $t_c + r_b$, the first component representing the direct effect of C and the other, the carryover effect of B. Similarly, the total treatment effect on the second observation on subject 6 is $t_b + r_c$, and so on. As far as the observations in the first period are concerned, the treatment effect will consist of only direct effects, assuming that all subjects have been treated alike before starting the experiment. If, in the analysis of the data, the above specification of the treatment effects of the different observations is made use of, it would be possible to obtain estimates of direct effects free from the influence of carryover effects, and vice versa.

The estimation of direct effects will be easier if the experiment is designed in such a way that each treatment is preceded by all other treatments equally often. It can be easily verified that the design shown in Table 9.17 has this property. Such arrangements in which each treatment is preceded by each other treatment equally often are sometimes called *balanced designs*.

Balancing in Table 9.17, however, cannot be said to be complete, since no treatment is allowed to follow itself. This creates certain problems. As Lucas (1957) and Cochran and Cox (1957, pp. 139–140) have pointed out, the estimates of the carryover effects yielded by the designs of the type shown in Table 9.17 would be less precise than the corresponding estimates of the direct effects and the variance of the sum of the direct and carryover effects of any treatment would be greater than the sum of the variances of the direct and carryover effects on account of the likely positive association between the two kinds of effects.

A procedure that makes the direct and carryover effects orthogonal has been suggested by Lucas (1957); the suggestion is to make use of what is called extra-period designs that can be easily constructed from the basic changeover designs of the type shown in Table 9.17 by repeating the treatments used in the last period in an extra period. The extra-period design derived from the design shown in Table 9.17 is shown in Table 9.18.

Table 9.17 **A CROSSOVER DESIGN THAT PERMITS ESTIMATION OF DIRECT AND CARRYOVER EFFECTS**

Period	Subject					
	1	2	3	4	5	6
1	A	B	C	A	B	C
2	B	C	A	C	A	B
3	C	A	B	B	C	A

Extra-period designs are not without their own drawbacks, however. Referring to the design presented in Table 9.18, it can be shown that the normal equations lead to the estimating equations

$$T_A - T_B = \tfrac{15}{2}(\hat{t}_a - \hat{t}_b) + \tfrac{1}{4}(B_2 + B_6 - B_3 - B_4)$$

for direct effects and

$$R_A - R_B = 6(\hat{r}_a - \hat{r}_b)$$

for first residual effects, i.e., carryover effects in the period immediately following the one of application,

where T_k = total of observations on treatment K

R_k = total of observations immediately following treatment K

B_s = total of observations on sth subject

$\hat{t}_k$ = estimate of direct effect of treatment K

$\hat{r}_k$ = estimate of first residual effect of treatment K

From these equations, it should be clear that estimation of contrasts involving direct effects, for example, $\hat{t}_a - \hat{t}_b$, is not as simple a matter as when carryover effects are absent and the successive observations can be assumed to be independent. The question therefore arises whether designs could be constructed that further simplify the estimation of direct effects. The changeover design shown in Table 9.19, first proposed by Quenouille (1953, p. 196), achieves that simplification.

The design in Table 9.19 permits the calculation of the sums of squares between subjects, between periods, and between direct effects in the same way as one would calculate them if carryover effects were absent. (The calculation of the SS between carryover effects, adjusted for subject effects, however, is not as simple as the calculation of the other SS's.)

Berenblut (1964) has pointed out that the Quenouille design shown in Table 9.19 is a particular case of a general arrangement involving v treatments, v^2 subjects, and $2v$ occasions for each subject.

While Berenblut's design permits estimation of contrasts involving direct effects in the same simple way as when carryover effects are absent, his design does not provide for such a simple estimation procedure for contrasts involving carryover effects. Lucas' design (Table 9.17), on the other hand, permits easy estimation of contrasts

Table 9.18   EXTRA-PERIOD DESIGN INVOLVING THREE TREATMENTS†

Period	Subject					
	1	2	3	4	5	6
1	A	B	C	A	B	C
2	B	C	A	C	A	B
3	C	A	B	B	C	A
4	C	A	B	B	C	A

† This design is derived from the basic changeover design shown in Table 9.17.

involving first residual effects while making the estimation of contrasts involving direct effects somewhat complicated. Apart from this difference between the two types of designs, there is the fact that Berenblut's design requires more subjects and more occasions per subject for a given number of treatments than Lucas' design (Table 9.17); if the number of treatments is three, for example, Berenblut's design requires nine subjects and six occasions per subject, whereas Lucas' designs requires six subjects and four occasions per subject.

It is also worth noting, before concluding the comparison of the two types of designs, that both types suffer from a common drawback. In neither does the precision of estimated contrast of first residual effects equal the precision of estimated contrast of direct effects.

So far, our remarks have been confined to designs that aim primarily at estimating direct and/or first residual effects. It should be noted that the procedures reviewed above are not strictly appropriate for estimating direct and first residual effects if it cannot be assumed that the effect of a treatment would not extend over periods beyond the first one following the one in which the treatment has been applied. For if the effect of any treatment extends over several periods, the estimates obtained in any of the ways discussed so far—estimates of contrasts involving direct effects or of those involving first residual effects—will be affected by disturbances due to the effects of treatments that persist in second, third, etc., periods following the one of application of the treatments. It is, however, possible to construct designs that balance treatments preceding as well as those applied two periods back. Williams (1949, 1950), for example, showed how to construct such designs using a complete orthogonal set of Latin squares.

Another point worth mentioning here concerns the construction of two-period designs that permit the estimation of treatment × period interaction. Balaam (1968) has shown that a two-period design for this purpose can be obtained by taking the first two periods of an appropriate Berenblut design. Obviously, such a design would require v^2 subjects if the number of treatments is v. It should be emphasized that Balaam's approach is applicable only if carryover effects are absent and the error terms can be assumed to be independently distributed. (By error term we mean here whatever is left in the observation after taking out the subject effect, the treatment effect, the period effect, and the interaction between treatment and period.)

A third point that may be mentioned here concerns the situation in which the observations show a marked trend over successive periods. When this happens, the

Table 9.19 QUENOUILLE DESIGN

Period	Subject			
	1	2	3	4
1	A	A	B	B
2	A	B	B	A
3	B	B	A	A
4	B	A	A	B

procedure described so far is not strictly applicable to estimating the direct and carry-over effects. Patterson (1950) has described procedures that can be used in such situations. Patterson's procedure involves making use of somewhat complex mathematical models. If carryover effects are not present, the problem stemming from the presence of a trend over time in the observations can be handled without too much difficulty. The next subsection is devoted to methods applicable to such situations.

Situations in Which the Error Component Exhibits a Time Trend While Carryover Effects Are Absent

Let us confine our attention to experiments in which only one subject is used for the entire experiment. (If there are two or more subjects, the procedure applied to one subject may be applied to the others and the results combined.) Let there be K treatments, and let the number of replications be R. The following is a simple example ($K = 2; R = 8$):

$$BAABABBAABBABAAB$$

where A and B stand for the two treatments.

Suppose the observations can be initially thought of as having two parts, the treatment effect and the remainder. Suppose further that what constitutes the remainder in the above decomposition of each observation can be thought of as the sum of two parts, a polynomial trend in time and an error term. In practical terms, the assumptions underlying the above model may be described as follows:

1 There is an aging effect which is responsible for the time trend in the data.
2 The aging effect can be represented as a polynomial in time (linear, quadratic, cubic, etc.).
3 The aging effect is statistically independent of the treatment effect.
4 The variations unaccounted for, i.e., the variance left unaccounted for by the treatment effect and the aging effect, are represented by random residuals.
5 The random components of the observations just described are identically distributed.

Whatever design is used, i.e., irrespective of the order in which the treatments are applied to the subject, analysis of the data by least squares is valid. But to apply the least-squares analysis, the degree of the polynomial representing the aging effect must be known beforehand. Ordinarily, one may not know beforehand the degree of the polynomial representing the aging effect. Under those circumstances, the estimation procedure becomes too complicated. No wonder, therefore, that researchers have been motivated to look for designs that permit estimation of treatment effects, eliminating aging effects, in situations like the one described above, without having to go through extensive calculations. Cox (1952) has given such a design for the two-variable case involving aging effect that can be represented by a polynomial of degree less than or equal to 3, there being no requirement that the exact degree of the polynomial be known beforehand. Cox's (1952) design permits estimation of treatment contrast in the same simple way as it would be estimated if aging effect were absent. Unfortunately, it is not possible to arrive at exactly orthogonal designs such as the one given by Cox (1952) in many cases. Reference may be made to Cox (1951) for methods to arrive at approximately orthogonal designs when exactly orthogonal ones are not possible. If the aging effect is linear, however, exactly orthogonal designs can be constructed in a variety of situations using what are known as magic squares and magic rectangles (Phillips, 1964, 1968a, b). Phillips (1968b) has also given some designs in which the aging effect, represented by a quadratic trend, has been balanced.

A Before-After Design Using A Single Subject

In the beginning of this section, reference was made to social reform experiments. The designs of reform experiments may be described as

$$\cdots AAA \cdots ABB \cdots B \cdots$$

In this design, B stands for the reform condition and A for the condition that prevailed prior to the institution of the reform. The question that the analyst may be asked to answer is: Has the change of treatment from A to B had any impact on the phenomenon under study, e.g., has liberalization of abortion laws produced any increase in abortion rate? The data yielded by reform experiments like the above can be put in the form of a time series consisting of, say, n observations $(y_1, y_2, \ldots, y_n)$ prior to the introduction of the reform and m observations $(y_{n+1}, y_{n+2}, \ldots, y_{n+m})$ afterward. We may say that the first n observations are on treatment A and the last m observations on treatment B. The research question then becomes: Has there been a shift in the level of the series as a result of the change of treatment? Note that even if there has been a shift in the level of the series, that shift need not necessarily be attributable to the change of treatment. Campbell (1969) described a number of possible rival hypotheses, any one of which may explain the shift in the level of the series in question. It is the responsibility of the analyst to examine each of the possible rival hypotheses. If the analyst becomes convinced that the shift, if any, in the level of the series is attributable to the change of treatment and is not due to any other plausible reason, then the model described below may prove useful in testing whether there has been any difference attributable to the change of treatment.

It is perhaps interesting to note that if the observations can be assumed to be statistically independent of each other, the above hypothesis can be tested using the

Table 9.20   STRUCTURE OF OBSERVATIONS IN THE BEFORE-AFTER EXPERIMENT USING A SINGLE SUBJECT

Occasion number	Observation	Additive components
1	y_1	$g + t_A + \varepsilon_1$
2	y_2	$g + t_A + b\varepsilon_1 + \varepsilon_2$
3	y_3	$g + t_A + b\varepsilon_1 + b\varepsilon_2 + \varepsilon_3$
$\cdots$		
n	y_n	$g + t_A + b\varepsilon_1 + \cdots + b\varepsilon_{n-1} + \varepsilon_n$
$n+1$	y_{n+1}	$g + t_B + b\varepsilon_1 + \cdots + b\varepsilon_n + \varepsilon_{n+1}$
$\cdots$		
$n+m$	y_{n+m}	$g + t_B + b\varepsilon_1 + \cdots + b\varepsilon_{n+m-1} + \varepsilon_{n+m}$

traditional t statistic. In most situations, however, it may not be logical to assume that the successive observations $y_1, y_2, \ldots$, are statistically independent of each other. Described below is one method of specifying the statistical dependence of successive observations in situations such as the one under discussion.

We assume that an additive model is applicable. Let g represent a component common for all the observations. Let the second component of each observation be the corresponding treatment effect. Let the observed value minus the sum of g and the corresponding treatment effect be called the *remainder*, for the moment. The model described below differs from the ones mentioned in the preceding sections essentially with respect to the specification of the structure of the *remainder* component.

Let us consider the remainder component of the first observation as the effect of a random shock. Another way of viewing it is as the effect of the uncontrolled events. We assume that once the subject experiences a random shock, he would undergo a permanent change. In other words, the random shock produces a permanent impact on the subject—permanent in the sense that it will last throughout the entire length of the time series taken for analysis. The permanent impact will be assumed to be a constant multiple of the amount of the shock actually received. Thus if ε_i is the shock effect in the ith period, $b\varepsilon_i$ will be taken to be the permanent effect resulting therefrom. (In many practical situations in which the above model applies, b seems to assume a value between 0 and 1. Hence, under those circumstances, b may be interpreted as that fraction of the random shock which becomes a permanent effect.) Obviously, the permanent effect will be carried over to the subsequent periods and therefore will form part of the observations obtained in later periods. We are thus led to the structure in Table 9.20.

From the structure shown in Table 9.20, it should be clear that there are four parameters to be estimated, g, t_A, t_B, and b. The main interest of the experimenter is likely to be centered in estimating and testing the significance of the treatment contrast $t_A - t_B$. Insofar as the interest is in drawing inference about $t_A - t_B$, there is no loss of generality if we set $t_A = 0$ and replace t_B by $t_B - t_A$ ($= d$, say). With these substitutions, the model described above reduces to the one suggested by Box and Tiao (1965). Except for the shift in level of the series represented by the parameter d, the Box and Tiao model is exactly the same as the exponentially weighted moving-average model widely used in economic research. The model is strictly applicable only if trend and seasonal fluctuations are absent. In other words, if aging effect is suspected to be present, or if some kind of periodic fluctuation accounts for some of the variations in the observations, these variations should be removed before applying the model.

Interestingly enough, the estimation of d is relatively an easy matter, as Box and Tiao (1965) have shown, if the value of b is known. Glass (1968) has applied the above model to a situation in which the value of b is unknown.

9.8 CONCLUDING REMARKS

Most of this chapter has been devoted to modifications of full factorials. Designs such as balanced incomplete blocks, quasi-Latin squares, etc., have been referred to only briefly. The review of the recent works on crossover designs can be described only as a cursory treatment. The references cited should be consulted for more information about these and other related designs. The exercises appended to the chapter are to be regarded as complements to the material presented in the text.

EXERCISES

9.1 One investigator wanted to compare teacher characteristics in team-teaching situations. The characteristics of interest were sex and experience, both having two levels (male, female; and more experienced, less experienced). Because of limitations of the availability of teacher teams and because of administrative problems, it was found impractical to try out all four treatment combinations in every school (block). If only two treatment combinations are to be tried in each school, suggest a suitable design for conducting the experiment.

9.2 Construct a 2^3 factorial arrangement in four blocks, each of size 2, confounding interactions (AB) and (AC). Which other factorial effect(s) is (are) confounded with blocks?

9.3 The following is a quarter replicate of a 2^3 factorial:

ac	ab

Which factorial effects can be estimated from this quarter replicate, under what assumptions, and how would you interpret your estimates?

9.4 With respect to the half replicate of a 2^6 factorial, in which the six-factor interaction is the defining contrast, state whether the following statements are true. If any statement is false, state why you think it is false.

(a) Every two-factor interaction has a four-factor interaction as alias.

(b) Every three-factor interaction has a two-factor interaction as alias.

(c) To obtain valid estimates of the two-factor interactions, it is enough to assume that the four-factor interactions are all negligible.

9.5 In Exercise 9.1 let A and B denote the two factors involved. Obtain, using blocks of size 2, a partially confounded factorial arrangement in which (A), (B), and (AB) are each confounded with blocks once and only once. Does the arrangement look like a balanced incomplete block design?

9.6 For the 2^3 factorial if you are asked to confound each main effect and each interaction once and only once with blocks (of size 4), would the arrangement you obtain look like a balanced incomplete block design?

9.7 Plans for balanced incomplete block designs are generally described in terms of the following parameters:

t = number of treatments (treatment combinations)
b = number of blocks
k = block size, i.e., number of treatments per block
r = number of replications of each treatment
λ = number of times each pair of treatments (treatment combinations) occurs together within some block

Show that $bk = rt$. Verify that in the design obtained in Exercise 9.6 $\lambda = 3$ and $\lambda(t - 1) = r(k - 1)$. This latter relationship holds for all balanced incomplete block designs.

9.8 Identify the interactions and/or main effects confounded in the following factorial arrangement. The columns are the blocks.

(1)	ab	a	b
abc	c	bc	ac
abd	d	bd	ad
cd	abcd	acd	bcd

[*Ans.* (*AB*), (*ACD*), and (*BCD*)]

9.9 Consider the tabulated experiment involving four experimental groups. Imagine that you have an aggregate measurement for each group at the end of the experiment. Treat the four observations as constituting those from a half replicate of a 2^3 factorial and

Experimental group	Whether the group is exposed to:		
	P Pretest	E Experimental stimulus	U Set of uncontrolled events
1	Yes	Yes	Yes
2	Yes	No	Yes
3	No	Yes	Yes
4	No	No	Yes

describe how you would estimate the main effects of P and E, assuming that all interactions involving U are negligible (see Namboodiri, 1970). Notice that the use of groups 3 and 4 in the above experiment actually involves enlarging the two-group design consisting of groups 1 and 2. The major reason for enlarging the two-group design to the four-group one (see, for example, Campbell and Stanley, 1963) is to guard against the possibility that the pretest itself may have an influence on the scores observed at the end of the experiment. To give an example, suppose the experiment involves an education program aimed to improve family health practices. The initial questionnaire may alert the families to deficiencies in their health practices and consequently may serve as a stimulus to remedy the deficiencies. Some authors feel that this is not a sufficient justification for enlarging the two-group design to one involving four groups (see, for example, Cochran, 1965). State and defend your views on the matter.

9.10 In many textbooks (e.g., Dayton, 1970) the following design is described as a repeated-measures design. Suppose there are n subjects and t treatments. To each subject all the treatments are applied, the order of application of treatments being randomized independently for each subject. The results of a simple experiment of this type may be symbolically represented as follows:

Subject	Treatment			
	1	2	$\cdots$	t
1	y_{11}	y_{12}	$\cdots$	y_{1t}
2	y_{21}	y_{22}	$\cdots$	y_{2t}
n	y_{n1}	y_{n2}	$\cdots$	y_{nt}

Is it valid to analyze this data set in the usual fashion, by partitioning the total SS into between-subjects SS, between-treatments SS, and residual SS and then completing the analysis-of-variance table? The F test associated with the analysis-of-variance table in this case is valid only if the conditions of homogeneity of variance and of covariance are satisfied by the data set at hand. To understand what this means, imagine that the above data set has been drawn at random from a t-variate universe. If this universe is characterized by a variance-covariance matrix whose diagonal elements are all σ^2 and whose off-diagonal elements are all $\rho\sigma^2$, we say the conditions of homogeneity referred to above prevail. How do we test this on the basis of the data set at hand? Let s_{ij} be the covariance (using $n - 1$ as the divisor) between the ith and jth columns of y values in the above data set, and let s_{ii} be the variance of the y values in the ith column (using $n - 1$ as the divisor). Let S be the matrix whose diagonal elements are $s_{ii}, i = 1, 2, \ldots, t$, and off-diagonal (i,j) elements are $s_{ij}, i \neq j = 1, 2, \ldots, t$. Let $\bar{s}$ be the average of the variances $s_{ii}, i = 1, 2, \ldots, t$, and let $\bar{c}$ be the average of the co-variances s_{ij}. Let S_0 be a $t \times t$ matrix whose diagonal elements are all $\bar{s}$ and whose off-diagonal elements are all $\bar{c}$. Now calculate

$$M = -(n - 1) \ln \frac{|S|}{|S_0|}$$

where $|S|$ and $|S_0|$ are the determinants of S and S_0 respectively, and

$$C = \frac{t(t + 1)^2(2t - 3)}{6(n - 1)(t - 1)(t^2 + t - 4)}$$

The statistic $B = (1 - C)M$ can be regarded as a chi square with $(t^2 + t - 4)/2$ degrees of freedom. This is a generalization of Bartlett's test for homogeneity of variance described in Chap. 5.

9.11 Studies conducted within the paradigm of operant behavior employ the *intrasubject-replication design* (often known as the reversal design) in which various treatments are successively applied to and removed from the same subject (see Sidman, 1960). For example, the $ABAB \cdots$ design in which A stands for "no reinforcement for a certain response," and B for a "reinforcing stimulus" following that response, is widely used to demonstrate that if a certain reinforcement is made available contingent upon a response, the effect is to increase the frequency of that response above operant level. What are some of the disadvantages of this design (see, for example, Bandura, 1969, pp. 242–244)? Those who have been engaged in applying the above design have often held that statistical tests of significance are of no relevance in making inferences from the results of their experiments. Do you agree with this position? Why or why not?

PART THREE

Simultaneous-Equation
Models

10

RECURSIVE MODELS AND ONE-WAY CAUSATION

Thus far it has been presumed that the choice of a dependent variable is not problematic and that each equation for a dependent variable can be handled separately. The independent variables X_i may be fixed by the investigator, as is commonly assumed in regression theory and for some types of experimental designs. Alternatively, they may be stochastic variables that, by assumption, are uncorrelated with the random error or disturbance terms. In this and the following chapter we shall consider much more complex though more realistic situations that are especially appropriate for nonexperimental research and for providing causal interpretations of empirical relationships.

10.1 INTRODUCTION TO SIMULTANEOUS EQUATIONS

In Chap. 1 we noted that many kinds of interpretive problems do not arise if one is willing to think in terms of fixed population parameters (or relationships between parameters) about which one wishes to obtain estimates based on sample data. The problem of inference can be conceived rather simply in terms of the investigator's ignorance of the parameter values or of certain pieces of information, e.g., values of Y, that are to be estimated from other pieces of information, for example, X_i. As long as the notion of prediction is conceived in terms of estimation of fixed values rather than prediction of actual *changes* that can be expected, very few theoretical problems are encountered. The dependent variable simply becomes the variable that one wishes to estimate, and the issue is confined to that of finding estimation procedures with optimal properties.

Unfortunately, the social scientist is seldom really interested in stopping at this point. It is obvious that real-world populations do not conform to the ideal. They are not fixed, being subject to at least two kinds of factors producing changes: (1) the population membership changes owing to births, deaths, and migration, and (2) the operation of outside or environmental forces from time to time produces changes either in the levels of the variables or the parameters themselves. It therefore becomes necessary to introduce greater realism into the underlying model to make it possible to link theoretical interpretations more systematically with one's data.

In particular, it is common practice in nonexperimental research to take the independent variables X_i in any given equation as random variables.[1] This implies that each of these variables, in its own right, might be considered as a dependent variable to be explained by a different set of variables including, perhaps, some or all of the remaining X_i in the original equation. Of course not *all* variables can be taken as dependent without becoming involved in an infinite regress, and in the next chapter we shall see more precisely how this difficulty can be handled. In brief, one may consider a set of what are termed *endogenous variables*, or variables to be explained by the theory under consideration, each of which may be dependent upon some or all of the remaining endogenous variables. But added to this set there must also be a set of *predetermined variables*, consisting of two types of variables, either previous (lagged) values of the endogenous variables or what are referred to as *exogenous variables*. The latter variables are, by the assumptions of one's theory, *not* dependent upon any of the endogenous variables, though they may be intercorrelated with each other for unknown reasons. These exogenous variables are the givens which the particular theory at hand cannot explain. They may have fixed values during the period of study or across all observations, but it will be most helpful if they are also subject to variation. If they can be brought explicitly into the equations and measured, they will contribute to the explanatory power of the model. If they cannot, we conceive of them as contributing to the error terms by producing unknown disturbances in certain of the variables.

Thus as soon as we admit that our populations do not remain fixed, we allow for stochastic or probabilistic elements *even in instances where the entire population has been sampled.* The stochastic terms therefore generally include both sampling error and the effects of unknown or unmeasured outside influences. As long as one confines oneself to the assumption of fixed populations and the notion that probabilistic terms enter only as a result of sampling error, it is merely the method of sampling that is at issue. But when one also allows for unknown disturbing influences that may affect both the independent and dependent variables, this particular kind of *ceteris paribus* assumption becomes much more problematic.

The choice of dependent variable becomes problematic once we allow for the

---

[1] In experimental designs, as we have seen, the independent variables (treatments, blocks) may be taken as either fixed or stochastic. Formal regression theory is usually developed for fixed X_i but can also be extended rather easily to random variables (see Johnston, 1972). In practice, however, the independent variables in nonexperimental research are practically never fixed by the investigator.

possibility of mutual influence of each variable on the others, referred to variously as reciprocal causation, causal circles, or feedback. It then turns out that in general it will not be realistic to make the kinds of assumptions that are necessary to justify the use of ordinary least squares. Furthermore, we may encounter situations in which estimation of the effects of one variable on another becomes impossible, even in the extreme case where there may be no sampling error or stochastic element in the equations. These problems of reciprocal causation will be postponed until the next chapter, but it is possible to indicate intuitively the nature of the difficulty.

Suppose, for example, that we are concerned with only two variables X and Y but suspect that there is reciprocal causation involved. That is, if one were to produce an actual change in X, this would result in a change in Y, with this change in Y also producing change in X. How could one estimate the effects of the one variable on the other? If one arbitrarily selected Y as the dependent variable, writing a linear equation of the form $Y = \alpha + \beta X + \varepsilon$ and estimating the parameters using ordinary least squares, would it be legitimate to estimate the effects of X on Y by b_{yx}?[1] If so, then by symmetry it would also be possible to estimate the effects of Y on X by b_{xy}. That something is wrong with such a procedure can be inferred from the fact that the product $b_{yx}b_{xy} = r_{xy}^2$ is (1) never negative, (2) never greater than unity, and (3) a function of the magnitude of the error variances of the variables that have been left out of the equation. Furthermore, if Y has no effect on X whereas X does affect Y, we know that it is impossible to obtain a zero value for b_{xy} and a nonzero value for b_{yx}, since both have the same numerators. It turns out that ordinary least squares is generally inappropriate for models involving reciprocal causation because error or disturbance terms will be correlated with the independent variables in their equations, contrary to assumption. In the next chapter we shall consider this difficulty in more detail and provide a reasonably simple alternative to ordinary least squares.

But we shall also encounter a more fundamental difficulty that forces a greater concern with clarifying the theoretical causal assumptions underlying the model. One can reasonably ask: If X and Y affect each other, how can we possibly justify the assumption of a stable population? For example, if an increase in X is produced by some exogenous agent, and if this results in an increase (or decrease) in Y, which in turn increases (or decreases) X, thereby further increasing Y, and so on, perhaps the changes will occur indefinitely. Strictly speaking, both X and Y might continue to change, but it may well be that the changes are of decreasing magnitude so that before very long they become negligible for all practical purposes. That is, the system may approach stability in the sense that both X and Y become exceedingly close to some equilibrium level. But perhaps the changes will *increase* in magnitude, so that the system explodes or at least the original model no longer holds.

In order to illustrate some of these rather abstract remarks, as well as the discussion that follows, let us consider a type of situation that occurs rather often in microlevel analyses, where the focus is on explaining the behaviors of individuals in

---

[1] For the sake of simplicity, in Parts Three and Four we shall omit all subscripts referring to individual cases. We shall, however, use double subscripts for the regression coefficients and their estimators, following the convention of representing the dependent variable by the first subscript and the independent variable by the second. Whenever conventional (ordinary) least-squares estimates are used, we shall also follow the convention of putting control variables after the dots in the appropriate subscripts.

social settings. We imagine two stimulus variables X_1 and X_2 that impinge on the individual from the social environment. We assume in this simplified model that the individual is unable to affect these stimuli, so that there is no feedback from behavior to these variables. But the stimulus variables X_1 and X_2 may be intercorrelated for unknown reasons, and we shall represent this possibility by connecting them by a curved double-headed arrow. We refer to variables such as these as exogenous to the system of variables we wish to explain.

Next we introduce two variables X_3 and X_4 to represent internal states of the individual, perhaps attitudes, motives, values, or expectancies that may be affected by X_1 and X_2. Here we admit the possibility that there may be reciprocal causation between these internal states, representing this by two separate arrows with heads pointing in opposite directions. Perhaps one kind of attitude affects another, or perhaps one of the internal states is an expectancy that affects and is affected by a motivational state that is aroused by X_1 and X_2. Finally, we let X_5 represent some form of behavior that we take to be influenced by either or both of the internal states X_3 and X_4, but not *directly* by the stimulus variables X_1 and X_2. We also allow for the possibility that there may be a sequence of behaviors such that previous behaviors may affect either or both of the two internal states X_3 and X_4. We thus have drawn double arrows between X_5 and both X_3 and X_4. The model is diagrammed in Fig. 10.1.

Any number of specific illustrations of this type of model can be noted, three of which are given in schematic form in Table 10.1. In the first case, the behavior involves a child's performances on examinations; the two stimulus variables are degree of familial support and peer-group influences; and the internal states are degree of knowledge and degree of motivation. In the second illustration we are concerned with the congressional voting behavior of a politician influenced by the exogenous variables X_1 = constituency demands, and X_2 = party competition. The internal states are X_3 = degree of desire for reelection and X_4 = the politician's attitudes on the specific policy questions. The third example involves the number of delinquent acts as X_5. The exogenous variables are police behavior X_1, on the one hand, and

Table 10.1 SOME SUBSTANTIVE VARIABLES APPROPRIATE FOR THE MODEL OF FIG. 10.1

Kind of individual	Stimuli		Internal states		Behavior
	X_1	X_2	X_3	X_4	X_5
Child taking examinations	Familial supports	Peer influences	Knowledge levels	Motivation levels	Performance levels
Politician voting in Congress	Constituency demands	Party competition	Desire for reelection	Attitudes on policy questions	Actual votes
Youth committing delinquent acts	Police behavior	Peer influences	Guilt levels	Self-respect	Delinquent acts

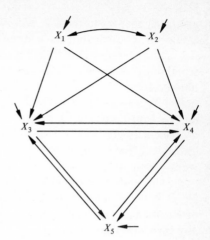

FIGURE 10.1

peer pressures X_2, on the other. The two internal states are level of guilt X_3 and degree of self-respect X_4.

In all three illustrations we allow for the possibility that the individuals' behaviors at earlier times may influence their internal states, which in turn may affect their later behaviors. For example, a poor test performance may lower the student's motivation on successive tests and may interfere with the learning process so as to affect adversely the degree of knowledge for the next test. This may lower test performance still further, which may again reduce both knowledge and motivation. Similarly, the commission of delinquent acts may lower both one's self-respect and level of guilt, which may in turn make further delinquent acts more probable. We see that under such circumstances it is entirely possible that the behavioral levels may set into motion internal processes that result in ever increasing or decreasing levels of these behaviors at later points in time. Perhaps these behaviors will eventually level off, but perhaps not.

It turns out that in order to handle this kind of problem one needs a *dynamic model* that involves the time element in a nontrivial way. For example, if one can specify the time lags involved and incorporate them into an explicit theory, it becomes possible to study *stability conditions*, which involve the relative magnitudes of the parameters and which may or may not be met with in any particular instance. We shall briefly consider the nature of such dynamic theories in the next chapter, though for the most part we shall be concerned with static simultaneous-equation models that do not specify time lags or precise temporal sequences.

Before we become concerned with these complex kinds of simultaneous-equation models, however, we shall discuss in the present chapter a much simpler kind of situation that is intermediate between the single-equation approaches dealt with in the previous chapters and the full simultaneous-equation model that allows for reciprocal causation. Suppose a finite set of variables can be hierarchically arranged in a causal sequence and labeled in such a way that if $i < j$, then X_i may cause X_j but X_j cannot cause X_i. That is, we may locate an X_1 such that a change in X_1 produced by an exogenous agent may result in changes in all of the remaining X_i but that exogenously produced changes in these X_i will *not* produce any change in X_1. Similarly, we may locate an X_2 such that an exogenous change in X_2 may produce changes in X_3, $X_4, \ldots, X_k$ but not in X_1. Likewise, a change in X_3 may affect the levels of all the remaining X_i with higher subscripts, but not X_1 and X_2, and so forth. Notice that it is not *necessary* that X_1 affect all the remaining variables, but in a reasonably complete model X_1 is likely to have direct effects on some of these X_i, indirect effects on others, and perhaps no effects on several others. We are assuming, however, that there are no feedback effects, direct or indirect, from any X_j to any X_i whenever $i < j$. In Table 10.1, for example, one attitude, say X_3, may affect the other X_4 but not vice versa. Both attitudes may affect behavior X_5, but behaviors cannot affect attitudes.

This kind of causal model is referred to as a *recursive system* because it is possible to build up the system, equation by equation. Intervention by an experimenter producing parameter changes at any point (say in the equation for X_4) will *not* have any effects on the levels of previous variables but may have effects on all subsequent variables. For example, suppose an experimenter could modify the situation so as to change β_{41}. This would affect the level of X_4, which in turn could affect X_5, $X_6, \ldots$, but not X_1, X_2, or X_3. A proper arrangement of the variables and their subscripts would result in a (linear and additive) equation system as follows:

$$X_1 = \varepsilon_1$$
$$X_2 = \beta_{21}X_1 + \varepsilon_2$$
$$X_3 = \beta_{31}X_1 + \beta_{32}X_2 + \varepsilon_3 \qquad (10.1)$$
$$\cdots\cdots\cdots\cdots\cdots\cdots\cdots\cdots\cdots\cdots\cdots$$
$$X_k = \beta_{k1}X_1 + \beta_{k2}X_2 + \cdots + \beta_{k,k-1}X_{k-1} + \varepsilon_k$$

where, for simplicity, constant terms have been omitted by assuming that each variable has been measured in terms of deviations about its own (population) mean and where the ε_i are disturbance terms which will be discussed below. Before considering the general k-variable recursive system, let us first examine the three-variable case for illustrative purposes.

10.2 THE THREE-VARIABLE CASE

The equation $X_1 = \varepsilon_1$ is merely a heuristic device to indicate that X_1 (say, an environmental stimulus) is taken as completely unexplained by any of the other variables in the system and hence is exogenous. There are two endogenous variables in the three-equation model, namely, X_2 (say, an attitude) and X_3 (say, a behavior) since both are functions of X_1. Therefore we really have the two-equation system

$$X_2 = \beta_{21}X_1 + \varepsilon_2$$
$$X_3 = \beta_{31}X_1 + \beta_{32}X_2 + \varepsilon_3$$

Suppose we focus on the equation for X_3, presuming that we are primarily concerned with a single-equation setup in which X_3 is taken as dependent upon both X_1 and X_2. We have of course allowed for intercorrelations between our independent variables X_1 and X_2, and we have controlled for one variable in studying the partial relationship of the remaining variable to X_3. In interpreting the partial correlation $r_{31 \cdot 2}$, for example, one can think of correlating the residuals of X_3 and X_1 after adjusting for X_2. But we now are taking X_2 as causally dependent upon X_1 and might naturally ask what entitled us to make these adjustments in X_1 in terms of one of its presumed effects X_2. Clearly, we admit the possibility that some strange things might be taking place. For example, there would also be nothing in the *mathematical* setup to stop us from relating X_1 and X_2, partialling for their presumed common effect X_3. The implication is that we shall need to have some guiding principles to tell us when controlling does and does not make sense theoretically, and it will turn out that these principles will require us to make a priori assumptions about the causal structure of the model.

Suppose we substitute for X_2 in the equation for X_3, getting

$$X_3 = \beta_{31}X_1 + \beta_{32}(\beta_{21}X_1 + \varepsilon_2) + \varepsilon_3$$
$$= (\beta_{31} + \beta_{32}\beta_{21})X_1 + (\beta_{32}\varepsilon_2 + \varepsilon_3)$$

We see that we have expressed X_3 in terms of X_1 alone and that the slope relating these two variables, ignoring X_2, is a composite of the *direct* effect of X_1, namely, β_{31}, and the *indirect* effect via X_2 as measured by the product of the slopes β_{21} and β_{32}. This result is entirely consistent with common sense. If, for example, $\beta_{31} = 2$, $\beta_{21} = -3$, and $\beta_{32} = .5$, then a unit increase in X_1 would be expected to increase X_3 by 2.0 units through the direct impact of X_1. But this unit increase in X_1 would also decrease X_2 by 3.0 units, which in turn would decrease X_3 by 1.5 units, so that the net result on X_3 would be an increase of .5 unit. The total slope relating X_1 and X_3 would thus be $2.0 - (3.0)(.5) = .5$, whereas the partial slope, with X_2 controlled, would be 2.0.

In the special case where $\beta_{31} = 0$ and where there is therefore no direct effect of X_1, there is only the indirect effect through X_2, for example, $-(3.0)(.5) = -1.5$. Here, a control for X_2, if this could be accomplished while still varying X_1, would produce a zero relationship between X_1 and X_3. In the special case where X_1 and X_2 are not causally connected in any way, our very simple causal system specifies that $\beta_{21} = 0$, so that there can be only a direct effect of X_1 on X_3. If $\beta_{32} = 0$, X_2 is not a direct cause of X_3, although there will be a spurious correlation between the two variables owing to their common cause X_1, assuming that $\beta_{21} \neq 0$ and $\beta_{31} \neq 0$. It should be made clear that the notion of "direct" relationship is always relative to the variables that have been explicitly included in the system of equations. In this example there are only three variables, and therefore the model does not allow for an indirect relationship between the two endogenous variables X_2 and X_3.

Thus the β_{ij} can be interpreted very simply in terms of the changes that would be produced in an experimental setting or in a natural situation in which exogenous factors were operating to produce (unit) changes in one of the X_i in the system. We shall see how this idea generalizes to the k-variable case, but before doing so we need to examine how the β_{ij} are estimated and the assumptions that will be necessary in order to justify the use of ordinary least squares. It is convenient to do so in the case of the three-variable model since the principles involved readily generalize.

Least-squares estimators will be unbiased and also more efficient, i.e., have a smaller standard error, than any other linear estimators provided that (1) the true structural equations are actually linear in form; (2) there is no measurement error in any of the independent variables in any given equation and only strictly random error in the measurement of the dependent variable; (3) the expected value of ε_i is zero, that is, $E(\varepsilon_i) = 0$; (4) the independent variables in any equation are uncorrelated in the population with the disturbance term of that equation [$\text{cov}(X_i, \varepsilon_j) = 0$, for $i < j$]; (5) each disturbance term has a constant variance $\sigma_{\varepsilon_i}{}^2$; and (6) sampling has been random.[1] If the only factor responsible for the disturbance term were sampling error, the crucial assumption 4 would be approximately met whenever the sample size is reasonably large. But when we allow for the effects of variables that have been omitted from the equations, the assumption of no correlations between independent variables and disturbance terms becomes much more problematic and therefore of crucial importance. If we could assume that the disturbance term is produced by a very large number of minor influences that operate more or less independently of each other, it would be plausible to assume that $\text{cov}(X_i, \varepsilon_j) = 0$ (if $i < j$) and also that the disturbance term has a normal distribution. The latter assumption of course justifies the usual F and t tests, though the normality assumption may be relaxed for reasonably large samples, e.g., if $n \geq 100$.

Since we are dealing with *simultaneous* equations, we must also be concerned about the covariances of the disturbance terms of the different equations. For recursive systems we shall assume that the expected values $E(\varepsilon_i \varepsilon_j) = 0$ for all $i \neq j$, meaning that the covariances between all pairs of disturbance terms are zero. In causal terms, this assumption would be justified if either of two conditions held: (1) each disturbance term is a resultant of a very large number of independently operating factors, only a few of which are common to any pair of ε_i, or (2) in instances where there are a small number of factors influencing each disturbance term, those factors affecting one disturbance term are uncorrelated with the factors affecting the other disturbances. Whether or not such assumptions about the disturbances are realistic in any given instance will depend upon the completeness of the causal system. Whenever common causes of the disturbance terms for two or more equations can be located or measured, they should be explicitly introduced into the equations as additional variables. For example, in any of the illustrations of the model for Fig. 10.1, it might be the case that there is some other cause of one of the internal states (say, X_3) that also directly affects the behavior in question. If so, this variable would have to be measured and brought into the system as a sixth variable. At some point, however, the theoretical system must be taken as theoretically closed or complete except for what are assumed to be strictly random sources of error in the equations.

Although we shall relax this assumption somewhat in the next chapter, we shall presently utilize the assumption that $E(\varepsilon_i \varepsilon_j) = 0$ for all $i \neq j$. It can be shown that this assumption of zero covariances between the disturbance terms implies that $E(X_i \varepsilon_j) = 0$ for all $i < j$, though not necessarily for $i > j$. In other words, we can

---

[1] This list of assumptions is really the same as those discussed in Part Two. Here we shall assume sample sizes sufficiently large not to require the normality of disturbance terms. Even where conventional F tests cannot be made, it can be shown that the assumptions listed here are sufficient to assure us that ordinary least-squares estimates will both be unbiased and have greater efficiency than any other linear estimators based on the same set of variables. Assumptions 2 and 4 will receive most of our attention in the remaining chapters. The assumption of random sampling applies if there is some finite population with (temporarily) fixed parameter values, e.g., the city of Chicago, from which one has sampled.

show that the disturbance term for any given equation, for example, ε_j, will be uncorrelated with the *independent* variables appearing in that equation but will generally be correlated with variables that depend upon X_j. This is consistent with the notion that exogenous factors affecting X_j will not affect any of the causally prior variables but may very well affect the variables that appear as dependent variables in later equations. We shall later see that nonrecursive systems, which allow for simultaneous effects of each variable on the others, are incompatible with the assumption required by ordinary least squares that $E(X_i \varepsilon_j) = 0$ for $i < j$.

Confining our attention to the three-variable case, we first note that since $X_1 = \varepsilon_1$, it immediately follows that the assumptions that $E(\varepsilon_1 \varepsilon_2) = E(\varepsilon_1 \varepsilon_3) = 0$ are equivalent to $E(X_1 \varepsilon_2) = E(X_1 \varepsilon_3) = 0$. But we need to show that $E(\varepsilon_2 \varepsilon_3) = 0$ implies that $E(X_2 \varepsilon_3) = 0$. If we rewrite our equations as

$$X_1 = \varepsilon_1 \qquad (10.2a)$$

$$X_2 - \beta_{21} X_1 = \varepsilon_2 \qquad (10.2b)$$

$$X_3 - \beta_{31} X_1 - \beta_{32} X_2 = \varepsilon_3 \qquad (10.2c)$$

multiply the equations pair by pair, and then take expected values, we get

$$E(X_1 X_2) - \beta_{21} E(X_1^2) = E(\varepsilon_1 \varepsilon_2) = 0 \qquad (10.3a)$$

$$E(X_1 X_3) - \beta_{31} E(X_1^2) - \beta_{32} E(X_1 X_2) = E(\varepsilon_1 \varepsilon_3) = 0 \qquad (10.3b)$$

$$E(X_2 X_3) - \beta_{31} E(X_1 X_2) - \beta_{32} E(X_2^2)$$
$$- \beta_{21} [E(X_1 X_3) - \beta_{31} E(X_1^2) - \beta_{32} E(X_1 X_2)] = E(\varepsilon_2 \varepsilon_3) = 0 \qquad (10.3c)$$

But the term within the brackets in Eq. (10.3c) is identical to the left-hand side of Eq. (10.3b), which is zero, and hence (10.3c) becomes

$$E(X_2 X_3) - \beta_{31} E(X_1 X_2) - \beta_{32} E(X_2^2) = 0$$

If we now return to Eq. (10.2c), multiply through by X_2 and take expected values, we get

$$E(X_2 X_3) - \beta_{31} E(X_1 X_2) - \beta_{32} E(X_2^2) = E(X_2 \varepsilon_3)$$

Since the left-hand side has just been shown to be zero under the assumption that $E(\varepsilon_2 \varepsilon_3) = 0$, it follows that $E(X_2 \varepsilon_3) = 0$. It should be noted that the above proof, although focusing on Eq. (10.2c), involved results which utilized the previous equations. One can similarly build up results for an equation involving X_4 as the dependent variable, showing that the assumption $E(\varepsilon_i \varepsilon_j) = 0$ for all $i \neq j$ implies that

$$E(X_1 \varepsilon_4) = E(X_2 \varepsilon_4) = E(X_3 \varepsilon_4) = 0$$

One can proceed recursively, equation by equation, to develop proofs using ordinary algebra, though the general proof for the k-equation case can be handled much more efficiently using matrix algebra.

Thus we see that the assumption that the disturbance terms are all mutually uncorrelated, together with the assumption that the recursive model has been correctly specified as linear and additive and that there are no measurement errors in the independent variables, implies that $E(X_i \varepsilon_j) = 0$ for all $i < j$, thus satisfying all the conditions necessary for ordinary least squares. It must be emphasized, however, that this result is dependent upon the recursive nature of the system, which thus permits one to break apart the system to study each equation separately. In effect, then, the single-equation methods considered up to this point all presuppose that the real world can be adequately modeled in recursive terms. That is, we may successively study each dependent variable without worrying about the possibility that some of the independent variables in a given equation are actually dependent, directly or indirectly, on the dependent variable in that equation.

Since ordinary least squares may legitimately be used for recursive systems, provided the crucial assumptions are satisfied, we shall introduce the customary dot notation to indicate which variables are being controlled. In the three-variable model our least-squares equations can therefore be written

$$X_1 = e_1$$
$$X_2 = b_{21}X_1 + e_2$$
and
$$X_3 = b_{31 \cdot 2}X_1 + b_{32 \cdot 1}X_2 + e_3$$

where the b_{ij} are unbiased estimates of the β_{ij} and the residuals e_i about the least-squares equations are taken as estimates of the unknown disturbance terms ε_i. It will be recalled that it is a property of ordinary least squares that the residual terms e_i are automatically uncorrelated with the independent variables in each equation. Therefore if each X_i is measured in terms of deviations about its (sample) mean, then $\sum X_i = \sum e_j = 0$ for all i and j and $\sum X_i e_j = 0$ for all $i < j$. This result readily generalizes to k variables, where it becomes necessary to introduce higher-order partials beginning with the equation for X_4.

10.3   THE GENERAL CASE: TESTING AMONG ALTERNATIVE MODELS

In the general (linear) k-equation model the assumption that the disturbance terms are mutually uncorrelated justifies one in singling out any specific equation for attention and applying ordinary least-squares theory to this equation separately. Thus the *structural equation*[1]

$$X_i = \beta_{i1}X_1 + \beta_{i2}X_2 + \cdots + \beta_{i,i-1}X_{i-1} + \varepsilon_i \qquad (10.4)$$

can be estimated by the least-squares equation

$$X_i = b_{i1 \cdot 23, \ldots, i-1}X_1 + b_{i2 \cdot 13, \ldots, i-1}X_2 + \cdots + b_{i, i-1 \cdot 12, \ldots, i-2}X_{i-1} + e_i$$

where the familiar dot notation has been introduced to indicate which variables are being controlled. As already noted, the number of control variables will generally increase with i if we follow the convention of labeling the causally prior variables with the lowest subscripts. For the sake of concreteness let us first focus on the five-variable

---

[1] We shall henceforth use the terms "structural equations" or "structural parameters" to refer to the representations of the true structural or causal properties of real-world phenomena, as contrasted with equations that are merely used for prediction or estimation purposes.

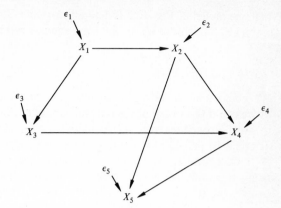

FIGURE 10.2

model of Fig. 10.2. The reader may illustrate this model concretely in terms of any of the examples given in Table 10.1. In Fig. 10.2, one of the stimulus variables affects the other, e.g., police behaviors affect peer behaviors. Also, one of these stimulus variables (X_2) affects behavior (X_5) directly. This could involve a number of mechanisms or intervening variables that have not (yet) been specified by the theory, necessitating our treating the X_2-to-X_5 relationship as direct.

Notice that the model is recursive since the variables have been ordered such that there are no arrows directed from any X_j to any X_i for which $i < j$. In this model there are $k(k-1)/2 = 10$ possible arrows, four of which have been omitted. This is the diagrammatic representation of the assumption that there are no *direct* causal linkages between four pairs of variables, namely, X_2 and X_3, X_1 and X_4, X_1 and X_5, and X_3 and X_5. We have imposed the a priori *assumption* that the structural parameters β_{32}, β_{41}, β_{51}, and β_{53} are all zero. In the next chapter when we consider what is referred to as the *identification problem* in some detail, we shall see that it is always necessary to make a sufficient number of untestable assumptions about any equation system to permit the estimation of the parameters involved. If we make just the right number of such assumptions, we obtain an *exactly identified* system, whereas if we then make additional assumptions, the system becomes *overidentified*. It turns out that a recursive system with all arrows retained is exactly identified provided we make the assumption that all disturbance terms are mutually uncorrelated. If we then erase certain of the arrows, assuming that the corresponding β_{ij} are zero, the system becomes overidentified. This means that there will be fewer unknown parameters than pieces of empirical data that can be used to estimate these parameters. Some of this excess information may then be used to test the adequacy of the model, since not all sets of empirical data will satisfy the model.

In the context of the model of Fig. 10.2 the four omitted arrows represent the assumption that four of the β_{ij} are assumed to be zero. Actually, if we had not made these assumptions, we could have *estimated* these particular β_{ij} using the comparable least-squares unstandardized slope estimates since the full recursive model, with all arrows drawn in, would have been exactly identified. We then might have *predicted* that these particular estimates of the β_{ij} would be close to zero under the assumption that the model of Fig. 10.2 is actually correct. That is, we would make the following empirical predictions about the least-squares estimates:[1]

$$b_{32 \cdot 1} = 0 \qquad b_{51 \cdot 234} = 0$$
$$b_{41 \cdot 23} = 0 \qquad b_{53 \cdot 124} = 0$$

We could then make the usual tests of significance of these regression coefficients, or we could place confidence intervals about each of them in order to assess the degree to which departures from zero could be attributed to sampling error or random shocks or could be judged minor enough to be practically insignificant.

Of course the unstandardized regression coefficients have the disadvantage of being functions of one's choice of units of measurement, so that it is often more convenient to state the null hypotheses in terms of the disappearance of partial correlation coefficients. Obviously, since the numerators of correlation and regression coefficients are identical, and are therefore both zero at the same time, the assumptions that particular β_{ij} are zero can also be tested by examining the partial correlation coefficients, which have the advantage of always varying between ± 1.0. In the above model, our empirically verifiable predictions therefore become

$$r_{23 \cdot 1} = 0 \qquad r_{15 \cdot 234} = 0$$
$$r_{14 \cdot 23} = 0 \qquad r_{35 \cdot 124} = 0$$

In general, assumptions that certain of the X_i do *not* directly affect some of the remaining variables can always be tested in terms of predictions to the effect that the comparable *partial* correlation coefficients will be approximately zero. It should be emphasized, however, that if *other* specific values for the β_{ij} are assumed, for example, $\beta_{35} = 6.32$ or $\beta_{35} = \beta_{45}$, it is not possible to work with partial correlations in this simple manner.

Clearly, the more arrows we omit, i.e., the more β_{ij} we assume equal to zero, the more empirical predictions we can make and therefore the easier it will be to reject the theory in question. In the recursive system we have already omitted half the possible arrows, namely, those directed from X_j to X_i, where $i < j$. If no additional arrows are omitted, the only rejectable predictions we can make concern possible nonlinearities and signs of the direct relationships. In other words, regardless of the data, it will always be possible to *estimate* each of the parameters in the model. As we erase arrows, we add a test criterion, namely, the disappearance of some partial correlation, for each arrow that has been erased. This implies that the simpler one's theory, in the sense of the fewer the presumed causal linkages, the easier it will be to reject the model.

Tests of null hypotheses can be considered in the light of equation systems of this sort. Each time we state a null hypothesis we are asserting that a certain relationship (either a partial or a total association) is nonexistent or negligible. The two-variable

---

[1] In the conventional statistical literature null hypotheses are stated in terms of certain parameters set equal to zero, for example, $\beta_{ij} = 0$, but we shall here designate our empirical predictions in terms of their estimates, for example, $b_{ij \cdot km} = 0$, so as to emphasize that there is always a problem of evaluating the goodness of fit of the model to the data. Thus we shall predict that certain ordinary least-squares estimates of the β_{ij} will be zero and then compare these predicted values with those actually obtained from the data.

case, e.g., a difference-of-means test or a one-way analysis-of-variance test, is therefore a very simple special case of the more general multivariate situation. Whether one chooses to formulate the problem in terms of tests of hypotheses or in terms of point and interval estimation is partly a matter of one's preferences and partly a question of the status of the theory at hand. Duncan (1966b) has argued that it is unnecessary to proceed to formulate a series of alternative models each of which is tested in terms of the degree to which a set of predictions concerning the disappearance of partial correlations tend to be satisfied by the data at hand. He points out that one can always draw in the full recursive model with no arrows omitted, though of course assuming a causal ordering among the variables. If certain of the partial correlations or regression coefficients then turn out to be approximately zero, one simply erases the corresponding arrows. In this manner one avoids a tedious search operation necessitated by posing alternative models.

Such a procedure, however, makes it possible to take advantage of sampling errors in an ex post facto manner. Unless one's theory is well accepted in advance, our position is that it is preferable to state and test a relatively small number of alternative models, choosing among these on the basis of the goodness of fit of the data to the model. We shall consider a concrete illustration of this procedure in the next section. In this way, certain of the alternatives can be clearly rejected. Whenever the sample size is sufficiently large, one may of course divide the sample in half, using the first half to develop the model by any means one can, while saving the second half to test the adequacy of the best-fitting model developed from the first set of data.

A number of important points can be illustrated in terms of the model of Fig. 10.2. The ordinary least-squares estimating equations would be

$$\begin{aligned}
X_2 &= b_{21}X_1 + e_2 \\
X_3 &= b_{31 \cdot 2}X_1 + e_3 \\
X_4 &= b_{42 \cdot 13}X_2 + b_{43 \cdot 12}X_3 + e_4 \\
X_5 &= b_{52 \cdot 134}X_2 + b_{54 \cdot 123}X_4 + e_5
\end{aligned} \tag{10.5}$$

As has been discussed elsewhere, it turns out that commonsense rules of thumb for deciding when and when not to control for "relevant" variables can be given a statistical justification, provided that one is willing to assume the appropriateness of recursive models and assumptions (see Blalock, 1962, 1964). Several very simple illustrations can be discussed briefly.

We note that in the equation for X_3 the coefficient for X_1 involves a control for X_2 even though X_2 does not appear explicitly in that equation. Of course sampling error alone will imply that $b_{32 \cdot 1}$ will never be exactly zero, but let us assume that, in fact, the data satisfy the model exactly so that $b_{32 \cdot 1}$ is zero. Is it then necessary to control for X_2? In other words, will $b_{31 \cdot 2}$ and b_{31} be identical if the data exactly fit the model? From the formula for $b_{31 \cdot 2}$ we have

$$b_{31 \cdot 2} = \frac{b_{31} - b_{32}b_{21}}{1 - b_{12}b_{21}}$$

But if $b_{32 \cdot 1} = 0$, as implied by the model, we have

$$b_{32 \cdot 1} = 0 = \frac{b_{32} - b_{31}b_{12}}{1 - b_{12}b_{21}}$$

and therefore $b_{32} = b_{31}b_{12}$, from which it follows that

$$b_{31 \cdot 2} = b_{31} \frac{1 - b_{12}b_{21}}{1 - b_{12}b_{21}} = b_{31}$$

Therefore there is no need to control for X_2 in studying the relationship between X_1 and X_3. Given the model, this should be intuitively obvious. We do not want to control for the effects of X_2 *unless* these effects are in other respects causally connected to X_3. If there had been either a direct link from X_2 to X_3 or any indirect causal connection through other variables in the model, we would have wanted to control for X_2 in studying the direct effect on X_3 of X_1. We have already noted that direct and indirect causal relationships are always relative to the variables explicitly included in the causal system. It would always be possible to insert intervening variables, e.g., between X_1 and X_3, so as to turn a direct relationship into an indirect one or to delete variables so as to turn an indirect relationship into a direct one.

If enough arrows have been omitted from a model, simplifications may occur that make it possible to omit certain control variables while still predicting the disappearance of the partial correlations. The reader may wish to check the assertion that the model of Fig. 10.2 implies that the partial correlation $r_{35 \cdot 24} = 0$ (except for sampling error), so that it is not necessary to control for X_1, though it would not be misleading to do so. Had there been a direct arrow between X_1 and X_5, however, the partial $r_{35 \cdot 24}$ would *not* be expected to vanish, since X_1 remains to produce a spurious relationship between X_3 and X_5.[1] As a general rule, however, it is safest to control for all independent variables with subscripts lower than those of the dependent variable under consideration unless the model clearly implies that this is not necessary.

There are several extremely important practical points that this example illustrates. First, it would be clearly an unsatisfactory state of affairs if it were necessary to control for *all* antecedent variables in any given situation. Suppose, for example, that there were a string of variables Z_i which produced variation in X_1 but which were uncorrelated with the remaining X_j except through their impact on X_1. We would discover that the resulting model would imply that it would *not* be necessary to control for any of these antecedent variables in studying the remainder of the system. This is in line with our practical interests, as we would otherwise become involved with an infinite regress of causes. We are of course conceptualizing the situation in such a way that the effects of these unknown prior causes are summarized in ε_1, and we have found it necessary to make similar assumptions about the remaining disturbance terms ε_j. The point is that if we had explicitly introduced a set of Z_i as additional

---

[1] One way to find out which partials can be expected to vanish is to use a water-flow analogy. We can imagine water flowing through pipes, downward to the ultimate dependent variable. If valves are placed at each intervening variable, and if the flow can be cut off completely by closing a certain combination of the valves, the remaining variables need not be controlled in order for the appropriate partial to disappear. In the illustration under consideration, if we closed off the valves at X_2 and X_4, it would not be necessary to close the valve at X_1 in order to prevent flow from X_3 to X_5. Strictly speaking, however, the water-flow analogy does not hold, because we can allow water to flow uphill and then down, e.g., from X_3 to X_2 to X_5, but it cannot flow downhill and then up, since this would involve flowing through a variable that is dependent upon both the variables under consideration. Thus water flowing from X_2 to X_3 could not get there by going through X_5 (or X_4) but could go through X_1 if the latter variable were connected to both X_2 and X_3.

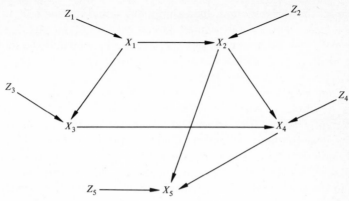

FIGURE 10.3

variables, as in the model of Fig. 10.3, we would have discovered that the resulting revised model would have implied that it was unnecessary to control for any of the Z_i *provided that* there were no additional arrows connecting any particular Z_i with more than one X_j. If, for example, Z_3 affected X_2 as well as X_3, the omission of Z_3 from the equation for X_3 would have resulted in a disturbance term ε_3 (containing Z_3) which would have been correlated with X_2, contrary to assumption. We see once more the importance of the assumptions about the disturbance terms.

A second point to note is that the model implies the necessity of *simultaneous* controls, as contrasted with controls applied sequentially. The model of Fig. 10.2 implies that $r_{14 \cdot 23} = 0$. It does *not* imply that a control for X_2 alone should reduce the relationship between X_1 and X_4 to zero. The correlation *may* move closer to zero with a single control, but it may not, according to the signs of the relationships. Suppose, for example, that all causal links involving these four variables are positive except that between X_2 and X_4. In this case, the indirect effect of X_1 on X_4 through X_2 would be negative, whereas that through X_3 would be positive. The two indirect paths might result in a very weak total association between X_1 and X_4. A control for X_2 alone, by taking out one of the two indirect influence paths, would result in a positive partial correlation $r_{14 \cdot 2}$ because effects through X_3 would not be removed. On the other hand, a control for X_3 alone would result in a negative value for $r_{14 \cdot 3}$, giving the investigator the impression that the association between X_1 and X_4 was subject to erratic behavior. The point is that the partial is predicted to disappear only when *both* intervening variables are controlled simultaneously. A similar principle holds in more complex models.

But it is also possible to overcontrol unless one is careful to specify the model clearly. When we say that all relevant variables should be controlled, we certainly do not mean that one should control for variables that are assumed to be *dependent* upon both variables under consideration. The recursive model implies that one does not introduce any variables with higher-numbered subscripts than that of the dependent variable with which one is concerned. Again, common sense suggests that we would become involved with an infinite regress if it were necessary to bring into the picture all the variables that, by assumption, would appear only in later equations. For example, suppose one wanted to control for X_3 in studying the relationship between X_1 and X_2. Except in very special cases it can readily be shown that $b_{21 \cdot 3} \neq b_{21}$, and if $r_{12} = 0$, it is obvious from the formula for $r_{12 \cdot 3}$ that $r_{12 \cdot 3} \neq 0$ unless either r_{13} or r_{23} also equals zero.

Confronted with a simple recursive model, an investigator may grant that these points are obvious. However, there are many practical situations in which one may inadvertently control for a dependent variable or manipulate the variation in dependent variables without recognizing this fact. Illustrations of such inadvertent statistical manipulations have been given elsewhere and can be noted here only in passing (see Blalock, 1964, 1967b). Sometimes dependent variables are manipulated in peculiar ways in the sample design, particularly when the ultimate dependent variable being studied represents a relatively rare phenomenon, e.g., suicide, drug addiction, or other forms of deviant behavior. The investigator may be forced to sample the cases by using a listing of these rare cases and may then decide to compare such a sample with a matched group of "normal" individuals. For example, a sample of inmates of a penitentiary may be matched with a group of normal persons, controlling for age, sex, race, socioeconomic status, and so forth. It may then not be recognized that relationships between the so-called independent variables may have been distorted in the process, and this, in turn, may affect the validity of one's inferences.

A second kind of situation in which dependent variables may inadvertently be used as control variables is one in which there may be reciprocal causation or feedback in the system and where the recursive model (in this simple form) is therefore inappropriate. In the model under consideration, suppose that X_3 also affects X_1 and that X_5 affects X_2. In studying the relationship between X_2 and X_4 should we control for X_5? Most of our commonsense rules of thumb for controlling break down unless recursive models can be assumed since the clearcut distinction between independent and dependent variables is misleading.

A more subtle and difficult problem involving possible inadvertent controls or manipulations of dependent variables arises whenever individual data are aggregated according to complex criteria. Very commonly in sociology and political science it is desirable to shift levels of aggregation from, say, counties to states or city blocks to census tracts. Individuals are aggregated according to geographic proximity, which is likely to be a very complex function of many of the variables in a causal system. Some of these, e.g., those relating to early socialization or contemporary influences, will tend to be causally prior to many dependent variables. But to the degree that residential location is a function of differential migration patterns, one's location is also likely to be *dependent* on factors such as income, education, race, and occupation. It has been shown (Blalock, 1964) that aggregating by dependent variables tends to have the opposite effect of controlling for dependent variables, but the essential point is that this kind of aggregation may tend to distort relationships between the causally prior variables. Unfortunately, when we aggregate by proximity, we ordinarily have no clearcut conception of exactly how we are affecting interrelationships, and therefore it becomes exceedingly difficult to generalize or compare results across different levels of aggregation (see Hannan, 1971).

One further note of caution is necessary before providing an example. Although not readily apparent from the model of Fig. 10.2, it will always be the case that several alternative models predict exactly the same empirical results. Assuming that one's tests are confined to simple covariance statements and that temporal sequences are not known or assumed a priori, there will be many ways in which a reasonably complex model can be altered by merely reordering the variables. In the simplest of cases involving just X_1 and X_2, we may always assume that X_1 causes X_2, and not vice versa, but we may assign either concrete variable, e.g., education or income, to the symbol X_1. Likewise, the abstract model of Fig. 10.2 contains five variables, and therefore it will be possible to assign five concrete variables to each of the X_i in $5! = 120$ different ways. This will be the case even without redirecting any of the arrows.

Obviously, then, we can proceed only by eliminating unsatisfactory models that do not fit the data, rather than establishing any specific model as the correct one. Put another way, there must be a large number of a priori and untestable assumptions made before one can test any given model. Sometimes there will be available additional information such as a knowledge of temporal sequences. But it will *always* be possible to challenge the required assumptions about the disturbance terms by suggesting variables that have been omitted from the system that jointly affect two or more of the variables. These points have been discussed extensively elsewhere but are nevertheless sufficiently important to require comment in the present context.

Example   Let us consider some empirical data and a series of models given by Goldberg (1966), whose study is based on a sample of 645 respondents interviewed by the Survey Research Center at the University of Michigan. The six variables used in the model are as follows:

1   Father's sociological characteristics (FSC)
2   Father's party identification (FPI)
3   Respondent's sociological characteristics (RSC)
4   Respondent's party identification (RPI)
5   Respondent's partisan attitudes (RPA)
6   Respondent's vote for president in 1956 (RV)

Three models considered by Goldberg, together with their implied predictions and the empirical results, are given in Figs. 10.4 to 10.6.

Goldberg's model I (Fig. 10.4) is seen to fit the data reasonably well except for the prediction that the partial correlation $r_{46 \cdot 1235}$ between respondent's party identification (RPI) and respondent's vote (RV) should be approximately zero. Actually, this correlation turned out to be .365 implying that there is some additional linkage between these two variables. A very simple modification of model I would therefore involve drawing in an additional arrow directly connecting X_4 to X_6. But Goldberg also considers an alternative model (model II) for which the direction of the causal relationship between RPI and RPA has been reversed. This reversal necessitates a labeling change, namely, the interchange of X_4 and X_5, to preserve the convention of taking variables with lower-numbered subscripts as causes of variables with higher-numbered subscripts. But models I and II also differ in two more respects: (1) model I contains arrows from FSC and RSC to RPA, whereas in model II RPA is strictly exogenous, and (2) model II contains arrows from FPI to RSC and from RSC to RPI.

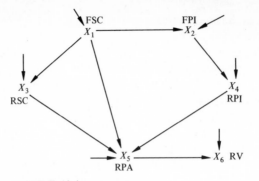

Predicted values	Actual values
$r_{23\cdot1} = 0$	.101
$r_{14\cdot23} = 0$	.017
$r_{34\cdot12} = 0$	.130
$r_{25\cdot134} = 0$	.032
$r_{16\cdot2345} = 0$	$-.019$
$r_{26\cdot1345} = 0$	.053
$r_{36\cdot1245} = 0$	$-.022$
$r_{46\cdot1235} = 0$	.365

FIGURE 10.4
Goldberg's model I.

These latter additions to model II may have been motivated by the nonzero partial correlations $r_{23\cdot1} = .101$ and $r_{34\cdot12} = .130$.

In any given instance it will be necessary to evaluate the goodness of fit of the model by using some rather arbitrary criterion. For small samples, this criterion might be that of the statistical significance of the partial correlations. In instances like Goldberg's illustration, especially where the sample size is large, it would be more appropriate to select some relatively small absolute magnitude, say .10. For example, it might be decided that modifications of the model will be necessary whenever a predicted zero partial correlation exceeds .10 in absolute magnitude. This would justify the addition of arrows from FPI to RSC and from RSC to RPI, as in model II, plus the addition of an arrow between RPI and RV, as in models II and III.

As we have already implied, the addition of arrows to a model means fewer remaining predictions that specific partial correlations will disappear. Had we drawn in all possible arrows from X_i to X_j, where $i < j$, we would have been reduced to no specific predictions at all, and the model would not have been rejectable except on grounds of nonlinearity, nonadditivity, or the incorrect prediction of signs. Therefore we wish to simplify the model as much as possible, and an obvious way is to erase arrows that are associated with very weak partial correlations. Although we have not supplied all the necessary information, the reader who calculates from Goldberg's original matrix of correlations the value of $r_{15\cdot234}$ (for model I) will discover that this partial has a very low numerical value. Presumably for this reason, Goldberg has deleted from model II the arrow between FSC and RPA, as well as that between RSC and RPA. Thus both models contain only seven arrows, thereby permitting predictions that eight partial correlations will disappear. Of course the reasoning is ex post facto in this kind of exploratory approach. As previously indicated, one may always hold out a portion of the data for more definitive tests of whatever best-fitting model has been derived from the exploratory analysis.

Model II is clearly unsatisfactory with respect to two of its predictions, namely, the partials involving the relationships between RPA (X_4) and both FPI (X_2) and the dependent variable RV (X_6). These partials are .357 and .470, respectively. Goldberg therefore constructs model III, which we shall not discuss further here; it is a modification of I and II and yields seven predictions that are reasonably well satisfied by the

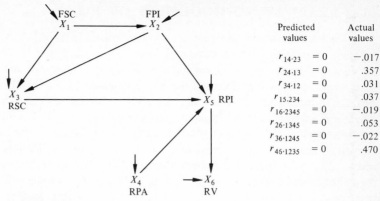

Predicted values	Actual values
$r_{14\cdot23} = 0$	$-.017$
$r_{24\cdot13} = 0$	$.357$
$r_{34\cdot12} = 0$	$.031$
$r_{15.234} = 0$	$.037$
$r_{16\cdot2345} = 0$	$-.019$
$r_{26\cdot1345} = 0$	$.053$
$r_{36\cdot1245} = 0$	$-.022$
$r_{46\cdot1235} = 0$	$.470$

FIGURE 10.5
Goldberg's model II.

data. Obviously, there are a number of alternative models that could also satisfy the data about as well. We have already referred to one of these, namely, a modified version of model I with an additional arrow between X_4 and X_6. This illustrates the general point made earlier that it will never be possible to eliminate all competing alternatives without making supplementary assumptions, many of which will not be testable. Knowledge of temporal sequences will provide a rationale for introducing certain simplifications and ruling out some of the alternatives. For example, the two variables referring to the father's characteristics, namely FSC and FPI, have been assumed to be causally prior to the respondent's characteristics, though one may certainly imagine instances in which a respondent may have influenced his father's characteristics. But we have already implied the possibility that RPI and RPA may mutually influence each other, since their causal roles were reversed between models I and II. In such instances, a nonrecursive model would of course be more appropriate. Here, however, we are restricting our attention to models involving one-way causation.

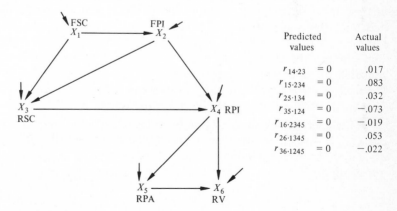

Predicted values	Actual values
$r_{14\cdot23} = 0$	$.017$
$r_{15.234} = 0$	$.083$
$r_{25\cdot134} = 0$	$.032$
$r_{35\cdot124} = 0$	$-.073$
$r_{16\cdot2345} = 0$	$-.019$
$r_{26\cdot1345} = 0$	$.053$
$r_{36\cdot1245} = 0$	$-.022$

FIGURE 10.6
Goldberg's model III.

10.4   EVALUATING GOODNESS OF FIT

The Goldberg illustration provides an excellent example through which to discuss the problem of evaluating the goodness of fit of any specific model. We have seen that his models I and II both contain some predictions that are not well satisfied by the data, with the result that a revised theory, represented by model III, was constructed. This last model's predictions are all reasonably well satisfied, and so we may stop with this model without further comment, recognizing, however, that a number of alternative models also yield predictions that cannot be rejected by the data. If we were interested in exploring this particular substantive model still further, we should of course attempt to specify additional alternatives that seem both reasonably plausible theoretically and compatible with the data. This would leave us with several versions that could not be rejected, and we would then presumably want to elaborate on or modify these models so that we could derive differential predictions from them that could be subject to testing. Or perhaps we could obtain additional kinds of empirical information such as the temporal sequences involved.

What overall criteria do we use to decide upon the adequacy of a model? If we relied on traditional tests of significance of null hypotheses (here to the effect that certain partials are zero), then if the sample size were sufficiently large, we might always reject the model as being at least slightly incompatible with the data. Applying this single criterion to large samples would certainly be unreasonable in view of the fact that, even if the model were perfectly correct, measurement errors in the variables would produce sufficient distortions to lead to the rejection of almost all null hypotheses. Therefore one must rely on some combination of results of significance tests and an inspection of the magnitudes and patterning of the partial correlations. For example, most of the departures from zero may involve a single variable which is suspected to have been measured rather poorly. Or none of the partials may be extremely close to zero, and yet none may be sufficiently large to cast doubt on the validity of the overall model. Also, some judgment must be made whether the model as a whole is to be rejected because of one or two deviations from predicted values or whether it is to be treated as modifiable, piecemeal and ex post facto, on the basis of an inspection of the data. Presumably, if there were a logically tight theoretical derivation that was presumed to stand or fall on the accuracy of a set of predictions taken together, we would want a single overall test and evaluation of goodness of fit. For example, the entire model might be rejected if a single test statistic yielded a probability smaller than a specified small quantity α. Such overall tests do exist in the literature, and the reader is referred to Jöreskog (1970, 1973), Goldberger (1973), and Goodman (1973) for discussions of these procedures.

The approach we shall stress in the present context is more exploratory and less efficient. It is based on the assumption that the investigator is much less interested in evaluating the overall goodness of fit of a model than in pinpointing specific defects and modifying the model on a piecemeal basis. This approach is of course much more vulnerable to the deficiencies of ex post facto analyses, since particular departures from predicted values may arise as a result of sampling errors. In effect, we are presuming that sample sizes are rather large, so that what are called specification

errors are the predominant sources of deviations from predicted values. These specification errors include random and nonrandom measurement errors and neglect of
many relevant variables in the model and errors resulting from a misspecification
of the correct form of the equations, e.g., nonlinearity, nonadditivity. If these types of
errors predominate, it seems to make little sense to rely heavily on significance tests,
whether these are tests of the entire model or of specific predictions taken individually.
Furthermore, we presume that the model has been generated by a very loose form
of theoretical reasoning that is not particularly sensitive to a relatively small number
of specific defects. In other words, we assume that it will be rather rare for one to be
able to pinpoint crucial tests of alternative theories such that one theory may be
unambiguously retained in favor of another. Thus our focus is on flexibility and the
importance of theory *building* as well as theory testing.

In view of all this, it seems wise to rely on multiple criteria of evaluation rather
than a single overall test or measure of goodness of fit. Our preference, here, would
be to look at each prediction separately and evaluate its adequacy in terms of whatever
assumptions one is making concerning the quality of one's measures and possible
distortions produced by measurement errors. If one does find it necessary to utilize a
general criterion for use in deciding on arrows to be added or subtracted in the process
of modifying a model, it seems sensible to adopt some degree of departure from zero
(or whatever value has been predicted), rather than relying on a test of significance
which puts the premium on sampling errors alone. Thus we could adopt the criterion
that whenever a partial correlation (predicted to be zero) is either greater than $+.10$
or less than $-.10$, this indicates that the model should be modified. Applying this
criterion in the case of the Goldberg illustration would then lead us to leave model III
unchanged.

No such criterion should be rigidly applied, however. First, we have noted that
the more arrows included in the model, the fewer the specific predictions that can be
made. Therefore if we adhered rigidly to the criterion, we would be more likely to
reject a model having only a few arrows, i.e., a simpler theory, than one that omits
only a very small number. Thus if we continue to modify a model by adding arrows,
we may expect to reach a point where the model very easily satisfies the data. Unfortunately, this will be true of many alternative models. Second, the criterion can be
rather easily satisfied if *all* correlations are rather weak. We would tend to prefer a
model that predicts a sufficient number of zero partials but for which the data yield
other partial and total correlations that are numerically large. All of this presupposes
that the models contain variables that are conceptually and empirically distinct, so
that problems of multicollinearity do not arise (see Gordon, 1968).

Once one has settled on a particular model as being sufficiently compatible with
the data, there remains the question of the best estimators to use to estimate the
structural parameters, or the β_{ij} in this instance. This is a technical question which
has a rather simple answer in the case of recursive systems with no measurement errors
but which is not so easily answered for nonrecursive systems or models containing
imperfectly measured variables. It can be shown in the recursive case that ordinary
least squares will produce unbiased and most efficient estimators. It might be thought
that some kind of weighted averaging procedure, as suggested by Boudon (1968),

might be more efficient, but Goldberger (1970) has shown that this is not true. However, the reader is referred to Goldberger (1973) and Jöreskog (1970, 1973) for discussions of the more general case. In the present chapter, then, we may rely on ordinary least squares. In the following chapter we shall discuss a modified procedure referred to as two-stage least squares that will answer our purposes very well for nonrecursive systems. For models involving measurement errors the reader is referred to a number of papers that appear in Goldberger and Duncan (1973) and also an alternative approach discussed by Wold (1975).

10.5   PATH ANALYSIS

The term *path analysis* was introduced by the biologist Sewall Wright (1934, 1960) to refer to a type of causal analysis that has much in common with the simultaneous-equation approaches of the econometricians. Apart from the work of Wright himself, the literatures on path analysis and econometric approaches seem to have developed along parallel lines without any extensive bibliographical references in common. Although most models involving path analysis have used recursive systems, nothing in principle restricts their application to one-way causation. The econometrics literature, however, is considerably richer and more developed for nonrecursive models, a fact which may merely reflect a difference in the substantive problems to which they have been applied or the theoretical perspectives taken. Our own discussion of path analysis will be confined to the recursive case.

The basic model underlying most path-analytic applications in the social sciences is the same as the one we have been discussing, namely, linear, additive, recursive equations with disturbance terms that are by assumption mutually uncorrelated. Therefore, we can anticipate that path analysis will yield exactly the same conclusions and inferences as the recursive-modeling approach. In contrast with the literature within econometrics, however, the path-analytic literature has tended to place a heavier emphasis on the heuristic use of visual diagrams and algorithms (or rules of thumb) for tracing paths, the explicit labeling of disturbance terms, and the use of standardized rather than unstandardized regression coefficients. Perhaps the best way to indicate the differences in approach, which are also partly notational, is to proceed by defining what is meant by a path coefficient and by stating a major theorem in path analysis for decomposing the total correlation between any two variables in a causal system.

A path coefficient may be defined as a standardized regression coefficient and interpreted as a ratio of two standard deviations, as indicated below (see Li, 1955; Land, 1969). Suppose we transform our original recursive system into an equivalent system involving normalized variables x_i all of which have unit variances and zero means. Unfortunately, many discussions of path analysis do not clearly distinguish between sample estimates and the population or structural parameters they are intended to estimate, and the most common literature therefore contains mixtures of parameters, for example, σ_i, and sample estimates, for example, p_{ij} and r_{ij}. Presumably, there are structural parameters π_{ij} which sample path coefficients are intended to estimate, but the sociological literature on the subject does not make this distinction.

We shall avoid this notational difficulty by confining ourselves to sample data, it being known that with reasonably large samples it will be possible to obtain un-biased and efficient estimates provided the usual least-squares assumptions have been met. Keeping this in mind, we transform each of the original X_i into standard form using the appropriate *sample* means and standard deviations. That is, we set each $x_i = (X_i - \bar{X}_i)/s_i$. Each of our least-squares equations can then be written in the form

$$x_i = p_{i1}x_1 + p_{i2}x_2 + \cdots + p_{i,i-1}x_{i-1} + v_i$$

where the residual term is also expressed in normalized form. The p_{ij} are then the *sample* path coefficients and are equivalent to standardized regression coefficients or *beta weights*, as discussed in the general statistical literature. It should be specifically noted that these p_{ij} are, in general, *partial* coefficients even though the customary dot notation is not retained in most discussions of path coefficients.

Since each variable has been expressed in standard form with unit (sample) variance, the path coefficient can be given a very simple interpretation. Instead of thinking in terms of unit changes in each variable that depend upon the units of measurement selected, e.g., pennies versus dollars, feet versus inches, we now think in terms of standard-deviation units, which are of course functions of the sample (or population) under investigation. A p_{ij} is a partial coefficient in the sense that it gives changes in the dependent variable for given changes in the appropriate indepen-dent variable, with all of the remaining variables controlled or held constant.

Since the standard deviation in the dependent variable x_i is unity, we may interpret p_{ij} as the proportion of the change (in standard-deviation units) in x_i for which x_j is directly responsible. That is, if we imagine that the independent variable x_j retains the same standard deviation (of unity) whereas all other variables in the equation (including the residual term) are held constant, then p_{ij} can be interpreted as the "new" standard deviation that would be produced in x_i. Ordinarily, p_{ij} will be less than unity in absolute value, though we can imagine situations in which x_i could vary *more* with all the remaining causes held constant than when all have standard deviations of unity. By convention we affix a positive or negative sign to p_{ij} to indicate the sign of the direct causal relationship, so that we may alternatively define p_{ij} as

$$p_{ij} = \pm \frac{s_{i \cdot j}}{s_i}$$

where $s_{i \cdot j}$ represents the standard deviation in x_i that would have resulted if x_j retained a standard deviation of unity and all the other causes of x_i remained constant.

The use of standardized coefficients, though forcing one to confine oneself to a single sample or population, permits certain simplifications that make it possible to write out a very simple formula for decomposing the total correlation between any two variables, including the correlation of a variable with itself. Since for standardized variables we have $r_{ij} = (1/n) \sum x_i x_j$, we may take the equation for x_i, multiply through by x_j and sum as follows:

$$x_i = p_{i1}x_1 + p_{i2}x_2 + \cdots + p_{i,i-1}x_{i-1} + v_i$$

Therefore

$$
\begin{aligned}
r_{ij} &= \frac{1}{n} \sum x_i x_j \\
&= p_{i1} \frac{\sum x_1 x_j}{n} + p_{i2} \frac{\sum x_2 x_j}{n} + \cdots + p_{i,i-1} \frac{\sum x_{i-1} x_j}{n} + \frac{\sum v_i x_j}{n} \\
&= p_{i1} r_{1j} + p_{i2} r_{2j} + \cdots + p_{i,i-1} r_{i-1,j} + 0
\end{aligned}
$$

Thus we obtain the following formula for the decomposition of a total correlation:

$$
r_{ij} = \sum_k p_{ik} r_{kj} \qquad \text{for } k < i \qquad (10.6)
$$

where we have utilized the fact that, for any $j < i$, $\sum v_i x_j = 0$. Thus, for recursive systems, we may decompose a total correlation into a sum of component parts, each of which is a product of a (partial) path coefficient p_{ik} and the total correlation r_{kj} between the appropriate x_k and the independent variable x_j, where we sum over all k for which there is a direct path from x_k to the dependent variable x_i.

Path Analysis Applied to Recursive Models

Let us now apply this basic formula for decomposing a total correlation to simple recursive models. Consider first the two-variable model for which x_1 is assumed to have a linear effect on x_2 and x_2 is also determined by a residual variable x_a that we previously labeled as a disturbance term v_2.[1] The model is given Fig. 10.7a. We assume that the independent variable x_1 and the unmeasured residual x_a are uncorrelated in the sample as well as the population, so that we may replace the assumption that $E(x_1 x_a) = 0$ with the assumption that $\sum x_1 x_a = r_{1a} = 0$, with all variables having been standardized so as to have zero (sample) means and standard deviations of unity. For this very simple model we can write two equations, one for r_{12} and a second for the correlation of x_2 with itself (which is of course unity). We get

$$
r_{12} = p_{21} r_{11} + p_{2a} r_{a1} = p_{21}
$$

since r_{a1} is zero, by assumption, and

$$
r_{22} = 1 = p_{21} r_{12} + p_{2a} r_{a2} = p_{21}{}^2 + p_{2a}{}^2
$$

since $p_{2a} = r_{a2}$.

The second of these equations gives the result that $p_{2a}{}^2 = 1 - p_{21}{}^2 = 1 - r_{12}{}^2$. Thus even though x_a is an unknown and we therefore cannot directly obtain the value of $p_{2a} = r_{a2}$, we can solve for the residual path coefficient in terms of the remaining quantities. We see that p_{2a} represents the square root of the proportion of the variance in x_2 that is unexplained by x_1. Numerical values of the residual path coefficients are sometimes inserted into path diagrams to indicate the magnitude of the unexplained

---

[1] This notational change is made in order to follow Wright (1934), who used literal subscripts to refer to unmeasured residual variables. Since much of the sociological literature follows Wright in this practice the reader may find it necessary to become familiar with both notations.

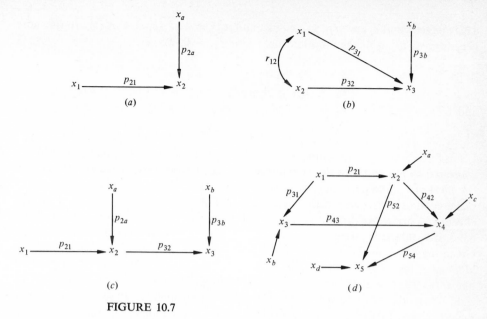

FIGURE 10.7

variance, though the computation of these coefficients does not shed any new light on the situation. In general, it turns out that for simple recursive models for which disturbances are assumed uncorrelated with independent variables in each equation, these residual path coefficients, when squared, are always $1 - R^2$, where R^2 represents the square of the multiple correlation of the explicit independent variables with the dependent variable under consideration.

Next let us consider the somewhat more complex three-variable model of Fig. 10.7b, which involves two explicit causes of the dependent variable x_3 plus a residual variable x_b. The explicit causes of x_3 may be intercorrelated for unknown reasons, as indicated by the curved double-headed arrow connecting x_1 and x_2. A path coefficient is of course asymmetric, and its use therefore implies that the direction of causation has been assumed. Had we been willing to assume that x_1 caused x_2, we would have so indicated by a straight arrow with a single head, in which case the path coefficient from x_1 to x_2 would have been labeled p_{21}. Likewise, if we had been willing to assume that x_2 caused x_1, we would have inserted p_{12} in the diagram. In either case the value of the path coefficient (p_{21} or p_{12}) would have been equivalent to r_{12}, as we have just seen. As we shall also see below, even if the relationship between x_1 and x_2 were spurious (say, due to x_0), we could have let r_{12} replace the compound path connecting x_1 and x_2, that is, $r_{12} = p_{10}p_{20}$. Thus regardless of the reason for the correlation between x_1 and x_2 we can utilize this correlation even though we cannot explain or decompose it. Customarily, the value of r_{12} is inserted in the path diagram next to the double-headed arrow. If $r_{12} = 0$ (or is very close to zero), the curved arrow is omitted.

If $r_{12} = 0$, we have a very simple extension of the previous model, and it can easily be seen that $r_{13} = p_{31}, r_{23} = p_{32}$, and $p_{3b}{}^2 = 1 - (r_{13}{}^2 + r_{23}{}^2)$. In the more general case, using the assumption that $r_{1b} = r_{2b} = 0$, we get

$$r_{13} = p_{31}r_{11} + p_{32}r_{12} + p_{3b}r_{1b} = p_{31} + p_{32}r_{12}$$
$$r_{23} = p_{31}r_{12} + p_{32}r_{22} + p_{3b}r_{2b} = p_{31}r_{12} + p_{32}$$

and
$$r_{33} = 1 = p_{31}r_{13} + p_{32}r_{23} + p_{3b}r_{3b}$$

or
$$p_{3b}{}^2 = 1 - p_{31}r_{13} - p_{32}r_{23}$$

Let us focus first on the decomposition of r_{13}. The direct effect of x_1 on x_3 is obviously measured by p_{31}, but how do we interpret the second component $p_{32}r_{12}$? If x_2 were the effect of x_1 (and not vice versa), we could replace r_{12} by p_{21}, in which case it would make good sense to refer to the compound path $p_{32}p_{21}$ as a measure of the indirect effect of x_1 just as we did in the case of the unstandardized coefficients. On the other hand, if x_2 affects x_1, we would not want to credit x_1 with indirect effects through x_2. Given the fact that the causal connection between x_1 and x_2 is assumed to be ambiguous, it makes no sense to think in terms of the indirect effects of x_1 in a model like this. The correlation r_{13} is thus decomposed, but only one component of the right-hand side can be given an unambiguous causal interpretation. Unfortunately there has been some confusion over this point in the sociological literature (Duncan, 1966*b*; Land, 1969), subsequently corrected by Duncan, (1971). Duncan (1971) attempts to provide an interpretation of sorts, leaving an overlap portion of the explanation as ambiguous, but this in no way resolves the difficulty.

The principle readily generalizes. If one begins with a set of exogenous variables that are intercorrelated for unknown reasons, it will be possible to assess their direct effects or their indirect effects through intervening (endogenous) variables but not the indirect effects of the exogenous variables through each other. Thus the feature of path analysis that permits the use of double-headed arrows cannot be utilized to provide interpretations in instances where the underlying theory is too weak to permit the assumption of the direction of causation among the exogenous variables. This is of course as expected.

If we take the equations for r_{13} and r_{23}, multiply the latter by r_{12}, and subtract from the former, we get

$$r_{13} - r_{12}r_{23} = p_{31} - p_{31}r_{12}{}^2$$

or
$$p_{31} = \frac{r_{13} - r_{12}r_{23}}{1 - r_{12}{}^2} = b_{31 \cdot 2}\frac{s_1}{s_3}$$

Similarly,

$$p_{32} = \frac{r_{23} - r_{12}r_{13}}{1 - r_{12}{}^2} = b_{32 \cdot 1}\frac{s_2}{s_3}$$

Thus when we solve for the p_{ij} in terms of the r_{ij}, we obtain the usual textbook formulas for the beta weights or standardized regression coefficients. There are therefore two ways of obtaining the path coefficients p_{ij} in simple recursive models. The first is to write down equations for the total correlations in terms of the path coefficients, then solve for the latter in terms of the former. The second is to utilize the formulas for standardized regression coefficients directly by multiplying the formulas for the unstandardized coefficients by the appropriate ratio s_j/s_i or the ratio of the standard deviation of the independent variable to that of the dependent variable.

As our final very simple model consider the three-variable causal chain in which x_1 affects x_2, which, in turn, affects x_3 as indicated in Fig. 10.7c. This model of course is identical to that of Fig. 10.7a except for the addition of x_3 and the unmeasured residual variable x_b, and therefore we may merely add to the equations for x_{12} and x_{22} the three equations

$$r_{13} = p_{32}r_{12} + p_{3b}r_{1b} = p_{32}r_{12} = p_{32}p_{21} = r_{12}r_{23}$$
$$r_{23} = p_{32}r_{22} + p_{3b}r_{2b} = p_{32}$$

and

$$r_{33} = 1 = p_{32}r_{23} + p_{3b}r_{b3}$$

or

$$p_{3b}{}^2 = 1 - p_{32}{}^2 = 1 - r_{23}{}^2$$

As we might expect, we see that $r_{23} = p_{32}$, meaning that if there is no path connecting x_1 to x_3 other than the path through x_2, there is no point in considering the antecedent cause x_1 when studying the relationship between x_2 and x_3.

The main point to note about this chain model, however, is that the correlation between the two end variables x_1 and x_3 can be expressed either as the product of the two path coefficients p_{21} and p_{32} or as the product of the correlations $r_{12}r_{23}$. Since the model predicts that, apart from sampling error, $r_{13} = r_{12}r_{23}$, this of course implies the prediction that $r_{13 \cdot 2} = 0$. In effect, we may write down three equations for the correlations r_{12}, r_{13}, and r_{23}, from which we need estimate only two path coefficients p_{21} and p_{32}. The latter coefficients could be estimated from the (completely trivial) equations for r_{12} and r_{23}, leaving the equation for r_{13} as a redundant equation supplying the test criterion $r_{13 \cdot 2} = 0$, or $r_{13} = p_{21}p_{32}$. In general, it will always be possible to locate one redundant equation for each direct path that has been omitted, and this can be written either in terms of some partial correlation set equal to zero or a total correlation set equal to a function of certain of the path coefficients and remaining (nonzero) correlations. The result for the three-variable simple chain also obviously generalizes to the k-variable case, $x_1 \rightarrow x_2 \rightarrow x_3 \rightarrow \cdots \rightarrow x_k$, the correlation between x_1 and x_k being

$$r_{1k} = p_{21}p_{32}p_{43} \cdots p_{k,k-1} = r_{12}r_{23}r_{34} \cdots r_{k-1,k}$$

A five-variable model Let us see how these results can be extended to the five-variable model of Fig. 10.2, which has been repeated as Fig. 10.7d with the residual terms relabeled as x_a, x_b, x_c, and x_d and the nonzero path coefficients indicated on the diagram. We have already noted that the four omitted direct arrows imply the predictions

$$r_{23 \cdot 1} = 0 \qquad r_{15 \cdot 234} = 0$$
$$r_{14 \cdot 23} = 0 \qquad r_{35 \cdot 124} = 0$$

We did not write down any additional equations but moved directly to the estimation of the six nonzero unstandardized regression coefficients β_{ij}, using ordinary least squares.

In the path-analytic approach we can write down equations for the 10 total correlations r_{ij}, $i \neq j$, from which we can solve for the six unknown p_{ij}. We accomplish this by utilizing a proper combination of six of the equations, saving the remaining four as redundant equations that may be used to test the goodness of fit of the model. As we shall see, we can select as our excess equations the four equations involving those r_{ij} for which the direct paths have been omitted, that is, r_{23}, r_{14}, r_{15}, and r_{35}. It will then turn out that these four equations are algebraically equivalent to the above predictions that the corresponding *partial* correlations are zero. Thus we shall see that the very simple result obtained above in the causal chain model is readily generalizable. For the sake of simplicity, we shall omit any further discussion of the equations for the correlations of each variable with itself, from which expressions for the residual path coefficients can be obtained. As already noted, these residual path coefficients turn out to be equivalent to $\sqrt{1 - R^2}$, or the square root of 1 minus the square of the multiple correlation with the appropriate dependent variable. The 10 equations for the r_{ij} are

$$r_{12} = p_{21} + p_{2a}r_{a1} = p_{21} \tag{10.7a}$$

$$r_{13} = p_{31} + p_{3b}r_{b1} = p_{31} \tag{10.7b}$$

(R) $\quad r_{23} = p_{31}r_{12} + p_{3b}r_{b2} = p_{31}r_{12} = p_{31}p_{21}$ $\tag{10.7c}$

(R) $\quad r_{14} = p_{42}r_{21} + p_{43}r_{31} + p_{4c}r_{c1} = p_{42}p_{21} + p_{43}p_{31}$ $\tag{10.7d}$

$$r_{24} = p_{42} + p_{43}r_{32} + p_{4c}r_{c2} = p_{42} + p_{43}p_{31}p_{21} \tag{10.7e}$$

$$r_{34} = p_{42}r_{23} + p_{43} + p_{4c}r_{c3} = p_{42}p_{31}p_{21} + p_{43} \tag{10.7f}$$

(R) $\quad r_{15} = p_{52}r_{21} + p_{54}r_{41} + p_{5d}r_{d1} = p_{52}p_{21} + p_{54}(p_{42}p_{21} + p_{43}p_{31})$ $\tag{10.7g}$

$$r_{25} = p_{52} + p_{54}r_{42} + p_{5d}r_{d2} = p_{52} + p_{54}(p_{42} + p_{43}p_{31}p_{21}) \tag{10.7h}$$

(R) $\quad r_{35} = p_{52}r_{23} + p_{54}r_{43} + p_{5d}r_{d3} = p_{52}p_{31}p_{21} + p_{54}(p_{43} + p_{42}p_{31}p_{21})$ $\tag{10.7i}$

$$r_{45} = p_{52}r_{24} + p_{54} + p_{5d}r_{d4} = p_{54} + p_{52}(p_{42} + p_{43}p_{31}p_{21}) \tag{10.7j}$$

Notice that we may proceed recursively through the equation system, ignoring x_4 and x_5 when studying x_1 through x_3, and ignoring x_5 when studying the first four variables. In Eqs. (10.7a) to (10.7c) the first two trivial equations are sufficient to give p_{21} and p_{31}, and therefore the third equation may be taken as redundant (marked R) and used for testing purposes. That is, the numerical value of r_{23} should be approximately equal to the product $p_{31}p_{21}$, which is to say that $r_{23 \cdot 1}$ should be approximately zero.

Next examine Eqs. (10.7d) to (10.7f) involving x_4. We see that r_{14}, r_{24}, and r_{34} have each been decomposed into relatively simple compound paths that are consistent with common sense. In the case of r_{14} we have the sum of the two compound paths $p_{42}p_{21}$ and $p_{43}p_{31}$ representing the indirect effects of x_1 on x_4 via the two (otherwise unconnected) intervening variables x_2 and x_3. The total correlation between x_2 and x_4 (and similarly that between x_3 and x_4) is decomposed into the direct path p_{42} plus a compound path $p_{43}p_{31}p_{21}$ representing the effect of x_1 on the relationship between x_2

and x_4. In this three-equation subsystem we again have only two path coefficients to estimate, and we may select one of these equations as redundant. We have selected Eq. (10.7d) involving r_{14} as the redundant equation, and it can easily be shown algebraically that this equation and Eq. (10.7c) together imply that $r_{14 \cdot 23} = 0$. Thus we obtain a result that is entirely consistent with the argument that an omitted arrow (or structural parameter) implies that the corresponding partial correlation should be zero except for sampling error.

The same kinds of results obtain for Eqs. (10.7g) to (10.7j) involving x_5 as the dependent variable. We now have only two new path coefficients p_{52} and p_{54} to estimate, and if we select Eqs. (10.7g) and (10.7i) as redundant, these two equations plus our earlier restrictions imply that $r_{15 \cdot 234}$ and $r_{35 \cdot 124}$ should be approximately zero. We also obtain expressions for the decompositions of the total correlations with x_5 that are straightforward extensions of the previous results. For example, r_{15} is decomposed into three compound paths, $p_{52}p_{21}$, $p_{54}p_{42}p_{21}$, and $p_{54}p_{43}p_{31}$, the latter two paths being through x_4. In the decomposition of r_{45} we have the direct path from x_4 to x_5 plus two compound paths involving x_2, which is a common cause of x_4 and x_5. Because certain of the arrows have been removed, these compound expressions are relatively more simple than would be necessary in more complete five-variable models. Of course if any of the conditions implied by the four redundant equations (R) are not met, it may be desirable to add arrows ex post facto, in which case some of Eqs. (10.7) would have to be modified.

It is possible to write down the equations for each of the r_{ij} directly in terms of the p_{ij} by inspection of the diagram through the use of a simple algorithm or rule of thumb for tracing paths, but use of such an algorithm is not recommended except as a heuristic device in instances where the model is at all complex. The reason is that it is very easy to miss certain of the more complex or indirect paths or to duplicate certain of the other paths. In brief, the algorithm states that one may trace forward only or back and then forward, but it is not permissible to trace forward and then back. In tracing forward only, one is of course obtaining the various direct and indirect effects of a variable. For example, in going from x_1 to x_4 one may move through either x_2 or x_3 via the compound paths $p_{42}p_{21}$ and $p_{43}p_{31}$ respectively. In going from x_1 to x_5 there is the compound path $p_{52}p_{21}$ through x_2 plus the path through x_4 which is itself compound.

When we trace backward and then forward, we take care of causally prior variables that may be producing partly spurious relationships. In the case of the relationship between x_2 and x_4 we decompose r_{24} into the direct effect of x_2 on x_4, namely p_{42}, plus the term $p_{43}p_{31}p_{21}$ representing the effect of x_1 on the relationship between x_2 and x_4. But if we were permitted to trace backward after first having gone forward, we would be moving through a common *effect* of the two variables. Thus we do not go forward from x_4 to x_5 and back to x_2 since the common effect x_5 does not enter into the relationship between x_2 and x_4, as we have already indicated. The algorithm for tracing paths provides a visual device for writing down the decomposition of each total correlation, and this may serve as a check on the algebraic equations.

Whenever any of the exogenous variables are connected by curved double-headed arrows, one may still proceed in terms of either the algebraic or algorithmic approach. For example, if one were unable to commit oneself on the direction of

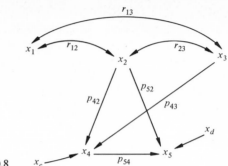

FIGURE 10.8

causation between x_1 and x_2, one would have to replace p_{21} by r_{12} throughout. Notice that the algorithmic rule for tracing paths would not have to be altered in this case. One must only be concerned about tracing forward and then back, and this will never be necessary in the case of exogenous variables, which are by assumption *not* dependent on any of the endogenous variables in the system. If there are more than two exogenous variables that are intercorrelated for unknown reasons, each *pair* of such variables must be linked by a separate curved arrow, as in Fig. 10.8, and the rule for tracing paths must be modified to exclude tracing through more than a single curved arrow in any one product expression in a compound path.

For example, in Fig. 10.8 in decomposing r_{14} we could use the path $p_{42}r_{12}$ and the path $p_{43}r_{13}$ but not the expression $p_{43}r_{12}r_{23}$, which would overlap the path $p_{43}r_{13}$ because the curved arrows cannot be interpreted as actual causal paths but merely as representations for the total correlations. If there were no correlation between x_1 and x_3, we would simply omit the curved arrow between these two variables, in which case it would *not* be legitimate to trace between them by moving through x_2. Because of the ambiguity in the causal relationships between these variables, this might entail tracing through a common effect, that is, x_1 and x_3 might be common causes of x_2 though unrelated to each other. Once more it should be emphasized that although p_{42} may be interpreted as a measure of the direct effects of x_2 on x_4, it is meaningless to talk about indirect effects through the other exogenous variables in the model of Fig. 10.8 without further assumptions about the causal relationships between the exogenous variables. One could, however, talk meaningfully about the indirect effects of x_2 on x_5 through the intervening variable x_4, since the direction of causation between x_2 and x_4 and between x_4 and x_5 is assumed to be known.

10.6 PATH ANALYSIS WITH UNSTANDARDIZED COEFFICIENTS

We have observed that path analysis, as a heuristic device, involves basically the same models and assumptions as are appropriate for recursive systems and that the basic arguments do not depend on the use of the standardized path coefficients p_{ij}. Wright (1960) refers to the unstandardized coefficients as *path regression coefficients* and notes that it is often appropriate to utilize both types of coefficients. Yet sociological applications of path analysis have involved almost exclusively the standardized p_{ij}, just as our predominant focus has been on the magnitudes of correlation coefficients. All such measures are population-specific, as we have already noted. As long as one thinks only in terms of fixed populations and generalizations from sample data, no special difficulties arise, but as soon as one shifts to comparisons *across* populations or allows for *changes* over time within a single population, the p_{ij} become noncomparable.

We have seen that the path coefficient, in the case of recursive systems with uncorrelated disturbance terms, is just a beta weight or standardized regression coefficient. This of course means that one may very simply transform from the standardized to the unstandardized regression coefficients by multiplying by a ratio of two standard deviations. For this reason, the distinction between the two types of measures is sometimes considered trivial. From the mathematical standpoint this may be true, but we must remember that the transformation cannot be made unless the standard deviations are *known*. Two implications of this obvious fact are immediately apparent: (1) Unless it becomes the common practice to report standard deviations along with the standardized coefficients, a reader cannot make the necessary transformation. Unfortunately it is the present practice within sociology and political science to omit this crucial information from publications. (2) Since there will rarely be theoretical or a priori grounds for saying much about the relative magnitudes of standard deviations prior to the collection of data, it becomes difficult to make very specific predictions concerning the relative magnitudes of the p_{ij}.

Suppose that one is comparing two path diagrams, one for whites and the other for blacks. Perhaps there may be available data from both the 1960 and 1970 censuses so that a comparison across time is also possible. The same model and structural parameter values may be appropriate for all four sets of data. Exactly the same sets of coefficients may be approximately zero. Assume that one is interested in explaining present occupation and is attempting to assess the relative contributions of father's occupation, respondent's education, and first job. The magnitudes of the structural parameters β_{ij} may be exactly the same for blacks and whites and for 1960 and 1970, and yet the path coefficients may differ because of unequal ratios of the standard deviations. For example, the fathers' occupations of blacks may be much more homogeneous than those of whites. Or perhaps there have been changes in several of the standard deviations over time. If one merely looks at the relative magnitudes of the p_{ij}, one may be led astray in inferring basic differences with respect to the causal processes operating. If father's occupation appears to be "more important" (in the sense of having larger path coefficients) for whites than blacks, this may merely be due to the fact that there is relatively greater dispersion in this variable for the white sample. One should not then automatically infer discrimination or basically different causal processes at work.

Rarely do our theories enable us to predict relative magnitudes of standard deviations, and this is especially true with respect to exogenous variables. We are interested, however, in assertions concerning the effects of *hypothetical* changes in certain of the variables. For example, are the increments in occupational status credited to an additional year of schooling the same for whites and blacks? Have these increments changed between 1960 and 1970? In other words, our theoretical interests are directly centered on hypothetical "if then" propositions even though our actual data may come to us in the form of real differences or changes appropriate to particular populations at two specific points in time. We need to keep this distinction in mind whenever we wish to link our data with our theories. We may need to use data about specific populations to *estimate* the structural parameters, but our basic concern is with the causal laws, rather than their manifestations in concrete populations.

Standardized measures have an important and convenient property in that they are not influenced by our choices of units of measurement, e.g., pennies versus dollars. This means that *if* one is willing to take as givens the various standard deviations that have been obtained for a concrete population or estimated from sample data, it is meaningful to use path coefficients to make comparisons across *variables*. One may thus say that, for whites, father's occupation explains more variance than respondent's education. But it must be recognized that such a result depends not only on the magnitudes of the unstandardized parameters but the empirical variability that pertains to that particular population.

For these reasons we recommend that investigators report *both* standardized p_{ij} and unstandardized regression coefficients for the benefit of those who may find it possible to replicate the study using approximately the same measures and for those who wish to make comparisons across subpopulations with different standard-deviation ratios. It is therefore desirable to develop comparable heuristic devices for decomposing unstandardized total regression coefficients and tracing direct and indirect effects. We turn next to such an analogous development of path diagrams, using slightly different causal models for the purpose of making additional interpretive remarks.

Let us return to our original notation, in which disturbance terms are represented as ε_i estimated empirically by the e_i which are the residuals about the ordinary least-squares equations. Thus we shall drop the notation involving a representation of the disturbances by x's with literal subscripts. To emphasize the fact that the least-squares coefficients involve controls for the other independent variables in each equation, we shall utilize the familiar dot notation of ordinary least squares. This is necessary in order to distinguish certain total coefficients not involving controls from partial coefficients that do involve the appropriate controls. Our recursive set of estimating equations is once again

$$X_1 = e_1$$
$$X_2 = b_{21}X_1 + e_2$$
$$X_3 = b_{31 \cdot 2}X_1 + b_{32 \cdot 1}X_2 + e_3$$
$$X_4 = b_{41 \cdot 23}X_1 + b_{42 \cdot 13}X_2 + b_{43 \cdot 12}X_3 + e_4 \qquad (10.8)$$
$$\cdots\cdots\cdots\cdots\cdots\cdots\cdots\cdots\cdots\cdots\cdots\cdots\cdots\cdots\cdots\cdots$$
$$X_k = b_{k1 \cdot 23,\ldots,k-1}X_1 + b_{k2 \cdot 13,\ldots,k-1}X_2$$
$$+ \cdots + b_{k,k-1 \cdot 23,\ldots,k-2}X_{k-1} + e_k$$

For simplicity we have again omitted the constant terms by assuming that each variable has been measured about its (sample) mean, but we have *not* standardized so that each variable has a (sample) standard deviation of unity. In other words, we are allowing for the possibility that samples and populations may differ with respect to variation in each variable. To emphasize the difference between the standardized and unstandard-ized variables we have returned to our original notation involving capital X_i.

We now proceed to derive a decomposition of b_{ij} representing the total slope estimate relating the dependent variable X_i to one of its (possible) causes X_j, with $j < i$. If we consider the equation

$$X_i = b_{i1 \cdot 23, \ldots, i-1} X_1 + b_{i2 \cdot 13, \ldots, i-1} X_2 + \cdots + b_{i, i-1 \cdot 12, \ldots, i-2} X_{i-1} + e_i$$

and multiply by X_j, then summing over all observations gives

$$\sum X_i X_j = b_{i1 \cdot 23, \ldots, i-1} \sum X_1 X_j + b_{i2 \cdot 13, \ldots, i-1} \sum X_2 X_j$$
$$+ \cdots + b_{ij \cdot 12, \ldots, (j-1)(j+1), \ldots, i-1} \sum X_j^2$$
$$+ \cdots + b_{i, i-1 \cdot 12, \ldots, i-2} \sum X_{i-1} X_j + \sum X_j e_i$$

But we know that the residuals about a least-squares equation are automatically uncorrelated with the independent variables in that equation, so that $\sum X_j e_i = 0$, and therefore the last term may be neglected. If we now divide by $\sum X_j^2$, we get

$$b_{ij} = \frac{\sum X_i X_j}{\sum X_j^2} = b_{i1 \cdot 23, \ldots, i-1} \frac{\sum X_1 X_j}{\sum X_j^2} + b_{i2 \cdot 13, \ldots, i-1} \frac{\sum X_2 X_j}{\sum X_j^2} + \cdots$$

$$+ b_{ij \cdot 12, \ldots, (j-1)(j+1), \ldots, i-1}(1) + \cdots + b_{i, i-1 \cdot 12, \ldots, i-2} \frac{\sum X_{i-1} X_j}{\sum X_j^2}$$

But we can represent expressions of the form $\sum X_i X_j / \sum X_j^2$ by total unstandardized slopes b_{ij}; therefore

$$b_{ij} = b_{i1 \cdot 23, \ldots, i-1} b_{1j} + b_{i2 \cdot 13, \ldots, i-1} b_{2j}$$
$$+ \cdots + b_{ij \cdot 12, \ldots, (j-1)(j+1), \ldots, i-1}$$
$$+ \cdots + b_{i, i-1 \cdot 12, \ldots, i-2} b_{i-1, j}$$

or

$$b_{ij} = \sum_k b_{ik \cdot 12, \ldots, (k-1)(k+1), \ldots, i-1} b_{kj}$$
$$j, k < i \qquad (10.9)$$

where the summation is over k. Thus we may represent a total slope as a sum of products of partial slopes and the total slope linking X_j with X_k. This formula is directly analogous to the formula for the decomposition of r_{ij} though the use of dot notation makes the expression on the right-hand side of Eq. (10.9) look more cumbersome.

Illustrations and Interpretations

Let us illustrate this rather abstract formulation in terms of simple models involving three and four variables. First consider the model of Fig. 10.9, where the estimates b_{21}, $b_{31 \cdot 2}$, and $b_{32 \cdot 1}$ have been inserted into the figure as least-squares estimates of the structural parameters β_{21}, β_{31}, and β_{32} representing the respective direct effects. Applying the above formula for the decomposition of b_{ij}, we obtain

$$b_{21} = b_{21} \qquad (10.10a)$$

$$b_{31} = b_{31 \cdot 2} + b_{32 \cdot 1} b_{21} \qquad (10.10b)$$

$$b_{32} = b_{32 \cdot 1} + b_{31 \cdot 2} b_{12} \qquad (10.10c)$$

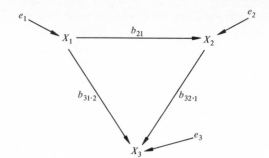

FIGURE 10.9

Although Eqs. (10.10b) and (10.10c) are analogous, as can be seen by interchanging X_1 and X_2, their causal interpretations are fundamentally different owing to the assumed causal asymmetry between X_1 and X_2 and the fact that since we are assuming that $\beta_{12} = 0$, the coefficient b_{12} should not be interpreted as an estimate of a parameter with causal meaning.

The interpretation of (10.10b) is straightforward intuitively. The coefficient b_{31} is the estimate of the *total* effect of X_1 on X_3, ignoring X_2 as an intervening variable. We have decomposed this total effect of X_1 into two paths, the direct effect being estimated by $b_{31.2}$ and the indirect effect estimated by the product $b_{32.1}b_{21}$. Assuming no sampling error, this means that if we changed X_1 by 1 unit, we would expect to change X_2 by b_{21} units, which would produce a change of $b_{21}b_{32.1}$ units in X_3 via this indirect path.

Suppose we now look at Eq. (10.10c) involving a decomposition of b_{32}, the estimate of the total regression of X_3 on X_2, ignoring their common cause X_1. Assume that we use b_{32} as our estimate of the effect of X_2 on X_3. Since this relationship is partly spurious, we expect a bias that can be estimated by the difference between b_{32} and $b_{32.1}$. The latter measures the direct effect of X_2, and since there are no indirect effects postulated in this particular model, this coefficient also measures the total effect on X_3 of a unit change in X_2. From (10.10c) we see that this bias in our estimate b_{32} (for large samples) will be approximately

$$b_{32} - b_{32.1} = b_{31.2}b_{12}$$

But since b_{12} cannot be given a simple causal interpretation, we replace it by its equivalent, namely $b_{21}(s_1{}^2/s_2{}^2)$, obtaining

$$b_{32} - b_{32.1} = b_{31.2}b_{21}\frac{s_1{}^2}{s_2{}^2}$$

Thus the approximate large-sample bias is a function not only of $b_{31.2}$ (the direct effect of X_1 on X_3) and b_{21} (the direct effect of X_1 on X_2) but also of the ratio of the two sample variances $s_1{}^2$ and $s_2{}^2$. If X_1 is either unknown to the investigator or uncontrolled in the analysis, and if it varies considerably relative to the other causes of X_2, then $s_1{}^2/s_2{}^2$ will be large and b_{32} will have a large bias unless either b_{21} or $b_{31.2}$ is negligible. This of course implies that unknown sources of spuriousness must be relatively constant if substantial biases are to be avoided. If the difference between b_{32} and $b_{32.1}$ were *not* a function of $s_1{}^2$, we would indeed be in difficulty, since this would mean that all possible sources of spuriousness (of which X_1 is merely an example) would have to be held *absolutely* constant.

We have seen that even though it is possible to decompose a total b_{ij} into components, this decomposition does not always lend itself to a simple causal interpretation. In the three-variable case we can only begin to see the general principles involved. The interpretation in the case of the effects of the causally prior variable X_1 was straightforward because X_2 was assumed not to affect X_1. The decomposition amounted to tracing forward from X_1 to X_3 by both the direct path and the indirect path through X_2. In the case of X_2, however, the effect of X_2 on X_3 should be estimated by $b_{32 \cdot 1}$ rather than b_{32}. In tracing paths we would want to count only the direct path from X_2 to X_3, considering the path via X_1 as a source of spuriousness rather than as a contribution of X_2 to X_3. These considerations suggest the following algorithm for tracing paths. The estimated total (direct and indirect) effect of X_j on X_i can be written as the sum of the simple and compound paths from X_j to X_i obtained by tracing forward only from X_j, provided that no paths are retraced and that we do not pass through any curved double-headed arrows.

The above algorithm makes sense intuitively in that in tracing effects we certainly do not want to trace back through common causes. Yet we must control for these common causes through the use of partial least-squares coefficients such as $b_{32 \cdot 1}$. The algorithm differs from that suggested by Wright for the standardized path coefficients, which permits tracing backward and then forward as well as tracing through curved arrows. The difference stems from the fact that Wright's algorithm provides an expression for the decomposition of a total *correlation*, whereas the above algorithm applies to the *effects* (direct and indirect) of one variable on another. Like Wright's algorithm, the above rule of thumb may be difficult to apply in complex models involving multiple paths because of the likelihood of either omitting or duplicating certain paths.

The final phrase in the algorithm referring to curved arrows implies that whenever the direction of causation between exogenous variables is ambiguous, we should not trace back and forth through paths connecting these variables as though their indirect effects could be inferred. For example, suppose we replaced the arrow from X_1 to X_2 in Fig. 10.9 by a curved double-headed arrow. Although we still could estimate $b_{31 \cdot 2}$ and $b_{32 \cdot 1}$ and empirically decompose the total slopes, we could not infer the indirect effects of X_1 via X_2 since we would not even know whether a change in X_1 would produce any change at all in X_2.

In the simple three-variable case, we cannot allow for both direct and indirect effects of the intervening variable X_2 on X_3, and therefore we cannot readily see how the principle generalizes. Let us therefore turn next to a four-variable recursive model.

Consider the model of Fig. 10.10 and the following system of least-squares equations estimating the corresponding structural equations:

$$X_1 = e_1 \tag{10.11a}$$

$$X_2 = b_{21}X_1 + e_2 \tag{10.11b}$$

$$X_3 = b_{31 \cdot 2}X_1 + b_{32 \cdot 1}X_2 + e_3 \tag{10.11c}$$

$$X_4 = b_{41 \cdot 23}X_1 + b_{42 \cdot 13}X_2 + b_{43 \cdot 12}X_3 + e_4 \tag{10.11d}$$

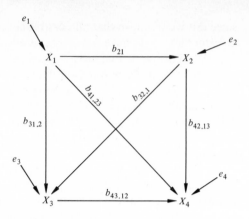

FIGURE 10.10

Multiplying (10.11d) by X_1 and summing, we get

$$\sum X_1 X_4 = b_{41\cdot 23}\sum X_1^2 + b_{42\cdot 13}\sum X_1 X_2 + b_{43\cdot 12}\sum X_1 X_3 + \sum X_1 e_4$$

and therefore since $\sum X_1 e_4 = 0$,

$$\frac{\sum X_1 X_4}{\sum X_1^2} = b_{41} = b_{41\cdot 23} + b_{42\cdot 13}b_{21} + b_{43\cdot 12}b_{31}$$

and we have again decomposed the total slope b_{41} into the component paths that could also have been obtained by tracing forward (only) in the diagram. This follows since from the three-equation model we have

$$b_{31} = b_{31\cdot 2} + b_{32\cdot 1}b_{21}$$

and therefore

$$b_{41} = b_{41\cdot 23} + b_{42\cdot 13}b_{21} + b_{43\cdot 12}(b_{31\cdot 2} + b_{32\cdot 1}b_{21})$$

Now suppose we want to estimate the effects of X_2 on X_4. If we controlled for X_1, as is appropriate, and if we followed the rule of tracing forward only, we would get

$$b_{42}^* = b_{42\cdot 13} + b_{32\cdot 1}b_{43\cdot 12}$$

where b_{42}^* is to be distinguished from b_{42}, the least-squares slope estimate that would be obtained if we *ignored* both X_1 and X_3. An expression for the latter is given by using the equation for the decomposition of b_{ij} as follows:

$$b_{42} = b_{42\cdot 13} + b_{41\cdot 23}b_{12} + b_{43\cdot 12}b_{32}$$

The approximate large-sample bias in b_{42} is therefore

$$b_{42} - b_{42}^* = b_{41\cdot 23}b_{12} + b_{43\cdot 12}(b_{32} - b_{32\cdot 1})$$

Again substituting $b_{21}(s_1^2/s_2^2)$ for b_{12} and using the equation for $b_{32} - b_{32\cdot 1}$, we get

$$b_{42} - b_{42}^* = b_{41\cdot 23}b_{21}\frac{s_1^2}{s_2^2} + b_{43\cdot 12}b_{31\cdot 2}b_{21}\frac{s_1^2}{s_2^2}$$

$$= (b_{41\cdot 23} + b_{43\cdot 12}b_{31\cdot 2})b_{21}\frac{s_1^2}{s_2^2}$$

Therefore the expected large-sample bias is a function of (1) the effect of X_1 on X_4 through paths that do not go through X_2, (2) b_{21}, the effect of X_1 on X_2, and (3) the ratio s_1^2/s_2^2 reflecting the relative amounts of variation in X_1 and X_2.

That the above results are all consistent with commonsense ways of estimating the direct and indirect effects of one variable on another does not make them trivial. In comparative research we often want to see whether the same laws hold up across populations or time periods, and therefore we want to estimate the direct and indirect effects of hypothetical changes (of, say, 1 unit) in a given variable, knowing that the actual changes or variations will be unique to each population or time interval. The algorithm presented above gives us a simple way of writing down such coefficients b_{ij}^* for the combined direct and indirect effects, and we have also given a method for decomposing the total b_{ij}. Provided that the system is recursive, the various direct effects can be properly estimated by the partial regression coefficients using ordinary least squares. Thus we have readily interpretable measures of both direct and indirect effects, as well as a decomposition of the b_{ij}, without reference to standardized measures such as correlation coefficients or beta weights.

10.7 CAUSAL INFERENCES WITH ORDINAL SCALES

A frequent practical question that often arises in the course of data analysis is: Can I still use the causal-modeling approach if my data have been measured only at the ordinal or nominal levels? A growing body of literature is beginning to supply an affirmative answer, though we must at the outset introduce a major note of caution before mentioning some of this literature very briefly. If one can assume that measurement at the ordinal and nominal levels does not mean that one's measuring instruments have been extremely crude and sloppy, and if such weak measurement is not a clue to conceptual ambiguities and highly indirect measurement, it appears safe to proceed along lines that have recently been proposed. But if one has obtained an ordinal scale with four or five levels that is to be related to a nominal variable with three levels, this may very well be due to the fact that one's research operations have not been sufficiently refined to reduce measurement errors below the point where safe inferences can be made. Of course the mere fact that one has obtained a ratio-scale indicator (say, years of formal schooling) does not automatically imply adequate measurement. But if one has used a simple dichotomy (high versus low), one may be almost certain that measurement errors will be substantial.

Thus before attempting to answer the general question of the appropriateness of causal inferences in the presence of ordinal and nominal data, one must ask the prior question: How good are my measures, and what kinds of distortions are being produced through simple classification or the resultant large numbers of ties that may appear in the data? If one honestly believes, for example, that ties do not reflect the crudity of one's measuring instruments but are due to natural groupings that are inherent in the real world, then one may utilize hypotheses that explicitly distinguish how the ties should be treated, choosing the appropriate ordinal measure on the basis of these hypotheses (Wilson, 1974). But if the ties are thought to have resulted from crudities in the measuring instrument, this will have very different implications for one's choice of a measure. Similarly, one must ask: Why are my marginal totals distributed as they are? Perhaps the marginals are totally a function of distributions that exist in the real world, but perhaps they are also functions of arbitrary decisions regarding cut points, or decisions, e.g., cutting at the median, that force the marginals to be approximately equal.

We shall comment briefly on some of these matters below, though we cannot engage in an extensive discussion in view of our present lack of knowledge concerning the implications of differing answers to these questions. Suffice it to say here that definitive answers to the general question of the appropriateness of causal inferences with ordinal and nominal data cannot be given until these prior measurement questions have been answered. We shall be concerned with the implications of measurement errors in Part Four. At present it appears easier to handle measurement-error models in the case of interval and ratio scales, and our later discussion will be almost totally concerned with measurement at this level. However, it should be clear at the outset that measurement-error problems cannot be avoided by the simple device of resorting to nominal and ordinal procedures. In many instances this will probably merely make matters worse rather than better, at least to the extent that scores are collapsed into a very small number of categories and the investigator is not sensitized to the implications of different marginal distributions (Carter, 1971).

Let us consider several concrete examples. Many so-called background and group-membership variables are measured as nominal scales, e.g., region of birth, religious denomination, types of formal organizations to which one belongs, or one's race-ethnicity category. Some of these may be roughly ordered. For example, one's occupational category may be placed in a prestige hierarchy; ethnic groups may be ordered according to the amount of discrimination they face; religious denominations may be ordered according to certain characteristics of their rituals, and so forth. In all these instances one would want to consider carefully the question: Is this nominal scale or ordered set of categories really the variable I am trying to measure, or is it being used as a very crude indicator of something else? Exactly what conceptual variable is being measured by "region" or "religious denomination," for instance? Is the biological dichotomy "sex" the causal variable of interest, or is it being taken as an indicator of some experience variable that does not take on identical values for all males or for all females? If exposure to discrimination or to a particular kind of socialization is the variable of interest, the simple dichotomies of sex and race will involve very high degrees of measurement error under some circumstances—and in fact the *degree* of measurement error is likely to be a function of other variables in the theoretical system.

In the case of ordered categories, such as occupational prestige categories, we assume that all individuals within each category are tied with respect to the dimension concerned. For example, suppose our occupational classes consist of the categories professional and managerial, clerical and sales, foremen and skilled labor, semiskilled and unskilled labor. If we are using only these categories, we must in effect assume that all professionals and managers have the same prestige scores. This will often be an unrealistic assumption. The presence of large numbers of ties may also be a clue to multidimensionality. For example, suppose a judge is trying to compare two individuals A and B. He may decide that A is higher than B in some respects but lower in others. Rather than recognizing that there may be more than one dimension involved, however, he may simply rate A and B as tied. We are suggesting that large numbers of ties, per se, may not invalidate an analysis, but they should serve as a warning that measurement errors may be serious.

In the discussion that follows we shall assume that all these difficulties have been resolved, that classification errors have been judged to be negligible, and that cutting points have been selected according to a carefully rationalized a priori theory. We can then set out to investigate just how much practical difference it makes if one modifies the categories in specified ways by using different cut points or perhaps collapsing them arbitrarily. The hope would be that ordinal and nominal measures of association are relatively robust with respect to such modifications. We may in fact imagine the existence of a true metric unknown to the investigator. There will then be a true set of structural equations that might have been estimated had this metric information been supplied.

Suppose, however, that the investigator uses only the ordinal information. Will the inferences made coincide with those that would have been possible if the metric information had been given? This is a meaningful kind of question that can be answered by means of simulated data having known properties. Ideally, it would be much better to obtain a mathematical solution of a more general nature since simulation provides empirical support only under specified conditions that may not always hold true. We shall first summarize briefly some results of simulation studies, commenting later on an analytic or mathematical approach that presupposes a particular kind of scoring system.

Reynolds (1971, 1974) has conducted a series of investigations that imply rather optimistic conclusions. These involved several different structural models; linear and nonlinear monotonic relations; normal, rectangular, and skewed population distributions; several methods of controlling; differing numbers of categories and choices of cut points; and a variety of ordinal measures of association. With the exception of gamma, which excludes ties altogether and is highly sensitive to the skewed marginals that often occur in the controlling process, Reynolds found that the ordinal measures commonly in use [Spearman's rho corrected for ties, tau_a, tau_b, tau_c, and Somers' (1962) d_{yx} and d_{xy}] all reduced very nearly to zero in situations where the value of a partial correlation was predicted to vanish. For example, when X and Y were constructed to be spuriously related due to a common cause W, the partial ordinal coefficients linking X to Y, controlling for W, were all reduced nearly to zero. The proportional reductions were all very high (in the neighborhood of 80 to 90 percent) except in instances where there were only two or three categories for the control variable. In general, it was found that if one uses as few as five levels for each variable, one obtains very good approximations to zero partials even where the original associations are quite high. This was also true of certain nonlinear but monotonic relationships. On the basis of Reynolds' rather extensive results we may tentatively conclude that for *tests* of causal models, where we predict the disappearance of certain partials, we may safely make inferences using ordinal measures of association provided that we do not collapse down to only two or three categories.

A much more difficult problem in the case of ordinal data, however, is how to make comparisons across samples or populations for which there are differing marginal totals or dispersions. We have argued that this is an important type of problem for interval and ratio scales since differences in dispersion in X relative to that in the disturbance term may very well account for differences in correlation or path coefficients even where the unstandardized regression coefficients are identical. In fact this is the essential reason why we introduced the discussion of path analysis with unstandardized coefficients. In Chap. 12 we shall also see that populations that differ with respect to the dispersion in one or more independent variables may appear to have different slope coefficients merely because of differential attenuation due to measurement errors. How can this kind of problem be conceptualized when we are dealing with ordinal data for which the notion of variance is undefined?

One way to study this problem is to look at it from the standpoint of developing a scoring system for ordinal variables that can be translated into a metric system, so that regression analysis can then be applied in a routine fashion. If one is willing to make a priori assumptions concerning the form of underlying frequency distributions, such a scoring system can be constructed rather easily. For example, if the underlying frequency distribution of X were rectangular, the ranks themselves could be used as scores. Or the "true" distribution of X might be assumed to be normal, as would be the case if X were caused by a large number of weakly intercorrelated independent variables, no two or three of which predominated. If so, one could then convert the ranks into normalized scores and proceed as though the data were at the interval level. Conceivably, one might be willing to postulate a particular form of a skewed distribution, such as chi square, and similarly convert the ranks into scores that followed this particular distribution.

These approaches have rarely been applied in sociological research, undoubtedly because of our unwillingness to make such strong assumptions about true frequency distributions. Another reason may be that a large percentage of our ordinal data come in the form of ordered categories that contain very large numbers of tied scores. Instead, we have tended to follow the lead of Kendall (1948), who developed a series of ordinal measures based on comparisons of ordered *pairs* of individuals, rather than assigning ranks to the individuals themselves. It can be shown that if one shifts from individuals to pairs of individuals as units, and if one adopts a very simple three-valued scoring system, then Kendall's tau becomes a special case of the Pearsonian product-moment coefficient and Somers' d_{yx} becomes a special case of the least-squares coefficient b_{yx}. The scoring system adopted is as follows: for every pair (i,j) of individuals we give X (and also Y) the scores

$$
\begin{array}{ll}
+1 & \text{if } X_i > X_j \\
0 & \text{if } X_i = X_j \\
-1 & \text{if } X_i < X_j
\end{array}
$$

Using these scores of $+1$, 0, and -1 for both X and Y, and taking as the total number of cases the number of pairs, we can then show that tau_b is equivalent to r and d_{yx} to b_{yx}. Furthermore, as Ploch has shown, one may then apply the basic tools of multivariate analysis to ordinal data, including not only partial and multiple correlational analysis but covariance analysis as well. Following the lead of Kendall (1948), Hawkes (1971) and Ploch (1974) have in effect argued that there is really no need to consider ordinal data as posing any special problems for the analyst and one may apply all our previous results for recursive equations to ordinal data.

But this argument seems to rely heavily on one's willingness to accept the above scoring system and its possible implications. It should be noted in this connection that the three-valued system implies some rather undesirable results that seem incompatible with other properties that one might like to hold for ordinal data. Hawkes (1971) has noted that the above scoring system yields the result that the variance in X (or Y), as conventionally defined, is only a function of the number of categories and the marginal totals, regardless of the ordering of these totals. Thus the set of ordered marginals (500,300,100,100,300,500) yields the same measure of dispersion as the ordered set (100,300,500,500,300,100), and both of these sets have smaller variances than the set of equal frequencies (300,300,300,300,300,300). Yet the first set contains most of its cases in the two extreme categories, whereas the second more closely approximates what we would expect for an approximately normal distribution of scores. Had the categories been assigned scores equally far apart, for example, 1, 2, 3, 4, 5, and 6, we would have obtained a substantially larger variance for the first set than the second, with the third set having an intermediate variance.

This scoring system also does not distinguish between pairs formed from adjacent columns (or rows) and those farther apart. Even though the distances between categories are unknown, transitivity alone impels one to take the position that a pair formed by selecting one individual from the first column and another from the second should receive a smaller score than a pair containing one member from the first column and the other from the fifth column. If the transitivity property for ordinality is presumed, and if strict inequality between A, B, C holds, with $A > B > C$, then the distance $\overline{AC}$ must be greater than either $\overline{AB}$ or $\overline{BC}$. The fact that distances are unknown would not, from this viewpoint, justify assigning exactly the same scores to *all* pairs that are ordered alike. This method of scoring seems to result in empirical behaviors of the tau measures and d_{yx} that do not correspond very well to their product-moment counterparts.

The behavior that we are especially concerned with in this connection is that which occurs when the dispersion in the X marginals is altered while the proportions or estimates of conditional probabilities remain unchanged. Consider the two sets of frequencies shown in Tables 10.2 and 10.3, where X is taken as independent and Y as dependent. The two tables differ with respect to the X marginals in that the first and third columns of Table 10.2 contain 500 cases each whereas the corresponding columns of Table 10.3 contain 2,000 cases each. But the proportions (as estimates of

Table 10.2

	X			
Y	Low	Med.	High	Total
Low	300	300	50	650
Med.	150	400	150	700
High	50	300	300	650
Total	500	1,000	500	2,000

$$d_{yx} = .41$$

Table 10.3

	X			
Y	Low	Med.	High	Total
Low	1,200	300	200	1,700
Med.	600	400	600	1,600
High	200	300	1,200	1,700
Total	2,000	1,000	2,000	5,000

$$d_{yx} = .50$$

conditional probabilities) are the same in each table. Under the circumstances we would want the two slope analogs of b_{yx} to be the same, but we find that the numerical values of d_{yx} are .41 and .50, respectively. Differences between the other tau-type measures are of comparable magnitude, and there is nothing in the formulas for these coefficients that would lead one to expect invariance.

If two populations differ with respect to their X marginals, one would naturally want to ask how this could occur. One possibility is that the research operations were noncomparable, but if so, then any kind of comparison would be difficult. With strictly ordered variables and no ties, the scores will of course be 1, 2, 3, ..., n, so that the dispersions for all populations must be identical. But in many instances we presume that although measurement has been crude, the variables can be theoretically conceived at a higher-than-ordinal level, perhaps an ordered metric if not an interval or ratio scale. If so, then we may fully expect the marginals to differ, and this may be given a *substantive* interpretation, as contrasted with being taken as an artifact of the scoring procedure used. Under these circumstances we would like to find a measure of association that will be invariant with shifts in the marginals of the independent variable.

One apparently simple device is to construct a standardized table for which all the X marginals are equal (say to 100) and *then* to calculate d_{yx} (or some other measure of ordinal association). But the matter does not appear to be this simple because we must also account for the Y marginals. Returning to the case of the linear model

$$Y = \alpha + \beta X + \varepsilon$$

we know that if we impose the assumptions that $E(\varepsilon) = 0$ and cov $(X,\varepsilon) = 0$, the variance of Y may be taken as a function of the *three* factors σ_x^2, β^2, and σ_ε^2 according to the equation

$$\sigma_y^2 = \beta^2 \sigma_x^2 + \sigma_\varepsilon^2$$

But in the case of our categorized ordinal (or nominal) data we have only three quantities: (1) the X marginals (some function of which corresponds to σ_x^2), (2) the Y marginals (some function of which corresponds to σ_y^2), and (3) the estimates of conditional probabilities derived from the cell frequencies. In effect, it appears that the variance of the disturbance term becomes hopelessly confounded with the slope.

If one is willing to assume that the omitted variables in one population have exactly the same variance as the omitted variables in the second population, one may then infer the relative magnitudes of slope estimates from the sizes of the correlation analogs. If one is unwilling to make this kind of assumption but is willing to impose a metric on the data according to some criterion, e.g., the normality of the frequency distribution, the slope analogs may be meaningfully compared. But if one does not want to make either of these assumptions, it appears to be impossible to obtain a genuine slope analog. This of course implies that comparisons of the relative magnitudes of ordinal measures across populations or time periods will not easily yield to simple substantive interpretations.

In summary, then, it appears safe to make *tests* of causal models on the basis of ordinal data, provided that only negligible measurement errors have been introduced through the research operation and categorization process. This is essentially because our tests involve predictions that certain partials will be approximately zero. But when we wish to make comparisons of nonzero values of ordinal measures of association, we cannot reach unambiguous conclusions unless we are willing to impose additional a priori assumptions, either about the metrics or about the relative magnitudes of the disturbance terms. For further discussion of this problem, see Blalock (1976).

Computations

In order to compute ordinal partials we may follow either of two general procedures according to the nature of the computer programs readily available. The first is to convert the original data for the n individuals into a new set of data for pairs of individuals. It will be convenient to distinguish the pair (i,j) from the pair (j,i), so that the total number of pairs will be $n(n-1)$. We then assign scores of -1, 0, and $+1$ to *each pair* for all variables. Because of the symmetry involved, this will mean that each variable will sum to zero and the sums of squares will equal the total number of pairs that are *not* tied on that particular variable. We may then compute all covariances and proceed as though we had interval-scale variables, using a regular regression program. We must remember, however, that the usual F tests will not be valid, in part because normality assumptions will be violated but primarily because the $n(n-1)$ pairs do not constitute independent replications. Discussions of significance tests are available in Goodman and Kruskal (1963) and in Quade (1971, 1974).

The other approach is to retain the original data for the n individuals and to compute the preferred ordinal measure by procedures that are described in most general texts; i.e., one may compute each two-variable ordinal measure by comparing the number of concordant and discordant pairs and using the appropriate formula. In computing partials one may then use either of two procedures. If all variables are at least at the ordinal level, and if all bivariate associations are monotonic, one may utilize formulas for partial ordinal measures that are exact counterparts of the product-moment formulas. For example, if we wish to relate X_2 to X_3, controlling for X_1, we may use the formula

$$\tau_{23 \cdot 1} = \frac{\tau_{23} - \tau_{12}\tau_{13}}{\sqrt{1 - \tau_{12}^2} \sqrt{1 - \tau_{13}^2}}$$

Kendall (1948) discussed this formula for the special case where there are no ties, but Hawkes (1971) has shown that the same formula is appropriate for the tied case as well. Furthermore, the formulas generalize to higher-order partials.

If some of the control variables are nominal, or if there are a large number of ties in these control variables, it may be preferable to use a weighted averaging procedure, as discussed by Davis (1967) and generalized by Quade (1971, 1974). The Quade approach has the advantage that one may use a completely flexible matching procedure that is not restricted to exact ties. For example, if IQ is to be used as a control, one may decide that individuals will be declared to be matched if their scores differ by no more than 10 points. This allows for the possibility that A may be matched with B and B with C whereas A and C may be too far apart to be considered matched.

Quade provides a very thorough discussion of this general approach, as well as the tradeoff that exists between the objective of precise matching, which may lead to rather large sampling errors, and rather loose matching criteria which introduce biases but reduce sampling errors.

10.8 CAUSAL INFERENCES WITH NOMINAL SCALES

With nominal scales many of the same kinds of problems exist in connection with crude categorizations that produce substantial measurement errors. Practically all discussions of nominal data presuppose that errors of classification are minimal and that the category boundaries are unproblematic. *If* this is in fact the case, then it appears safe to proceed, but if the nominal categories mask conceptual ambiguities and measurement errors, we shall be in difficulty. Again we need to ask the same kinds of questions as before. Why are our marginal totals as they are? What if two studies produce different marginals? What substantive and/or artifactual factors could have produced these differences? And why do we get large numbers of ties? Not the least of our problems will be that of finding a reasonably simple interpretation of results for any pairing of variables each of which has three or more levels. Suppose, for example, that we use five religious denominations and relate these to seven unordered occupational categories. We can say that our "variables" are "religion" and "occupation," but if the pattern of percentages is at all complex, it may become almost mandatory to infer some underlying dimensions that account for this patterning. If so, we shall be returning to a problem involving ordinal scales.

If all variables in the theoretical system are simple dichotomies, they may simply be scored as 1's and 0's and inserted into regression programs, as previously discussed. Certainly, if most of the endogenous or mutually dependent variables are interval scales whereas a few of the exogenous variables are dichotomies, this very simple approach will make a good deal of sense. Or the nominal variables may be taken to define covariance partitions. If certain of the exogenous variables are nominal scales with multiple categories, one may insert these as dummy variables and proceed without complication. It is when the *dependent* variable (for any given equation) is nominal that difficulties arise. Of course if one is faced with a large number of nominal dependent variables with numerous categories, there will be formidable interpretive problems, regardless of the mode of statistical analysis. In this sense, it appears wise to adopt the habit of always looking with suspicion on any dependent variable that cannot be dimensionalized, arguing that such a variable is more likely a resultant of poor conceptualization than anything else. There may, however, be a few instances, e.g., complex choice behavior, where one can do no better than nominal dependent variables.

Dichotomies can be approached from a number of different perspectives, e.g., information theory, that, in principle, are generalizable (with conceptual discomfort, to be sure) to complex multiple classifications. One such perspective explicitly focuses on causal interpretations of conditional probabilities. The reader is referred to a rather elementary discussion of a probabilistic discussion of causality given by Wilber (1967), which is basically compatible with a more formal and detailed treatment of the subject by Suppes (1970). The essential idea is that if there are two events A and B, with B preceding A, then if B causes A, the probability of A, given B, will not be the same as the unconditional probability of A. Suppes distinguishes positive causes, for which $P(A \mid B) > P(A)$, from negative causes, for which $P(A \mid B) < P(A)$. As an example of a negative cause we might say that the presence of an inoculation (B) lowers the probability of disease (A). This is simply a way of referring to a negative relationship between the two variables, of course.

We must allow for the existence of other possible causes of A and in particular for the possibility that A and B are spuriously related due to any number of C_i. For this reason, Suppes refers to B as a *prima facie* cause of A. If C precedes both B and A, and if $P(A \mid BC) = P(A \mid C)$, then B is inferred to be a *spurious cause* of A or what we have been referring to as spuriously correlated with A. Finally, if it could ever be shown that there are no C_i for which $P(A \mid BC) = P(A \mid C)$, then we could infer that B is a *direct* cause of A. The argument is much more complex and refined than this, but for our present purposes it will be sufficient to illustrate the ideas briefly and comment on their interpretation.

When we say that $P(A \mid B) \neq P(A)$, we of course imply that knowing something about B helps us modify our prediction about A. This says nothing about causation and must therefore be supplemented by further information and assumptions. One of these is the presumed temporal ordering, which places B prior to A. Another is the absence of C_i that would show the relationship between A and B to be spurious. This latter assumption is, of course, completely untestable. In fact, even if we were to find such a C that led us to infer a spurious relationship between A and B, it remains possible that had we introduced additional C_i, this relationship might have been modified. Thus we are basically faced with the same kind of situation as in connection with continuous variables, where we found it necessary to postulate a causal ordering among the variables and to make certain assumptions about the unknown disturbance terms. We have no explicit disturbance term in this instance, but we do have an explicit reference to probabilities. Just as we were not in a position to say anything a priori about the absolute magnitude of the variance of the disturbance term, we also cannot predict the actual magnitudes of the conditional and unconditional probabilities in the absence of additional knowledge. Our "test" of the causal assertion comes in the form of a prediction that if certain variables are not directly linked (here A and B), then one probability estimate will be (approximately) equal to another, for example, $P(A \mid BC) = P(A \mid C)$. In this instance, the implied equality asserts that if we are given C, then additional knowledge of B will in no way affect our prediction of A. Put causally, if C occurs, then A will occur with the same probability regardless of whether or not B occurs. Notice, however, that if we do not know whether C has occurred, a knowledge that B has occurred will generally modify our prediction about A, that is, $P(A \mid B) \neq P(A)$. We are not explicitly talking about controlling for C, but there is an obvious connection that can be clarified by referring to a specific illustration. Labeling the absence of A, B, and C as $\bar{A}$, $\bar{B}$, and $\bar{C}$, respectively, suppose we are given the following set of frequencies:

$$
\begin{array}{ll}
A\ B\ C = 40 & \bar{A}\ B\ C = 20 \\
A\ B\ \bar{C} = 30 & \bar{A}\ B\ \bar{C} = 10 \\
A\ \bar{B}\ C = 10 & \bar{A}\ \bar{B}\ C = 30 \\
A\ \bar{B}\ \bar{C} = 20 & \bar{A}\ \bar{B}\ \bar{C} = 40
\end{array}
$$

Let us first construct the three bivariate relationships as 2×2 tables (Tables 10.4 to 10.6). We immediately see that A and C are unrelated, whereas both A and B, and B and C, are associated, the former pair more strongly than the latter. If we assume that C occurs before B, which in turn occurs before A, and if we wish to allow for both indirect and direct links between A and C, we may postulate the following model:

In this case we are not predicting a spurious relationship between B and A, but our previous argument has been that if there is no direct causal link between the two, then if we control for C, the relationship between A and B should vanish. If it does not, we shall need to obtain a measure for the strength of the direct link between these two variables. Similarly, it is entirely possible that the direct link between A and C might be absent, in which case we would have expected the partial between A and C, controlling for B, to vanish. If it does not, we again want a measure of the degree of relationship. In relating B and C, however, we have argued that we should *not* control for A, which is presumed to be dependent upon both B and C. Thus we would want to take the total table relating B and C, testing this for significance and then measuring the degree of association in some way. Since it is well known that a difference of proportions is a special case of a slope, and since we have been relying on unstandardized regression coefficients as measures of the direct relationship between two variables, we may simply measure this association between B and C by the difference of proportions

$$d_{bc} = .6 - .4 = .2$$

Now let us display the partial tables relating A to B controlling for C and A to C controlling for B (Tables 10.7 to 10.10). We see that the data have been constructed so as to eliminate statistical interactions from consideration. That is, the relationship $d_{ab \cdot c}$ between A and B within C (given C) is exactly the same as the relationship $d_{ab \cdot \bar{c}}$ between A and B within $\bar{C}$ (given not C). The same applies to the partial between A and B.

We noted that the total association between A and C is zero, whereas the partial is $-\frac{1}{12} = -.083$. Thus C has a small negative direct effect on A that is exactly

Table 10.4

	C	$\bar{C}$	
B	60	40	100
$\bar{B}$	40	60	100
	100	100	200

Table 10.5

	C	$\bar{C}$	
A	50	50	100
$\bar{A}$	50	50	100
	100	100	200

Table 10.6

	B	$\bar{B}$	
A	70	30	100
$\bar{A}$	30	70	100
	100	100	200

Table 10.7

	C		
	B	$\bar{B}$	
A	40	10	50
$\bar{A}$	20	30	50
	60	40	100

$$d_{ab \cdot c} = \tfrac{40}{60} - \tfrac{10}{40} = \tfrac{5}{12}$$

Table 10.8

	$\bar{C}$		
	B	$\bar{B}$	
A	30	20	50
$\bar{A}$	10	40	50
	40	60	100

$$d_{ab \cdot \bar{c}} = \tfrac{30}{40} - \tfrac{20}{60} = \tfrac{5}{12}$$

counterbalanced by the positive effect on B which in turn has a positive effect on A. We note that the total effect (an analog to the total regression coefficient) may be decomposed as follows:

$$d_{ac} = d_{ac \cdot b} + d_{bc}d_{ab \cdot c}$$

or numerically,

$$0 = -\tfrac{1}{12} + (.2)(\tfrac{5}{12})$$

Since the interaction is exactly zero in this simple illustration, the same results hold in the case of the unused control categories $\bar{B}$ and $\bar{C}$, respectively. Had there been a slight interaction, we could have used a weighted average. If the interactions had been large, however, we would have needed to reformulate the model in more complex terms.

This probabilistic orientation to causal models may be extended to nominal classifications with any number of categories, although the picture is likely to become confusing if one simultaneously allows for numerous variables (say, more than five) each of which has multiple categories. For two variables A and B having multiple levels $(A_1, A_2, \ldots, A_m)$ and $(B_1, B_2, \ldots, B_n)$, we may form a series of conditional and unconditional probabilities. For example, we may take B as a prima facie cause of A if B precedes A and if *any* of the conditional probabilities $P(A_i \mid B_j)$ differ from the comparable unconditional probabilities $P(A_i)$. It is of course possible for all these to be equal except for one, for example, $P(A_4 \mid B_5) \neq P(A_4)$. If so, it would probably make sense to distinguish the fourth category of A from all the rest, and similarly for the fifth category of B, and then to proceed ex post facto with the two variables re-defined as dichotomies. Of course the picture will rarely be this simple, and we must

Table 10.9

	B		
	C	$\bar{C}$	
A	40	30	70
$\bar{A}$	20	10	30
	60	40	100

$$d_{ac \cdot b} = \tfrac{40}{60} - \tfrac{30}{40} = -\tfrac{1}{12}$$

Table 10.10

	$\bar{B}$		
	C	$\bar{C}$	
A	10	20	30
$\bar{A}$	30	40	70
	40	60	100

$$d_{ac \cdot \bar{b}} = \tfrac{10}{40} - \tfrac{20}{60} = -\tfrac{1}{12}$$

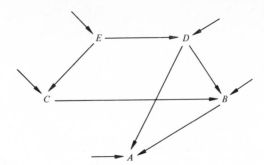

FIGURE 10.11

also recognize the possibility of sampling errors in view of the fact that empirical frequencies are being used as estimates of the probabilities.

Whenever the various interactions seem to be negligible, interests of simplicity will dictate that these interactions be neglected. In this case we may proceed to control for whatever variables are implied by the causal model through a process of standardization (see Rosenberg, 1962) involving the use of weighted averages. Suppose, for example, we have the model of Fig. 10.11, which is identical to that of Fig. 10.2 except for a change in symbols. The model implies that there should be no relationship between A and C if simultaneous controls are introduced for B, D, and E. This may be translated into the prediction that $P(A \mid CBDE) = P(A \mid BDE)$. That is, given that we are in the subcategory BDE, our prediction about A will not be affected by a knowledge of C. If all five variables are dichotomies, we shall have similar predictions within the subcategories $BD\bar{E}$, $B\bar{D}E$, $B\bar{D}\bar{E}$, and so forth. If the interactions are very small, then within each of these subcategories there will be only negligible associations between A and C and we may use a weighted average as an overall estimate. The model of Fig. 10.11 also predicts that $P(C \mid DE) = P(C \mid E)$, $P(B \mid ECD) = P(B \mid CD)$, and $P(A \mid EBCD) = P(A \mid BCD)$.

Multiple interactions Whenever there are numerous interactions as well as a large number of variables or attributes in the causal model, the complexity of the situation can quickly get out of hand. Most discussions of interaction in the case of nominal scales are deceptively simple in this respect in that they either deal with dichotomies or if they are applied to nominal scales with more than three categories, they will be confined to only two or three independent variables. Suppose, however, that we were concerned with a ten-variable model and that in the ninth equation there were five nominal-scale independent variables, each with four or more categories. The number of possible interaction terms or parameters would become extremely large. In such instances it is especially important to conduct a preliminary screening and to include only those interaction terms which explain a reasonable amount of the variance. The Automatic Interaction Detector (AID) program developed by Sonquist and Morgan (1964) may be used for this purpose, using, it is hoped only a portion of one's total sample. Having located these specific interaction terms, and having provided a theoretical rationale for their existence, one may then use the remainder of one's data to make a series of specific tests on an a priori basis.

Having delimited the number of interaction terms somewhat, one may add these specific terms to a regression equation in the form of cross-product terms, testing for their significance as indicated in Part One. This procedure will be especially appropriate whenever the mix of variables favors the interval scales and particularly when the dependent variables in the equations of interest are interval or ratio scales. However, whenever most of the variables are nominal, or whenever a dependent variable consists of nominal categories, procedures described by Goodman (1970, 1973) may be much more appropriate. Goodman's approach basically permits one to evaluate the goodness of fit of alternative models which contain varying numbers of interaction and main-effects parameters. It does not rely on the normality and homoscedasticity assumptions required by the F and t tests used in connection with regression analyses.

The problems of reconciling the statistical literature stemming from regression and least-squares analysis, on the one hand, and from the analysis of multidimensional contingency tables, on the other, seem to be primarily notational and conceptual. Recent work by Jöreskog (1970), among others, stresses that on the most general level the problem can be conceived as that of making inferences about underlying covariance structures, subject to certain given or hypothesized constraints. We have described these in terms of postulated zero values for certain of the parameters or in terms of the nonexistence of certain connecting paths in causal diagrams. In contingency analysis, however, we are much more familiar with notions such as pairwise independence, conditional independence, and expected and actual frequency distributions. The present efforts of Goodman, especially, seem directed at making translations between these two kinds of statistical perspectives. Once the translation is made, from the standpoint of the multiple contingency-table approach the problem becomes that of estimating the expected frequencies in the various cells, using the method of maximum likelihood under the null hypothesis, and then testing the goodness of fit of the model to the data. But these purely *statistical* considerations, which seem readily resolvable, may turn out to be far less problematic than the substantive *measurement* problems that underlie one's decisions whether to treat the data as nominal, ordinal, interval, or ratio scales.

Causal models of any reasonable degree of complexity will generally contain *mixtures* of interval-ratio, ordinal, and nominal scales, so that it will not be easy to decide which kind of analysis to employ. It might appear safest to drop back to procedures that make the weakest assumptions, namely, those requiring only nominal scales. But in doing so one is likely to sacrifice two very important objectives: (1) the goal of reducing measurement errors as much as possible and (2) that of making meaningful comparisons of unstandardized coefficients across populations or time periods. In view of the difficulty of assessing the relative advantages and disadvantages for each alternative, when in doubt, it is advisable to analyze one's data by several different methods. Usually, one will reach basically the same conclusions in each case, but whenever this does not happen, one may gain additional insights that might have been lost if only a single method had been used. For example, it may turn out that a model fits very well when interval-level assumptions have been made but when all variables have been treated as nominal scales, the fit is much less satisfactory. Perhaps this may be due to measurement errors introduced in the process of collapsing control categories. Or perhaps the model is really inadequate, but this has been disguised by nonlinearity somewhere in the (incorrectly specified) regression equations. The general implication here, as elsewhere, is that whenever one is dealing with fallible data and imperfect knowledge, it is best to utilize a very flexible approach to data analysis and to rely on as many cross checks as possible.

10.9   CONCLUDING REMARKS

This chapter has not introduced any really new procedures or computing routines but has dealt almost exclusively with matters of interpretation. We have seen that the recursive causal model justifies the use of ordinary least squares applied to single equations. But we have had to make a number of rather major restrictive assumptions in order to justify this use. It is therefore extremely important to study the implications for data analysis whenever these simplifying assumptions are relaxed in the interest of greater realism. Unfortunately, the simultaneous relaxation of several assumptions usually leads to situations in which there are too many unknowns for solution, in which case the estimation problem becomes hopeless. But it is nevertheless fruitful to examine what happens whenever these assumptions are relaxed one at a time.

The key assumptions required for ordinary least squares to yield unbiased estimates with maximum efficiency are (1) that the model has been correctly specified as linear and additive, (2) that the disturbance terms are mutually uncorrelated and therefore uncorrelated with each of the "independent" variables in their respective equations, and (3) that there are no measurement errors involved. In the next chapter we shall discuss nonrecursive models involving reciprocal causation for which assumption 2 cannot in general be satisfied. In Part Four we shall relax the assumption of perfect measurement and study the implications of both random and nonrandom measurement errors, though confining ourselves to recursive models. Certain other kinds of complications must be discussed much more briefly.

As noted at several points, the assumption that the disturbance terms are mutually uncorrelated is absolutely crucial in justifying the use of ordinary least squares. There are a number of ways in which this assumption may be violated. First, there may be important common causes that have been omitted from the model. In fact, a skeptic may *always* point to this possibility and never be denied since the claim is basically nonrejectable. Provided that the investigator has made a reasonable effort to anticipate the existence of such potential disturbing influences, measure them, and introduce them explicitly into the equation system, the burden of proof then falls on the skeptic to identify specific additional variables and to replicate the study using a more complete causal model. Scientific knowledge thereby accumulates. Put another way, we attempt to construct models of increasing complexity even though any particular model will always be incomplete because of the necessity of allowing for disturbance terms with unknown properties.

There is a second sense in which the disturbance terms may be intercorrelated. Looking at any particular equation, one may suspect that the disturbance terms for certain of the individuals may be intercorrelated across these individuals. If the cases represent observations of a single individual over time, as is frequently the case in the econometrics literature, it will often be true that the disturbance term at time t will be correlated with the disturbances for the adjacent times $t - 1$ and $t + 1$. This is the familiar problem of autocorrelation (or self-correlation) of the disturbance terms, a subject which will be discussed in the next chapter. The problem does not seem so simple, however, when one encounters autocorrelation across *spatial* units. For example, adjacent census tracts are likely to be subjected to unmeasured or unknown disturbances that are positively correlated. But since it is more difficult to order spatial units unambiguously along a single spatial dimension, the extension of procedures utilized for time-series data may not prove to be completely straightforward. Nevertheless, the difficulties do not appear insurmountable.

Whenever the independent variables in any particular equation are highly intercorrelated, a problem known as *multicollinearity*, sampling errors for the regression coefficients and partial correlations will be large, thereby necessitating the use of extremely large samples. Furthermore, since there are likely to be measurement errors superimposed on sampling errors, the existence of multicollinearity between subsets of independent variables will result in the inability to separate out the individual effects of these variables. As we shall see in the next chapter, a practical expedient in such instances is to treat such highly intercorrelated variables as a single block which can then be distinguished from other mutually intercorrelated blocks of variables. For discussions of this multicollinearity problem see Blalock (1963), Christ (1966), Gordon (1968), and Johnston (1972).

We have also seen how the general linear model can be adapted to handle many kinds of nonlinearity, e.g., logarithmic relationships, polynomials, or exponentials, and nonadditive relationships, e.g., multiplicative models. In principle, these same adaptations can be utilized in the case of recursive models, although one's interpretations may have to be modified. For example, suppose the least-squares equation for X_4 is

$$X_4 = b_{41 \cdot 23}X_1 + b_{42 \cdot 13}X_2 + b_{43 \cdot 12}X_3 + e_4$$

Technically there is no reason why X_3 cannot be the product or some more complex function of X_1 and X_2. If we take $X_3 = kX_1X_2$, we in effect have an additional equation in the system, one that is exact but nonlinear. Exact *linear* functions create identification problems involving too many unknowns in the recursive models. We shall see why this is the case in the next chapter. Exact nonlinear equations create similar difficulties, as can be seen from the obvious fact that if X_3 is the product of X_1 and X_2, it will be highly linearly correlated with either or both of these variables and we shall therefore be confronted with a multicollinearity problem. A similar difficulty would arise if X_3 were a higher power of either X_1 or X_2.

Given the fact that measurement errors will practically always be involved, the use of nonlinear or nonadditive models in simultaneous equations may remain only a theoretical possibility. Of course if *all* relationships can be expressed in the same form, certain relatively simple transformations may be utilized. Thus if all variables are linearly related to log X, for example, log population, it makes sense to utilize log X instead of X throughout. Likewise, if all equations could be written as multiplicative relationships, then simple logarithmic transformations would convert each equation into one that is in additive form. But if some equations involve multiplicative relationships and others additive ones, and if some variables are linearly related to log X and others to X itself, such simple transformations will not automatically resolve the difficulties.

These remarks illustrate the major point that there are many distinct ways in which complexities may arise and that if complexities of one particular type are introduced, it will be advisable to compensate by simplifying the system in other respects. In the present chapter we are using multiple equations and are allowing for the possibility that a large number of variables and equations may be necessary in order to achieve a satisfactory theoretical explanation. But the forms of these equations are of the simplest possible type, as are our assumptions about disturbance terms and the nonexistence of measurement errors.

EXERCISES

10.1 Using a four-variable recursive model for which $E(X_i) = E(\varepsilon_i) = 0$, for all $i = 1, 2,$ 3, 4, show that if cov $(\varepsilon_i,\varepsilon_j) = 0$, for all $i \neq j$, then cov $(X_i,\varepsilon_4) = 0$, for $i = 1, 2, 3$.

10.2 (a) Show that apart from sampling error, $r_{15} = r_{12}r_{23}r_{34}r_{45}$ in the simple causal chain $X_1 \rightarrow X_2 \rightarrow X_3 \rightarrow X_4 \rightarrow X_5$.

(b) Practically speaking, what does this imply about correlations between pairs of variables that are remotely linked to each other through such simple chains?

(c) Show that this model implies that a control for any *one* of the variables X_2, X_3, or X_4 will be sufficient to reduce the partial correlation between X_1 and X_5 to zero, except for sampling errors.

(d) Suppose the above chain were modified by adding a direct causal link between X_2 and X_4. How would this modify your answer to part (c)? Can you find a simple expression for r_{15} that is analogous to that in part (a)?

10.3 For the model given below in Fig. E10.3:

(a) Write out the structural equation for each of the X_i.

(b) Which partial or total r's are predicted to be zero?

(c) Are there any alternative models involving these same six variables and having exactly the same predictions as in part (b)?

(d) Assuming that the data satisfy the model exactly, show that the ordinary least-squares formulas for partial correlations and slopes imply that

$$\text{(i)}\ b_{31 \cdot 2} = b_{31} \qquad \text{and} \qquad \text{(ii)}\ |r_{31 \cdot 2}| \geq |r_{31}|$$

(e) Give an intuitive explanation of why it is unnecessary to control for either X_2 or X_4 when inferring the correctness of the submodel involving X_1, X_3, and X_5.

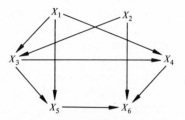

FIGURE E10.3

10.4 Using the model of Exercise 10.3:

(a) Write out the path equations for each of the r_{ij} and express each r_{ij} in terms of path coefficients alone.

(b) Suppose X_1 and X_2 are correlated for unknown reasons. How would each of the above equations be modified?

(c) Assuming that $r_{12} = 0$, solve for p_{31}, p_{32}, p_{41}, and p_{43} in terms of the r_{ij}. Compare your results with the formulas for standardized beta weights.

(d) Write out a similar set of equations for decomposing the unstandardized b_{ij} in terms of partial and total b's.

(e) From part (d) give an expression for the direct and indirect effects of X_1 on X_4, X_5, and X_6.

(f) Do the same for the direct and indirect effects of X_3 on X_5 and X_6. Also, give an expression representing the approximate bias if you had used the total b_{63} to estimate the effects of X_3 on X_6. Give an interpretation.

10.5 Selecting one of the three substantive examples used to illustrate the model of Fig. 10.1, construct a realistic recursive model that would allow for reciprocal causation between the internal states and behavior. This can be done by distinguishing between time periods t_1, t_2, t_3, and so forth. Renumber your variables, if necessary, and then answer parts (a) and (b) of Exercise 10.3 in terms of this new model.

11

NONRECURSIVE MODELS AND RECIPROCAL CAUSATION

In Chap. 10 we confined our attention to recursive systems of equations, where it is possible to consider equations one at a time and to apply ordinary least squares under the assumption that the disturbance terms are uncorrelated with each independent variable in a given equation. We discovered that such recursive systems provide the rationale for most if not all of our simple rules of thumb for interpreting results whenever control variables are introduced and for tracing direct and indirect effects of changes in any particular variable.

11.1 SUBSTANTIVE EXAMPLES OF RECIPROCAL CAUSATION

As we have already implied, however, many situations are not this simple. Consider several parties in mutual interaction, with the responses of one serving as stimuli for the others. Sometimes it will be possible to write down an equation representing the behavior of each party as a dependent variable but with this same variable representing an independent variable in the equation for another party. For example, one party may be the suppliers of goods, a second the customers, and a third the distributors who adjust prices in accord with supply and demand. Or there may be several nations involved in an arms race, with the levels of arms in each nation being functions of the levels in the other nations. Or an equation might be written to represent the behavior of each member of a small group, perhaps a nuclear family. In all such instances, if one had available a detailed and accurate record of the temporal sequences involved, it would be possible to turn the system into a recursive one through the device of lagging the variables and assigning the proper temporal sequences. For example, X_1 might represent the actions of party 1 at time 1; X_2 the response of party 2 at time 2; X_3 the reply of party 1 at time 3, and so forth. Often, however, we lack such detailed information and yet wish to estimate the effects of each party's behavior on that of the others.

Perhaps a more common type of situation, at least within sociology and political science where time-series data are usually unavailable, is one in which we have a number of simultaneous measures on particular units that are not considered to be in mutual interaction but are taken as independent replications of causal processes at work within each unit. For example, a social psychologist may wish to study rather complicated relationships between the set of perhaps five distinct attitudinal dimensions and several forms of behavior. The attitudes may have developed over time as a result of gradual shifts or a series of rather minor events. It would therefore be out of the question to attempt to identify discrete points in time $t_1, t_2, \ldots, t_k$ that could be associated with these continuous rates of change. To be sure, one might *study* the individuals at discrete time periods, but it would be utterly unrealistic to assume that changes in each variable could be associated with these particular points in time. It is not as though the individual experienced a major trauma that made him more authoritarian, which later made him more prejudiced, and which still later made him favor the Republican party. It would be much more realistic to assume that all three kinds of attitudes were gradually shifting, with perhaps a strain toward consistency or balance.

We noted in the previous chapter that a number of social psychological models involve combinations of exogenous variables, internal states, and behaviors. In the model of Fig. 10.1 we used X_1 through X_5 to designate these variables. In the present chapter we shall follow the practice of econometricians and designate the exogenous variables by a separate letter Z_i, renumbering the three endogenous variables as X_1, X_2, and X_3, the last of which represents the behavior in question. If we use as our example the child taking a series of examinations, then the variables are

$$Z_1 = \text{familial support} \quad X_1 = \text{knowledge level}$$
$$Z_2 = \text{peer influences} \quad X_2 = \text{motivation level}$$
$$X_3 = \text{performance level}$$

Since Z_1 and Z_2 are taken as exogenous, or given, we write only three equations, one for each of the X_i. The coefficients of the X_i are designated by β_{ij} whereas those of the exogenous Z_i are designated by γ_{ij}. The linear equations are

$$X_1 = \alpha_1 + \beta_{12}X_2 + \beta_{13}X_3 + \gamma_{11}Z_1 + \gamma_{12}Z_2 + \varepsilon_1$$
$$X_2 = \alpha_2 + \beta_{21}X_1 + \beta_{23}X_3 + \gamma_{21}Z_1 + \gamma_{22}Z_2 + \varepsilon_2 \qquad (11.1)$$
$$X_3 = \alpha_3 + \beta_{31}X_1 + \beta_{32}X_2 + \varepsilon_3$$

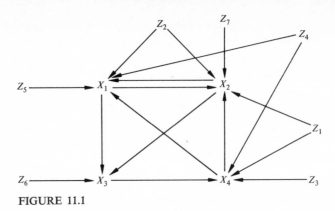

FIGURE 11.1

Clearly, the equations for X_1 and X_2 contain more unknown parameters than in the recursive systems.[1] The equation for performance X_3 omits the two exogenous variables Z_1 and Z_2, and we shall later see that this fact will permit us to estimate the remaining coefficients in this equation. But the first two equations will be *underidentified*, meaning that there are too many unknowns for solution. This implies that if all variables were perfectly measured, we could separate out the effects of the two internal states on behavior but we could not distinguish the effects of the two exogenous variables on the internal states unless the equations were modified in some essential way. It will also turn out that we shall not be able to assume that the disturbance terms ε_i are uncorrelated with any of the "independent" X_i that appear in their respective equations. As we shall see, this will require us to modify the ordinary least-squares procedure that was appropriate for single equations and for recursive models.

As an alternative illustration, suppose an investigator is doing a macrolevel study of national development. Although the data points may occur at more than a single time owing to differences in reporting practices and the availability of data, the investigator may not wish to assume that these data-collection periods correspond with any abrupt changes in the variables, as though the systems were responding to major exogenous events that occurred precisely at those times. More likely, there will have been continual changes in variables such as population size, per capita GNP, industrialization, urbanization, literacy levels, and so forth. There will be a few variables, e.g., those related to outcomes of political elections or major acts of violence or disasters, that may be assumed to change levels at discrete points in time, but ordinarily there will be many others for which changes will be continuous. In effect, everything will be changing at once, making it virtually impossible to say very much about temporal sequences, apart from plotting these changes descriptively in relation to a temporal axis.

Not all variables will be reciprocally interrelated to a major extent, so that in practice one will ordinarily be concerned with a relatively small subset of mutually

---

[1] In this chapter we shall reinsert the constant terms α_i in each equation since the computational procedures will not be the same as those used for ordinary least squares. We shall see that these constant terms pose no special difficulties for analysis or interpretation, however.

interrelated endogenous variables along with a number of other variables that do not depend on these variables. At least we hope that the real world can be adequately approximated by models of this type. Otherwise, if all variables affected each other, it would be difficult to know where to begin, and (as we shall see) we would encounter formidable problems that would make estimation impossible except under very special circumstances. Before considering the general k-equation case in some detail, let us provide a concrete illustration of a model involving four endogenous, or mutually interdependent, variables and seven exogenous variables that are assumed not to depend on any of these endogenous variables.

The study by Mason and Halter (1968) involved data collected on 195 Oregon farmers concerning the diffusion of certain agricultural innovations and the resultant productivity of their farms. The model (Fig. 11.1) contains four endogenous variables:

X_1 = farmer's social influence
X_2 = his prestige
X_3 = his innovation-adoption score
X_4 = his level of production

As can be seen, there is assumed to be direct reciprocal causation between influence and prestige, with productivity affecting both these variables. But productivity is itself affected by adoption practices, which are in turn affected by influence and prestige. Clearly, there is no single variable in this block that can be taken as dependent, since each of the four is dependent upon at least one of the others. This fact immediately implies that disturbance terms in each equation will ordinarily be correlated with the independent variables in that equation, thereby invalidating the use of ordinary least squares. We shall return to this point once we have discussed the nature of the assumptions we shall make about these disturbances terms.

The model also contains seven additional variables labeled as Z_i in order to distinguish them from the endogenous X_i. Notice that there are no arrows going from any of the endogenous X_i to any of the exogenous Z_i. The exogenous variables may be intercorrelated (though no curved double-headed arrows have been drawn in the diagram), but these intercorrelations cannot be explained by the present theory. The seven exogenous Z_i are

Z_1 = farmer's type of residence
Z_2 = his educational level
Z_3 = his age
Z_4 = his level of control over production resources
Z_5 = his level of exposure to mass media appropriate for exercise of influence
Z_6 = his level of use of technical information sources
Z_7 = his level of exposure to mass media appropriate for recognition

It is readily apparent, even from this single example, that what are exogenous variables for one investigator may be taken as endogenous by another. One would certainly expect Z_5 and Z_7 to be causally interdependent, with both being affected by education Z_2. Similarly, Z_1 may affect Z_6, with age Z_3 possibly affecting all of the remaining Z_i through a string of other variables not contained in the model. Thus in a more complete theory one might attempt to specify how the Z_i are interrelated by putting them into blocks of reciprocally interrelated variables and perhaps by locating additional causes of these Z_i. We shall later see that it will be advantageous to develop *block-recursive models* that are characterized by recursive (one-way) relationships *between* blocks but that allow for reciprocal causation *within* blocks.

Any particular model may be inadequate in a number of ways, but the approach itself permits flexibility with respect to modifications. For example, Mason and Halter decided to use the two mass-media exposure variables Z_5 and Z_7 as exogenous. An alternative formulation might have involved taking these variables as functions of influence and prestige, in which case Z_5 and Z_7 would have to be brought into the set of endogenous variables and relabeled as X_5 and X_6. Why not play it safe, then, and allow for effects of all of the X_i on all of the Z_i, turning the system into one large set of 11 mutually interdependent endogenous variables? This would necessitate allowing for 10 unknown slope coefficients (plus 1 intercept term) in each equation, however, and it turns out that this is far too many unknowns. Even with no disturbance terms in the equations and with all variables perfectly measured, we could not estimate the structural parameters from the empirical data. In effect, we must impose a number of a priori assumptions that usually take the form of assuming zero values for some of the parameters. But of course this means that certain of the variables do not directly cause some of the other variables.

In Sec. 11.3 we shall consider the necessary and sufficient conditions for identifying the structural parameters of a k-equation system, which in practical terms means the conditions under which it is at least *possible* to estimate the parameters from empirical data. Whenever these conditions are met, the next question then becomes that of *how* to estimate the parameters by statistical means, a topic that will be discussed in the following section.

11.2   THE IDENTIFICATION PROBLEM

As we have already implied, the identification problem refers to the possibility that there may be too many unknown parameters in one or more of the equations, so that it becomes impossible to estimate the magnitudes of the effects of independent variables that appear in this equation. Why should this difficulty arise? In Part Four we shall encounter identification problems stemming from a different source, namely, the fact that certain of the variables in the system have not been measured. In such a situation it is rather easy to see why there may be too many unknowns, since the more unmeasured variables there are in a system the fewer the pieces of empirical information with which to obtain our estimates. But with all variables perfectly measured and with only one parameter in each equation associated with each of these variables, how can there be too many unknowns? In most general terms, the identification problem arises because we are using a static simultaneous-equation model to try to separate out the effects of each variable on the others, which implies that for each pair of variables X_i and X_j we may need to estimate *both* the effects of X_i on X_j (as represented by β_{ji}) and of X_j on X_i (as represented by β_{ij}). Furthermore, we are attempting to do this without having to say anything about the time lags.

Until we come to Sec. 11.7, the equation system we use is static in the sense that it presupposes an equilibrium-type situation. In effect we assume that the interrelationships between the *endogenous* variables are such that any changes that may have been produced by an exogenous agent in any one of these variables have worked rather rapidly through the endogenous variables, so that the level of each of these variables has stabilized by the time the observations have been taken. That is, we assume that our observations involve equilibrium levels of the endogenous variables. Yet we wish to infer from these equilibrium levels the causal processes that have produced them.

In economics the classic example of this problem involves the attempt to infer the underlying supply and demand functions that have generated equilibrium prices and quantities. Empirically, the analyst is presented with the *intersection* of the supply and demand curves, representing the equilibrium levels, from which the task is to infer the entire curves. Because of the operation of exogenous factors, these equilibrium levels will shift over time. It is this empirical fact, *together with a series of untestable a priori assumptions*, that enables the investigator to infer the underlying functions. This example of supply and demand functions has been thoroughly discussed in the econometrics literature and will be of little interest to sociologists, social psychologists, or political scientists. But the principles involved are very general and, surprisingly enough, apply not only to cross-sectional data but to time-series data as well. In fact, most of the data available to economists are of the second variety, indicating that longitudinal data will not in themselves enable us to resolve the difficulty as simply as one might imagine.

Very commonly the sociologist or political scientist is posed with the following kind of problem. The data are available at a single point in time, or at most two or three points in time that do not coincide at all well with the lag periods that may exist in the real world. For instance, two parties may be in interaction and may be responding to each other at monthly intervals, whereas the data may be collected on an annual basis. Let us suppose, however, that the data have been collected at a single point in time. A child is given an examination, at which time an effort is made to assess his or her "true knowledge" and motivation, as well as parental encouragement and peer-group norms. It may be argued naïvely that the performance level cannot possibly affect any of the latter variables and that the two internal states cannot affect anything other than the behavior. This may be true in this *specific* instance, but it may also be the case that *past* performances, which have not been measured, have affected present levels of knowledge and motivation. If an investigator had first given the exam and then perhaps a month later attempted to measure motivation change, it might have been inferred that the behavior in fact influenced the internal state. As this example clearly indicates, one should be careful not to equate the time at which the data were gathered with the time periods during which the variables were mutually influencing each other.

In situations like this, where one cannot infer anything very definite about the temporal sequences because of the unavailability of the data, it does not follow that one must use a recursive model. In this illustration the behavior X_3 may be conceptualized as a set of behavior *sequences*, some of which may have affected the internal states (though not the two exogenous variables in this very simple example). *If* we are willing to assume that these sequences have resulted in a stabilized pattern such that motivation, knowledge, and performance—all of which may have mutually affected each other in the past—are now approximately constant, *then* we may go ahead to analyze the data and to try to infer·the effects of these variables on each other.

For the remainder of this chapter we shall concentrate illustratively on the Mason and Halter example, where we assume a similar kind of stabilization process to be at work. We imagine that the levels of the exogenous variables are temporarily fixed for a given farmer, though these levels will of course vary across individual cases, thereby providing essential empirical information that will help us sort out the various effects. But for each individual we imagine that the four endogenous variables that mutually affect each other produce changes that are rather rapid. When the observer comes onto the scene, perhaps after one of the exogenous variables has shifted its level, we assume that the mutual effects between the endogenous variables have completely worked themselves out, so that the farmer's levels of these variables are in equilibrium. We then rely on *comparisons* across farmers, plus our a priori assumptions, to infer the *processes* that have been at work within each farmer.

Let us begin to formalize the argument by returning to a very simple two-variable model, using for illustrative purposes the variables influence X_1 and prestige X_2 that are interrelated by means of the pair of linear structural equations

$$X_1 = \alpha_1 + \beta_{12}X_2 + \varepsilon_1$$
$$X_2 = \alpha_2 + \beta_{21}X_1 + \varepsilon_2$$

(11.2)

Although the identification problem results from a *mathematical* difficulty that does not involve the disturbance terms, we have retained the disturbance terms ε_1 and ε_2 to represent the effects of the omitted exogenous variables Z_1 through Z_7 as well as those which were not explicitly considered by Mason and Halter. The structural parameters β_{12} and β_{21}, which need not have the same sign, give us the actual changes that would be produced in the one variable with a unit change in the other. Conceivably, each successive change in one variable could become greater and greater, in which case an equilibrium level would not be reached. This would be the case, in this simple example, if the product $\beta_{12}\beta_{21}$ were greater than unity in absolute value.[1] We shall suppose, however, that this is not the case and that we wish to establish the equilibrium levels of the two variables.

To be specific suppose that $\beta_{12} = 2.0$ and $\beta_{21} = .25$. Then if the system were in equilibrium, where we arbitrarily set the initial levels of X_1 and X_2 at zero, and if one of the exogenous variables produced a unit increase in X_2, we could trace out the changes that would occur in the system. If the change in X_2 were to cause an (almost) immediate change in X_1 whereas the change in X_1 produced a change in X_2 with a one-period delay, we could diagram the process as in Fig. 11.2, where the time periods are designated as $t_1, t_2, t_3, \ldots$. The unit increase in X_2 would result in a change of 2.0 units in X_1. This, in turn, would result in a change of $(2.0)(.25) = .50$ unit in X_2. The next change in X_1 would then be $(.50)(2.0) = 1.0$, which would be followed by a change in X_2 of .25 unit. We see that before very long the changes in both variables

---

[1] For discussions of these stability conditions see Baumol (1959), Blalock (1969), Samuelson (1947), and Simon (1957).

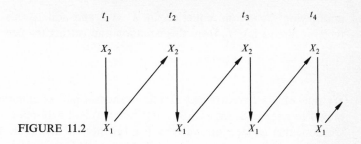

FIGURE 11.2

would become negligible, and it can be shown using our knowledge of infinite series that the limiting values representing the total changes in each variable would be as follows:

$$X_2: 1 + \tfrac{1}{2} + \tfrac{1}{4} + \tfrac{1}{8} + \tfrac{1}{16} + \tfrac{1}{32} + \cdots = 2.0$$
$$X_1: 2 + 1 + \tfrac{1}{2} + \tfrac{1}{4} + \tfrac{1}{8} + \tfrac{1}{16} + \cdots = 4.0$$

These results were derived knowing the values of the structural parameters. But what if we began with only the equilibrium values, as obtained from the data, and wished to estimate the structural parameters from these data? This is the situation in which we are practically always interested. If we had been able to observe each of the successive changes in X_1 and X_2, as just described, we could of course determine the parameter values. But suppose we were merely able to observe the equilibrium result that $X_1 = 4$ and $X_2 = 2$. Could we have arrived at the values of the structural parameters β_{12} and β_{21}? No, we could not. One possibility is of course that $\beta_{12} = \beta_{21} = 0$ and that there were two exogenous changes, one affecting X_1 and the other X_2. But it turns out that there are infinitely many combinations of possible values of the two parameters that could have produced exactly these same equilibrium values.

A good way to see why we are facing this difficulty, from the mathematical point of view, is to note that we are dealing with a pair of equations in two unknowns and that we can derive mathematically equivalent sets, i.e., ones with the same solution, by making *linear transformations*. We take the first equation, multiply it by an arbitrary constant λ_1 and add the new equation to the second. Similarly, we take the original second equation, multiply it by another constant λ_2 and add it to the first. The resulting pair of equations will be mathematically equivalent to the first pair. The (static) equations we would use to represent the equilibrium relationships between X_1 and X_2 for the above hypothetical process would be

$$X_1 = 2X_2$$
$$X_2 = 1 + .25X_1 \tag{11.3}$$

where we imagine that both variables begin with the value zero but where X_2 is given an initial impetus of 1 unit from the exogenous source. That is, when X_1 is (temporarily) zero, $X_2 = 1$, with the process continuing until the equilibrium solution is approximated to a very close degree.

Equations (11.3) of course have the "solution" $X_1 = 4$, $X_2 = 2$, which would be the observed values at equilibrium. Algebraically, we can obtain this solution by multiplying the second equation by $\lambda_2 = 2$ and adding this to the first equation, thereby eliminating X_2 from this equation and getting the new equation system

$$X_1 = 4 + 0X_2 = 4$$
$$X_2 = 2 + 0X_1 = 2$$
(11.4)

We thus obtain the extremely simple equivalent pair of equations that we have been taught to refer to as the solution and that we shall here refer to as a *reduced form* in which neither endogenous variable is a function of the other. (See Sec. 11.5.)

But suppose we had multiplied the original first equation by $\lambda_1 = -.5$, adding it to the second, and also multiplied the original second equation by $\lambda_2 = 1$, adding it to the first original equation. It can easily be verified that, on simplification, we would have obtained

$$X_1 = \tfrac{4}{3} + \tfrac{4}{3}X_2$$
$$X_2 = \tfrac{1}{2} + \tfrac{3}{8}X_1$$
(11.5)

which also has the solution $X_1 = 4$, $X_2 = 2$. We now have produced three mathematically equivalent sets of equations, and it is obvious that there are indefinitely many such sets that can be derived from the first by using arbitrary values of λ_1 and λ_2. One particular choice of the lambdas will result in the solution or reduced form. But if we are merely given the solution, how can we recover the original pair of equations and identify it uniquely from all the rest?

Perhaps it will be asked: Why do we need to identify the original pair, since all of the remainder are mathematically equivalent? We must remember that we are trying to say something about real-world *processes*, namely, the effects of each variable on the other, when we are given only the equilibrium solution or reduced forms. We shall presently see that these solutions or reduced forms are merely means for estimating the structural parameters. The latter enable us to answer important questions about the causal processes involved, namely, what the effects on the entire system of variables (here only X_1 and X_2) would be if one were able to produce a unit change in any one of the variables. We see that we would arrive at different conclusions according to which of the possible equivalent sets of equations we used. Put another way, although the sets are *mathematically* equivalent, in the sense of having the same solutions, they are not *substantively* identical because they imply different dynamic processes at work among the variables.

Our problem arises in this simple two-variable case because the model does not put any restrictions or side conditions on the parameters α_i and β_i. If we can impose a sufficient number of a priori assumptions on the model, we can then select among the indefinite number of pairs of equations that have the same solution or reduced form to find the single pair that simultaneously satisfies these additional a priori conditions. Let us look at the algebra of the two-variable case. If we begin with the (error-free) equations

$$X_1 = \alpha_1 + \beta_{12}X_2$$
$$X_2 = \alpha_2 + \beta_{21}X_1$$

multiply the first equation by λ_1, and add it to the second equation, we get

$$(\lambda_1\alpha_1 + \alpha_2) + (\beta_{21} - \lambda_1)X_1 + (\lambda_1\beta_{12} - 1)X_2 = 0$$

This, upon solving for X_2, is equivalent to

$$X_2 = \frac{\lambda_1 \alpha_1 + \alpha_2}{1 - \lambda_1 \beta_{12}} + \frac{\beta_{21} - \lambda_1}{1 - \lambda_1 \beta_{12}} X_1 \qquad (11.6)$$

which is of the form $X_2 = A_2 + B_{21} X_1$. But unless we have specified ahead of time a value for the coefficient of X_1, we shall not be able to distinguish between β_{21} and B_{21}, which will not be equal unless $\lambda_1 = 0$. And since we may assign λ_1 any value we wish, the situation is hopeless.

But now suppose we were to impose the a priori assumption that X_1 is not a direct cause of X_2, so that we know that the true coefficient of X_1 must be zero. If someone were to present us with the "false" equation for X_2 involving the coefficient $B_{21} = (\beta_{21} - \lambda_1)/(1 - \lambda_1 \beta_{12})$, we would be able to discover the hoax. If $\beta_{21} = 0$, then in order for B_{21} also to be zero we must have $\lambda_1 = 0$, which means that the second equation has not been modified at all. Had we been able to specify some other value for the coefficient of X_1, say 6.3541, we would likewise be able to discover the hoax if someone were to present us with a mathematically equivalent set of equations involving a different coefficient for X_1. But we have turned the problem into a trivial one by presuming that we already know the effects of X_1 on X_2, whereas our original problem treated *both* β_{21} and β_{12} as unknowns! How can we get around the difficulty?

Let us now introduce some of the exogenous variables into the system, though still confining ourselves to the two endogenous variables X_1 and X_2 in order to keep the algebra simple. First let us bring in Z_2, which is assumed to affect both variables. Our two (error-free) equations now become

$$X_1 = \alpha_1 + \beta_{12} X_2 + \gamma_{12} Z_2$$
$$X_2 = \alpha_2 + \beta_{21} X_1 + \gamma_{22} Z_2 \qquad (11.7)$$

Suppose someone were to multiply the first equation by λ_1 and add it to the second equation, pretending to us that the coefficients so obtained were actually the true structural or causal parameters. Could we now discover the hoax? The new second equation would become

$$X_2 = \frac{\lambda_1 \alpha_1 + \alpha_2}{1 - \lambda_1 \beta_{12}} + \frac{\beta_{21} - \lambda_1}{1 - \lambda_1 \beta_{12}} X_1 + \frac{\lambda_1 \gamma_{12} + \gamma_{22}}{1 - \lambda_1 \beta_{12}} Z_2 \qquad (11.8)$$

which is of the general form $X_2 = A_2 + B_{21} X_1 + C_{22} Z_2$.

We shall not be in any better position than before, since the first two terms are the same as those previously obtained and the last term involves Z_2, whose coefficient has not been specified. Similarly, the second equation of (11.7) could have been multiplied by an arbitrary λ_2 and added to the structural equation for X_1, resulting in a false first equation of exactly the same form as the true structural equation, namely, $X_1 = A_1 + B_{12} X_2 + C_{12} Z_2$. Thus if we allow for the possibility that Z_2 affects both X_1 and X_2 without providing a priori values for either of the coefficients we cannot use Z_2 to help us identify the true structural equations.

Mason and Halter, however, provide additional exogenous variables, including Z_5 (exposure to mass media appropriate for the exercise of influence) and Z_7 (exposure to media appropriate for recognition). Furthermore, and this is most important, they specify a priori values for two of the coefficients, since they assume that Z_7 does not directly affect X_1 and Z_5 does not directly affect X_2. The equations for X_1 and X_2 therefore become

$$X_1 = \alpha_1 + \beta_{12}X_2 + \gamma_{12}Z_2 + \gamma_{15}Z_5$$
$$X_2 = \alpha_2 + \beta_{21}X_1 + \gamma_{22}Z_2 + \gamma_{27}Z_7 \tag{11.9}$$

Suppose the first equation is multiplied through by λ_1 and added to the second equation. The new equation for X_2 will now contain a term involving Z_5, namely the coefficient $\lambda_1\gamma_{15}/(1 - \lambda_1\beta_{12})$. This will not be zero unless either $\lambda_1 = 0$, in which case the second equation will not have been altered at all, or $\gamma_{15} = 0$, which would mean that Z_5 is not a direct cause of X_1. Thus if Z_5 does legitimately belong in the equation for X_1, any alternative second equation that is presented to us can immediately be spotted as false because Z_5 will appear in it with a nonzero coefficient, contrary to our a priori assumption. The same point obviously holds in connection with Z_7 and the first equation, which can be distinguished from any false candidate by virtue of the fact that the latter would contain a nonzero coefficient for Z_7, which we are assuming does not directly affect X_1.

Notice that it is possible for the coefficients in one equation to be identified whereas the coefficients in other equations are not. Suppose, for example, that we had available only Z_5 but not Z_7. The second equation would still be identified, whereas the first would not, since identification of the first equation depends on the presence of Z_7 in the system. From this simple illustration we can infer that an equation is more likely to be identified the more variables in the remainder of the system that have been omitted from this particular equation or that can be assigned specific nonzero values on some a priori basis. Also, we might anticipate *statistical* estimation problems whenever the Z_i are only very minor causes of some of the endogenous variables, and therefore appear with very small coefficients, or whenever the Z_i are highly intercorrelated so that multicollinearity results. Since the identification problem results from the *mathematical* fact of having too many unknowns in the system, however, these statistical estimation problems need not concern us yet.

11.3 NECESSARY AND SUFFICIENT CONDITIONS FOR IDENTIFICATION IN THE GENERAL CASE

It would be difficult to study the identification problem for the general k-equation case through the use of ordinary algebra. Fortunately the topic has been well studied within the econometrics literature, where matrix-algebraic proofs are readily available (see, especially, Fisher, 1966; Johnston, 1972). It will be sufficient for our purposes to give the results of this work by stating the necessary and sufficient conditions for identifying the parameters in any given equation. We shall confine our attention to linear additive systems with the reminder that we are making the (often unrealistic) assumption that each variable has been perfectly measured. We shall reintroduce disturbance terms ε_i into the system and make the fundamental assumption that the ε_i are uncorrelated with all of the exogenous Z_i, though not necessarily with the remaining X_i in each equation. We shall return to a more detailed discussion of these assumptions after stating the necessary and sufficient conditions for identification.

Suppose we have k mutually interdependent endogenous variables along with m exogenous variables. We write down the k-equation system as follows:

$$X_1 = \alpha_1 + \beta_{12}X_2 + \cdots + \beta_{1k}X_k + \gamma_{11}Z_1 + \gamma_{12}Z_2 + \cdots + \gamma_{1m}Z_m + \varepsilon_1$$
$$X_2 = \alpha_2 + \beta_{21}X_1 + \cdots + \beta_{2k}X_k + \gamma_{21}Z_1 + \gamma_{22}Z_2 + \cdots + \gamma_{2m}Z_m + \varepsilon_2$$

$$\cdots\cdots\cdots\cdots\cdots\cdots\cdots\cdots\cdots\cdots\cdots\cdots\cdots\cdots\cdots\cdots\cdots\cdots \quad (11.10)$$

$$X_k = a_k + \beta_{k1}X_1 + \cdots + \beta_{k,k-1}X_{k-1} + \gamma_{k1}Z_1 + \gamma_{k2}Z_2$$
$$+ \cdots + \gamma_{km}Z_m + \varepsilon_k$$

where cov $(Z_i, \varepsilon_j) = 0$ for all i and j. Since each equation contains $k + m$ unknown parameters, there will be a total of $k(k + m)$ parameters in the system, which is far too many to be estimated from the empirical data. Therefore certain restrictions or side conditions must be put on these parameters; as already implied, the most common assumptions are those which specify zero values for certain of the parameters, meaning that some of the variables are assumed not to be direct causes of certain of the remaining ones. Other kinds of restrictive assumptions are of course possible. Certain of the coefficients might be set equal to each other or might be assumed to have known ratios. An intuitively appealing kind of restriction on the model would be to specify only the *signs* of certain of the coefficients, but it can be shown that such very imprecise restrictions will generally not be of much help. Therefore we shall confine our attention to *zero assumptions* that involve setting some of the parameter values equal to zero.

A *necessary* (but not sufficient) condition (referred to as the *order condition*) for identifying the parameters in any particular equation can now be stated very simply. If there are k endogenous variables and therefore k equations, one must leave out at least $k - 1$ variables from any equation that is to be identified. That is, one must set $k - 1$ of the coefficients equal to zero. If precisely $k - 1$ have been set equal to zero, we say that the equation is *exactly identified*. If more than $k - 1$ have been set equal to zero, the equation is overidentified, in contrast to the underidentified equation, which has fewer than $k - 1$ coefficients set equal to zero. Underidentified equations are hopeless. The distinction between exactly and overidentified systems is primarily useful in considering alternative methods of parameter estimation.

Let us write out the equations for the Mason and Halter model (Fig. 11.1) and verify that the necessary condition for identification has been met in all equations. Since there are four endogenous variables, we see that we must omit at least $k - 1 = 3$ variables from each equation. The equations are

$$X_1 = \alpha_1 + \beta_{12}X_2 + \beta_{14}X_4 + \gamma_{12}Z_2 + \gamma_{14}Z_4 + \gamma_{15}Z_5 + \varepsilon_1$$
$$X_2 = \alpha_2 + \beta_{21}X_1 + \beta_{24}X_4 + \gamma_{21}Z_1 + \gamma_{22}Z_2 + \gamma_{27}Z_7 + \varepsilon_2$$
$$X_3 = \alpha_3 + \beta_{31}X_1 + \beta_{32}X_2 + \gamma_{36}Z_6 + \varepsilon_3 \quad (11.11)$$
$$X_4 = \alpha_4 + \beta_{43}X_3 + \gamma_{41}Z_1 + \gamma_{43}Z_3 + \gamma_{44}Z_4 + \varepsilon_4$$

Since there are eleven variables in all, only three of which need be eliminated from each equation, we immediately see that the necessary condition will be met by any equation that contains eight or fewer variables (including the "dependent" variable).

The first two equations contain six variables, the third four variables, and the final equation five variables, so that all four equations are overidentified. In the abbreviated two-endogenous-variable model previously considered, it is necessary to omit $k - 1 = 1$ variable from each equation. Before we introduced any of the Z_i, we had only two variables in the system and both equations were underidentified. The addition of Z_2 into the model did not help, since this exogenous variable appeared in both equations and there were still no variables omitted from either equation. But when we introduced Z_5 and Z_7, each of which was omitted from one of the equations, the system became exactly identified.

Thus we can merely count the number of variables in the equations or even more simply the number of arrowheads at each variable in order to determine whether the necessary condition for identification has been met. Usually if this necessary condition has been met, it will also mean that the sufficient condition has also been met, but since this will not automatically be the case, mention should be made of a condition that is *both* necessary and sufficient. Unfortunately, this more restrictive condition is not as easy to state simply or to apply in practice. This both necessary and sufficient condition (referred to as the *rank condition* for identification) can be stated as follows. Given a linear system of k equations, the coefficients in any particular equation can be identified if and only if it is possible to form at least one nonzero determinant of $k - 1$ rows and columns by proceeding as follows:

1 Rewrite each equation with all terms except the disturbance term on the left-hand side.
2 Form a matrix of the coefficients on the left-hand side, including the constant terms (if any).
3 Delete from this matrix all columns that do not have a zero in the prescribed position for the particular equation in question; also delete the row corresponding to this equation.
4 Try to find at least one $(k - 1) \times (k - 1)$ nonzero determinant from the above matrix. If this can be done, the equation in question can be identified.

Let us apply this necessary and sufficient condition to the Mason and Halter illustration. We rewrite the equations leaving blank spaces corresponding to omitted variables as follows:

$$-\alpha_1 + \quad X_1 - \beta_{12}X_2 \qquad - \beta_{14}X_4 \qquad -\gamma_{12}Z_2 \qquad -\gamma_{14}Z_4 - \gamma_{15}Z_5 \qquad\qquad\qquad = \varepsilon_1$$
$$-\alpha_2 - \beta_{21}X_1 + \quad X_2 \qquad - \beta_{24}X_4 - \gamma_{21}Z_1 - \gamma_{22}Z_2 \qquad\qquad\qquad\qquad\qquad -\gamma_{27}Z_7 = \varepsilon_2$$
$$-\alpha_3 - \beta_{31}X_1 - \beta_{32}X_2 + \quad X_3 \qquad\qquad\qquad\qquad\qquad\qquad -\gamma_{36}Z_6 \qquad\quad = \varepsilon_3$$
$$-\alpha_4 \qquad\qquad - \beta_{43}X_3 + \quad X_4 - \gamma_{41}Z_1 \qquad -\gamma_{43}Z_3 - \gamma_{44}Z_4 \qquad\qquad\qquad = \varepsilon_4$$

Next we form an array of the coefficients:

Const	X_1	X_2	X_3	X_4	Z_1	Z_2	Z_3	Z_4	Z_5	Z_6	Z_7
$-\alpha_1$	1	$-\beta_{12}$	0	$-\beta_{14}$	0	$-\gamma_{12}$	0	$-\gamma_{14}$	$-\gamma_{15}$	0	0
$-\alpha_2$	$-\beta_{21}$	1	0	$-\beta_{24}$	$-\gamma_{21}$	$-\gamma_{22}$	0	0	0	0	$-\gamma_{27}$
$-\alpha_3$	$-\beta_{31}$	$-\beta_{32}$	1	0	0	0	0	0	0	$-\gamma_{36}$	0
$-\alpha_4$	0	0	$-\beta_{43}$	1	$-\gamma_{41}$	0	$-\gamma_{43}$	$-\gamma_{44}$	0	0	0

In the first equation the omitted variables are X_3, Z_1, Z_3, Z_6, and Z_7. We therefore use only these columns in the matrix, striking out the first row, getting the matrix

$$\begin{bmatrix} 0 & -\gamma_{21} & 0 & 0 & -\gamma_{27} \\ 1 & 0 & 0 & -\gamma_{36} & 0 \\ -\beta_{43} & -\gamma_{41} & -\gamma_{43} & 0 & 0 \end{bmatrix}$$

from which we must be able to form at least one nonvanishing 3×3 determinant.

We immediately encounter a difficulty since the structural parameters will practically always be unknown. Whenever the necessary (order) condition has been met, there will be at least $k - 1$ columns for the matrix. If we take any combinations of $k - 1$ columns together with the $k - 1$ rows, we shall generally expect to find a nonvanishing determinant except under unusual circumstances, e.g., when some variables are exact linear functions of the others. For example, we might use the first three columns of the above matrix to form the determinant

$$\begin{vmatrix} 0 & -\gamma_{21} & 0 \\ 1 & 0 & 0 \\ -\beta_{43} & -\gamma_{41} & -\gamma_{43} \end{vmatrix} = -\gamma_{21}\gamma_{43}$$

which will not be zero unless either Z_1 does not directly affect X_2 or Z_3 does not directly affect X_4. Similarly, in order to check on the identification of the second equation one would strike out the second row and all columns except those corresponding to X_3, Z_3, Z_4, Z_5, and Z_6.

There is one special type of situation worth noting in which the necessary (order) condition may be met whereas the necessary and sufficient condition will not. With two equations that contain exactly the same combination of variables, when one selects only those columns corresponding to the omitted variables for the first of these equations, the row corresponding to the second of these equations will also contain only zeros, making it impossible to find a nonzero determinant of greater than $(k - 2) \times (k - 2)$ dimensionality. For example, suppose the equation for X_2 in the Mason and Halter model contained exactly the same variables as the first equation. The second row of the reduced matrix would then consist entirely of zeros. If the second equation, in addition, contained fewer variables than the first, the first equation could not be identified whereas the second would.

11.4   A NOTE ON THE ASSUMPTIONS OF RECURSIVE MODELS

The foregoing discussion of identification in linear systems assumes only that the (population) disturbances are uncorrelated with the Z_i. It has been pointed out that the ε_i can generally be expected to be correlated with the endogenous variables in each equation, since the disturbance terms represent expressions for the residual causes of the X_i which mutually affect each other. For example in the two-equation case involving influence and prestige, omitted variables that affect prestige will be correlated with influence, since prestige is assumed to be one of the causes of influence. It is this correlation between the disturbance terms and the X_i that produces biases in ordinary least squares, and we shall see that one of the alternatives to ordinary least squares (OLS), namely, two-stage least squares (2SLS), involves an explicit procedure for replacing the troublesome X_i with estimates that get around this difficulty.

In the case of recursive systems we also made the important assumption that the disturbance terms are *mutually* uncorrelated, i.e., that cov $(\varepsilon_i, \varepsilon_j) = 0$ for all $i \neq j$. We then showed that this assumption implied that each disturbance term is uncorrelated with the X_i that appear as independent variables in its equation. This assumption has seemed unduly restrictive to many econometricians, though, as we shall presently see, the notion of block recursiveness, which is necessary to justify the exclusion of variables from a theoretical system, implies that *some* disturbance terms must be assumed to be mutually uncorrelated (see Fisher, 1966).

Are recursive systems of equations identified? Obviously we have acted as though they are, since we proceeded to obtain estimates of the parameters in each of the equations. Yet if we apply the necessary or order condition for identifiability, we immediately see that only the equation for X_2 is identified. For example, the full equation for X_3, namely, $X_3 = \alpha_3 + \beta_{31}X_1 + \beta_{32}X_2 + \varepsilon_3$, contains all the variables in the system (up to that point), so that the necessary condition is not met. However, if we add to the zero assumptions (to the effect that half of the β_{ij} have been set equal to zero) the assumption that the covariances between the disturbances are zero, it turns out that the full recursive model is exactly identified (Fisher, 1966).

Thus we see the crucial role played by the assumption of a lack of correlation between the disturbance terms. Not only is this assumption needed in order to justify the use of ordinary least squares, but it is needed to identify the coefficients as well. For example, if any of the variables are *exact* (linear) functions of the remaining X_i, then the equations for these particular variables will not contain any disturbance terms and the covariance assumption will break down. Practically, this means that if we *define* one or more "independent" variables in terms of the others, we shall not be able to separate out their effects on subsequent dependent variables as long as we confine ourselves to linear systems. A case in point is the identification problem arising in studies that attempt to assess the impact of status inconsistency or occupational mobility when we also allow for the separate effects of the status variables themselves (Blalock, 1967c).

In the less extreme case, where one of the X_i is highly correlated with a linear combination of other X_i, the disturbance term in its equation will have a very small variance, with the result that sampling errors will be exceedingly large. Suppose, for example, that X_3 is linearly dependent on X_1 and X_2 and that the multiple correlation of these variables with X_3 is of the order of magnitude of .90 or larger. Then the disturbance term in the equation for X_3 will have a small variance, and its covariance with other disturbance terms is likely to be so close to zero that it cannot be distinguished from sampling error. Strictly speaking, the subsequent equations can be identified, but the sampling errors of the parameter estimates will be so large as to be practically useless. The question of multicollinearity is thus directly related to the identification problem in the sense that one can think of the identification problem as involving the limiting case of extreme multicollinearity where there are minor disturbance terms that approach zero, with the sampling errors of the parameter estimates approaching infinity.

In a sense it seems completely unreasonable to have to pin our hopes on such assumptions about disturbance terms, but we cannot relax these assumptions too far without encountering a hopeless situation. As we have noted, the fact that we find it necessary to introduce exogenous Z_i that are assumed to be uncorrelated with the disturbance terms in the equations for the X_i implies that we must assume the existence of sets of variables that do not depend on these X_i. We imagine that the Z_i, in turn, depend upon other variables and that the equations for these Z_i also involve disturbance terms, say η_i. Then the assumption that the Z_i and the ε_i are uncorrelated basically requires us to assume that the η_i from previous sets are uncorrelated with the ε_i. In other words, we must assume that *some* pairs of disturbance terms are mutually uncorrelated.

The major general implication of these remarks is of course that *some* assumptions must always be made about the disturbance terms and their relationships with the variables in the structural system. In experimental designs, randomization makes it possible to use extremely simple causal models while still getting by with reasonably plausible assumptions about the disturbances—though sometimes not quite as plausible as the notion of the ideal experiment would lead one to believe. But in nonexperimental research it becomes necessary to compensate for the inability to randomize and rigidly control certain variables by bringing more variables explicitly into the equation system. In any particular instance, of course, it will be necessary to compromise between the need for relative simplicity and the need for realistic assumptions. The practicing social scientist need not understand all the proofs and derivations behind the statistical estimating procedures, but he or she needs to be very much aware of the required assumptions.

11.5 REDUCED FORMS AND INDIRECT LEAST SQUARES

Given the linear system of k equations

$$X_1 = \alpha_1 + \beta_{12}X_2 + \beta_{13}X_3 + \cdots + \beta_{1k}X_k + \gamma_{11}Z_1$$
$$+ \gamma_{12}Z_2 + \cdots + \gamma_{1m}Z_m + \varepsilon_1$$

$$X_2 = \alpha_2 + \beta_{21}X_1 + \beta_{23}X_3 + \cdots + \beta_{2k}X_k + \gamma_{21}Z_1$$
$$+ \gamma_{22}Z_2 + \cdots + \gamma_{2m}Z_m + \varepsilon_2$$

$$\cdots\cdots\cdots\cdots\cdots\cdots\cdots\cdots\cdots\cdots\cdots\cdots\cdots\cdots\cdots\cdots\cdots$$

$$X_k = \alpha_k + \beta_{k1}X_1 + \beta_{k2}X_2 + \cdots + \beta_{k,k-1}X_{k-1} + \gamma_{k1}Z_1$$
$$+ \gamma_{k2}Z_2 + \cdots + \gamma_{km}Z_m + \varepsilon_k$$

we have seen that a necessary condition for identification is that at least $k - 1$ variables be left out of any equation that is to be identified. *If* this condition (or some other conditions amounting to at least $k - 1$ independent restrictions on the parameters) is met, we know that estimation of the structural parameters may be possible. But we have also noted that ordinary least squares (OLS) produces biased estimates except in recursive systems.

In this section and Sec. 11.6 we shall discuss two alternative estimation techniques that have the property of yielding *consistent* estimators, which have negligible large-sample biases. In addition to these techniques, several other approaches have been discussed in the econometrics literature but are too technical for the present work. In general, these other approaches use more information about the entire system of equations and are slightly more efficient. On the other hand, they require stronger assumptions about overidentifying restrictions and appear to be more sensitive to specification error.

In order to introduce the simultaneous-equation alternatives to ordinary least squares, we must first consider the *reduced forms* of the structural equations. A reduced form contains a single endogenous variable expressed as a function of all of the exogenous variables but not of the remaining endogenous variables. We can obtain the reduced forms from the original structural system by algebraically eliminating all but one of the endogenous variables, as will be illustrated presently. This means that $k - 1$ variables will be eliminated from each equation, and since the rank condition will also be met in this case, this assures that the reduced-form coefficients can be identified and thus estimated. However, this does not mean that given the estimates of the reduced-form parameters, one can then obtain unique estimates of the underlying structural parameters. As we have already noted, there may be infinitely many sets of structural parameters that all yield the same reduced-form parameters, and the identification problem can be conceptualized as that of inferring the structural parameters from the reduced-form parameters. Unless the necessary and sufficient conditions for identification are met, it will be impossible to follow this reverse route. Put another way, we can always estimate the reduced-form parameters from the data, but since there may be infinitely many sets of structural parameters that are observationally equivalent, the underlying structure that has generated the reduced form may not be recoverable.

The reduced-form parameters are conventionally represented as π_{ij}. In order to avoid possible confusion with path coefficients we shall use the symbol $\hat{\pi}_{ij}$, rather than p_{ij}, to represent the OLS estimator of these π_{ij}. Following this notation, and representing the constant terms by π_{i0}, we can write out the following expressions for the reduced-form equations:

$$X_1 = \pi_{10} + \pi_{11}Z_1 + \pi_{12}Z_2 + \cdots + \pi_{1m}Z_m + \eta_1$$
$$X_2 = \pi_{20} + \pi_{21}Z_1 + \pi_{22}Z_2 + \cdots + \pi_{2m}Z_m + \eta_2$$
$$\cdots\cdots\cdots\cdots\cdots\cdots\cdots\cdots\cdots\cdots\cdots\cdots\cdots\cdots\cdots\cdots\cdots \tag{11.12}$$
$$X_k = \pi_{k0} + \pi_{k1}Z_1 + \pi_{k2}Z_2 + \cdots + \pi_{km}Z_m + \eta_k$$

It turns out that if we make the assumptions previously made for the equation system (11.9)—that $E(\varepsilon_i) = 0$; that cov $(Z_i,\varepsilon_j) = 0$, for all i and j; and that there are no specification or measurement errors—then ordinary least squares yields unbiased estimators of the reduced-form parameters π_{ij}. In addition, if the disturbance terms η_i are normally distributed with constant variances, then conventional F and t tests can be made provided random sampling has been used.

Before looking at the algebra it seems useful to return to the substantive examples of the external stimuli (Z_1 and Z_2), the internal states (X_1 and X_2), and some form of behavior (X_3) as represented by the Eqs. (11.1). The reduced-form equations for X_1, X_2, and X_3 would involve only Z_1 and Z_2 as independent variables and would be interpretable as giving us the direct and indirect effects of these two exogenous variables without explicitly bringing in any of the intervening endogenous variables. These reduced forms could be used to compare the relative effects of the Z_i, but the *mechanisms* through which they operate, namely, the values of the β_{ij} and γ_{ij}, would not be known. As long as these structural parameters remained unchanged, and provided that they were of no theoretical interest to the investigator, these reduced-form equations could be used to assess the relative contributions of the Z_i. But if any of the β_{ij} or γ_{ij} were to be modified, or if their values differed across several populations, we would have no idea how to predict changes in the reduced-form parameters π_{ij}. In terms of one of the examples we have been using, if the congressmen were to change either their sensitivities to constituency preferences or their goals regarding reelection, the reduced-form parameters would be useless in enabling us to predict their votes. As long as their motivational structure remained the same, however, the π_{ij} would also remain unchanged. In effect, the nonrecursive system (11.1) becomes transformed into a much simpler recursive one but at the price of a loss in theoretical richness and explanatory power.

Let us now work out the algebra for a relatively simple two-equation model so that we can see the principles involved more clearly. Suppose that we are given two endogenous and two exogenous variables related by the system

$$X_1 = \alpha_1 + \beta_{12}X_2 + \gamma_{11}Z_1 + \gamma_{12}Z_2 + \varepsilon_1$$
$$X_2 = \alpha_2 + \beta_{21}X_1 + \gamma_{21}Z_1 + \gamma_{22}Z_2 + \varepsilon_2 \tag{11.13}$$

Since all four variables appear in both equations, we immediately see that neither equation can be identified unless additional assumptions are made. But let us see how expressions for the reduced-form parameters can be obtained. We must eliminate X_2 from the first equation and X_1 from the second. This can be accomplished by multiplying by the appropriate coefficients and adding or subtracting one equation to or from the other. In order to eliminate X_2 from the first equation we multiply the second equation by β_{12} and add it to the first. This gives

$$(1 - \beta_{12}\beta_{21})X_1 = (\alpha_1 + \beta_{12}\alpha_2) + (\gamma_{11} + \beta_{12}\gamma_{21})Z_1$$
$$+ (\gamma_{12} + \beta_{12}\gamma_{22})Z_2 + (\varepsilon_1 + \beta_{12}\varepsilon_2)$$

or

$$X_1 = \frac{\alpha_1 + \beta_{12}\alpha_2}{1 - \beta_{12}\beta_{21}} + \frac{\gamma_{11} + \beta_{12}\gamma_{21}}{1 - \beta_{12}\beta_{21}} Z_1 + \frac{\gamma_{12} + \beta_{12}\gamma_{22}}{1 - \beta_{12}\beta_{21}} Z_2 + \frac{\varepsilon_1 + \beta_{12}\varepsilon_2}{1 - \beta_{12}\beta_{21}} \tag{11.14a}$$

By a similar process we obtain

$$X_2 = \frac{\alpha_2 + \beta_{21}\alpha_1}{1 - \beta_{12}\beta_{21}} + \frac{\gamma_{21} + \beta_{21}\gamma_{11}}{1 - \beta_{12}\beta_{21}} Z_1 + \frac{\gamma_{22} + \beta_{21}\gamma_{12}}{1 - \beta_{12}\beta_{21}} Z_2 + \frac{\varepsilon_2 + \beta_{21}\varepsilon_1}{1 - \beta_{12}\beta_{21}} \tag{11.14b}$$

Thus we have reduced the original equations to an algebraically equivalent set containing only one endogenous variable in each equation. By examining the coefficients we see that

$$\pi_{10} = \frac{\alpha_1 + \beta_{12}\alpha_2}{1 - \beta_{12}\beta_{21}} \qquad \pi_{11} = \frac{\gamma_{11} + \beta_{12}\gamma_{21}}{1 - \beta_{12}\beta_{21}} \qquad \pi_{12} = \frac{\gamma_{12} + \beta_{12}\gamma_{22}}{1 - \beta_{12}\beta_{21}}$$

$$\pi_{20} = \frac{\alpha_2 + \beta_{21}\alpha_1}{1 - \beta_{12}\beta_{21}} \qquad \pi_{21} = \frac{\gamma_{21} + \beta_{21}\gamma_{11}}{1 - \beta_{12}\beta_{21}} \qquad \pi_{22} = \frac{\gamma_{22} + \beta_{21}\gamma_{12}}{1 - \beta_{12}\beta_{21}} \tag{11.15}$$

and the new error terms are

$$\eta_1 = \frac{\varepsilon_1 + \beta_{12}\varepsilon_2}{1 - \beta_{12}\beta_{21}} \quad \text{and} \quad \eta_2 = \frac{\varepsilon_2 + \beta_{21}\varepsilon_1}{1 - \beta_{12}\beta_{21}}$$

Since the new error terms η_1 and η_2 are functions of the fixed parameters and ε_1 and ε_2, which are by assumption uncorrelated with the exogenous variables Z_1 and Z_2, we see that this assumption required by ordinary least squares is automatically met. It can also be demonstrated that under the assumptions of the model, ordinary least squares provides the most efficient linear unbiased estimates of all reduced-form parameters.

The expressions for the π_{ij} in terms of the structural parameters show that if the structural parameters were known a priori, the reduced-form parameters could be obtained algebraically. But we notice that there are six π_{ij} whereas there are eight structural parameters, and therefore it is apparent that if we knew the reduced-form parameters π_{ij}, we could not solve for the underlying structural parameters. This is the identification problem in a nutshell. We shall be able to obtain estimates of the π_{ij} from the data, but these data alone will not be sufficient to estimate the structural parameters. By a simple process of counting unknowns we can see that we must reduce the number of unknown structural parameters to six, although it might be more difficult to determine in this case whether *any* two of the unknowns could be assigned arbitrary values.

From our earlier consideration of the identification problem we know that in order for the first equation to be identified, at least one variable must be left out of the equation (or assigned some known coefficient). The same holds for the second equation, and this implies that for all the coefficients in *both* equations to be identified, we must remove at least one variable from each equation. Let us suppose that Z_1 affects only X_1 directly whereas Z_2 affects only X_2 directly. This means that $\gamma_{12} = \gamma_{21} = 0$. Our expressions for the π_{ij} now reduce to

$$\pi_{10} = \frac{\alpha_1 + \beta_{12}\alpha_2}{1 - \beta_{12}\beta_{21}} \quad \pi_{11} = \frac{\gamma_{11}}{1 - \beta_{12}\beta_{21}} \quad \pi_{12} = \frac{\beta_{12}\gamma_{22}}{1 - \beta_{12}\beta_{21}}$$

$$\pi_{20} = \frac{\alpha_2 + \beta_{21}\alpha_1}{1 - \beta_{12}\beta_{21}} \quad \pi_{21} = \frac{\beta_{21}\gamma_{11}}{1 - \beta_{12}\beta_{21}} \quad \pi_{22} = \frac{\gamma_{22}}{1 - \beta_{12}\beta_{21}} \qquad (11.16)$$

Given these simplifications, we can now solve *uniquely* for the structural parameters, obtaining

$$\alpha_1 = \frac{\pi_{10}\pi_{22} - \pi_{20}\pi_{12}}{\pi_{22}} \quad \alpha_2 = \frac{\pi_{20}\pi_{11} - \pi_{10}\pi_{21}}{\pi_{11}}$$

$$\beta_{12} = \frac{\pi_{12}}{\pi_{22}} \quad \beta_{21} = \frac{\pi_{21}}{\pi_{11}} \qquad (11.17)$$

$$\gamma_{11} = \frac{\pi_{11}\pi_{22} - \pi_{12}\pi_{21}}{\pi_{22}} \quad \gamma_{22} = \frac{\pi_{11}\pi_{22} - \pi_{12}\pi_{21}}{\pi_{11}}$$

We see that each of the structural parameters is a *nonlinear* function of the π_{ij}. This means that even though ordinary least squares applied to the reduced-form equations will yield unbiased estimates, the resulting solutions for the structural parameters will not in general give unbiased estimates. But it can be shown that the resulting estimates are *consistent*, which means that for large samples their biases will be negligible.

The procedure we have exemplified here is referred to as *indirect least squares* and involves two basic steps: (1) we estimate the reduced-form parameters π_{ij} by ordinary least squares, and (2) then we solve *algebraically* for the structural parameter estimates. The $\hat{\pi}_{ij}$ estimators of the π_{ij} are unbiased, whereas the estimates obtained by substitution of the $\hat{\pi}_{ij}$ for the appropriate π_{ij} in expressions (11.17) are consistent.

The algebraic solution in step 2 depends, however, on there being exactly the right numbers of both kinds of parameters. We have seen that in underidentified systems the number of unknown structural parameters will be greater than the number of estimable reduced-form parameters, and step 2 cannot be carried out. But where the system is overidentified, there will be fewer structural parameters than reduced-form parameters and the algebraic equations will therefore not ordinarily be mathematically consistent. In other words, we may get different results depending on which of Eqs. (11.15) we choose to use in the solution, and there will be no *unique* algebraic solution for the unknown structural parameters. The method of indirect least squares therefore breaks down for overidentified systems. We shall return to this point after considering an alternative approach. Before doing so, however, let us briefly discuss the matrix generalization of indirect least squares.

A Matrix-Algebra Generalization of Indirect Least Squares[1]

We have considered only a two-equation illustration, where the algebra required to obtain solutions was relatively simple. As can be imagined, the algebra for four- or five-equation systems becomes much more tedious, and the general principles are exceedingly difficult to extract. It is therefore much more convenient to represent the situation in matrix notation and to develop proofs for the general case in terms of matrix algebra. Since we are assuming only a minimal acquaintance with matrix algebra, however, we shall merely outline the nature of the arguments, which are developed more systematically in textbooks on econometrics.

In order to represent the general linear system of equations (11.10) in a more convenient matrix form we can move all the terms except the error term to the left-hand side and rewrite the constant terms α_i as γ_{i0} times a dummy variable Z_0, which is always unity. We thus have the system

$$
\begin{aligned}
X_1 - \beta_{12}X_2 - \beta_{13}X_3 - \cdots - \beta_{1k}X_k - \gamma_{10}(1) - \gamma_{11}Z_1 - \cdots \\
- \gamma_{1m}Z_m = \varepsilon_1 \\[6pt]
-\beta_{21}X_1 + X_2 - \beta_{23}X_3 - \cdots - \beta_{2k}X_k - \gamma_{20}(1) - \gamma_{21}Z_1 - \cdots \\
- \gamma_{2m}Z_m = \varepsilon_2 \quad (11.18) \\[6pt]
\cdots\cdots\cdots\cdots\cdots\cdots\cdots\cdots\cdots\cdots\cdots\cdots\cdots\cdots\cdots \\[6pt]
-\beta_{k1}X_1 - \beta_{k2}X_2 \qquad - \cdots + X_k - \gamma_{k0}(1) - \gamma_{k1}Z_1 - \cdots \\
- \gamma_{km}Z_m = \varepsilon_k
\end{aligned}
$$

[1] This section may be skimmed or omitted on first reading.

which can be represented by the matrix equation

$$\boldsymbol{\beta}\mathbf{X} + \boldsymbol{\Gamma}\mathbf{Z} = \boldsymbol{\varepsilon} \qquad (11.19)$$

where $\mathbf{X}$, $\mathbf{Z}$, and $\boldsymbol{\varepsilon}$ are column vectors with dimensions $k \times 1$, $(m + 1) \times 1$, and $k \times 1$, respectively:

$$\mathbf{X} = \begin{bmatrix} X_1 \\ X_2 \\ X_3 \\ \vdots \\ X_k \end{bmatrix} \qquad \mathbf{Z} = \begin{bmatrix} Z_0 \\ Z_1 \\ Z_2 \\ Z_3 \\ \vdots \\ Z_m \end{bmatrix} \qquad \text{and} \qquad \boldsymbol{\varepsilon} = \begin{bmatrix} \varepsilon_1 \\ \varepsilon_2 \\ \varepsilon_3 \\ \vdots \\ \varepsilon_k \end{bmatrix}$$

and where $\boldsymbol{\beta}$ and $\boldsymbol{\Gamma}$ are, respectively, $k \times k$ and $k \times (m + 1)$ matrices:

$$\boldsymbol{\beta} = \begin{bmatrix} 1 & -\beta_{12} & -\beta_{13} & \cdots & -\beta_{1k} \\ -\beta_{21} & 1 & -\beta_{23} & \cdots & -\beta_{2k} \\ -\beta_{31} & -\beta_{32} & 1 & \cdots & -\beta_{3k} \\ \cdots & \cdots & \cdots & \cdots & \cdots \\ -\beta_{k1} & -\beta_{k2} & -\beta_{k3} & \cdots & 1 \end{bmatrix} \qquad \boldsymbol{\Gamma} = \begin{bmatrix} -\gamma_{10} & -\gamma_{11} & -\gamma_{12} & \cdots & -\gamma_{1m} \\ -\gamma_{20} & -\gamma_{21} & -\gamma_{22} & \cdots & -\gamma_{2m} \\ \cdots & \cdots & \cdots & \cdots & \cdots \\ -\gamma_{k0} & -\gamma_{k1} & -\gamma_{k2} & \cdots & -\gamma_{km} \end{bmatrix}$$

The dimensions of $\mathbf{Z}$ and $\boldsymbol{\Gamma}$ involve $m + 1$ instead of m because the constant term has been rewritten as $\gamma_{i0}Z_0 = \gamma_{i0}$ and thereby incorporated with the exogenous-variable terms.

By recalling that in matrix multiplication one takes the rows of the first matrix times the successive columns of the second matrix, the reader should verify that the products $\boldsymbol{\beta}\mathbf{X}$ and $\boldsymbol{\Gamma}\mathbf{Z}$ are each *vectors* with k rows:

$$\boldsymbol{\beta}\mathbf{X} = \begin{bmatrix} X_1 - \beta_{12}X_2 - \beta_{13}X_3 - \cdots - \beta_{1k}X_k \\ -\beta_{21}X_1 + X_2 - \beta_{23}X_3 - \cdots - \beta_{2k}X_k \\ \cdots \cdots \cdots \cdots \cdots \cdots \\ -\beta_{k1}X_1 - \beta_{k2}X_2 - \beta_{k3}X_3 - \cdots + X_k \end{bmatrix}$$

$$\boldsymbol{\Gamma}\mathbf{Z} = \begin{bmatrix} -\gamma_{10} - \gamma_{11}Z_1 - \gamma_{12}Z_2 - \cdots - \gamma_{1m}Z_m \\ -\gamma_{20} - \gamma_{21}Z_1 - \gamma_{22}Z_2 - \cdots - \gamma_{2m}Z_m \\ \cdots \cdots \cdots \cdots \cdots \cdots \\ -\gamma_{k0} - \gamma_{k1}Z_1 - \gamma_{k2}Z_2 - \cdots - \gamma_{km}Z_m \end{bmatrix}$$

and we see that each of the terms in the matrix equation (11.19) is of dimension $k \times 1$.

Let us next represent the reduced-form system in matrix form as

$$\mathbf{X} = \boldsymbol{\Pi}\mathbf{Z} + \boldsymbol{\eta} \qquad (11.20)$$

where we can again absorb the constant terms by writing them as $\pi_{i0}Z_0$, where Z_0 is always equal to unity. The dimensions of $\boldsymbol{\Pi}$ are of course the same as those of $\boldsymbol{\Gamma}$, and the dimensions of the new error vector are $k \times 1$.

If we now premultiply both sides of the equation

$$\boldsymbol{\beta}\mathbf{X} + \boldsymbol{\Gamma}\mathbf{Z} = \boldsymbol{\varepsilon}$$

by $\boldsymbol{\beta}^{-1}$, we get

$$\boldsymbol{\beta}^{-1}\boldsymbol{\beta}\mathbf{X} + \boldsymbol{\beta}^{-1}\boldsymbol{\Gamma}\mathbf{Z} = \boldsymbol{\beta}^{-1}\boldsymbol{\varepsilon}$$

Therefore transferring $\boldsymbol{\beta}^{-1}\boldsymbol{\Gamma}\mathbf{Z}$ to the right-hand side gives

$$(\boldsymbol{\beta}^{-1}\boldsymbol{\beta})\mathbf{X} = \mathbf{IX} = \mathbf{X} = -\boldsymbol{\beta}^{-1}\boldsymbol{\Gamma}\mathbf{Z} + \boldsymbol{\beta}^{-1}\boldsymbol{\varepsilon}$$

Comparing this result with the reduced-form equation (11.20), we see that

$$\boldsymbol{\Pi} = -\boldsymbol{\beta}^{-1}\boldsymbol{\Gamma} \quad \text{and} \quad \boldsymbol{\eta} = \boldsymbol{\beta}^{-1}\boldsymbol{\varepsilon} \quad (11.21)$$

which are general matrix formulas for the reduced-form parameters in terms of the structural parameters.

We illustrate these results in terms of the two-equation system (11.13), with $\gamma_{12} = \gamma_{21} = 0$, for which we derived expressions for the π_{ij} using ordinary algebra. In this particular case we have

$$\boldsymbol{\beta} = \begin{bmatrix} 1 & -\beta_{12} \\ -\beta_{21} & 1 \end{bmatrix} \quad \boldsymbol{\Gamma} = \begin{bmatrix} -\alpha_1 & -\gamma_{11} & 0 \\ -\alpha_2 & 0 & -\gamma_{22} \end{bmatrix}$$

$$\boldsymbol{\beta}^{-1} = \begin{bmatrix} \dfrac{1}{1 - \beta_{12}\beta_{21}} & \dfrac{\beta_{12}}{1 - \beta_{12}\beta_{21}} \\ \dfrac{\beta_{21}}{1 - \beta_{12}\beta_{21}} & \dfrac{1}{1 - \beta_{12}\beta_{21}} \end{bmatrix}$$

$$\boldsymbol{\Pi} = \begin{bmatrix} \pi_{10} & \pi_{11} & \pi_{12} \\ \pi_{20} & \pi_{21} & \pi_{22} \end{bmatrix} = -\boldsymbol{\beta}^{-1}\boldsymbol{\Gamma}$$

$$= \begin{bmatrix} \dfrac{\alpha_1 + \alpha_2\beta_{12}}{1 - \beta_{12}\beta_{21}} & \dfrac{\gamma_{11}}{1 - \beta_{12}\beta_{21}} & \dfrac{\beta_{12}\gamma_{22}}{1 - \beta_{12}\beta_{21}} \\ \dfrac{\alpha_2 + \alpha_1\beta_{21}}{1 - \beta_{12}\beta_{21}} & \dfrac{\beta_{21}\gamma_{11}}{1 - \beta_{12}\beta_{21}} & \dfrac{\gamma_{22}}{1 - \beta_{12}\beta_{21}} \end{bmatrix}$$

We see that the terms within the $\boldsymbol{\Pi}$ matrix are exactly the same as those previously derived. Thus we have a more general way of arriving at expressions for the reduced-form parameters, which is of course also much more convenient in more complex linear systems.

11.6 TWO-STAGE LEAST SQUARES

A closely related approach to simultaneous-equation estimation that also involves the reduced-form equations is two-stage least squares (2SLS). Unlike indirect least squares, this second technique can be applied to overidentified equations. In the special case where an equation is exactly identified, it can be shown that indirect least squares and two-stage least squares are mathematically equivalent. If all equations are exactly identified, both these techniques turn out to be equivalent to the more sophisticated *full-information methods* that are too technical to be discussed in the present work. These latter methods, as previously implied, are somewhat more efficient than two-stage least squares whenever it is possible to use overidentifying restrictions from equations other than the one being estimated.

The general idea behind two-stage least squares is basically that of purifying the endogenous variables that appear in the equation to be estimated in such a way that they become uncorrelated with the disturbance term in that equation. Without loss of generality, let us suppose that we wish to estimate the structural parameters in the first equation of the system

$$X_1 = \alpha_1 + \beta_{12}X_2 + \beta_{13}X_3 + \cdots + \beta_{1k}X_k + \gamma_{11}Z_1 + \cdots + \gamma_{1m}Z_m + \varepsilon_1$$

where at least $k - 1$ of the coefficients have been set equal to zero. Since the endogenous variables $X_2, X_3, \ldots, X_k$ will be correlated with ε_1, whereas by assumption the exogenous Z_i are not, we may replace each of the empirically obtained X_i by its *estimated* value derived from the appropriate reduced-form equation. This will make each of these estimated X_i *exact* functions of the exogenous Z_i and therefore uncorrelated with the error terms. To be specific, we calculate estimated values $\hat{X}_2, \hat{X}_3, \ldots, \hat{X}_k$ according to the formulas

$$\hat{X}_2 = \hat{\pi}_{20} + \hat{\pi}_{21}Z_1 + \hat{\pi}_{22}Z_2 + \cdots + \hat{\pi}_{2m}Z_m$$
$$\hat{X}_3 = \hat{\pi}_{30} + \hat{\pi}_{31}Z_1 + \hat{\pi}_{32}Z_2 + \cdots + \hat{\pi}_{3m}Z_m$$
$$\cdots\cdots\cdots\cdots\cdots\cdots\cdots\cdots\cdots\cdots\cdots\cdots\cdots\cdots\cdots\cdots \quad (11.22)$$
$$\hat{X}_k = \hat{\pi}_{k0} + \hat{\pi}_{k1}Z_1 + \hat{\pi}_{k2}Z_2 + \cdots + \hat{\pi}_{km}Z_m$$

where the $\hat{\pi}_{ij}$ are the ordinary least-squares estimators of the π_{ij}. This constitutes the first stage of the procedure. We now *replace* the observed value X_2 by its predicted value $\hat{X}_2$ and do similarly for all of the remaining X_i except the dependent variable X_1 whose equation is being estimated. In effect, this does away with all of the correlations between the endogenous variables and disturbance terms and thus the complications that produce the bias in ordinary least squares.

Having inserted these new derived or estimated X_i, we again use ordinary least squares to compute a set of coefficients $\tilde{a}_1$ and $\tilde{b}_{1j}$ in the equation

$$X_1 = \tilde{a}_1 + \tilde{b}_{12}\hat{X}_2 + \cdots + \tilde{b}_{1k}\hat{X}_k + c_{11}Z_1 + \cdots + c_{1m}Z_m + \tilde{e}_1 \quad (11.23)$$

Since the exogenous Z_i are assumed uncorrelated with the disturbance term ε_1, there is no need to replace them by any other values, and in fact since they have been taken as exogenous, we have not had occasion to write down any equations for them. The coefficients of Eq. (11.23) obtained by applying least squares in this second step can be shown to be consistent estimators of the structural parameters under the assumptions of the model.

Example   Using this same structural model

$$X_1 = \alpha_1 + \beta_{12}X_2 + \gamma_{11}Z_1 + \varepsilon_1$$
$$X_2 = \alpha_2 + \beta_{21}X_1 + \gamma_{22}Z_2 + \varepsilon_2$$

let us suppose that we have the following very simple data for an extremely small sample $n = 6$:

Individual number	X_1	X_2	Z_1	Z_2
1	0	1	2	3
2	2	4	1	5
3	4	2	4	0
4	6	3	8	5
5	8	6	6	10
6	10	5	9	6
Total	30	21	30	29

We first write down the reduced-form equations which omit X_2 from the equation for X_1 and X_1 from the equation for X_2. We obtain ordinary least-squares estimates $\hat{\pi}_{ij}$ for the coefficients in the two reduced-form equations as follows:

$$X_1 = \hat{\pi}_{10} + \hat{\pi}_{11}Z_1 + \hat{\pi}_{12}Z_2 + v_1 = -1.071 + .9224Z_1 + .3019Z_2 + v_1$$

and

$$X_2 = \hat{\pi}_{20} + \hat{\pi}_{21}Z_1 + \hat{\pi}_{22}Z_2 + v_2 = .8504 + .0970Z_1 + .4479Z_2 + v_2$$

Thus the numerical values of the $\hat{\pi}_{ij}$ are

$$\begin{aligned}
\hat{\pi}_{10} &= -1.071 & \hat{\pi}_{20} &= .8504 \\
\hat{\pi}_{11} &= .9224 & \hat{\pi}_{21} &= .0970 \\
\hat{\pi}_{12} &= .3019 & \hat{\pi}_{22} &= .4479
\end{aligned}$$

Since both structural equations are exactly identified, we can obtain the indirect least-squares estimates from these $\hat{\pi}_{ij}$ by means of the following *algebraic* equations:

$$\hat{\alpha}_1 = \frac{\hat{\pi}_{10}\hat{\pi}_{22} - \hat{\pi}_{20}\hat{\pi}_{12}}{\hat{\pi}_{22}} = \frac{(-1.071)(.4479) - (.8504)(.3019)}{.4479} = -1.644$$

$$\hat{\beta}_{12} = \frac{\hat{\pi}_{12}}{\hat{\pi}_{22}} = \frac{.3019}{.4479} = .6739$$

$$\hat{\gamma}_{11} = \frac{\hat{\pi}_{11}\hat{\pi}_{22} - \hat{\pi}_{12}\hat{\pi}_{21}}{\hat{\pi}_{22}} = \frac{(.9224)(.4479) - (.3019)(.0970)}{.4479} = .8570$$

$$\hat{\alpha}_2 = \frac{\hat{\pi}_{20}\hat{\pi}_{11} - \hat{\pi}_{10}\hat{\pi}_{21}}{\hat{\pi}_{11}} = \frac{(.8504)(.9224) - (-1.071)(.0970)}{.9224} = .9629$$

$$\hat{\beta}_{21} = \frac{\hat{\pi}_{21}}{\hat{\pi}_{11}} = \frac{.0970}{.9224} = .1051$$

$$\hat{\gamma}_{22} = \frac{\hat{\pi}_{11}\hat{\pi}_{22} - \hat{\pi}_{12}\hat{\pi}_{21}}{\hat{\pi}_{11}} = \frac{(.9224)(.4479) - (.3019)(.0970)}{.9224} = .4162$$

Thus our indirect least-squares estimates of the structural equations are

$$X_1 = -1.644 + .6739X_2 + .8570Z_1 + \hat{e}_1$$

$$X_2 = .9629 + .1051X_1 + .4162Z_2 + \hat{e}_2$$

In order to obtain the two-stage least-squares estimates we need to return to the reduced-form estimates $\hat{\pi}_{ij}$ in order to calculate *predicted* values of X_1 and X_2. The predicted values of X_2, represented by $\hat{X}_2$, will be inserted into the equation for X_1, whereas the predicted values of X_1, represented by $\hat{X}_1$, will be used in obtaining the parameter estimates in the equation for X_2. We calculate each of the $\hat{X}_1$ and $\hat{X}_2$ values directly from the reduced-form equations by inserting the appropriate values of Z_1 and Z_2. For example, in the first observation, the Z_1 and Z_2 values are 2 and 3, respectively. Therefore the $\hat{X}_1$ score for this first individual is calculated as follows:

$$\hat{X}_1 = -1.071 + .9224(2) + .3019(3) = 1.679$$

When we obtain the 2SLS estimators in the equation for X_2, we must use this predicted X_1 value, which is an exact function of the Z's, rather than the true value of 0, which is also a function of the disturbance term ε_1. We now calculate the other predicted values of $\hat{X}_1$ and $\hat{X}_2$, which are displayed below along with the actual values of X_1, X_2, Z_1, and Z_2.

Individual number	X_1	$\hat{X}_1$	X_2	$\hat{X}_2$	Z_1	Z_2
1	0	1.679	1	2.388	2	3
2	2	1.361	4	3.187	1	5
3	4	2.619	2	1.238	4	0
4	6	7.818	3	3.866	8	5
5	8	7.482	6	5.911	6	10
6	10	9.042	5	4.410	9	6
Total	30	30.001	21	21.000	30	29

Notice, incidentally, that except for rounding errors, $\sum X_1 = \sum \hat{X}_1$ and similarly $\sum X_2 = \sum \hat{X}_2$.

Finally, we replace all the X_2 scores by the predicted $\hat{X}_2$ scores in the equation for X_1 and use ordinary least squares to estimate the structural equation for X_1. Similarly, we use the predicted $\hat{X}_1$ scores, along with Z_2, in the equation for X_2. In order to distinguish these estimates from those used earlier in the case of indirect least squares, namely, the $\hat{\alpha}_{ij}$, $\hat{\beta}_{ij}$, and $\hat{\gamma}_{ij}$, we may label the 2SLS estimators as $\tilde{a}_i$, $\tilde{b}_{ij}$, and c_{ij}. Our results are

$$X_1 = \tilde{a}_1 + \tilde{b}_{12}\hat{X}_2 + c_{11}Z_1 + \tilde{e}_1 = -1.644 + .6739\hat{X}_2 + .8571Z_1 + \tilde{e}_1$$

$$X_2 = \tilde{a}_2 + \tilde{b}_{21}\hat{X}_1 + c_{22}Z_2 + \tilde{e}_2 = .9629 + .1051\hat{X}_1 + .4162Z_2 + \tilde{e}_2$$

which we see to be identical to those obtained using indirect least squares, except of course for slight rounding errors. As a practical expedient, since indirect least squares can be used only with exactly identified equations, and since canned 2SLS programs are readily available, it will usually be more convenient to use the 2SLS procedure.

11.7 OVERIDENTIFICATION AND TESTING HYPOTHESES

As previously noted, there are somewhat more efficient estimation techniques for overidentified systems that take full advantage of the additional information in all the equations at once. The two approaches we have discussed are essentially based on modified single-equation methods which use only the knowledge of the exogenous variables appearing in the remaining equations. For example, we saw in the case of two-stage least squares that in using the reduced forms we needed only the information that particular Z_i omitted from the first equation did appear somewhere in the remaining structural equations. But we did not use any a priori information about these equations, nor did we require estimates of the parameters in these other equations.

The reader has probably already reached the conclusion, however, that these relatively more complex simultaneous-equation procedures may be beyond the practical stage for most social science research. They require rather strong a priori assumptions regarding exogenous variables and negligible measurement errors. Therefore it seems reasonable to conclude that for some time to come the use of 2SLS will be sufficient for most purposes. In fact, it appears to be the case that the more efficient full-information methods are more sensitive to specification errors than 2SLS.[1] The general rule seems to be that the more a priori assumptions we can safely make, the more information we can legitimately use and the more efficient our estimating procedure. But if we use false information or assumptions, we can expect to pay the price in terms of increased sensitivity to specification errors.

Furthermore, we have been assuming that a priori knowledge about specific parameters is either given or not given. Actually, in practical situations we shall have varying degrees of faith in these assumptions. Some may be based on reasonably sound theoretical grounds, whereas others may be made merely for the sake of convenience, e.g., to produce enough identifying restrictions to make estimation possible. As soon as we admit to varying degrees of faith in these a priori assumptions, we introduce a further complication requiring subjective probability distributions for the parameters. Econometricians have, in fact, developed certain statistical testing and estimating procedures based on such subjective probabilities, but they are far too technical for this text (see Christ, 1966). In addition, they presume a great deal more theoretical knowledge than is yet attainable in the other social sciences. Let us therefore consider a much simpler though less sophisticated alternative approach.

We have seen that the indirect least-squares approach requires exact identification. In this case, two-stage least squares and the remaining more sophisticated approaches all become mathematically equivalent to indirect least squares. If all equations are exactly identified, no empirical *tests* of the model will be possible (apart from predictions concerning the signs of the coefficients), though parameter estimates may of course be obtained. But if one or more equations are overidentified, we can use the excess of empirical information over the number of unknowns to make predictions that will not be satisfied unless the data conform to the model. As we have noted for recursive systems, the greater the excess of empirical information over the

---

[1] King (1974), after a careful review of previous simulation studies as well as a series of his own simulations, concludes that 2SLS is at least as good as, if not superior to, two of these alternatives, three-stage least squares and limited-information maximum likelihood. The advantages of 2SLS over OLS increase with sample size. The only difficulties with 2SLS seem to occur when the Z_i are highly intercorrelated.

number of unknowns or the more overidentifying restrictions we impose, the more difficult it will be to satisfy the conditions of the model.

These considerations suggest the following strategy. We should always attempt to impose as many reasonable identifying restrictions as possible, so as to increase the number of empirical predictions that can be tested by the data. Ordinarily these will take the form of assumptions that some variables do not appear in particular equations. In effect, these then imply that certain of the β_{ij} or γ_{ij} are zero. This, in turn, would mean that if just the right number of these supposedly zero parameters were reinserted into the model to make all equations exactly identified, our empirical estimates of these coefficients should turn out to be close to zero. If they are not, then something is wrong with the model. Our strategy, then, is to reinsert certain of the coefficients back into the equations so as to make all of the equations exactly identified and then to compare the estimates of these parameters with the hypothesized values (ordinarily zero). This strategy of making all equations exactly identified also does away with the need for considering the more sophisticated estimation procedures.

But if an equation is overidentified, which means that more than $k - 1$ of the parameters have been specified in advance, how do we go about making it exactly identified? This requires us to reinsert certain of these "known" parameters into the equation as though they were unknown. But which ones do we reinsert? Here we can take into consideration the fact that we have varying degrees of faith in these assumptions. If it is possible to order our assumptions in each equation by the degree of faith we have in them, we may proceed as follows. In each equation we select the $k - 1$ (zero) assumptions in which we have most faith to be used in order to achieve exact identification. The remaining overidentifying restrictions are merely noted and stated as hypotheses to be tested by the data. That is, the X_i and Z_i corresponding to the latter overidentifying restrictions are inserted back into the equation system as though the parameter values were unknown, these being estimated either by 2SLS or by indirect least squares.

For illustrative purposes let us return once more to the Mason and Halter model of Fig. 11.1. In this overidentified model the following parameters have been assumed equal to zero:

Equation for X_1	$\beta_{13}, \gamma_{11}, \underline{\gamma_{13}}, \underline{\gamma_{16}}, \gamma_{17}$
Equation for X_2	$\beta_{23}, \gamma_{23}, \underline{\gamma_{24}}, \gamma_{25}, \gamma_{26}$
Equation for X_3	$\beta_{34}, \gamma_{31}, \gamma_{32}, \gamma_{33}, \underline{\gamma_{34}}, \gamma_{35}, \gamma_{37}$
Equation for X_4	$\beta_{41}, \beta_{42}, \underline{\gamma_{42}}, \gamma_{45}, \gamma_{46}, \underline{\gamma_{47}}$

In order to achieve exact identification we set $k - 1 = 3$ coefficients equal to zero by selecting the three zero assumptions from each equation in which we have the most faith or the highest degree of certainty. Let us say that these are the underlined coefficients in each of the four equations. The remaining coefficients are reinserted, giving the following exactly identified equations:

$$X_1 = \alpha_1 + \beta_{12}X_2 + \beta_{14}X_4 + \gamma_{11}Z_1 + \gamma_{12}Z_2 + \gamma_{14}Z_4 + \gamma_{15}Z_5 + \gamma_{17}Z_7 + \varepsilon_1$$

$$X_2 = \alpha_2 + \beta_{21}X_1 + \beta_{23}X_3 + \beta_{24}X_4 + \gamma_{21}Z_1 + \gamma_{22}Z_2 + \gamma_{25}Z_5 + \gamma_{27}Z_7 + \varepsilon_2$$

$$X_3 = \alpha_3 + \beta_{31}X_1 + \beta_{32}X_2 + \gamma_{32}Z_2 + \gamma_{33}Z_3 + \gamma_{35}Z_5 + \gamma_{36}Z_6 + \gamma_{37}Z_7 + \varepsilon_3$$

$$X_4 = \alpha_4 + \beta_{42}X_2 + \beta_{43}X_3 + \gamma_{41}Z_1 + \gamma_{43}Z_3 + \gamma_{44}Z_4 + \gamma_{45}Z_5 + \gamma_{46}Z_6 + \varepsilon_4$$

The coefficients in the above equations are estimated using indirect least squares or 2SLS, with the specific hypotheses that the estimated values of the following parameters will be approximately zero:

Equation for X_1 $\hat{\gamma}_{11}, \hat{\gamma}_{17}$
Equation for X_2 $\hat{\beta}_{23}, \hat{\gamma}_{25}$
Equation for X_3 $\hat{\gamma}_{32}, \hat{\gamma}_{33}, \hat{\gamma}_{35}, \hat{\gamma}_{37}$
Equation for X_4 $\hat{\beta}_{42}, \hat{\gamma}_{45}, \hat{\gamma}_{46}$

We thus have 11 specific predictions that are directly analogous to the kinds of predictions that were made in the case of recursive models. The parameters are of course "partial" coefficients in the sense that other variables in the system are held constant through statistical adjustments. But the conventional dot notation of ordinary least squares has not been used and has been replaced by the convention of designating estimators by placing a hat over the appropriate Greek letter.

Once more, we encounter the problem of evaluating the goodness of fit of the data to the model. Unfortunately, only the asymptotic properties of the 2SLS estimators are known, and this information is of little value in connection with samples of finite size. There have been a number of Monte Carlo simulation studies of the behavior of various estimators, including 2SLS, for different models, sample sizes, levels of intercorrelations between the Z_i, and specification errors. The results of these studies are by no means definitive, however, and not easily summarized (see King, 1974; Johnston, 1972). They do show that 2SLS and the more sophisticated alternatives all have larger standard errors than OLS, though smaller estimated biases and mean square errors, i.e., deviations about the true parameter values. One might therefore utilize OLS formulas for standard errors, with the realization that the true standard errors will almost certainly be greater than those based on OLS calculations. If samples under investigation are small, for example, $n \leq 50$, it may be wise to estimate standard errors by a series of Monte Carlo experiments based on models that approximate the presumed causal structure(s). In the case of large samples where sampling errors are considered much less problematic than measurement and specification errors, judgments as to goodness of fit might be more appropriately based on the absolute magnitudes of the departures from zero values.

11.8 THE USE OF LAGGED ENDOGENOUS VARIABLES

Thus far we have treated exogenous variables as though they were completely distinct variables from the set of endogenous variables. We shall now expand the discussion to include what are referred to as lagged endogenous variables or previous values of the endogenous variables. For example, one might use a child's previous test scores or a youth's previous delinquent behaviors as causes of present attitudes or behaviors. We shall see that these lagged endogenous variables can play exactly the same role as exogenous variables in the statistical analysis and in helping to identify the structural parameters, though they involve a number of different kinds of practical problems, as might be expected. Actually, it is conventional within the econometrics literature to refer to the Z_i as *predetermined* variables and to recognize two types of predetermined variables, truly exogenous variables and lagged endogenous variables. Exactly the same assumptions are made about the two kinds of predetermined variables, though we shall see that the assumptions of zero correlation with the disturbance terms becomes much more problematic in the case of lagged endogenous variables.

Many econometric models contain combinations of exogenous and lagged endogenous variables for a number of reasons. Obviously, the use of lagged variables (endogenous or exogenous) requires data at two or more points in time and preferably time-series data over a prolonged period. Equally obvious is the point that very inclusive models involving numerous equations require large numbers of Z_i in order to achieve identification, and there may not be a sufficient number of truly exogenous variables to go around. This is particularly likely in the case of macrolevel models involving entire nations or large regions, since outcome variables for these large macro units are likely to affect so many variables that there are very few exogenous variables for which the required assumptions are realistic (Fisher, 1965). On the other hand, for individual persons or small groups, or even many bureaucratic organizations, there will generally be a large number of environmental influences at work with negligible feedback from the microsystem involved. Of course many of the potential exogenous variables will be either unknown or unmeasured, whereas others will be so highly intercorrelated that they cannot be used as distinct Z_i without producing unusually large sampling errors. For all these reasons it may be desirable to increase the number of Z_i by using lagged endogenous variables.

Other reasons for introducing lagged variables pertain to the desirability of working with dynamic models which involve explicit recognition of temporal sequences, which permit one to study the time paths of each variable, and which are formulated in such a way that stability conditions can be used to ascertain whether the system will stabilize (unless subjected to exogenous shocks) or explode. The subject of dynamic theories is beyond the scope of a textbook on statistics except insofar as temporal data create special problems for statistical analysis. It is sufficient to point out that the use of lagged endogenous variables may introduce greater realism into a theoretical system, quite apart from the additional Z_i that are supplied. For example, many policy decisions involve expectations that may be taken as functions of past as well as present values of certain of the variables. Strotz and Wold (1960) have cogently argued that most static simultaneous-equation models, which do not incorporate lagged relationships, involve specification errors that should ideally be corrected by allowing for the lags that inevitably occur in stimulus-response or behavioral interactions between parties such as buyers and sellers or warring nations.

The use of lagged variables has certain disadvantages, however. First, it is necessary to provide a realistic model specifying the appropriate intervals. Sometimes it makes sense to treat the temporal dimension as though it were discrete. That is, changes in all the variables are conceptualized as occurring at the discrete points in time $t, t + 1, \ldots$, where the intervals between the t's are not necessarily equal. For example a social psychologist may study small-group processes at 20 consecutive meetings of the group. In the model one may be willing to make the assumption that there are no intervening meetings and that values of variables pertaining to the meeting at time t may be affected by what occurred at the meeting at time $t - 1$ or at previous meetings.

In other instances it may be reasonable to assume that decisions are made at a particular time t and they cannot be revised until a later time $t + k$. Suppose, for example, that an automobile manufacturer decides to produce a given number of bodies of a particular type during the months of July, August, and September and then to revise these production figures in October in the light of sales estimates. In other words, the decision remains in effect for a fixed period of time, and it is not possible to make continual adjustments on a day-to-day basis. Similarly, political parties may develop campaign strategies, revise their policies, or appoint new officers only at specified intervals, decisions being based on the outcome of the previous election or perhaps an assessment of changes that have occurred between the two previous elections.

Provided that data can be collected at intervals that are at least as frequent as these natural periods between decisions, it becomes possible to distinguish X_i at time t (X_{it}) from the same variable X_i at time $t - 1 (X_{i,t-1})$, the latter variable being treated as predetermined at time t. However, if changes are occurring more or less continuously, or if observations are possible only at much less frequent intervals, such as every fifth election or group meeting, it may not be possible to obtain the necessary data to estimate the structural parameters. For example, if individual data with respect to decisions to migrate have been aggregated, then even though the individual decisions may occur at discrete points in time, the aggregated migration rates will be continuous functions of time. When time is treated as a continuous variable, theoretical models are usually represented in terms of differential equations in which the rate of change in X_i is designated by the symbol dX_i/dt. Empirically, such equations may be approximated by what are termed *difference equations* involving discrete time periods t, $t + 1,\ldots$ and *distributed lags* that allow for the effects not only of the immediate past but of more remote periods as well (see Rao and Miller, 1971). Sociologists and political scientists have generally lacked adequate time-series data over more than two or three time periods, but if and when such data become available, we shall be in a position to utilize the econometrics literature on the subject without great difficulty. As we shall presently see, however, the use of change data depends rather heavily on the assumption that measurement errors are negligible, and herein lies our principal difficulty.

In order to see some of the basic issues in very simplified terms, let us return to the simple two-variable system of Fig. 11.2, in which we assumed that changes in X_2 will have (almost) immediate effects on X_1 but that changes in X_1 that occur at time $t - 1$ will not produce changes in X_2 until the next time period t. A linear model representing these assumptions is

$$X_{1t} = \alpha_1 + \beta_{12}X_{2t} + \varepsilon_{1t}$$
$$X_{2t} = \alpha_2 + \beta_{21}X_{1,t-1} + \varepsilon_{2t} \tag{11.24}$$

where we assume that the disturbance terms ε_{1t} and ε_{2t} represent omitted variables that have (almost) instantaneous effects on X_1 and X_2, respectively. At time t the value of $X_{1,t-1}$ is taken as predetermined and might have been labeled Z_1. That is, X_{1t} and $X_{1,t-1}$ are treated as though they were distinct variables in the model. With adequate time-series data they could be measured separately, though we would of course expect them to be intercorrelated.

Notice that Eqs. (11.24) are recursive and are exactly identified. $X_{1,t-1}$ (or Z_1) has been omitted from the first equation, and X_{1t} has been omitted from the second. If someone were to multiply the first equation by λ_1 and add it to the second, we would immediately spot the hoax since the new second equation would contain X_{1t} as well as the lagged value $X_{1,t-1}$, whereas by assumption there is no immediate effect of X_1 on X_2. Thus, with only two distinct variables, we have found a way of resolving the identification problem. But we have had to pay the price of specifying additional assumptions concerning the time lags involved.

We encounter a further difficulty with the recursive system (11.24). Previously we assumed in the case of a recursive system that the ε_i were mutually uncorrelated, and this in turn justified the assumption, required by ordinary least squares, that each disturbance term is uncorrelated with all the independent variables in its equation. In the present context we would have to assume that ε_{1t} is uncorrelated with X_{2t}, which is itself a function of $X_{1,t-1}$ (as well as ε_{2t}). But of course X_1 at the earlier time $t - 1$ is a function of $\varepsilon_{1,t-1}$, and this means that ε_{1t} and $\varepsilon_{1,t-1}$ must be uncorrelated. That is, ε_1 must be uncorrelated with its own values at earlier time periods! Usually, of course, this will not be the case; the disturbance term will be self-correlated or autocorrelated. We turn next to a consideration of this problem especially likely to occur whenever the intervals between observations are short.

11.9 AUTOCORRELATION OF THE DISTURBANCE TERMS

Problems of autocorrelation of course may occur in single-equation models for which there is a single dependent variable but observations are made on the same individual over time. An analogous difficulty occurs whenever spatial units, e.g., contiguous counties or states, have been distinguished but are being used to represent supposedly independent replications. The smaller the spatial or temporal units, i.e., the closer they are together, the more likely the disturbance terms will be to contain components that do not shift their values quickly enough to justify the assumption that they can be treated as a truly random variable. Let us consider both a temporal and a spatial illustration.

Suppose a social psychologist is conducting a classroom experiment, taking observations each day over a period of several months. For example, he or she may be attempting to manipulate the environment, either systematically or according to a randomized scheme, while observing the changes in behavior that occur. Certain of the variables that are omitted from the explanatory system will be true constants, but others will shift their values more gradually than the changes in the experimental manipulations. Thus one omitted variable may take on high values during the entire first week, gradually shifting to low values during the second and third weeks, becoming high again during the next 2-week period. Contributions to the disturbance term from this particular source will therefore be positively associated. That is, if the disturbance term takes on a high (positive) value on Tuesday of the first week, it is also likely to be high on Wednesday and Thursday of that same week. Unknown disturbances that have only temporary effects (say, of 1 or 2 min) will of course tend to produce uncorrelated disturbance terms, but to the degree that the investigator has been unable to measure and explicitly introduce or control for the disturbances that have more prolonged effects, one may expect positive serial correlations between the disturbances. If one looked at pairs of disturbances ε_t and ε_{t+1} for all possible values of t, treating the first member of each pair as X and the second member as Y, one would obtain a positive value for r_{xy}.

Autocorrelation in the case of contiguous spatial units has not received the attention it deserves, but the basic issues are the same. Two adjacent census tracts (or counties) are likely to have similar values for many variables such as socioeconomic characteristics, labor-force characteristics, political variables, delinquency rates, and so forth. To the degree that such variables constitute major components of the disturbance terms, one can expect autocorrelation between adjacent units to result. But unlike temporal data, spatial locations cannot be ordered along a single continuum, so that somewhat arbitrary decisions would have to be made concerning the "adjacency" of, say, pairs of counties. Given a decision rule for ordering spatial units according to adjacency, one may then proceed to test for autocorrelation in the manner to be described below.[1]

A very simple test for autocorrelated disturbances, the Durbin-Watson d test, is based on the idea that in the case of positive (negative) autocorrelation the average differences between ε_t and ε_{t-1} will be smaller (larger) than one would expect if nonadjacent pairs had been compared. Of course the true disturbances for the structural equations will not be available, but they can be estimated (in recursive systems) by the residuals e_t and e_{t-1} about the least-squares equations. We form the statistic d as follows:

$$d = \frac{\sum\limits_{t=2}^{n} (e_t - e_{t-1})^2}{\sum\limits_{t=1}^{n} e_t^2} \qquad (11.25)$$

where of course the sample mean of the $e_t = 0$ and there are n observations, one each at times $1, 2, 3, 4, \ldots, n$. Note that the summation for the numerator is necessarily from $t = 2$ to n, whereas the summation in the denominator runs from 1 to n. The distribution of d is symmetrical about a mean of 2.0. Although the exact distribution of d is unknown, Durbin and Watson provide a pair of limiting values d_l and d_u that can be used as follows. In the usual case of tests of the null hypothesis H_0 of no autocorrelation against the alternative hypothesis H_1 of *positive* serial correlation, we reject H_0 in favor of H_1 if $d < d_l$, and we retain H_0 if $d > d_u$. If the calculated value of d, which of course can never be negative, comes out between d_l and d_u, the decision is indeterminate. If negative autocorrelation has been predicted, we reject H_0 if $d > 4 - d_l$ and retain H_0 if $d < 4 - d_u$, with intermediate values of d again implying an indeterminate decision. The critical values of d_l and d_u depend not only on sample size and significance level used but also on the number of independent variables in the equation. Tables for d_l and d_u can be found in Durbin and Watson (1951), Christ (1966), or Rao and Miller (1971). Although the test was designed for the single-equation approach, and therefore for recursive systems as well, Durbin (1957) concludes that it holds approximately more generally for nonrecursive models.

---

[1] For example, such a decision rule might involve placing a grid or concentric-circles pattern over the map of a city or other territorial unit and then numbering the subunits, e.g., census tracts, sequentially according to a prearranged scheme.

The test is biased against discovering either positive or negative serial correlations, however, if the equation contains lagged values of the *dependent* variable.

What if the results of the Durbin-Watson test are such that the null hypothesis of no autocorrelation must be rejected? Is there any way the autocorrelation can be removed? To do so requires one to postulate the nature of what is termed the autoregressive equation, or the equation for ε_t in terms of disturbances at earlier times. In principle, such an equation might involve a number of complexities including the possibility that ε_t is a function of disturbance terms in the remote past. In practice, however, the simple linear equation

$$\varepsilon_t = \rho \varepsilon_{t-1} + \eta_t \qquad (11.26)$$

is most commonly used in econometric research. It is assumed that the disturbance component η_t is independent of all other variables in the system and is not autocorrelated. Thus the disturbance term ε_t is taken as a linear function of the immediate past disturbance ε_{t-1} (but *not* of previous disturbances at $t-2$, $t-3$, and so forth) plus a strictly random component η_t that can be conceptualized as having been produced by a multitude of random events the effects of which are of very short duration.

The coefficient ρ cannot be greater than unity, or otherwise the disturbance terms would not have a finite variance and would explode by having increasingly greater effects as time passes. It can be shown that if $|\rho| < 1$, then

$$\sigma_\varepsilon{}^2 = \frac{\sigma_\eta{}^2}{1 - \rho^2} \qquad \text{for all } t$$

Therefore, the greater the value of ρ^2 the greater the variance of ε relative to the variance of η. Unfortunately, ρ will not be known, and therefore its numerical value must be either estimated or assumed a priori. An iterative process may be used to estimate ρ, which of course will be positive in the case of positive serial correlation (see Christ, 1966; Johnston, 1972). More simply, however, it is often assumed that ρ takes on its maximum value (assuming stability) of unity. In this case the equation for ε_t becomes $\varepsilon_t = \varepsilon_{t-1} + \eta_t$. This assumption amounts to postulating that each shift produced by a change in ε_t moves the dependent variable to a new level and that there is no return to the original level until the next shift takes place at time $t + 1$.

We can now take out the autocorrelation between disturbance terms by *transforming* all the variables in the system in accord with the postulated autoregressive structure. In this simple case, noting that $\varepsilon_t - \varepsilon_{t-1} = \eta_t$, we see that we can get rid of the troublesome ε's by taking *first differences*. That is, for each $t \geq 2$, we form

$$\Delta X_{it} = X_{it} - X_{i,t-1}$$

thus sacrificing one observation. Having done this for all variables in the system (both the X_i and the Z_j in the case of a nonrecursive system), we form a new set of equations in the ΔX_i having exactly the same structural parameters as the original system.

To see how the procedure works, consider the very simple single equation involving only X_1 and X_2, namely,

$$X_{2t} = \alpha_2 + \beta_{21} X_{1t} + \varepsilon_t \qquad (11.27)$$

We are making the fundamental assumption that the same basic causal laws and parameter values hold at all points in time, so that at time $t - 1$ the equation would be

$$X_{2,t-1} = \alpha_2 + \beta_{21}X_{1,t-1} + \varepsilon_{t-1} \qquad (11.28)$$

Therefore, subtracting Eq. (11.28) from (11.27), we get

$$X_{2t} - X_{2,t-1} = (\alpha_2 - \alpha_2) + \beta_{21}(X_{1t} - X_{1,t-1}) + (\varepsilon_t - \varepsilon_{t-1})$$

or $\qquad\qquad \Delta X_{2t} = \beta_{21}\Delta X_{1t} + \eta_t \qquad\qquad\qquad\qquad (11.29)$

Since the disturbance term of (11.29) is just η_t, we have eliminated the autocorrelation, on the assumption that $\rho = 1.0$. Of course it is likely that ρ is somewhere between 0 and 1, and we shall have overcorrected for the autocorrelation. If so, when we form first differences, the new disturbance terms are likely to have a *negative* serial correlation, and therefore the Durbin-Watson test should be used again, this time to test for negative serial correlation. Should it prove necessary to reject the null hypothesis, one might try an intermediate value for ρ (say, .50 or .75) and repeat the process by forming $\Delta X_{it} = X_{it} - .5X_{i,t-1}$ (for the assumption that $\rho = .50$).

Most generally, then, we may follow three basic steps: (1) postulate an autoregressive structure by writing an equation for ε_t, (2) transform the original variables in accord with the postulated autoregressive structure so as to leave only the η_t, and (3) apply the appropriate estimating procedure, e.g., OLS or 2SLS, to the transformed variables. It can be shown that for single equations with no lagged variables the presence of autocorrelation does not systematically bias the ordinary least-squares estimators. However, it makes them less efficient than estimators based on the transformed equations, e.g., first differences. Also, if least-squares formulas are mistakenly used to estimate the variances of the estimators, in order to make tests of significance or to construct confidence intervals, they will *underestimate* the true variances. This means that one may gain a false sense of security in one's estimates. In *nonrecursive* systems involving lagged endogenous variables, autocorrelation will result in estimators that are not consistent in instances where the estimators otherwise would be consistent. This means that there will be nonnegligible large-sample biases, the magnitudes of which will depend on many factors, including, of course, the seriousness of the autocorrelation effect.

There is also what may amount to a much more serious practical drawback to the procedure of working with the transformed variables, as in the above instance of the use of first differences. The differences between the X_i at two successive times are likely to be relatively small compared with the total variation in the variables before transformation. This implies that difference scores are likely to be hypersensitive to the effects of random *measurement* errors. As we shall see in the next chapter, random measurement errors in independent variables produce slope attenuations to the degree that the measurement-error variances are large relative to the variances in the true scores. Therefore it is to our advantage to produce as much variation in the X_i as possible, so that the effects of random measurement error will not predominate. But by working with transformed scores we may obtain differences that are more functions of measurement error than real changes, this being an important practical stumbling block in many studies involving change data. Therefore we find it necessary to correct for autocorrelation by using a procedure that increases sensitivity to measurement errors! If the latter can be assumed to be negligible, then transformations make sense. If autocorrelations are negligible, then the original scores may safely be utilized. But if *both* autocorrelation and random measurement errors are present to unknown degrees, there may be no way to resolve the difficulty without making a considerable effort to improve the measurement or to bring the sources of autocorrelation explicitly into the model.

11.10   THEORY BUILDING AND BLOCK-RECURSIVE MODELS

At least in sociology there has been a tendency to confine the use of causal models to variables that are either almost perfectly measured or for which measurement-error models are relatively straightforward. This implies that many potentially relevant variables are omitted from these models, not because our theories indicate that they are of minor importance but primarily because they cannot be easily measured. All models of reality must of course be incomplete and oversimplified, but it does not follow from this that practical matters of research convenience should dictate one's choice of variables to be included. Whenever this occurs, one is putting the cart before the horse, as it were. This in turn means that justifications for untested assumptions cannot easily be made on strictly theoretical grounds. In effect, one must say to one's readers: I am having to assume that this variable has a negligible effect because I cannot measure it. It would be fortunate if nature were so kind to us in this respect, but it is unlikely. In such cases it would be much more honest and fruitful to say: I found it necessary to omit the following variables because I could not measure them. I believe that the omission of these variables is likely to bias my results in the following specified ways. Since there was no way to check on these biases, however, my study cannot provide definitive results, and I strongly urge others to attempt a careful measurement of these variables.

We have not placed any great emphasis on the subject of theory construction in this text, concentrating instead on theory testing and the estimation of parameters under the presumption that reasonably complete theories have already been formulated. The remarks in this concluding section are intended to correct in part for this one-sided emphasis. In the context of the comments made in the previous paragraph, the implication is that one should attempt to construct as complete a theory as possible *before* deciding which variables to omit on the basis of the limitations of one's research design or data-collecting capabilities. Compromises will always be necessary, but if the theory has been formulated in much more complete terms than the more delimited model to be used with one's data, it will then become possible to make *explicit* decisions about the simplifications that are being made.

This implies, then, that the theorist-investigator should provide the reader with a variety of models that range in degree of complexity as well as in specific details, e.g., the reversal of arrows or a specification of nonlinearity. For instance there may be both a long-term model and a short-term one, the latter involving somewhat more restrictive assumptions regarding feedbacks that may be neglected in the short run but not over a longer period. Likewise, the more complex models may involve as many as 50 to 100 variables, whereas the empirical investigation may have to be confined to a small subset of these. The more complex models may then be used to help rationalize the process of introducing successive simplifications justifying the delimitation of the empirical study, provided that certain assumptions about the remaining variables are deemed appropriate. Such more inclusive theories may also be used to enable readers of different empirical studies to put the latter together in a cumulative fashion.

It is obviously impossible to indicate precisely what we have in mind in such highly general terms, but perhaps a specific illustration will help. Suppose that we are dealing with two or more actors in interaction. One of the first tasks will be to delineate a list of relevant actors, which may of course be organized groups as well as single individuals. There will then be a list of variables appropriate for each actor. Perhaps the same list can be used for all actors, with the differences in the values of the structural parameters representing the differential effects on the several parties. For example, by setting some of the parameters for party A equal to zero, we may introduce simplifications for this party that we do not wish to introduce for the remaining actors. In the case of two persons in interaction—perhaps a delinquent youth and a police officer—we shall have two sets of variables referring to the behaviors of the two parties, two different sets to their internal states, e.g., goals, motives, expectations, or attitudes, and still other sets referring to a diversity of environmental stimuli, some of which will be truly exogenous and some of which may be affected by the behaviors of either or both of the parties.

Once one has delineated sets of conceptually interrelated variables, one may begin to construct complex models by a process of explicitly examining all pairs of variables and asking whether or not there is presumed to be any direct causal link between them. This process may uncover additional intervening variables that are thought to interpret these linkages that were originally taken to be direct relative to the other variables previously included in the model. Finally, of course, one must close the theoretical system by making assumptions about the residual causes of each variable that has been explicitly included. Let us say, for illustrative purposes, that one has constructed a general model with the following variables:

Behaviors of parties A and B
　　Five variables for A, six for B
Internal states of parties A and B
　　Eight variables for A and the same eight for B
Exogenous environmental variables
　　Fifteen variables affecting A
　　Ten variables affecting B (eight of which overlap those affecting A)
Outcomes of behaviors of A and B
　　Six variables, four of which feed back to affect internal states of A and B

Such an overall model would undoubtedly look quite complex if expressed in terms of causal diagrams and even more complex if written down in equation form. But it will often be possible to group these variables into blocks of variables that may coincide reasonably well with the sets of variables as originally grouped conceptually. We may aim to construct a *block-recursive model* characterized by reciprocal causation *within* the individual blocks but recursive connections (and uncorrelated disturbances) *between* blocks (see Fisher, 1966). An example of a five-block recursive system is given in Fig. 11.3. Such a schematic diagram not only involves a reduction in the total number of arrows, since we now have only one arrow connecting each block of variables, but it also makes it easier to visualize the overall picture with a view to making further simplifications or subdividing the overall model for a more careful look at the separate blocks.

It should be specifically noted that certain of the blocks may contain previous or lagged values of variables that appear in the other blocks. For example, certain of the variables that affect the internal states of party A at time t may be past behaviors of B at times $t - 1$, $t - 2$, or earlier, or perhaps *changes* in the levels of certain exogenous variables as well as their present states. Outcomes of the previous behaviors of both A and B may be placed together in a single block, which may be taken to influence present levels of certain but not all of the environmental factors. The essential feature of the block-recursive model is that there can be no reciprocal causation or feedback between blocks. If such reciprocal causation is thought to exist, e.g., between blocks 3 and 4, the model must be reconceptualized. This can be done either by combining the variables in these blocks into a single larger block or by introducing specific assumptions about the time lags and constructing a new block involving lagged values of the variables in one of these blocks. For example, one may form a block $2A$ consisting of previous values of block 4 variables, with block $2A$ variables affecting block 3 and block 3 variables in turn affecting block 4.

This kind of scheme makes it possible to see at a glance which sets of variables shall definitely be dropped from a given empirical analysis, which must be retained, and which may be used on a selective basis. Suppose, for example, that an investigator wishes to focus on the k variables in block 4. Since all block 5 variables are dependent upon the block 4 variables and variables in earlier blocks, block 5 should be dropped from the analysis. To introduce these variables as controls, for example, would be misleading. Since all the variables in block 4 are assumed to form a mutually interdependent set, this implies that the investigator must include all these variables in the submodel to be empirically investigated. If any of these variables were omitted (or very poorly measured), unknown biases would be introduced because all k equations should be handled simultaneously. Of course if certain of the feedbacks within block 4 were assumed to be negligible, the model might be simplified so as to take out those variables in the block that depend upon the remaining variables but have only negligible effects on them. Suppose, for example, that there were three such variables in block 4. Then the model could be reconceptualized by placing these "dependent" variables in a new block $4A$ and then confining one's attention to the remaining $k - 3$ variables in the old block 4.

One need not include all exogenous variables, however. This is indeed fortunate since it would require the impossible, namely, that one use all the causes of each variable in one's model. We know that in order to identify the coefficients in each equation (say, in block 4) we need only bring in a sufficient number of exogenous variables on a selective basis. We saw that a number of criteria can be used as guide-

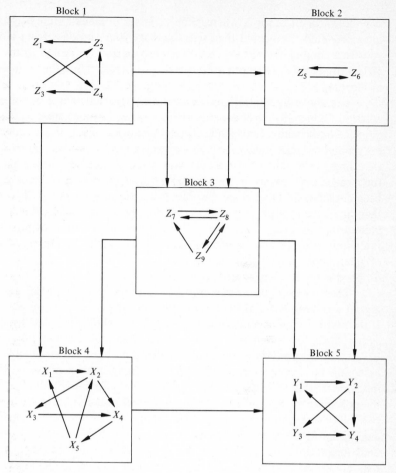

FIGURE 11.3

lines here, though we did not mention the obvious fact that ease and accuracy of measurement may be one such criterion. In particular, we found that it is advantageous to bring in those exogenous variables which are assumed to affect directly only a relatively small number of the endogenous variables. These exogenous variables must be important enough causes and must vary sufficiently to be at least moderately correlated with the endogenous variables they are presumed to affect.

We also do not want our exogenous variables to be too highly intercorrelated, which often means that they should come from different blocks whenever possible. Suppose, for example, that block 3 contains a number of highly interrelated socio-economic-status variables that are thought to affect block 4 variables but in complex ways. Since it will be difficult to disentangle their individual effects on block 4 variables, it will therefore be advantageous to select only one variable from this block. Or if measurement of the remaining variables in block 3 is not too difficult, it may be preferable to construct a combined index to represent this block. Perhaps the same may hold true of block 2, whereas the variables in block 1 may be only very weakly intercorrelated. We could then select perhaps four variables from block 1 and also construct combined indices to obtain single measures for blocks 2 and 3. This would yield a total of six exogenous variables for use in estimating the structural parameters in block 4.

The above illustration points up the obvious advantage of constructing initial models of a high degree of complexity *before* selecting one's final set of variables to be included in the study. It also indicates the advantage of selecting a sufficient number of exogenous variables to achieve highly overidentified equations. Suppose, for example, that three of the equations in block 4 are exactly identified. It may then turn out that the measures of certain of the exogenous variables are much less adequate than had been anticipated. Or perhaps the intercorrelations in block 1 will be much stronger than had been expected. In either case, the equations concerned will become under-identified (if the exogenous variables have to be dropped altogether) or subject to large sampling errors (if the exogenous variables are highly intercorrelated). But if the original model is highly overidentified, these particular exogenous variables may be dropped without complication.

These considerations may also be useful in deciding on the appropriateness of lagged endogenous variables as supplements to the exogenous variables in the model. If the number of exogenous variables on the original list is not sufficient to make the system highly overidentified, and if it is expected that a certain number of them will have to be dropped from the analysis, then one must either simplify the model by assuming zero values for some of the remaining coefficients or specify a dynamic model with well-defined lag periods. The second alternative will of course necessitate the collection of data at several times. It will also require an examination of the possibility of autocorrelation, which, in turn, may be removable only by making the assumption that the changes are real and only negligibly due to measurement errors. Once again, the important point is that such decisions must be thought through before data are collected rather than being brought in as ex post facto rationalizations for these decisions.

EXERCISES

11.1 Suppose it is known that a unit increase in X_1 produces an increase in X_2 of 5 units, that a unit increase in X_2 produces a decrease in X_3 of .3 unit, and that a unit increase in X_3 produces an increase in X_1 of .4 unit. Also, X_1 and X_3 begin with values of zero, but a change in an exogenous variable shifts the level of X_2 from 0 to 10 units.

(a) Write out a system of equations that represents these facts and obtain the algebraic solution.

(b) Assuming values at t_0 of 0, 10, and 0 for X_1, X_2, and X_3, respectively, and allowing one time period for each shift to occur, write out the set of numerical values that each variable would have over the first 24 time periods; i.e., the change in X_2 produces a change in X_3 at t_1; this change produces a change in X_1 at t_2, which changes X_2 at t_3, and so forth. Do these values appear to be converging toward the solution found in part (a)?

(c) Now suppose a unit change in X_3 produces a change in X_1 of 4 units, rather than .4 unit. What will happen to the three variables over time? From these two sets of coefficients can you see a condition that must hold among the coefficients in order for the system to stabilize?

11.2 Take the model for Exercise 10.3. Suppose that X_6 is also a cause of X_5, so that there is reciprocal causation between these two variables. How does this affect the various predictions made for the model and the method of data analysis? Will all equations be identified? Explain.

11.3 Suppose you have four variables interrelated according to Fig. E.11.3 below.
    (a) Add a minimum number of Z_i such that each equation becomes exactly identified and insert them into the diagram.
    (b) Write out the set of linear equations implied by this revised model.
    (c) Show that both the necessary (order) and necessary and sufficient (rank) conditions for identification have been satisfied.

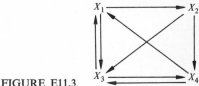

FIGURE E11.3

11.4 For the model given below in Fig. E.11.4:
    (a) Show that the necessary and sufficient conditions for identification are met.
    (b) Select a hypothetical ordering of the zero assumptions of this model according to the degree of faith you have in their reasonableness. Show how you could modify the equation system for this model so as to make each equation exactly identified.
    (c) State the empirical predictions that you would then make in order to test the adequacy of the model.

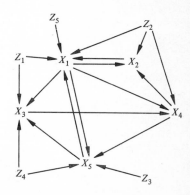

FIGURE E11.4

11.5 Use the exactly identified model in Fig. E.11.5 on page 532.
    (a) Write out both the structural and reduced-form equations and solve for the π_{ij} in terms of the β_{ij} and the γ_{ij}.
    (b) Solve for the β_{ij} and γ_{ij} in terms of the π_{ij}.
    (c) From this model indicate how you would use 2SLS to estimate the coefficients in each structural equation.

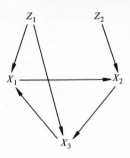

FIGURE E11.5

11.6 Take the numerical example that was used to illustrate the computation of indirect least squares and 2SLS (page 515). Suppose the equation system is respecified so that $\gamma_{11} = \gamma_{22} = 0$, whereas γ_{12} and γ_{21} are no longer equal to zero. Recalculate the indirect least-squares and 2SLS estimates, verifying that these two procedures again yield identical results.

PART FOUR

Models Involving Measurement Errors

12

RANDOM MEASUREMENT ERRORS
AND THEIR IMPLICATIONS

Whenever one wishes to test explicitly formulated theories, one is made very much aware of the numerous assumptions required. Among the most important of these assumptions are those involving measurement errors. It is still the common practice in social research to make the unrealistic assumption that measurement errors are negligible and therefore can be completely ignored in the data analysis. The usefulness of statistical analysis obviously depends on our being able to introduce somewhat less restrictive assumptions regarding measurement errors, and this of course should apply to nonstatistical reasoning as well. The purpose of this chapter is to discuss the nature of several specific approaches to handling strictly random measurement errors and to examine their implications. In the following chapter we shall be concerned with specific kinds of nonrandom or systematic measurement errors.

12.1 GENERAL REMARKS ABOUT MEASUREMENT-ERROR MODELS

It is well to begin by realizing that *all* variables are measured with some degree of error, so that it does not make sense to attempt to dichotomize between a set of measured variables and another set of "unmeasurable" ones. Nevertheless, in any particular piece of research an investigator may have to take certain of the variables as unmeasured, recognizing that a second investigator may attempt to obtain at least indirect measures of some of these variables. Social scientists have differing tastes with respect to including in their theories variables that are likely to be extremely difficult to measure or require highly indirect forms of measurement. There is no need for us to take sides on this matter, though it is important to recognize the nature of the issues that are involved. However, we must avoid any possible tendency to assume away the problem by pretending that all variables in the model have been perfectly measured. Another methodological stance that seems dysfunctional from the standpoint of the accumulation of knowledge is that which takes poor measurement for granted and which weakens this measurement still further in the process of adjusting the mode of data analysis to this low quality of measurement. For example, we do not recommend the practice of dichotomizing or trichotomizing variables merely because they have not been measured as interval scales. This simply adds further measurement error to that already existing and lends support to the notion that there is little or no point in striving for greater accuracy of measurement.

The general strategy that is applied in this and the following chapter involves a number of basic steps. First, one must attempt to define one's theoretical variables as clearly as possible so that some sort of highly specific measurement-error model can be constructed. This is usually the most difficult but also the most crucial step in the whole process. Here one must be especially careful not to let the existence of a reasonably simple metric dictate a definition of the variable that is not really intended. For example, the variable of theoretical interest may be "quality of education," or "total amount learned," both of which may be rather poorly indexed by years of formal schooling. Many of our most important variables are not easily associated with such simple metrics, however, and even those which are often do not imply a unique measure. Thus if one wanted to measure the degree of "urbanization" across a number of societies, one could employ a variety of very similar measures, no one of which is theoretically more appropriate than the others. Suppose investigator A selected as the measure of urbanization the proportion of the total population living in cities of over 100,000 population. Investigator B might use the proportion living in cities of over 80,000, and investigator C might use the logarithm of the proportion of the population between the ages of 15 and 65 living in cities over 500,000. Presumably, all three of these measures would be highly intercorrelated, but obviously this will not always be the case with respect to all the indicators one might choose. If not, then how can we begin to conceptualize the problem?

This example illustrates the more general point that we cannot begin to develop measurement-error models until this first step has been taken. Our second step will consist of writing down an equation linking the "true" value to the measured indicator. But in this example what would we take as the "true" amount of urbanization? How could we assess whether or not investigator A's measure involved only random measurement error whereas B's contained a specific bias? The more general point, of course, is that data analysis presupposes that one knows what one is doing. In this case, it is assumed that we have a rigorous theoretical definition of our variable. Of course if one were to be an extreme perfectionist in this respect, one could never proceed beyond the first step except in the simplest of cases. This in turn would tend to confine one's analysis to a very small subset of the variables usually found in social science theories. Nevertheless, it is worth stressing that we can come to grips with measurement errors only to the degree that we are able to improve our theoretical conceptualizations. An explicit attention to such measurement errors, it is hoped, may help to focus our energies in this direction.

This being a text in statistics, we shall henceforth (with tongue in cheek) assume that all conceptual variables have been clearly defined, so that it is theoretically meaningful to speak about "true" values of all variables in the causal system. As noted above, we can then take the second step of writing down an equation linking true and measured scores. But such equations must be based on substantive grounds. In certain instances we may *define* one variable to be an exact function of others that have previously been defined and linked with operational measures. Thus one may define density as mass divided by volume, acceleration as a change in velocity per unit of time, or status inconsistency as the difference between two statuses. Such derived concepts pose no special problems with respect to measurement, though they often lead to complications in data *analyses* unless one is very careful.

Most of our measured variables are assumed to be linked with theoretical constructs by *empirical* processes motivating our equations that interrelate them. Thus we do not measure mass directly but infer a value on the basis of pointer readings which we assume are *effects* of the mass of the body operating as one among several forces on a spring scale or balance. Similarly, we assume that the income that gets punched onto a computer card is caused by a respondent's answers based on a knowledge of the true income, but perhaps also on some deliberate or unconscious distortions. Or the recorded income may be partly caused by processes within the interviewer or coder that produced an error. Sometimes, in contrast, we measure variables in terms of their presumed *causes*, e.g., when we try to measure hunger by the number of hours since the last feeding or frustration by the manipulations of the experimenter.

Thus this second step of writing equations linking measured and true values often (though not always) involves a type of causal modeling based on exactly the same kind of reasoning as discussed in Part Three. There are two important differences, however: (1) We must now take some of the variables as measured and others (the true values) as unmeasured. This will in general introduce a large number of unknowns into the system, which will of course lead to problems of identification. (2) Because of the identification problems produced by these unmeasured variables our causal models will have to be much simpler. In particular, we shall confine ourselves to recursive models and relatively small numbers of "true scores," or variables that are considered to be theoretically important. In principle, the approach can be extended routinely to more complex models provided that the equations do not become underidentified.

As a general rule the more complex we make these measurement-error causal models the greater must be the ratio of measured to unmeasured variables. We are in effect augmenting whatever causal models are deemed appropriate for the basic theory with what we might refer to as an *auxiliary theory* involving the linkages between true scores and measured variables. Unfortunately there must be a tradeoff, here, because of the need to construct highly overidentified models that are easily rejectable. If our general theories are very complex, in the sense of containing many unknown coefficients, then our measurement-error models must be simple. (In this connection we should note that in Part Three our measurement-error models involved the extreme limit of simplicity, namely, all true and measured values being identical.) But if our measurement-error models are highly complex, perhaps because of the indirectness of many measurements, then the basic theory being tested must be rather simple. The implication is clear, if not disheartening. If we believe that reality is complex and that our substantive theories must be much more complex than those with which we are accustomed to dealing, then we must have extremely good measures of almost every variable in the theoretical system!

There is a final third step in addition to the obvious one of providing substantive interpretations for one's findings. This is the deductive task of extracting the implications of the measurement-error models for testing and for estimating the relevant parameters. It is this third step that will occupy most of our attention in the remainder of this and the final chapter. This step depends upon the prior two and in particular on the validity of the assumptions made in constructing the measurement-error models. Given the uncertainties that will almost inevitably be encountered in taking these first two steps, the obvious implication is that one will be well advised to take a very flexible approach to the final testing and estimation stage, cautioning one's readers about the many assumptions that have been made along the way.

With these preliminary remarks in mind, we turn to a more detailed examination of the implications of measurement errors of various kinds. Several kinds of questions will concern us in this and the following chapter:

1 The nature of the distortions introduced by random and nonrandom measurement errors and how these affect one's inferences in various kinds of situations
2 Possible corrections that can be made and the nature of the assumptions they require
3 The risks one takes in making these corrections in instances where the required assumptions are not completely justified

It would be comforting to find that more or less cut-and-dried methods could always be used and that these procedures required minimal assumptions. As we shall see, however, it will always be necessary to make certain untestable a priori assumptions that will require good judgment as to the relative seriousness of different kinds of possible distortions. The better one's theory the easier it will be to justify such assumptions. But accurate measurement will usually be necessary to establish these theories in the first place. Herein lies a major dilemma for the social scientist.

12.2   RANDOM MEASUREMENT ERRORS AND SLOPE ATTENUATIONS

If we were to define random measurement error as involving those errors of measurement which result from the operation of random processes, we would be left with the question of exactly what this means. In the real world we can certainly imagine situations in which measurement errors are produced by a very large number of minor causes that are only weakly intercorrelated. For example, there may be minor coding errors, lapses of memory on the part of respondents, punching errors, rounding errors, and so forth, that have about the same aggregate effect on the measured variables as would have been obtained if one had added a component e to the true score X by entering a table of normally distributed random numbers. In practice, we are likely to have systematic errors superimposed upon such strictly random errors of measurement, but in the present chapter we are assuming that there are no such systematic errors.

Let us consider random measurement errors in the following way. Let X be the true score, let X' be the measured score, and let e represent the measurement-error component according to the equation $X' = X + e$.[1] We assume that $E(e) = 0$, as would be the case in real-life situations if half of the measurement-error distortions were in the positive direction and half in the negative, with the probability distribution of the measurement-error term being symmetrical. If the expected value of e is *not* zero, the distortions are systematic. If $E(e) = K \neq 0$, these distortions are constants that affect only the intercept terms in one's equations. If $X' = cX + e$, where $c \neq 1$, there is an error of scale that will also result in incorrect slope estimates. But the most fundamental (and perhaps most risky) assumption we shall make is that the measurement-error component e is uncorrelated with all other variables in the theoretical system. In particular, we shall assume that cov $(X,e) = 0$ and that cov $(Y,e) = 0$, where Y is dependent on X. Note, however, that the *measured* value X' will of necessity be correlated with e since we are taking X' to be a function of X and e. We shall later need to allow for random measurement errors in more than a single variable, in which case we shall assume that cov $(e_i,e_j) = 0$, for all $i \neq j$, where e_i and e_j represent the measurement-error components for X_i and X_j respectively. In other words, we assume that the factors that produce measurement errors in one variable are not systematically related in the aggregate to factors producing measurement errors in the other variables.

Our definition of random measurement errors will thus be as follows. If X_i' is an indicator of the true value X_i and is related to X_i by the equation $X_i' = a + cX_i + e_i$, then measurement error in X_i will be said to be random if and only if

$$a = 0 \qquad (12.1a)$$

$$c = 1 \qquad (12.1b)$$

$$E(e_i) = 0 \qquad (12.1c)$$

$$\text{cov } (X_i,e_i) = 0 \qquad (12.1d)$$

$$\text{cov } (X_j,e_i) = 0 \qquad (12.1e)$$

where X_j represents any other variable in the causal system

$$\text{cov } (e_i,e_j) = 0 \qquad i \neq j \qquad (12.1f)$$

$$\sigma_{e_i}^2 = \text{const} \qquad (12.1g)$$

Conditions (12.1a) and (12.1b) may be relaxed whenever the conceptual variable is not defined in such a way as to imply a definite zero point or metric unit, in which case (12.1c) will also be unimportant. Conditions (12.1d) and (12.1e), as already noted, are especially important and will distinguish the random measurement-error models

---

[1] In this chapter and the next we shall use e_i terms to represent *measurement* errors, whereas in Part III we used the e_i to represent residual terms in least-squares estimating equations.

from most of those discussed in the next chapter. Condition (12.1*f*) states that there must not be any correlations between the measurement-error components of the variables in the theoretical system, and (12.1*g*) assures us that the variance of e_i is not related to any other variables. In effect, our definition of random measurement error is intended to specify a set of conditions that would be very well approximated if the measurement-error terms were produced from a table of random numbers with a mean of zero. They would be approximated if measurement errors resulted from the operation of an extremely large number of weakly interrelated minor sources of error, in which case the e_i would also be approximately normally distributed.

Random measurement errors in a *dependent* variable may be conceived as being absorbed into the error term and therefore create no special difficulties apart from increasing the error variance and therefore attenuating correlation coefficients. Consider the equation

$$Y = \alpha + \beta X + \varepsilon_y$$

If $Y' = Y + e_y$, then

$$Y' = \alpha + \beta X + (\varepsilon_y + e_y)$$

where the two terms within the parentheses are empirically indistinguishable unless, of course, Y as well as Y' is known. The new error variance will be $\sigma_{\varepsilon_y}^2 + \sigma_{e_y}^2$, and therefore to the degree that $\sigma_{e_y}^2$ is large relative to $\sigma_{\varepsilon_y}^2$ it will be difficult to explain a high percentage of the variance in the measured value Y'. But since, by assumption, e_y is uncorrelated with X (and with any other causes of Y), the introduction of the measurement-error component e_y will not systematically bias the ordinary least-squares slope estimate. Therefore we shall not be concerned with random measurement errors in dependent variables until we turn our attention to multiple-indicator approaches later in the chapter. It should be kept in mind, however, that in recursive systems a variable that is dependent in one equation may be independent in another equation, so that in the k-equation case we can ignore random measurement errors only in the ultimate dependent variable X_k.

In the case of the independent variable X, if we let the measured value $X' = X + e$, we know that in general

$$\text{var } X' = \text{var } X + \text{var } e + 2 \text{ cov } (X,e)$$

For strictly random measurement errors we assume that cov $(X,e) = 0$ for very large samples, although for finite samples there will always be some slight correlation between measurement errors and the true values of X. In terms of sigma notation we may write

$$\sigma_{x'}^2 = \sigma_x^2 + \sigma_e^2$$

It is difficult to specify an exact but simple general formula for the slope attenuation, but if the remaining assumptions required for ordinary least squares are met, it can be shown (Johnston, 1972) that in the case of the equation $Y = \alpha + \beta X + \varepsilon_y$ the large-sample expected value of $b_{yx'}$ is given by the approximate formula

$$E(b_{yx'}) = \frac{\beta}{1 + \sigma_e^2/\sigma_x^2} = \beta \frac{\sigma_x^2}{\sigma_x^2 + \sigma_e^2} = \beta \frac{\sigma_x^2}{\sigma_{x'}^2} \qquad (12.2)$$

Thus the least-squares slope estimator will be attenuated to the degree that $\sigma_e{}^2$ is large relative to $\sigma_x{}^2$, the variance in the true X scores. Since in most practical applications neither of these quantities will be known, it will be difficult to assess the actual degree of attenuation. Before discussing the question of how one may attempt to correct for or remove these effects of random measurement errors in X, let us consider a number of situations in which such measurement error can lead to faulty causal inferences.

Because we now need to distinguish between true values and their measured indicators, we shall change our notation somewhat by referring to the true scores as W, X, Y, and Z. When there is only a single indicator of each variable, we shall use primes to designate the respective measured values. Later in the chapter, when we study multiple-indicator approaches, we shall refer to the several indicators of X as $X_1, X_2, \ldots, X_k$, and similarly for W, Y, and Z. Thus capital letters without primes or subscripts will designate unmeasured variables or "true scores," whichever terminology the reader prefers to adopt.

12.3 ILLUSTRATIONS AND IMPLICATIONS OF SLOPE ATTENUATIONS[1]

It might be thought that purely random errors usually result in conservative inferences in view of the fact that correlations will generally be attenuated in magnitude. For example, we know that in the two-variable case involving X and Y, random measurement errors in either variable will attenuate the correlation between them. Therefore, an investigator who finds a relationship which is statistically significant and moderately large, even with measurement error, may have added confidence that the true relationship is even more pronounced. Hence one is being conservative in inferring a relationship. What about more complex multivariate analyses, however? The same sort of situation might be thought to occur in instances where one is introducing control variables and making inferences on the basis of the behavior of the partial coefficients. But the picture is no longer so simple.

Basically, the problem reduces to the fact that random measurement errors introduce unknowns into the system and the values of these unknowns cannot be easily estimated. As a result, there will be additional models or causal explanations that can also account for the data. These alternative explanations cannot be merely ignored, and an important task is to develop methodological tools which will enable us to become explicitly aware of these alternatives so that we can deal with them systematically.

Spuriousness and Intervening Variables

A simple way to introduce the problem will be to discuss models in which measurement error occurs in an intervening variable or in a variable producing a spurious relationship between two other variables. Suppose we are given either of the models

---

[1] This section is taken, with minor modifications, from H. M. Blalock, Some Implications of Random Measurement Error for Causal Inferences, *American Journal of Sociology*, **71**: 37–47 (July 1965), by permission of The University of Chicago Press.

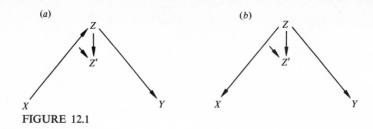

FIGURE 12.1

of Fig. 12.1, where Z has been measured imperfectly. The measured value Z' is taken as being caused by the true value plus a number of factors which are unrelated to either X or Y and which are assumed to create purely random disturbances (as indicated by side arrows, Fig. 12.1). We are assuming no measurement errors in X or Y, although it has been shown elsewhere that purely random errors in either of these variables will not lead to faulty causal inferences in the case of the models of Fig. 12.1, provided Z is perfectly measured.[1]

In both these models, a control for Z (if perfectly measured) would completely wipe out the association between X and Y, except for sampling errors. The investigator could then infer, correctly, that the relationship between X and Y is not direct, although one would need additional information to choose between models (a) and (b).

If Z is measured with error, however, the investigator is actually controlling not for Z but for Z'. It can easily be shown that in the case of random measurement error, a control for Z' will reduce but not eliminate the association between X and Y in both models of Fig. 12.1. Furthermore, the degree of reduction in this association depends upon the actual amount of measurement error involved. This can be seen as follows. We know that, except for sampling error, $r_{xy \cdot z} = 0$, or

$$r_{xy} = r_{xz}r_{yz}$$

Also,
$$r_{xz'} = r_{xz}r_{zz'} \quad \text{and} \quad r_{yz'} = r_{yz}r_{zz'}$$

and therefore
$$r_{xz'}r_{yz'} = r_{zz'}^2 r_{xz}r_{yz} = r_{zz'}^2 r_{xy}$$

Hence
$$|r_{xy}| = \left| \frac{r_{xz'}r_{yz'}}{r_{zz'}^2} \right| \geq |r_{xz'}r_{yz'}|$$

If we were to control for Z' instead of Z, we would obtain $r_{xy \cdot z'}$ as follows:

$$r_{xy \cdot z'} = \frac{r_{xy} - r_{xz'}r_{yz'}}{\sqrt{1 - r_{xz'}^2}\sqrt{1 - r_{yz'}^2}} = \frac{r_{xy}(1 - r_{zz'}^2)}{\sqrt{1 - r_{xz'}^2}\sqrt{1 - r_{yz'}^2}}$$

This partial correlation will be zero if and only if $r_{zz'}^2 = 1$, excluding the trivial case when $r_{xy} = 0$. That is, the measurement of Z must be perfect except for an intercept and/or a scale factor.

---

[1] Random measurement errors in X and Y (measured as X' and Y', respectively) will attenuate the correlations with Z in such a way that the expected value of the partial correlation $r_{x'y' \cdot z}$ will be zero (see Blalock, 1964, chap. 5).

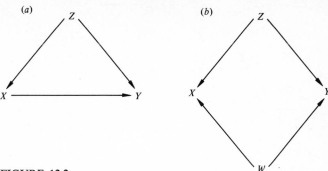

FIGURE 12.2

These results are of course consistent with common sense, but we should note that a number of alternative explanations now become rival hypotheses to the models of Fig. 12.1. One possibility is that Z has, in fact, been accurately measured but that there is also a direct causal link between X and Y. Another is that the relationship between X and Y is spurious but that an additional variable W would also need to be controlled. These alternatives are given in Fig. 12.2, where we have assumed perfect measurement in all variables.

Correlated Independent Variables

Situations frequently arise in which the investigator wishes to compare the supposed effects of two independent variables which are themselves highly correlated. As already noted, such comparisons are made difficult by the phenomenon of multi-collinearity, or the fact that sampling errors of partial correlations become unusually large. But even where the sample size is sufficiently large to assure accurate estimates, random measurement errors can also lead to faulty inferences. Suppose we wished to choose among the three alternative models represented in Fig. 12.3. The curved double-headed arrows connecting X and Z in each instance merely indicate that there is no concern with explaining the correlation between X and Z, although it is assumed that these variables are highly related.

First suppose that both X and Z have been perfectly measured. Assuming that the sample size is sufficiently large to ignore sampling error, one could choose among alternatives (a), (b), and (c) by obtaining the partial correlations $r_{xy \cdot z}$ and $r_{zy \cdot x}$. If the former disappeared but not the latter, we would infer that X is not a direct cause of Y but that Z is a direct cause. Likewise, we would select alternative (c) if $r_{zy \cdot x}$ but not $r_{xy \cdot z}$ disappeared. If neither disappeared or approached zero, we would prefer the second model.

Now suppose that alternative (a) is actually correct but that Z has been measured with random error. The appropriate causal model would then be as indicated in Fig. 12.4, where the measured value Z' is again assumed to be composed of the actual Z score plus a random element. It can be seen that if the magnitude of the correlation between Z and Z' is less than that between X and Z, $r_{xy \cdot z'}$ will be numerically larger

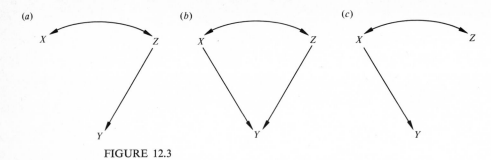

FIGURE 12.3

than $r_{z'y \cdot x}$. Since the investigator will ordinarily not be in a position to estimate the magnitude of $r_{zz'}$, and in fact may ignore the measurement error altogether, one may be led to infer that either of the alternatives (*b*) or (*c*) is preferable to (*a*). In other words, one may infer a direct linkage between X and Y when it does not exist.

For example, the correlations between the variables may be as indicated in Table 12.1. These data are consistent with the model of Fig. 12.4 since, except for sampling errors, we would have

$$r_{xz'} = r_{xz}r_{zz'} \quad \text{or} \quad .40 = (.80)(.50)$$

$$r_{xy} = r_{xz}r_{yz} \quad \text{or} \quad .48 = (.80)(.60)$$

and $\qquad r_{yz'} = r_{yz}r_{zz'} \quad \text{or} \quad .30 = (.60)(.50)$

Since Z will of course not be measured, the only correlations that will be obtained empirically are those which fall to the right of and below the vertical and horizontal dashed lines, respectively. In this particular illustrative example, it is obvious by inspection of this submatrix that the variables X, Y, and Z' are related by a common factor Z and that $r_{xy \cdot z'}$ is numerically greater than $r_{z'y \cdot x}$. In fact the values of these two partials are .41 and .13, respectively. Thus the correlation between X and Y would be reduced only from .48 to .41 with a control for Z', whereas that between Z' and Y would be reduced from .30 to .13 with a control for X. One might then infer that alternative (*c*) rather than (*a*) of Fig. 12.3 is actually correct.

Here the correlation between Z and its measured value Z' was rather low (.50) compared with an r_{xz} of .80. Had $r_{zz'}$ also been .80, we can see by symmetry that the two partials would have been identical and there would have been no reason to prefer (*c*) over (*a*). Perhaps alternative (*b*) would have been taken as most plausible. In any given empirical instance, one may always make use of different sets of presumed correlations with Z or other variables assumed to be measured imperfectly. If

Table 12.1

	Z	X	Y	Z'
Z	—	.80	.60	.50
X	.80	—	.48	.40
Y	.60	.48	—	.30
Z'	.50	.40	.30	—

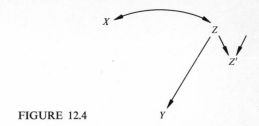

FIGURE 12.4

reasonable upper limits on the magnitudes of measurement errors can be estimated, these values can be used to appraise the compatibility of the obtained data with any given model.

The above example is obviously extreme in several respects. A correlation of .8 between X and Z may be unusually high, although with measurement errors in either or both of the variables there may be no way of accurately estimating this "true" correlation. Likewise, we are apt to have measurement errors in X as well as Z, and these may be of a comparable magnitude. Unfortunately, however, the investigator will seldom be in a position to estimate the actual magnitudes of such errors. There will be too many unknowns for a definitive result. Since a high correlation between independent variables will also lead to considerable sampling error superimposed on measurement error, it may prove almost impossible in most instances to disentangle their separate effects on a given dependent variable.

The effects of random measurement errors in one or more highly intercorrelated variables can be seen from a slightly different perspective. Suppose Y is related to two independent variables X and W according to the equation

$$Y = \alpha + \beta_1 X + \beta_2 W + \varepsilon_y$$

If there is random measurement error in X but not W, then if X' replaces X in the equation, the estimate of β_1 will be biased toward zero. But to the degree that X and W are highly intercorrelated, the estimate of β_2 will be biased *away from zero*. That is, W will receive credit for the effects of X. This principle readily generalizes to the multivariate case, with the most accurately measured variables receiving credit for the effects of those which have been measured with the greatest measurement-error variances relative to the variances in the true values. Since these measurement-error variances will ordinarily be unknown, it thus becomes exceedingly risky to attempt to separate out the individual effects of variables that are highly interrelated. Instead, one should either obtain a single score for each block of variables or make use of multiple-partial measures that treat these highly correlated variables as a block, rather than individually (see Gordon, 1968; Sullivan, 1974).

Developmental Sequences

As Hyman (1955) has pointed out, it may be very difficult in practice to distinguish between a developmental sequence, in which a background factor W causes X, which in turn causes Y, and a situation in which W is producing a spurious relationship between X and Y. In their simplest forms, these two situations can be represented as in Fig. 12.5. Notice that in both instances W is assumed to cause X, but the real

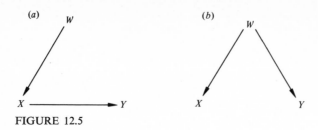

FIGURE 12.5

question is whether or not X is a direct cause of Y. If there were no measurement error in any of the variables, and if outside variables did not disturb the picture, it would be possible to choose between these two alternatives by noting what happens to $r_{xy \cdot w}$. In the case of a developmental sequence a control for W will *reduce* the partial correlation between X and Y, although not to zero. But $r_{xy \cdot w}$ will vanish almost completely if the relationship between X and Y is spurious because of W. We have already seen, however, that random measurement error in W will mean that the partial will not be zero in the model of Fig. 12.5b, nor would it vanish if another uncontrolled variable Z were also producing a spurious relationship. In short, if the value of $r_{xy \cdot w}$ is less in magnitude than r_{xy} but is not reduced to zero, it will be difficult to distinguish a developmental sequence from other alternatives.

If we examine the behavior of the partial slope $b_{yx \cdot w}$, however, we note that under the alternative of Fig. 12.5a, the expected value of $b_{yx \cdot w}$ should be exactly the same as the expected value of b_{yx}. In other words, we do not expect the estimate of the slope to be affected by a control for W under alternative (a), and this will hold true even with random measurement errors in W. This is consistent with one's research aims and with causal expectations. A control for the antecedent variable W would reduce the variation in X, thereby reducing the value of $r_{xy}{}^2$, or the proportion of the total variation in Y explained by X. But the causal law connecting X and Y remains the same. In the case of spuriousness, however, the magnitude of the slope would be reduced when W is controlled. Thus by noting whether or not the value of the slope b_{yx} changes with a control for W, we can distinguish between alternatives (a) and (b) in Fig. 12.5 even with random measurement error in W.

Now, suppose there is also random measurement error in X, the "independent" variable under investigation. Alternative (a) can now be represented as in Fig. 12.6. We have already noted that such random measurement error in the independent variable X will produce attenuation in the estimates of the slope b_{yx} as well as in the correlation r_{xy}. Furthermore, the degree of downward bias will increase with the amount of measurement error, relative to the total variation in X. When we now control for W, the expected value of the slope estimate will no longer remain constant. The control for W reduces the variation in X, relative to measurement error, and this will generally (except for sampling error) decrease $b_{yx' \cdot w}$.[1] In other words, a control

---

[1] Using the predicted relationships $r_{wy \cdot x} = r_{wx' \cdot x} = r_{x'y \cdot x} = 0$, it can be shown that except for sampling error

$$b_{yx' \cdot w} = b_{yx'} \frac{1 - r_{wx}{}^2}{1 - r_{wx}{}^2 r_{xx'}^2}$$

Hence the magnitude of $b_{yx' \cdot w}$ will be less than that of $b_{yx'}$ except when $r_{xx'}^2 = 1.0$, that is, when there is no random measurement error in X.

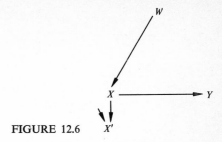

FIGURE 12.6

for W will reduce *both* the correlation and regression coefficients, and the developmental sequence can no longer be easily distinguished from other situations.

Generally speaking, the reduction in the magnitude of $r_{x'y}$ and $b_{yx'}$ will not be very pronounced in a developmental sequence, compared with situations in which X and Y are spuriously related. But the relative magnitudes of the reductions of course depend on a number of parameters, many of which will have unknown values. For example, X and Y may be spuriously related because of W, but measurement of W may be extremely poor.

Factor Analysis

Since factor analysis is basically a special type of causal analysis involving inferences about unknown factors, it is not immune to difficulties introduced by measurement error. Multiple-factor analysis usually leads to difficult problems of interpretation for reasons similar to those presently under consideration. The use of postulated unmeasured factors means that a large number of plausible alternative models can satisfy the data equally well. When we add to the unknown factors the possibility of unknown measurement errors, these difficulties become even more pronounced. We shall illustrate only in the case of a single-factor model, although the kinds of issues involved can readily be generalized.

In the single-factor model, a factor F is postulated in order to account for the intercorrelations between a number of measured variables X_i. The causal model can be diagrammed as in Fig. 12.7. Notice, first, that even with perfect measurement of the X_i, there are other models which could also account for these same relationships. For example, X_1 may cause F, which in turn causes the remaining X_i, as in Fig. 12.8. Therefore, any one of the measured variables may be an indirect cause of the remainder through F as an intervening variable. This should immediately put one on guard against assuming that F is a common cause of the X_i merely because the results of a factor analysis give a single-factor solution involving a factor that can be readily identified.

Now suppose there is random measurement error in some of the X_i. This will of course attenuate the correlation of F with those X_i which have been poorly measured. But it will not affect the inference that a single factor is needed to account for the intercorrelations. For example, if X_1 and X_2 have been measured with random measurement error, r_{12} may be small numerically but $r_{12 \cdot F}$ should be approximately zero. Thus a control for the factor should still produce a correlation matrix of residuals which are approximately zero.

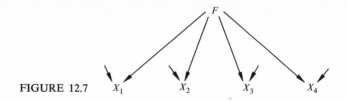

FIGURE 12.7

The question arises, however, how the factor will be named or identified. Let us imagine two research situations. In the first, X_1 and X_2 have been very poorly measured, whereas in the second the major random measurement errors are in X_3 and X_4. In the first instance, then, the factor loadings are apt to be highest on X_3 and X_4, whereas the reverse will hold in the second situation. Even though two investigators may obtain similar unknown F's, there is likely to be the psychological tendency to give a name to this common F which reflects the highest numerical loadings. For example, the investigator in the first situation may very well decide to ignore X_1 and X_2 in deciding what to call the factor.

Furthermore, we again encounter a situation in which alternative models may equally well account for the results at hand. For example, instead of F being a cause of all the X_i, suppose that X_1 caused all the remaining variables, with F dropping out of the picture. But X_1 might be measured with random measurement error. We would then be dealing with the model of Fig. 12.9. If it were possible to measure X_1 exactly, factor analysis would produce a factor with extremely high loadings on X_1. Furthermore, a control for X_1 would wipe out all the associations between the remaining X_i. But with random measurement error in X_1 this would not be the case. We would have a causal model empirically indistinguishable from that of Fig. 12.7, where F is the cause of all of the X_i including X_1.

Again, difficulties would arise in giving a name to the supposed factor in Fig. 12.9. If X_1 were amount of education, it is not too likely that the investigator would give the unknown factor that name, unless measurement errors in X_1 were minimal. Because any of the X_i could be in a position of X_1 in Fig. 12.9, one could not rule out the possibility that a given X_i is the common cause of the other variables but has been imperfectly measured. Since in one sense this would represent a simpler explanation of the pattern of intercorrelations than a theory which required one to postulate the existence of an unknown factor, some investigators might actually prefer it. Unfortunately, however, there would be no easy way of deciding which of the X_i should be singled out as the cause of all the others.

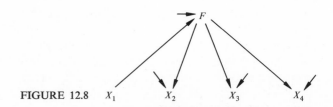

FIGURE 12.8

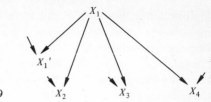

FIGURE 12.9

Inferring Interaction

We have noted that random measurement error in X produces attenuation in the slope $b_{yx'}$ in the sense that least-squares estimates will tend to underestimate the magnitude of the population regression coefficient. Furthermore, the degree of attenuation increases with the magnitude of the random measurement errors in X relative to the actual variation in X. Now suppose an investigator finds that two subsamples differ significantly with respect to values of $b_{yx'}$. One might then wish to infer an interaction or nonadditive effect on the dependent variable Y.

But one of two alternative possibilities is quite plausible. Perhaps there is more random measurement error in X in one of the subsamples than in the other. For example, responses of blacks may have been less accurate than those of whites. Or, even supposing the same measurement accuracy in the two subsamples, there may be less variation in X in one of the subsamples. Given the measurement error, this would of course have the effect of producing greater attenuation in the subsample having the small relative variation in X. In both these situations, there may actually be no interaction effect, but it will be difficult to distinguish these cases from instances where a nonadditive model is more appropriate.

12.4 GROUPING APPROACHES TO ESTIMATING RANDOM MEASUREMENT ERRORS[1]

Given the assumption that random measurement error exists, so that $\sigma_e^2 > 0$, how can the situation be handled in the data *analysis* stage of research? Of course the preferred and obvious resolution to the problem is to eliminate or minimize the sources of random error at the data *collection* stage, but even the most careful and expensive piece of research will result in some errors of measurement. Therefore it is important to consider what can and cannot be done once the data are in hand. We shall see that not all proposed resolutions actually work, and each has its disadvantages in terms of required a priori assumptions or additional data that must be collected.

Since measurement-error attenuations are a function of the relative sizes of the measurement-error variances and the variances in the true values of the independent variables, one obvious possibility is to design the study to maximize the variation in all X_i that are to be used as independent variables. To the degree that one may choose among populations with varying degrees of homogeneity with respect to these X_i, this consideration would suggest selecting very heterogeneous populations for study.

---

[1] Portions of Secs. 12.4, 12.5, and 12.6 parallel closely the discussion in H. M. Blalock, Estimating Measurement Errors in Causal Models, *Revue Internationale de Sociologie*, (2) **5**: 300–318 (September 1969).

Unfortunately, however, populations that are heterogeneous with respect to the X_i immediately under investigation are also likely to be heterogeneous with respect to unmeasured variables that contribute to the disturbance terms in the equations for the dependent, or endogenous, variables of interest. Usually it is not possible to estimate the relative magnitudes of the respective variances in advance of data collection, so that the principle of selecting populations that are heterogeneous with respect to the X_i but homogeneous with respect to unmeasured variables is not a very useful one in practice.

One way of reducing the measurement-error variances that is commonly used whenever strictly random measurement errors are assumed and whenever it may safely be assumed that measurement, per se, does not affect the true X_i values is to take repeated measurements and use the arithmetic mean of these scores as the estimate of the true scores. Thus one may weigh a block of wood five times or measure its length with calipers five times, averaging both sets of figures. One assumes in doing this that the wood is not reacting to the measurement process so that the properties involved (mass and length) remain constant during the measurement process. Furthermore, the sources of measurement error are assumed to stem from processes, e.g., the senses of sight and touch on the part of the observer, that are in no way systematically related to the properties being measured. This averaging operation reduces σ_e^2/n according to the law of large numbers without systematically affecting the true scores. The larger the number of readings being averaged, the greater the reduction in the measurement-error variance.

With human subjects several sources of difficulty make such repeated measurements less useful. First, if they are carried out consecutively over a time span that is not negligible, the property being studied may be changing as a result of natural processes, quite apart from the effects of measurement per se. Therefore it becomes difficult to distinguish real changes from differences due to measurement error. We shall return to this particular difficulty in a later section of this chapter. Second, individuals react to the process of measurement: they recall their previous responses, they lose motivation, or they attempt to appear overly consistent. Third, the linkage between the research operations and the variable of theoretical interest is often sufficiently indirect to make a strictly random measurement-error model unrealistic. For all these reasons, investigators often prefer to substitute multiple measures involving somewhat *different* operations for repeated measures involving basically the same operations. This of course opens up the added possibility that the different measures may not all be equivalent in the sense that some may involve systematic errors. Even if this possibility is ruled out, the error variances will not all be the same. In effect, the use of diverse measures introduces additional unknowns into the system, though perhaps guarding against certain other kinds of systematic distortions.

It must be anticipated, therefore, that analysis procedures involving multiple indicators will be relatively complex and also require additional a priori assumptions, only a few of which can be empirically tested. If there are multiple measures X_i, all related to X (the common factor) by equations of the simple type $X_i = X + e_i$, where the e_i are strictly random according to Eqs. (12.1), then combining these into a single index will reduce the relative magnitude of the measurement-error variance. But if the conditions (12.1) are *not* met, there may be systematic distortions produced by this averaging process.

What can be done if only a single measure is available for each variable? One obvious possibility is to make an a priori assumption about either the absolute magnitude of σ_e^2 or the ratio σ_e^2/σ_x^2. When random measurement errors are neglected altogether, the implicit assumption is that $\sigma_e^2 = 0$. But if we consider this latter assumption unrealistic, can we replace it by some specific alternative value? Of course if previous research involving repeated measurements has provided us with such an estimate, we can utilize it to good advantage. However, this implies a much higher degree of standardization of measurement and coordination of research efforts than is yet evident in either sociology or political science. A somewhat more realistic possibility is to make several estimates of the measurement-error variance, indicating to the reader the implications of each value for the subsequent analysis. In the absence of any other data such a practice is certainly to be recommended over the present one of ignoring measurement errors altogether. The difficulty is that in multivariate analyses there will be so many different combinations of possible measurement-error estimates for each variable that the results are likely to depend very heavily on the particular choice of estimates used. If so, the reader would at least be sensitized to ambiguities in the data that would otherwise be ignored.

One very simple and intuitively appealing procedure has been suggested in the statistical literature. The Wald-Bartlett grouping procedure, discussed below, relies on the notion that one may form averages of scores for individuals who appear to be similar with respect to their X scores, thereby substituting averages of *similar* individuals for repeated measurements on the *same* individual. However, the procedure breaks down for a reason that is highly instructive, namely, the difficulty of locating individuals who are similar with respect to true scores when we know only their measured scores. In brief, it has been proposed by Wald (1940), Bartlett (1949), and others that if we can find a method of grouping together cases according to their true scores on the independent variable, computing the mean X and Y scores for each group, we can estimate β_{yx} by the equation

$$\tilde{b}_{yx} = \frac{\overline{Y}_2 - \overline{Y}_1}{\overline{X}_2 - \overline{X}_1}$$

where $\overline{X}_1$ and $\overline{Y}_1$ are the respective means for group 1 and $\overline{X}_2$ and $\overline{Y}_2$ are the means for group 2. Wald originally proposed subdividing the total sample into two groups by ranking cases according to the measured X' value and placing all cases with X' scores below the median in group 1 and the remainder in group 2. Bartlett altered this procedure by omitting cases with intermediate values of X', using only the more extreme scores. Additional modifications and refinements to the procedure need not concern us here.

One of the basic assumptions of these procedures is that the method be independent of the measurement errors in X. This will of course be violated since it is the X' values, instead of the true X scores, which must be used in the grouping operation. The practical value of the procedure thus depends on its robustness under departures from this assumption. Using a number of data sets, with several parameter values and differing underlying population distributions (normal, rectangular, and highly skewed), we have found that the Wald-Bartlett estimates behave almost the same as the least-squares estimates (see Carter and Blalock, 1970).

For instance, for one set of data where the true value of $\beta_{yx} = 1.0$ for which σ_e^2 was substantial relative to σ_x^2, the least-squares estimates over 10 replications (with $n = 300$) had a mean of .257, whereas four different versions of Wald-Bartlett estimators had means of .275, .269, .260, and .275. Thus for these simulated data, which involved underlying rectangular distributions, the Wald-Bartlett estimates turned out to be only slightly superior to the least-squares estimates. For normal populations the mean values of least-squares estimates, using 100 replications of samples of size 60, were almost identical to the Wald-Bartlett estimates, though for highly skewed populations the least-squares estimates were slightly superior. In all cases studied the biases were estimated to be of the same order of magnitude.

In retrospect, it does not seem possible that we can correct for unknown random measurement errors by any simple statistical manipulations of this kind, although perhaps some such device may ultimately be discovered. The remaining approaches to be discussed in this chapter rely heavily on a priori assumptions about the causal structure of the model. It appears these assumptions can be tested only indirectly by utilizing multiple estimates and applying consistency criteria to evaluate their adequacy.

12.5 THE INSTRUMENTAL-VARIABLES APPROACH

In the econometrics literature we find discussions of an instrumental-variables approach used in parameter estimation in simultaneous-equation systems. This approach may also be used to obtain consistent estimators of β that are approximately unbiased in large samples. The basic idea can be very simply illustrated for the two-variable model and can readily be generalized to more complex models involving several independent variables. Consider the equation

$$Y = \alpha + \beta X + \varepsilon_y$$

If we can find additional variables Z_i that are direct or indirect causes of X but that do not appear in the equation for Y, we can use these Z_i as *instrumental variables* by taking ratios of covariances between these variables and Y and X. In particular, the estimator

$$b_{yx'}^* = \frac{\sum yz_i}{\sum x'z_i}$$

where $x' = X' - \bar{X}'$, $y = Y - \bar{Y}$, and $z_i = Z_i - \bar{Z}_i$, will be a consistent estimator under the usual required assumptions for ordinary least squares *provided* that Z_i is uncorrelated with ε_y. In other words this implies that the exogenous variables that are being used as instrumental variables Z_i must not be connected to Y by any path aside from that through X.

We have also contrasted the relative adequacy of the instrumental-variables and least-squares approaches using simulated data, and it is possible to summarize these results briefly (see Blalock, Wells, and Carter, 1970). Where the required assumptions were actually met, the instrumental-variables estimates performed much better in the case of samples of size 500. Using four sets of replications at each of three levels of random measurement error in X, we found that the means of the ordinary least-squares estimates of $\beta = 2.0$ attenuated to 1.31, .34, and .04 as the ratio of σ_e^2/σ_x^2 was increased, but the means of four replications each for the corresponding instrumental-variables estimates were 2.07, 2.19, and 2.37.

If the instrumental-variables approach were also superior in situations where the required assumptions were not met, one would be very fortunate. If there is actually specification error in the equations, as well as measurement error, then in general neither ordinary least squares nor the instrumental-variables procedure will provide unbiased or consistent estimates. The question then arises which approach one would generally prefer. We have studied only one important kind of specification error, namely, the case where one or more of the Z_i in fact belong in the equation for Y contrary to assumption. To be specific consider the following equations

$$Y = \alpha_1 + \beta_1 X + \gamma_1 Z + \varepsilon_y \quad \text{and} \quad X = \alpha_2 + \gamma_2 Z + \varepsilon_x$$

where ε_x and ε_y are random disturbances that are uncorrelated with the independent variables in their respective equations. However, suppose that in effect we have omitted Z from the first equation by erroneously assuming that $\gamma_1 = 0$. Since Z is a cause of X, if it has been incorrectly omitted from the first equation, this will create a correlation between X and the revised disturbance term ε_y', which now contains the effects of Z. This produces a bias in the ordinary least-squares estimate of β_1.

It can be shown that the approximate large-sample biases in the least-squares and instrumental-variables estimators will be:

$$\text{Ordinary least-squares bias} = \frac{\gamma_1}{\gamma_2}\left(1 - \frac{\sigma_{\varepsilon_x}^2}{\sigma_x^2}\right)$$

$$\text{Instrumental-variables bias} = \frac{\gamma_1}{\gamma_2}$$

Thus in general there will be a greater systematic bias in the instrumental-variables estimator whenever in fact no measurement error is superimposed on the specification error. Put differently, if we suspect that specification error, estimated by the ratio γ_1/γ_2, is large compared to the measurement error in X, we would prefer ordinary least squares. The ratio of the relative biases in the two estimates depends, however, on the factor $1 - \sigma_{\varepsilon_x}^2/\sigma_x^2$. Notice that if Z is an important cause of the variation in X, this factor will be almost unity and the ordinary least-squares bias will then be approximately the same as that for the instrumental-variables estimator. But if the correlation between Z and X is very weak, this factor may be considerably less than unity. In other words, it is to our advantage to have instrumental variables that are important causes of X but that have negligible direct effects on Y.

Therefore in any particular situation one will be faced with the problem of making a priori assumptions concerning the relative magnitudes of measurement and specification errors, a difficult task indeed. If one lacks an adequate theory—almost inevitable in sociology—one should attempt to find several different instrumental variables that are not too highly interrelated. If these all yield approximately the same estimates, which differ from those obtained from ordinary least squares, then one has more assurance that there is attenuation in the latter due to random measurement error. One may also wish to combine this use of multiple instrumental variables with the multiple-indicator approach discussed in Sec. 12.6.

Example   Suppose the correlations (or path coefficients) between Z_1, X, X', and Y are as indicated below and that the model holds true exactly (without sampling errors):

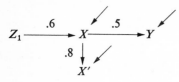

If we take $X' = X + e$ and make the assumptions necessary for the measurement error in X to be strictly random, the following set of (hypothetical) numerical values will exactly produce the above correlations, as can easily be verified:

$$\sum z_1^2 = 900 \qquad \sum z_1 x = 360 \qquad \sum xx' = 400$$
$$\sum x^2 = 400 \qquad \sum z_1 x' = 360 \qquad \sum z_1 y = 450$$
$$\sum y^2 = 2{,}500 \qquad \sum xy = 500 \qquad \sum x'y = 500$$
$$\sum x'^2 = 625$$

In addition, we see that $r_{z_1 x'} = .48$, $r_{x'y} = .40$, and $r_{z_1 y} = .30$, as predicted for this very simple model.

If we had the true X values available, we could readily calculate the ordinary least-squares (OLS) value for b_{yx} as $500/400 = 1.25$. We might have also used the true X's to obtain the instrumental-variable estimate $b_{yx}^* = \sum z_1 y / \sum z_1 x = 450/360 = 1.25$. This latter estimate would of course have a larger standard error than b_{yx} and would therefore never be preferred over the OLS estimator if the true X scores were known. But now suppose the true X values are not available and we are forced to use the measured X' scores instead. In this case our OLS estimator becomes $b_{yx'} = 500/625 = .80$, which involves a substantial bias in spite of the reasonably high correlation between X and X'. However, had we used the instrumental-variable estimator $b_{yx'}^* = \sum z_1 y / \sum z_1 x' = 450/360 = 1.25$, we would have obtained the correct slope estimate in this simple case of a perfect fit between model and data.

If we had been able to find a second instrumental variable Z_2, again exactly satisfying the model but with a smaller variance and weaker correlation with X, we might have obtained the following results:

$$\sum z_2^2 = 400 \qquad r_{z_2 x} = .40$$
$$\sum z_2 x = 160 \qquad r_{z_2 y} = .20$$
$$\sum z_2 y = 200 \qquad r_{z_2 x'} = .32$$
$$\sum z_2 x' = 160$$

Here our instrumental-variable estimator would be $b_{yx'}^* = \sum z_2 y / \sum z_2 x' = 200/160 = 1.25$, which is again the correct value. But although this ratio is correct and is equal to the value of the OLS estimator based on the true value of X, its standard error is larger than that of the estimator based on the instrumental variable Z_1 because of the weaker correlations between Z_2 and the remaining variables, compared with those involving Z_1.

This numerical example illustrates the possibility of finding multiple instrumental variables. Of course in actuality, even if the model were correctly specified, the estimates based on these instrumental variables would never be exactly the same. This suggests the desirability of taking some kind of average of the results, the simplest of which would be their arithmetic mean. Hauser and Goldberger (1971) and Goldberger (1973), however, show that maximum efficiency can be achieved by weighting the estimates so that those having the smallest standard errors receive the greatest weights. In the above example, the estimate based on Z_1 would receive a greater weight than that based on Z_2. The reader is referred to the sources cited for details and further illustrations.

It seems advisable to inject a note of caution in connection with such procedures that stress the maximization of efficiency. It is very possible that there may be specification errors in the model, in which case we would prefer to rely most heavily on estimates derived from models in which we have the most faith. In our hypothetical illustration, for example, we may strongly suspect that Z_1 affects Y by some path that does not go through X, whereas this is thought to be much less likely for Z_2. In other words, we may have less faith in the adequacy of the simple model for the instrumental variable Z_1 than for Z_2. Procedures that stress maximizing efficiency or minimizing standard errors typically presume that specification errors are negligible, and they do not allow for varying degrees of faith in the adequacy of one's models. *If* one had no reason to prefer one instrumental variable over another in terms of the likelihood of specification errors, it makes sense to use the Hauser-Goldberger procedure. But in many practical instances it seems highly likely that certain instrumental variables will be more suspect than others, in which case it makes sense to give more weight to results based on the preferred Z_i. One must also keep in mind the possibility that some of these Z_i may be measured with systematic errors that are correlated with other variables in the model. The general point, of course, is that rarely can one safely use simple rules of thumb or statistical formulas without carefully considering whether other factors may invalidate their use.

12.6 THE MULTIPLE-INDICATOR APPROACH

Work with repeated measures or composite indices is extremely common in the social sciences, and considerable systematic knowledge of the subject has accumulated in both psychology and economics. Yet surprisingly little of this systematic knowledge, beyond the use of factor analysis and related techniques, has diffused into the fields of sociology and political science. In these fields, as well as in the remaining social sciences, measurement decisions involving the combination of indicators have for the most part been ad hoc in nature. As a result, there is considerable conceptual confusion and a definite lack of clear-cut guidelines for handling multiple indicators. The extensive though technical literature within econometrics on aggregation has been virtually ignored by sociologists until recently. Factor analysis has been utilized to good advantage in social psychological research but has also been applied uncritically in many measurement situations for which the underlying model has been inappropriate. Fortunately, the growing sociological literature on multiple-indicator approaches is beginning to correct for some of these deficiencies, though at the same time pointing up many of the complications that arise whenever measurement is highly indirect or measurement errors substantial.

Basically, the multiple-indicator approach relies on the fact that whenever there are several measures of each variable, there may be enough intercorrelations between these measured variables to yield overidentified systems and thus a number of rejectable empirical predictions. The measures of each variable are kept distinct, rather than being combined into a single index, until the predictions of the measurement model have been tested. Estimates of the true values of the variables, and their intercorrelations, are based on these indicators once the predictions of the model have been tested. In effect, the use of multiple indicators provides a kind of insurance against both random and nonrandom measurement errors of various kinds. But the price one pays is that of greater complexity of the model, as well as the necessity of making explicit assumptions concerning the linkages of the indicators with the unmeasured variables and, possibly, with each other. We shall see in the next chapter that whenever the model allows for nonrandom measurement errors, the number of unknowns may increase to a point where the system becomes underidentified. In general, the greater the complexity and realism we permit in the model, the more measured indicators we need relative to the total number of variables in the system.

Let us begin with a very simple two-variable model, as formulated in path-analysis terminology by Costner (1969). Consider the model of Fig. 12.10, in which X and Y (without subscripts) are taken as the underlying variables whose true relationship is to be estimated, the measured values of these variables being indicated by subscripts rather than primes. Hereafter, we shall use the visual device of representing the measurement-error terms by short arrows. The lack of connection between these residual arrows and the remaining variables in the system indicates that we are taking the measurement errors to be strictly random, so that $E(e_i) = 0$ for all indicators. The covariances of all measurement-error terms with each other, with all true values, and with the remaining indicator variables are also all assumed to be zero. In the next chapter we shall selectively relax certain of these assumptions in order to allow for particular kinds of nonrandom errors.

Following Costner, we shall make use of the convenient path analysis notation, though our argument could also be presented in terms of correlation coefficients. In fact, in the very simple model of Fig. 12.10 each path coefficient is a total correlation. These path coefficients will be labeled $a, b, c, \ldots, k$, rather than as p_{ij}, in order to avoid cumbersome subscripting notation. If we are dealing with strictly random measurement errors, each path coefficient that connects an unmeasured variable to one of its indicators will be an unmeasured correlation coefficient, but this will not always be true of the path coefficients connecting the several unmeasured variables.

Since X and Y are both unmeasured and all the path coefficients connect some variable to either X or Y, none of the path coefficients can be directly computed from the data. But we will be able to obtain six correlation coefficients involving the X_i and Y_j. We will have four cross correlations using the pairs (X_1, Y_1), (X_1, Y_2), (X_2, Y_1), and (X_2, Y_2) plus two correlations involving pairs of indicators of the same variable

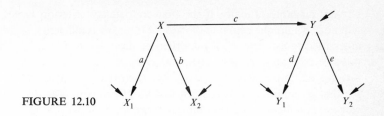

FIGURE 12.10

(X_1,X_2) and (Y_1,Y_2). These six correlations may be used to estimate the path co-efficients and to test for the consistency between the data and the model, this test being possible because of the excess of empirical information (the six r's) over un-known parameters (the five unknown path coefficients). We obtain the following equations for the six correlations:

$$r_{x_1x_2} = ab \qquad r_{x_1y_2} = ace$$

$$r_{y_1y_2} = de \qquad r_{x_2y_1} = bcd \qquad (12.3)$$

$$r_{x_1y_1} = acd \qquad r_{x_2y_2} = bce$$

We now note that if we pair certain of the cross correlations and multiply them together, the results should be equal (except for sampling error). Thus

$$r_{x_1y_1}r_{x_2y_2} = r_{x_1y_2}r_{x_2y_1} = abc^2de \qquad (12.4)$$

This result provides an empirical test of the model, since not all data will satisfy this condition. Following Costner, we shall refer to Eq. (12.4) as the *first consistency criterion*. If the data in fact satisfy the criterion to within some predetermined degree of accuracy (say, the right-hand side is within $\pm.05$ of the left-hand side), then we may use either of these products to estimate the true correlation between X and Y as follows:

$$c^2 = \frac{abc^2de}{abde} = \frac{r_{x_1y_1}r_{x_2y_2}}{r_{x_1x_2}r_{y_1y_2}} = \frac{r_{x_1y_2}r_{x_2y_1}}{r_{x_1x_2}r_{y_1y_2}} \qquad (12.5)$$

The remaining path coefficients, or correlations, can also be estimated by similar operations. Thus there will be two estimates each of a^2, b^2, d^2, and e^2, one of which is

$$a^2 = \frac{a^2bce}{bce} = \frac{r_{x_1x_2}r_{x_1y_2}}{r_{x_2y_2}} \qquad d^2 = \frac{acd^2e}{ace} = \frac{r_{x_1y_1}r_{y_1y_2}}{r_{x_1y_2}}$$

$$(12.6)$$

$$b^2 = \frac{ab^2ce}{ace} = \frac{r_{x_1x_2}r_{x_2y_2}}{r_{x_1y_2}} \qquad e^2 = \frac{acde^2}{acd} = \frac{r_{x_1y_2}r_{y_1y_2}}{r_{x_1y_1}}$$

Several cautions deserve careful notice. First, the fact that we are using ratios of products of correlations means that unless one has very large samples, resulting sampling errors are likely to be substantial. Also, if the correlations obtained are very weak, these products will be close to zero. In particular, if either correlation in the denominator of the expression for c^2 is close to zero, at least one term in the numerator will also be small (assuming the correctness of the model) and the resulting fraction will be very unstable. Finally, if the first consistency criterion is not met exactly, there will not be a unique estimate of each of the path coefficients. As with the two numerators used to estimate c^2, it may then be desirable to form an average of the various possible estimates. In this particular example there will be only two estimates of c^2, but in situations where there are more than two indicators of each variable we shall see that a number of different estimators of c^2 (as well as the remaining coefficients) can be obtained. A very simple procedure that will work satisfactorily in many instances is to form an arithmetic mean of the several estimates, though Hauser and Goldberger (1971) have shown that a weighted average that favors indicators having the smallest error variances is maximally efficient. Jöreskog (1970) has developed a very general confirmatory factor-analysis procedure that provides maximum-likelihood estimators for such overidentified systems.

Now suppose we have three measures of both X and Y, as in Fig. 12.11. If strictly random errors can be assumed for all indicators, we see that we have markedly increased the number of empirical correlations to $(6)(5)/2 = 15$, while increasing the number of unknown path coefficients only to 7. Therefore we in effect have eight excess equations that may be used to test the adequacy of the model. The two X indicators (X_1, X_2) may be paired with any of the three Y-indicator combinations (Y_1, Y_2), (Y_1, Y_3), and (Y_2, Y_3), providing three consistency criteria. But there will also be three additional criteria involving the X-indicator combination (X_1, X_3) and three more involving the pair (X_2, X_3). Not all of these criterion equations will be independent, but it is clear that we shall have multiple tests utilizing this first kind of consistency criterion. When we consider nonrandom measurement errors, we shall see that certain kinds of nonrandom errors will produce results that satisfy some of these equations but not others, so that a careful examination of those equations which are and are not satisfied may enable one to infer possible sources of nonrandom measurement error. Unfortunately, however, there are a number of sources of nonrandom measurement error that also imply these very same consistency criteria; i.e., there are alternatives to the strictly random measurement-error model that cannot be distinguished empirically from it on the basis of this first type of consistency criterion, which is basically the tetrad-differences approach familiar to students of factor analysis.

Costner therefore proposes a *second* consistency criterion that can be used whenever there are more than two indicators of either X or Y. Basically, this second criterion involves estimating the true correlation between X and Y in several different ways. If the random-error model is correct, these alternative estimates of c^2 should all be approximately equal. We have two estimates of c^2 from the pairs (X_1, X_2) and (Y_1, Y_2), but there are eight additional pairings of X and Y indicators, as we have already noted. With so many possible estimates we naturally expect that sampling error alone will produce certain differences, especially since each estimate involves a ratio of products of correlations. Perhaps a series of simulation studies will provide useful guidelines to the order of magnitude of differences between estimators that can be attributed to sampling error alone under the random-error model. At present we cannot provide any really definitive guidelines or rules of thumb, though the interested reader is referred to Hauser and Goldberger (1971) for a more technical discussion and set of references. In the following chapter on nonrandom error we shall illustrate the application of this second test criterion to some relatively simple models.

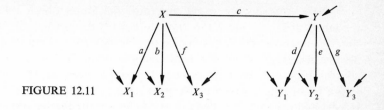

FIGURE 12.11

Example Let us suppose that Y represents a form of behavior and X an internal state. To be specific let Y be the true rate of delinquency for a set of individuals, as measured by the two indicators Y_1 = self-reported rate and Y_2 = officially recorded rate. We of course recognize that both these measured values will contain nonrandom sources of error, but for the present chapter we shall suppose that the errors are strictly random. Let X represent the true level of self-respect, which is imperfectly measured by X_1, a self-reported rate, and X_2, a score obtained from a projective test of some kind. Again we recognize that there may be a number of biasing factors at work and in particular a tendency for the two self-report measures (X_1 and Y_1) to be spuriously correlated as a result of the operation of unmeasured personality variables or perhaps a degree of distrust of the interviewer. Nevertheless, we shall presume that the model of Fig. 12.10 holds exactly and without even any sampling errors. If the correlation between the true X and Y scores were .5, if the correlations between X and X_1 and X_2 were .8 and .7, respectively, and if the correlations between Y and Y_1 and Y_2 were .9 and .6, respectively, we would then obtain the following set of correlations between the measured variables:

$$r_{x_1 x_2} = (.8)(.7) = .56 \qquad r_{x_1 y_2} = (.8)(.5)(.6) = .24$$

$$r_{y_1 y_2} = (.9)(.6) = .54 \qquad r_{x_2 y_1} = (.7)(.5)(.9) = .315$$

$$r_{x_1 y_1} = (.8)(.5)(.9) = .36 \qquad r_{x_2 y_2} = (.7)(.5)(.6) = .21$$

Applying the first consistency criterion, we would get

$$r_{x_1 y_1} r_{x_2 y_2} = r_{x_2 y_1} r_{x_1 y_2} \qquad \text{or} \qquad (.36)(.21) = (.24)(.315) = .0756$$

which of course holds exactly in this perfect-fitting hypothetical example. We could then estimate c^2, or the true correlation between X and Y, as follows:

$$c^2 = \frac{r_{x_1 y_1} r_{x_2 y_2}}{r_{x_1 x_2} r_{y_1 y_2}} = \frac{.0756}{(.56)(.54)} = .25$$

Likewise our estimates of a^2, b^2, d^2, and e^2 would also yield back the "correct" values, no matter which combinations of indicators we used. For example we could estimate a^2 by either

$$a^2 = \frac{r_{x_1 x_2} r_{x_1 y_2}}{r_{x_2 y_2}} = \frac{(.56)(.24)}{.21} = .64$$

or

$$a^2 = \frac{r_{x_1 x_2} r_{x_1 y_1}}{r_{x_2 y_1}} = \frac{(.56)(.36)}{.315} = .64$$

which are of course identical since we are assuming away any possible sampling errors.

Now let us suppose we can find third indicators of both X and Y in accord with the model of Fig. 12.11. As we try to add really distinct indicators, we may expect to find it increasingly difficult to locate further indicators about which we are willing to assume relatively simplified measurement-error models, in this case strictly random sources of measurement errors. Let us imagine, however, that we have been able to find a set of judges who are able to score each individual with respect to both self-respect and delinquency. Perhaps the judges in the former instance are parents and teachers and in the second instance a group of peers. Let these additional indicators be X_3 and Y_3, respectively, and the corresponding paths to the true values be f and g, both having the numerical value of .7. Of course, in a real research situation we would not have available any of these correlations with the true values but only the set of intercorrelations between the indicators. If the model of Fig. 12.11 held exactly, without sampling error, the nine additional correlations between the indicators would be

$$r_{x_1x_3} = (.8)(.7) = .56 \qquad r_{x_3y_1} = (.7)(.5)(.9) = .315$$

$$r_{x_2x_3} = (.7)(.7) = .49 \qquad r_{x_3y_2} = (.7)(.5)(.6) = .21$$

$$r_{y_1y_3} = (.9)(.7) = .63 \qquad r_{x_3y_3} = (.7)(.5)(.7) = .245$$

$$r_{y_2y_3} = (.6)(.7) = .42 \qquad r_{x_1y_3} = (.8)(.5)(.7) = .28$$

$$r_{x_2y_3} = (.7)(.5)(.7) = .245$$

We immediately see the richness of the model in terms of the excess of empirical correlations over unknown path coefficients (here, also correlations). This provides a large number of consistency checks of two types. First, we can use the nine pairings of two X and two Y indicators to form products of correlations to check the first consistency criterion. Three of these pairings are

$$r_{x_1y_1}r_{x_2y_2} = r_{x_1y_2}r_{x_2y_1} \qquad (.36)(.21) = (.24)(.315) = .0756$$

$$r_{x_1y_1}r_{x_2y_3} = r_{x_1y_3}r_{x_2y_1} \qquad (.36)(.245) = (.28)(.315) = .0882$$

$$r_{x_1y_2}r_{x_3y_3} = r_{x_1y_3}r_{x_3y_2} \qquad (.24)(.245) = (.28)(.21) = .0588$$

If the first consistency criterion is considered satisfied, one may then proceed to estimate the several "unknown" path coefficients by a number of different means. As we have already noted, and as we shall see in the next chapter, it is possible for the first consistency criterion to be satisfied in a number of different kinds of models involving nonrandom measurement errors. But the second criterion—involving the comparison of the estimates of c^2—usually will not be satisfied in such instances. In other words, in the presence of nonrandom errors we can expect that the several estimates of c^2 will differ substantially.

We again encounter the problem of evaluating the goodness of fit of a model that is highly overidentified. We also may wish to obtain the most efficient estimators of each parameter by means of a weighting of the separate estimates. Once more we refer the reader to Hauser and Goldberger (1971) and to Goldberger (1973), but with the same kinds of reservations as before. If one has more or less equal faith in all the indicators, in terms of the correctness of the model as specified, then it makes good sense to use a single test criterion to evaluate the overall fit and then to obtain estimators with maximum efficiency. For example, in a study of performance levels of children one might have three separate tests of performance, all deemed equally valid on a priori grounds. If three measures of X could also be found that were all equally plausible on substantive grounds, then a single overall goodness-of-fit test would make very good sense.

But in much sociological and political science research our measures are likely to be based on diverse research operations or to tap different "aspects" of a poorly understood variable such as industrialization, political alienation, antiminority prejudice, or deviant behavior. In these instances an investigator may have strong a priori reasons for preferring one indicator over the others and may merely be using the others for the purpose of making consistency checks. In situations like this it may be wise to proceed in an exploratory fashion, indicator by indicator, rather than using a single overall consistency check. This latter approach is more in the spirit of suggestions by Sullivan (1974) and Costner and Schoenberg (1973). We shall explore these avenues further in the next chapter when we consider specific kinds of nonrandom measurement-error models.

A very new approach developed by Wold and his colleagues (Wold, 1975) appears to be intermediate between these two strategies, the first of which presumes that the model is correctly specified and which stresses efficiency, and the second of which is far more exploratory. The Wold procedure, NIPALS, is a *n*onlinear *i*terative *pa*rtial *l*east-*s*quares procedure involving the successive applications of ordinary least squares and requiring fewer assumptions about the covariances of the disturbance terms and exogenous variables. In situations where the model has been correctly specified and the assumptions required for maximum likelihood methods met, NIPALS turns out to be less efficient than the type of approach suggested by Hauser and Goldberger (1971), but if certain of these assumptions are not met, it appears that the NIPALS estimates may be superior. At present, the question of the most appropriate strategy of analysis in highly overidentified systems (with and without measurement errors) is very much an open one, so that it seems wise to apply several different approaches and compare the results.

The above numerical examples vividly illustrate the point that products of correlations that are only moderate to begin with are likely to be very small numerically. This merely emphasizes the importance of the remark made earlier that sampling and specification errors are likely to be crucial. It might be argued that the correlation between the "true" values X and Y was taken to be too small in these examples and that this accounts for many of the numerically low coefficients. But if reality is complex, and if multiple causation prevails, we are likely to discover that our empirical correlations are not much higher than this. If this is the kind of reality that actually exists, we are going to be in difficulty unless we can obtain extremely large samples and much better indicators of the very large numbers of variables that will be needed in our explanatory models. Suppose, for example, we were dealing not with a single X but with five intercorrelated independent variables, each of which is rather poorly measured by four or five indicators. It would indeed be difficult to use the multiple-indicator approach in such situations. One or two high-quality measures of each variable would be much more useful.

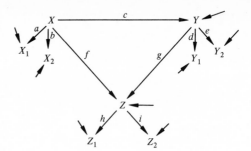

FIGURE 12.12

12.7 EXTENSION TO THE MULTIVARIATE CASE[1]

The above approach can easily be extended to k-equation recursive systems, though the nonrecursive case appears to be much more complicated. Let us first consider the three-variable model represented in Fig. 12.12. For simplicity, we shall confine our attention to situations where there are exactly two indicators of each variable since additional indicators will merely provide additional test equations. We have noted that it is an important feature of recursive systems that any variable clearly dependent on two (or more) variables may be ignored in studying the relationship between these variables. Therefore Z may be ignored in studying the relationship between X and Y, and thus the situation involving these two variables is exactly the same (including notation) as in Fig. 12.10. We can therefore confine our attention to the relationships between X and Z and between Y and Z.

Looking first at the relationship between X and Z, we can write the correlation between X_1 and Z_1 as the sum of two separate paths, the first being afh and the second being the indirect path through Y, namely, $acgh$. Since a and h (the two measurement-error paths) are common to both these longer paths, we can factor them out, obtaining the expression $ah(f + cg)$. If we do the same for the other combinations of X_i and Z_i, we get

$$
\begin{array}{ll}
r_{x_1 x_2} = ab & r_{x_1 z_2} = ai(f + cg) \\
r_{z_1 z_2} = hi & r_{x_2 z_1} = bh(f + cg) \qquad (12.7) \\
r_{x_1 z_1} = ah(f + cg) & r_{x_2 z_2} = bi(f + cg)
\end{array}
$$

The compound path represented by $f + cg$ appears in each of the pairings involving an X_i and Z_i and can be shown by the method of path analysis to equal r_{xz}, which, although not directly given, can be estimated in a straightforward manner. We form the products $r_{x_1 z_1} r_{x_2 z_2}$ and $r_{x_1 z_2} r_{x_2 z_1}$ and note that they should both equal $abhi(f + cg)^2 = abhir_{xz}^2$. This predicted relationship again provides a test of the random-error assumption, and we can estimate r_{xz}^2 by dividing either of these products or their (weighted) average by

$$
r_{x_1 x_2} r_{z_1 z_2} = abhi
$$

---

[1] Portions of this section have been taken, with modifications, from H. M. Blalock, Multiple Indicators and the Causal Approach to Measurement Error, *American Journal of Sociology*, **75**: 264–272 (September 1969), by permission of The University of Chicago Press.

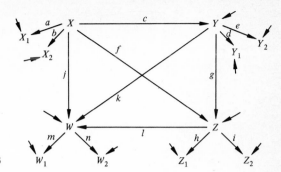

FIGURE 12.13

Turning to the relationship between Y and Z, we see that a similar result holds. If we correlate Y_1 with Z_1, the result can be expressed as the sum of the two paths dgh and $dcfh$, the latter indirect path involving tracing backward from Y_1 to Y to X and then forward to Z and Z_1. This compound path can also be factored into $dh(g + cf)$ and similarly for the remaining combinations. We thus get

$$r_{y_1y_2} = de \qquad\qquad r_{y_1z_2} = di(g + cf) = dir_{yz}$$

$$r_{z_1z_2} = hi \qquad\qquad r_{y_2z_1} = eh(g + cf) = ehr_{yz} \qquad (12.8)$$

$$r_{y_1z_1} = dh(g + cf) = dhr_{yz} \qquad r_{y_2z_2} = ei(g + cf) + eir_{yz}$$

We see that again we can obtain both a test for randomness of the measurement error and an estimate of r_{yz}^2. All the various measurement-error paths can likewise be estimated.

The extension to additional variables is straightforward, as can be illustrated with the four-variable model of Fig. 12.13. Since W is dependent on each of the other variables, we may ignore it in estimating all the paths involving X, Y, and Z. We then focus on the relationships of each of these variables with W. We can see how the procedure works by confining our attention to X and W, since the pairings of Y and Z with W involve similar processes of tracing paths. There are four paths connecting X_1 and W_1, one being via the direct link between X and W, two being through a single intervening variable, and the last involving the three-step path from X to Y, from Y to Z, and from Z to W. Thus

$$r_{x_1w_1} = am(j + ck + fl + cgl) = amr_{xw} \qquad (12.9)$$

and similarly $\qquad r_{x_1w_2} = anr_{xw} \qquad r_{x_2w_1} = bmr_{xw} \qquad r_{x_2w_2} = bnr_{xw} \qquad (12.10)$

and therefore $\qquad\qquad r_{x_1w_1}r_{x_2w_2} = r_{x_1w_2}r_{x_2w_1} = abmnr_{xw}^2$

This procedure can obviously be extended to any number of recursively related variables.

The Use of Single Measures

Every variable in the models we have thus far considered has been measured by at least two indicators. Are there any conditions under which a single measure can be used to represent one (or more) of the variables? We shall see that in very special

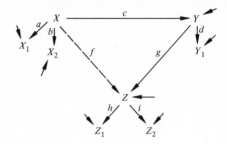

FIGURE 12.14

circumstances a single estimator can be used, provided one is willing to make rather strong a priori assumptions about the model. Let us consider the model of Fig. 12.14 in which the intervening variable Y has been measured by the single indicator Y_1. The remaining paths have been designated as in Fig. 12.12, with the dashed arrow between X and Z indicating that we shall want to consider the special case where $f = 0$, that is, where there is no direct link between X and Z. Since Y_2 and e have been removed from the model of Fig. 12.12, we are more restricted in the relationships we can use. However, in considering the relationship between X and Y we have

$$r_{x_1 y_1} r_{x_2 y_1} = ab(cd)^2 = r_{x_1 x_2}(cd)^2$$

and therefore

$$(cd)^2 = \frac{r_{x_1 y_1} r_{x_2 y_1}}{r_{x_1 x_2}}$$

From this relationship alone we cannot disentangle c and d, the total path connecting X to Y_1. But we can combine this result with a similar one obtained in relating Y_1 to the two indicators of Z. We have

$$r_{y_1 z_1} r_{y_1 z_2} = hi(dg)^2 = r_{z_1 z_2}(dg)^2$$

and therefore

$$(dg)^2 = \frac{r_{y_1 z_1} r_{y_1 z_2}}{r_{z_1 z_2}}$$

We now eliminate d, the coefficient for the measurement error in Y, by dividing one of these expressions by the other. We thus obtain an expression for the ratio c to g:

$$\left(\frac{c}{g}\right)^2 = \frac{r_{x_1 y_1} r_{x_2 y_1} r_{z_1 z_2}}{r_{x_1 x_2} r_{y_1 z_1} r_{y_1 z_2}}$$

Next consider the paths between X and Z as represented by the compound expression $f + cg$. We see that if there were no direct link between X and Z, so that $f = 0$, we would be in a position to obtain an expression for the product cg, which together with the above expression for their ratio, would make it possible to solve for either of them separately. If in fact $f = 0$, it would be possible to treat Y as a completely unmeasured-intervening variable, with the result that

$$(cg)^2 = \frac{r_{x_1 z_1} r_{x_2 z_2}}{r_{x_1 x_2} r_{z_1 z_2}}$$

This expression can be obtained by merely replacing Z for Y in the model of Fig. 12.10. Had the direct link between X and Z not been zero, it would have been necessary to use the compound path $f + cg$ instead of the simple product cg. This would have meant that the presence of f in the expression would have made it impossible to solve for either c or g. With f set equal to zero, however, we can take the product of the (squared) ratio of c to g and that of their (squared) product, obtaining

$$c^4 = \left(\frac{c}{g}\right)^2 (cg)^2 = \frac{r_{x_1y_1}r_{x_2y_1}r_{x_1z_1}r_{x_2z_2}}{r_{x_1x_2}^2 r_{y_1z_1}r_{y_1z_2}} \qquad (12.11)$$

and similarly

$$g^4 = \frac{(cg)^2}{(c/g)^2} = \frac{r_{y_1z_1}r_{y_1z_2}r_{x_1z_1}r_{x_2z_2}}{r_{z_1z_2}^2 r_{x_1y_1}r_{x_2y_1}} \qquad (12.12)$$

These results are of course highly complex and very much subject to the vagaries of sampling error. Furthermore, we have had to make the assumption that $f = 0$. This amounts to assuming that Y is the only intervening link between X and Z. Since we are allowing for the possibility of measurement error in Y, this particular assumption cannot be tested by the simple device of computing $r_{xz \cdot y}$. With random measurement error in Y, this partial cannot be expected to vanish, and we shall not be in a position to decide the degree to which the nonvanishing partial was due to measurement error or to the existence of f. Therefore we have to rely completely on the a priori assumption that $f = 0$.

It can similarly be shown that if we had two measures for both X and Y but a single measure for Z, it would not have been possible to separate the component paths by any such simplifying assumptions. In particular, the paths g and h would always be found together as a product, so that the device of multiplying a product by a ratio could not be used. The same holds if X had been measured by a single indicator, with Y and Z having two measures each. It therefore appears as though the only relatively simple kind of model for which a single indicator can be used is one in which the single indicator is linked with an intervening variable in a simple causal chain of the form $X \rightarrow Y \rightarrow Z$.

12.8   MEASUREMENTS AT SEVERAL POINTS IN TIME[1]

Rather than using several distinct indicators that have been measured at a single point in time, it is often desirable or practically necessary to use the same or nearly equivalent measures at two or more points in time. The *test-retest method* of estimating reliability is of course a very familiar procedure in educational testing, as are the pre- and postmeasurement procedures commonly utilized in social psychological experiments. Though beset with many practical difficulties of maintaining respondent cooperation, assuring anonymity, and locating respondents who have moved, the panel design of survey research is becoming increasingly common in social research. The basic difficulty posed by all such measurement designs is that of separating real change

[1] Portions of this section have been taken, with modifications, from H. M. Blalock, Estimating Measurement Error Using Multiple Indicators and Several Points in Time, *American Sociological Review*, **35**: 101–111 (February 1970), by permission of the American Sociological Association.

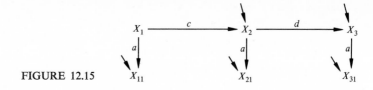

FIGURE 12.15

from measurement error. In panel studies the objective is often that of inferring real change, say with respect to voting intentions. The purpose of obtaining premeasurements as well as postmeasurements in an experimental design is similarly that of inferring the amount of change that has occurred. But on other occasions it may be desirable to obtain several measures under the assumption that there are no real changes and that any differences obtained can be used to assess reliability or the extent of random measurement error.

Provided that there are a sufficient number of measures over time *and* provided that one is willing to make certain a priori untestable assumptions, it is indeed possible to infer the degree to which differences are due to measurement errors, on the one hand, and real changes, on the other. In general, the more measures one has available, the more realistic assumptions one can make concerning nonrandom sources of error while still having overidentified models. But of course there will always be practical limitations imposed because of the need to retain the cooperation of respondents, to economize, and to obtain measures of many different kinds of variables. Therefore it will often be difficult to obtain more than three or four measures of each variable, and we shall see in the next chapter that this will restrict the complexity of one's measurement models. In the present chapter we are concerned with strictly random sources of measurement error and therefore use only very simple models.

Let us consider the model of Fig. 12.15, as discussed by Heise (1969). We represent the unmeasured variable X at three time periods by X_1, X_2, and X_3, while using double subscripts for the indicators. Since there is only a single indicator at each point in time in this model, the indicators have been labeled X_{11}, X_{21}, and X_{31}, respectively.[1] Sources of stability in X are designated as c and d, thus allowing for possible differences in stability over time.

Notice that we are explicitly assuming that the path coefficient a linking X to its indicator remains constant over time. This basic assumption permits important simplifications compared with models in which the unmeasured variables are distinct, thereby reducing the number of unknowns and permitting the introduction of more complex assumptions regarding measurement errors. However, this kind of assumption can be disarmingly simple, as has been noted by Wiley and Wiley (1970). A path coefficient (or correlation), being a standardized measure, will not be invariant if there are changes in the variances of either variable. In this instance, the path coefficient a would not be expected to remain unchanged if there were changes in the measurement-error variances or in the variances of the true X_i over time. Wiley and Wiley note that although it may be reasonable to assume that measurement-error variances remain

---

[1] In the next chapter, when we introduce models with two or three indicators at each time period, the first subscript will refer to the time and the second to the indicator number.

constant over time, we cannot so easily assume that the variances of the true (but unmeasured) X_i will remain constant. Therefore, one should attempt to assess the plausibility of this particular simplifying assumption compared with others that might be made.[1] Unfortunately, it will always be necessary to make a certain number of not too realistic assumptions in order to obtain definite solutions. The important feature of causal models is that such assumptions are made explicit, so that they are open to challenge.

Heise calls our attention to the fact that if we have data for only two time periods, we shall be able to compute only a single correlation and it will be impossible to solve for the two unknowns, a and c. For example, if we lacked data at time 3, we would have only the equation

$$r_{x_{11}x_{21}} = a^2 c$$

If it were reasonable to assume no change in the true value of X, so that $c = 1$, we could then obtain a^2, which is the square of the reliability coefficient discussed in connection with split-half reliability. Thus, if a correction for unreliability were used in test-retest situations, it would require the assumption that X had remained stable over time. But we encounter difficulties as soon as we allow for both real change and measurement error. However, if we also had data at time 3, we would be able to solve for the three unknowns a, c, and d, using the three empirical correlations

$$r_{x_{11}}r_{x_{21}} = a^2 c \qquad r_{x_{11}x_{31}} = a^2 cd \qquad r_{x_{21}x_{31}} = a^2 d$$

from which
$$a^2 = \frac{r_{x_{11}x_{21}}r_{x_{21}x_{31}}}{r_{x_{11}x_{31}}} = \frac{a^4 cd}{a^2 cd} \qquad (12.13)$$

$$c = \frac{r_{x_{11}x_{31}}}{r_{x_{21}x_{31}}} \qquad \text{and} \qquad d = \frac{r_{x_{11}x_{31}}}{r_{x_{11}x_{21}}}$$

These results show why it is necessary to be cautious in applying simple formulas for correcting correlations between indicators for attenuations due to measurement error. The formula for attenuation corrections generally used in the educational testing literature requires the assumption that the two (or possibly more) measures of each variable are equivalent or parallel forms in a very specific sense. The measures of each variable must be equally good in the sense that (1) they involve only random measurement errors and that $E(e_i) = 0$ for each indicator and (2) the measurement-error variances $\sigma_{e_i}^2$ for all measures of X (and similarly for the remaining variables) must be equal. In the model of Fig. 12.10 this would of course imply that $a = b$ and also that $d = e$, with the result that the formula for c^2 in (12.5) simplifies. The formula for the true correlation between X and Y thus becomes

$$c^2 = r_{xy}^2 = \frac{r_{x_1y_1}^2}{r_{x_1x_2}r_{y_1y_2}} \qquad (12.14)$$

---

[1] This will not be easy, though some sort of rough approximation will always be desirable. In particular, if the variance of the *measured* value of X remains roughly constant over time, it may seem reasonable to assume that its components (the variance of the true value and the variance of the measurement-error term) are also both constant. Of course, one might be increasing and the other decreasing by exactly the same amount, but in the absence of a prediction to this effect, we might rule out this particular possibility as being rather implausible.

where any of the three remaining pairs of (X_i, Y_j) indicators (or a weighted average of the four correlations) might have been used in the numerator. Thus the above correction requires not only large samples (as will also be necessary for all formulas in this section) but also the use of parallel or equivalent forms for each of the variables.

Equivalent forms are rarely used in sociological or political science research, partly because we have failed to take measurement-error problems seriously but also because of practical problems of expediency. We have seen that the administration of the same test or set of items at several points in time opens up the possibility that real change has occurred in the interim. So-called split-halves testing procedures would seem to have useful analogs in attitudinal research, provided that a careful item analysis is used to produce nearly equivalent halves. This of course permits one to obtain two or more (nearly) identical measures at the same point in time. It has long been recognized, however, that there will be many potential sources of nonrandom measurement error introduced by such simultaneous measurements. Some of these are peculiar to the general type of measuring instrument being used, e.g., questionnaire versus interview, but others may be due to the mood of the respondent at the time, possible distractions, the way the instructions are given, and so forth. Therefore as a guard against such nonrandom sources of error it is recommended that the investigator use diverse types of measuring techniques. But of course this means that indicators are unlikely to be equivalent in the sense that their measurement-error variances are all approximately equal. Thus the most desirable strategy from the standpoint of handling strictly random measurement errors is not necessarily the most realistic procedure for guarding against nonrandom errors.

12.9 UNSTANDARDIZED COEFFICIENTS AND MEASUREMENT ERRORS

For the sake of simplicity, we have discussed measurement-error problems and multiple indicators in terms of standardized measures, namely, correlation and path coefficients, even though we previously noted that comparisons across samples, populations, or time periods become difficult when only the standardized measures are used and reported. It is not possible in this general text to treat this problem in any detail, and in fact there seem to be a number of complications that have not as yet been adequately discussed in the literature. The reader is referred, however, to the discussions by Wiley and Wiley (1970), Wiley (1973), Schoenberg (1972), Werts, Linn, and Jöreskog (1971), and Jöreskog (1973).

In brief, these authors have commented on the fact that where cross-sample comparisons are desired, it will be preferable to work with the variance-covariance matrices rather than matrices of correlation coefficients. Unfortunately, one cannot obtain estimates of the unstandardized parameters without imposing some kinds of additional assumptions on the model. One such possibility is that the measurement-error variances remain constant over time or that they are the same from one population to the next. The latter kind of assumption seems much more reasonable for subpopulations (or subsamples) studied by the same investigator than for studies conducted by different investigators (see Schoenberg, 1972). But it may be more reasonable in the case of one set of indicators than another, a possibility that may further complicate the analysis. If one is willing to select one indicator of each unmeasured variable as a *reference indicator*, considering the remainder to be supplementary indicators, then one may specify the units for the unmeasured variables by arbitrarily setting at unity the regression coefficients connecting the reference indicators to the true values (see Wiley, 1973; Schoenberg, 1972). This of course cannot be done for all indicators simultaneously, though it could be approximated in the case of parallel forms of the same research instrument. It remains a possibility, however, that indicator 1 for X may have a greater measurement-error variance in sample A than in sample B, whereas indicator 2 has a greater measurement-error variance in sample B. If so, a procedure like Jöreskog's (1973) will still yield a solution, but one's interpretations of such results may depend upon one's relative degrees of faith in the model's assumptions regarding each indicator.

Thus although formal procedures exist for dealing with measurement-error models in terms of unstandardized coefficients, there appear to be a sufficient number of potential complications to require the reader to exercise a good deal of caution in connection with this problem. Certainly, however, it makes sense to compare the variances of indicators across populations (samples) or across time periods. If these variances in the comparable indicators are approximately equal, it will make little practical difference whether one uses the standardized or unstandardized versions. If these variances are very different, one must theorize why this is happening *in the case of each indicator*, since the reasons may not be identical across all indicators. Perhaps it will then be plausible to introduce simplifying assumptions and to select out a reference indicator for each variable. If so, it would seem appropriate to use the unstandardized measures.

12.10 INDICATORS AS CAUSES OF UNMEASURED VARIABLES

Thus far our discussion of multiple-indicator approaches has been confined to what can be termed *effect indicators* that are assumed to be linked with true values through causal processes in which the indicators have been taken as effects of the unmeasured variables. Measurement errors involving response errors, recording errors, or data-processing errors can all be conceptualized in these terms. But there are occasions where the measured indicators involve manipulations by experimenters or the exposure to natural events or group memberships, whereas the theoretical variables of interest are taken as dependent upon these situational variables. For example, an experimenter may withhold sleep or food from an individual and measure the degree of frustration by the number of hours during which the deprivation has taken place. Or one may attempt to create several distinct levels of motivation in one's subjects by using different experimental manipulations, inferring motivation from the kind of manipulation received. Similarly, the theorist may be interested in measuring the degree of threat posed by a minority group through such indicators as its relative size or its competitive position in the labor force. "Education," or total amount learned, may be indirectly measured by total exposure to schooling as indicated by the number of years of formal education. The basic assumption is that amount of knowledge absorbed is an effect of the length of time exposed to formal educational processes. Similarly, a man's economic influence may be measured by his economic resources or by the number of individuals he employs or manages.

In most of these situations we note that the theoretical variable of interest is some postulated property such as frustration, motivation, threat, knowledge, or influence. The objective indicators, on the other hand, are likely to involve completely different dimensions such as time exposed to some treatment, financial resources, or situational factors, e.g., the size or competitive position of a minority. Therefore it may not make very good sense to conceptualize the linkage as involving strictly random measurement error since, at the very least, there would have to be a change of scale as well as a stochastic element contributing to the measurement error. We shall return to some of these same illustrations in discussing nonrandom errors. Nevertheless it is instructive to begin with the simplest possible model in which we take the *true* value X as being equal to the measured value X' plus a strictly random component e. Let us see how this model for what we shall term *cause indicators* differs from those involving effect indicators.

If we write the equation $X = X' + e$, making the assumption that $E(e) = 0$, it is apparent that the sign of the error component e makes no difference. Therefore, we might just as well have written the equation as $X = X' - e$, from which we have $X' = X + e$, which is the basic measurement-error equation used in the case of effect indicators. But we also previously assumed that cov $(X,e) = 0$, whereas we are now taking e to represent all the remaining causes of X other than X'. Therefore for cause indicators X and e will be correlated. But if we make the assumption that the remaining causes of X represented by e are uncorrelated with X' the cause indicator of X, then we obtain the result that

$$\sigma_x^2 = \sigma_{x'}^2 + \sigma_e^2$$

since cov $(X',e) = 0$, or
$$\sigma_{x'}^2 = \sigma_x^2 - \sigma_e^2 \qquad (12.15)$$

Thus the variance in the true scores will be greater than the variance in the measured scores X'. In this case if we use the slope estimator $b_{yx'}$ to estimate β_{yx}, it turns out that we do not produce any systematic bias. But let us consider some modifications in the multiple-indicator models that are necessitated by the use of cause indicators.

Consider models I to IV of Fig. 12.16. The notion of random measurement error, in ordinary parlance, is not sufficiently specific to enable one to decide whether or not each of these models should be subsumed under the heading, but in all four models it is being assumed that the residual causes of X and Y that create measurement errors are uncorrelated (in the aggregate) with all other variables in the system, except by virtue of the fact that they may be indirect causes of some of the variables through either X or Y. For example, in model I the unnamed causes of X represented by the side arrow are assumed to be uncorrelated with X_1 and X_2, but they will of course be correlated with Y, Y_1, and Y_2 because of the causal linkage between X and Y. The residual causes of Y_1 and Y_2 are assumed to be uncorrelated with all other variables in the system. The same assumptions hold for model II, though this model allows for an empirical correlation between the two cause indicators X_1 and X_2. In model III the residual causes of X represented by the side arrow are assumed to be

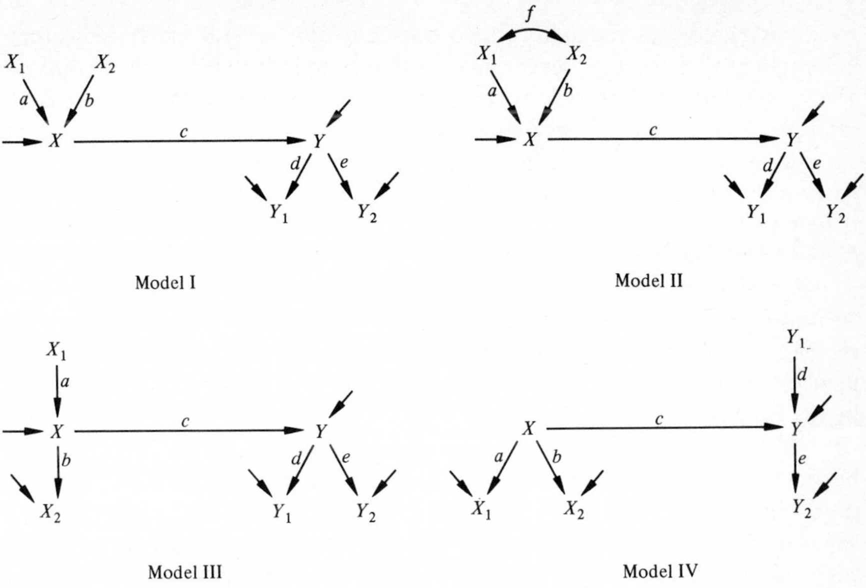

Model I   Model II

Model III   Model IV

FIGURE 12.16

uncorrelated with X_1 but will be correlated with X_2 because of the causal linkage between X and X_2. In model IV, Y_1 is assumed to be uncorrelated with all other variables except Y, whereas Y_2 will of course be correlated with the X variables. Even if one does not choose to refer to these models as random-error models, they are at least sufficiently similar to those previously discussed to warrant a discussion in the present chapter.

Notice that the predictions of model I are identical to those made in connection with the effect-indicator model of Fig. 12.10 except for the lack of implied association between X_1 and X_2. The first consistency criterion will therefore be satisfied, since it involves only the last four equations of the set (12.3). However, the procedure used to estimate the true correlation between X and Y now breaks down because we cannot divide by the (zero) correlation between X_1 and X_2. More generally we would expect the two cause indicators of X to be correlated, as in model II. The first consistency criterion is also satisfied in this model, but if we were to use the formula for estimating c^2 that was appropriate for Fig. 12.10, we would obtain

$$\frac{r_{x_1y_1}r_{x_2y_2}}{r_{x_1x_2}r_{y_1y_2}} = \frac{c^2de(a + bf)(b + af)}{fde} = c^2\frac{(a + bf)(b + af)}{f} \neq c^2 \quad (12.16)$$

We thus encounter the first of several models for which the first consistency criterion is satisfied but the previous formulas for estimating the coefficients should not be used. Without the addition of other variables to the model, this particular model would be empirically indistinguishable from that of Fig. 12.10.

The predictions for model III are identical to those for the model of Fig. 12.10, however, and therefore c^2 would be correctly estimated by using the equation (12.5) appropriate for Fig. 12.10. These first three models all seem reasonably common in social psychological research, where one may use several cause indicators of an independent variable, e.g., several manipulations to induce frustration, combined with several response indicators, e.g., measures of aggression, to measure Y. As another example, X_1 might be mother's education, X_2 father's education, and X the conceptual variable "exposure to knowledge within the family setting." The assumption would be that other sources of exposure will be idiosyncratic and therefore unrelated to either parent's education. Or X_1 may be race, X_2 sex, and X exposure to discrimination. In the latter example, X_1 and X_2 are likely to have a negligible mutual correlation, whereas mother's and father's educations will undoubtedly be positively related. As illustrations of model III, we might find an investigator using a single manipulation combined with an effect indicator of the induced internal state X. Perhaps X_2 may be a physiological response of some kind or a measure of perceived frustration, as indicated by the subject.

Models of type IV, in which the *dependent* variable has a cause indicator, may be much less common in empirical research. It is to be noted that cause indicators of the dependent variable cannot be assumed to have any easily explainable or predictable correlations with the X variables (see the predictions involving Y_1). The procedures suggested in connection with effect indicators therefore break down completely. This fact should be carefully noted in connection with general recursive systems in which a particular variable may appear as an independent variable in some equations but as the dependent variable in one of the equations in the set. In the next chapter we shall consider modifications of models I to III to allow for nonrandom errors, dropping model IV from further consideration. We shall also briefly discuss the possibility that variables such as X_1 and X_2 in model II may have joint effects on X that are non-additive.

EXERCISES

12.1 Referring to Exercise 10.2, suppose there is random measurement error in X_2.
   (*a*) How would each of your answers in this exercise have to be modified?
   (*b*) What general implications do you see in this example for the use of control variables whenever you suspect that relatively simple causal chain models will be appropriate?

12.2 Assume that the model in Fig. E12.2 is correct and that there is only random measurement error in X. The correlations are given on the diagram, and the variances of W, X, X', and Y are 100, 400, 625, and 225, respectively.
   (*a*) Calculate the covariances. What is the true value of b_{yx}?
   (*b*) Suppose that X is unmeasured and that X' is used instead. What is $b_{yx'}$? What is the instrumental-variable estimate based on X'?

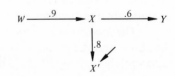

FIGURE E12.2

12.3   Suppose that X_1, X_2, and X_3 are effect indicators of X and that Y_1, Y_2, and Y_3 are effect indicators of Y. You wish to estimate the true correlation between X and Y as part of a submodel in which X causes Y. You begin with a random measurement-error model and the following matrix of intercorrelations between the indicators:

	X_1	X_2	X_3	Y_1	Y_2	Y_3
X_1		.70	.67	.40	.32	.50
X_2			.61	.35	.28	.44
X_3				.33	.27	.41
Y_1					.41	.62
Y_2						.55

(a) Ignoring X_3 and Y_3, and therefore dropping back to a two-indicator model, check on the adequacy of the random-error model by using the first consistency criterion. Then proceed to estimate all the correlations involving X and Y, the true values.

(b) Using all three indicators of both X and Y, write out a number of test equations for evaluating the goodness of fit of the random-error model, using the first consistency criterion.

(c) Obtain a series of estimates of the true correlation between X and Y. Do the same for each of the path coefficients linking the true values to their indicators.

(d) On the basis of your results evaluate the overall adequacy of the model and obtain single estimates of the path coefficients, using arithmetic means of the estimates obtained in part (c).

12.4   Suppose you have two measures of X at each of three points in time and are willing to assume that measurement errors are random.

(a) What kinds of simplifying assumptions might seem plausible, and what kinds of practical conditions would be necessary for these assumptions to be approximated?

(b) Represent these assumptions in your path diagram.

(c) How would you estimate the stability coefficients between t_1 and t_2 and between t_2 and t_3? What consistency checks could be made?

13

NONRANDOM MEASUREMENT ERRORS[1]

As soon as one allows for the possibility of nonrandom measurement errors, one must begin to face a number of difficult problems, both conceptually and methodologically. Strictly random errors are generally discussed under the heading of *reliability*, and, as we have seen, there are definite ways of estimating reliability coefficients and correcting for unreliability provided that multiple indicators are available. But the possibility that errors are nonrandom opens up the whole question of *validity* and the problem of assessing the degree to which one is measuring what one intends to measure. How can these nonrandom errors be represented in mathematical equations in a realistic way? Obviously, one must have certain ideas about the sources of such nonrandom errors and the kinds of distortions being produced; otherwise the situation is too vague to handle. But as we have already noted, in many instances it is even difficult to specify exactly what one means by the "true" value of X.

Let us consider two examples. On the one hand, we can readily imagine a "true" percentage of persons who voted for President Nixon in 1968 or even the true percentage who "intended" to vote for him. If a survey consistently underestimated this true percentage in every state, or if it underestimated it for males but overestimated it for females, we would say that measurement error was nonrandom.

[1] The introduction and Secs. 13.1 and 13.2 are taken, with modifications, from H. M. Blalock, A Causal Approach to Nonrandom Measurement Errors, *American Political Science Review*, **64**: 1099–1111 (December 1970), by permission of the American Political Science Association.

On the other hand, consider a variable such as the "true" amount of political alienation in the United States. Lacking a theoretical definition of alienation that clearly specifies the research operations, it is even difficult for us to think in terms of measurement biases or nonrandom errors. Yet we would want to design an instrument that does not systematically exaggerate the alienation of blacks while underestimating that of whites (see Carr, 1971). The notion of nonrandom measurement errors makes some sense in this instance, but it is difficult to pin down because of inadequate theoretical conceptualization. Unfortunately, many of our most important concepts in the behavioral sciences are more like that of alienation than that of a voting percentage.

In the case of alienation, if we could imagine a "true" value X, we might be pleased if we could obtain a measure X' that is an exact linear function of X, say $X' = a + bX$, with no error term. The coefficient b would measure an error of scale, whereas a would give the intercept error. If there were some meaningful zero point (no alienation), we would prefer that the intercept error be zero, so that we could use the properties of a ratio scale. But in tests of causal models, or in estimating path coefficients, we would encounter no special difficulties if $a \neq 0$. We would have measured X perfectly except for an exact linear transformation. Of course if we wished to report a measure of central tendency, such as the proportion intending to vote for Nixon, such errors of scale and zero point could be disastrous practically, and we would want to be assured that $X' = X$.

Even these two very simple examples illustrate that there may be many different sources and kinds of measurement error. Some types involve the possibility that the measurement errors in X may be functions of X itself, whereas others stem from the influence of disturbing factors Z_i that may be correlated with X or with other variables in the theoretical system. In many particular instances, measurement errors may be partly a function of X, partly due to Z_i, and partly due to sources that produce approximately random variation. We shall first examine situations in which measurement error in X is taken as a function of X itself.

13.1   MEASUREMENT ERRORS AS A FUNCTION OF X

Errors of intercept, scale, and functional form can all be considered to be *nonstochastic* errors in which the measured and true values of X are not identical. Some such possibilities are the following:

Intercept error: $X' = a + X \qquad a \neq 0$

Scale error: $X' = bX \qquad b \neq 1$

Intercept and scale error: $X' = a + bX$

Nonlinear errors: $X' = a + b \ln X \qquad X' = ke^{bX}$

$$X' = a + b_1 X + b_2 X^2 + \cdots + b_k X^k$$

Obviously, unless a concept is defined sufficiently clearly to imply a definite unit of measurement, it will be impossible to conceptualize the problem of nonrandom measurement error in terms of equations like the above. As already implied, it is perhaps true that most of the important concepts in fields such as sociology and political science have not been refined to a sufficient degree to make such distinctions meaningful. In many instances, the specification of the concept will not permit one to decide between linear transformations or even monotonic nonlinear transformations. Thus if "industrialization" is measured in one study by X' and in a second by $X'' = a + bX'$ (with $b > 0$) or by $X''' = a + b \ln X'$ (with $b > 0$), there may be no possible way of saying which measure is the more "valid" in terms of the theoretical definition being used. Many discussions that rely on assessing validity in terms of empirical intercorrelations, for example, completely ignore the possibility that nonlinear though monotonic transformations will alter the magnitudes of product-moment correlations. Clearly, if the theoretical definitions of concepts such as industrialization, urbanization, discrimination, or anomie do not imply a metric, then "validity" cannot be assessed in terms of magnitudes of correlation coefficients.[1]

In the remainder of the chapter we shall suppose that the theoretical concept has been defined clearly enough to permit consensus on the ideal metric that is to be used. The fact that this assumption places a severe limitation on the utility of the statistical analysis of nonrandom measurement errors merely indicates the distance we shall have to travel before the degree of measurement error can be assessed in any truly systematic way. Nevertheless, it seems useful to discuss a number of specific sources of nonrandom errors if only to sensitize the investigator to the problem and to indicate how difficult it will be in any particular instance to assess the validity of one's measuring instruments in the absence of careful theoretical conceptualization.

In the present context, let us simplify our analysis by ignoring the problem of errors of scale and intercept by taking $X' = X + e$. However, we shall permit the error term e to be some function of the true X and a random component u, and we shall find that we thereby reintroduce a scale and intercept factor as a result of this error component. This seems to be a relatively straightforward way of conceptualizing the measurement-error problem without introducing undue simplicity. Although the error component may be a nonlinear function of X, as we shall illustrate below, we shall confine our actual analysis to the linear situation, where we may write

$$X' = X + e \quad \text{and} \quad e = c + dX + u \qquad (13.1)$$

where u is a random variable with expected value zero, with variance σ_u^2 independent of X, and with cov $(X,u) = 0$.[2] We may then write

$$X' = c + (1 + d)X + u \qquad (13.2)$$

Since X and u are uncorrelated, we have

$$\sigma_{X'}^2 = (1 + d)^2 \sigma_X^2 + \sigma_u^2 \qquad (13.3)$$

---

[1] We do not wish to equate validity with the absence of nonrandom measurement errors, since if random measurement errors exist, an indicator cannot be perfectly valid. The notion of validity is a notoriously slippery concept in the social science literature, and we shall therefore avoid the term.

[2] There is another possibility that we shall not discuss further, namely, that the error *variance* σ_e^2 is a function of X. Sometimes it is reasonable to take $\sigma_e^2 = a + bX$, where $b > 0$, in which case it may be advisable to transform X to $\ln X$. However, this may produce nonlinear relationships with other variables in the causal system.

Notice that if d is negative, it is entirely possible for $\sigma_{X'}{}^2$ to be less than $\sigma_X{}^2$.*
This means that if ordinary least squares is used to estimate β in the equation
$Y = \alpha + \beta X + \varepsilon$, we shall generally obtain a biased estimate but we can no longer
count on this bias being in the downward direction. Also, it turns out that the
method of instrumental variables cannot be used to obtain a consistent estimator.
The factor $1 + d$ will appear in this estimate, and since d will be unknown, there
appears to be no purely statistical device for eliminating its influence in the derived
estimate. In effect, then, the d coefficient will be confounded in slope estimates
relating X to other variables.

Before examining the implications of this kind of complication for path analysis,
let us consider some substantive examples for which we might anticipate that the
measurement-error component is a function of X. We shall later discuss instances
where the error is a *correlate* of X due to the operation of extraneous variables, but
we are now considering instances where the level of X itself has a direct or indirect
effect on the measurement-error component.

Some Examples

One obvious possibility is that the true value of X may be consistently underestimated
or overestimated by a constant proportion of the actual value. For example, persons
may report their ages to be .9 of their true ages or their incomes to be 1.25 times their
true incomes. Somewhat more realistically, such fractional representations might
be appropriate for certain subpopulations, e.g., people over 21, or white-collar
workers. Whereas differential proportions applied to various subpopulations imply
nonlinear relationships between the measurement-error component and X, the situation
might reasonably be approximated by a series of linear models applied to diverse
subpopulations.

Perhaps a more common phenomenon is that of regression effects due to
measurement error. Of course if one is dealing empirically with fallible measures
and purely random measurement errors, one will encounter a regression effect of a
somewhat different nature. For example, if at time 1 we measure individuals with
random error, then certain persons with intermediate true values will appear extreme
because of the random errors. But with the next measure, they will be likely to regress
toward the mean, leaving us with the impression that there is a homogenizing influence
at work. However, there will be other individuals who appeared intermediate at
the first measurement whose second scores will have diverged from the mean, again
due entirely to random measurement error. A true homogenizing influence can be
distinguished from random measurement-error effects, in this instance, by a comparison of the variances of the measured values at both points in time. If the regression
effect is an artifact of random measurement errors, these variances should be approximately equal, whereas if there has been a homogenizing effect, the variance at time 1
will be greater than that at time 2 (see Campbell and Stanley, 1963).

We are concerned with a different kind of regression effect, however, namely,

---

* An empirical example where this occurs in connection with floor and ceiling effects has been discussed
by Siegel and Hodge (1968).

that due to nonrandom errors that are negatively related to the true X scores. Suppose there is a norm operating to the effect that it is desirable to appear "moderate" or "normal." If respondents are given a series of questions designed to tap political conservatism, one might then expect that ultraconservatives will try to appear only moderately conservative, whereas radicals may wish to appear as liberals and liberals as moderates. If increasing values of X represent increasing conservatism, the measured scores of conservatives will be less than the true scores and measured scores of liberals will be greater than the true values. In other words, the slope d linking e to X in (13.1) will be negative.

Of course the norm might not operate equally in all subpopulations and might even favor polarization, in which case liberals might wish to appear radical and conservatives as ultraconservatives. In this case the slope linking e to X would be positive, and $1 + d$ would be greater than unity. Again, if different norms operated in different subgroups, a more complex model would be required, and measurement error would be taken as a joint (nonadditive) function of X and other variables. As a first approximation, however, linear models with different error slopes could be used. Our principal difficulty in these and other instances is a lack of theory concerning specific sources of nonrandom errors.

Another source of errors producing regression effects may arise in the process of aggregating microlevel data, although once more the mechanisms involved seem poorly understood. Suppose, for example, that one has a social psychological theory that on the individual level the number of blacks to whom a white individual is exposed will affect the individual's political attitudes concerning the role of the federal government in school-desegregation issues. But the measurement of exposure to blacks would necessarily be very indirect and perhaps indexed by the percentage of blacks in the local area. But how big an area? Living within a mile radius? Within the same county? The same state? The larger the unit selected the less adequate the measure of exposure, and also the less the variation in minority percentage. That is, whereas counties may range in minority percentage from 0 up to almost 100, if states are used as the aggregate units, the largest percentage may be in the neighborhood of 45 percent. Likewise, at least in the South, there are no states with under 5 percent black. In effect, the range of variation in X, percent black, will be reduced whenever larger aggregates are used. As larger and less homogeneous aggregates are used, one would thus expect the slope $1 + d$ to approach zero unless other mechanisms were simultaneously at work. Obviously, sources of nonrandom error such as those produced by aggregation have not been adequately studied by political scientists and sociologists.

As an example of measurement errors that are more complex functions of X, let us consider possible distortions produced by so-called floor and ceiling effects and by measurement instruments that are relatively insensitive to variations in extreme scores. Obviously, if there is a true floor or ceiling effect, then substantial measurement errors can appear only in a single direction or they would be recognized as nonsensical and thereby modified. If "percent urban" were measured as negative or greater than 100, an error would immediately be noted. In fact, it would be impossible to obtain such scores by conventional means of measuring percent urban. But if the true percent urban (by whatever definition) in a county were very close to

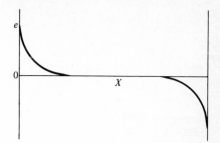

FIGURE 13.1

zero, measurement errors would usually be in the positive direction. Similarly, these errors would be in the negative direction at the upper limit. Assuming approximately random errors in intermediate positions, we might expect that the function relating e to X might be approximately as in Fig. 13.1. Such a function might be approximated by a cubic equation or by three straight lines the middle one of which has a horizontal slope. Where the floor and ceiling effects merge into each other imperceptibly, this measurement-error situation might be approximated by a single straight line with a slightly negative value for the error slope d.

Measurement Errors Due to Classification

A very common practice whenever measurement is clearly recognized as being crude is to resort to a relatively small number of ordered categories. In the extreme case the analyst may use dichotomies in order to simplify the analysis. It may not be recognized that such very simple procedures produce both random and nonrandom measurement errors that become increasingly serious as the number of categories is reduced. Of course it will ordinarily be difficult if not impossible to estimate the extent of such measurement errors, but this fact does not miraculously make these errors disappear. It merely makes it impossible to assess their distorting effects. In order to understand these effects it will ·undoubtedly be necessary to conduct a series of simulation studies on data with known properties, using various methods of categorization in diverse causal situations (see Reynolds, 1971; Labovitz, 1967). Here it will be possible to provide only a single illustration.

Suppose that the true underlying distribution of X is approximately normal, with a mean of 50 and a standard deviation of 10. We then know the proportion of cases expected to fall within any particular interval. Now suppose that a social scientist, ignorant of the true distribution, obtains an approximate measure using a scale that enables one to divide the population into six ordered categories, as indicated in Fig. 13.2. Ordinarily there will also be random errors of measurement that can result in misclassification superimposed upon effects produced by the fact that individuals in the tails of the distribution will all be lumped together in the extreme categories.

Ignoring the random errors of classification, let us examine the nature of the measurement errors that would be produced if the analyst simply assigned scores of

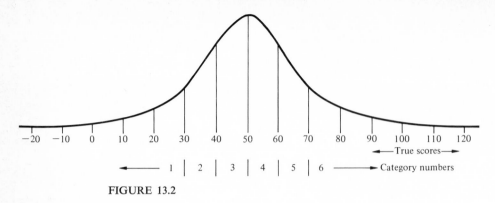

FIGURE 13.2

1 to 6 to the categories and then proceeded to analyze the data as though no measure-ment errors existed and as though one had a legitimate interval scale.[1] The X' scores therefore range from 1 to 6. Let us say that the true X scores range from -20 to 120. We thus have a change of scale in addition to any measurement errors, but this can be handled by dividing each X by 10, the width of the analyst's intervals, and then subtracting 1.5. Thus a true score of 50 is converted to an X' score of 3.5, which is equivalent to the average score of the two middle categories. We can then obtain the measurement-error components at selected levels of X from the equation $e = X' - X$ as shown in Table 13.1.

For example, an individual with a true score of -20 would be placed in the first category and would receive an X' score of 1, whereas the adjusted X score would

[1] It might seem absurd for an analyst to treat this kind of data as an interval scale, but we must remember that ordinal data also involve equally difficult problems of interpretation. In this particular illustration we are assuming no additional distortions produced by the numerical values assigned to the categories; i.e., the true distance between category limits is always 10. If the analyst used unequal intervals that were scored 1 to 6, there would be the added complication of nonlinearity. Data given by Labovitz (1967) and Reynolds (1971) suggest, however, that such distortions generally may not be serious.

Table 13.1

True X	$\dfrac{X}{10} - 1.5$	X'	e	True X	$\dfrac{X}{10} - 1.5$	X'	e
-20	-3.5	1	4.5	$50+$	3.5	4	.5
-10	-2.5	1	3.5	$60-$	4.5	4	$-.5$
0	-1.5	1	2.5	$60+$	4.5	5	.5
10	$-.5$	1	1.5	$70-$	5.5	5	$-.5$
20	.5	1	.5	$70+$	5.5	6	.5
$30-$	1.5	1	$-.5$	80	6.5	6	$-.5$
$30+$	1.5	2	.5	90	7.5	6	-1.5
$40-$	2.5	2	$-.5$	100	8.5	6	-2.5
$40+$	2.5	3	.5	110	9.5	6	-3.5
$50-$	3.5	3	$-.5$	120	10.5	6	-4.5

FIGURE 13.3

be $-20/10 - 1.5 = -3.5$. Therefore the measurement-error component e for this individual would be $X' - X = 4.5$. As we proceed down the columns we note that e decreases as the true X increases until we reach the boundary of the second class interval at $X = 30$. For any individual having a score slightly less than 30 (symbolized by $30-$ in the table), the X' score will remain 1. Individuals whose scores are slightly over 30, however, will be placed in the second category (assuming the absence of random errors), so that for these individuals the error component $e = 2 - (30/10 - 1.5) = +.5$. This jump in X' values near the category limits produces a sawtooth effect, as indicated in Fig. 13.3. This effect continues until we reach the upper boundary of category 5, at which point the values of e decrease linearly as X increases. When we plot the measurement error e against X, as in Fig. 13.3, we note that except for the sawtooth effect the relationship is very similar to that indicated in Fig. 13.2 for floor and ceiling effects.

In practical terms, this kind of distortion can be reduced by using a large number of categories and by making sure that the extreme categories contain relatively few cases. If this has been done, and if one is willing to consider distortions producing the sawtooth effect as approximately random, one may safely ignore the extreme effects and use a random measurement-error model. But if only a small number of categories have been used or if there are numerous cases in the extreme categories, the measurement error should be taken as a negative function of X. A linear approximation, it is hoped, will be reasonably adequate.

Path analyses with linear models Let us assume that a linear approximation is appropriate and that we wish to assess the implications of measurement errors that are functions of X. We have already seen that if $X' = c + (1 + d)X + u$, we must anticipate a change in scale that will affect our estimates of regression coefficients. But what about path coefficients or correlations? Will our causal inferences be affected, apart from the matter of scale? Common sense would suggest that the same tools appropriate to random measurement error may be used, but common sense is a notoriously poor guide in complex situations like this. If we represent the nonrandom component as a function of X, we can construct a causal diagram in which X' is taken as a function of two variables, X and e, with e in turn being a function of X, as indicated in Fig. 13.4. Furthermore, we can construct relatively

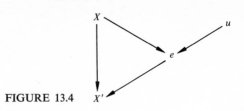

FIGURE 13.4

complex models allowing for multiple indicators and several imperfectly measured variables.

Let us consider the model of Fig. 13.5 involving two indicators of X, both of which have been measured with errors proportional to X, and a dependent variable Y likewise measured with nonrandom errors by indicators Y_1 and Y_2. We may ask whether it is possible to estimate the path coefficient c connecting the true values of X and Y in the presence of so many unknowns. Clearly, there are more unknowns than equations, but it does not therefore follow that some of these unknowns cannot be estimated.

Since there are four measured variables, there will be six intercorrelations, the path equations for which are as follows:

$$
\begin{aligned}
r_{x_1x_2} &= ab + ahi + bfg + fghi = (a + fg)(b + hi) = p_1^* p_2^* \\
r_{y_1y_2} &= de + dlm + ejk + jklm = (d + jk)(e + lm) = p_3^* p_4^* \\
r_{x_1y_1} &= c(a + fg)(d + jk) = cp_1^* {}_3^* \\
r_{x_1y_2} &= c(a + fg)(e + lm) = cp_1^* p_4^* \\
r_{x_2y_1} &= c(b + hi)(d + jk) = cp_2^* p_3^* \\
r_{x_2y_2} &= c(b + hi)(e + lm) = cp_2^* p_4^*
\end{aligned}
\tag{13.4}
$$

where we have been able to utilize the component paths

$$
\begin{aligned}
p_1^* &= a + fg & p_2^* &= b + hi \\
p_3^* &= d + jk & p_4^* &= e + lm
\end{aligned}
$$

Notice that each of these compound paths p_i^* involves three unknowns that always appear together in identical expressions. In effect, there are two excess unknowns in each of these compound expressions, thus accounting for eight of the unknowns in the system. Although none of these four compound paths can be decomposed without additional data or assumptions, it is possible to estimate c and each of the four p_i^* from the six equations. Of particular interest is the estimate of c obtained as follows:

$$
c^2 = \frac{p_1^* p_2^* c^2 p_3^* p_4^*}{p_1^* p_2^* p_3^* p_4^*} = \frac{r_{x_1y_1} r_{x_2y_2}}{r_{x_1x_2} r_{y_1y_2}} = \frac{r_{x_1y_2} r_{x_2y_1}}{r_{x_1x_2} r_{y_1y_2}}
\tag{13.5}
$$

This estimate is of course identical with the estimate obtained earlier in connection with the random measurement-error model. Furthermore, we note that we obtain the same redundant equation for testing purposes, namely, the prediction

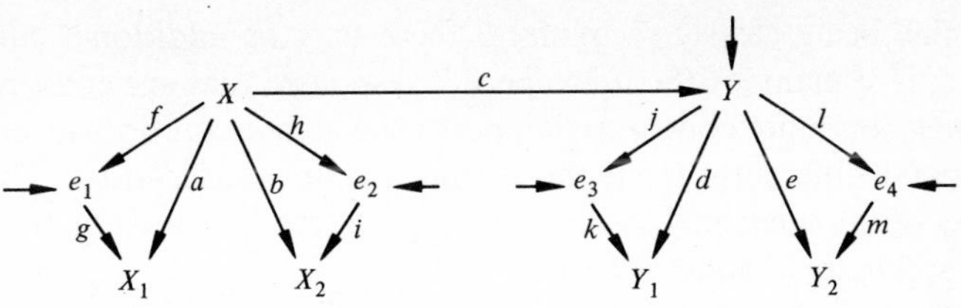

FIGURE 13.5

that $r_{x_1y_1}r_{x_2y_2} = r_{x_1y_2}r_{x_2y_1}$. However, this means that the first consistency criterion cannot be used to ascertain whether measurement errors are purely random or whether there is an additional error component that is proportional to X. It can be shown that we could also have estimated c if we had had a perfect measure of Y instead of two indicators, but if there had been only a single measure of Y involving measurement error, we could not have estimated c without additional assumptions.

The caution should be repeated that path coefficients and correlations will not be invariant across populations or time periods, even where the same fundamental laws are operative, i.e., where the same unstandardized coefficients hold. Therefore where one wishes to compare coefficients in the presence of variances that differ from one population to the next, the above procedure will not be appropriate. Also, we must once more emphasize that in order to be assured of reasonable sampling errors with these rather complex expressions, one must be dealing with large samples and correlations of at least moderate size.

13.2   MEASUREMENT ERRORS DUE TO A SINGLE EXTRANEOUS FACTOR

Perhaps a more common type of nonrandom measurement error, and certainly one that is more difficult to handle, stems from the influence of extraneous factors that may also be correlated with X and other variables in the causal system. We can think of numerous examples. When we deal with official data, e.g., those compiled by national censuses or government bodies, we are likely to find common biases affecting more than a single variable. For example, a government may wish to appear more "advanced" economically and politically than it actually is. Size of population, gross national product, and degree of urbanization may all be inflated, whereas indexes of crime and internal disruption may be biased in the opposite direction. Census underenumeration of blacks may have varying effects on a number of different economic and political indicators.

In survey interviews or any other situation where attitudes, knowledge, or ability are being measured it is well recognized that motivational factors are likely to produce nonrandom errors. Respondents with low motivation or interest may tend to check items randomly. On ability or knowledge tests this is likely to lower their performance scores; on attitudinal items it may give them intermediate scores on many different scales. If the questions are tapping several underlying dimensions, without this being clearly recognized, there may be additional confounding effects. For example if many of the items permit responses that are conventional or popular while others are more controversial, persons who are conformists or "other directed" may respond differently from those who are nonconformists. This may hold true regardless of the questions being asked, so that there is a common component across several supposedly distinct scales.

Unfortunately, unless these extraneous factors are recognized and controlled in the data-collection stage, they are likely to remain as unknowns. If they occur to an investigator as an afterthought, the chances are that they themselves will have been only indirectly measured. Furthermore, if numerous such extraneous factors are operative, it will be difficult to bring them all into the model in such a way that their relationships with the other variables are made explicit.

In order to illustrate the general principle that the number of indicators must be large relative to the number of unknowns in the system, let us begin by considering a model where there is only a single source of nonrandom errors in the measurement of X but we have only two measures of X. We shall later take up models in which there are three or more measures of X but possibly several sources of nonrandom errors in these indicators. In the model of Fig. 13.6 we have X, which is again measured by two indicators; Y, which is dependent on X and assumed perfectly measured (though similar results would be obtained if we had two indicators of Y with random errors or errors proportional to Y); W, which is taken as a cause of both X and Y; and an extraneous factor Z, taken as a cause of X_1 and X_2 and as possibly correlated with W, X, and Y. In the model of Fig. 13.6, Z is assumed to cause W, X, and Y. Obviously there are many other possibilities, even confining our attention to the links between W, X, Y, and Z, but we may use this particular model as illustrative of both the procedure that can be followed and the difficulties one can expect to encounter.

If we were to treat X as unmeasured but the remaining variables as measured, we would have five measured variables and therefore ten equations, one for each of the correlations that can be obtained from the data. There are also ten unknown path coefficients that could be found by solving these ten equations, which would be rather complex nonlinear equations that could not be easily solved algebraically. However, as previously noted, Z will generally be either unknown or poorly measured. It is therefore important to examine the implications of the model assuming that Z is unknown and that one has available only the six intercorrelations between W, X_1, X_2, and Y. There are of course now four more unknowns than equations, so that additional simplifications will be necessary. Keep in mind that our objective is to estimate the causal links h, i, and j involving only W, X, and Y. Therefore it is conceivable that some of these path coefficients can be estimated even though there are too many unknowns in the total system.

In introducing simplifications it is undesirable to place many restrictions on the relationships between W, X, and Y, other than linearity and one-way causation, since these are the variables of fundamental interest. Of course it is possible to assume away the links between Z and X_1 and X_2, but these are the very sources of nonrandom measurement error of concern to us. We might wish to consider a model

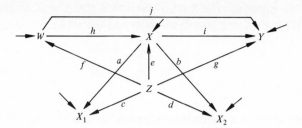

FIGURE 13.6

in which Z affected X_1 but not X_2, which might be subject only to random measurement error. Unless there are two or more X_i unrelated to Z, however, this kind of simplification will not really help. Let us therefore consider the kinds of simplifications introduced if Z is unrelated to W, X, or Y.

It turns out that the important simplifications will occur if we set $f = 0$, thereby taking out the link between Z and the independent variable W. In this case the six equations can be written down by inspection as follows:

$$r_{x_1x_2} = ab + cd + e(ad + bc) \qquad r_{x_1y} = a(i + eg + hj) + c(g + ei)$$

$$r_{x_1w} = ah \qquad\qquad\qquad r_{x_2y} = b(i + eg + hj) + d(g + ei) \qquad (13.6)$$

$$r_{x_2w} = bh \qquad\qquad\qquad r_{wy} = hi + j$$

The last of these equations contains only h, i, and j, which are of major interest, and the second and third equations contain only a, b, and h. But the remaining equations contain too many compound paths to be of any help. If equivalent forms had been used in the measurement of X, so that it would be reasonable to assume that $a = b$ and $c = d$, the system would contain two redundant equations for testing purposes (third and fifth equations) but there would still be too many unknowns for solution.

If we take out the link g between Z and Y, while retaining e, we remove two terms each from the expressions for r_{x_1y} and r_{x_2y} but leave the remaining equations unchanged. Similarly, if we remove e but not g, we achieve somewhat greater simplifications, but it still remains impossible to solve for h, i, and j. Let us therefore consider the case where we take out all the links e, f, and g of Z with the variables of interest. Even with this very restrictive model the equations reduce to

$$r_{x_1x_2} = ab + cd \qquad r_{x_1y} = a(i + hj)$$

$$r_{x_1w} = ah \qquad\qquad r_{x_2y} = b(i + hj) \qquad (13.7)$$

$$r_{x_2w} = bh \qquad\qquad r_{wy} = hi + j$$

We now have six equations and seven unknowns. But the equations are not all independent since the ratio a/b can be obtained by dividing the second equation by the third or the fourth by the fifth. This gives us a redundant equation for testing purposes, but even in this very restricted model we still cannot solve for the path coefficients.

Finally, if we introduce the further restriction that $j = 0$, so that W does not affect Y except through X, the equations reduce to the set

$$r_{x_1 x_2} = ab + cd \qquad r_{x_1 y} = ai$$
$$r_{x_1 w} = ah \qquad r_{x_2 y} = bi \qquad (13.8)$$
$$r_{x_2 w} = bh \qquad r_{wy} = hi$$

Although this set contains a redundant equation, as before, and although the compound path cd cannot be decomposed, it becomes possible to solve for the remaining paths. For example,

$$h^2 = \frac{h}{i} hi = \frac{r_{x_1 w}}{r_{x_1 y}} r_{wy} = \frac{r_{x_2 w}}{r_{x_2 y}} r_{wy}$$

and

$$i^2 = \frac{i}{h} hi = \frac{r_{x_1 y}}{r_{x_1 w}} r_{wy} = \frac{r_{x_2 y}}{r_{x_2 w}} r_{wy}$$

Thus in order to estimate the path coefficients of interest we have had to impose a large number of restrictions on the model. In effect, we have had to assume that the source Z of nonrandom measurement error is uncorrelated with any of the remaining variables in the system. Not only this, but we have had to use a very simple model connecting W, X, and Y. If we were primarily interested in the path i connecting X to its effect Y, we would be obliged, in this instance, to locate a cause W of X that does not affect Y by any other path. Recall that the method of instrumental variables required a similar kind of assumption.

Since we are clearly in difficulty if we wish to allow for nonrandom measurement errors in X whenever only two indicators of X are available, let us see to what extent the situation is improved when three or more indicators have been used.

13.3   MIXED RANDOM AND NONRANDOM ERRORS[1]

Whenever there are more than two measures of each variable, it becomes possible to handle certain kinds of nonrandom errors without encountering identification problems. In the present section we shall deal with selected models that are characterized by assumptions to the effect that most of the indicators of X and Y involve strictly random errors whereas a few involve nonrandom errors. Of course, as we extend the number of indicators and variables, the number of possible models becomes too great to permit systematic coverage of all likely alternatives. The general strategy of approach, however, can be applied to many concrete examples. Obviously, it will be advantageous whenever possible to anticipate the most plausible alternative models *in advance of* data collection, so as to guide the choice of indicators to be used.

It turns out that a good many nonrandom-error models imply that the first consistency criterion, as proposed by Costner, will be met within the limits of sampling error. However, Costner also proposed a second criterion, namely, that the several

---

[1] Portions of this section closely parallel the discussion in H. M. Blalock, Estimating Measurement Errors in Causal Models, *Revue Internationale de Sociologie*, (2)**5**: 300–318 (September 1969).

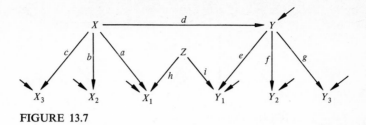

FIGURE 13.7

estimates of the same path coefficients be approximately equal. In the present section we shall see how the two kinds of criteria can be used in exploratory fashion to help locate troublesome indicators that may involve nonrandom errors of various kinds. The strategy depends, however, on the assumption that there are relatively few indicators subject to these nonrandom errors. Whenever most or all indicators involve nonrandom errors, we must make a series of simplifying assumptions about the sources of these errors. This latter kind of possibility will be considered in the next section.

Let us first take up two somewhat different models in which one of the indicators of X is correlated with one of the indicators of Y not only because of the correlation between X and Y but through some additional path. In Fig. 13.7, X_1 and Y_1 are partly spuriously related owing to a disturbing factor Z. The model of Fig. 13.8, in contrast, includes the assumption that there is a direct causal arrow running from X_1 to Y_1. In the example of attitude measurement, Z might be some disturbance that jointly affected one's answers to two adjacent items in the test or questionnaire. Or perhaps the response to the item X_1 provided a clue or in some other way directly influenced the answer to a second item (Y_1). The numbering of the paths in Figs. 13.7 and 13.8 corresponds reasonably well to previous notation. Notice that there are three indicators for each variable and the ordering of the X_i in the figures has been reversed so as to simplify the diagrams.

The reader can easily verify that in both models the pairs of indicators that do *not* involve the pair (X_1, Y_1) will not produce any problems. For example, if we combine the indicators X_2 and X_3 with indicators Y_2 and Y_3, we obviously have the same situation as in Fig. 12.10. If X_1 appears with either X_2 or X_3 in conjunction with Y_2 and Y_3, there will likewise be no problem since the disturbances that affect the relationship between X_1 and Y_1 are not involved. For example, the only path connecting X_1 and Y_2 is adf. Of the nine test comparisons that are possible, five should hold as predicted previously under a strictly random model. The four remaining pair combinations will all have identical structures in both Figs. 13.7 and 13.8, and therefore we may confine our attention to the combination that pairs X_1 and X_2 with Y_1 and Y_2. In the model of Fig. 13.7 involving Z we obtain

$$
\begin{array}{ll}
r_{x_1x_2} = ab & r_{x_1y_2} = adf \\[4pt]
r_{y_1y_2} = ef & r_{x_2y_1} = bde \qquad (13.9) \\[4pt]
r_{x_1y_1} = ade + hi & r_{x_2y_2} = bdf
\end{array}
$$

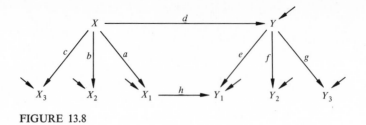

FIGURE 13.8

If we now take the usual paired products of correlations, we get

$$r_{x_1y_1}r_{x_2y_2} = (ade + hi)bdf = abd^2ef + bdfhi \qquad (13.10)$$

and

$$r_{x_1y_2}r_{x_2y_1} = (adf)bde = abd^2ef \qquad (13.11)$$

which are obviously not equal. Numerically, if we used the hypothetical data for the model of Fig. 12.11, and if $h = .6$ and $i = .5$, we would get

$$r_{x_1y_1} = (.8)(.5)(.9) + (.6)(.5) = .66$$

and therefore

$$r_{x_1y_1}r_{x_2y_2} = (.66)(.21) = .1386$$

which is clearly not equal to the value of .0756 obtained previously in connection with Fig. 12.11.

Thus we fail to pass the first consistency criterion for random errors proposed by Costner. This will also be the case for the remaining three sets of pairs involving the X_1 and Y_1 combination. An examination of the several combinations of pair sets that pass and fail the test criterion might be sufficient to help find the source of nonrandom error. But there is a supplementary test in this type of model. It is possible to obtain an estimate of the path hi that will be identical (except for sampling error) for all four sets of pair combinations that involve the nonrandom error. We compute this estimate by *subtracting* the second product of correlations (13.11) from the first (13.10) and then dividing by $r_{x_2y_2} = bdf$:

$$r_{x_1y_1}r_{x_2y_2} - r_{x_1y_2}r_{x_2y_1} = bdfhi = r_{x_2y_2}hi$$

and therefore

$$hi = \frac{r_{x_1y_1}r_{x_2y_2} - r_{x_1y_2}r_{x_2y_1}}{r_{x_2y_2}} \qquad (13.12)$$

In our particular numerical example we get

$$hi = \frac{(.66)(.21) - (.24)(.315)}{(.7)(.5)(.6)} = \frac{.0630}{.21} = .30$$

which of course is consistent with the known a priori values $h = .6$ and $i = .5$.

Thus for the causal model of Fig. 13.7 there ought to be five combinations that satisfy the first consistency criterion. Each of the remaining four should produce similar estimates of hi, the nonrandom-error component. This information will be enough to estimate all coefficients except for h and i, which cannot be separated empirically in this particular model. Indeed, the models of Figs. 13.7 and 13.8 cannot be empirically distinguished unless some additional information is given. We can

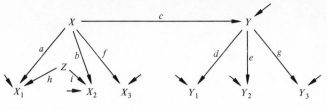

FIGURE 13.9

see this by noting that the algebra for Fig. 13.8 is the same as that of Fig. 13.7 except that the simple path h replaces the compound path hi. This follows since in Fig. 13.8 $r_{x_1y_1} = ade + h$ whereas all of the other correlations are the same as those for Fig. 13.7. This further illustrates the general point that there are always certain causal models that are indistinguishable empirically from each other. As we have just observed, however, each may be readily distinguished from a number of alternatives that do not satisfy the conditions of the model. For instance, it would be rather unusual to have five of the nine comparisons satisfying the first test criterion and the remaining four all involving the combination X_1 and Y_1 and yielding the same estimates for hi (or h).

The two apparently very different models of Figs. 13.9 and 13.10 have implications which differ from those of Figs. 13.7 and 13.8. In Fig. 13.9 there is a source of spuriousness Z producing an additional causal path between two indicators of a *single* unmeasured variable (in this case X). A similar situation would arise if there were two or more sources of spuriousness, say Z_1 and Z_2, relating X_1 to X_2 and X_2 to X_3. The principal feature of this type of model is that no causal path connecting any of the X_i to any of the Y_j is subject to these sources of nonrandom error. Since the (X_i, Y_j) correlations are the only ones that appear in the first test equalities, each of the tests will be passed in spite of the existence of nonrandom measurement errors.

The *estimates* of the path c between X and Y will differ, however. Since in Fig. 13.9 $r_{x_1x_2} = ab + hi$, whereas the other correlations among the X's or among the Y's will be unaffected by Z, we will get a biased estimator of c whenever X_1 and X_2 are used as a pair. It will be recalled that the *denominator* of this estimate contains the product of the correlations between X_i and X_j and between Y_i and Y_j and that the former product will contain the term hi whenever the (X_1, X_2) pair is involved.

Using the same data as for Fig. 12.11, letting $h = .5$ and $i = .4$, and assuming no sampling errors whatsoever, we get

$$r_{x_1x_2} = (.8)(.7) + (.5)(.4) = .76$$

If we used as our estimate of c^2 the value

$$c^2 = \frac{r_{x_1y_1}r_{x_2y_2}}{r_{x_1x_2}r_{y_1y_2}}$$

we would get

$$c^2 = \frac{(.36)(.21)}{(.76)(.54)} = \frac{.0756}{.4104} = .184$$

which underestimates the correct value of .25.

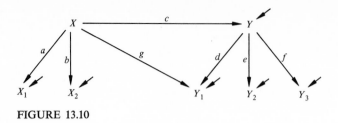

FIGURE 13.10

In the model of Fig. 13.9, the six estimates that do not contain the (X_1, X_2) pair will not be systematically distorted by Z, whereas the three that do contain this pair will produce the same extent of bias. Unfortunately, it will not be possible to predict the direction of this bias unless the signs of h and i are known.

The model of Fig. 13.10 contains a very different type of nonrandom error. As the figure has been drawn, the variable X is itself a cause of measurement error in one of the indicators of Y (in addition to being a direct cause of Y). Put another way, Y_1 is actually an indicator of both X and Y. In this kind of model the pairings of X_1 and X_2 with Y_2 and Y_3 will be uninfluenced by this disturbance, but any pairs involving Y_1 may be affected. For example.,

$$r_{x_1y_1} = acd + ag = a(cd + g) \qquad r_{x_1y_2} = ace$$
$$r_{x_2y_1} = bcd + bg = b(cd + g) \qquad r_{x_2y_2} = bce \tag{13.13}$$

and therefore

$$r_{x_1y_1}r_{x_2y_2} = abce(cd + g) = r_{x_1y_2}r_{x_2y_1}$$

We thus have an additional example of nonrandom measurement error in which the first test criterion can be expected to be satisfied (within the limits of sampling error). Once more the estimate of the causal path between X and Y will be biased because when we divide either $r_{x_1y_1}r_{x_2y_2}$ or $r_{x_1y_2}r_{x_2y_1}$ by the product $r_{x_1x_2}r_{y_1y_2}$, the resulting fraction will contain the unknown g.

The implication of these last two illustrations is that there are causal models involving nonrandom errors that cannot be detected by applying the first consistency criterion alone, but if one combines this test with the computation of the several estimates of the path linking X and Y, one may obtain possible clues as to the source of the measurement bias. However, there will always be several models that imply the same empirical predictions, so a systematic effort should be made to examine the implications of various alternative models.

The above illustrations involve only a single source of nonrandom error at a time. Because with three indicators of each variable the models are highly over-identified, it is not too difficult to locate almost by inspection the indicator or pair of indicators that is the source of the nonrandom errors. But suppose we had two different sources, perhaps one involving X_1 and X_2 and the second linking X_3 to Y_1. It would then be much more difficult to pinpoint the problem without a systematic search procedure. Costner and Schoenberg (1973) suggest the following general strategy in the presence of several such sources of nonrandom errors. First, form all possible two-indicator submodels and use them to search for cross-variable sources of error. One may use the first consistency criterion to check for nonrandom errors of this particular type. If any are found, they can be inserted into the completely random-error model. Perhaps the pairs (X_1, Y_4) and (X_3, Y_2) will need to be connected by paths as a result of this first step in the search procedure, though it may not be known why they are so connected. Perhaps X_1 is a direct cause of Y_4, whereas X_3 and Y_2 are spuriously related owing to some common feature in the way they have been measured.

Having inserted these cross-variable connections among certain of the indicators, and having assured oneself that the model has not become underidentified in doing so, one may then look at all possible three-indicator submodels. In the models we have been considering there is of course only a single three-indicator model, but if we had three indicators of X and four of Y, there would be four such submodels. If we had had four indicators of each variable, there would have been sixteen three-indicator submodels that could be examined in sequence to look for within-variable sources of nonrandom errors, e.g., a common source between X_2 and X_3, and also for the possibility that one of the true values influences an indicator of the other variable, for example, X directly affects Y_1. Costner and Schoenberg provide a much more detailed discussion of this search procedure, as well as several concrete examples. It seems to be the case, however, that if there are a large number of complications relative to the number of indicators, even this type of search will uncover so many alternative models that additional a priori assumptions must be imposed. If certain of the indicators have been obtained by a single means (say, through interviewing) whereas others have a possibly different single source of error, it may become possible to impose a sufficient number of reasonable restrictions on the parameters to achieve a solution in spite of a relatively large proportion of indicators that are subject to nonrandom as well as random errors. We next turn our attention to this kind of possibility.

13.4 MULTIPLE SOURCES OF NONRANDOM ERRORS[1]

Obviously, if we allow for too many sources of nonrandom error at once, we shall encounter identification problems. Yet there are many empirical situations in which there will be multiple sources of error, so that there may not be any indicators that are subject to only random measurement errors. If so, simplifications will have to be introduced in some manner. One rather plausible kind of assumption is that whenever the same method of measurement has been applied to several variables, it will have similar effects on the respective indicators.

One of the first major efforts to systematize this kind of working assumption

---

[1] Portions of this section have been taken, with modifications, from H. M. Blalock, Estimating Measurement Error Using Multiple Indicators and Several Points in Time, *American Sociological Review*, **35**: 101–111 (February 1970), by permission of the American Sociological Association.

was by Campbell and Fiske (1959), who made the important suggestion that several variables be measured by two or more distinct procedures. For example, each of four variables might be tapped by interviews and also by questionnaires. If there were a general methods effect from interviewing common to all variables and a different common effect of the questionnaire procedure, then certain empirical checks might be made to assess the validity of the measures. The basic strategy has been elaborated and improved by Althauser and Heberlein (1970), Althauser, Heberlein, and Scott (1971), and Alwin (1973) and will not be discussed further here.

Instead, we shall return to a consideration of measurements at several points in time, where it is assumed that X has been measured in either two or three distinct ways at each of two or three points in time. Such an assumption makes it reasonable to suppose that the distortions produced by method 1 will be similar at each point in time, and likewise for the remaining methods. Of course such simplifying assumptions will not be completely realistic, particularly if the true variance in X is not constant, but, as we have seen on numerous occasions, certain simplifications will always be necessary in any given piece of research. If one is unhappy about any *particular* set of assumptions, however, it may be possible to relax these assumptions by paying the price of collecting additional indicators. Let us begin by considering a model involving two indicators of X at each of three points in time, comparing the results of this model with one involving three indicators at two points in time.

In the model of Fig. 13.11, a single letter f is utilized to represent the correlations connecting the sources of measurement error at the same point in time. This means that we are assuming that whatever sources, e.g., the ordering of items, are producing spurious covariation between the indicators at time 1 are operating to the same extent at times 2 and 3, relative to the extent of covariation produced by X. As we shall see, it will be possible to test for this particular assumption of equal f's, provided one is willing to accept the correctness of the remaining assumptions.

We are also assuming that it is appropriate to use the same correlation coefficient g linking the sources of measurement error in the first indicator during the interval between times 1 and 2 and between times 2 and 3. Such an assumption would seem reasonably plausible only if the two intervals are roughly comparable in duration. A different coefficient g' has been used for the first indicator between times 1 and 3, thus allowing for attenuation in nonrandom measurement error due to memory loss or other factors. If one were willing to assume some specific relationship between g and g' (say, $g' = g^2$), then it would be possible to reduce the number of unknowns by 1. Similarly, we have introduced h and h' to represent the comparable sources of nonrandom errors for the second indicator between the three points in time.

It should be specifically noted that we are assuming *no* measurement-error correlations connecting different indicators at different points in time. If such additional correlations were introduced, we would have too many unknowns for solution, unless our assumptions about these additional disturbances were unusually simplistic, e.g., all identical. As a general principle, we wish to keep the number of unknowns well below the number of equations so that we shall have a substantial number of redundant equations for testing purposes.

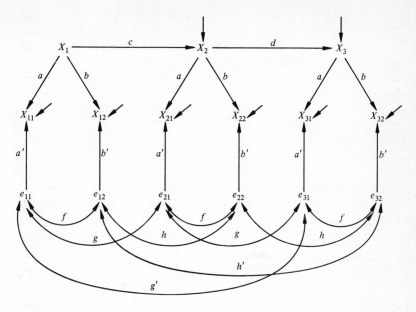

FIGURE 13.11

Since there are a total of six measured values, we may write down the following fifteen equations, one for each correlation:

$$r_{x_{11}x_{12}} = ab + a'b'f \qquad (13.14a)$$

$$(R) \quad r_{x_{21}x_{22}} = ab + a'b'f \qquad (13.14b)$$

$$(R) \quad r_{x_{31}x_{32}} = ab + a'b'f \qquad (13.14c)$$

$$r_{x_{11}x_{21}} = a^2c + (a')^2g \qquad (13.14d)$$

$$r_{x_{21}x_{31}} = a^2d + (a')^2g \qquad (13.14e)$$

$$r_{x_{11}x_{31}} = a^2cd + (a')^2g' \qquad (13.14f)$$

$$r_{x_{12}x_{22}} = b^2c + (b')^2h \qquad (13.14g)$$

$$r_{x_{22}x_{32}} = b^2d + (b')^2h \qquad (13.14h)$$

$$r_{x_{12}x_{32}} = b^2cd + (b')^2h' \qquad (13.14i)$$

$$r_{x_{11}x_{22}} = abc \qquad (13.14j)$$

$$r_{x_{11}x_{32}} = abcd \qquad (13.14k)$$

$$(R) \quad r_{x_{12}x_{21}} = abc \qquad (13.14l)$$

$$(R) \quad r_{x_{12}x_{31}} = abcd \qquad (13.14m)$$

$$r_{x_{21}x_{32}} = abd \qquad (13.14n)$$

$$(R) \quad r_{x_{22}x_{31}} = abd \qquad (13.14o)$$

Looking first at the simplest of these equations, (13.14j) to (13.14o), we see that three of them can be taken as redundant (R) and thus can be used as conditions which the data must satisfy if the model is to be retained. For example, from (13.14j) and (13.14l) we have the implied result that $r_{x_{11}x_{22}} = r_{x_{12}x_{21}}$, which should hold within the limits of sampling error. The three equations of this set that have been taken as nonredundant yield

$$c = \frac{r_{x_{11}x_{32}}}{r_{x_{21}x_{32}}} = \frac{abcd}{abd} \qquad d = \frac{r_{x_{11}x_{32}}}{r_{x_{11}x_{22}}} = \frac{abcd}{abc}$$

(13.15)

$$ab = \frac{r_{x_{11}x_{22}}r_{x_{21}x_{32}}}{r_{x_{11}x_{32}}} = \frac{(abc)(abd)}{abcd}$$

If we now subtract Eq. (13.14e) from (13.14d), we can eliminate $(a')^2 g$ as follows:

$$a^2(c - d) = r_{x_{11}x_{21}} - r_{x_{21}x_{31}} = \delta_1 \qquad (13.16)$$

and similarly, subtracting (13.14h) from (13.14g) gives

$$b^2(c - d) = r_{x_{12}x_{22}} - r_{x_{22}x_{32}} = \delta_2 \qquad (13.17)$$

Provided that $c - d \neq 0$, we can divide the first of these equations by the second, getting

$$\frac{a^2}{b^2} = \frac{\delta_1}{\delta_2} = \frac{r_{x_{11}x_{21}} - r_{x_{21}x_{31}}}{r_{x_{12}x_{22}} - r_{x_{22}x_{32}}} \qquad (13.18)$$

If we then multiply both sides by $(ab)^2$, we get using (13.15)

$$a^4 = \frac{\delta_1}{\delta_2}\left(\frac{r_{x_{11}x_{22}}r_{x_{21}x_{32}}}{r_{x_{11}x_{32}}}\right)^2$$

and therefore

$$a^2 = \pm\sqrt{\frac{\delta_1}{\delta_2}}\frac{r_{x_{11}x_{22}}r_{x_{21}x_{32}}}{r_{x_{11}x_{32}}} \qquad (13.19)$$

Similarly,

$$b^2 = \pm\sqrt{\frac{\delta_2}{\delta_1}}\frac{r_{x_{11}x_{22}}r_{x_{21}x_{32}}}{r_{x_{11}x_{32}}} \qquad (13.20)$$

where we would use either the plus or the minus sign to give positive values for a^2 and b^2. Finally, we may use these estimates of a^2, b^2, c, and d to solve for the composite paths $a'b'f$, etc. Since Eqs. (13.14a) to (13.14c) involve only ab and $a'b'f$, we could have allowed for three different disturbance terms f_1, f_2, and f_3, although this would not permit our using Eqs. (13.14b) and (13.14c) for testing purposes.

This procedure depends heavily on the assumption that $c \neq d$. If, in fact, the two stability coefficients are nearly identical in the population, then even though their sample counterparts may be slightly different, there will be very large sampling errors for estimates of the ratio a^2/b^2 and also for all of the estimates dependent upon this ratio. Therefore, for all practical purposes, the procedure will be useful only if the stability coefficients c and d are very different. Of course, if the three observations were unequally spaced temporally, we might expect this condition to hold, but then

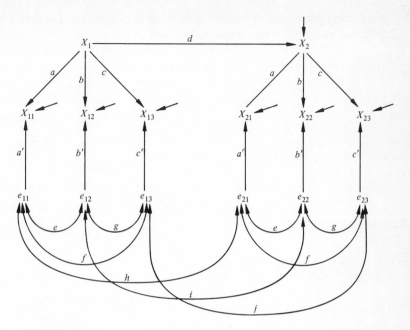

FIGURE 13.12

it would be unrealistic to assume that g and h remain constant from one interval to the next. Therefore, this procedure for estimating coefficients under conditions where nonrandom measurement errors are allowed depends on our finding situations of a rather peculiar nature, such as a pronounced curvilinear trend in the stability coefficients. Hannan, Rubinson, and Warren (1974) have noted a similar difficulty in connection with more complex panel-data models.

In experimental designs where an experimental variable has been introduced between times 1 and 2 and where individuals are measured under other conditions at time 3, it may be possible to create such situations. Or, in natural settings, where some major event has intervened between times 1 and 2 and where a third observation has been taken to imply more long-run consequences of this event, there may also be a sufficient difference between c and d. But if we merely have three observations taken during "normal" time periods, the stability coefficients may be too close together. Of course, Eqs. (13.14j) to (13.14o) may be used to estimate c and d so as to decide whether or not to proceed with the estimates of the remaining coefficients. It would be advantageous to know more about the sampling errors of these rather complex expressions.

Let us now consider Fig. 13.12, in which we have three indicators at each of two points in time. If we wish to allow for the possibility that the three indicators are not equally "good" (by some criterion), not only must we use the three different symbols a, b, and c to represent their direct links with X but we must also allow for different correlations between the sources of measurement error at the same points in time (e, f, and g) and across time (h, i, and j). We shall again assume no correlations connecting the sources of measurement error in one indicator at time 1 with a different indicator at time 2. Our equations are now

$$r_{x_{11}x_{12}} = ab + a'b'e \tag{13.21a}$$

$$r_{x_{11}x_{13}} = ac + a'c'f \tag{13.21b}$$

$$r_{x_{12}x_{13}} = bc + b'c'g \tag{13.21c}$$

$$\text{(R)} \quad r_{x_{21}x_{22}} = ab + a'b'e \tag{13.21d}$$

$$\text{(R)} \quad r_{x_{21}x_{23}} = ac + a'c'f \tag{13.21e}$$

$$\text{(R)} \quad r_{x_{22}x_{23}} = bc + b'c'g \tag{13.21f}$$

$$r_{x_{11}x_{21}} = a^2d + (a')^2h \tag{13.21g}$$

$$r_{x_{12}x_{22}} = b^2d + (b')^2i \tag{13.21h}$$

$$r_{x_{13}x_{23}} = c^2d + (c')^2j \tag{13.21i}$$

$$r_{x_{11}x_{22}} = abd \tag{13.21j}$$

$$r_{x_{11}x_{23}} = acd \tag{13.21k}$$

$$\text{(R)} \quad r_{x_{12}x_{21}} = abd \tag{13.21l}$$

$$r_{x_{12}x_{23}} = bcd \tag{13.21m}$$

$$\text{(R)} \quad r_{x_{13}x_{21}} = acd \tag{13.21n}$$

$$\text{(R)} \quad r_{x_{13}x_{22}} = bcd \tag{13.21o}$$

We immediately see that since there are six equations that can be considered redundant, there are only nine independent equations that can be used to solve for the unknowns. Therefore certain simplifications will be necessary. If we begin with Eqs. (13.21j) to (13.21o) (three of which are redundant), we find that we can no longer estimate the stability coefficient d as was possible when we had three points in time. We can, however, use the three nonredundant equations of this set to obtain

$$\frac{a}{b} = \frac{r_{x_{11}x_{23}}}{r_{x_{12}x_{23}}} \qquad \frac{a}{c} = \frac{r_{x_{11}x_{22}}}{r_{x_{12}x_{23}}} \qquad \frac{b}{c} = \frac{r_{x_{11}x_{22}}}{r_{x_{11}x_{23}}}$$

$$a^2d = \frac{r_{x_{11}x_{22}}r_{x_{11}x_{23}}}{r_{x_{12}x_{23}}} \qquad b^2d = \frac{r_{x_{11}x_{22}}r_{x_{12}x_{23}}}{r_{x_{11}x_{23}}} \tag{13.22}$$

$$c^2d = \frac{r_{x_{11}x_{23}}r_{x_{12}x_{23}}}{r_{x_{11}x_{22}}}$$

We immediately see that the last three expressions can be used in conjunction with Eqs. (13.21g) to (13.21i) to solve for the three compound paths representing the disturbances over time, $(a')^2h$, $(b')^2i$, and $(c')^2j$. However, since Eqs. (13.21d) to (13.21f) are redundant with Eqs. (13.21a) to (13.21c), we cannot estimate a, b, and c if we know only the ratios a/b, a/c, and b/c.

If we assume that $(a')^2h = (b')^2i = (c')^2j$, we merely make two of the three equations (13.21g) to (13.21i) redundant. Let us therefore examine the implications of the assumption that $a'b'e = a'c'f = b'c'g$. If this is the case, we may subtract

(13.21*b*) from (13.21*a*), obtaining an expression for $a(b - c)$. Utilizing the expressions for a/b and a/c, we obtain the result that

$$a^2 = \frac{r_{x_{11}x_{22}}r_{x_{11}x_{23}}}{r_{x_{12}x_{23}}} \frac{r_{x_{11}x_{12}} - r_{x_{11}x_{13}}}{r_{x_{11}x_{22}} - r_{x_{11}x_{23}}} \tag{13.23}$$

Similar expressions can of course be obtained for b^2 and c^2. Also we get

$$d = \frac{r_{x_{11}x_{22}} - r_{x_{11}x_{23}}}{r_{x_{11}x_{12}} - r_{x_{11}x_{13}}} \tag{13.24}$$

It should be noted that each of these expressions contains a ratio of *differences* between two correlations, and we can therefore expect considerable sampling error whenever the comparable population differences are small. If we examine the nature of the pairs of correlations involved in these differences, we see that the estimating procedure depends rather heavily on our being able to find three indicators that are related rather differently to X. For example, if the second and third indicators are more or less interchangeable, it would seem rather unlikely that their correlations with the first indicator would be very different, and thus we would expect both pairs of differences between correlations to be rather small. This result is somewhat analogous to the practical restriction obtained in connection with three time periods, namely, the requirement that the two stability coefficients be different in value.

13.5 CAUSE INDICATORS AND NONRANDOM ERRORS

It was pointed out in the previous chapter that it is sometimes necessary or feasible to use some indicators that are taken as causes, rather than effects, of certain of the theoretical variables. We noted that in many experimental designs the investigator manipulates the setting, the instructions to subjects, or perhaps the behavior of stooges in order to produce certain motivational states in the subjects. These motivational states, rather than the specific manipulations, may be the theoretical variables of interest, the prediction being that changes in these internal states will, in turn, affect behavior. In natural settings we often take group memberships, e.g., family status, religious denomination, or union membership, as indicators of socialization variables assumed to affect psychological states. We also examined the implications of several random-error models, one of which involved a combination of one cause and one effect indicator of X.

Let us now consider the nature of certain complications that may be introduced whenever the cause and effect indicators of X can be interrelated through paths other than the individual paths connecting each indicator to X. Obviously there will be numerous possibilities, only a few of which can be examined. First we note that a cause indicator, such as X_1 in Fig. 13.13, may be causally linked to the dependent variable Y and to X's effect indicators by one or more paths that do not go through X. If X is not directly measured but only X_1, X_2, and X_3 are known, it will clearly be impossible to estimate the direct effect of X on Y if we allow for this degree of complexity. We shall confine our attention to situations in which X_1, the cause indicator of X, is assumed not to affect Y or any of its indicators except through X. This is

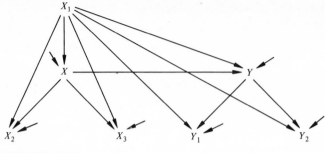

FIGURE 13.13

of course the crucial simplifying assumption that is usually made in experimental designs. It amounts to assuming that the manipulation has been simple enough to activate the single motivational factor X while not influencing a wide variety of additional potential intervening variables.

Miller (1971) has discussed this model in some detail, noting that multiple cause indicators may be needed if one is unwilling to make this particular kind of simplifying assumption. Costner (1971a) has pointed out that certain kinds of flaws in experimental designs can be discovered, provided that one has a combination of cause and effect indicators of X. In particular, he notes that in the model in which X_1 may directly affect Y or its indicators Y_1 and Y_2, the indicators X_2 and X_3 can be used to infer the existence of these complications provided that the direct arrows between X_1 and both X_2 and X_3 are removed.

We shall consider a somewhat different set of simplified models in which the cause and effect indicators may be interrelated via paths other than through X but in which there are no further complications. Turning to Fig. 13.14, let us first assume that the second cause indicator X_3 is unavailable and that we are attempting to infer the effects of X on Y on the basis of X_1 and X_2. The model predicts the following equations:

$$r_{x_1x_2} = ab + f \qquad r_{x_1y_2} = ace$$
$$r_{y_1y_2} = de \qquad r_{x_2y_1} = cd(b + af) \qquad (13.25)$$
$$r_{x_1y_1} = acd \qquad r_{x_2y_2} = ce(b + af)$$

from which we see that the first consistency criterion is satisfied, whereas if we attempt to estimate c^2 from the formula

$$\frac{r_{x_1y_1}r_{x_2y_2}}{r_{x_1x_2}r_{y_1y_2}} = \frac{ac^2de(b + af)}{(ab + f)de} = c^2 \frac{ab + a^2f}{ab + f} \qquad (13.26)$$

we obtain a biased estimate. If all correlations are positive, the estimate of c^2 is biased downward unless $a^2 = 1$, with the degree of attenuation increasing with f. This result implies that if the experimental manipulation or natural event affects X_2 by any mechanism other than through X, this fact will affect our inferences about the supposed effects of X on Y. Furthermore, the first test criterion will not permit one to distinguish this situation from the random-error model in which $f = 0$.

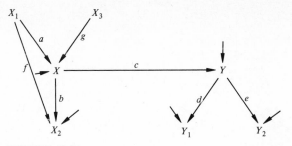

FIGURE 13.14

Suppose, instead, that we happened to select the cause indicator X_3 rather than X_1. In this very simplified model we are assuming no correlation whatsoever between X_1 and X_3, as well as no additional linkage between X_3 and the effect indicator X_2. Under these conditions it might be thought that if X_1 were unknown or ignored, the pair of indicators X_2 and X_3 would produce results similar to those for the random model. But it can easily be seen that although the first consistency criterion will once more be satisfied, the usual estimate of c^2 will be

$$\frac{r_{x_3y_1}r_{x_2y_2}}{r_{x_2x_3}r_{y_1y_2}} = \frac{c^2 deg(b + af)}{bdeg} = c^2 \left(1 + \frac{af}{b}\right) \qquad (13.27)$$

The implication is that if the effect indicator is influenced by *any* cause of X through *any* mechanism that does not involve X as the only intervening variable, we are in difficulty. Clearly, this is a discouraging result which places a severe limitation on the use of combined cause and effect indicators.

Next consider the model of Fig. 13.15 in which there are two effect indicators and a single cause indicator, the latter being linked to one of the effect indicators X_2 via at least one path that does not go through X. The equations for this model are

$$\begin{array}{ll} r_{x_1x_2} = ab + f & r_{x_1y_2} = ace \\ r_{x_1x_3} = ag & r_{x_2y_1} = cd(b + af) \\ r_{x_2x_3} = g(b + af) & r_{x_2y_2} = ce(b + af) \qquad (13.28) \\ r_{y_1y_2} = de & r_{x_3y_1} = gcd \\ r_{x_1y_1} = acd & r_{x_3y_2} = gce \end{array}$$

The first consistency criterion is satisfied for all three of the pairs (X_1, X_2), (X_1, X_3), and (X_2, X_3). Applying the second criterion in estimating c^2, we get

$$\frac{r_{x_1y_1}r_{x_2y_2}}{r_{x_1x_2}r_{y_1y_2}} = \frac{ac^2de(b + af)}{(ab + f)de} = c^2 \frac{ab + a^2f}{ab + f} \neq c^2$$

$$\frac{r_{x_1y_1}r_{x_3y_2}}{r_{x_1x_3}r_{y_1y_2}} = \frac{ac^2deg}{adeg} = c^2 \qquad (13.29)$$

$$\frac{r_{x_2y_1}r_{x_3y_2}}{r_{x_2x_3}r_{y_1y_2}} = \frac{c^2 deg(b + af)}{deg(b + af)} = c^2$$

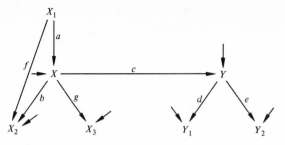

FIGURE 13.15

Thus, two of the three estimates of c^2 should be approximately correct except for sampling error; and provided one is willing to assume the correctness of the model in other respects, this would enable the investigator to infer which of the two effect indicators is directly linked with the cause indicator X_1. Obviously, if there were additional effect indicators of X_1, most of which were assumed to involve only random error, it would then be easier to locate the one or two faulty effect indicators. In instances where it is necessary to allow for nonrandom sources of error in most of the indicators, one must introduce simplifications of the same nature as were made in connection with models in which all measures were taken as effect indicators.

Although these particular models do not sufficiently illustrate the point, cause indicators guard against different kinds of sources of nonrandom error than effect indicators guard against (Curtis and Jackson, 1962). Therefore, whenever possible, it will be desirable to use multiple indicators of both types. It may well turn out to be the case, however, that really useful cause indicators will be difficult to find in many practical applications. The investigator should nevertheless be sensitive to the fact that cause and effect indicators do not always behave similarly with respect to their predicted intercorrelations with other indicators. One should be especially on guard against procedures that supposedly permit one to appraise the "validity" of an indicator on the basis of magnitudes of correlation coefficients, without the benefit of a specific theoretical model.

13.6 NONADDITIVE MEASUREMENT-ERROR MODELS

Thus far all the measurement-error models we have considered, with the exception of the brief discussion of errors due to classification, have involved the assumption that all equations are of the linear and additive form. Given the fact that the unmeasured variables in these equations typically create situations involving relatively large numbers of unknowns, such additive relationships will often serve as the only approximations that do not produce underidentified equations. Nevertheless, there are many instances in which it is reasonable to modify these additive equations so as to take into consideration certain kinds of complications that are frequently noted in the literature. In particular, one often finds situations where objective indicators are substituted for internal psychological states and where simple multiplicative models appear to be more realistic than additive ones. The basic reason is that persons may not all react to objective situations in the same way. They may define these situations differently or attach very different meanings to them.

Suppose we have a single dependent variable Y assumed to be perfectly measured or measured only with random error. Suppose, also, that we have two independent variables X_1 and X_2 representing two postulated internal states (perhaps two attitudes) that are imperfectly measured by the cause indicators X_1' and X_2', respectively. Let us further assume that the true equation for Y involving X_1 and X_2 is actually linear and additive in form as given by

$$Y = \alpha + \beta_1 X_1 + \beta_2 X_2 + \varepsilon$$

However, suppose we wish to allow for the fact that individuals may react differently to the objective variables X_1' and X_2'. For example, X_1' may be the percentage of a minority in the local area, whereas the theoretical variable X_1 may be the degree of threat experienced by members of the dominant group. Not all such individuals will be equally threatened by the presence of the minority. Similarly, X_2' may be the individual's income level, whereas the theoretical variable X_2 may be degree of economic security. Not all persons having the same objective incomes may be equally secure economically for a number of reasons.

When we use additive equations with constant coefficients, we implicitly assume homogeneity with respect to the causal processes that are operative. That is, we assume that a given change in X_1 (holding X_2 constant) will always produce the same change in Y because of the constancy of β_1. But the more heterogeneous our population the greater the likelihood that there will be different sets of β's for different subpopulations within the larger population. For example, males may react differently from females, younger persons from older ones, highly educated from those with less education, and so forth. Therefore the coefficients that have thus far been treated as constants should perhaps more realistically be taken as *variables U, V,* and *W.* These variables, in their own right, may require theoretical explanation and may very well be causally linked with some of the remaining variables in the theoretical system. For example, the value of β_1 may be systematically related to the level of income X_2' or to economic security X_2.

In general, we might anticipate that the more heterogeneous the population under investigation the less reasonable it is to assume constant coefficients for the X_i. Thus as one attempts to extend the scope of one's generalizations, realism dictates that what may have previously been taken as constants for relatively homogeneous subpopulations must now be treated as variables. Suppose we let $\alpha = U$, $\beta_1 = V$, and $\beta_2 = W$. The equation linking Y to the true values of X_1 and X_2 then becomes

$$Y = U + VX_1 + WX_2 + \varepsilon \qquad (13.30)$$

and we have two multiplicative terms VX_1 and WX_2 to consider. All five of the "independent" variables in Eq. (13.30) may of course be interrelated with other variables in the system.

In the present context we shall confine our attention to linear and additive models in the case of the *true* X_i while allowing for multiplicative terms involving the measured variables X_1' and X_2'. That is, we shall assume homogeneity with respect to the coefficients β_1 and β_2, as well as α, which will be taken as constants. But when we substitute the indicators for X_1 and X_2, we shall allow for slippage in the form of different coefficients connecting X_1' with X_1 and X_2' with X_2. That is, we may wish to allow for the fact that different individuals may interpret, react to, or give different meanings to the same objective phenomena, the X_i'. This will introduce multiplicative terms into the equations involving the *measured* variables. Thus if people react differently to the objective variable X_1', we shall allow for this possibility by replacing the constant β_1' by the variable WV', and similarly for other terms in the equation.

Obviously, there are numerous possible models that might be discussed. We shall confine our attention, however, to two illustrative kinds of models, those involving cause indicators of an independent variable and those involving effect indicators of a dependent variable.

Cause Indicators of Independent Variables

We have already noted the illustration where an objective situational factor, such as the percentage of a minority X', is thought to influence one or more postulated internal states of individuals, e.g., perceived threats, and where one's theory is primarily concerned with the latter unmeasured variables as they relate to dependent variables such as aggressive behavior. Let us therefore focus on a somewhat different example involving the measurement of education. The most common measures of "education" in the sociological literature involve the number of years of formal schooling. Conceptually, however, the interest may be in "total amount learned," "amount of knowledge retained," "prestige accruing from having received formal education," "educational maturity," or some other similar notion. More generally, we often find it desirable to substitute a measure of time or money invested in an activity, e.g., watching television, for an internal psychological state that is difficult to define conceptually or to measure operationally. We assume that amount of knowledge absorbed or the influence of an activity is roughly proportional to the time invested.

Yet we clearly recognize that knowledge absorbed is not only a function of time invested but also of the quality of the experience, in this case quality of education. It is also a function of the retention of previous knowledge, which, in turn, may be a function of quality. If quality were held constant, it would be proper to use the time-based measure under the assumption that there would be only a constant scale factor to be applied to all individuals. That is, if amount learned X were a constant function of time invested X', we could write $X = kX'$, or perhaps some nonlinear but monotonic function such as $X = k \ln X'$ or $X = k_1 e^{k_2 X'}$. But if quality actually varies across individuals, and particularly if quality is a function of other variables in the theoretical system, the problem is more difficult.

One very simple multiplicative model that takes quality into consideration is the equation $X = kWX'$, where W is a quality coefficient that may be a function of another variable in the theoretical system, say father's income Z. If we wish to relate education to a dependent variable Y, which may also be a function of Z, we may then use the equation system

$$Y = \alpha_1 + \beta_1 X + \beta_2 Z + \varepsilon_y$$
$$X = kWX' \qquad\qquad (13.31)$$
$$W = \alpha_2 + \beta_3 Z + \varepsilon_w$$

where for convenience the second equation has been taken as exact. If years of education X' replaces the unmeasured "true education," which is also a function of quality, the equation for Y in terms of X' and Z becomes

$$Y = \alpha_1 + \beta_1(kWX') + \beta_2 Z + \varepsilon_y$$
$$= \alpha_1 + \beta_1 k(\alpha_2 + \beta_3 Z + \varepsilon_w)X' + \beta_2 Z + \varepsilon_y$$

which is of the form

$$Y = A + B_1 X' + B_2 Z + B_3 X'Z + U \qquad\qquad (13.32)$$

Thus the theoretical system implies that there should be an interaction (in the form of the product term $X'Z$) between years of schooling X' and father's income Z even though the joint effects of true education and father's income on Y are strictly additive. In substantive terms, if quality and father's income are positively related, as one would expect, then education will appear to have a greater payoff in terms of Y for children of wealthier parents. This is of course true for education as measured in terms of years of formal schooling X', but is not the case with respect to true education X according to the above model.

This type of model does not create special difficulties, theoretically, as long as one is careful to distinguish between X and X'. In practice, however, it is common to report empirical results in terms of years of schooling X' (neglecting the quality aspect) but to discuss one's results as though $X' = X$. In other words, it is common to slip back and forth between one's measured variables and the theoretical constructs by making the implicit assumption that variables such as quality W are either constant or unrelated to the remaining variables in the theoretical system.

It would be desirable if one could estimate the parameters α_1, α_2, β_1, β_2, and β_3 from the empirical coefficients A, B_1, B_2, and B_3, but it will be seen that the former set contains one more unknown than the latter and that the important coefficient β_1 giving the effects of true education on Y cannot be estimated without additional information. This is what one would expect since X is unmeasured. Of course if quality W were measured, we could obtain X, apart from a scale and random-error factor, through the equation $X = kWX'$. Given the imperfect measurement of all variables in the system, one may at least take advantage of the prediction that X' and Z should have nonadditive effects on Y to provide a weak test of the model. If there were several indicators of Y and Z and several variables in the role of X', multiple tests would be possible. In general, whenever one must resort to weak tests of this sort, it is advisable to rely on multiple tests, since there will of course be many alternative explanations for any single nonadditive relationship. If multiple tests all lend plausibility to the model, this would justify the effort to obtain more precise measures of quality W and to study more systematically its relationship with the remaining variables in the theoretical system.

Effect Indicators of Dependent Variables

Although we might also focus our attention on illustrations involving effect indicators of independent variables, we have selected for our final example a model in which a dependent variable can be measured by effect indicators that are multiplicative functions of the true values and some other variable W that may be related to some of the remaining variables in the system. One example of this kind of situation may occur in attitudinal research, where the measured value Y' is a function of the true attitude and a component W that involves the respondent's resistance or unwillingness to state his or her true position. Perhaps the item is a controversial or threatening one, with some respondents being inhibited from stating a position that appears too extreme. But the degree of threat may be a function of education or economic status, so that the "threat coefficient" W needs to be brought explicitly into the theoretical system.

Let us consider a somewhat different example in which the dependent variable Y is conceptualized to be the frequency or total number of behavioral acts of some type, whereas Y' involves a nonrandom sampling of these acts. If the investigator relies on self-reporting, the respondent may underreport the true frequency, either because of faulty memory or because of the controversial nature of the behavior. Or one may systematically overestimate the frequency. If the investigator relies on official records that are incomplete or collected with ambiguous guidelines as criteria, the recorded frequency may depart from the true frequency in either direction but in some nonrandom fashion. Recorded crime and delinquency rates are an obvious example of a nonrandom sampling of behavioral acts. It is well known that the number of recorded crimes, arrests, and convictions represents but a fraction of the actual number of criminal acts, but the proportions are not constant for all sub-populations or all types of offenses.

If Y represents the actual frequency of the behavior in question, whether that of a single individual or a rate characteristic of a census tract or larger geographic territory, then the recorded frequency Y' can be taken as the product of Y times W, where W represents the probability (or proportion of times) that the behavior is recorded. Obviously, then, W depends upon a number of factors including the skills of the individuals in avoiding detection (if the act concerned is a deviant act) and the motivation and ability of the recording agents to produce a value of W close to unity. In the case of criminal behavior, the latter would depend on such factors as the number of police in the area, the degree of cooperation they receive from witnesses and victims, the degree of bias in their recording system, their physical resources, the total number of crimes in the area, the degree to which efficiency and accuracy are demanded by the public at large and by supervisory personnel, and so forth. Obviously, then, the probability of being detected W will be a function of many of the variables that are also likely to be included in the theoretical system. Let us assume that W is a function of X (say, socioeconomic status), where we are attempting to study the effects of this same X on the true rates Y.

If $Y' = WY$, where for simplicity we neglect any possible stochastic error term, and if both Y and W depend upon X according to the linear equations

$$Y = \alpha_1 + \beta_1 X + \varepsilon_y \quad \text{and} \quad W = \alpha_2 + \beta_2 X + \varepsilon_w \quad (13.33)$$

then the measured rate Y' will be a nonlinear function of socioeconomic status X according to the formula

$$Y' = WY = (\alpha_1 + \beta_1 X + \varepsilon_y)(\alpha_2 + \beta_2 X + \varepsilon_w)$$

which is a parabola of the general form

$$Y' = A + B_1 X + B_2 X^2 + U \qquad (13.34)$$

Therefore even though there is a linear relationship between X and the true rate Y, the relationship between X and the measured value Y' is predicted to be parabolic. In the special case where Y and X are *not* linearly related, so that $\beta_1 = 0$, it can be seen that we would expect a *linear* relationship between the measured value Y' and X.

Of course it would be desirable to estimate the separate coefficients α_1, α_2, β_1, and β_2, but this will not be possible without additional information. Again, if there are multiple indicators in the role of Y', or several measures of X, we can perform a series of weak tests of the model. Each such test will take the form of predicting that the relationships between measured values of Y and measures of X will be nonlinear and parabolic in form. Should such predictions prove correct, the investigator would be motivated to attempt to measure W more directly and to construct alternative theories relating W to other variables in the system.

13.7 CONCLUDING REMARKS

The general point of this chapter has been the argument that whenever nonrandom sources of measurement error are anticipated, the investigator should attempt to state explicit theories linking the sources of these measurement errors to the remaining variables in the theoretical system. Furthermore, because of identification problems resulting from the large number of unknowns, one will of necessity have to rely on multiple measures of as many variables as possible. But the use of multiple indicators, alone, is not sufficient. There must be an accompanying theory that permits one to link both measured and unmeasured variables in terms of specific equations. In the interest of simplicity, most of these equations will probably have to be linear and additive in form, although we have just examined several situations in which multiplicative relationships may provide a more reasonable alternative.

One should attempt to develop as realistic an *auxiliary theory* of measurement error as possible, taking into consideration whatever confounding factors seem necessary. It will undoubtedly be discovered, however, that allowances for all plausible sources of error will produce underidentified systems. To the degree that this can be anticipated *in advance* of data collection, the investigator may then make a series of important decisions that will ordinarily be closed off once the data have been collected. One may either simplify the theory by removing arrows from the model, thereby assuming that certain of the coefficients are zero, or one may search for additional indicators to increase the proportion of measured to unmeasured variables. In either case, the objective should be to achieve systems that are *overidentified*, so that consistency checks may be made. Since these checks are basically weak tests of the theory in the sense that there will ordinarily be many plausible rival hypotheses, it will as a general rule be advisable to make the theoretical system highly overidentified so that multiple tests can be made. When one is reasonably satisfied that a given model is appropriate, it will be possible to proceed to the estimation of the parameters whenever they are estimable. For this purpose, simple arithmetic means may be used as rough approximations although weighted averages will give estimates of greater efficiency.

We have examined a range of different kinds of measurement-error models, including those in which indicators may be taken as causes as well as effects of the unmeasured variables. But we have merely scratched the surface in terms of the possible alternatives. We also have had to imply at several points that for many kinds of variables of theoretical interest to social scientists our theoretical concep-tualizations are simply not clear enough to suggest how we should proceed. The major point in this connection is that a choice of statistical estimating procedures must be predicated upon a series of theoretical assumptions. If the latter are not made explicit, one can hardly expect to develop definitive guidelines for analysis.

In many instances we simply lack the resources to collect more than a handful of highly indirect indices of a very complex theoretical construct such as the notion of industrialization or that of political instability. In studying individuals we cannot possibly work with detailed and noncomparable "life histories" and therefore find it necessary to use group-membership categories or rough measures of characteristics of peers or friends in order to attempt to measure so-called structural effects or the impacts of social environmental factors. But these group memberships and peer characteristics may also be partly dependent on the variables of interest, e.g., when an individual is self-selected into a voluntary organization or informal network of peers. In all these instances there then may be complex reciprocal causation between indicators and unmeasured variables, whereas we have confined our attention to recursive measurement-error models. In such situations we may be dealing with a set or block of variables, some measured and others not, that are complexly inter-related in unknown ways. If we use a simple effect-indicator model in this kind of situation, we may introduce serious distortions, though perhaps not. Until we study a sufficient number of specific models of this nature, perhaps through a series of simulation studies, we simply shall not know. In the meantime, how can we proceed in such cases?

If the theoretical concept is an important one for the analysis, this kind of situation calls for careful conceptual clarifications based on a series of exploratory research projects before the main investigation. Empirical procedures such as factor analysis, latent-structure analysis, or smallest-space analysis that permit one to examine the question of underlying dimensionality are important tools for this exploratory phase, particularly if the discovery of multidimensionality can lead to conceptual clarifications making it possible to distinguish between a number of (unidimensional) variables that can be more carefully measured during subsequent investigations. The general principle of analysis in this connection is that whenever one anticipates that basic concepts are likely to involve more than single underlying dimensions, it is wise to conduct methodological investigations of these kinds of complications *before* the major study gets underway, while there is still time to reconceptualize one's variables and decide upon possible modifications in the measurement procedure.

Assuming that one has done as much preliminary work of this nature as feasible and that one is left with relationships between unmeasured and measured variables sufficiently ambiguous to preclude constructing simple causal models, it may become necessary to construct a single index for the entire block. Of course if all of the indicators for a block are highly intercorrelated, it is not likely to make much difference how one obtains such a measure. In the limiting case where all indicators are perfectly intercorrelated, one may merely use that single indicator which is easiest and cheapest to obtain. When the intercorrelations are of the order of .8 or .9, a simple arithmetic mean of two or three of the cheapest indicators will undoubtedly produce results that are virtually equivalent to those obtained by more sophisticated weighting schemes. At the opposite extreme, where indicators are only very weakly intercorrelated, the choice among weighting schemes will make a much greater difference, but in these situations there is likely to be such a high degree of uncertainty that *no* procedure can be relied upon to yield unambiguous results. Here, it would generally seem wise to treat each indicator as tapping a single theoretical variable, provided that the resulting causal model makes sense theoretically. The interested reader is referred to Jacobson and Lalu (1974) for a discussion of the relative advantages and disadvantages of single indicators, multiple indicators, and combined indices under a variety of circumstances.

When the measured values of variables in a conceptual block are moderately intercorrelated, it is especially difficult to know how to proceed in the absence of a theory. In view of this, the best strategy is probably a flexible one that involves analyzing the data in several different ways. If the measures within a block are to be combined into a single score, there appear to be two general approaches that are likely to yield very similar results under most circumstances, assuming that there is a single clustering of scores implying unidimensionality. The first is to use a strictly internal criterion for selecting the weights, such as the factor loadings of each indicator with the single factor representing the block; i.e., we may factor-analyze the set of scores for the block, and *if* a single-factor model seems appropriate, we obtain an overall factor score for the block *before* examining its relationship with other variables in the system. Strictly speaking, since the causal interrelationships within the block may be much more complex than presumed by the factor-analysis model, we would be using factor analysis merely as a rationale for reducing the block of variables to a single measure rather than as a tool for investigating the causal structure within the block. We may also measure other blocks in a similar fashion and then treat each block as though it were a single variable in the model.

The second kind of approach involves selecting weights for the indicators within a block according to their predictions to other variables *outside* this block. The latter may be treated either individually or collectively. In the first instance, let us suppose that we have two blocks A and B causally prior to the indicators X_1, X_2, and X_3, which are taken to be separate indicators of the variable X. We might weight the indicators in block A (and separately those in block B) so as to maximize the explained variance in X_1. We could then repeat the same process for X_2 and X_3, making separate tests of causal models if we wished (see Sullivan, 1971, 1974). This allows for the possibility that the weights for the indicators using X_1 as a criterion may be very different from those using X_2 as a criterion, and it also permits one to make more than one test of the overall model taking advantage of this flexibility. If, however, we preferred to combine the X_i beforehand, we could again do so by factor-analyzing these variables. Or we could use canonical correlation to obtain optimal weights for *both* the independent variable (here block A) and the dependent variable (here X) taken simultaneously (see Hauser and Goldberger, 1971). Once again, if we shifted to a different dependent variable Y (perhaps measured by four indicators), the weights given to the block A indicators might be very different from those obtained using X as a dependent variable. Furthermore, if we added several more indicators to X, say X_4 and X_5, this would be expected to change the weights given to the block A indicators.

There appears to be very little point in attempting to compare the relative merits of these alternative approaches in the abstract, for reasons that have already been indicated. If one has no idea what is going on within the various blocks, one may choose to rely on some such notion as maximizing the unexplained variance, either in a single indicator of a dependent variable or in a combined index. But this is a purely statistical argument that cannot readily be given a substantive justification unless and until one is willing to postulate a causal structure between the indicators within each of the relevant blocks. Thus, if one finds it necessary to fall back upon such strictly statistical criteria of maximization (or minimization) of some quantity, this should be taken as a sign of ignorance of the theoretical structure of the model rather than as a stopping point of the analysis.

Finally, before concluding our discussion of measurement errors, we should note that we have not discussed any of the complications that may result when one or more variables is defined to be some mathematical function of certain of the others. We may briefly note three kinds of illustrations of this kind of situation. First, one variable may be taken as a difference (or sum) of two others, or as some more complex function involving differences. For example, occupational mobility may be measured as the difference between occupational statuses at times 1 and 2; status inconsistency as a difference between two statuses occupied simultaneously; or cognitive imbalance as the difference between two attitudinal levels. In situations like this, if we wish to separate out the effects of such a derived variable from those of the two (or more) components, we shall not be able to do so unless we specify some particular kind of nonadditive model. Thus one cannot separate out the (additive) effects of first occupation, second occupation, *and* occupational mobility on some dependent variable (Blalock, 1967c; Taylor, 1973).

A second kind of complication that is especially likely to occur in macrolevel studies frequently arises where variables are defined as ratios or products of other variables. Often the same variable, e.g., population size, appears in the denominators of many other variables, e.g., per capita measures or proportions, but it may also appear in the numerators of others. For example, one may wish to relate the size of an organization to the proportion of its members who are in managerial positions. In such situations we may anticipate that correlations will exist that are merely artifacts of the measures used, and we should make explicit note of the fact that ratios (or products) introduce nonlinearities into what might otherwise be taken as linear relationships. This is a rather old problem, which has been recently called to the attention of sociologists by Freeman and Kronenfeld (1973) and Schuessler (1973).

A third and rather difficult type of problem involving exact functional relationships is encountered when we attempt to aggregate individual data to obtain macrolevel measurements or when we wish to reverse the process and disaggregate to smaller units. Here we may be concerned with the "consistency" of results across different levels of aggregation, and it turns out that we must make rather restrictive linearity assumptions if we ever hope to achieve reasonably simple results. This issue was originally posed to a sociological audience in a classic paper on ecological correlations by Robinson (1950) and has also been discussed by Goodman (1953, 1959), Duncan and Davis (1953), Duncan et al. (1961), and Blalock (1964, 1971b). The reader is especially referred to Hannan (1971) for a discussion and integration of the sociological literature on the subject with the much more extensive and technical treatments of aggregation in the econometrics literature.

Measurement problems are highly technical and require an understanding of a number of different approaches to multivariate analysis. But it would be a mistake for the reader to infer from this that technical discussions alone can resolve the tough conceptual problems stemming from a lack of a major effort to clarify the theoretical definitions of the basic concepts that appear in the social science literature. It is in the area of measurement especially that the integration of theory and research is absolutely essential. One's statistical analyses of data must take measurement decisions as a starting point, although we have just seen that many of the same kinds of statistical tools that are applied to multivariate analyses of complex causal models may also be used to help resolve certain kinds of measurement decisions and to rationalize the process of estimating structural parameters in the face of imperfect measurement. In effect, all this means that we can have faith in the results of these analyses only to the degree that we are comfortable with the quality of our data and the linkages that have been established between our theoretical definitions and research operations. Thus we see a knowledge of the tools of multivariate analysis as providing what amounts to a necessary condition for real progress whenever we must deal with a complex reality. But, by the same token, this knowledge cannot provide the sufficient conditions as well. The latter will also require good substantive theories.

EXERCISES

13.1   Using the data of Exercise 12.3:
　　　(a) Suppose there is a common cause W which affects the indicators Y_1 and Y_2, with $r_{wy_1} = .40$ and $r_{wy_2} = .50$. What modifications will this make in the calculations for Exercise 12.3? Comment on your results.
　　　(b) Now suppose that there is no such W but there is a Z which is a common cause of X_3 and Y_1, with $r_{x_3 z} = .60$ and $r_{y_1 z} = .50$. How does this modify the results of Exercise 12.3?
　　　(c) Suppose that the revised model contains *both* W and Z. How does this affect the results?
13.2   Given Fig. E13.2 (page 610):
　　　(a) Write out a set of equations expressing the correlations between indicators in terms of the various path coefficients.
　　　(b) Can you find a way to estimate c^2 from this model? If not, what simplifications will make it possible to estimate c^2?

(c) Suppose there were also a correlation between X_1 and X_2. How would this affect your answer to part (b)?

(d) For the original model, with no correlation between X_1 and X_2, check to see if the first consistency criterion is satisfied for all combinations of indicators.

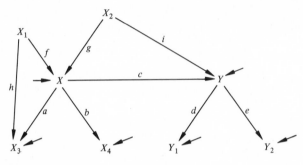

FIGURE E13.2

13.3 Suppose you have reason to suspect that in a simple model in which X causes Y, with two indicators of both X and Y, there is measurement error in Y that is proportional to X. Construct a set of equations representing this possibility and assess the implications from the standpoint of (a) whether or not the first consistency criterion would be approximately satisfied and (b) whether or not the estimate of the true correlation between X and Y would be affected. Now do the same for the case where measurement error in X is proportional to Y. Assume random measurement errors elsewhere in the system.

13.4 Two stimuli X_1 and X_2 are assumed to influence an internal state X, which in turn affects Y in accord with a linear equation. But the coefficients linking X to its cause indicators X_1 and X_2 are assumed to be each linear functions of W, which is also a cause of Y. Write out a set of equations to represent these possibilities and discuss their implications for data analysis under the assumption that X cannot be directly measured.

Appendixes

PROGRAM CARTROCK

A.1 DESCRIPTION[1]

Instructions for using program CARTROCK are embedded in the appropriate places in the listing; also, see Secs. 2.15 and 2.16. When properly instructed, the program will provide solutions for all single-equation ordinary-least-squares regression models. Since it has the unusual feature of requiring no explicit unit vector, it will perform all analyses for Part One of this book. The job-control statements given here are only illustrative, as they will vary from installation to installation.

A.2 TECHNICAL REQUIREMENTS

Region Size

The program requires approximately 61 to 75 kilobytes of computer space. (The size is approximate because the user-supplied SUBROUTINE DATA will vary.)

---

[1] Program CARTROCK was written by Richard C. Rockwell, of the University of North Carolina, with minor program and documentation changes by the author of Part One. We are grateful to Dr. Rockwell for permission to use his program.

Compiler

The program can be compiled by any Fortran compiler permitting use of logical IF statements, logical operators, logical relations, implied DO loops, and COMMON statements.

Input and Output

The program is written to use a card reader (unit 5) as input, a line printer (unit 6) as output, and a magnetic-tape drive (unit 4) as a scratch tape for intermediate results. (The scratch tape must accommodate storage of all analysis-variable values for all cases.)

Maximum Number of Cases

The maximum of 500 can be changed by increasing the dimensions of Y and XINV in the main program. Where N cases are desired, the dimension of Y should be N or greater and the dimension of XINV should be N^2 or greater.

Maximum Number of Input Variables

This maximum is 25, but it can be changed by increasing the dimension of XIN in the main program to any number.

Maximum Number of Analysis Variables

This maximum of 100 can be changed by increasing the dimensions of XUSE, Z, B, and IND to the desired maximum and the dimension of XIND to the square of the desired maximum.

A.3 PROGRAM ORGANIZATION

Job-control statements (user-supplied, see local programmer)

MAIN PROGRAM
SUBROUTINE LOC
SUBROUTINE MINV
SUBROUTINE GTPRD
SUBROUTINE GMPRD
SUBROUTINE DATA

(Subroutine is to be user-modified; see SUBROUTINE DATA, below, in program listing)
Job-control statements (user-supplied, see local programmer)
Main control card (user-supplied, see Sec. A.5)
Data cards (user-supplied)
Analysis-control cards (user-supplied, see Sec. A.6)
Job-control statements (user-supplied, see local programmer)

A.4 PROGRAM LISTING

```
      DIMENSION XIN(25), XUSE(100), XIND(30000), Y(1200), XINV(2500),
     1  Z(50), B(50), IND(100)
      COMMON NVAR, NVEC, XIN, XUSE
      INTEGER*2 NAL1(3), NAL2(2), NAL3(3), NAL4(2)
100   WRITE (6,1)
  1   FORMAT  ('1INSTRUCTIONAL REGRESSION PACKAGE FOR THE SOCIAL SCIENCE
     1ES')
      REWIND 4
      READ (5,2,END=210) NAL1, NAL2, N, NVAR, NVEC, NREG, MIKE
  2   FORMAT (5A2,5I5)
      IF (N * NVAR * NVEC .NE. 0) GO TO 110
      WRITE (6,3) N, NVAR, NVEC
  3   FORMAT ('0CONTROL CARD ERROR'/' N = ',I5/' NVAR = ',I5/' NVEC = ',
     1I5/' NONE OF THE ABOVE MAY BE ZERO.')
      RETURN
110   IF (NREG .EQ. 0) NREG = 1
      WRITE (6,4) NAL1, NAL2, N, NVAR, NVEC, NREG
  4   FORMAT ('0',5A2/' N = ',I5,5X,'NVAR = ',I5,5X,'NVEC = ',I5,5X,
     1'NREG = ',I5)
      DO 140 I = 1, N
      DO 120 J = 1, NVAR
120   XIN(J) = -9999999.
      DO 130 J = 1, NVEC
130   XUSE(J) = -9999999.
      CALL DATA
      IF(I-MIKE) 131,131,132
131   WRITE (6,5) I, (XIN(J), J = 1,NVAR)
  5   FORMAT ('0INPUT VARIABLES FOR CASE ',I5/(1X,10G11.4))
      WRITE (6,6) (XUSE(J), J = 1,NVEC)
  6   FORMAT (' VECTORS FOR LINEAR MODELS'/(1X,10G11.4))
132   WRITE (4) (XUSE(K), K = 1,NVEC)
140   CONTINUE
      DO 200 I = 1, NREG
      REWIND 4
      READ (5,7,END=210) NAL3,NAL4,MY,NP,(IND(J),J=1,12)
  7   FORMAT(5A2,14I5)
      IF (NP-12) 144,144,142
142   READ(5,13,END=210) (IND(J),J=13,NP)
 13   FORMAT(10X,14I5)
144   WRITE (6,8) NAL1, NAL2, N, NVAR, NVEC, I, NREG, NAL3, NAL4, MY,
     1 NP, (IND(J), J = 1,NP)
  8   FORMAT ('1',5A2/' N = ',I5,5X,'NVAR = ',I5,5X'NVEC = ',I5,5X,
     1' REGRESSION NUMBER ',I5,' OF ',I5,' REGRESSIONS'/1X,5A2/
     2' DEPENDENT VARIABLE IS VECTOR ',I5/' THERE ARE ',I5,' PREDICTORS'
     3/' THEY ARE'/(20I6))
      IF (MY .GT. NVEC .OR. MY .LE. 0) GO TO 220
      IF (NP .GT. NVEC .OR. NP .LE. 0) GO TO 220
      DO 145 J = 1, NP
      IF (IND(J) .EQ. MY .OR. IND(J) .GT. NVEC .OR. IND(J) .LE. 0)
     1 GO TO 220
145   CONTINUE
      XY = 0
      DO 160 J = 1, N
```

(continued overleaf)

```
      READ (4) (XUSE(K), K = 1,NVEC)
      Y(J) = XUSE(MY)
      XY = XY + Y(J)
      DO 150 K = 1, NP
      L = IND(K)
      CALL LOC (J, K, IJ, N, NP, 0)
150   XIND(IJ) = XUSE(L)
160   CONTINUE
      CALL GTPRD (XIND, XIND, XINV, N, NP, NP)
      CALL MINV (XINV, NP, D, Z, B)
      IF (D) 180, 170, 180
170   WRITE (6,9)
  9   FORMAT ('0CANNOT SOLVE FOR A UNIQUE SET OF COEFFICIENTS FOR THESE
     1PREDICTORS'/' AM GOING ON TO NEXT PROBLEM')
      GO TO 200
180   CALL GTPRD (XIND, Y, Z, N, NP, 1)
      CALL GMPRD (XINV, Z, B, NP, NP, 1)
      CALL GMPRD (XIND, B, XINV, N, NP, 1)
      FN = N
      XY = XY/FN
      SSR = 0
      SSYM = 0
      DO 190 J = 1, N
      SSR = (Y(J) - XINV(J)) **2 + SSR
190   SSYM = (Y(J) - XY) **2 + SSYM
      RSQ = (SSYM - SSR) / SSYM
      WRITE (6,10) XY, SSR, SSYM, RSQ, (IND(K), B(K), K = 1,NP)
 10   FORMAT ('0MEAN OF DEPENDENT VARIABLE = ',G11.4/' ERROR SUM OF SQUA
     1RES = ',G15.4/' TOTAL SUM OF SQUARES = ',G15.4/' COEFFICIENT OF DE
     2TERMINATION = ',F8.4/'0COEFFICIENTS OF PREDICTOR VECTORS'/' VECTOR
     3      COEFFICIENT'/(I7,F16.4))
200   CONTINUE
      RETURN
210   WRITE (6,11)
 11   FORMAT ('0UNEXPECTED END OF CONTROL CARDS')
      RETURN
220   WRITE (6,12)
 12   FORMAT ('0ERROR ON ANALYSIS CONTROL CARD')
      RETURN
      END


      SUBROUTINE LOC(I,J,IR,N,M,MS)                              LOC   440
      IX=I                                                      LOC   460
      JX=J                                                      LOC   470
      IF(MS-1) 10,20,30                                         LOC   480
 10   IRX=N*(JX-1)+IX                                           LOC   490
      GO TO 36                                                  LOC   500
 20   IF(IX-JX) 22,24,24                                        LOC   510
 22   IRX=IX+(JX*JX-JX)/2                                       LOC   520
      GO TO 36                                                  LOC   530
 24   IRX=JX+(IX*IX-IX)/2                                       LOC   540
      GO TO 36                                                  LOC   550
 30   IRX=0                                                     LOC   560
      IF(IX-JX) 36,32,36                                        LOC   570
 32   IRX=IX                                                    LOC   580
```

```
36 IR=IRX                               LOC  590
   RETURN                               LOC  600
   END                                  LOC  610


   SUBROUTINE GMPRD(A,B,R,N,M,L)        GMPR 370
   DIMENSION A(1),B(1),R(1)             GMPR 380
   IR=0                                 GMPR 400
   IK=-M                                GMPR 410
   DO 10 K=1,L                          GMPR 420
   IK=IK+M                              GMPR 430
   DO 10 J=1,N                          GMPR 440
   IP=IR+1                              GMPR 450
   JI=J-N                               GMPR 460
   IB=IK                                GMPR 470
   R(IR)=0                              GMPR 480
   DO 10 I=1,M                          GMPR 490
   JI=JI+N                              GMPR 500
   IB=IB+1                              GMPR 510
10 R(IR)=R(IR)+A(JI)*B(IB)              GMPR 520
   RETURN                               GMPR 530
   END                                  GMPR 540


   SUBROUTINE GTPRD(A,B,R,N,M,L)        GTPR 360
   DIMENSION A(1),B(1),R(1)             GTPR 370
   IR=0                                 GTPR 390
   IK=-N                                GTPR 400
   DO 10 K=1,L                          GTPR 410
   IJ=0                                 GTPR 420
   IK=IK+N                              GTPR 430
   DO 10 J=1,M                          GTPR 440
   IB=IK                                GTPR 450
   IR=IR+1                              GTPR 460
   R(IR)=0                              GTPR 470
   DO 10 I=1,N                          GTPR 480
   IJ=IJ+1                              GTPR 490
   IB=IB+1                              GTPR 500
10 R(IR)=R(IR)+A(IJ)*B(IB)              GTPR 510
   RETURN                               GTPR 520
   END                                  GTPR 530


   SUBROUTINE MINV(A,N,D,L,M)           MINV 330
   DIMENSION A(1),L(1),M(1)             MINV 340
   D=1.0                                MINV 560
   NK=-N                                MINV 570
   DO 80 K=1,N                          MINV 580
   NK=NK+N                              MINV 590
   L(K)=K                               MINV 600
   M(K)=K                               MINV 610
   KK=NK+K                              MINV 620
   BIGA=A(KK)                           MINV 630
   DO 20 J=K,N                          MINV 640
   IZ=N*(J-1)                           MINV 650
   DO 20 I=K,N                          MINV 660
```

(*continued overleaf*)

```
         IJ=IZ+I                                                    MINV 670
10    IF( ABS(BIGA)- ABS(A(IJ))) 15,20,20                          MINV 680
15    BIGA=A(IJ)                                                    MINV 690
      L(K)=I                                                        MINV 700
      M(K)=J                                                        MINV 710
20    CONTINUE                                                      MINV 720
      J=L(K)                                                        MINV 760
      IF(J-K) 35,35,25                                              MINV 770
25    KI=K-N                                                        MINV 780
      DO 30 I=1,N                                                   MINV 790
      KI=KI+N                                                       MINV 800
      HOLD=-A(KI)                                                   MINV 810
      JI=KI-K+J                                                     MINV 820
      A(KI)=A(JI)                                                   MINV 830
30    A(JI) =HOLD                                                   MINV 840
35    I=M(K)                                                        MINV 880
      IF(I-K) 45,45,38                                              MINV 890
38    JP=N*(I-1)                                                    MINV 900
      DO 40 J=1,N                                                   MINV 910
      JK=NK+J                                                       MINV 920
      JI=JP+J                                                       MINV 930
      HOLD=-A(JK)                                                   MINV 940
      A(JK)=A(JI)                                                   MINV 950
40    A(JI) =HOLD                                                   MINV 960
45    IF(BIGA) 48,46,48                                             MINV1010
46    D=0.0                                                         MINV1020
      RETURN                                                        MINV1030
48    DO 55 I=1,N                                                   MINV1040
      IF(I-K) 50,55,50                                              MINV1050
50    IK=NK+I                                                       MINV1060
      A(IK)=A(IK)/(-BIGA)                                           MINV1070
55    CONTINUE                                                      MINV1080
      DO 65 I=1,N                                                   MINV1120
      IK=NK+I                                                       MINV1130
      HOLD=A(IK)                                                    MINV1140
      IJ=I-N                                                        MINV1150
      DO 65 J=1,N                                                   MINV1160
      IJ=IJ+N                                                       MINV1170
      IF(I-K) 60,65,60                                              MINV1180
60    IF(J-K) 62,65,62                                              MINV1190
62    KJ=IJ-I+K                                                     MINV1200
      A(IJ)=HOLD*A(KJ)+A(IJ)                                        MINV1210
65    CONTINUE                                                      MINV1220
      KJ=K-N                                                        MINV1260
      DO 75 J=1,N                                                   MINV1270
      KJ=KJ+N                                                       MINV1280
      IF(J-K) 70,75,70                                              MINV1290
70    A(KJ)=A(KJ)/BIGA                                              MINV1300
75    CONTINUE                                                      MINV1310
      D=D*BIGA                                                      MINV1350
      A(KK)=1.0/BIGA                                                MINV1390
80    CONTINUE                                                      MINV1400
      K=N                                                           MINV1440
100   K=(K-1)                                                       MINV1450
      IF(K) 150,150,105                                             MINV1460
105   I=L(K)                                                        MINV1470
```

```
      IF(I-K) 120,120,108                                    MINV1480
108 JQ=N*(K-1)                                               MINV1490
    JR=N*(I-1)                                               MINV1500
    DO 110 J=1,N                                             MINV1510
    JK=JQ+J                                                  MINV1520
    HOLD=A(JK)                                               MINV1530
    JI=JR+J                                                  MINV1540
    A(JK)=-A(JI)                                             MINV1550
110 A(JI) =HOLD                                              MINV1560
120 J=M(K)                                                   MINV1570
    IF(J-K) 100,100,125                                      MINV1580
125 KI=K-N                                                   MINV1590
    DO 130 I=1,N                                             MINV1600
    KI=KI+N                                                  MINV1610
    HOLD=A(KI)                                               MINV1620
    JI=KI-K+J                                                MINV1630
    A(KI)=-A(JI)                                             MINV1640
130 A(JI) =HOLD                                              MINV1650
    GO TO 100                                                MINV1660
150 RETURN                                                   MINV1670
    END                                                      MINV1680


    SUBROUTINE DATA
    DIMENSION VAR(25),VEC(100),M(30)
    COMMON NVAR,NVEC,VAR,VEC
    .
    .
    .
```

In this location the user must supply a set of Fortran statements which will read a row vector, VAR(I), containing all variables to be read from cards for one case. Additionally, statements must be provided which make any necessary transformations and construct a row vector, VEC(I), containing all variables to be analyzed for one case. The dimension of VAR(I) (the number of variables to be read) is stored in location NVAR. The dimension of VEC(I) (the number of vectors to be constructed) is stored in NVEC.

```
    RETURN
    END
```

```
//GO.FT04F001 DD UNIT=TAPE
//GO.SYSIN DD *
    (Main control card, explained below, goes here.)
    (All data cards follow the main control card here.)
    (Analysis-control cards, explained below, go here.)
/*
```

A.5 MAIN CONTROL CARD

The main control card must contain the following information:

Columns 1–10   A label comprising any combination of letters, numbers, and blanks which will serve to identify the computer output for you.

Columns 11–15   The number of cases to be processed. This must be an integer, right-justified in the field. It is stored in location N for use by the program. $N \le 500$. (The limit can be increased by changing the maximum dimension of Y in the first statement in the main program, above.)

Columns 16–20   The number of variables to be read in from cards. This must be an integer, right-justified in field. (This must be exactly the number of variables specified in your READ and FORMAT statements in SUBROUTINE DATA.) It is stored in location NVAR for use by the program. NVAR $\leq$ 25.

Columns 21–25   The number of vectors (VEC's) created by SUBROUTINE DATA. This must be an integer, right-justified in field. It is stored in location NVEC for use by the program. NVEC $\leq$ 100.

Columns 26–30   The number of separate models (analyses, regressions) to be performed by the program during this run. This must be an integer, right-justified in field. It is stored in location NREG for use by the program. There is no limit for NREG.

A.6   ANALYSIS-CONTROL CARDS

These cards immediately follow data cards. There are as many of these cards as you have specified in columns 26 to 30 of the main control card.

Columns 1–10   Any combination of letters, numbers, or blanks which will serve to identify this model to you, e.g., FULL MODEL.

Columns 11–15   The identification of the vector in which the dependent variable is stored; i.e., if the vector numbered 6 is to be a dependent variable, punch 6 in these columns, right-justified in field.

Columns 16–20   The number of independent variables or predictors in this model. Punch right-justified.

Successive Five-Character Fields

Those identifying vectors which are to be your independent, or predictor, vectors. Use exactly as many five-character fields as you have independent vectors in your model. Use one field per vector, punched right-justified. If the number of vectors is too large to be punched on the first card, continue to succeeding cards, skipping columns 1 to 10 and beginning in column 11, with five-character fields. For example, if your model involves 16 predictor vectors, you might punch

FULL MODEL   1   16   2   3   4   5   6   7   8   9   10   11   12   13

                       14   15   16   17

where vector 1 is your dependent variable and vectors 2 to 17 are your predictor variables or vectors.

SIMULATED DATA SET

Simulated data follow which should be punched on IBM cards and used with the illustrations in Chap. 6 and with exercises in Chaps. 5 and 6. The data can be interpreted according to the schedule in Table B.1.

B.1 CODING SCHEDULE

Variable No.	Column	Description of variable	Maximum range
	1–3	Serial family identification number	001–200
1	4	Region of residence for husband and wife: 1 = Northeast 2 = Midwest 3 = Mountain 4 = Pacific 5 = South	1–5
2	5	Urbanness of residence for husband and wife: 1 = urban 2 = suburban 3 = rural	1–3
3	6–7	Husband's age, years	00–99
4	8–9	Husband's years of education	00–20
5	10–12	Husband's occupational prestige: 000 = low 100 = high	000–100
6	13–17	Husband's income, dollars	00000–40000
7	18	Husband's political party preference: 0 = Democratic 1 = Republican	0–1

(*continued overleaf*)

B.1 (*continued from previous page*)

Variable No.	Column	Description of variable	Maximum range
8	19–21	Husband's liberalism-conservatism 000 = ultraliberal 100 = ultraconservative	000–100
9	22–23	Wife's age, years	00–99
10	24–25	Wife's years of education	00–20
11	26	Wife's political party preference: 0 = Democratic 1 = Republican	0–1
12	27–29	Wife's liberalism-conservatism: 000 = ultraliberal 100 = ultraconservative	000–100
13	30–31	Husband's father's years of education	00–20
14	32–34	Husband's father's occupational prestige: 000 = low 100 = high	000–100
15	35–39	Husband's father's income, dollars	00000–40000
16	40	Husband's father's political party preference: 0 = Democratic 1 = Republican	0–1
17	41–43	Husband's father's liberalism-conservatism: 000 = ultraliberal 100 = ultraconservative	000–100
18	44–45	Husband's mother's years of education	00–20
19	46	Husband's mother's political party preference: 0 = Democratic 1 = Republican	0–1
20	47–49	Husband's mother's liberalism-conservatism: 000 = ultraliberal 100 = ultraconservative	000–100

B.2 READING THE DATA

The user should employ the following standard READ and FORMAT statements when using these data. This will permit us to refer to specific variables VAR(I)s by their variable number designated above.

```
      READ (5,11) (VAR(I),I=1,20)
   11 FORMAT (3X,2F1.0,2F2.0,F3.0,F5.0,F1.0,F3.0,2F2.0,F1.0,F3.0,F2.0,
   1 F3.0,F5.0,F1.0,F3.0,F2.0,F1.0,F3.0)
```

B.3 DATA

The following data should be punched on cards. The vertical line separates the identification number from the first variable. The identification number should be right-justified in columns 1 to 3. Thus, the line separates columns 3 and 4.

```
 1|1258 7 18 24370 6257 61 69 6 13    00 29 51 91
 2|2141 8 15 37550 4847120 37 6 14 26800 18 70 44
 3|5360 7 10 37150 3571 91 72 6  2   840 52 51 91
 4|112512 83138221 7726160 92 8  4 37690 67120 31
 5|125112 56208811 5353101 45 8  2 18460 28120 49
 6|234511 42 96190 3646130 58 8  9    00 61110 72
 7|335910 43 68850 3866130 50 8  4 25940 51100 36
 8|424112  5   570 8839 71 77 8 14  5890 64121 98
 9|5334 9 43 39201 7633111 68 8 15 41150 25 91 80
10|134510 39 70000 5252140 70 9 29 42360 71100 54
11|233010 19 28730 8031131 92 9 29 41391 85101 88
12|335213 31 59690 4657120 48 9 30 16350 25130 66
13|414812 87181451 5252121 45 9 27 38731 80121 91
14|535110 22 16480 4553120 30 9 31 62951 75101 82
15|114211 36111810 6945140 8510 39 38750 44110 59
16|212512 27 52470 6826170 6010 16 41270 55110 65
17|225414 72229351 3754101 4010 25 26420 28140 46
18|312613 55 68200 6327140 7310 15 32370 69130 37
19|334111 18 52500 6245150 6910 23 41160 30111 90
20|532513 32 45320 5022100 4910 41 83600 45130 66
21|533112 62 83101 6628 71 6010 27 31360 55120 39
22|534014 47104361 6946150 4410 43 34491 85140 81
23|535811 25 72510 2562130 4510 23 30030 50110 56
24|534011 43 58790 3546150 4210 35 41670 51110 60
25|113612 52137660 7339140 8111 24 25710 40120 57
26|223812 23 67690 4943170 9311 30 44520 34120 35
27|234412 39150361 7046111 7011 40 54020 39120 41
28|415712  4 37000 6159 90 5811 23 32480 61120 74
29|423913 84215041 3140111 3011 25 55920 55130 80
30|432613 76119420 6722 90 7311 15 28940 70130 40
31|435512 89144861 4159110 4611 37 63750 48120 49
32|513914 69192261 5738111 5611 35 31660 68140 47
33|524215 65186110 5544140 6511 40 45440 71150 69
34|532613 26 61050 4827140 6311 35 26970 37130 35
35|533114 61 95061 6928101 5911 18 56840 42140 71
36|535314 32 92211 7352101 6011 22 26131 88141 92
37|112813 72151171 7430140 7912 11 15180 77100 60
38|113112 97142541 6932140 7312 22 63380 79 90 51
39|113414 27 61020 8030100 6312 68 93751 70130 67
40|113416 55161000 8337160 7812 36 29580 73160 65
41|113614 83180971 6740150 8912 27 24600 45120 61
42|114011 32 48340 6238120 6512 56 98160 62 70 58
43|114111 15 16080 7640120 7212 23 40720 49 71 85
44|114211 44113431 8639121 8412 81125991 56 61 55
45|122612 49 68240 8027170 9912 21 59460 67 80 23
46|122715 61119600 8925130 8512 14 35840 55140 66
47|122813 28 86390 8226110 7412 63 68990 76100 56
48|123117 90229911 4933180 9912 91149241 50160 81
49|123112 16 55910 7930110 7712 29 27810 78 80 63
50|123216 75167161 5932140 9412 22 38080 44170 62
```

```
 51 123214 30  92450 7530110 7812 42 94160 69120 51
 52 123811 22  86300 7741140 7612 73107341 61 61 61
 53 124315 59191860 8040110 6512 53 97610 76140 38
 54 125413 85267571 4164141 4012 51 59801 79101 72
 55 132513 45  39720 7426140 7512 37 38920 32100 52
 56 133316 70139351 7936161 7612 67 75281 65171 81
 57 133511 32  49341 9536140 7512 91121321 59 71 49
 58 134011 44103330 5939130 7412 56 68800 76 70 51
 59 135016 85302721 4555151 4812 48 68610 71160 60
 60 135113 79182220 5361140 5812 40 45290 79100 49
 61 135811 38  70091 8359121 7512 33 75181 81 61 70
 62 212615 61129621 7822111 7212  9 24110 32150 74
 63 212615 36  89790 6622110 9112 65128741 69120 43
 64 212711 19  13300 6124130 8812 49 80800 54 61 72
 65 212715 39  98580 8428160 7412 76100861 66131 66
 66 212914 73142980 5126100 4212 91161191 55110 43
 67 213016 80173601 7128131 6912 68128971 58141 65
 68 214814 47141440 3955140 5012 22 66180 67130 65
 69 222615 94142511 4924130 9912 72110670 44130 57
 70 222812 35  33400 7328130 7012 74120761 69 81 61
 71 222911 34  51890 5431150 8612 41 45130 61 70 69
 72 223017 95240341 4030160 9612 96130761 53160 43
 73 223116 75157381 6134171 5412 83108801 68151 66
 74 223913 65132801 5637 91 5412  2  4940 44100 63
 75 224311 22  38840 8348150 7812 29 78601 74 71 72
 76 224311 41115800 4246140 8712 43 68040 74 60 39
 77 224412 40135751 6241 61 6012 93138891 51 91 58
 78 224912 33  25150 7048 71 7012 38 91921 73 91 68
 79 225413 40  91661 6860121 6212 43 52491 78101 74
 80 225712 81262641 2867160 7012 14  1200 36 80 48
 81 232615 41  63660 5124130 8112 89130871 62120 43
 82 232716 88129841 6327150 6512 37 70490 47160 72
 83 232814 78134691 7325101 6912 42 51601 81131 77
 84 233116 58109301 8129131 7112 53 69561 79170 57
 85 235812 59163771 6458 71 4812 78 93911 68 81 57
 86 312512 90143161 7025141 7312 96145571 57 91 49
 87 312711 27  38440 7223120 6212 59104641 68 61 57
 88 313116 93207241 5833160 8412 37 42470 48170 58
 89 315511 38  90700 4160140 6512 32 84340 35 70 59
 90 322514 41  82761 7823121 7712 41 91751 70121 71
 91 322615 27  73160 7626140 7112 60 75270 56150 77
 92 324112 97225591 3239 71 3612  7 44710 50 90 36
 93 324415 57211301 5551161 4912 73122951 66131 60
 94 325512 84160181 4456 81 3312 48 53211 74 91 71
 95 335516 87235511 5062151 5012 55 93481 68170 67
 96 332514 63101110 6123120 5612 21 21620 34130 42
 97 332615 93133851 6523121 5712 32 57490 33140 75
 98 332711 40  58580 6529150 5912 86110411 67 61 57
 99 332811 33  27300 5128140 6512 66 94130 73 70 28
100 332917 100176701 5626130 7612 96141351 51160 63
```

```
101333012 89148411  5929101  6312  51 87890  76 91 67
102333119 81174031  6334190  6712  90164751  56160 75
103333618 67221601  6137170  7012  90164041  52151 57
104333712 18 26980  6741160  6112  32 33851  86 91 76
105334312 49129770  5345120  4512  88116511  66 91 56
106334417100335291  3448161  3812  80132731  62160 85
107334611 36 71871  8048130  6412  24 42880  65 71 81
108334711 30 70191  8354151  7912  60 98371  65 71 67
109412615 87135131  5923120  7412  20 11190  42150 44
110412712 73122081  6725 91  6012  23 43120  51 91 83
111413013 46 79280  8126 90  6812   4   690  55110 77
112413412 36 98160  7735120  6812  17 28890  54 80 31
113413712 76176831  5635 71  4312  13   00  51 81 94
114413814 54144690  7938120  6412   9   00  71130 45
115413915 33 70121  8539131  7712  60101581  67121 72
116414315 85249931  4249160  6512  83109171  59130 81
117422917 58141431  5728150  9212  75147181  62171 70
118423515 33 74900  8838150  9212  41 50580  66140 59
119424015 54186350  8245160  7212   8 20510  52140 55
120424015 87231891  3141131  2812  13  4240  39140 54
121424011 22 39480  8438120  8112  90164461  59 61 43
122424111 23 70350  6942130  8012  28 23160  49 60 24
123424216 72249941  3142140  7612  30 37160  35170 83
124425312  2 28670  6553 80  5112  17 37120  64 90 58
125425615 88236731  1955110  7812  21 23060  50140 39
126425916 93297351  2164141  1012  35 55981  87160 79
127432715 91146071  5925131  4912  81 94071  60141 68
128432813 35 44550  7226110  7412   6 34930  51100 64
129434012 32105441  7246170  7612  45 55231  75 81 70
130434911 39 71731  7258150  6812  33 53951  79 61 76
131512712 69 93201  7328151  7512  91123781  58 81 53
132512917 80140651  6825131  6312  83139601  61161 64
133513016 63103860  6628130  5712   8 24300  65170 61
134513612 20 19420  4738140  5512  93118111  62 81 50
135514115 69182251  6245151  5612  87127771  61131 62
136524315 40111470  5846140  6612  43 98090  32150 75
137523315 68177920  6532130  5812  92148531  50130 59
138525411 15 69710  4457130  4012  38 74660  59 70 59
139532611 34 56560  5426140  7112  72 96671  62 61 57
140532819 91196371  5427180  7712  95158261  55171 63
141533113 65 82591  6629110  4412  56 91091  77111 66
142533613 93179951  4139130  4012  95133511  59100 51
143533812 71128121  5640120  3712  36 32660  37 81 76
144534212 17 31310  4146150  4812  95150941  60 81 47
145535215 90268961  3556141  2712  94114581  60121 56
146113113 81180431  7531121  7213  45 76311  78101 75
147132613 65 79730  7023100  8313  56 95300  72100 50
148322715 58128351  6728151  6713  72 96481  63131 64
149413313 48 99031  7731101  6813  31 55741  85101 84
150524313 38122200  6248130  5813  37 63370  63101 74
```

```
151533215  45 93441  7534150  4913  94137861  62121  53
152112918  58159461  8829171  8314  75106741  65151  67
153113913  64181401  7537 91  6914  46 91031  79101  71
154214317  79269551  5748171  5314  75133801  65141  69
155225116  69264991  3559170  6014  82130971  60130  79
156312614  38 89510  7622101·8814  26 24860  67100  37
157313917  72237691  5942170  6314  54101580  64130  55
158412917  97208081  5127150  8914  34 47150  55150  44
159433615  46144170  6833110  7314  94163071  55110  72
160512618  75128200  6622140  6014  70100800  49150  85
161524513  35105020  5253140  6414  45 78980  75100  67
162524017100291571  2344170  7514  55 86431  71150  58
163532518  77118141  6925180  6214  59109491  74151  72
164115118  91369791  4060190  8615  56124310  42140  55
165322515  38 79500  8523130  6815  34 74460  57120  49
166412516  58 92650  9121120  7315  53 85151  72140  51
167423716  55159121  5336131  5015  31 48090  51140  82
168112717  65138030  8226150  8116  88208741  49140  73
169122519  59152360  9222160  9916  80159581  56150  58
170124519  66260031  5045160  7316  72168991  66170  77
171212516  88120141  7623141  7816  79162141  57131  59
172213317  81172291  7431141  6616  76123781  64141  63
173222518  79184691  5923161  4816  91171801  51141  52
174222619  70134730  7324170  8116  65123460  86150  53
175233516  66143421  6736150  4216  74161371  57130  55
176312519  67108551  8923171  8016  91183251  55161  52
177322816  63128111  6228150  7416  80193511  52121  54
178324419  73286601  4348181  4116  70115551  68151  63
179332615  84121820  6124130  6216  75157370  79120  74
180333017  68131871  7727141  6616  66108061  68141  66
181333616  66137891  7534131  6716  81140991  64121  56
182412619  57144960  9424170  8816  54116120  48170  90
183412619  79171141  6224170  8916  85198231  57160  80
184413515  35102770  8834121  6816  64125781  71121  59
185422517  84133920  8925170  8516  60112900  60140  47
186423116  67175121  4827121  4616  83139691  65121  54
187512519  88153841  6824181  5916  74139881  60151  64
188513416  56155050  5933140  7216  68128290  47120  44
189525319  82307171  2063190  5016  63109010  43170  78
190523115  38 72050  7629120  5416  87202351  50131  52
191222519  99212581  5124181  4317  86153401  55161  65
192313019  96223681  5728160  9017100185481  54180  57
193512718  90195221  6127171  6117  59162461  70141  62
194313319  97285501  5430151  5018  90182611  59171  60
195422519  91174490  8926190  9918  80189760  67180  44
196112619  77136781  8622151  7819  96233101  48171  47
197212519  69121420  7225190  9919  86188831  64150  67
198322919  71199681  5927161  4819  69198651  63171  63
199412519  65106150  9125190  9919  78156151  60190  83
200523619  76205881  4338180  7219  65159311  72180  67
```

C.1 LINEAR FORMS AND ELEMENTARY PROPERTIES OF MATRICES

In elementary mathematics one is introduced to functions like $3x + 5y - 2z$, in which x, y, and z are variables. In this function all variables occur with degree 1, and furthermore no term contains more than one variable. Such a function is known as a *linear form*. It is customary to represent such a function by a single letter such as w or by the symbol $f(x,y,z)$ and to write

$$w = 3x + 5y - 2z \quad \text{or} \quad f(x,y,z) = 3x + 5y - 2z \qquad \text{(C.1)}$$

Sometimes our interest centers on the coefficients in the linear form, especially when we are comparing two or more linear forms in the same variables. We may therefore find it useful to write the coefficients without associating them explicitly with the variables as in (C.1). In so doing it may be convenient to follow the convention of writing the coefficient of x first, followed by those of y and z in that order. This helps us compare the coefficients of linear forms in the same set of variables. Following this convention, let us write the coefficients in the linear form (C.1) as

$$(3, 5, -2) \qquad \text{(C.2)}$$

This array of numbers may be considered as an entity in itself. In secondary school mathematics we were introduced to the notion that a number can be represented by a point on a line, or vice versa, and that a pair of numbers can be represented by a point in a plane, or vice versa. In the same manner we may see a correspondence between triplets like (C.2) and a point in a three-dimensional space and in general between an n-tuple and a point in an n-dimensional space. Let us write two n-tuples and label them **A** and **B**:

$$\mathbf{A} = (a_1, a_2, \ldots, a_n) \qquad \mathbf{B} = (b_1, b_2, \ldots, b_n) \qquad (C.3)$$

The term *vector* is used to refer to n-tuples such as the ones shown in (C.3). Obviously a vector may contain one *element* only, two elements, or any number of elements. Treating vectors as entities in themselves, we shall define the operation addition as follows: *the sum of the two vectors **A** and **B** in (C.3) is defined as a vector **C** formed by adding the corresponding elements of **A** and **B**.* Thus $\mathbf{C} = \mathbf{A} + \mathbf{B} = (a_1 + b_1, a_2 + b_2, \ldots, a_n + b_n)$.

EXERCISE C.1

Let $\mathbf{A} = (2,3)$ and $\mathbf{B} = (1,0)$. Find **C** the sum of **A** and **B**. Plot the points **A**, **B**, and **C** in a two-dimensional space. Verify that the points corresponding to **A**, **B**, and **C** and the origin $(0,0)$ give the vertices of a parallelogram.

We shall now define multiplication of a vector by a constant. If $\mathbf{A} = (a_1, a_2, \ldots, a_n)$ and c is a constant, then $\mathbf{A} \times c = c \times \mathbf{A} = (ca_1, ca_2, \ldots, ca_n)$. Thus, if $\mathbf{A} = (1,2,3,4)$ and we want the product of **A** with the number 2, the required product is $2\mathbf{A} = (2 \times 1, 2 \times 2, 2 \times 3, 2 \times 4) = (2,4,6,8)$. In describing above the addition of vectors and the multiplication of a vector by a constant we have implicitly defined the equality between two vectors; we shall now give a formal definition of equality: *two vectors **A** and **B** are equal if each element of **A** is equal to the corresponding element of **B**.* Thus **A** and **B** shown in (C.3) are equal if $a_1 = b_1, a_2 = b_2, \ldots, a_n = b_n$.

We now define the *inner, or dot, product of the two vectors **A** and **B** in (C.3)* to be $a_1 b_1 + a_2 b_2 + \cdots + a_n b_n$. Note that this product is a *number*. For instance, if $\mathbf{A} = (1,3,-1)$ and $\mathbf{B} = (-1,-3,2)$, then the inner product $\mathbf{A} \cdot \mathbf{B} = 1(-1) + 3(-3) + (-1)2 = (-1) + (-9) + (-2) = -12$. If the inner product of two vectors is zero, the vectors are said to be *orthogonal*.

EXERCISE C.2

Using numerical examples, verify the following properties:

1. $\mathbf{A} \times \mathbf{B} = \mathbf{B} \times \mathbf{A}$, where **A** and **B** are two vectors of the same dimension, i.e., with the same number of elements. (In other words, **A** and **B** are 1-tuples, or both are 2-tuples, or both are 3-tuples, ...)

2. If **A**, **B**, and **C** are three vectors of the same dimension, $\mathbf{A} \times (\mathbf{B} + \mathbf{C}) = \mathbf{A} \times \mathbf{B} + \mathbf{A} \times \mathbf{C} = (\mathbf{B} + \mathbf{C}) \times \mathbf{A} = \mathbf{B} \times \mathbf{A} + \mathbf{C} \times \mathbf{A}$.

3. If a is a number, then

$$(a \times \mathbf{A}) \times \mathbf{B} = a \times (\mathbf{A} \times \mathbf{B}) \qquad \text{and} \qquad \mathbf{A} \times (a \times \mathbf{B}) = a \times (\mathbf{A} \times \mathbf{B})$$

4. If **A** is a vector with all its elements zero and **B** is any other vector of the same dimension, then $\mathbf{A} \times \mathbf{B} = 0$ and $\mathbf{A} \times \mathbf{A} = 0$.

5. If **A** is a vector with at least one element not equal to zero, then $\mathbf{A} \times \mathbf{A}$ is greater than zero.

6. Find the inner product of the two vectors $(1,1)$ and $(-1,1)$. Plot the points corresponding to these vectors in a two-dimensional space; join each point to the origin. Verify that the angle between the two lines is $90°$. This illustrates the geometric definition of orthogonality: Two vectors **A** and **B** are orthogonal if the angle between them is $90°$, in other words, if one is perpendicular to the other. Try the same with the vectors $(1,1)$ and $(1,-1)$.

Now let us consider two linear forms:

$$\begin{aligned} w_1 &= 2x + 3y + 4z \\ w_2 &= 3x + 3y + 2z \end{aligned} \qquad (C.4)$$

The coefficients in the two linear forms may be written as an array of numbers as follows:

$$\begin{bmatrix} 2 & 3 & 4 \\ 3 & 3 & 2 \end{bmatrix} \tag{C.5}$$

An array of numbers of the type (C.5) is called a *matrix*. The above matrix has two rows and three columns; the rows are [2 3 4] and [3 3 2], and the columns are $\begin{bmatrix} 2 \\ 3 \end{bmatrix}, \begin{bmatrix} 3 \\ 3 \end{bmatrix}, \begin{bmatrix} 4 \\ 2 \end{bmatrix}$. The rows of this matrix may be viewed as 3-tuples and the columns as 2-tuples.

Any matrix with m rows and n columns is called an m by n or $m \times n$ matrix, *always mentioning the number of rows first*. It may be noted that a row vector with n elements is a $1 \times n$ matrix; similarly the columns of any $m \times n$ matrix may be viewed as column vectors each with m elements. If the number of rows equals the number of columns, the matrix is called a *square matrix*. Thus

$$\begin{bmatrix} 1 & 2 \\ 0 & -1 \end{bmatrix} \quad \text{and} \quad \begin{bmatrix} 1 & 1 & 1 \\ 2 & 1 & 0 \\ 0 & 1 & -1 \end{bmatrix}$$

are both square matrices.

We shall now define addition and multiplication of matrices. Addition is defined only for matrices of the same dimension, i.e., matrices of the same specification, $m \times n$. Thus addition is defined for the matrices

$$\mathbf{A} = \begin{bmatrix} 1 & 1 & 0 \\ 2 & 1 & -1 \end{bmatrix} \quad \text{and} \quad \mathbf{B} = \begin{bmatrix} 3 & 1 & 2 \\ 1 & -1 & 0 \end{bmatrix}$$

but not for the matrices

$$\mathbf{C} = \begin{bmatrix} 1 & 1 \\ 2 & 1 \end{bmatrix} \quad \text{and} \quad \mathbf{B} = \begin{bmatrix} 3 & 1 & 2 \\ 1 & -1 & 0 \end{bmatrix}$$

Addition of two matrices is performed by adding the corresponding elements in the two matrices. Thus if $\mathbf{C}$ is the sum of the matrices $\mathbf{A}$ and $\mathbf{B}$ given above, then

$$\mathbf{C} = \begin{bmatrix} 1 & 1 & 0 \\ 2 & 1 & -1 \end{bmatrix} + \begin{bmatrix} 3 & 1 & 2 \\ 1 & -1 & 0 \end{bmatrix} = \begin{bmatrix} 1+3 & 1+1 & 0+2 \\ 2+1 & 1+(-1) & (-1)+0 \end{bmatrix}$$

$$= \begin{bmatrix} 4 & 2 & 2 \\ 3 & 0 & -1 \end{bmatrix}$$

In the description given above we have used the definition of the equality of two matrices: *two matrices are said to be equal if the corresponding elements are equal.*

It can easily be verified that

$$\begin{bmatrix} a_1 & a_2 & a_3 \\ b_1 & b_2 & b_3 \end{bmatrix} + \begin{bmatrix} 0 & 0 & 0 \\ 0 & 0 & 0 \end{bmatrix} = \begin{bmatrix} a_1 & a_2 & a_3 \\ b_1 & b_2 & b_3 \end{bmatrix}$$

or more generally that the addition of a *zero matrix*, or *null matrix*, i.e., a matrix with all its elements zero, to a given matrix leaves the latter unaltered.

EXERCISE C.3

If

$$A = \begin{bmatrix} 1 & -1 & 0 \\ 2 & 2 & 4 \end{bmatrix} \qquad B = \begin{bmatrix} 4 & 3 & 1 \\ 1 & 2 & -1 \end{bmatrix} \qquad \text{and} \qquad 0 = \begin{bmatrix} 0 & 0 & 0 \\ 0 & 0 & 0 \end{bmatrix}$$

verify that

$$A + B = B + A \qquad A + 0 = 0 + A \qquad B + 0 = 0 + B = B$$

The multiplication of a matrix with a number is defined as follows: *if c is a number and A a matrix, then c times A is the matrix whose elements are c times the corresponding elements of A.* Thus

$$4 \times \begin{bmatrix} 2 & 1 & 3 \\ 5 & 7 & -1 \end{bmatrix} = \begin{bmatrix} 8 & 4 & 12 \\ 20 & 28 & -4 \end{bmatrix}$$

and

$$-5 \times \begin{bmatrix} 1 & -1 & 0 \\ 2 & 1 & 7 \\ 7 & -1 & -5 \end{bmatrix} = \begin{bmatrix} -5 & 5 & 0 \\ -10 & -5 & -35 \\ -35 & 5 & 25 \end{bmatrix}$$

EXERCISE C.4

Verify that for any matrix A, $A + (-1) \times A = 0$, where 0 is a null matrix.

We define one more notion related to a matrix before considering the multiplication of one matrix by another. *The matrix obtained by changing the rows of a given matrix A into columns and vice versa is called the transpose of A and is denoted by the symbol A′.* Thus

$$\begin{bmatrix} 1 & 2 \\ 2 & 2 \\ 3 & 5 \end{bmatrix} \quad \text{is the transpose of} \quad \begin{bmatrix} 1 & 2 & 3 \\ 2 & 2 & 5 \end{bmatrix}$$

and

$$\begin{bmatrix} 3 & 1 & 5 \\ 5 & 7 & 8 \\ 2 & 1 & -1 \end{bmatrix} \quad \text{is the transpose of} \quad \begin{bmatrix} 3 & 5 & 2 \\ 1 & 7 & 1 \\ 5 & 8 & -1 \end{bmatrix}$$

EXERCISES C.5

1. Let

$$A = \begin{bmatrix} 1 & 2 & 1 \\ -1 & 0 & 3 \end{bmatrix} \qquad \text{and} \qquad B = \begin{bmatrix} -1 & 5 & 7 \\ 3 & 7 & 8 \end{bmatrix}$$

Find $A + B$, $3B$, $-7B$, $A - B$, and $2A - B$.
2. Using the matrices A and B in Exercise 1, find $A′$ and $B′$.
3. Verify that the transpose of $A′$ is A.
4. Verify that the transpose of $A′ + B′$ is $A + B$ and that the transpose of $A + B$ is $A′ + B′$.
5. Write down the row vectors and the column vectors of A and B.

We shall now define multiplication of one matrix by another. Like addition, multiplication is defined only in special cases. Let us illustrate how multiplication is performed; then we shall specify when it is valid. Consider

$$A = \begin{bmatrix} a_{11} & a_{12} & a_{13} \\ a_{21} & a_{22} & a_{23} \end{bmatrix} \qquad \text{and} \qquad B = \begin{bmatrix} b_{11} & b_{12} \\ b_{21} & b_{22} \\ b_{31} & b_{32} \end{bmatrix}$$

We shall denote the row vectors of A by a_1 and a_2 and the column vectors of B by b_1 and b_2. The transpose of the column vectors of B is denoted by b_1' and b_2'. Then $b_1' = [b_{11} \quad b_{21} \quad b_{31}]$, and $b_2' = [b_{12} \quad b_{22} \quad b_{32}]$. We shall also use the notation AB for the product of A on the *left* and B on the *right*. The product AB can now be defined in terms of the product of the vectors just defined. Recall that the product of vectors is a *number*.

$$AB = \begin{bmatrix} a_1b_1' & a_1b_2' \\ a_2b_1' & a_2b_2' \end{bmatrix}$$

$$= \begin{bmatrix} a_{11}b_{11} + a_{12}b_{21} + a_{13}b_{31} & a_{11}b_{12} + a_{12}b_{22} + a_{13}b_{32} \\ a_{21}b_{11} + a_{22}b_{21} + a_{23}b_{31} & a_{21}b_{12} + a_{22}b_{22} + a_{23}b_{32} \end{bmatrix}$$

Example Consider the matrices

$$A = \begin{bmatrix} 1 & 2 & -1 \\ 0 & 1 & 1 \end{bmatrix} \quad \text{and} \quad B = \begin{bmatrix} 1 & 2 \\ 3 & 1 \\ 1 & 1 \end{bmatrix}$$

Then

$$AB = \begin{bmatrix} [1 & 2 & -1] \times [1 & 3 & 1] & [1 & 2 & -1] \times [2 & 1 & 1] \\ [0 & 1 & 1] \times [1 & 3 & 1] & [0 & 1 & 1] \times [2 & 1 & 1] \end{bmatrix}$$

$$= \begin{bmatrix} 6 & 3 \\ 4 & 2 \end{bmatrix}$$

The first element in the first row of the product is obtained as

$$1 \times 1 + 2 \times 3 + (-1) \times 1 = 1 + 6 + (-1) = 6$$

Similarly, the second element in the first row is obtained as

$$1 \times 2 + 2 \times 1 + (-1) \times 1 = 2 + 2 + (-1) = 3$$

and so on.

It should be clear, from what has been said above, that *the mechanics involved in obtaining the product* AB, *with* A *on the left and* B *on the right, is to multiply each column of the matrix on the right with each row of the matrix on the left.* This may be remembered as the *row-by-column rule for multiplication.*

Also it should be clear from the description given above that multiplication AB is not defined if the number of columns in A is not the same as the number of rows in B. Thus if

$$A = \begin{bmatrix} 1 & 2 \\ 3 & 4 \end{bmatrix} \quad \text{and} \quad B = \begin{bmatrix} 1 & 5 & 7 \\ 3 & 5 & 7 \\ 1 & 1 & 2 \end{bmatrix}$$

then neither the product AB nor BA is defined. But if

$$C = \begin{bmatrix} 1 & 2 \\ 3 & 4 \\ 5 & 7 \end{bmatrix} \quad \text{and} \quad D = \begin{bmatrix} 1 & 5 & 7 \\ 7 & 5 & 8 \\ 1 & 1 & 1 \end{bmatrix}$$

then the product DC is defined but not CD. Repeating what was just mentioned, to see whether a product is defined you ask: Is the number of columns of the matrix on the left equal to the number of rows of the matrix on the right? If the answer is yes, the matrices are *conformable* and the product is defined; otherwise not.

EXERCISES C.6

1. Find the products AB and BA, where

$$A = \begin{bmatrix} a & 0 & 0 \\ 0 & b & 0 \\ 0 & 0 & c \end{bmatrix} \quad \text{and} \quad B = \begin{bmatrix} p & 0 & 0 \\ 0 & q & 0 \\ 0 & 0 & r \end{bmatrix}$$

Note The first element in the first row (column), the second element in the second row (column), the third element in the third row (column), etc., form the *principal diagonal* of a matrix. If in a matrix all elements other than those in the principal diagonal are zero, the matrix is called a *diagonal matrix*. As may be easily verified, the product of two diagonal matrices is itself a diagonal matrix with the elements in the principal diagonal equal to the product of the corresponding elements in the principal diagonals of the two matrices being multiplied.

2. Find the product AB, where

$$A = \begin{bmatrix} 1 & 1 & 1 \\ 2 & 1 & 2 \\ 3 & 0 & 3 \end{bmatrix} \quad \text{and} \quad B = \begin{bmatrix} 2 & 1 \\ 1 & 2 \\ 2 & 1 \end{bmatrix}$$

Is the product BA defined in this case?

3. Find the products AI and IA, where

$$A = \begin{bmatrix} a_{11} & a_{12} & a_{13} \\ a_{21} & a_{22} & a_{23} \\ a_{31} & a_{32} & a_{33} \end{bmatrix} \quad \text{and} \quad I = \begin{bmatrix} 1 & 0 & 0 \\ 0 & 1 & 0 \\ 0 & 0 & 1 \end{bmatrix}$$

Note A diagonal matrix with all the elements in the principal diagonal equal to 1 is called a *unit matrix* or *identity matrix* and is denoted by the symbol I. As may be easily verified, the multiplication by an identity matrix leaves the original matrix unaltered. It may also be verified that $AI = IA$.

4. Find the product AB and BA where

$$A = \begin{bmatrix} 3 & 5 \\ 1 & 0 \end{bmatrix} \quad \text{and} \quad B = \begin{bmatrix} 1 & 0 \\ 0 & 0 \end{bmatrix}$$

Verify that $AB \neq BA$.

5. Verify that $AB = BA$, where

$$A = \begin{bmatrix} t & 0 \\ 0 & t \end{bmatrix} \quad \text{and} \quad B = \begin{bmatrix} a_{11} & a_{12} \\ a_{21} & a_{22} \end{bmatrix}$$

6. Prove that if

$$B = \begin{bmatrix} b_{11} & 0 \\ 0 & b_{22} \end{bmatrix} \quad b_{11} \neq b_{22} \quad \text{and} \quad A = \begin{bmatrix} a_{11} & a_{12} \\ a_{21} & a_{22} \end{bmatrix}$$

then $AB = BA$ only if a_{12} and a_{21} are both equal to zero.

7. If

$$\begin{bmatrix} x & a \\ b & y \end{bmatrix} \begin{bmatrix} 1 & 2 \\ 3 & 6 \end{bmatrix} = \begin{bmatrix} 1 & 2 \\ 3 & 6 \end{bmatrix}$$

determine x and y in terms of a and b.

8. Given

$$A = \begin{bmatrix} 2 & 1 \\ 1 & 2 \end{bmatrix} \quad B = \begin{bmatrix} 1 & 2 \\ 2 & 1 \end{bmatrix} \quad \text{and} \quad C = \begin{bmatrix} 3 & 4 \\ 2 & 1 \end{bmatrix}$$

verify that $(AB)C = A(BC)$. Verify also that $2AB = A2B$; $(AB)aC = (AaB)C = aA(BC)$.

9. Using the matrix A in Exercise 8, verify that $AA^2 = A^2A = A^3$, where $A^2 = AA$ and $A^3 = AAA$. Also verify that $AA^3 = A^2A^2 = A^3A = A^4$.

10. With

$$V = \begin{bmatrix} 0 & 3 \\ 0 & 0 \end{bmatrix} \quad \text{and} \quad W = \begin{bmatrix} 0 & 4 \\ 0 & 0 \end{bmatrix}$$

calculate VW and WV. Notice that both products have all their elements zero, although this is not true of either factor.

11. With

$$T_1 = \begin{bmatrix} 1 & 0 & 0 \\ 1 & 1 & 0 \\ 1 & 1 & 1 \end{bmatrix} \quad \text{and} \quad T_2 = \begin{bmatrix} 1 & 2 & 3 \\ 0 & 1 & 2 \\ 0 & 0 & 1 \end{bmatrix}$$

calculate $(T_1)^2$, $(T_2)^2$, T_1T_2, and T_2T_1.

Note A matrix with only zeros above the principal diagonal is called a *lower triangular matrix* and one with all the elements below the principal diagonal zero is called an *upper triangular matrix*.

12. Verify that the product of any matrix with its transpose is always a matrix which is square and symmetrical. (A *symmetrical matrix* is one whose elements below the principal diagonal are the mirror reflections of the elements above the principal diagonal.) Thus in a symmetrical matrix, the first element in the second row is the same as the second element in the first row, and in general the ith element in the jth row is the same as the jth element in the ith row.

13. Verify using T_1 and T_2 in Exercise 11 that the transpose of a product is equal to the product of the transposes written in reverse order; i.e., the transpose of the product T_1T_2, in which T_1 is on the left and T_2 is on the right, is equal to the product of T_2' on the left and T_1' on the right. In other words, $(T_1T_2)' = T_2'T_1'$. In general, to get the transpose of a product AB we first interchange the factors, i.e., write B on the left and A on the right, replace each factor by its transpose, and then obtain the product.

C.2 SYSTEMS OF LINEAR EQUATIONS

We are now ready to talk about how to represent a system of simultaneous equations as an equation in matrices. Consider the following system of equations:

$$\begin{aligned} 2x + 3y - z &= 1 \\ x + 2y + z &= -2 \\ 3x + 5y + z &= 2 \end{aligned} \qquad \text{(C.6)}$$

The first equation in (C.6) can be written as a matrix equation as follows:

$$\begin{bmatrix} 2 & 3 & -1 \end{bmatrix} \begin{bmatrix} x \\ y \\ z \end{bmatrix} = 1 \qquad \text{(C.7)}$$

Similarly the second and the third equations in (C.6) can be written in matrix form

$$[1 \quad 2 \quad 1] \begin{bmatrix} x \\ y \\ z \end{bmatrix} = -2 \qquad (C.8)$$

and

$$[3 \quad 5 \quad 1] \begin{bmatrix} x \\ y \\ z \end{bmatrix} = 2 \qquad (C.9)$$

All three equations, (C.7), (C.8), and (C.9), can be written in one matrix equation:

$$\begin{bmatrix} 2 & 3 & -1 \\ 1 & 2 & 1 \\ 3 & 5 & 1 \end{bmatrix} \begin{bmatrix} x \\ y \\ z \end{bmatrix} = \begin{bmatrix} 1 \\ -2 \\ 2 \end{bmatrix} \qquad (C.10)$$

The 3×3 matrix on the left of (C.10) is called the *coefficient matrix*. The matrix obtained by augmenting the coefficient matrix with the column on the right-hand side of (C.10) is called the *augmented matrix*; the augmented matrix in (C.10) is thus:

$$\begin{bmatrix} 2 & 3 & -1 & 1 \\ 1 & 2 & 1 & -2 \\ 3 & 5 & 1 & 2 \end{bmatrix}$$

When discussing systems of simultaneous equations, it is emphasized that the equations lead to a unique set of solutions only if there are as many consistent and independent equations as there are unknowns. These requirements can now be translated in terms of the properties of the coefficient matrix and the augmented matrix. First let us recall what is meant by independence and consistency when these terms are applied to simultaneous equations. Consider the following two equations in two unknowns x and y:

$$\begin{aligned} 2x + 3y &= 5 \\ 4x + 6y &= 10 \end{aligned} \qquad (C.11)$$

In this case, the second equation is not independent of the first since it can be derived from the first by multiplying both sides by 2. *If in a system of simultaneous equations, any one equation can be obtained as a linear combination of one or more of the others, the former is said to be linearly dependent on the others.* Now consider the following two equations in x and y:

$$\begin{aligned} x + y &= 5 \\ x + y &= 10 \end{aligned} \qquad (C.12)$$

These two equations are not consistent, since for no values of x and y can both be satisfied; for any pair of values for x and y, the left-hand sides of the equations will be identical while the right-hand sides will be different.

The requirement that there must be as many independent equations as there are unknowns can be expressed as the requirement that the *rank* of the coefficient matrix must be the same as the number of unknowns; and the requirement that the equations must be consistent can be expressed as the condition that the rank of the coefficient matrix must be equal to the rank of the augmented matrix. This gives us the reason for studying the *rank* of a matrix.

As already mentioned, any matrix can be thought of as composed of a number of column vectors or of a number of row vectors. The *column rank* of a matrix is defined as the maximum number of linearly independent column vectors in the matrix. Similarly the *row rank* of a matrix is defined as the maximum number of linearly independent row vectors in the matrix. One of the main theorems concerning rank of a matrix is that *the column rank equals the row rank*.

In order to compute the rank of a matrix easily, we make use of the property that the following operations (known as *elementary operations*) on the columns of a matrix do not change its rank:

1 Multiplying or dividing one column by a nonzero number
2 Interchanging two columns
3 Adding one column to another
4 Deleting a column consisting entirely of zeros

Since the row rank equals the column rank, the same operations, just mentioned, applied to rows instead of columns do not affect the rank of the matrix either. By applying the above operations we shall be able to obtain a matrix whose rank can be calculated more easily. To demonstrate, let us calculate the rank of the matrix:

$$A = \begin{bmatrix} 1 & 1 & 2 & 0 & 4 \\ 2 & 1 & 3 & 1 & 7 \\ 3 & 1 & 4 & 2 & 10 \\ 4 & 1 & 5 & 3 & 13 \\ 5 & 1 & 6 & 4 & 16 \end{bmatrix}$$

STEP 1 Leave column 1 as it is and subtract suitable multiples of column 1 from the other columns so that the first element in each of them becomes zero. In this case we subtract column 1 from column 2 and twice column 1 from column 3; leave the fourth column as it is; and subtract 4 times column 1 from column 4. These operations lead to

$$A_1 = \begin{bmatrix} 1 & 0 & 0 & 0 & 0 \\ 2 & -1 & -1 & 1 & -1 \\ 3 & -2 & -2 & 2 & -2 \\ 4 & -3 & -3 & 3 & -3 \\ 5 & -4 & -4 & 4 & -4 \end{bmatrix}$$

STEP 2 Leave the first two columns as they are in A_1. Subtract appropriate multiples of column 2 of A_1 from the third, fourth, etc., columns of A_1 so that the second row of the third, fourth, etc., columns of A_1 becomes zero. Note that this operation will not affect the first row of these columns. In this case we need to subtract column 2 from column 3; -1 times column 2 from column 4; and column 2 from column 5. That leads to

$$A_2 = \begin{bmatrix} 1 & 0 & 0 & 0 & 0 \\ 2 & -1 & 0 & 0 & 0 \\ 3 & -2 & 0 & 0 & 0 \\ 4 & -3 & 0 & 0 & 0 \\ 5 & -4 & 0 & 0 & 0 \end{bmatrix}$$

STEP 3   Leave out all columns which entirely consist of zeros. The remaining number of columns (in this case 2) is the column rank of the matrix.

The two columns that remain after the operation in step 3 form a 5×2 matrix:

$$A_3 = \begin{bmatrix} 1 & 0 \\ 2 & -1 \\ 3 & -2 \\ 4 & -3 \\ 5 & -4 \end{bmatrix}$$

It can be easily verified that the two columns of A_3 are linearly independent. That is, it is *not possible* to give a general formula of the following type:

$$\begin{bmatrix} 1 \\ 2 \\ 3 \\ 4 \\ 5 \end{bmatrix} = a \begin{bmatrix} 0 + b \\ -1 + b \\ -2 + b \\ -3 + b \\ -4 + b \end{bmatrix}$$

for any a and b.

EXERCISES C.7

1. Find the rank of the following matrices:

$$A = \begin{bmatrix} 2 & 1 & 3 \\ 5 & 2 & 1 \end{bmatrix} \qquad B = \begin{bmatrix} -1 & 1 & -2 \\ 3 & 2 & 11 \end{bmatrix}$$

$$C = \begin{bmatrix} 1 & 2 & -3 \\ -1 & -2 & 3 \\ 5 & 3 & 1 \\ 0 & 0 & 0 \end{bmatrix} \qquad D = \begin{bmatrix} 1 & 0 & 1 & 0 & 0 \\ 1 & 1 & 0 & 0 & 0 \\ 0 & 1 & 1 & 0 & 0 \\ 0 & 0 & 1 & 1 & 0 \\ 0 & 1 & 0 & 1 & 1 \end{bmatrix}$$

2. Examine the equations

$$3x - 2y + 2z = 1$$
$$4x - 2y + 3z = 2$$
$$2x - 3y + z = 3$$

Verify that the rank of the coefficient matrix is equal to the rank of the augmented matrix. Write the above equations as a matrix equation. Premultiply the matrix equation with the following matrix:

$$\begin{bmatrix} 7 & -4 & -2 \\ 2 & -1 & -1 \\ -8 & 5 & 2 \end{bmatrix}$$

Verify that premultiplication leads to the equation:

$$\begin{bmatrix} 1 & 0 & 0 \\ 0 & 1 & 0 \\ 0 & 0 & 1 \end{bmatrix} \begin{bmatrix} x \\ y \\ z \end{bmatrix} = \begin{bmatrix} -7 \\ -3 \\ 8 \end{bmatrix}$$

and hence to the solution $x = -7$, $y = -3$, and $z = 8$.

3. Using

$$A = \begin{bmatrix} 3 & -2 & 2 \\ 4 & -2 & 3 \\ 1 & 0 & 1 \end{bmatrix}$$

verify that the rank of A is the same as the rank of AA' and of $A'A$; that is, the rank of a matrix is equal to the rank of the product of that matrix and its transpose.

4. Given $x + y = 5$ and $x + y = 10$, are the ranks of the coefficient and augmented matrices the same?

C.3 DETERMINANTS

Before we proceed to consider some of the other important properties of matrices, we should become familiar with the definition and elementary properties of *determinants*. Consider the following pair of equations:

$$\begin{aligned} ax + by &= p \\ cx + dy &= q \end{aligned} \tag{C.13}$$

Multiply the first equation with c and the second with a and then subtract one from the other; we get $(ad - bc)y = aq - cp$, and hence

$$y = \frac{aq - cp}{ad - bc} \tag{C.14}$$

Similarly, if we multiply the first equation by d and the second one by b and subtract one from the other, we get $(ad - bc)x = dp - bq$, or

$$x = \frac{dp - bq}{ad - bc} \tag{C.15}$$

Notice that the solution for x is a ratio of two quantities, the denominator, involving the coefficients a, b, c, and d, and the numerator, involving the coefficients of y and the constant terms on the right-hand side of (C.13). A similar feature is true of the solution for y too. The quantity occurring in the denominator of the solutions for x and y involves all the elements of the coefficient matrix and only those. This shows the usefulness of stipulating a procedure which assigns a unique numerical value to the arrangement of numbers in the coefficient matrix. We assign the numerical value $ad - bc$ to the arrangement

$$\begin{bmatrix} a & b \\ c & d \end{bmatrix}$$

and use the term *determinant* for that numerical value. The notation used for the determinant of the above arrangement of numbers is

$$\begin{vmatrix} a & b \\ c & d \end{vmatrix}$$

in which the arrangement is enclosed within two parallel vertical lines (instead of square brackets). It should be emphasized that this notion is applied only to arrangements in the form of square matrices.

The quantity $aq - cp$ occurring in the numerator of the ratio in (C.14) can be represented as the determinant

$$\begin{vmatrix} a & p \\ c & q \end{vmatrix}$$

and similarly the quantity $dp - bq$ as the determinant

$$\begin{vmatrix} p & b \\ q & d \end{vmatrix}$$

Thus the solutions (C.14) and (C.15) can be rewritten

$$x = \frac{\begin{vmatrix} p & b \\ q & d \end{vmatrix}}{\begin{vmatrix} a & b \\ c & d \end{vmatrix}} \quad \text{and} \quad y = \frac{\begin{vmatrix} a & p \\ c & q \end{vmatrix}}{\begin{vmatrix} a & b \\ c & d \end{vmatrix}} \tag{C.16}$$

Note that the determinant in the numerator for x in (C.16) is obtained from the determinant of the coefficient matrix by replacing the coefficients of x by the corresponding constant terms on the right of (C.13). Similarly, the determinant in the numerator of y in (C.16) is obtained from the determinant of the coefficient matrix by replacing the coefficients of y by the corresponding constant terms on the right of (C.13).

Generalizing the above procedure to the case of three equations in three unknowns,

$$\begin{aligned} ax + by + cz &= p \\ dx + ey + fz &= q \\ lx + my + nz &= r \end{aligned} \tag{C.17}$$

we may write the solution as

$$x = \frac{\begin{vmatrix} p & b & c \\ q & e & f \\ r & m & n \end{vmatrix}}{\begin{vmatrix} a & b & c \\ d & e & f \\ l & m & n \end{vmatrix}} \quad y = \frac{\begin{vmatrix} a & p & c \\ d & q & f \\ l & r & n \end{vmatrix}}{\begin{vmatrix} a & b & c \\ d & e & f \\ l & m & n \end{vmatrix}} \quad \text{and} \quad z = \frac{\begin{vmatrix} a & b & p \\ d & e & q \\ l & m & r \end{vmatrix}}{\begin{vmatrix} a & b & c \\ d & e & f \\ l & m & n \end{vmatrix}} \tag{C.18}$$

with the understanding of course that the 3×3 (third-order) determinants appearing in the numerator and denominator are defined in such a way that they provide the same solution as the one we would have arrived at if we had solved (C.17) by any other method. The following is such a definition. Consider the determinant in the denominators in (C.18),

$$\begin{vmatrix} a & b & c \\ d & e & f \\ l & m & n \end{vmatrix} = a \begin{vmatrix} e & f \\ m & n \end{vmatrix} - b \begin{vmatrix} d & f \\ l & n \end{vmatrix} + c \begin{vmatrix} d & e \\ l & m \end{vmatrix} \tag{C.19}$$

The definition given in (C.19) involves the following rules:

1 The value of the determinant on the left of (C.19) is the sum of three products each of which is a product of an element in the first row of the determinant and an associated factor.

2 The factor associated with a, the first element in the first row of the determinant, is $+1$ times the determinant obtained by discarding the row and column containing a.

3 The factor associated with b, the second element in the first row of the determinant, is -1 times the determinant obtained by discarding the row and column containing b.

4 The factor associated with *c*, the third element in the first row of the determinant, is +1 times the determinant obtained by discarding the row and column containing *c*.

The factor associated with *a* on the right-hand side of (C.19) is called the *cofactor* of *a* in the determinant on the left of (C.19). Similarly the factor associated with *b* is called the cofactor of *b* and that associated with *c* is called the cofactor of *c*.

The cofactor is defined not only for the elements of the first row but for all elements of the determinant. The following schema tells the sign to be attached to the determinant obtained by discarding the row and column containing a given element in order to get the corresponding cofactor:

$$\begin{bmatrix} + & - & + \\ - & + & - \\ + & - & + \end{bmatrix} \qquad \text{(C.20)}$$

Thus to get the cofactor of *f* in the determinant on the left of (C.19) we first write down the determinant obtained by discarding the row and column containing *f*; then attach to that determinant the appropriate sign, in this case a minus sign since the schema (C.20) shows a minus sign in the place occupied by *f* in the determinant. Thus the cofactor of *f* is

$$- \begin{vmatrix} a & b \\ l & m \end{vmatrix}$$

and similarly the cofactor of *n* is

$$+ \begin{vmatrix} a & b \\ d & e \end{vmatrix}$$

and so on.

The process of writing down the value of the determinant in the fashion shown in (C.19) is called the *expansion* of the determinant in terms of the elements of the first row and the corresponding cofactors. A determinant can be expanded in terms of the elements of any row (or column) and the corresponding cofactors simply by summing the products of the elements of the given row (or column) and the corresponding cofactors. Thus,

$$\begin{vmatrix} a & b & c \\ d & e & f \\ l & m & n \end{vmatrix} = -d \begin{vmatrix} b & c \\ m & n \end{vmatrix} + e \begin{vmatrix} a & c \\ l & n \end{vmatrix} - f \begin{vmatrix} a & b \\ l & m \end{vmatrix} \qquad \text{(C.21)}$$

It is also equal to

$$l \begin{vmatrix} b & c \\ e & f \end{vmatrix} - m \begin{vmatrix} a & c \\ d & f \end{vmatrix} + n \begin{vmatrix} a & b \\ d & e \end{vmatrix} \qquad \text{(C.22)}$$

and so on.

The generalization of the foregoing to determinants of higher order is straightforward. Thus if we have a fourth-order determinant,

$$|\mathbf{A}| = \begin{vmatrix} a_1 & b_1 & c_1 & d_1 \\ a_2 & b_2 & c_2 & d_2 \\ a_3 & b_3 & c_3 & d_3 \\ a_4 & b_4 & c_4 & d_4 \end{vmatrix} \qquad \text{(C.23)}$$

we write down its expansion in terms of the elements of the third row as $a_3A_3 + b_3B_3 + c_3C_3 + d_3D_3$, where A_3 is the cofactor of a_3, B_3 is the cofactor of b_3, and so on. Note that A_3 itself is a third-order determinant which can be similarly expanded, and ultimately we arrive at a single number as the value of the determinant $|\mathbf{A}|$.

EXERCISES C.8

1. Let

$$A = \begin{bmatrix} 2 & 3 \\ 1 & 4 \end{bmatrix} \quad \text{and} \quad B = \begin{bmatrix} 1 & 5 \\ 2 & 3 \end{bmatrix}$$

be two matrices. Verify the theorem that the determinant of the product of two matrices is equal to the product of the determinants of the two matrices.

2. Try to verify the above theorem using the matrices

$$A = \begin{bmatrix} a_1 & a_2 \\ b_1 & b_2 \end{bmatrix} \quad \text{and} \quad B = \begin{bmatrix} c_1 & c_2 \\ d_1 & d_2 \end{bmatrix}$$

3. Verify that

$$\begin{vmatrix} 1 & 0 & k \\ 0 & 1 & 0 \\ 0 & 0 & 1 \end{vmatrix} = 1$$

4. Prove that

$$\begin{vmatrix} 1 & 0 & k \\ 0 & 1 & 0 \\ 0 & 0 & 1 \end{vmatrix} \begin{vmatrix} a_1 & b_1 & c_1 \\ a_2 & b_2 & c_2 \\ a_3 & b_3 & c_3 \end{vmatrix} = \begin{vmatrix} a_1 + ka_3 & b_1 + kb_3 & c_1 + kc_3 \\ a_2 & b_2 & c_2 \\ a_3 & b_3 & c_3 \end{vmatrix}$$

and hence verify that the addition of a multiple of the elements of any row of a determinant to the corresponding elements of any other row leaves the value of the determinant unaltered.

5. Verify that

$$\begin{vmatrix} a_1 & b_1 & c_1 \\ a_2 & b_2 & c_2 \\ a_3 & b_3 & c_3 \end{vmatrix} \begin{vmatrix} 1 & 0 & k \\ 0 & 1 & 0 \\ 0 & 0 & 1 \end{vmatrix} = \begin{vmatrix} a_1 & b_1 & ka_1 + c_1 \\ a_2 & b_2 & ka_2 + c_2 \\ a_3 & b_3 & ka_3 + c_3 \end{vmatrix}$$

and hence that the addition of a multiple of any column to another column of a determinant does not change the value of the determinant.

6. Verify that

$$\begin{vmatrix} k & 0 & 0 \\ 0 & 1 & 0 \\ 0 & 0 & 1 \end{vmatrix} = k$$

hence that

$$\begin{vmatrix} ka_1 & b_1 & c_1 \\ ka_2 & b_2 & c_2 \\ ka_3 & b_3 & c_3 \end{vmatrix} = k \begin{vmatrix} a_1 & b_1 & c_1 \\ a_2 & b_2 & c_2 \\ a_3 & b_3 & c_3 \end{vmatrix}$$

and hence that multiplying a column of a determinant by a constant number amounts to multiplying the value of the determinant by that constant.

7. Similarly verify that multiplying the elements in a row of a determinant by a constant amounts to multiplying the value of that determinant by that constant.

8. Verify that

$$\begin{vmatrix} a_1 & 0 & 0 \\ a_2 & b_2 & 0 \\ a_3 & b_3 & c_3 \end{vmatrix} = a_1 b_2 c_3$$

i.e., the product of the diagonal elements.

Note A determinant of the above type, in which all elements above the principal diagonal are zero, is called a *triangular determinant*. Determinants with zeros all below the principal diagonal are also called *triangular determinants*. Evaluation of triangular determinants is quite simple, as you have just seen. This property is exploited in evaluating other determinants by reducing them to triangular form using interchange of rows or addition of multiples of one row to another, operations which do not affect the value of the determinant.

9. Using the results of Exercise 4, prove that if two rows of a determinant are identical, the value of the determinant is zero.

10. Using the results of Exercise 5, prove that if two columns of a determinant are identical, the value of the determinant is zero.

11. It has been shown that a determinant can be expanded in terms of the elements of any given row and the corresponding cofactors. Suppose you have the determinant

$$\begin{vmatrix} a_1 & b_1 & c_1 \\ a_2 & b_2 & c_2 \\ a_3 & b_3 & c_3 \end{vmatrix}$$

Let us denote the cofactor of any element of the determinant by the corresponding capital letter; thus A_1 is the cofactor of a_1, C_2 is the cofactor of c_2, and so on. Using the result of Exercise 9, prove that $a_2A_1 + b_2B_1 + c_2C_1 = 0$. *Hint:* Show that $a_2A_1 + b_2B_1 + c_2C_1$ is the expansion of the determinant

$$\begin{vmatrix} a_2 & b_2 & c_2 \\ a_2 & b_2 & c_2 \\ a_3 & b_3 & c_3 \end{vmatrix}$$

which has identical first and second rows.

12. Using the notation of Exercise 11, show that $a_3A_1 + b_3B_1 + c_3C_1 = 0$.

Note If you take the elements of any one row in any determinant and multiply them with the cofactors of the corresponding elements in another row, the sum of such products will be zero. Similar results hold true for columns.

13. Using the notations of Exercise 11 and the results of Exercises 11 and 12, show that

$$\begin{vmatrix} a_1 & b_1 & c_1 \\ a_2 & b_2 & c_2 \\ a_3 & b_3 & c_3 \end{vmatrix} \begin{vmatrix} A_1 & A_2 & A_3 \\ B_1 & B_2 & B_3 \\ C_1 & C_2 & C_3 \end{vmatrix} = \begin{vmatrix} a_1 & b_1 & c_1 \\ a_2 & b_2 & c_2 \\ a_3 & b_3 & c_3 \end{vmatrix} \begin{vmatrix} a_1 & b_1 & c_1 \\ a_2 & b_2 & c_2 \\ a_3 & b_3 & c_3 \end{vmatrix} \begin{vmatrix} a_1 & b_1 & c_1 \\ a_2 & b_2 & c_2 \\ a_3 & b_3 & c_3 \end{vmatrix}$$

Hint: The product of the two determinants on the left of the above equation is equal to

$$\begin{vmatrix} a_1A_1 + b_1B_1 + c_1C_1 & 0 & 0 \\ 0 & a_2A_2 + b_2B_2 + c_2C_2 & 0 \\ 0 & 0 & a_3A_3 + b_3B_3 + c_3C_3 \end{vmatrix}$$

14. Verify that a square matrix is singular if its rank is less than the number of rows.

Note A matrix whose determinant is not equal to zero is referred to as *nonsingular*, and one whose determinant is equal to zero is called *singular*.

C.4   SOLUTION OF SIMULTANEOUS EQUATIONS: THE FULL-RANK CASE

Among the methods of solving simultaneous equations that are consistent, the procedure involving elimination of the unknowns one by one may be the most familiar, but we shall describe a procedure that involves the use of the *inverse of the coefficient matrix.* Let us consider a simple case first. Suppose we want to solve the following equations:

$$x + 3y = 6$$
$$3x + y = 6 \qquad \text{(C.24)}$$

In matrix notation this equation system can be written

$$\begin{bmatrix} 1 & 3 \\ 3 & 1 \end{bmatrix} \begin{bmatrix} x \\ y \end{bmatrix} = \begin{bmatrix} 6 \\ 6 \end{bmatrix} \qquad \text{(C.25)}$$

If we can find a matrix

$$\begin{bmatrix} a & b \\ c & d \end{bmatrix}$$

such that its product with the coefficient matrix in (C.25) is an identity matrix, i.e.,

$$\begin{bmatrix} a & b \\ c & d \end{bmatrix} \begin{bmatrix} 1 & 3 \\ 3 & 1 \end{bmatrix} = \begin{bmatrix} 1 & 0 \\ 0 & 1 \end{bmatrix} \qquad \text{(C.26)}$$

then by premultiplying both sides of (C.25) with

$$\begin{bmatrix} a & b \\ c & d \end{bmatrix}$$

we can reduce it to

$$\begin{bmatrix} 1 & 0 \\ 0 & 1 \end{bmatrix} \begin{bmatrix} x \\ y \end{bmatrix} = \begin{bmatrix} a & b \\ c & d \end{bmatrix} \begin{bmatrix} 6 \\ 6 \end{bmatrix}$$

This is the same as

$$\begin{bmatrix} x \\ y \end{bmatrix} = \begin{bmatrix} a & b \\ c & d \end{bmatrix} \begin{bmatrix} 6 \\ 6 \end{bmatrix} \qquad \text{(C.27)}$$

and hence immediately gives the solutions for x and y in the given equation system. The question then is: How can we find a matrix

$$\begin{bmatrix} a & b \\ c & d \end{bmatrix}$$

given the coefficient matrix

$$\begin{bmatrix} 1 & 3 \\ 3 & 1 \end{bmatrix}$$

so as to satisfy (C.26)?

A matrix

$$\begin{bmatrix} a & b \\ c & d \end{bmatrix}$$

which satisfies (C.26) is called the *inverse* of

$$\begin{bmatrix} 1 & 3 \\ 3 & 1 \end{bmatrix}$$

It may be worth noting that the concept of the inverse of a matrix is an extension of the concept of the reciprocal of a number. The process of obtaining the inverse of a matrix is called inverting the matrix. We shall describe two methods for inverting a matrix. One uses the cofactors and the determinant of the given matrix, and the other follows more or less closely the routine for solving simultaneous equations by eliminating the unknowns one by one.

METHOD 1 Suppose we want to find the inverse of the matrix

$$A = \begin{bmatrix} a_{11} & a_{12} & \cdots & a_{1n} \\ a_{21} & a_{22} & \cdots & a_{2n} \\ \cdots & \cdots & \cdots & \cdots \\ a_{n1} & a_{n2} & \cdots & a_{nn} \end{bmatrix}$$

We first write the matrix of cofactors of each element of **A**. Let the matrix of cofactors be

$$\begin{bmatrix} A_{11} & A_{12} & \cdots & A_{1n} \\ A_{21} & A_{22} & \cdots & A_{2n} \\ \cdots & \cdots & \cdots & \cdots \\ A_{n1} & A_{n2} & \cdots & A_{nn} \end{bmatrix}$$

where A_{ij} is the cofactor of a_{ij}, $i = 1, 2, \ldots, n$; $j = 1, 2, \ldots, n$.

We divide each element of the transpose of the matrix of cofactors by the determinant $|A|$. The resulting matrix will be the inverse of **A**. To illustrate, suppose

$$A = \begin{bmatrix} 1 & 3 \\ 3 & 1 \end{bmatrix}$$

The matrix of cofactors is

$$\begin{bmatrix} 1 & -3 \\ -3 & 1 \end{bmatrix}$$

which when transposed becomes

$$\begin{bmatrix} 1 & -3 \\ -3 & 1 \end{bmatrix}$$

When divided by the determinant of **A** ($= -8$), this becomes

$$\begin{bmatrix} -\frac{1}{8} & \frac{3}{8} \\ \frac{3}{8} & -\frac{1}{8} \end{bmatrix}$$

which can be easily verified to be the inverse of **A**:

$$\begin{bmatrix} -\frac{1}{8} & \frac{3}{8} \\ \frac{3}{8} & -\frac{1}{8} \end{bmatrix} \begin{bmatrix} 1 & 3 \\ 3 & 1 \end{bmatrix} = \begin{bmatrix} 1 & 0 \\ 0 & 1 \end{bmatrix}$$

To give another illustration, consider

$$A = \begin{bmatrix} 3 & 5 & 4 \\ 1 & 2 & -4 \\ 2 & 3 & 1 \end{bmatrix}$$

It is easily verified that the matrix of cofactors of **A** is

$$\begin{bmatrix} 14 & -9 & -1 \\ 7 & -5 & 1 \\ -28 & 16 & 1 \end{bmatrix}$$

which when transposed becomes

$$\begin{bmatrix} 14 & 7 & -28 \\ -9 & -5 & 16 \\ -1 & 1 & 1 \end{bmatrix}$$

Dividing each element of this matrix by the determinant of A ($= -7$), we get

$$\begin{bmatrix} -2 & -1 & 4 \\ +\frac{9}{7} & +\frac{5}{7} & -\frac{16}{7} \\ +\frac{1}{7} & -\frac{1}{7} & -\frac{1}{7} \end{bmatrix}$$

which can be easily verified to be the inverse of **A**.

Since one of the steps in inverting a given matrix involves division by the determinant of the matrix, it is obvious that when we talk about inverting a matrix, we are referring to square matrices which are nonsingular. Another way of stating this is that only square matrices of *full rank* admit inverses. By full rank is meant the maximum possible rank. We shall later on generalize the notion of inverse to singular and rectangular matrices. For the moment we confine our attention to nonsingular square matrices.

METHOD 2   One can invert a given matrix using another routine. We write the given matrix on the left and a corresponding identity matrix on the right. Then we perform on *both* a sequence of operations involving either subtracting a constant multiple of one row from another or multiplying a row by a constant. The object is to reduce the matrix on the left to an identity matrix. Once this object is accomplished, the matrix that emerges on the right is the inverse of the given matrix. This procedure is illustrated in Table C.1.

Notice that operation 1 described in Table C.1 is aimed at reducing all elements in the first column of the given matrix to 1. By the end of the second operation the first column of the given matrix is reduced to that of a corresponding identity matrix. By the end of the third operation all elements except the first one in column 2 are made equal to 1. Operation 4 makes the diagonal and the below-diagonal elements in column 2 correspond to those of an identity matrix. By the end of the fifth operation the original matrix is reduced to a triangular matrix with all its diagonal elements equal to 1. The remaining operations are aimed at reducing the elements above the principal diagonal to zero.

EXERCISES C.9

1. Invert each of the following matrices:

$$\mathbf{B} = \begin{bmatrix} 1 & 2 & -3 \\ 0 & 1 & 2 \\ 0 & 0 & 1 \end{bmatrix} \qquad \mathbf{C} = \begin{bmatrix} 1 & 1 & -1 & -1 \\ 1 & -1 & -1 & 1 \\ 1 & 1 & 1 & 1 \\ 1 & -1 & 1 & -1 \end{bmatrix}$$

2. Given the matrix **C** in Exercise 1, compute **C′C** and invert it.
3. Solve the following systems of equations:

(a) $x + y + z = 3$ (b) $x + y + z + w = 10$

   $x - y + 2z = 2$ $x + 2y + 2z + 3w = 23$

   $2x + 3y - 3z = 2$ $x + 2y + 3z + 2w = 22$

                                $x + 2y + 3z + 4w = 30$

4. Using the coefficient matrix of part (a) in Exercise 3, verify that the inverse of the transpose of a matrix is the transpose of the inverse of that matrix. That is, $(\mathbf{A}^{-1})' = (\mathbf{A}')^{-1}$.

5. Using the coefficient matrix of part (a) of Exercise 3 and the matrix inverted in Table C.1, verify that the inverse of the product of two matrices is equal to the product of the inverses of the matrices in the reverse order. That is, $(AB)^{-1} = B^{-1}A^{-1}$.

6. The routine described in Table C.1 is translated into a computer program using Fortran language in Golden (1965 p. 114). Obtain the inverse of the following matrices using any available program:

$$(a)\ \begin{bmatrix} 3 & 5 & 4 \\ 1 & 2 & -4 \\ 2 & 3 & 1 \end{bmatrix} \qquad (b)\ \begin{bmatrix} 22 & 7 & 7 & 8 & 7 \\ 7 & 7 & 0 & 2 & 3 \\ 7 & 0 & 7 & 3 & 2 \\ 8 & 2 & 3 & 8 & 0 \\ 7 & 3 & 2 & 0 & 7 \end{bmatrix}$$

Table C.1 A ROUTINE FOR INVERTING A MATRIX

Operation† (1)	Left (2)	Right (3)
	$\begin{bmatrix} 3 & 5 & 4 \\ 1 & 2 & -4 \\ 2 & 3 & 1 \end{bmatrix}$	$\begin{bmatrix} 1 & 0 & 0 \\ 0 & 1 & 0 \\ 0 & 0 & 1 \end{bmatrix}$
1. Divide row 1 by 3 and row 3 by 2	$\begin{bmatrix} 1 & \frac{5}{3} & \frac{4}{3} \\ 1 & 2 & -4 \\ 1 & \frac{3}{2} & \frac{1}{2} \end{bmatrix}$	$\begin{bmatrix} \frac{1}{3} & 0 & 0 \\ 0 & 1 & 0 \\ 0 & 0 & \frac{1}{2} \end{bmatrix}$
2. Subtract row 1 from rows 2 and 3	$\begin{bmatrix} 1 & \frac{5}{3} & \frac{4}{3} \\ 0 & \frac{1}{3} & -\frac{16}{3} \\ 0 & -\frac{1}{6} & -\frac{5}{6} \end{bmatrix}$	$\begin{bmatrix} \frac{1}{3} & 0 & 0 \\ -\frac{1}{3} & 1 & 0 \\ -\frac{1}{3} & 0 & \frac{1}{2} \end{bmatrix}$
3. Divide row 2 by $\frac{1}{3}$ and row 3 by $-\frac{1}{6}$	$\begin{bmatrix} 1 & \frac{5}{3} & \frac{4}{3} \\ 0 & 1 & -16 \\ 0 & 1 & 5 \end{bmatrix}$	$\begin{bmatrix} \frac{1}{3} & 0 & 0 \\ -1 & 3 & 0 \\ 2 & 0 & -3 \end{bmatrix}$
4. Subtract row 2 from row 3	$\begin{bmatrix} 1 & \frac{5}{3} & \frac{4}{3} \\ 0 & 1 & -16 \\ 0 & 0 & 21 \end{bmatrix}$	$\begin{bmatrix} \frac{1}{3} & 0 & 0 \\ -1 & 3 & 0 \\ 3 & -3 & -3 \end{bmatrix}$
5. Divide row 3 by 21	$\begin{bmatrix} 1 & \frac{5}{3} & \frac{4}{3} \\ 0 & 1 & -16 \\ 0 & 0 & 1 \end{bmatrix}$	$\begin{bmatrix} \frac{1}{3} & 0 & 0 \\ -1 & 3 & 0 \\ \frac{1}{7} & -\frac{1}{7} & -\frac{1}{7} \end{bmatrix}$
6. Subtract -16 times row 3 from row 2	$\begin{bmatrix} 1 & \frac{5}{3} & \frac{4}{3} \\ 0 & 1 & 0 \\ 0 & 0 & 1 \end{bmatrix}$	$\begin{bmatrix} \frac{1}{3} & 0 & 0 \\ \frac{9}{7} & \frac{5}{7} & -\frac{16}{7} \\ \frac{1}{7} & -\frac{1}{7} & -\frac{1}{7} \end{bmatrix}$
7. Subtract $\frac{5}{3}$ times row 2 and $\frac{4}{3}$ times row 3 from row 1	$\begin{bmatrix} 1 & 0 & 0 \\ 0 & 1 & 0 \\ 0 & 0 & 1 \end{bmatrix}$	$\begin{bmatrix} -2 & -1 & 4 \\ \frac{9}{7} & \frac{5}{7} & -\frac{16}{7} \\ \frac{1}{7} & -\frac{1}{7} & -\frac{1}{7} \end{bmatrix}$

† The operations described in column 1 of the table are to be performed on the matrices obtained just above. Thus operation 1 is to be performed on the first matrix in column 2 and the corresponding matrix in column 3, operation 2 is to be performed on the second matrix in column 2 and the corresponding matrix in column 3, and so on.

C.5   SOLUTION OF SIMULTANEOUS EQUATIONS: THE CASE OF LESS THAN FULL RANK

Let us now consider the case of systems of equations in which the equations are consistent but the number of independent equations falls short of the number of unknowns. Consider, for example, the following system of equations:

$$\begin{aligned} x + 3y + 2z &= 7 \\ -2x + 2y - 3z &= -4 \\ -x + 5y - z &= 3 \end{aligned} \qquad \text{(C.28)}$$

It is easily seen that the third equation is obtained by adding the first two. Hence we have fewer than three independent equations in the system. The question is: Can we arrive at unique solutions for the unknowns in these equations? In order to examine this question let us proceed as if we had not noticed that there are fewer than three independent equations in the system.

Let us try to solve these equations by eliminating the unknowns one by one. Dividing the second equation throughout by -2 and the third equation throughout by -1, we get

$$\begin{aligned} x - y + 1.5z &= 2 \\ x - 5y + z &= -3 \end{aligned} \qquad \text{(C.29)}$$

If we now subtract the first equation in (C.28) from each of the equations in (C.29), we get the following two equations which do not contain x:

$$\begin{aligned} -4y - 0.5z &= -5 \\ -8y - 1.0z &= -10 \end{aligned} \qquad \text{(C.30)}$$

Now let us divide the first equation in (C.30) throughout by -4 and the second one in (C.30) throughout by -8, getting

$$\begin{aligned} y + 0.125z &= 1.25 \\ y + 0.125z &= 1.25 \end{aligned} \qquad \text{(C.31)}$$

The equations in (C.31) are identical to each other. Consequently we cannot eliminate any more of the unknowns. But we may arbitrarily fix any one of the unknowns at any value and convert the whole problem to one involving only two unknowns. Let us fix z at some arbitrarily chosen value, say 0. This will lead to the following solutions for y and x from (C.31) and (C.29):

$$y = 1.25 \qquad x = 3.25$$

Thus $x = 3.25$, $y = 1.25$, $z = 0$ is seen to be one solution set satisfying (C.28). By giving z some other arbitrary value we can get another solution set. In fact there is an unlimited number of solution sets possible for (C.28). Notice that this procedure is equivalent to setting z equal to an arbitrarily chosen value in the original set of equations (C.28) and then solving for x and y from any two of the equations.

In general if we have a consistent system of n equations in n unknowns but only r of them are linearly independent (r being less than n), we may give arbitrary values to $n - r$ unknowns and solve for the remaining r unknowns using any set of r equations. The solutions so obtained together with the arbitrary values given to the $n - r$ unknowns provide us with one possible set of solutions to the given system of equations.

At this point it is useful to introduce the notion of *generalized inverse* of a matrix.

Consider a system of consistent equations of the form

$$\begin{bmatrix} a_{11} & a_{12} & \cdots & a_{1n} \\ a_{21} & a_{22} & \cdots & a_{2n} \\ \multicolumn{4}{c}{\cdots\cdots\cdots\cdots\cdots} \\ a_{m1} & a_{m2} & \cdots & a_{mn} \end{bmatrix} \begin{bmatrix} x_1 \\ x_2 \\ \vdots \\ x_n \end{bmatrix} = \begin{bmatrix} c_1 \\ c_2 \\ \vdots \\ c_m \end{bmatrix}$$

or in shorter notation

$$Ax = c$$

where A = coefficient matrix whose (i,j) element is a_{ij}
   x = column vector of unknowns whose ith element is x_i
   c = column vector of constants with c_i as ith element

Notice that A is an $m \times n$ matrix. Let A be of any rank. A generalized inverse (or g-inverse) of A is an $n \times m$ matrix denoted by A_g^{-1}, the subscript indicating that the reference is to a generalized inverse, such that the solution for x obtained as

$$x = A_g^{-1}c$$

satisfies $Ax = c$.

It is possible to calculate g-inverses of any given matrix in a number of ways. We shall describe only one method. First find out the redundancies among the rows of the given matrix. In

$$A = \begin{bmatrix} 1 & 3 & 2 \\ -2 & 2 & -3 \\ -1 & 5 & -1 \end{bmatrix}$$

we notice that the third row is the sum of the first two. We may therefore treat the third row as redundant. (This is an arbitrary decision. We may treat either the first or the second row as redundant.) Omit the rows thus identified as redundant and omit the corresponding columns. This process reduces A to

$$\begin{bmatrix} 1 & 3 \\ -2 & 2 \end{bmatrix}$$

If all redundancies have been removed, the resulting reduced matrix should have a true inverse in the sense of Sec. C.4. In the present case we notice that

$$\begin{bmatrix} \frac{2}{8} & -\frac{3}{8} \\ \frac{2}{8} & \frac{1}{8} \end{bmatrix}$$

is the true inverse of the reduced matrix

$$\begin{bmatrix} 1 & 3 \\ -2 & 2 \end{bmatrix}$$

We now insert rows and columns of zeros from where redundant rows and columns were originally removed. This yields a g-inverse of the original matrix. Applying this to the present case, we have

$$\begin{bmatrix} \frac{2}{8} & -\frac{3}{8} & 0 \\ \frac{2}{8} & \frac{1}{8} & 0 \\ 0 & 0 & 0 \end{bmatrix}$$

as a g-inverse of

$$\begin{bmatrix} 1 & 3 & 2 \\ -2 & 2 & -3 \\ -1 & 5 & -1 \end{bmatrix}$$

A solution set satisfying (C.28) can be obtained as

$$\begin{bmatrix} x \\ y \\ z \end{bmatrix} = \begin{bmatrix} \frac{2}{8} & -\frac{3}{8} & 0 \\ \frac{2}{8} & \frac{1}{8} & 0 \\ 0 & 0 & 0 \end{bmatrix} \begin{bmatrix} 7 \\ -4 \\ 3 \end{bmatrix} = \begin{bmatrix} \frac{26}{8} \\ \frac{10}{8} \\ 0 \end{bmatrix}$$

which is easily verified to satisfy (C.28).

Had we removed the first row and first column of $\mathbf{A}$, we would have obtained as a g-inverse of $\mathbf{A}$

$$\begin{bmatrix} 0 & 0 & 0 \\ 0 & -\frac{1}{3} & \frac{3}{13} \\ 0 & -\frac{5}{13} & \frac{2}{3} \end{bmatrix}$$

leading to the solution set

$$x = 0 \qquad y = 1 \qquad z = 2$$

The point to remember is that the method described above for calculating g-inverses will always provide a g-inverse and will lead to a corresponding solution set satisfying the original equations. In the case of a rectangular matrix, say m by n, of rank r, we omit suitable rows and columns so as to identify a square submatrix of order r and rank r, whose true inverse is then augmented below by $m - r$ rows of zero and on the right by $n - r$ columns of zero. The resulting matrix is a g-inverse of the given matrix.

EXERCISES C.10

1. Verify that if $\mathbf{A}_g{}^{-1}$ is a g-inverse of $\mathbf{A}$,

$$\mathbf{A}\mathbf{A}_g{}^{-1}\mathbf{A} = \mathbf{A}$$

2. Verify that if $\mathbf{A}_g{}^{-1}$ is a g-inverse of $\mathbf{A}$, and $\mathbf{A}_g{}^{-1}\mathbf{A} = \mathbf{H}$, then $\mathbf{H}^2 = \mathbf{H}$.
3. Find a g-inverse of

$$\begin{bmatrix} 1 & 3 & 0 & 5 \\ -1 & 1 & 1 & 0 \\ 2 & -1 & 2 & 5 \\ -1 & 2 & -1 & -1 \end{bmatrix}$$

4. Given the matrix equation

$$\begin{bmatrix} 1 & 3 & 5 \\ 2 & -1 & 4 \\ 4 & -9 & 2 \end{bmatrix} \begin{bmatrix} x \\ y \\ z \end{bmatrix} = \begin{bmatrix} 8 \\ 11 \\ 17 \end{bmatrix}$$

(a) Verify that in the coefficient matrix

$$2(\text{row } 1) + (\text{row } 3) = 3(\text{row } 2)$$

(b) Treat row 1 and column 1 as redundant and obtain a g-inverse.
(c) Treat row 2 and column 2 as redundant and obtain a g-inverse.
(d) Treat row 3 and column 3 as redundant and obtain a g-inverse.

C.6 MISCELLANEOUS TOPICS

Partitioned Matrices

Sometimes it is useful to represent a given matrix as a juxtaposition of two or more matrices. Such a representation is called *partitioning* a given matrix. A matrix is partitioned by drawing dashed or dotted lines between any two rows and/or columns. For example

$$A = \begin{bmatrix} a & b & \vdots & c & d \\ e & f & \vdots & g & h \\ \cdots & \cdots & \vdots & \cdots & \cdots \\ i & j & \vdots & k & l \\ m & n & \vdots & o & p \end{bmatrix} = \begin{bmatrix} A_1 & \vdots & A_2 \\ \cdots & \vdots & \cdots \\ A_3 & \vdots & A_4 \end{bmatrix}$$

where $A_1 = \begin{bmatrix} a & b \\ e & f \end{bmatrix}$ $A_2 = \begin{bmatrix} c & d \\ g & h \end{bmatrix}$ $A_3 = \begin{bmatrix} i & j \\ m & n \end{bmatrix}$ $A_4 = \begin{bmatrix} k & l \\ o & p \end{bmatrix}$

Obviously if we have

$$A = \begin{bmatrix} P & \vdots & Q \\ \cdots & \vdots & \cdots \\ R & \vdots & S \end{bmatrix}$$

P and **Q** must have the same number of rows and so must **R** and **S**. Similarly, **P** and **R** must have the same number of columns and so must **Q** and **S**.

The product of two partitioned matrices is obtained in the usual manner: we treat each part as an ordinary element and apply the row-by-column multiplication rule. Thus if

$$A = \begin{bmatrix} P & \vdots & Q \\ \cdots & \vdots & \cdots \\ R & \vdots & S \end{bmatrix} \quad \text{and} \quad B = \begin{bmatrix} T & \vdots & U \\ \cdots & \vdots & \cdots \\ V & \vdots & W \end{bmatrix}$$

then

$$AB = \begin{bmatrix} P & \vdots & Q \\ \cdots & \vdots & \cdots \\ R & \vdots & S \end{bmatrix} \begin{bmatrix} T & \vdots & U \\ \cdots & \vdots & \cdots \\ V & \vdots & W \end{bmatrix} = \begin{bmatrix} PT + QV & \vdots & PU + QW \\ \cdots & \vdots & \cdots \\ RT + SV & \vdots & RU + SW \end{bmatrix}$$

provided **A** and **B** have been partitioned in such a way that products like **PT**, **QV**, **RT**, etc., exist. This means that the number of columns in **P** must be equal to the number of rows in **T**, and so on.

If **A** has been partitioned as

$$A = \begin{bmatrix} P & \vdots & Q \\ \cdots & \vdots & \cdots \\ R & \vdots & S \end{bmatrix}$$

it is easy to verify that

$$A' = \begin{bmatrix} P' & \vdots & R' \\ \cdots & \vdots & \cdots \\ Q' & \vdots & S' \end{bmatrix}$$

EXERCISES C.11

1. You are given a partitioned matrix

$$A = \begin{bmatrix} a & \vdots & b & c \\ \cdots & \vdots & \cdots & \cdots \\ d & \vdots & e & f \\ g & \vdots & h & i \end{bmatrix} = \begin{bmatrix} A_1 & \vdots & A_2 \\ \cdots & \vdots & \cdots \\ A_3 & \vdots & A_4 \end{bmatrix}$$

If

$$B = \begin{bmatrix} m & n \\ o & p \\ q & r \end{bmatrix}$$

how would you partition B into

$$\begin{bmatrix} B_1 \\ -- \\ B_2 \end{bmatrix}$$

so that the product AB can be written as a partitioned matrix involving products such as A_1B_1 and so on?

2. Suppose

$$X_* = \begin{bmatrix} x_{11} & x_{12} & \cdots & x_{1m} & q_{11} & q_{12} & \cdots & q_{1k} \\ x_{21} & x_{22} & \cdots & x_{2m} & q_{21} & q_{22} & \cdots & q_{2k} \\ \cdots\cdots\cdots\cdots\cdots\cdots\cdots\cdots\cdots\cdots\cdots \\ x_{n1} & x_{n2} & \cdots & x_{nm} & q_{n1} & q_{n2} & \cdots & q_{nk} \end{bmatrix}$$

and we partition X_* by drawing a dotted line between columns m and $m + 1$. If we designate the partitioned X_* as

$$X_* = [X \mid Q]$$

obtain X'_*X_*.

Quadratic Forms

Functions such as

$$Q = 5x_1{}^2 + 14x_1x_2 + 9x_2{}^2$$

are called *quadratic forms*. A quadratic form in the n variables $x_1, x_2, \ldots, x_n$ can be written

$$Q = \sum_i \sum_j a_{ij}x_ix_j$$

where the a_{ij}'s are constants.

All quadratic forms can be conveniently represented in matrix notation. Thus $Q = 5x_1{}^2 + 14x_1x_2 + 9x_2{}^2$ can be expressed as

$$[x_1 \quad x_2] \begin{bmatrix} 5 & 7 \\ 7 & 9 \end{bmatrix} \begin{bmatrix} x_1 \\ x_2 \end{bmatrix}$$

In general a quadratic form involving n variables may be written as

$$Q = X'AX$$

where $X' = [x_1 \quad x_2 \quad \cdots \quad x_n]$, and A is an $n \times n$ matrix of constants. A is called the *matrix of the quadratic form Q*.

Without loss of generality we may assume the matrix of a quadratic form to be symmetric, that is, A in $X'AX$ is such that $A' = A$. To prove this let us suppose that we have a quadratic form $X'BX$ in which B is not symmetric. We notice that $X'BX$ is a 1×1 matrix and consequently $X'BX = X'B'X$, since the transpose of a 1×1 matrix is the matrix itself. Hence we may write

$$Q = \tfrac{1}{2}X'BX + \tfrac{1}{2}X'B'X = X'AX$$

where $A = \tfrac{1}{2}B + \tfrac{1}{2}B'$. But A is then symmetric. Thus if we are given a quadratic form whose matrix is not symmetric, we can always rewrite it in such a way that the matrix of the quadratic form is symmetric.

The simplest quadratic form is one in which the matrix of the quadratic form is an identity matrix. When this is the case, we notice that the quadratic form is simply a sum of squares

$$[x_1 \quad x_2 \quad \cdots \quad x_n] \begin{bmatrix} 1 & 0 & \cdots & 0 \\ 0 & 1 & \cdots & 0 \\ \cdots\cdots\cdots\cdots \\ 0 & 0 & \cdots & 1 \end{bmatrix} \begin{bmatrix} x_1 \\ x_2 \\ \vdots \\ x_n \end{bmatrix} = \sum_{i=1}^{n} x_i^2 = X'X$$

where $X' = [x_1 \quad x_2 \quad \cdots \quad x_n]$. All sums of squares are thus quadratic forms. Of particular interest to us is the sum of squares

$$\sum \varepsilon^2 = \sum (y - \beta_0 x_0 - \beta_1 x_1 - \cdots - \beta_k x_k)^2$$

which we meet in regression analysis. Writing Y for the column vector of y values, X for the matrix with its first column consisting of the x_0 vector, second column of the x_1 vector, and so on, we have

$$\sum \varepsilon^2 = (Y - X\beta)'(Y - X\beta) \qquad (C.32)$$

where $\beta' = [\beta_0 \quad \beta_1 \quad \cdots \quad \beta_k]$.

It is possible to prove that in the above situation any quadratic form with $X'X$ as its matrix can be expressed as a sum of squares of r quantities, where r is the rank of $X'X$. The proof of this interesting result will not be attempted here. It is stated here because it has an important application worth mentioning.

Consider the equations

$$(X'X)\hat{\beta} = X'Y \qquad (C.33)$$

In the regression theory this equation system is known as the *normal equations*. It can be shown that any solution set satisfying the normal equations will lead to the minimum value of the sum of squares $\sum \varepsilon^2 = (Y - X\beta)'(Y - X\beta)$. To prove this we notice that if $\hat{\beta}$ is any solution set satisfying (C.33), we may write $\sum \varepsilon^2$ as

$$\sum \varepsilon^2 = [Y - X\hat{\beta} + X(\hat{\beta} - \beta)]'[Y - X\hat{\beta} + X(\hat{\beta} - \beta)]$$
$$= (Y - X\hat{\beta})'(Y - X\hat{\beta}) + (\hat{\beta} - \beta)'X'X(\hat{\beta} - \beta) \qquad (C.34)$$

The question is: For what values of β is (C.34) minimum, and is that minimum value unique? We notice that the part of $\sum \varepsilon^2$ that changes with β is the quadratic form

$$Q = (\hat{\beta} - \beta)'X'X(\hat{\beta} - \beta)$$

Hence the minimum value of $\sum \varepsilon^2$ corresponds to the minimum value of Q. But Q is a quadratic form with $X'X$ as its matrix. Applying the result mentioned earlier, we notice that Q can be expressed as the sum of squares of r quantities, where r is the rank of $X'X$. Now a sum of squares cannot be negative. Hence the minimum value of Q is 0; and Q attains this minimum when $\beta = \hat{\beta}$. When Q attains this minimum, the value of $\sum \varepsilon^2$ becomes minimum and is

$$(Y - X\hat{\beta})'(Y - X\hat{\beta})$$

Notice that this value is the one we would have got if we had substituted $\hat{\beta}$ for β in (C.32). Hence any solution set satisfying the normal equations when substituted in the sum of squares of the residuals, that is, $\sum \varepsilon^2$, gives the minimum value of the latter.

It may be worth emphasizing that this result applies to the full-rank case as well as the less than full-rank case of the normal equations.

EXERCISES C.12

1. Carry out the proof just given for the case involving one regressor x assuming that the (y,x) pairs are $(y_1,x_1), \ldots, (y_n,x_n)$.

2. Show algebraically that the minimum value of $\sum \varepsilon^2$ obtained when substituting $\hat{\hat{\beta}}$, that is, any solution set satisfying the normal equations $(X'X)\hat{\hat{\beta}} = X'Y$, in $(Y - X\beta)'(Y - X\beta)$ can be expressed as

$$Y'Y - \beta'X'Y$$

3. Is $Y'Y - \hat{\hat{\beta}}'X'Y$ referred to in Exercise 2 of the same value irrespective of the solution set for β used? In other words is the minimum of $\sum \varepsilon^2$ unique? *Hint:* From (C.34) we have

$$\sum \varepsilon^2 \geq (Y - X\hat{\hat{\beta}})'(Y - X\hat{\hat{\beta}})$$

APPENDIX D

TABLES

Table D.1   PERCENTAGE POINTS OF STUDENT'S t DISTRIBUTION†

df	Level of significance for one-tailed test					
	.10	.05	.025	.01	.005	.0005
	Level of significance for two-tailed test					
	.20	.10	.05	.02	.01	.001
1	3.078	6.314	12.706	31.821	63.657	636.619
2	1.886	2.920	4.303	6.965	9.925	31.598
3	1.638	2.353	3.182	4.541	5.841	12.941
4	1.533	2.132	2.776	3.747	4.604	8.610
5	1.476	2.015	2.571	3.365	4.032	6.869
6	1.440	1.943	2.447	3.143	3.707	5.959
7	1.415	1.895	2.365	2.998	3.499	5.408
8	1.397	1.860	2.306	2.896	3.355	5.041
9	1.383	1.833	2.262	2.821	3.250	4.781
10	1.372	1.812	2.228	2.764	3.169	4.587
11	1.363	1.796	2.201	2.718	3.106	4.437
12	1.356	1.782	2.179	2.681	3.055	4.318
13	1.350	1.771	2.160	2.650	3.012	4.221
14	1.345	1.761	2.145	2.624	2.977	4.140
15	1.341	1.753	2.131	2.602	2.947	4.073
16	1.337	1.746	2.120	2.583	2.921	4.015
17	1.333	1.740	2.110	2.567	2.898	3.965
18	1.330	1.734	2.101	2.552	2.878	3.922
19	1.328	1.729	2.093	2.539	2.861	3.883
20	1.325	1.725	2.086	2.528	2.845	3.850
21	1.323	1.721	2.080	2.518	2.831	3.819
22	1.321	1.717	2.074	2.508	2.819	3.792
23	1.319	1.714	2.069	2.500	2.807	3.767
24	1.318	1.711	2.064	2.492	2.797	3.745
25	1.316	1.708	2.060	2.485	2.787	3.725
26	1.315	1.706	2.056	2.479	2.779	3.707
27	1.314	1.703	2.052	2.473	2.771	3.690
28	1.313	1.701	2.048	2.467	2.763	3.674
29	1.311	1.699	2.045	2.462	2.756	3.659
30	1.310	1.697	2.042	2.457	2.750	3.646
40	1.303	1.684	2.021	2.423	2.704	3.551
60	1.296	1.671	2.000	2.390	2.660	3.460
120	1.289	1.658	1.980	2.358	2.617	3.373
∞	1.282	1.645	1.960	2.326	2.576	3.291

† Abridged from Table III of R. A. Fisher and F. Yates, *Statistical Tables for Biological, Agricultural and Medical Research*, published by Longman Group, Ltd., London, 1973 (previously published by Oliver & Boyd, Edinburgh) by permission of the authors and publishers.

Table D.2 UPPER PERCENTAGE POINTS OF THE χ^2 DISTRIBUTION†

Probability

df	.99	.98	.95	.90	.80	.70	.50	.30	.20	.10	.05	.02	.01	.001
1	.0³157	.0³628	.00393	.0158	.0642	.148	.455	1.074	1.642	2.706	3.841	5.412	6.635	10.827
2	.0201	.0404	.103	.211	.446	.713	1.386	2.408	3.219	4.605	5.991	7.824	9.210	13.815
3	.115	.185	.352	.584	1.005	1.424	2.366	3.665	4.642	6.251	7.815	9.837	11.341	16.268
4	.297	.429	.711	1.064	1.649	2.195	3.357	4.878	5.989	7.779	9.488	11.668	13.277	18.465
5	.554	.752	1.145	1.610	2.343	3.000	4.351	6.064	7.289	9.236	11.070	13.388	15.086	20.517
6	.872	1.134	1.635	2.204	3.070	3.828	5.348	7.231	8.558	10.645	12.592	15.033	16.812	22.457
7	1.239	1.564	2.167	2.833	3.822	4.671	6.346	8.383	9.803	12.017	14.067	16.622	18.475	24.322
8	1.646	2.032	2.733	3.490	4.594	5.527	7.344	9.524	11.030	13.362	15.507	18.168	20.090	26.125
9	2.088	2.532	3.325	4.168	5.380	6.393	8.343	10.656	12.242	14.684	16.919	19.679	21.666	27.877
10	2.558	3.059	3.940	4.865	6.179	7.267	9.342	11.781	13.442	15.987	18.307	21.161	23.209	29.588
11	3.053	3.609	4.575	5.578	6.989	8.148	10.341	12.899	14.631	17.275	19.675	22.618	24.725	31.264
12	3.571	4.178	5.226	6.304	7.807	9.034	11.340	14.011	15.812	18.549	21.026	24.054	26.217	32.909
13	4.107	4.765	5.892	7.042	8.634	9.926	12.340	15.119	16.985	19.812	22.362	25.472	27.688	34.528
14	4.660	5.368	6.571	7.790	9.467	10.821	13.339	16.222	18.151	21.064	23.685	26.873	29.141	36.123
15	5.229	5.985	7.261	8.547	10.307	11.721	14.339	17.322	19.311	22.307	24.996	28.259	30.578	37.697
16	5.812	6.614	7.962	9.312	11.152	12.624	15.338	18.418	20.465	23.542	26.296	29.633	32.000	39.252
17	6.408	7.255	8.672	10.085	12.002	13.531	16.338	19.511	21.615	24.769	27.587	30.995	33.409	40.790
18	7.015	7.906	9.390	10.865	12.857	14.440	17.338	20.601	22.760	25.989	28.869	32.346	34.805	42.312
19	7.633	8.567	10.117	11.651	13.716	15.352	18.338	21.689	23.900	27.204	30.144	33.687	36.191	43.820
20	8.260	9.237	10.851	12.443	14.578	16.266	19.337	22.775	25.038	28.412	31.410	35.020	37.566	45.315
21	8.897	9.915	11.591	13.240	15.445	17.182	20.337	23.858	26.171	29.615	32.671	36.343	38.932	46.797
22	9.542	10.600	12.338	14.041	16.314	18.101	21.337	24.939	27.301	30.813	33.924	37.659	40.289	48.268
23	10.196	11.293	13.091	14.848	17.187	19.021	22.337	26.018	28.429	32.007	35.172	38.968	41.638	49.728
24	10.856	11.992	13.848	15.659	18.062	19.943	23.337	27.096	29.553	33.196	36.415	40.270	42.980	51.179
25	11.524	12.697	14.611	16.473	18.940	20.867	24.337	28.172	30.675	34.382	37.652	41.566	44.314	52.620
26	12.198	13.409	15.379	17.292	19.820	21.792	25.336	29.246	31.795	35.563	38.885	42.856	45.642	54.052
27	12.879	14.125	16.151	18.114	20.703	22.719	26.336	30.319	32.912	36.741	40.113	44.140	46.963	55.476
28	13.565	14.847	16.928	18.939	21.588	23.647	27.336	31.391	34.027	37.916	41.337	45.419	48.278	56.893
29	14.256	15.574	17.708	19.768	22.475	24.577	28.336	32.461	35.139	39.087	42.557	46.693	49.588	58.302
30	14.953	16.306	18.493	20.599	23.364	25.508	29.336	33.530	36.250	40.256	43.773	47.962	50.892	59.703

For larger values of df, the expression $\sqrt{2\chi^2} - \sqrt{2\,df - 1}$ may be used as a normal deviate with unit variance, remembering that the probability for χ^2 corresponds with that of a single tail of the normal curve.

† From Table IV of R. A. Fisher and F. Yates, *Statistical Tables for Biological, Agricultural and Medical Research*, published by Longman Group, Ltd., London, 1973 (previously published by Oliver & Boyd, Edinburgh), by permission of the authors and publishers.

Table D.3 UPPER PERCENTAGE POINTS OF THE F DISTRIBUTION†

$p = .05$

n_1 / n_2	1	2	3	4	5	6	8	12	24	∞
1	161.4	199.5	215.7	224.6	230.2	234.0	238.9	243.9	249.0	254.3
2	18.51	19.00	19.16	19.25	19.30	19.33	19.37	19.41	19.45	19.50
3	10.13	9.55	9.28	9.12	9.01	8.94	8.84	8.74	8.64	8.53
4	7.71	6.94	6.59	6.39	6.26	6.16	6.04	5.91	5.77	5.63
5	6.61	5.79	5.41	5.19	5.05	4.95	4.82	4.68	4.53	4.36
6	5.99	5.14	4.76	4.53	4.39	4.28	4.15	4.00	3.84	3.67
7	5.59	4.74	4.35	4.12	3.97	3.87	3.73	3.57	3.41	3.23
8	5.32	4.46	4.07	3.84	3.69	3.58	3.44	3.28	3.12	2.93
9	5.12	4.26	3.86	3.63	3.48	3.37	3.23	3.07	2.90	2.71
10	4.96	4.10	3.71	3.48	3.33	3.22	3.07	2.91	2.74	2.54
11	4.84	3.98	3.59	3.36	3.20	3.09	2.95	2.79	2.61	2.40
12	4.75	3.88	3.49	3.26	3.11	3.00	2.85	2.69	2.50	2.30
13	4.67	3.80	3.41	3.18	3.02	2.92	2.77	2.60	2.42	2.21
14	4.60	3.74	3.34	3.11	2.96	2.85	2.70	2.53	2.35	2.13
15	4.54	3.68	3.29	3.06	2.90	2.79	2.64	2.48	2.29	2.07
16	4.49	3.63	3.24	3.01	2.85	2.74	2.59	2.42	2.24	2.01
17	4.45	3.59	3.20	2.96	2.81	2.70	2.55	2.38	2.19	1.96
18	4.41	3.55	3.16	2.93	2.77	2.66	2.51	2.34	2.15	1.92
19	4.38	3.52	3.13	2.90	2.74	2.63	2.48	2.31	2.11	1.88
20	4.35	3.49	3.10	2.87	2.71	2.60	2.45	2.28	2.08	1.84
21	4.32	3.47	3.07	2.84	2.68	2.57	2.42	2.25	2.05	1.81
22	4.30	3.44	3.05	2.82	2.66	2.55	2.40	2.23	2.03	1.78
23	4.28	3.42	3.03	2.80	2.64	2.53	2.38	2.20	2.00	1.76
24	4.26	3.40	3.01	2.78	2.62	2.51	2.36	2.18	1.98	1.73
25	4.24	3.38	2.99	2.76	2.60	2.49	2.34	2.16	1.96	1.71
26	4.22	3.37	2.98	2.74	2.59	2.47	2.32	2.15	1.95	1.69
27	4.21	3.35	2.96	2.73	2.57	2.46	2.30	2.13	1.93	1.67
28	4.20	3.34	2.95	2.71	2.56	2.44	2.29	2.12	1.91	1.65
29	4.18	3.33	2.93	2.70	2.54	2.43	2.28	2.10	1.90	1.64
30	4.17	3.32	2.92	2.69	2.53	2.42	2.27	2.09	1.89	1.62
40	4.08	3.23	2.84	2.61	2.45	2.34	2.18	2.00	1.79	1.51
60	4.00	3.15	2.76	2.52	2.37	2.25	2.10	1.92	1.70	1.39
120	3.92	3.07	2.68	2.45	2.29	2.17	2.02	1.83	1.61	1.25
∞	3.84	2.99	2.60	2.37	2.21	2.09	1.94	1.75	1.52	1.00

Values of n_1 and n_2 represent the degrees of freedom associated with the larger and smaller estimates of variance respectively.

† Abridged from Table V of R. A. Fisher and F. Yates, *Statistical Tables for Biological, Agricultural and Medical Research*, published by Longman Group, Ltd., London, 1973 (previously published by Oliver & Boyd, Edinburgh), by permission of the authors and publishers.

Table D.3 UPPER PERCENTAGE POINTS OF THE F DISTRIBUTION (CONTINUED)

$p = .01$

n_1 \ n_2	1	2	3	4	5	6	8	12	24	∞
1	4052	4999	5403	5625	5764	5859	5981	6106	6234	6366
2	98.49	99.01	99.17	99.25	99.30	99.33	99.36	99.42	99.46	99.50
3	34.12	30.81	29.46	28.71	28.24	27.91	27.49	27.05	26.60	26.12
4	21.20	18.00	16.69	15.98	15.52	15.21	14.80	14.37	13.93	13.46
5	16.26	13.27	12.06	11.39	10.97	10.67	10.27	9.89	9.47	9.02
6	13.74	10.92	9.78	9.15	8.75	8.47	8.10	7.72	7.31	6.88
7	12.25	9.55	8.45	7.85	7.46	7.19	6.84	6.47	6.07	5.65
8	11.26	8.65	7.59	7.01	6.63	6.37	6.03	5.67	5.28	4.86
9	10.56	8.02	6.99	6.42	6.06	5.80	5.47	5.11	4.73	4.31
10	10.04	7.56	6.55	5.99	5.64	5.39	5.06	4.71	4.33	3.91
11	9.65	7.20	6.22	5.67	5.32	5.07	4.74	4.40	4.02	3.60
12	9.33	6.93	5.95	5.41	5.06	4.82	4.50	4.16	3.78	3.36
13	9.07	6.70	5.74	5.20	4.86	4.62	4.30	3.96	3.59	3.16
14	8.86	6.51	5.56	5.03	4.69	4.46	4.14	3.80	3.43	3.00
15	8.68	6.36	5.42	4.89	4.56	4.32	4.00	3.67	3.29	2.87
16	8.53	6.23	5.29	4.77	4.44	4.20	3.89	3.55	3.18	2.75
17	8.40	6.11	5.18	4.67	4.34	4.10	3.79	3.45	3.08	2.65
18	8.28	6.01	5.09	4.58	4.25	4.01	3.71	3.37	3.00	2.57
19	8.18	5.93	5.01	4.50	4.17	3.94	3.63	3.30	2.92	2.49
20	8.10	5.85	4.94	4.43	4.10	3.87	3.56	3.23	2.86	2.42
21	8.02	5.78	4.87	4.37	4.04	3.81	3.51	3.17	2.80	2.36
22	7.94	5.72	4.82	4.31	3.99	3.76	3.45	3.12	2.75	2.31
23	7.88	5.66	4.76	4.26	3.94	3.71	3.41	3.07	2.70	2.26
24	7.82	5.61	4.72	4.22	3.90	3.67	3.36	3.03	2.66	2.21
25	7.77	5.57	4.68	4.18	3.86	3.63	3.32	2.99	2.62	2.17
26	7.72	5.53	4.64	4.14	3.82	3.59	3.29	2.96	2.58	2.13
27	7.68	5.49	4.60	4.11	3.78	3.56	3.26	2.93	2.55	2.10
28	7.64	5.45	4.57	4.07	3.75	3.53	3.23	2.90	2.52	2.06
29	7.60	5.42	4.54	4.04	3.73	3.50	3.20	2.87	2.49	2.03
30	7.56	5.39	4.51	4.02	3.70	3.47	3.17	2.84	2.47	2.01
40	7.31	5.18	4.31	3.83	3.51	3.29	2.99	2.66	2.29	1.80
60	7.08	4.98	4.13	3.65	3.34	3.12	2.82	2.50	2.12	1.60
120	6.85	4.79	3.95	3.48	3.17	2.96	2.66	2.34	1.95	1.38
∞	6.64	4.60	3.78	3.32	3.02	2.80	2.51	2.18	1.79	1.00

Values of n_1 and n_2 represent the degrees of freedom associated with the larger and smaller estimates of variance respectively.

Table D.3   UPPER PERCENTAGE POINTS OF THE F DISTRIBUTION (CONTINUED)

$p = .001$

n_2 \ n_1	1	2	3	4	5	6	8	12	24	∞
1	405284	500000	540379	562500	576405	585937	598144	610667	623497	636619
2	998.5	999.0	999.2	999.2	999.3	999.3	999.4	999.4	999.5	999.5
3	167.5	148.5	141.1	137.1	134.6	132.8	130.6	128.3	125.9	123.5
4	74.14	61.25	56.18	53.44	51.71	50.53	49.00	47.41	45.77	44.05
5	47.04	36.61	33.20	31.09	29.75	28.84	27.64	26.42	25.14	23.78
6	35.51	27.00	23.70	21.90	20.81	20.03	19.03	17.99	16.89	15.75
7	29.22	21.69	18.77	17.19	16.21	15.52	14.63	13.71	12.73	11.69
8	25.42	18.49	15.83	14.39	13.49	12.86	12.04	11.19	10.30	9.34
9	22.86	16.39	13.90	12.56	11.71	11.13	10.37	9.57	8.72	7.81
10	21.04	14.91	12.55	11.28	10.48	9.92	9.20	8.45	7.64	6.76
11	19.69	13.81	11.56	10.35	9.58	9.05	8.35	7.63	6.85	6.00
12	18.64	12.97	10.80	9.63	8.89	8.38	7.71	7.00	6.25	5.42
13	17.81	12.31	10.21	9.07	8.35	7.86	7.21	6.52	5.78	4.97
14	17.14	11.78	9.73	8.62	7.92	7.43	6.80	6.13	5.41	4.60
15	16.59	11.34	9.34	8.25	7.57	7.09	6.47	5.81	5.10	4.31
16	16.12	10.97	9.00	7.94	7.27	6.81	6.19	5.55	4.85	4.06
17	15.72	10.66	8.73	7.68	7.02	6.56	5.96	5.32	4.63	3.85
18	15.38	10.39	8.49	7.46	6.81	6.35	5.76	5.13	4.45	3.67
19	15.08	10.16	8.28	7.26	6.61	6.18	5.59	4.97	4.29	3.52
20	14.82	9.95	8.10	7.10	6.46	6.02	5.44	4.82	4.15	3.38
21	14.59	9.77	7.94	6.95	6.32	5.88	5.31	4.70	4.03	3.26
22	14.38	9.61	7.80	6.81	6.19	5.76	5.19	4.58	3.92	3.15
23	14.19	9.47	7.67	6.69	6.08	5.65	5.09	4.48	3.82	3.05
24	14.03	9.34	7.55	6.59	5.98	5.55	4.99	4.39	3.74	2.97
25	13.88	9.22	7.45	6.49	5.88	5.46	4.91	4.31	3.66	2.89
26	13.74	9.12	7.36	6.41	5.80	5.38	4.83	4.24	3.59	2.82
27	13.61	9.02	7.27	6.33	5.73	5.31	4.76	4.17	3.52	2.75
28	13.50	8.93	7.19	6.25	5.66	5.24	4.69	4.11	3.46	2.70
29	13.39	8.85	7.12	6.19	5.59	5.18	4.64	4.05	3.41	2.64
30	13.29	8.77	7.05	6.12	5.53	5.12	4.58	4.00	3.36	2.59
40	12.61	8.25	6.60	5.70	5.13	4.73	4.21	3.64	3.01	2.23
60	11.97	7.76	6.17	5.31	4.76	4.37	3.87	3.31	2.69	1.90
120	11.38	7.31	5.79	4.95	4.42	4.04	3.55	3.02	2.40	1.56
∞	10.83	6.91	5.42	4.62	4.10	3.74	3.27	2.74	2.13	1.00

Values of n_1 and n_2 represent the degrees of freedom associated with the larger and smaller estimates of variance respectively.

Table D.4 UPPER PERCENTAGE POINTS OF THE STUDENTIZED RANGE†

Error df	α	\multicolumn{10}{c}{r = number of means or number of steps between ordered means}									
		2	3	4	5	6	7	8	9	10	11
5	.05	3.64	4.60	5.22	5.67	6.03	6.33	6.58	6.80	6.99	7.17
	.01	5.70	6.98	7.80	8.42	8.91	9.32	9.67	9.97	10.24	10.48
6	.05	3.46	4.34	4.90	5.30	5.63	5.90	6.12	6.32	6.49	6.65
	.01	5.24	6.33	7.03	7.56	7.97	8.32	8.61	8.87	9.10	9.30
7	.05	3.34	4.16	4.68	5.06	5.36	5.61	5.82	6.00	6.16	6.30
	.01	4.95	5.92	6.54	7.01	7.37	7.68	7.94	8.17	8.37	8.55
8	.05	3.26	4.04	4.53	4.89	5.17	5.40	5.60	5.77	5.92	6.05
	.01	4.75	5.64	6.20	6.62	6.96	7.24	7.47	7.68	7.86	8.03
9	.05	3.20	3.95	4.41	4.76	5.02	5.24	5.43	5.59	5.74	5.87
	.01	4.60	5.43	5.96	6.35	6.66	6.91	7.13	7.33	7.49	7.65
10	.05	3.15	3.88	4.33	4.65	4.91	5.12	5.30	5.46	5.60	5.72
	.01	4.48	5.27	5.77	6.14	6.43	6.67	6.87	7.05	7.21	7.36
11	.05	3.11	3.82	4.26	4.57	4.82	5.03	5.20	5.35	5.49	5.61
	.01	4.39	5.15	5.62	5.97	6.25	6.48	6.67	6.84	6.99	7.13
12	.05	3.08	3.77	4.20	4.51	4.75	4.95	5.12	5.27	5.39	5.51
	.01	4.32	5.05	5.50	5.84	6.10	6.32	6.51	6.67	6.81	6.94
13	.05	3.06	3.73	4.15	4.45	4.69	4.88	5.05	5.19	5.32	5.43
	.01	4.26	4.96	5.40	5.73	5.98	6.19	6.37	6.53	6.67	6.79
14	.05	3.03	3.70	4.11	4.41	4.64	4.83	4.99	5.13	5.25	5.36
	.01	4.21	4.89	5.32	5.63	5.88	6.08	6.26	6.41	6.54	6.66
15	.05	3.01	3.67	4.08	4.37	4.59	4.78	4.94	5.08	5.20	5.31
	.01	4.17	4.84	5.25	5.56	5.80	5.99	6.16	6.31	6.44	6.55
16	.05	3.00	3.65	4.05	4.33	4.56	4.74	4.90	5.03	5.15	5.26
	.01	4.13	4.79	5.19	5.49	5.72	5.92	6.08	6.22	6.35	6.46
17	.05	2.98	3.63	4.02	4.30	4.52	4.70	4.86	4.99	5.11	5.21
	.01	4.10	4.74	5.14	5.43	5.66	5.85	6.01	6.15	6.27	6.38
18	.05	2.97	3.61	4.00	4.28	4.49	4.67	4.82	4.96	5.07	5.17
	.01	4.07	4.70	5.09	5.38	5.60	5.79	5.94	6.08	6.20	6.31
19	.05	2.96	3.59	3.98	4.25	4.47	4.65	4.79	4.92	5.04	5.14
	.01	4.05	4.67	5.05	5.33	5.55	5.73	5.89	6.02	6.14	6.25
20	.05	2.95	3.58	3.96	4.23	4.45	4.62	4.77	4.90	5.01	5.11
	.01	4.02	4.64	5.02	5.29	5.51	5.69	5.84	5.97	6.09	6.19
24	.05	2.92	3.53	3.90	4.17	4.37	4.54	4.68	4.81	4.92	5.01
	.01	3.96	4.55	4.91	5.17	5.37	5.54	5.69	5.81	5.92	6.02
30	.05	2.89	3.49	3.85	4.10	4.30	4.46	4.60	4.72	4.82	4.92
	.01	3.89	4.45	4.80	5.05	5.24	5.40	5.54	5.65	5.76	5.85
40	.05	2.86	3.44	3.79	4.04	4.23	4.39	4.52	4.63	4.73	4.82
	.01	3.82	4.37	4.70	4.93	5.11	5.26	5.39	5.50	5.60	5.69
60	.05	2.83	3.40	3.74	3.98	4.16	4.31	4.44	4.55	4.65	4.73
	.01	3.76	4.28	4.59	4.82	4.99	5.13	5.25	5.36	5.45	5.53
120	.05	2.80	3.36	3.68	3.92	4.10	4.24	4.36	4.47	4.56	4.64
	.01	3.70	4.20	4.50	4.71	4.87	5.01	5.12	5.21	5.30	5.37
∞	.05	2.77	3.31	3.63	3.86	4.03	4.17	4.29	4.39	4.47	4.55
	.01	3.64	4.12	4.40	4.60	4.76	4.88	4.99	5.08	5.16	5.23

† Abridged from Table 29 in E. S. Pearson and H. O. Hartley (eds.), *Biometrika Tables for Statisticians*, vol. 1, 2d ed., Cambridge University Press, New York, 1958, reproduced with the kind permission of the editors and trustees of *Biometrika*.

Table D.4 UPPER PERCENTAGE POINTS OF THE STUDENTIZED RANGE
(CONTINUED)

r = number of means or number of steps between ordered means										
12	13	14	15	16	17	18	19	20	α	Error df
7.32	7.47	7.60	7.72	7.83	7.93	8.03	8.12	8.21	.05	5
10.70	10.89	11.08	11.24	11.40	11.55	11.68	11.81	11.93	.01	
6.79	6.92	7.03	7.14	7.24	7.34	7.43	7.51	7.59	.05	6
9.48	9.65	9.81	9.95	10.08	10.21	10.32	10.43	10.54	.01	
6.43	6.55	6.66	6.76	6.85	6.94	7.02	7.10	7.17	.05	7
8.71	8.86	9.00	9.12	9.24	9.35	9.46	9.55	9.65	.01	
6.18	6.29	6.39	6.48	6.57	6.65	6.73	6.80	6.87	.05	8
8.18	8.31	8.44	8.55	8.66	8.76	8.85	8.94	9.03	.01	
5.98	6.09	6.19	6.28	6.36	6.44	6.51	6.58	6.64	.05	9
7.78	7.91	8.03	8.13	8.23	8.33	8.41	8.49	8.57	.01	
5.83	5.93	6.03	6.11	6.19	6.27	6.34	6.40	6.47	.05	10
7.49	7.60	7.71	7.81	7.91	7.99	8.08	8.15	8.23	.01	
5.71	5.81	5.90	5.98	6.06	6.13	6.20	6.27	6.33	.05	11
7.25	7.36	7.46	7.56	7.65	7.73	7.81	7.88	7.95	.01	
5.61	5.71	5.80	5.88	5.95	6.02	6.09	6.15	6.21	.05	12
7.06	7.17	7.26	7.36	7.44	7.52	7.59	7.66	7.73	.01	
5.53	5.63	5.71	5.79	5.86	5.93	5.99	6.05	6.11	.05	13
6.90	7.01	7.10	7.19	7.27	7.35	7.42	7.48	7.55	.01	
5.46	5.55	5.64	5.71	5.79	5.85	5.91	5.97	6.03	.05	14
6.77	6.87	6.96	7.05	7.13	7.20	7.27	7.33	7.39	.01	
5.40	5.49	5.57	5.65	5.72	5.78	5.85	5.90	5.96	.05	15
6.66	6.76	6.84	6.93	7.00	7.07	7.14	7.20	7.26	.01	
5.35	5.44	5.52	5.59	5.66	5.73	5.79	5.84	5.90	.05	16
6.56	6.66	6.74	6.82	6.90	6.97	7.03	7.09	7.15	.01	
5.31	5.39	5.47	5.54	5.61	5.67	5.73	5.79	5.84	.05	17
6.48	6.57	6.66	6.73	6.81	6.87	6.94	7.00	7.05	.01	
5.27	5.35	5.43	5.50	5.57	5.63	5.69	5.74	5.79	.05	18
6.41	6.50	6.58	6.65	6.73	6.79	6.85	6.91	6.97	.01	
5.23	5.31	5.39	5.46	5.53	5.59	5.65	5.70	5.75	.05	19
6.34	6.43	6.51	6.58	6.65	6.72	6.78	6.84	6.89	.01	
5.20	5.28	5.36	5.43	5.49	5.55	5.61	5.66	5.71	.05	20
6.28	6.37	6.45	6.52	6.59	6.65	6.71	6.77	6.82	.01	
5.10	5.18	5.25	5.32	5.38	5.44	5.49	5.55	5.59	.05	24
6.11	6.19	6.26	6.33	6.39	6.45	6.51	6.56	6.61	.01	
5.00	5.08	5.15	5.21	5.27	5.33	5.38	5.43	5.47	.05	30
5.93	6.01	6.08	6.14	6.20	6.26	6.31	6.36	6.41	.01	
4.90	4.98	5.04	5.11	5.16	5.22	5.27	5.31	5.36	.05	40
5.76	5.83	5.90	5.96	6.02	6.07	6.12	6.16	6.21	.01	
4.81	4.88	4.94	5.00	5.06	5.11	5.15	5.20	5.24	.05	60
5.60	5.67	5.73	5.78	5.84	5.89	5.93	5.97	6.01	.01	
4.71	4.78	4.84	4.90	4.95	5.00	5.04	5.09	5.13	.05	120
5.44	5.50	5.56	5.61	5.66	5.71	5.75	5.79	5.83	.01	
4.62	4.68	4.74	4.80	4.85	4.89	4.93	4.97	5.01	.05	∞
5.29	5.35	5.40	5.45	5.49	5.54	5.57	5.61	5.65	.01	

Althauser, R. P., and T. A. Heberlein. 1970. "Validity and the Multitrait-Multimethod Matrix." Chapter 9 of Borgatta (1970).

———, ———, and R. A. Scott. 1971. "A Causal Assessment of Validity: The Augmented Multitrait-Multimethod Matrix." Chapter 22 of Blalock (1974).

Alwin, D. F. 1973. "Approaches to the Interpretation of Relationships in the Multitrait-Multimethod Matrix." Chapter 4 in Costner (1973–1974).

Anderson, R. L., and T. A. Bancroft. 1952. *Statistical Theory in Research.* New York: McGraw-Hill.

Andrews, F. M., and R. C. Messenger. 1973. *Multivariate Nominal Scale Analysis: A Report on a New Analysis Technique and a Computer Program.* Ann Arbor, Mich.: Institute for Social Research.

Armitage, P. 1955. "Tests for Linear Trends in Proportions and Frequencies." *Biometrics,* 11:375–386.

Aronson, E., and J. Mills. 1959. "The Effect of Severity of Initiation on Liking for a Group." *Journal of Abnormal and Social Psychology,* 59:177–181.

Baker, B. O., C. D. Hardyck, and L. F. Petrinovich. 1966. "Weak Measurements vs. Strong Statistics: An Empirical Critique of S. S. Stevens' Proscriptions on Statistics." *Educational and Psychological Measurement,* 26:291–309.

Balaam, L. N. 1968. "A Two-Period Design with t^2 Experimental Units." *Biometrics,* 24:61–74.

Bandura, A. 1969. *Principles of Behavior Modification.* New York: Holt.

Barnard, G. A. 1947. "Significance Tests for 2 × 2 Tables." *Biometrika*, **34**:123–138.

Bartlett, M. S. 1937. "Some Examples of Statistical Methods of Research in Agriculture and Applied Biology." *Journal of the Royal Statistical Society*, Supplement **4**:137–170.

———. 1949. "The Fitting of Straight Lines If Both Variables Are Subject to Error." *Biometrics*, **5**:207–242.

Baumol, W. J. 1959. *Economic Dynamics.* New York: Macmillan.

Berenblut, I. I. 1964. "Change-over Designs with Complete Balance for First Residual Effects." *Biometrics*, **20**:707–712.

Blalock, H. M. 1962. "Four-Variable Causal Models and Partial Correlations." *American Journal of Sociology*, **68**:182–194.

———. 1963. "Correlated Independent Variables: The Problem of Multicollinearity." *Social Forces*, **42**:233–237.

———. 1964. *Causal Inferences in Nonexperimental Research.* Chapel Hill: University of North Carolina Press.

———. 1967a. "Causal Inferences, Closed Populations, and Measures of Association." *American Political Science Review*, **61**:130–136.

———. 1967b. "Causal Inferences in Natural Experiments: Some Complications in Matching Designs." *Sociometry*, **30**:300–315.

———. 1967c. "Status Inconsistency, Social Mobility, Status Integration and Structural Effects." *American Sociological Review*, **32**:790–801.

———. 1969. *Theory Construction.* Englewood Cliffs, N.J.: Prentice-Hall.

———. 1970. "Estimating Measurement Error Using Multiple Indicators and Several Points in Time." *American Sociological Review*, **35**:101–111.

——— (ed.). 1971a. *Causal Models in the Social Sciences.* Chicago: Aldine-Altherton.

———. 1971b. "Aggregation and Measurement Error." *Social Forces*, **50**:151–165.

———. 1972. *Social Statistics*, 2d ed. New York: McGraw-Hill.

——— (ed.). 1974. *Measurement in the Social Sciences.* Chicago: Aldine.

———. 1976. "Can We Find a Genuine Ordinal Slope Analogue?" To appear in Heise (1976), in press.

———, and A. B. Blalock (eds.). 1968. *Methodology in Social Research.* New York: McGraw-Hill.

———, C. S. Wells, and L. F. Carter. 1970. "Statistical Estimation with Random Measurement Error." Chapter 5 of Borgatta (1970).

———, A. Aganbegian, F. Borodkin, R. Boudon, V. Capecchi (eds.). 1975. *Quantitative Sociology.* New York: Academic Press.

Boneau, C. A. 1960. "The Effects of Violations of Assumptions Underlying the *t* Test." *Psychological Bulletin*, **57**:49–64.

Borgatta, E. (ed.). 1969. *Sociological Methodology.* San Francisco: Jossey-Bass.

———. 1970. *Sociological Methodology.* San Francisco: Jossey-Bass.

Borko, H. (ed.). 1962. *Computer Applications in the Behavioral Sciences.* Englewood Cliffs, N.J.: Prentice-Hall.

Bose, R. C., M. M. Clatworthy, and S. Shrikhande. 1954. "Tables of Partially Balanced Designs with Two Associate Classes." *North Carolina Agricultural Experimental Station Technical Bulletin* 102.

Bottenberg, R. A., and J. Ward, Jr. 1963. "Applied Multiple Linear Regression." *National Technical Information Service Publication* AD 413 128.

Boudon, R. 1968. "A New Look at Correlation Analysis." Chapter 6 of Blalock and Blalock (1968).

Box, G. E. P. 1953. "Nonnormality and Tests on Variances." *Biometrika*, **40**:318–335.

—— and G. S. Watson. 1962. "Robustness to Nonnormality of Regression Tests." *Biometrika*, **49**:93–106.

—— and G. C. Tiao. 1965. "A Change in Level of Nonstationary Time Series." *Biometrika*, **52**:181–192.

Bross, I. 1958. "How to Use Ridit Analysis." *Biometrics*, **14**:18–38.

Campbell, D. T. 1969. "Reforms as Experiments." *American Psychologist*, **24**:409–429.

—— and D. W. Fiske. 1959. "Convergent and Discriminant Validation by the Multitrait-Multimethod Matrix." *Psychological Bulletin*, **56**:81–105.

—— and J. C. Stanley. 1963. *Experimental and Quasi-experimental Designs for Research.* Chicago: Rand McNally.

Carr, L. G. 1971. "The Srole Items and Acquiescence." *American Sociological Review*, **36**:287–293.

Carter, L. F. 1971. "Inadvertent Sociological Theory." *Social Forces*, **50**:12–25.

—— and H. M. Blalock, Jr. 1970. "Underestimation of Error in Wald-Bartlett Slope Estimation." *Journal of the Royal Statistical Society*, (C)**19**:34–41.

Christ, C. F. 1966. *Econometric Models and Methods.* New York: Wiley.

Cochran, W. G. 1965. "The Planning of Observational Studies of Human Populations." *Journal of the Royal Statistical Society*, (A)**128**:234–265.

—— and G. M. Cox. 1957. *Experimental Designs.* New York: Wiley.

Cohen, J. 1968. "Multiple Regression as a General Data-Analytic System." *Psychological Bulletin*, **70**:426–443.

Costner, H. L. 1969. "Theory, Deduction, and Rules of Correspondence." *American Journal of Sociology*, **75**:245–263.

——. 1971*a*. "Utilizing Causal Models to Discover Flaws in Experiments." *Sociometry*, **34**:398–410.

—— (ed.). 1971*b*. *Sociological Methodology.* San Francisco: Jossey-Bass.

——. 1972. *Sociological Methodology.* San Francisco: Jossey-Bass.

——. 1973–1974. *Sociological Methodology.* San Francisco: Jossey-Bass.

—— and R. Schoenberg. 1973. "Diagnosing Indicator Ills in Multiple Indicator Models." Chapter 9 of Goldberger and Duncan (1973).

Cox, D. R. 1951. "Some Systematic Designs." *Biometrika*, **38**:312–323.

——. 1952. "Some Recent Work on Systematic Experimental Designs." *Journal of the Royal Statistical Society*, (B)**14**:211–228.

——. 1958. *Planning of Experiments.* New York: Wiley.

——. 1970. *The Analysis of Binary Data.* London: Methuen.

Curtis, R. F., and E. F. Jackson. 1962. "Multiple Indicators in Survey Research." *American Journal of Sociology*, **68**:195–204.

Davies, O. L. (ed.). 1956. *The Design and Analysis of Industrial Experiments.* New York: Hafner.

Davies, V., E. Gross, and J. F. Short, Jr. 1958. *Experiments in Teaching Effectiveness Applied to Introductory Sociology.* Department of Sociology, State College of Washington.

Davis, J. A. 1967. "A Partial Coefficient for Goodman and Kruskal's Gamma." *Journal of the American Statistical Association*, **62**:189–193.

Dayton, C. M. 1970. *The Design of Educational Experiments.* New York: McGraw-Hill.

Duncan, O. D. 1966*a*. "Methodological Issues in the Analysis of Social Mobility." Chapter 2 of Smelser and Lipset (1966).

——. 1966*b*. "Path Analysis: Sociological Examples." *American Journal of Sociology*, **72**:1–16.

———. 1971. "Addenda" to "Path Analysis: Sociological Examples." Chapter 7 in Blalock (1971a).

———, and B. Davis. 1953. "An Alternative to Ecological Correlation." *American Sociological Review*, **18**:665–666.

———, R. P. Cuzzort, and B. Duncan. 1961. *Statistical Geography*. Glencoe, Ill.: The Free Press.

Durbin, J. 1957. "Testing for Serial Correlation in Systems of Simultaneous Regression Equations." *Biometrika*, **44**:370–377.

——— and G. S. Watson. 1951. "Testing for Serial Correlation in Least Squares Regression II." *Biometrika*, **38**:159–178.

Edwards, A. L. 1941. "Political Frames of Reference as a Factor Influencing Recognition." *Journal of Abnormal and Social Psychology*, **36**:34–50.

Federer, W. T. 1955. *Experimental Design: Theory and Application*. New York: Macmillan.

Feldstein, M. S. 1966. "A Binary Variable Multiple Regression Method of Analysing Factors Affecting Perinatal Mortality and Other Outcomes of Pregnancy." *Journal of the Royal Statistical Society*, (A)**29**:61–73.

Finney, D. J. 1964. *Statistical Method in Biological Assay*, 2d ed. New York: Hafner.

——— and A. D. Outhwaite. 1956. "Serially Balanced Sequences in Bioassay." *Proceedings of the Royal Society*, **145**:493–507.

Fisher, F. M. 1965. "The Choice of Instrumental Variables in the Estimation of Economywide Econometric Models." *International Economic Review*, **6**:245–274.

———. 1966. *The Identification Problem in Economics*. New York: McGraw-Hill.

Fisher, R. A. 1947. *The Design of Experiments*, 4th ed. Edinburgh: Oliver & Boyd.

——— and F. Yates. 1973. *Statistical Tables for Biological, Agricultural and Medical Research*, 6th ed. London: Longman Group, Ltd.

Freeman, J. H., and J. E. Kronenfeld. 1973. "Problems of Definitional Dependency: The Case of Administrative Intensity." *Social Forces*, **52**:108–121.

Gart, J. J., and J. R. Zweifel. 1967. "On the Bias of Various Estimators of the Logit and Its Variance with Application to Quantal Bioassay." *Biometrika*, **54**:181–187.

Gerard, H. B., and G. C. Mathewson. 1966. "The Effects of Severity of Initiation on Liking for a Group: A Replication." *Journal of Experimental Social Psychology*, **2**:278–287.

Glass, G. V. 1968. "Analysis of Data on the Connecticut Speeding Crackdown as a Time Series Quasi-Experiment." *Law and Society Review*, **3**:55–76.

Goldberg, A. S. 1966. "Discerning a Causal Pattern among Data on Voting Behavior." *American Political Science Review*, **60**:913–922.

Goldberger, A. S. 1970. "On Boudon's Method of Linear Causal Analysis." *American Sociological Review*, **35**:97–101.

———. 1973. "Efficient Estimation in Overidentified Models: An Interpretive Analysis." Chapter 7 of Goldberger and Duncan (1973).

——— and O. D. Duncan (eds.). 1973. *Structural Equation Models in the Social Sciences*. New York: Seminar Press.

Golden, J. T. 1965. *Fortran IV: Programming and Computing*. Englewood Cliffs, N.J.: Prentice-Hall.

Goodman, L. A. 1953. "Ecological Regressions and Behavior of Individuals." *American Sociological Review*, **18**:663–664.

———. 1959. "Some Alternatives to Ecological Correlation." *American Journal of Sociology*, **64**:610–625.

———. 1968. "The Analysis of Cross-Classified Data: Independence, Quasi-Independence, and Interactions in Contingency Tables with or without Missing Entries." *Journal of the American Statistical Association,* **63**:1091–1131.

——. 1970. "The Multivariate Analysis of Qualitative Data: Interactions among Multiple Classifications." *Journal of the American Statistical Association,* **65**:226–256.

——. 1972*a*. "A Modified Multiple Regression Approach to the Analysis of Dichotomous Variables." *American Sociological Review,* **37**:28–46.

——. 1972*b*. "A General Model for the Analysis of Surveys." *American Journal of Sociology,* **77**:1035–1086.

——. 1973. "Causal Analysis of Data from Panel Studies and Other Kinds of Surveys." *American Journal of Sociology,* **78**:1135–1191.

—— and W. H. Kruskal. 1963. "Measurement of Association for Cross-Classifications, III: Approximate Sampling Theory." *Journal of the American Statistical Association,* **58**:310–364.

Gordon, R. A. 1968. "Issues in Multiple Regression." *American Journal of Sociology,* **73**:592–616.

Grizzle, J. E. 1965. "The Two-Period Change-over Design and Its Use in Clinical Trials." *Biometrics,* **21**:467–480.

——, C. F. Starmer, and G. G. Koch. 1969. "Analysis of Categorical Data by Linear Models." *Biometrics,* **25**:489–504.

Gujarati, D. 1970*a*. "Use of Dummy Variables in Testing for Equality between Sets of Coefficients in Two Linear Regressions: A Note." *American Statistician,* **24**(1):50–52.

——. 1970*b*. "Use of Dummy Variables in Testing for Equality between Sets of Coefficients in Linear Regressions: A Generalization." *American Statistician,* **24**(5):18–22.

Hannan, M. T. 1971. *Aggregation and Disaggregation in Sociology.* Lexington, Mass.: Heath.

——, R. Rubinson, and J. T. Warren. 1974. "The Causal Approach to Measurement Error in Panel Analysis: Some Further Contingencies." Chapter 10 of Blalock (1974).

Hauser, R. M., and A. S. Goldberger. 1971. "The Treatment of Unobservable Variables in Path Analysis." Chapter 4 of Costner (1971*b*).

Hawkes, R. K. 1971. "The Multivariate Analysis of Ordinal Measures." *American Journal of Sociology,* **76**:908–926.

Heise, D. R. 1969. "Separating Reliability and Stability in Test-Retest Correlation." *American Sociological Review,* **34**:93–101.

—— (ed.). 1976. *Sociological Methodology.* San Francisco: Jossey-Bass.

Herman, W. L., J. E. Potterfield, C. M. Dayton, and K. G. Amershek. 1969. "The Relationship of Teacher-centered Activities and Pupil-centered Activities to Pupil Achievement in 18 Fifth-Grade Social Studies Classes." *American Educational Research Journal,* **6**:227–239.

Hetherington, E. M., and M. Carlson. 1964. "Effects of Candidate Support and Election Results upon Attitudes to the Presidency." *Journal of Social Psychology,* **64**:333–338.

Hill, R., J. M. Stycos, and K. W. Back. 1959. *The Family and Population Control.* Chapel Hill: The University of North Carolina Press.

Hoel, P. G. 1962. *Introduction to Mathematical Statistics,* 2d ed. New York: Wiley.

Hunt, E. B. 1961. "Memory Effects in Concept Learning." *Journal of Experimental Psychology,* **62**:598–604.

Hyman, H. 1955. *Survey Design and Analysis.* Glencoe, Ill.: The Free Press.

Ireland, C. T., and S. Kullback. 1968. "Contingency Tables with Given Marginals." *Biometrika,* **55**:179–188.

Jacobson, A. L., and N. M. Lalu. 1974. "An Empirical and Algebraic Analysis of

Alternative Techniques for Measuring Unobserved Variables." Chapter 7 of Blalock (1974).

Jennings, E. 1967. "Fixed Effects Analysis of Variances by Regression Analysis." *Multivariate Behavioral Research,* **2**:95–108.

Johnston, J. 1972. *Econometric Methods,* 2d ed. New York: McGraw-Hill.

Jones, H. L. 1958. "Inadmissible Samples and Confidence Limits." *Journal of the American Statistical Association,* **53**:482–490.

Jöreskog, K. G. 1970. "A General Method for the Analysis of Covariance Structures." *Biometrika* **57**:239–251.

———. 1973. "A General Method for Estimating a Linear Structural Equation System." Chapter 5 of Goldberger and Duncan (1973).

Kempthorne, O. 1952. *The Design and Analysis of Experiments.* New York: Wiley.

———. 1969. *An Introduction to Genetic Statistics.* Ames, Iowa: The Iowa State University Press.

Kendall, M. G. 1948. *Rank Correlation Methods.* New York: Hafner.

Keuls, M. 1952. "The Use of Studentized Range in Connection with an Analysis of Variance." *Euphytica,* **1**:112–122.

Keyfitz, N. 1953. "A Factorial Arrangement of Comparisons of Family Size." *American Journal of Sociology,* **58**:470–480.

King, D. 1974. "Feedback and Reciprocal Causation: A Monte Carlo Study of the Relative Small Sample Properties of Simultaneous Equation Estimators." Unpublished Ph.D. dissertation in sociology. Chapel Hill: University of North Carolina.

Kirk, R. E. 1968. *Experimental Design: Procedures for the Behavioral Sciences.* Belmont, Calif.: Brooks/Cole.

Kish, L. 1959. "Some Statistical Problems in Research Design." *American Sociological Review,* **24**:328–338.

———. 1962. "Studies of Interviewer Variance for Attitudinal Variables." *Journal of the American Statistical Association,* **57**:92–115.

———. 1965. *Survey Sampling.* New York: Wiley.

Koch, G. G., W. D. Johnson, and H. D. Tolley. 1972. "A Linear Models Approach to the Analysis of Survival and Extent of Disease in Multidimensional Contingency Tables." *Journal of the American Statistical Association,* **67**:783–796.

——— and S. Lemeshow. 1972. "An Application of Multivariate Analysis to Complex Sample Survey Data." *Journal of the American Statistical Association,* **67**:780–782.

Labovitz, S. 1967. "Some Observations on Measurement and Statistics." *Social Forces,* **46**:151–160.

Land, K. C. 1969. "Principles of Path Analysis." Chapter 1 of Borgatta (1969).

Lerea, L., and B. Ward. 1966. "The Social Schema of Normal and Speech Defective Children." *Journal of Social Psychology,* **69**:87–94.

Li, C. C. 1955. *Population Genetics.* Chicago: University of Chicago Press.

Li, J. C. R. 1964. *Statistical Inference,* vol. 2. Ann Arbor, Mich.: Edwards Brothers.

Lucas, H. L. 1957. "Extra-Period Latin-Square Change-over Design." *Journal of Dairy Science,* **40**:225–239.

Mallios, W. S. 1970. "The Analysis of Structural Effects in Experimental Designs." *Journal of the American Statistical Association,* **65**:808–827.

Mason, R., and A. N. Halter. 1968. "The Application of a System of Simultaneous Equations to an Innovation Diffusion Model." *Social Forces,* **47**:182–195.

McGinnis, R. 1958. "Randomization and Inference in Sociological Research." *American Sociological Review,* **23**:408–414.

McNemar, Q. 1962. *Psychological Statistics*, 3d ed. New York: Wiley.

Mendenhall, W. 1968. *An Introduction to Linear Models and the Design and Analysis of Experiments*. Belmont, Calif.: Wadsworth.

Miller, A. D. 1971. "Logic of Causal Analysis: From Experimental to Nonexperimental Designs." Chapter 15 of Blalock (1971*a*).

Miller, N. E., and R. Bugelski. 1948. "Minor Studies in Aggression: the Influence of Frustrations Imposed by the In-Group on Attitudes Expressed toward Out-Groups." *Journal of Psychology*, **25**:437–442.

Mood, A. M. 1950. *Introduction to the Theory of Statistics*. New York: McGraw-Hill.

Morgan, J. N., and R. C. Messenger. 1973. *THAID: a Sequential Analysis Program for the Analysis of Nominal Scale Dependent Variables*. Ann Arbor, Mich.: Institute for Social Research.

Morrison, D. E., and R. E. Henkel. 1969. "Significance Tests Reconsidered." *The American Sociologist*, **4**:131–140.

Mosteller, F., and C. Youtz. 1961. "Tables of the Freeman-Tukey Transformations for the Binomial and Poisson Distributions." *Biometrika*, **48**:433–440.

Myers, R. H. 1971. *Response Surface Methodology*. Boston: Allyn and Bacon.

Namboodiri, N. K. 1970. "A Statistical Exposition of the 'Before-After' and 'After-Only' Designs and Their Combinations." *American Journal of Sociology*, **76**:83–102.

———. 1972. "Experimental Designs in Which Each Subject Is Used Repeatedly." *Psychological Bulletin*, **77**:54–64.

Newman, D. 1939. "The Distribution of the Range in Samples from a Normal Population, Expressed in Terms of an Independent Estimate of Standard Deviation." *Biometrika*, **31**:20–30.

Noble, B. 1969. *Applied Linear Algebra*. Englewood Cliffs, N.J.: Prentice-Hall.

Norton, D. W. 1952. "An Empirical Investigation of Some Effects of Nonnormality and Heterogeneity on the F-Distribution." Unpublished Ph.D. dissertation. Iowa State University.

Patterson, H. D. 1950. "The Analysis of Change-over Trials." *Journal of Agricultural Science*, **40**:375–380.

Paull, A. E. 1950. "On a Preliminary Test for Pooling Mean Squares in the Analysis of Variance." *Annals of Mathematical Statistics*, **21**:539–556.

Pearson, E. S., and H. O. Hartley. 1958. *Biometrika Tables for Statisticians*, 2d ed. Cambridge: Cambridge University Press.

Phillips, J. P. N. 1964. "The Use of Magic Squares for Balancing and Assessing Order Effects in Some Analysis of Variance Designs." *Applied Statistics*, **13**:67–73.

———. 1968*a*. "A Simple Method of Constructing Certain Magic Rectangles of Even Order." *Mathematical Gazette*, **379**:9–12.

———. 1968*b*. "Method of Constructing One-Way and Factorial Designs Balanced for Trend." *Applied Statistics*, **17**:162–170.

Ploch, D. R. 1974. "Ordinal Measures of Association and the General Linear Model." Chapter 12 of Blalock (1974).

Quade, D. 1971. *Nonparametric Partial Correlation*. Amsterdam: Mathematisch Centrum.

———. 1974. "Nonparametric Partial Correlation." Chapter 13 of Blalock (1974).

Quenouille, M. H. 1953. *The Design and Analysis of Experiments*. London: Charles Griffin.

Rao, C. R. 1965. *Linear Statistical Inference and Its Applications*. New York: Wiley.

Rao, P., and R. L. Miller. 1971. *Applied Econometrics*. Belmont, Calif.: Wadsworth.

Reynolds, H. T. 1971. "Making Causal Inferences with Ordinal Data." Chapel Hill, N.C.: Institute for Research in Social Science. Working Paper 5.

———. 1974. "Ordinal Partial Correlation and Causal Inferences." Chapter 14 of Blalock (1974).

Robinson, W. S. 1950. "Ecological Correlations and the Behavior of Individuals." *American Sociological Review*, **15**:351–357.

Rockwell, R. 1974. "Multiple Linear Regression: A Higher Common Denominator." Chapel Hill, N.C.: Institute for Research in Social Science. Unpublished manuscript.

Rosenberg, M. 1962. "Test Factor Standardization as a Method of Interpretation." *Social Forces*, **41**:53–61.

———. 1968. *The Logic of Survey Analysis*. New York: Basic Books.

Ross, J. A., and P. Smith. 1968. "Orthodox Experimental Designs." Chapter 9 of Blalock and Blalock (1968).

Samuelson, P. A. 1947. *The Foundations of Economic Analysis*. Cambridge, Mass.: Harvard University Press.

Savage, L. J. 1962. *The Foundations of Statistical Inference: A Discussion*. New York: Barnes & Noble.

Scheffé, H. 1959. *The Analysis of Variance*. New York: Wiley.

Schoenberg, R. 1972. "Strategies for Meaningful Comparison." Chapter 1 of Cosnert (1972).

Schuessler, K. 1971. *Analyzing Social Data: A Statistical Orientation*. Boston: Houghton Mifflin.

———. 1973. "Ratio Variables and Path Models." Chapter 10 of Goldberger and Duncan (1973).

Selvin, H. C. 1957. "A Critique of Tests of Significance in Survey Research." *American Sociological Review*, **22**:519–527.

Senders, V. L. 1958. *Measurement and Statistics*. New York: Oxford University Press.

Sheehe, P. R., and I. D. J. Bross. 1961. "Latin Squares to Balance Immediate Residual and Other Order Effects." *Biometrics*, **17**:405–414.

Sidman, M. 1960. *Tactics of Scientific Research*. New York: Basic Books.

Siegel, P. M., and R. W. Hodge. 1968. "A Causal Approach to the Study of Measurement Error." Chapter 2 of Blalock and Blalock (1968).

Simon, H. A. 1957. *Models of Man*. New York: Wiley.

Smelser, N. J., and M. S. Lipset (eds.). 1966. *Social Structure and Mobility in Economic Development*. Chicago: Aldine.

Snedecor, G. M., and W. G. Cochran. 1967. *Statistical Methods*, 6th ed. Ames, Iowa: The Iowa State University Press.

Somers, R. H. 1962. "A New Asymmetric Measure of Association for Ordinal Variables." *American Sociological Review*, **27**:799–811.

Sonquist, J. A., and J. N. Morgan. 1964. *The Detection of Interaction Effects*. Ann Arbor, Mich.: Institute for Social Research.

Staats, A. W., C. K. Staats, and W. G. Heard. 1959. "Language Conditioning of Meaning to Meaning Using a Semantic Generalization Paradigm." *Journal of Experimental Psychology*, **57**:187–192.

Strotz, R. H., and H. A. Wold. 1960. "Recursive versus Nonrecursive Systems." *Econometrica*, **28**:417–427.

Sullivan, J. L. 1971. "Multiple Indicators and Complex Causal Models." Chapter 18 of Blalock (1971a).

———. 1974. "Multiple Indicators: Some Criteria of Selection." Chapter 8 of Blalock (1974).

Suppes, P. 1970. *A Probabilistic Theory of Causality*. Amsterdam: North-Holland.

Taylor, H. F. 1973. "Linear Models of Consistency: Some Extensions of Blalock's Strategy." *American Journal of Sociology*, **78**:1192–1215.

Theil, H. 1970. "On the Estimation of Relationships Involving Qualitative Variables." *American Journal of Sociology*, **76**:103–154.

Tukey, J. W. 1949. "One Degree of Freedom for Nonadditivity." *Biometrics*, **5**:232–242.

———. 1953. "The Problem of Multiple Comparisons." Princeton, N. J.: Princeton University. Mimeo ms. of 396 pages.

———. 1955. "Queries." *Biometrics*, **11**:111–113.

Veldman, D. 1967. *Fortran Programming for the Behavioral Sciences*. New York: Holt.

Wald, A. 1940. "The Fitting of Straight Lines If Both Variables Are Subject to Error." *Annals of Mathematical Statistics*, **11**:284–300.

Walker, S. H., and D. B. Duncan. 1967. "Estimation of the Probability of an Event as a Function of Several Independent Variables." *Biometrika*, **54**:167–179.

Ward, J., Jr. 1962. "Multiple Linear Regression Models." Chapter 10 of Borko (1962).

———. 1969. "Synthesizing Regression Models: An Aid to Learning Effective Problem Analysis." *American Statistician*, **23**(2):14–20.

Werts, C. E., R. L. Linn, and K. G. Jöreskog. 1971. "Estimating the Parameters of Path Models Involving Unmeasured Variables." Chapter 23 of Blalock (1971*a*).

Wiggins, J. A., F. Dill, and R. D. Schwartz. 1965. "On 'Status-Liability.'" *Sociometry*, **28**:197–209.

Wilber, G. 1967. "Causal Models and Probability." *Social Forces*, **46**:81–89.

Wiley, D. E. 1973. "The Identification Problem for Structural Equation Models with Unmeasured Variables." Chapter 4 of Goldberger and Duncan (1973).

——— and J. A. Wiley. 1970. "The Estimation of Measurement Error in Panel Data." *American Sociological Review*, **35**:112–117.

Williams, D. E., M. Wark, and F. D. Minifie. 1963. "Ratings of Stuttering by Audio, Visual, and Audiovisual Cues." *Journal of Speech and Hearing Research*, **6**:91–100.

Williams, E. J. 1949. "Experimental Designs Balanced for Estimation of Residual Effects of Treatments." *Australian Journal of Scientific Research*, **2**:149–169.

———. 1950. "Experimental Designs Balanced for Pairs of Residual Effects." *Australian Journal of Scientific Research*, **3**:351–363.

Wilson, T. P. 1974. "Measures of Association for Bivariate Ordinal Hypotheses." Chapter 11 of Blalock (1974).

Winch, R. F., and D. T. Campbell. 1969. "Proof: No. Evidence? Yes. The Significance of Tests of Significance." *American Sociologist*, **4**:140–143.

Winer, B. J. 1962. *Statistical Principles in Experimental Design*. New York: McGraw-Hill.

Wold, H. A. 1975. "Path Models with Latent Variables That Have Several Indicators: Partings of the Ways in the Light of NIPALS Modelling." Chapter 6 of Blalock et al. (1975).

Wortham, A. W., and T. E. Smith. 1960. *Practical Statistics in Experimental Design*. Dallas: Dalls.

Wright, S. 1934. "The Method of Path Coefficients." *Annals of Mathematical Statistics*, **5**:161–215.

———. 1960. "Path Coefficients and Path Regressions: Alternative or Complementary Concepts?" *Biometrics*, **16**:189–202.

Yates, F. 1934. "The Analysis of Multiple Classification with Unequal Numbers in the Different Classes." *Journal of the American Statistical Association*, **29**:51–66.

———. 1960. *Sampling Methods for Census and Surveys*, 3d ed. London: Griffin.

SUBJECT INDEX